Handbook of Graphical Models

Chapman & Hall/CRC
Handbooks of Modern Statistical Methods

Series Editor

Garrett Fitzmaurice, *Department of Biostatistics, Harvard School of Public Health, Boston, MA, U.S.A.*

The objective of the series is to provide high-quality volumes covering the state-of-the-art in the theory and applications of statistical methodology. The books in the series are thoroughly edited and present comprehensive, coherent, and unified summaries of specific methodological topics from statistics. The chapters are written by the leading researchers in the field, and present a good balance of theory and application through a synthesis of the key methodological developments and examples and case studies using real data.

Longitudinal Data Analysis
Edited by Garrett Fitzmaurice, Marie Davidian, Geert Verbeke, and Geert Molenberghs

Handbook of Spatial Statistics
Edited by Alan E. Gelfand, Peter J. Diggle, Montserrat Fuentes, and Peter Guttorp

Handbook of Markov Chain Monte Carlo
Edited by Steve Brooks, Andrew Gelman, Galin L. Jones, and Xiao-Li Meng

Handbook of Survival Analysis
Edited by John P. Klein, Hans C. van Houwelingen, Joseph G. Ibrahim, and Thomas H. Scheike

Handbook of Mixed Membership Models and Their Applications
Edited by Edoardo M. Airoldi, David M. Blei, Elena A. Erosheva, and Stephen E. Fienberg

Handbook of Missing Data Methodology
Edited by Geert Molenberghs, Garrett Fitzmaurice, Michael G. Kenward, Anastasios Tsiatis, and Geert Verbeke

Handbook of Design and Analysis of Experiments
Edited by Angela Dean, Max Morris, John Stufken, and Derek Bingham

Handbook of Cluster Analysis
Edited by Christian Hennig, Marina Meila, Fionn Murtagh, and Roberto Rocci

Handbook of Discrete-Valued Time Series
Edited by Richard A. Davis, Scott H. Holan, Robert Lund, and Nalini Ravishanker

Handbook of Big Data
Edited by Peter Bühlmann, Petros Drineas, Michael Kane, and Mark van der Laan

Handbook of Spatial Epidemiology
Edited by Andrew B. Lawson, Sudipto Banerjee, Robert P. Haining, and María Dolores Ugarte

Handbook of Neuroimaging Data Analysis
Edited by Hernando Ombao, Martin Lindquist, Wesley Thompson, and John Aston

Handbook of Statistical Methods and Analyses in Sports
Edited by Jim Albert, Mark E. Glickman, Tim B. Swartz, Ruud H. Koning

Handbook of Methods for Designing, Monitoring, and Analyzing Dose-Finding Trials
Edited by John O'Quigley, Alexia Iasonos, Björn Bornkamp

Handbook of Quantile Regression
Edited by Roger Koenker, Victor Chernozhukov, Xuming He, and Limin Peng

Handbook of Environmental and Ecological Statistics
Edited by Alan E. Gelfand, Montserrat Fuentes, Jennifer A. Hoeting, Richard L. Smith

For more information about this series, please visit: https://www.crcpress.com/go/handbooks

Handbook of Graphical Models

Edited by

Marloes Maathuis

Mathias Drton

Steffen Lauritzen

Martin Wainwright

CRC Press is an imprint of the
Taylor & Francis Group, an **informa** business
A CHAPMAN & HALL BOOK

CRC Press
Taylor & Francis Group
6000 Broken Sound Parkway NW, Suite 300
Boca Raton, FL 33487-2742

First issued in paperback 2020

ISBN-13: 978-1-4987-8862-5 (hbk)
ISBN-13: 978-0-367-73260-8 (pbk)

Library of Congress Cataloging-in-Publication Data

Names: Drton, Mathias, editor. | Lauritzen, Steffen, editor. | Wainwright, Martin, editor.
Title: Handbook of graphical models / edited by Marloes Maathuis, Mathias Drton, Steffen Lauritzen and Martin Wainwright.
Description: Boca Raton, Florida : CRC Press, c2019. | Includes bibliographical references and index.
Identifiers: LCCN 2018010969| ISBN 9781498788625 (hardback : alk. paper) | ISBN 9780429463976 (e-book) | ISBN 9780429874246 (web pdf) | ISBN 9780429874239 (epub) | ISBN 9780429874222 (mobi/kindle)
Subjects: LCSH: Graphical modeling (Statistics)
Classification: LCC QA279 .H3435 2019 | DDC 519.5--dc23
LC record available at https://lccn.loc.gov/2018010969

Visit the Taylor & Francis Web site at
http://www.taylorandfrancis.com

and the CRC Press Web site at
http://www.crcpress.com

Contents

Preface

A graphical model is a statistical model that is associated to a graph. The nodes of the graph correspond to the random variables of interest, and the edges encode allowed conditional dependencies among the variables. The factorization properties underlying graphical models facilitate tractable computation with multivariate distributions, making the models a valuable tool in a plethora of applications. Furthermore, directed graphical models admit intuitive causal interpretations and have become a cornerstone for causal inference.

While there exist a number of excellent books on graphical models, the field has grown so much that individual authors can hardly cover its entire scope. Moreover, the field is interdisciplinary by nature, with important contributions from a range of disciplines, including statistics, computer science, electrical engineering, biology, mathematics and philosophy. Through chapters by leading researchers from these different areas, this handbook provides a broad and accessible overview of the state of the art.

The book contains a total of twenty-one chapters, grouped into five parts:

I. Conditional independencies and Markov properties

II. Computing with factorizing distributions

III. Statistical inference

IV. Causal inference

V. Applications

Part I reviews the foundations of graphical models. It discusses how graphs can encode conditional independencies between random variables, or equivalently, a factorization of the joint distribution of the variables. The main theme of Part II is how to perform efficient computations based on the joint distribution of a given graphical model, in particular by leveraging the associated factorization properties. In Part III, the focus of the book shifts to problems of statistical inference, such as learning the graph and estimating the associated parameters from available data. Part IV focuses on the causal interpretation of directed acyclic graphs. The corresponding chapters review fundamental concepts of graphical approaches to causal inference, and also treat statistical aspects such as learning a directed acyclic graph from data. Finally, Part V shows how graphical models are used in selected applied problems in forensic science and biology.

Part I forms the basis of the book. The remaining Parts II through V can be read independently, while cross-references between the chapters highlight connections. The topics of the chapters range from explanations of basic concepts at a level that is suitable to newcomers to descriptions of recent developments or original research. As such, the book targets a wide audience, including graduate students in statistics, mathematics and computer science, users of graphical models in applied research, as well as experts on graphical models. Most of all, we hope that the book will spark further research in this exciting field.

We would like to express our sincere thanks to all authors for their valuable contributions, and to Rob Calver, Cynthia Klivecka, and Lara Spieker for their help and guidance throughout the entire process.

Marloes Maathuis
ETH Zurich
Mathias Drton
University of Copenhagen and University of Washington
Steffen Lauritzen
University of Copenhagen
Martin Wainwright
University of California, Berkeley

Contributors

Jeffrey A. Bilmes
Department of Electrical Engineering
University of Washington
Seattle, WA, USA

A. Philip Dawid
Statistical Laboratory
University of Cambridge
Cambridge, UK

Vanessa Didelez
Department of Biometry and Data Management
Leibniz Institute for Prevention Research and Epidemiology – BIPS
Bremen, Germany

Arnaud Doucet
Department of Statistics
University of Oxford
Oxford, UK

Robin Evans
Department of Statistics
University of Oxford
Oxford, UK

Jana Janková
Statistical Laboratory
University of Cambridge
Cambridge, UK

Thomas Kahle
Faculty of Mathematics
Otto von Guericke University of Magdeburg
Magdeburg, Germany

Luca La Rocca
Department of Physics, Informatics and Mathematics
University of Modena and Reggio Emilia
Modena, Italy

John Lafferty
Department of Statistics and Data Science
Yale University
New Haven, CT, USA

Anthony Lee
School of Mathematics
University of Bristol
Bristol, UK

Hongzhe Li
Department of Biostatistics, Epidemiology and Informatics
University of Pennsylvania
Philadelphia, PA, USA

Han Liu
Department of Electrical Engineering and Computer Science
Northwestern University
Evanston, IL, USA

Po-Ling Loh
Department of Electrical and Computer Engineering
University of Wisconsin – Madison
Madison, WI, USA

Jing Ma
Biostatistics Program
Fred Hutchinson Cancer Research Center
Seattle, WA, USA

Hélène Massam
Department of Mathematics and Statistics
York University
Toronto, ON, Canada

Ofer Meshi
Google

Julia Mortera
Department of Economics
Università Roma Tre
Rome, Italy

Sach Mukherjee
German Center for Neurodegenerative Diseases
Bonn, Germany

Chris Oates
School of Mathematics, Statistics and Physics
Newcastle University
Newcastle, UK

Johannes Rauh
Max Planck Institute for Mathematics in the Sciences
Leipzig, Germany

Alberto Roverato
Department of Statistical Sciences
University of Bologna
Bologna, Italy

Nicholas Ruozzi
Erik Jonsson School of Engineering and Computer Science
University of Texas at Dallas
Richardson, TX, USA

Alexander G. Schwing
Department of Electrical and Computer Engineering
University of Illinois at Urbana-Champaign
Urbana, IL, USA

Ilya Shpitser
Department of Computer Science
Johns Hopkins University
Baltimore, MD, USA

Peter Spirtes
Department of Philosophy
Carnegie Mellon University
Pittsburgh, PA, USA

Johan Steen
Department of Intensive Care
Ghent University Hospital
Ghent, Belgium

Milan Studený
Institute of Information Theory and Automation
The Czech Academy of Sciences
Prague, Czech Republic

Seth Sullivant
Department of Mathematics
North Carolina State University
Raleigh, NC, USA

Caroline Uhler
Department of Electrical Engineering and Computer Science
Massachusetts Institute of Technology
Cambridge, MA, USA

Sara van de Geer
Seminar for Statistics
ETH Zurich
Zurich, Switzerland

Stijn Vansteelandt
Department of Applied Mathematics, Computer Science and Statistics
Ghent University
Ghent, Belgium

Kun Zhang
Department of Philosophy
Carnegie Mellon University
Pittsburgh, PA, USA

Piotr Zwiernik
Department of Economics and Business
Universitat Pompeu Fabra
Barcelona, Spain

Part I

Conditional independencies and Markov properties

1

Conditional Independence and Basic Markov Properties

Milan Studený

Institute of Information Theory and Automation, The Czech Academy of Sciences

CONTENTS

The aim of the chapter

In this chapter, the concept of *conditional independence* (CI) is recalled and an overview of both former and recent results on the description of CI structures is given. The traditional graphical models, namely those ascribed to *undirected graphs* (UGs) and *directed acyclic graphs* (DAGs), can be interpreted as special cases of statistical models of a CI structure. Therefore, an overview of Markov properties for these two basic types of graphs is also given. Markov properties for more general graphs are discussed in Chapter 2.

1.1 Introduction: Historical Overview and an Example

In this section, some earlier results on CI are recalled and an example is given to informally illustrate the concept of probabilistic CI.

1.1.1 Stochastic conditional independence

Loève [27] already defined the concept of CI in terms of σ-algebras in his book on probability theory in the 1950s. Phil Dawid [12] was probably the first statistician who explicitly formulated certain basic formal properties of stochastic CI. He observed that several statistical concepts, e.g., the one of a sufficient statistic, can equivalently be defined in terms of generalized CI and this observation allows one to derive many results in an elegant way with the aid of those formal properties. These basic formal properties of stochastic CI were later independently formulated in the context of philosophical logic by Spohn [50], who was interested in the interpretation of the concept of CI and its relation to causality. The same properties, this time formulated in terms of σ-algebras, were also explored by Mouchart and Rolin [39]. The author of this chapter was told that the conditional independence symbol $\perp\!\!\!\perp$ was proposed by Dawid and Mouchart in their discussion in the late 1970s.

The significance of the CI concept for probabilistic reasoning was later recognized by Pearl and Paz [43], who observed that the above-mentioned basic formal properties of CI are also valid for certain ternary separation relations induced by undirected graphs. This led them to an idea of describing such formal ternary relations by graphs and introducing an abstract concept of a *semi-graphoid.* The even more abstract concept of a *separoid* was later suggested by Dawid [13]. Pearl and Paz [43] also raised a conjecture that semi-graphoids coincide with probabilistic CI structures, which was refuted by myself in [52] using some tools of information theory.

A lot of effort and time has been devoted to the problem of characterizing all possible CI structures induced by four discrete random variables. The final solution to that problem was achieved by Matúš [36, 33, 34]; the number of these structures is 18478 [66] and they are decomposed into 1098 types.

1.1.2 Graphs and local computation method

The idea to use graphs whose nodes correspond to random variables in order to describe CI structures had appeared in statistics earlier than Pearl and Paz suggested this approach in the context of computer science. One can distinguish between two basic traditional trends, namely, using undirected and directed (acyclic) graphs. Note that statistical models described by such graphs can be understood as the models of (special) CI structures.

Undirected graphs (UGs) appeared in the 1970s in statistical physics as tools to describe relations among discrete random variables. Moussouris [40] introduced several Markov properties relative to an UG for distributions with positive density and showed their equivalence with a factorization condition. Darroch, Lauritzen, and Speed [10] realized that UGs can be used to describe statistical models arising in the theory of contingency tables, so they introduced a special class of (undirected) graphical models and interpreted them in terms of CI. At the same time, the use of UGs was considered in the area of multivariate statistical analysis. Dempster [15] introduced covariance selection models for continuous real random variables, which were interpreted in terms of CI by Wermuth [67].

In the 1980s, directed acyclic graphs (DAGs) found their applications in the decision-making theory in connection with influence diagrams. Smith [49] used the above-mentioned formal properties of CI to easily show the correctness of some operations with influence diagrams. Pearl's book [42] on probabilistic reasoning had a substantial impact on promotion of graphical methods in artificial intelligence; in the book, he defined a directional separation criterion (d-separation) for DAGs and pinpointed the role of CI.

The theoretical breakthrough leading to (graphical) probabilistic expert systems was the *local computation method.* Lauritzen and Spiegelhalter [26] offered a methodology to perform efficient computations of conditional probabilities for (discrete) measures which are Markovian with respect to a DAG.

1.1.3 Conditional independence in other areas

Probability theory and statistics are not the only fields in which the concept of CI was introduced and examined. An analogous concept of *embedded multivalued dependency* (EMVD) was studied in the 1970s in the theory of relational databases. Sagiv and Walecka [44] showed that there is no finite axiomatic characterization of EMVD structures. Shenoy [48] observed that one can introduce the concept of CI within various calculi for dealing with knowledge and uncertainty in artificial intelligence (AI), including Spohn's theory of natural (ordinal) conditional functions, Zadeh's possibility theory and the Dempster-Shafer theory of evidence.

This motivated several papers devoted to formal properties of CI within various uncertainty calculi in AI. For example, Vejnarová [63] studied the properties of CI in the frame of possibility theory and it was shown in [54] that there is no finite axiomatic characterization of the CI structures arising in the context of natural conditional functions. Various concepts of conditional irrelevance have also been introduced and their formal properties were examined in the theory of *imprecise probabilities*; let us mention the concept of epistemic irrelevance introduced by Cozman and Walley [9].

1.1.4 Geometric approach and methods of modern algebra

The observation that graphs cannot describe all possible discrete stochastic CI structures led me to proposing a linear-algebraic method of their description in [57]. In this approach, certain vectors whose components are integers and correspond to subsets of the set of variables, called (structural) *imsets*, are used to describe the CI structures. The approach allows one to apply geometric methods of combinatorial optimization to learning graphical models and to approaching the CI implication problem. Hemmecke et al. [21] answered two of the open problems related to the method of imsets and disproved a geometric conjecture from [57] about the cone corresponding to the (structural) imsets.

The application of methods of modern algebra and (polyhedral) geometry to problems arising in mathematical statistics has recently led to establishing a new field of *algebraic statistics*. Drton, Sturmfels and Sullivant [16] in their book on this topic devoted one chapter to advanced algebraic tools for describing statistical models of CI structure. The topic of probabilistic CI thus naturally became one of the topics of interest in that area.

1.1.5 A motivational example

This section contains a small story to explain the intuitive sense of CI. The section can be skipped without affecting the flow of the chapter.

Imagine that organizers of a conference entitled *Probabilistic Graphical Models*, to be held in September 2018 in Prague, have the task to organize a lunch for the participants in a student cafeteria during a lunch break. Because of time limitations, they give the participants a limited choice of drinks and dishes. Cruel organizers intentionally decide to ignore human rights of vegetarians and teetotalers; thus, 3 items are to be served consecutively:

a ... a drink (exclusive choice is either BEER or white WINE),
c ... the main course (the choice is PORK or FISH),
b ... a dessert (the choice is SALAD or CAKE).

The participants are asked to decide about their main courses c after obtaining their drinks a. This can substantially influence their decisions: a well-known fact is that white wine pairs with fish while beer is the best fit with a traditional Czech dish which consists of roasted pork, sauerkraut and dumplings. Thus, a typical participant already drinking wine chooses fish, while only a minority of wine-drinkers take pork. Analogously, a typical beer-drinker goes for pork and a minority of them take fish. An analogous decision problem for participants occurs when they finish their main courses. Since the pork with sauerkraut is fat, a typical pork-eater decides to compensate that by choosing a light salad; on the other hand, fish has low calories, which makes a majority of fish-eaters decide for a sweet cake. Assume for simplicity that the proportion of non-typical participants in each group is $\frac{1}{4}$ and that drinking preferences are equal. This leads to the following scheme, allowing us to compute the overall probabilities:

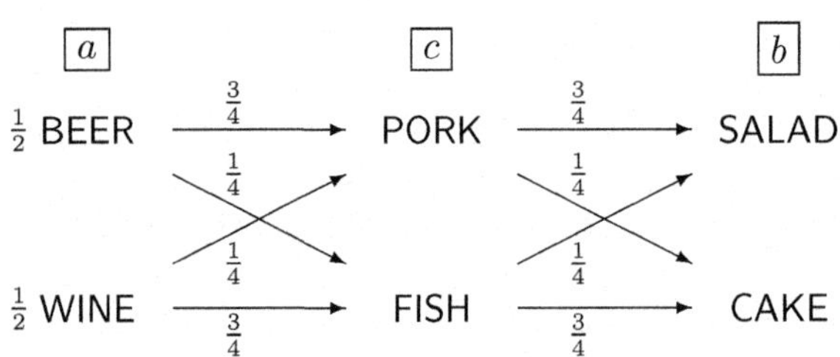

It is clear from the description of the situation that (the decision about) the last event b dominantly depends on (the previous decision about) the event c; in fact, it is independent of what the results of the event a was. This description characterizes the situation when *events* a *and* b *are conditionally independent given the values of the event* c, which is conventionally denoted by $a \perp\!\!\!\perp b \mid c$.

This example also illustrates the intuitive difference between conditional and unconditional independence of events. One can observe higher correlation between beer and salad,

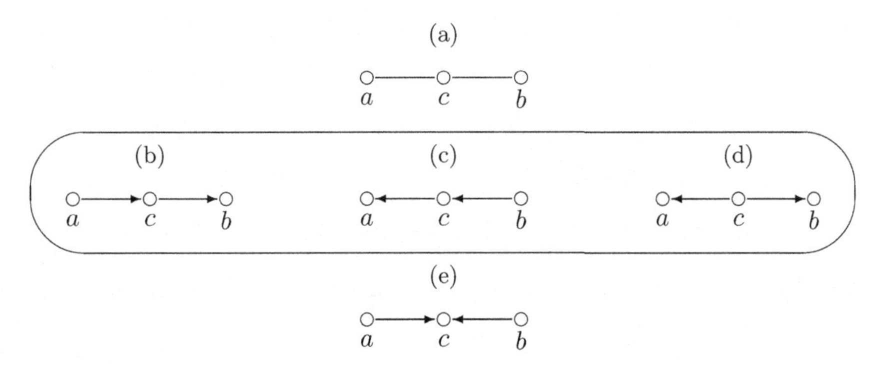

FIGURE 1.1: Examples of graphs for a description of a CI structure.

respectively between wine and cake:

$$
\begin{array}{lll}
\multirow{2}{*}{(\mathsf{BEER, SALAD})} & (\mathsf{BEER, PORK, SALAD}) \to \frac{1}{2}\cdot\frac{3}{4}\cdot\frac{3}{4}=\frac{9}{32} & \multirow{2}{*}{$\to \frac{10}{32},$} \\
 & (\mathsf{BEER, FISH, SALAD}) \to \frac{1}{2}\cdot\frac{1}{4}\cdot\frac{1}{4}=\frac{1}{32} & \\
\multirow{2}{*}{(\mathsf{BEER, CAKE})} & (\mathsf{BEER, PORK, CAKE}) \to \frac{1}{2}\cdot\frac{3}{4}\cdot\frac{1}{4}=\frac{3}{32} & \multirow{2}{*}{$\to \frac{6}{32},$} \\
 & (\mathsf{BEER, FISH, CAKE}) \to \frac{1}{2}\cdot\frac{1}{4}\cdot\frac{3}{4}=\frac{3}{32} & \\
\multirow{2}{*}{(\mathsf{WINE, SALAD})} & (\mathsf{WINE, PORK, SALAD}) \to \frac{1}{2}\cdot\frac{1}{4}\cdot\frac{3}{4}=\frac{3}{32} & \multirow{2}{*}{$\to \frac{6}{32},$} \\
 & (\mathsf{WINE, FISH, SALAD}) \to \frac{1}{2}\cdot\frac{3}{4}\cdot\frac{1}{4}=\frac{3}{32} & \\
\multirow{2}{*}{(\mathsf{WINE, CAKE})} & (\mathsf{WINE, PORK, CAKE}) \to \frac{1}{2}\cdot\frac{1}{4}\cdot\frac{1}{4}=\frac{1}{32} & \multirow{2}{*}{$\to \frac{10}{32}.$} \\
 & (\mathsf{WINE, FISH, CAKE}) \to \frac{1}{2}\cdot\frac{3}{4}\cdot\frac{3}{4}=\frac{9}{32} & \\
\end{array}
$$

Note that situations when a confounding variable c exists for correlated variables a and b often occurs in connection with the so-called *Simpson's paradox.*

Graphs can be used to depict such conditional in/dependence relations among (random) variables; for example, the undirected graph in Figure 1.1(a) is traditionally used to depict the above-described situation. Directed edges (= arrows) can then be used to give some additional information or interpretation. Thus, the directed graph in Figure 1.1(b), which brings the same CI information as that in Figure 1.1(a), can be used to show the time direction. Nevertheless, the role of variables a and b is exchangeable in the sense that the probability distribution remains the same if one swaps a and b. Indeed, one can consider an absurd situation when the cruel organizers decide the next day to reverse the order of courses in order to torture the participants with the thirst. The point is that the probability distribution will be the same but the graph in Figure 1.1(c) then better reflects the additional information about the time direction. The third option is that one decides to use arrows to pinpoint the central role of the variable c, which leads to the graph shown in Figure 1.1(d). All the directed graphs in the oval within Figure 1.1 give the same CI information (about the underlying distribution); in this case we say they are *independence* or *Markov equivalent.*

The directed graph in Figure 1.1(e) is, however, interpreted in another way. This graph is traditionally used to describe the situation when variables a and b are unconditionally independent but conditionally dependent given c; in notation, $a \perp\!\!\!\perp b$ and $a \not\perp\!\!\!\perp b \,|\, c$. Let us give another silly example of when such a probability distribution can occur. Imagine that a poor man has the last two coins to be spent this evening and he has to decide whether to buy some food or a bottle of beer he likes a lot. Thus, he has to decide exclusively between the hunger and the thirst; nevertheless, he slightly tends to avoid the hunger. Therefore, he decides to toss the coins and if two heads occur then he buys the beer, otherwise the food. Then the result a of tossing the first coin will be independent of the result b of tossing the second coin while the event c whether he buys food or beer depends on the joint configuration of values of a and b. In this case, a and b are *not* conditionally independent given the values of c. Some authors then say that variables a and b are *marginally independent* but not conditionally independent given (the values of) variable c.

The latter example also allows us to explain the difference between two traditional interpretations of graphical models. In this chapter we deal with the basic *CI interpretation* when the graph in Figure 1.1(e) is understood solely as a record of CI information about the distribution. Other authors have given an extended *causal interpretation* of graphs, where arrows are used to encode expected causal relationships among variables. Thus, the graph in Figure 1.1(e) can be understood as a pictorial representation of the fact that the variable c causally/functionally depends on (combination of) variables a and b. Note, however, that causal interpretation of graphs is based on very specific assumptions on data generation mechanism. The causal interpretation is discussed in Part IV of this handbook.

1.2 Notation and Elementary Concepts

In this section, notation is introduced and elementary notions are recalled. Throughout the chapter N is a finite non-empty index set whose elements correspond to random *variables* (and to nodes of graphs in graphical context). The symbol $\mathcal{P}(N) := \{A : A \subseteq N\}$ will denote the *power set* of N.

1.2.1 Discrete probability measures

This section mainly deals with the discrete case and does not require any special previous knowledge from the reader.

Definition 1.2.1. A *discrete probability measure over* N is defined as follows:

(i) For each $i \in N$ a non-empty finite set X_i is given, which is the *individual sample space* for the variable i. This defines a *joint sample space*, which is the Cartesian product $\mathsf{X}_N := \prod_{i \in N} \mathsf{X}_i$.

(ii) A probability measure P on X_N is given; it is determined by its *density*, which is a function $p : \mathsf{X}_N \to [0,1]$ such that $\sum_{x \in \mathsf{X}_N} p(x) = 1$. Then $P(\mathbb{A}) = \sum_{x \in \mathbb{A}} p(x)$ for any $\mathbb{A} \subseteq \mathsf{X}_N$.

A general *probability measure over* N is defined analogously, but instead of a finite set X_i, a measurable space $(\mathsf{X}_i, \mathcal{X}_i)$ is assumed for any $i \in N$. The joint sample space is endowed with the product σ-algebra $\bigotimes_{i \in N} \mathcal{X}_i$. Some measures on $(\mathsf{X}_N, \bigotimes_{i \in N} \mathcal{X}_i)$ cannot be determined by densities in the general case.

Given $A \subseteq N$, any list of elements $[x_i]_{i \in A}$ such that $x_i \in \mathsf{X}_i$ for $i \in A$ will be called a *configuration* for A. The set X_A of configurations for A is then the *sample space for* A. Given disjoint $A, B \subseteq N$, we will use concatenation AB as a shorthand for (disjoint) union $A \cup B$. Given disjoint configurations $a \in \mathsf{X}_A$ and $b \in \mathsf{X}_B$, the symbol $[a, b]$ will denote their *join*, i.e., the joint list. If the joint configuration is an argument of a function, say of a density $p : \mathsf{X}_{AB} \to \mathbb{R}$, then the brackets will be omitted and we will write $p(a, b)$ instead of $p([a, b])$; similarly in the case of the join of three or more disjoint configurations.

In the case of $A \subseteq B$ and $b \in \mathsf{X}_B$, the symbol b_A will denote the *restriction* of the configuration b for A, that is, the restricted list. The mapping from X_B to X_A ascribing b_A to $b \in \mathsf{X}_B$ is the corresponding marginal *projection.* In particular, the symbol $b_\emptyset$ will denote the *empty configuration*, that is, the empty list of elements.

Given $i \in N$, the symbol i will often be used as an abbreviation for the singleton $\{i\}$. In particular, if $i \in A \subseteq N$ and $a \in \mathsf{X}_A$ then the symbol a_i will be a simplified notation for the marginal configuration $a_{\{i\}}$; of course, it is nothing but the i-th component of the configuration a.

Given disjoint $A, B \subseteq N$ and configuration sets $\mathbb{A} \subseteq \mathsf{X}_A$, $\mathbb{B} \subseteq \mathsf{X}_B$, we introduce $\mathbb{A} \times \mathbb{B} := \{[a,b] \,:\, a \in \mathbb{A} \;\&\; b \in \mathbb{B}\}$. Note that $\mathbb{A} \times \mathbb{B}$ is typically the Cartesian product but if $A = \emptyset$ and $\mathbb{A} \neq \emptyset$, that is, if $\mathbb{A} = \{a_\emptyset\}$ only contains the empty configuration, then one has $\mathbb{A} \times \mathbb{B} = \mathbb{B}$; analogously in the case of $B = \emptyset \neq \mathbb{B}$.

Definition 1.2.2. Given $A \subseteq N$ and a probability measure P over N, the *marginal measure for A* is the measure P_A over A defined by the relation

$$P_A(\mathbb{A}) := P(\{x \in \mathsf{X}_N \,:\, x_A \in \mathbb{A}\}) \quad \text{for } \mathbb{A} \subseteq \mathsf{X}_A \quad (\mathbb{A} \in \textstyle\bigotimes_{i \in A} \mathcal{X}_i \text{ in general}).$$

In the discrete case, the *marginal density for A* is the density of P_A; it is given by the formula

$$p_A(a) = P(\{x \in \mathsf{X}_N \,:\, x_A = a\}) = \sum_{c \in \mathsf{X}_{N \setminus A}} p(a,c) \quad \text{for } a \in \mathsf{X}_A,$$

where p is the (joint) density of the probability measure P.

Note that a simple *vanishing principle* for marginal densities will be tacitly used in § 1.3.1: if $x \in \mathsf{X}_N$, $C \subseteq B \subseteq N$ then $p_C(x_C) = 0$ implies $p_B(x_B) = 0$. The next elementary concept in the discrete case is that of a conditional probability, where the conditioning objects are (marginal) configurations.

Definition 1.2.3. Given disjoint sets $A, C \subseteq N$ of variables and a discrete probability measure P over N, the *conditional probability on X_A given C* is a (partial) function of two arguments denoted by $P_{A|C}(*|*)$, where the asterisks stand for the respective arguments. Specifically,

$$P_{A|C}(\mathbb{A}|c) \;:=\; \frac{P_{AC}(\mathbb{A} \times \{c\})}{P_C(\{c\})} \equiv \frac{P_{AC}(\mathbb{A} \times \{c\})}{p_C(c)}$$

$$\text{where } \mathbb{A} \subseteq \mathsf{X}_A \text{ and } c \in \mathsf{X}_C \text{ with } p_C(c) > 0.$$

The *conditional density for A given C* is also a (partial) function, in this case both arguments are the respective marginal configurations:

$$p_{A|C}(a|c) := \frac{p_{AC}(a,c)}{p_C(c)} \equiv P_{A|C}(\{a\}|c) \quad \text{for } a \in \mathsf{X}_A,\ c \in \mathsf{X}_C \text{ with } p_C(c) > 0.$$

Observe that the marginal measure can be viewed as a special case of the conditional probability where the conditioning configuration is empty, that is, $C = \emptyset$. Another observation is that, for any *positive configuration*, that is, $c \in \mathsf{X}_C$ with $p_C(c) > 0$, the function $\mathbb{A} \subseteq \mathsf{X}_A \mapsto P_{A|C}(\mathbb{A}|c)$ is a probability measure over A. It is clear that $P_{A|C}(*|*)$ only depends on the marginal P_{AC}.

In the computer science community, the conditional density is sometimes called a *conditional probability table.* Let us emphasize that the ratio defining the conditional density is not defined for conditioning on *zero configurations* $c \in \mathsf{X}_C$ with $p_C(c) = 0$, an important detail which is, unfortunately, omitted or even ignored in some machine learning (text)books.

Note that the assumption that the density is *strictly positive*, that is, $p(x) > 0$ for all $x \in \mathsf{X}_N$, is too restrictive in the area of probabilistic expert systems because it does not allow for modeling functional dependencies between random variables.

In the discrete case, one does not need to extend the conditional probability to zero configurations in order to define the notion of CI; however, in the general case, one has to consider different versions of conditional probability, which makes the general definition of CI more technical (see § 1.3.2).

1.2.2 Continuous distributions

In this section, which can be skipped by beginners, we assume that the reader is familiar with the standard notions of measure theory. The meaning of the term *probability distribution* encountered in the literature depends on the field in which it is actually encountered. In probability theory, it usually means a (general) probability measure, while in statistics its meaning is typically restricted to measures given by densities, and in computer science it is often identified with the concept of a density function.

In statistics, one typically works with real continuous distributions and these are defined through densities. There is a quite wide class of probability measures for which the concept of density (function) makes good sense.

Definition 1.2.4. A probability measure over N is *marginally continuous* if it is absolutely continuous with respect to the product of its one-dimensional marginals, that is, if

$$(\bigotimes_{i\in N} P_i)(\mathbb{A}) = 0 \quad \text{implies} \quad P(\mathbb{A}) = 0 \qquad \text{for any } \mathbb{A} \in \textstyle\bigotimes_{i\in N} \mathcal{X}_i,$$

in notation $P \ll \bigotimes_{i\in N} P_i$, where the symbol $\otimes$ is used to denote both the product of (probability) measures and the product of σ-algebras.

An equivalent definition of a marginally continuous measure is that there exists a (dominating) system of σ-finite measures μ^i on $(\mathsf{X}_i, \mathcal{X}_i)$ for $i \in N$ such that $P \ll \bigotimes_{i\in N} \mu^i$ (see [57, Lemma 2.3]). It is easy to verify that every discrete probability measure over N is marginally continuous: the dominating system of measures is the system of counting measures, that is, $\mu^i(\mathbb{A}) = |\mathbb{A}|$ for any $i \in N$ and $\mathbb{A} \subseteq \mathsf{X}_i$. Another standard example is a *regular Gaussian measure* over N; in this case, for any $i \in N$, $\mathsf{X}_i = \mathbb{R}$ is the set of real numbers endowed with the Borel σ-algebra and μ^i is the Lebesgue measure.

Having fixed individual sample spaces and a dominating system of σ-finite measures, every marginally continuous measure P can be introduced through its *joint density* f, which is the Radon-Nikodym derivative of P with respect to $\mu := \bigotimes_{i\in N} \mu^i$. For all $A \subseteq N$, we put $\mathcal{X}_A := \bigotimes_{i\in A} \mathcal{X}_i$ and accept a convention that $\mathcal{X}_\emptyset := \{\emptyset, \mathsf{X}_\emptyset\}$ is the only (trivial) σ-algebra on $\mathsf{X}_\emptyset$.

The *marginal density for* $A \subseteq N$ is then defined as the Radon-Nikodym derivative f_A of the marginal P_A with respect to $\mu^A := \bigotimes_{i\in A} \mu^i$, where $\mu^\emptyset$ is the only probability measure on $(\mathsf{X}_\emptyset, \mathcal{X}_\emptyset)$ by a convention. Recall that it is an $\mathcal{X}_A$-measurable function satisfying $P_A(\mathbb{A}) = \int_{x\in\mathbb{A}} f_A(x)\, \mathrm{d}\mu^A(x)$ for any $\mathbb{A} \in \mathcal{X}_A$. The marginal density f_A can be understood as a function on the joint sample space X_N depending only on the marginal configuration x_A. The joint and marginal densities are determined uniquely in the sense of μ-everywhere.

1.3 The Concept of Conditional Independence

In this section, several equivalent definitions of probabilistic CI in the discrete case are presented; the general case is discussed in the end of the section.

1.3.1 Conditional independence in the discrete case

In this section, a number of equivalent definitions of probabilistic CI are given and illustrated by two examples. Our attention will intentionally be restricted to the discrete case in order to keep the text accessible to beginners.

The following symmetric definition of CI was chosen as the basic one because it is analogous to the definition of stochastic independence, which is characterized by the requirement that the joint distribution is the product of marginal ones.

Definition 1.3.1. Let $A, B, C \subseteq N$ be pairwise disjoint sets of variables and P a discrete probability measure over N. We say that *A and B are conditionally independent given C with respect to P* and write $A \perp\!\!\!\perp B \mid C\ [P]$ if

$$\forall\, \mathbb{A} \subseteq \mathsf{X}_A \quad \forall\, \mathbb{B} \subseteq \mathsf{X}_B \quad \forall\, c \in \mathsf{X}_C \ \text{ such that } \ p_C(c) > 0$$
$$P_{AB|C}(\mathbb{A} \times \mathbb{B}|c) = P_{A|C}(\mathbb{A}|c) \cdot P_{B|C}(\mathbb{B}|c). \tag{1.1}$$

It follows from the definition that the validity of $A \perp\!\!\!\perp B \mid C\ [P]$ only depends on the marginal measure P_{ABC}. Clearly, a modified formulation of (1.1) is that, for each positive configuration $c \in \mathsf{X}_C$, the conditional probability $P_{AB|C}(*|c)$ is the product of some measures over A and B. The condition (1.1) has a natural interpretation of *conditional irrelevance*: once the value $c \in \mathsf{X}_C$ for C is known, the variables in A and B do not influence each other, i.e., the occurrence of a value $b \in \mathsf{X}_B$ does not influence the probability of occurrence of $a \in \mathsf{X}_A$, and vice versa. Also, (1.1) can be extended to a general case, as explained in § 1.3.2. On the other hand, (1.1) is not suitable for verification.

Fortunately, there are elegant equivalent conditions given in terms of densities. Specifically, given pairwise disjoint $A, B, C \subseteq N$ and a discrete probability measure P over N, the CI statement $A \perp\!\!\!\perp B \mid C\ [P]$ has the following equivalent formulation in terms of *marginal densities*:

$$\forall\, x \in \mathsf{X}_{ABC} \qquad p_C(x_C) \cdot p_{ABC}(x) = p_{AC}(x_{AC}) \cdot p_{BC}(x_{BC}), \tag{1.2}$$

which easily implies a seemingly weaker condition

$$\forall\, x \in \mathsf{X}_{ABC} \text{ with } p_{ABC}(x) > 0 \qquad p_{ABC}(x) = \frac{p_{AC}(x_{AC}) \cdot p_{BC}(x_{BC})}{p_C(x_C)}. \tag{1.3}$$

Using the vanishing principle, the reader can easily see that (1.1)⇒(1.2)⇒(1.3); the implication (1.3)⇒(1.1) follows from the next fact.

Observation 1.3.1. There exists a probability measure $\bar{P}$ on X_{ABC} such that

$$\bar{P}_{AC} = P_{AC}, \quad \bar{P}_{BC} = P_{BC}, \text{ and } A \perp\!\!\!\perp B \mid C\ [\bar{P}].$$

The measure $\bar{P}$ is uniquely determined and satisfies $P_{ABC} \ll \bar{P}$.

Proof. We define the value $\bar{p}(x)$ of the density of $\bar{P}$ by the formula on the right-hand side of (1.3) for $x \in \mathsf{X}_{ABC}$ with $p_C(x_C) > 0$ and $\bar{p}(x) = 0$ in the case of $p_C(x_C) = 0$. The remaining statements are left to the reader as an exercise. □

Observation 1.3.1 even holds for any pair of discrete probability measures Q on X_{AC} and R on X_{BC} satisfying $Q_C = R_C$ in place of P_{AC} and P_{BC}. The measure $\bar{P}$ can then be called the *conditional product of Q and R*; this result implies that, for any such *consonant* pair of measures Q and R, a distribution P over ABC exists having them as marginals, namely $\bar{P}$.

To verify (1.3)⇒(1.1), we use the construction from the proof of Observation 1.3.1 and apply (1.3) to see that $\bar{p}(x) = p_{ABC}(x)$ in the case of $p_{ABC}(x) > 0$. Then we realize that the values of both $\bar{p}$ and p_{ABC} sum up to 1 to extend the equality $\bar{p}(x) = p_{ABC}(x)$ to the case of $p_{ABC}(x) = 0$.

Another CI characterization in terms of marginal densities appeared in [40]; it can be interpreted as a *cross-exchange condition* for configurations:

$$\forall\, a, \bar{a} \in \mathsf{X}_A,\ \ \forall\, b, \bar{b} \in \mathsf{X}_B,\ \ \forall\, c \in \mathsf{X}_C \ \text{ one has}$$
$$p_{ABC}(a, b, c) \cdot p_{ABC}(\bar{a}, \bar{b}, c) = p_{ABC}(a, \bar{b}, c) \cdot p_{ABC}(\bar{a}, b, c)\,. \tag{1.4}$$

To verify (1.2)⇒(1.4), we distinguish between the cases of $p_C(c) = 0$, when (1.4) is evident, and $p_C(c) > 0$. In the latter case, derive (1.4) whose sides are both multiplied by $p_C(c) \cdot p_C(c)$ from equalities (1.2) applied to $x = [a, b, c]$, $x = [\bar{a}, \bar{b}, c]$, $x = [\bar{a}, b, c]$, and $x = [a, \bar{b}, c]$. The implication (1.4)⇒(1.2) can be shown by summing over $\bar{a}$ and $\bar{b}$ in (1.4). The condition (1.4) is particularly easy to verify in the binary case, when $|\mathsf{X}_i| = 2$ for all $i \in N$.

Example 1.3.2. To illustrate the application of the above equivalent definitions of (discrete probabilistic) CI, let us take $N = \{a, b, c\}$, $\mathsf{X}_i = \{0, 1\}$ and introduce a binary probability measure P on X_N by its density p as follows:

$$p(0,0,0) = p(0,1,1) = p(1,0,1) = p(1,1,0) = \frac{1}{8} + \varepsilon\,,$$
$$p(0,0,1) = p(0,1,0) = p(1,0,0) = p(1,1,1) = \frac{1}{8} - \varepsilon\,,$$

for some $0 \le \varepsilon \le \frac{1}{8}$. We have $p_{ab}(0,0) = p_{ab}(0,1) = p_{ab}(1,0) = p_{ab}(1,1) = \frac{1}{4}$. No matter what the parameter ε is, the cross-exchange condition (1.4) holds:

$$p_{ab}(0,0) \cdot p_{ab}(1,1) = \frac{1}{4} \cdot \frac{1}{4} = \frac{1}{16} = \frac{1}{4} \cdot \frac{1}{4} = p_{ab}(0,1) \cdot p_{ab}(1,0)\,,$$

which means $a \perp\!\!\!\perp b \,|\, \emptyset\ [P]$ or, in a brief record, $a \perp\!\!\!\perp b\ [P]$. One can also test that using the condition (1.2): since one has $p_a(0) = p_a(1) = \frac{1}{2}$ and $p_b(0) = p_b(1) = \frac{1}{2}$, using the fact that $p_\emptyset(x_\emptyset) = 1$ for any $x \in \mathsf{X}_{ab}$ we have

$$p_\emptyset(x_\emptyset) \cdot p_{ab}(x_{ab}) = 1 \cdot \frac{1}{4} = \frac{1}{4} = \frac{1}{2} \cdot \frac{1}{2} = p_a(x_a) \cdot p_b(x_b)\,,$$

again showing $a \perp\!\!\!\perp b \,|\, \emptyset\ [P]$. On the other hand, for all $0 < \varepsilon \le \frac{1}{8}$, the cross-exchange condition (1.4) does not hold with $c = 0$:

$$p(0,0,0) \cdot p(1,1,0) = (\frac{1}{8} + \varepsilon)^2 \ne (\frac{1}{8} - \varepsilon)^2 = p(0,1,0) \cdot p(1,0,0)\,,$$

which means $a \not\perp\!\!\!\perp b \,|\, c\ [P]$. The density p is strictly positive except $\varepsilon = \frac{1}{8}$, which is one of the classic examples of interesting zero-admitting densities. Recall a traditional tale from probabilistic reasoning on how such a distribution can occur. Imagine that variables a and b describe the result of a simultaneous (independent) toss of two fair coins, with outcomes 0 and 1, and a witness rings a bell whenever different outcomes occur. The variable c has the value 1 if the bell rings, otherwise it has the value 0.

Factorization and other equivalent definitions

An elegant characterization of a CI statement is in terms of *factorization*:

$$\exists\, f : \mathsf{X}_{AC} \to \mathbb{R},\ \exists\, g : \mathsf{X}_{BC} \to \mathbb{R} \ \text{ such that} \quad \forall\, x \in \mathsf{X}_{ABC} \quad p_{ABC}(x) = f(x_{AC}) \cdot g(x_{BC}), \tag{1.5}$$

where the functions f and g are called *potentials*. To show (1.2)⇒(1.5), put $f = p_{AC}$ and $g(z) = \frac{p_{BC}(z)}{p_C(z_C)}$ in the case of $p_C(z_C) > 0$ and $g(z) = 0$ otherwise. To show (1.5)⇒(1.2) introduce marginal potentials $f_C(c) = \sum_{a \in \mathsf{X}_A} f(a,c)$, $g_C(c) = \sum_{b \in \mathsf{X}_B} g(b,c)$ for $c \in \mathsf{X}_C$ and observe by summing in (1.5) that $p_{AC} = f \cdot g_C$, $p_{BC} = f_C \cdot g$ and $p_C = f_C \cdot g_C$. Then substitute these equalities and (1.5) to both sides of (1.2). In comparison with the condition (1.3), the factorization condition (1.5) does not require the potentials to be expressed in terms of marginal densities, which makes (1.5) more suitable for verification.

The concept of CI is often introduced in terms of *conditional densities*. An elegant symmetric definition of CI in these terms is the following one:

$$\forall\, x \in \mathsf{X}_{ABC} \ \text{ such that } p_C(x_C) > 0, \text{ one has} \quad p_{AB|C}(x_{AB}|x_C) = p_{A|C}(x_A|x_C) \cdot p_{B|C}(x_B|x_C). \tag{1.6}$$

To see it is equivalent to the previous conditions observe that (1.2)⇒(1.6)⇒(1.3). Nevertheless, the most popular definition in the terms of conditional densities is the next asymmetric one, which basically says that the conditional distribution $P_{A|BC}$ *does not depend on the variables in* B:

$$\forall\, x \in \mathsf{X}_{ABC} \text{ with } p_{BC}(x_{BC}) > 0 \quad p_{A|BC}(x_A|x_{BC}) = p_{A|C}(x_A|x_C). \tag{1.7}$$

One can easily show that (1.2)⇒(1.7)⇒(1.3). The interpretation of the condition (1.7), which is common in the theory of Markov processes, is that the *future* A depends on the *past* B only through the *present* C. Of course, there are lots of modifications of this condition, for example that $p_{A|BC}(*|*)$ only depends on AC, but these modifications are omitted in this chapter.

Example 1.3.3. To illustrate the application of the factorization property, take another example of a discrete distribution with zero-admitting density. Put again $N = \{a,b,c\}$, $\mathsf{X}_i = \{0,1\}$ and introduce a probability measure P on X_N by its density p as follows:

$$p(0,0,0) = p(1,1,1) = \frac{1}{2}, \quad p(x) = 0 \text{ for remaining configurations } x \in \mathsf{X}_N.$$

To verify that $a \perp\!\!\!\perp b \,|\, c\ [P]$ holds using the condition (1.5), introduce functions $f : \mathsf{X}_{ac} \to \mathbb{R}$ and $g : \mathsf{X}_{bc} \to \mathbb{R}$ as follows:

$$f(0,0) = f(1,1) = \frac{1}{2}, \qquad f(0,1) = f(1,0) = 0,$$
$$g(0,0) = g(1,1) = 1, \qquad g(0,1) = g(1,0) = 0.$$

For $x \in \mathsf{X}_N$, one has $f(x_{ac}) \cdot g(x_{bc}) \neq 0$ iff either $x = (0,0,0)$ or $x = (1,1,1)$ and the value is $\frac{1}{2}$ then. Thus, (1.5) holds and, by symmetry argument, we observe that $i \perp\!\!\!\perp j \,|\, k\ [P]$ is true for any choice of distinct $i,j,k \in N$. On the other hand, one has $p_{ab}(0,0) = p_{ab}(1,1) = \frac{1}{2}$ and $p_{ab}(0,1) = p_{ab}(1,0) = 0$, which allows one to observe, using the cross-exchange condition (1.4), that $a \not\perp\!\!\!\perp b \,|\, \emptyset\ [P]$; hence, by symmetry, $i \not\perp\!\!\!\perp j \,|\, \emptyset\ [P]$ for any distinct $i,j \in N$.

1.3.2 More general CI concepts

This section, to be skipped by beginners, assumes that the reader is familiar with deeper notions of measure theory. Its aim is to explain how probabilistic CI is defined in terms of σ-algebras and how this abstract definition is reduced to the cases of general and marginally continuous probability measures.

A crucial concept is that of *conditional probability*, where the conditioning object is a σ-algebra. Let $\boldsymbol{P}$ be a probability measure on a measurable space $(\mathsf{X},\mathcal{X})$, $\mathcal{C}\subseteq\mathcal{X}$ a σ-algebra and $\tilde{\mathbb{A}}\in\mathcal{X}$ an event. A version of *conditional probability* of $\tilde{\mathbb{A}}$ given $\mathcal{C}$ (= conditioned by $\mathcal{C}$) is any $\mathcal{C}$-measurable function $h:\mathsf{X}\to[0,1]$, denoted by $\boldsymbol{P}[\tilde{\mathbb{A}}|\mathcal{C}]$, such that

$$\forall\,\tilde{\mathbb{C}}\in\mathcal{C}\qquad \boldsymbol{P}(\tilde{\mathbb{A}}\cap\tilde{\mathbb{C}}) = \int_{\tilde{\mathbb{C}}} h(x)\,\mathrm{d}\boldsymbol{P}(x) \equiv \int_{\tilde{\mathbb{C}}} \boldsymbol{P}[\tilde{\mathbb{A}}|\mathcal{C}](x)\,\mathrm{d}\boldsymbol{P}(x)\,. \tag{1.8}$$

It follows from the Radon-Nikodym theorem that such a function h exists and is unique in the sense of $\boldsymbol{P}_{\mathcal{C}}$-everywhere equality, where $\boldsymbol{P}_{\mathcal{C}}$ denotes the restriction of $\boldsymbol{P}$ to the measurable space $(\mathsf{X},\mathcal{C})$. One can introduce the concept of CI for σ-algebras as follows: given σ-algebras $\mathcal{A},\mathcal{B},\mathcal{C}\subseteq\mathcal{X}$, we say that $\mathcal{A}$ and $\mathcal{B}$ are *conditionally independent given* $\mathcal{C}$ and write $\mathcal{A}\perp\!\!\!\perp\mathcal{B}\,|\,\mathcal{C}$ if

$$\begin{aligned}&\forall\,\tilde{\mathbb{A}}\in\mathcal{A}\ \ \forall\,\tilde{\mathbb{B}}\in\mathcal{B}\\ &\qquad \boldsymbol{P}[\tilde{\mathbb{A}}\cap\tilde{\mathbb{B}}|\mathcal{C}](x) = \boldsymbol{P}[\tilde{\mathbb{A}}|\mathcal{C}](x)\cdot\boldsymbol{P}[\tilde{\mathbb{B}}|\mathcal{C}](x)\quad\text{for }\boldsymbol{P}_{\mathcal{C}}\text{-a.e }x\in\mathsf{X}\,.\end{aligned} \tag{1.9}$$

Note that the validity of (1.9) does not depend on the choice of particular versions of conditional probabilities; its equivalent formulation is the condition

$$\forall\,\tilde{\mathbb{A}}\in\mathcal{A}\quad\text{there exists }\mathcal{C}\text{-measurable version of }\boldsymbol{P}[\tilde{\mathbb{A}}|\mathcal{B}\vee\mathcal{C}]\,,$$

where $\mathcal{B}\vee\mathcal{C}$ is the σ-algebra generated by $\mathcal{B}\cup\mathcal{C}$; see [57, Lemma A.6]. This condition can be interpreted as an analogue of the discrete condition (1.7).

Let us now describe how the CI definition (1.9) works in the case of a (general) *probability measure P over N* mentioned in Definition 1.2.1. In this case, we put $(\mathsf{X},\mathcal{X}):=(\mathsf{X}_N,\bigotimes_{i\in N}\mathcal{X}_i)$, $\boldsymbol{P}:=P$. Recall from § 1.2.2 that, for $A\subseteq N$, $\mathcal{X}_A\equiv\bigotimes_{i\in A}\mathcal{X}_i$ denotes the product σ-algebra on X_A, with $\mathcal{X}_\emptyset\equiv\{\emptyset,\mathsf{X}_\emptyset\}$. It can be ascribed the respective *coordinate σ-algebra* $\mathcal{A}:=\{\mathbb{A}\times\mathsf{X}_{N\setminus A}\,:\,\mathbb{A}\in\mathcal{X}_A\}$ of subsets of $\mathsf{X}=\mathsf{X}_N$; one then has $\mathcal{A}\subseteq\mathcal{X}$.

Given disjoint $A,C\subseteq N$, let $\mathcal{C}$ denote the coordinate σ-algebra for $\mathcal{X}_C$. Any event $\mathbb{A}\in\mathcal{X}_A$ can be ascribed its cylindrical extension $\tilde{\mathbb{A}}:=\mathbb{A}\times\mathsf{X}_{N\setminus A}$; the conditional probability $x\in\mathsf{X}_N\mapsto\boldsymbol{P}[\tilde{\mathbb{A}}|\mathcal{C}](x)$ then depends on x_C and can be identified with an $\mathcal{X}_C$-measurable function on X_C, to be denoted by $c\in\mathsf{X}_C\mapsto P_{A|C}(\mathbb{A}|c)$. Thus, (1.8) allows one to introduce the concept of *conditional probability on X_A given C* as a function $P_{A|C}:\mathcal{X}_A\times\mathsf{X}_C\to[0,1]$ of two arguments such that, for any $\mathbb{A}\in\mathcal{X}_A$, the function $c\in\mathsf{X}_C\mapsto P_{A|C}(\mathbb{A}|c)$ is $\mathcal{X}_C$-measurable and satisfies

$$P_{AC}(\mathbb{A}\times\mathbb{C}) = \int_{\mathbb{C}} P_{A|C}(\mathbb{A}|c)\,\mathrm{d}P_C(c)\quad\text{for any }\mathbb{C}\in\mathcal{X}_C\,.$$

Observe that this is a natural generalization of the concept from Definition 1.2.3. Given pairwise disjoint $A,B,C\subseteq N$, the condition $\mathcal{A}\perp\!\!\!\perp\mathcal{B}\,|\,\mathcal{C}$ from (1.9) turns into the requirement

$$\begin{aligned}&\forall\,\mathbb{A}\in\mathcal{X}_A\ \ \forall\,\mathbb{B}\in\mathcal{X}_B\\ &\qquad P_{AB|C}(\mathbb{A}\times\mathbb{B}|c) = P_{A|C}(\mathbb{A}|c)\cdot P_{B|C}(\mathbb{B}|c)\quad\text{for }P_C\text{-a.e. }c\in\mathsf{X}_C\end{aligned}$$

which directly generalizes (1.1) and can be considered as a definition of the CI statement $A\perp\!\!\!\perp B\,|\,C\;[P]$ in the case of a (general) measure P over N.

In the case of a *marginally continuous* measure P over N (see § 1.2.2) one can introduce CI in terms of marginal densities. Specifically, it was shown in [57, Lemma 2.4] that, provided a dominating system of measures μ^i on $(\mathsf{X}_i, \mathcal{X}_i)$, $i \in N$, is fixed, one has $A \perp\!\!\!\perp B \,|\, C \;[P]$ for pairwise disjoint $A, B, C \subseteq N$ iff

$$f_C(x_C) \cdot f_{ABC}(x_{ABC}) \;=\; f_{AC}(x_{AC}) \cdot f_{BC}(x_{BC}) \quad \text{for } \mu\text{-a.e. } x \in \mathsf{X}_N,$$

where f_D, $D \subseteq N$, denotes the marginal density for D. This condition generalizes (1.2) and one can also generalize the other equivalent conditions from § 1.3.1 in terms of densities. For example, (1.5) takes the form: there exist $\mathcal{X}_{AC}$-measurable $h : \mathsf{X}_{AC} \to \mathbb{R}$ and $\mathcal{X}_{BC}$-measurable $g : \mathsf{X}_{BC} \to \mathbb{R}$ such that

$$f_{ABC}(x) = h(x_{AC}) \cdot g(x_{BC}) \quad \text{for } \mu\text{-a.e. } x \in \mathsf{X}_N.$$

1.4 Basic Properties of Conditional Independence

In this section, we introduce (probabilistic) CI structures and recall their basic formal properties. We also relate formal CI models to classic statistical models.

1.4.1 Conditional independence structure

A *disjoint triplet over* N is an ordered triplet $A, B, C \subseteq N$ of pairwise disjoint subsets of N. Notation $\langle A, B|C\rangle$ will be used to indicate the intended interpretation of such a triplet as a formal statement that the variables in A are independent of/dependent on the variables in B conditionally the variables in C. The system of all disjoint triplets over N will be denoted by $\mathcal{T}(N)$.

A *formal independence model over* N is a subset $\mathcal{M}$ of $\mathcal{T}(N)$, whose elements are interpreted as independence statements. We write $A \perp\!\!\!\perp B \,|\, C \;[\mathcal{M}]$ to indicate that $\langle A, B|C\rangle \in \mathcal{M}$ is interpreted as an independence statement and $A \not\perp\!\!\!\perp B \,|\, C$ if $\langle A, B|C\rangle$ is interpreted as a dependence statement.

The *conditional independence structure* induced by a probability measure P over N is a formal independence model (over N) composed of those triplets which represent valid CI statements with respect to P:

$$\mathcal{M}_P := \{\, \langle A, B|C\rangle \in \mathcal{T}(N) : \;\; A \perp\!\!\!\perp B \,|\, C \;[P] \,\}.$$

Not every formal independence model is a CI structure. The next proposition presents basic formal properties of CI structures.

Observation 1.4.1. Let P be a probability measure over N. Then one has for (pairwise disjoint) $A, B, C, D \subseteq N$:

(i) $\emptyset \perp\!\!\!\perp B \,|\, C \;[P]$,

(ii) $A \perp\!\!\!\perp B \,|\, C \;[P] \;\Leftrightarrow\; B \perp\!\!\!\perp A \,|\, C \;[P]$,

(iii) $A \perp\!\!\!\perp BD \,|\, C \;[P] \;\Leftrightarrow\; \{\, A \perp\!\!\!\perp D \,|\, C \;[P] \;\;\&\;\; A \perp\!\!\!\perp B \,|\, DC \;[P] \,\}$.

Moreover, if P has a strictly positive density then

(iv) $\{\, A \perp\!\!\!\perp B \,|\, DC \;[P] \;\;\&\;\; A \perp\!\!\!\perp D \,|\, BC \;[P] \,\} \;\Rightarrow\; A \perp\!\!\!\perp BD \,|\, C \;[P]$.

Recall that a discrete measure P on X_N has a (strictly) positive density if $p(x) > 0$ for all $x \in \mathsf{X}_N$. In the general case (see § 1.2.2) a measure P over N has a positive density if it is marginally continuous and a dominating system μ^i, $i \in N$, of σ-finite measures exists such that $\mu \equiv \bigotimes_{i\in N} \mu^i \ll P$.

The proof of Observation 1.4.1 given below is intentionally restricted to the discrete case so that beginners can understand it fully. A reader familiar with calculus of densities can modify the proof to cover the marginally continuous case (see § 1.2.2), provided that he/she is familiar with peculiarities of the almost-everywhere equality of densities. Nevertheless, in a general case of CI in terms of σ-algebras (see § 1.3.2), deeper measure-theoretical considerations are needed to derive the result; see [57, § A.7].

Proof. We assume the discrete case throughout the proof. To verify (i), we use (1.1) and realize that in the case of $A = \emptyset$ one has either $\mathbb{A} = \emptyset = \mathbb{A} \times \mathbb{B}$ or $\{\mathbb{A} \neq \emptyset \;\&\; \mathbb{A} \times \mathbb{B} = \mathbb{B}\}$. The condition (ii) is evident. To verify (iii), we combine (1.2) and (1.3). For the implication $A \perp\!\!\!\perp BD \,|\, C \;\Rightarrow\; A \perp\!\!\!\perp D \,|\, C$, we use (1.2): the summation over B-configurations in $p_C \cdot p_{ABDC} = p_{AC} \cdot p_{BDC}$ gives $p_C \cdot p_{ADC} = p_{AC} \cdot p_{DC}$. As concerns $A \perp\!\!\!\perp BD \,|\, C \Rightarrow A \perp\!\!\!\perp B \,|\, DC$, we multiply the above-derived equalities (the latter with swapped sides) to get

$$p_C \cdot p_{ABDC} \cdot p_{AC} \cdot p_{DC} = p_{AC} \cdot p_{BDC} \cdot p_C \cdot p_{ADC} \,.$$

Because canceling is possible here for positive $ABDC$-configurations, one gets

$$\forall p_{ABDC} > 0 \quad p_{ABDC} \cdot p_{DC} = p_{ADC} \cdot p_{BDC} \,,$$

which is, by (1.3), $A \perp\!\!\!\perp B \,|\, DC$. The proof of the converse, that is, the implication $\{A \perp\!\!\!\perp D \,|\, C \;\&\; A \perp\!\!\!\perp B \,|\, DC\} \Rightarrow A \perp\!\!\!\perp BD \,|\, C$, is analogous.

To verify $\{A \perp\!\!\!\perp B \,|\, DC \;\&\; A \perp\!\!\!\perp D \,|\, BC\} \Rightarrow A \perp\!\!\!\perp BD \,|\, C$ in (iv), we use (1.3) for both CI statements and get by canceling (because of $p_{BDC} > 0$):

$$\frac{p_{ADC} \cdot p_{BDC}}{p_{DC}} = p_{ABDC} = \frac{p_{ABC} \cdot p_{BDC}}{p_{BC}} \quad \Rightarrow \quad \frac{p_{ADC}}{p_{DC}} = \frac{p_{ABC}}{p_{BC}} \,.$$

Choose and fix a configuration $b \in \mathsf{X}_B$ and write

$$\forall\, [a,d,c] \in \mathsf{X}_{ADC} \qquad p_{A|DC}(a|d,c) = \frac{p_{ADC}(a,d,c)}{p_{DC}(d,c)} = \frac{p_{ABC}(a,b,c)}{p_{BC}(b,c)} \,,$$

which means that $p_{A|DC}$ does not depend on $d \in \mathsf{X}_D$. By condition (1.7), it follows that $A \perp\!\!\!\perp D \,|\, C \; [P]$. By (iii), this together with $A \perp\!\!\!\perp B \,|\, DC \; [P]$ implies $A \perp\!\!\!\perp BD \,|\, C \; [P]$. □

Note that the property in Observation 1.4.1(iv), called *intersection*, need not be valid for a discrete distribution P which is not strictly positive. Indeed, Example 1.3.3 shows that one can have $i \perp\!\!\!\perp j \,|\, k \; [P]$ for all distinct $i,j,k \in N$, while $i \not\perp\!\!\!\perp j \,|\, \emptyset \; [P]$; the latter implies $i \not\perp\!\!\!\perp \{j,k\} \,|\, \emptyset \; [P]$. Chapter 3 of this handbook analyzes the validity of the intersection property in more details.

Side-remark about relational databases

Formal independence models satisfying the conditions (i)-(iii) from Observation 1.4.1 occur also outside of statistics. There is one area in computer science where a concept analogous to the concept of probabilistic CI has been studied. It is the theory of *relational databases*, which is approximately 15 years older than probabilistic reasoning. The problem there is how to efficiently organize data in large data banks [7]. Researchers in this area became interested in special concepts of functional and multi-valued dependencies (in databases), which allowed them to reduce the memory demands, and tried to axiomatize them [3]. Moreover, there is a concept of *embedded multivalued dependency* (EMVD) [44], which is completely analogous to probabilistic CI and exhibits similar formal properties.

In this theory, the elements of N are called *attributes*, and each attribute $i \in N$ is ascribed a finite (individual) sample space X_i of possible values. A *relational database over* N is simply a set of configurations over N.

One can introduce natural operations with relational databases, some of which were already mentioned in § 1.2.1. Given $A \subseteq B \subseteq N$ and a relational database $\mathbb{D} \subseteq \mathsf{X}_B$ over B, the *projection* of $\mathbb{D}$ onto A is a relational database over A defined by $\mathbb{D}_A := \{b_A : b \in \mathbb{D}\}$. The second important operation is that of a combination, which is an analogue of the operation of conditional product for discrete probability measures from Observation 1.3.1. Specifically, given a disjoint triplet $\langle A, B|C\rangle$ over N and databases $\mathbb{D}^1 \subseteq \mathsf{X}_{AC}$, $\mathbb{D}^2 \subseteq \mathsf{X}_{BC}$ its *combination* is a relational database over ABC defined as follows:

$$\mathbb{D}^1 \bowtie \mathbb{D}^2 \;:=\; \{\, [a,b,c] \in \mathsf{X}_{ABC} \;:\; [a,c] \in \mathbb{D}^1 \;\&\; [b,c] \in \mathbb{D}^2 \,\}.$$

There is an analogy of the CI concept: given $\langle A, B|C\rangle \in \mathcal{T}(N)$ and a database $\mathbb{D}$ over N, we say that an *embedded multivalued dependency* (EMVD) statement $A \perp\!\!\!\perp B \,|\, C \; [\mathbb{D}]$ holds if $\mathbb{D}_{ABC} = \mathbb{D}_{AC} \bowtie \mathbb{D}_{BC}$, in words, if the projection of $\mathbb{D}$ onto ABC is the combination of its projections onto AC and BC.

We leave it to the reader to verify that the formal independence model induced by $\mathbb{D}$ satisfies the conditions (i)-(iii) from Observation 1.4.1.

1.4.2 Statistical model of a CI structure

The aim of this section is to explain that formal independence models can be interpreted as common statistical models. Recall that by a (mathematical) *statistical model* is meant a class of probability measures $\mathbb{M}$ on a prescribed sample space, which is a measurable space $(\mathsf{X}, \mathcal{X})$. In multivariate statistical analysis, one usually has a *joint sample space* $(\mathsf{X}_N, \mathcal{X}_N)$ in the place of $(\mathsf{X}, \mathcal{X})$.

Typically, a statistical model $\mathbb{M}$ is a parameterized class of measures and all of them are absolutely continuous with respect to some given σ-finite measure μ on $(\mathsf{X}, \mathcal{X})$, which is a product measure $\mu = \bigotimes_{i\in N} \mu^i$ in the case of $(\mathsf{X}_N, \mathcal{X}_N)$. Each probability measure in $\mathbb{M}$ is then determined by its density with respect to μ and, quite often, they are assumed to be mutually absolutely continuous. The parameters usually belong to a convex subset $\Theta \subseteq \mathbb{R}^n$ for some $n \geq 1$.

Assume that a *distribution framework* is specified, that is, a collection Ψ of probability measures on the sample space is determined from which the probability measures in $\mathbb{M}$ should be chosen. For example, in the discrete case, Ψ could be the class of all measures with positive density, while in the continuous case with $\mathsf{X}_i = \mathbb{R}$ for $i \in N$, one can have the class of regular Gaussian distributions on $\mathbb{R}^N$ in the place of Ψ. Then, every formal independence model $\mathcal{M} \subseteq \mathcal{T}(N)$ over N can be ascribed a class of probability measures

$$\mathbb{M} = \{\, P \in \Psi \;:\; A \perp\!\!\!\perp B \,|\, C \; [P] \quad \text{whenever } \langle A, B|C\rangle \in \mathcal{M} \,\},$$

which can be called the *statistical model of CI structure* given by $\mathcal{M}$.

This concept generalizes the classic concept of a *graphical model* [68, 24]. Indeed, the reader can learn in § 1.7.1 that every UG G over N induces the class $\mathbb{M}_G$ of Markovian measures over N through a formal independence model $\mathcal{M}_G$ induced by G. In general, statistical models of CI structures are very complicated; however, graphical models provide a subclass of nice models.

1.5 Semi-graphoids, Graphoids, and Separoids

The notions discussed in this section have been inspired by the research on stochastic CI, but they rather belong to the area of discrete mathematics. Pearl and Paz [43] introduced the following concept in 1987.

Definition 1.5.1. A *disjoint semi-graphoid over* N is a formal independence model $\mathcal{M}$ over N satisfying the following conditions/axioms:

$\emptyset \perp\!\!\!\perp B \,|\, C \; [\mathcal{M}]$ triviality,

$A \perp\!\!\!\perp B \,|\, C \; [\mathcal{M}] \;\Rightarrow\; B \perp\!\!\!\perp A \,|\, C \; [\mathcal{M}]$ symmetry,

$A \perp\!\!\!\perp BD \,|\, C \; [\mathcal{M}] \;\Rightarrow\; A \perp\!\!\!\perp B \,|\, DC \; [\mathcal{M}]$ weak union,

$A \perp\!\!\!\perp BD \,|\, C \; [\mathcal{M}] \;\Rightarrow\; A \perp\!\!\!\perp D \,|\, C \; [\mathcal{M}]$ decomposition,

$A \perp\!\!\!\perp D \,|\, C \; [\mathcal{M}] \;\&\; A \perp\!\!\!\perp B \,|\, DC \; [\mathcal{M}] \;\Rightarrow\; A \perp\!\!\!\perp BD \,|\, C \; [\mathcal{M}]$ contraction.

A disjoint semi-graphoid $\mathcal{M}$ will be called a *graphoid* (over N) if it satisfies

$A \perp\!\!\!\perp B \,|\, DC \; [\mathcal{M}] \;\&\; A \perp\!\!\!\perp D \,|\, BC \; [\mathcal{M}] \;\Rightarrow\; A \perp\!\!\!\perp BD \,|\, C \; [\mathcal{M}]$ intersection.

Given $\mathcal{M} \subseteq \mathcal{T}(N)$, its *semi-graphoid closure* is the smallest semi-graphoid over N containing $\mathcal{M}$. The *graphoid closure of* $\mathcal{M}$ can be introduced analogously.

Semi/graphoid closures are well defined because set intersection of semi/graphoids over N is a semi/graphoid over N. The CI implications in Definition 1.5.1 are nothing else but detailed conditions from Observation 1.4.1, which basically says that every probabilistic CI structure is a disjoint semi-graphoid, and even a graphoid if the distribution has a positive density.

There are areas different from probability theory in which semi-graphoids have occurred. We have seen in the side-remark from § 1.4.1 that every relational database can be ascribed a disjoint semi-graphoid. The undirected separation criterion from § 1.7.1 allows one to ascribe a graphoid to every UG over N and the same holds for the directional separation criterion from § 1.8.1. Let us give three more examples; their verification is left to the reader.

A class of subsets: take $\mathcal{T} \subseteq \mathcal{P}(N) \equiv \{A : A \subseteq N\}$ and define

$$A \perp\!\!\!\perp B \,|\, C \; [\mathcal{T}] \;:=\; \forall\, T \in \mathcal{T} \quad T \subseteq ABC \;\Rightarrow\; [\,T \subseteq AC \text{ or } T \subseteq BC\,].$$

A natural conditional function: given a finite joint sample space X_N, this is a function $\kappa : \mathsf{X}_N \to \mathbb{Z}$ such that $\min\{\,\kappa(x) \,:\, x \in \mathsf{X}_N\} = 0$. Introduce a marginal (function) for any $A \subseteq N$ by the formula: $\kappa_A(y) := \min\{\,\kappa(y,z) \,:\, z \in \mathsf{X}_{N\setminus A}\,\}$ for any $y \in \mathsf{X}_A$. Define

$$\begin{aligned} A \perp\!\!\!\perp B \,|\, C \; [\kappa] \;:=\; & \forall\, x \in \mathsf{X}_N \\ & \kappa_C(x_C) + \kappa_{ABC}(x_{ABC}) = \kappa_{AC}(x_{AC}) + \kappa_{BC}(x_{BC})\,. \end{aligned}$$

Note that this is a concept taken over from [51].

A supermodular function: this is a set function $m : \mathcal{P}(N) \to \mathbb{R}$ such that $m(D \cup E) + m(D \cap E) \geq m(D) + m(E)$ for all $D, E \subseteq N$. Define

$$A \perp\!\!\!\perp B \,|\, C \; [m] \;:=\; m(C) + m(ABC) = m(AC) + m(BC)\,.$$

Note that semi-graphoids defined in this way coincide with *structural semi-graphoids* mentioned in § 1.10.1.

Some authors do not regard the restriction to disjoint triplets over N as necessary and consider a *general semi-graphoid* over N, which is a set of ordered triplets $A \perp\!\!\!\perp B \,|\, C$ of (not necessarily disjoint) subsets of N, which satisfies the following three conditions:

- $B \subseteq C \;\Rightarrow\; A \perp\!\!\!\perp B \,|\, C$,
- $A \perp\!\!\!\perp B \,|\, C \;\Leftrightarrow\; B \perp\!\!\!\perp A \,|\, C$,
- $A \perp\!\!\!\perp B \cup D \,|\, C \;\Leftrightarrow\; \{ A \perp\!\!\!\perp D \,|\, C \;\&\; A \perp\!\!\!\perp B \,|\, D \cup C \}$.

A general semi-graphoid is induced by a discrete probability measure P over N through the condition (1.2) where non-disjoint triplets are allowed. Then $A \perp\!\!\!\perp A \,|\, C\; [P]$ means that $\forall\, p_{AC} > 0$ one has $p_{AC} = p_C$, which corresponds to *functional dependency of A on C*; note that an axiomatic characterization of probabilistic functional dependency structures was given by Matúš [30]. Thus, general semi-graphoids are broader than disjoint semi-graphoids because they involve functional dependency relation modeling.

Dawid took an even more general point of view and introduced an abstract concept of a separoid; below we describe a simplification of his definition [13].

Definition 1.5.2. Let $\mathbb{S}$ be a joint semi-lattice, that is, a partially ordered set in which every two elements a, b have a supremum (= a join), denoted by $a \vee b$. A set of ordered triplets $a \perp\!\!\!\perp b \,|\, c$ of elements of $\mathbb{S}$ will be called a *separoid* if

- $b \vee c = c \;\Rightarrow\; a \perp\!\!\!\perp b \,|\, c$,
- $a \perp\!\!\!\perp b \,|\, c \;\Leftrightarrow\; b \perp\!\!\!\perp a \,|\, c$,
- $a \perp\!\!\!\perp b \vee d \,|\, c \;\Leftrightarrow\; \{ a \perp\!\!\!\perp d \,|\, c \quad \& \quad a \perp\!\!\!\perp b \,|\, d \vee c \}$.

Of course, every general semi-graphoid over N is a separoid on the lattice $(\mathcal{P}(N), \subseteq)$. Another prominent example requires for the reader to be familiar with measure theory: given a probability measure $\boldsymbol{P}$ on a measurable space $(\mathsf{X}, \mathcal{X})$, let $\mathbb{S}$ be the set of all σ-algebras contained in $\mathcal{X}$, ordered by inclusion. Then the ternary relation $\mathcal{A} \perp\!\!\!\perp \mathcal{B} \,|\, \mathcal{C}$ introduced in § 1.3.2 is a separoid.

1.5.1 Elementary and dominant triplets

To represent a (disjoint) semi-graphoid over N in the memory of a computer, one does not need all $|\mathcal{T}(N)| = 4^{|N|}$ bits.

Definition 1.5.3. A disjoint triplet $\langle A, B|C \rangle$ over N will be called *trivial* if either $A = \emptyset$ or $B = \emptyset$; it will be called *elementary* if $|A| = 1 = |B|$. The system of elementary triplets over N will be denoted by $\mathcal{T}_\epsilon(N)$.

Clearly, the trivial triplets can always be excluded from considerations because they are contained in any semi-graphoid. On the other hand, the elementary triplets are substantial because of the following fact.

Observation 1.5.1. Let $\mathcal{M}$ be a disjoint semi-graphoid over N. Then, for every disjoint triplet $\langle A, B|C \rangle \in \mathcal{T}(N)$, one has $A \perp\!\!\!\perp B \,|\, C\; [\mathcal{M}]$ iff

$$\forall i \in A \quad \forall j \in B \quad \forall K \text{ with } C \subseteq K \subseteq ABC \setminus \{i, j\} \qquad i \perp\!\!\!\perp j \,|\, K\; [\mathcal{M}]\,. \tag{1.10}$$

In particular, for two disjoint semi-graphoids $\mathcal{M}^1$ and $\mathcal{M}^2$ over N, one has $\mathcal{M}^1 \subseteq \mathcal{M}^2$ iff $\mathcal{M}^1 \cap \mathcal{T}_\epsilon(N) \subseteq \mathcal{M}^2 \cap \mathcal{T}_\epsilon(N)$, which implies that any semi-graphoid $\mathcal{M}$ is uniquely determined by its elementary trace $\mathcal{M} \cap \mathcal{T}_\epsilon(N)$.

Proof. The necessity of (1.10) can be easily derived using the decomposition and weak union properties combined with the symmetry property. For the converse implication, suppose that $\langle A, B|C\rangle$ is not trivial and use induction on $|AB|$; the instance $|AB| = 2$ is evident. Supposing $|AB| > 2$ either A or B is not a singleton. Owing to the symmetry property, one can – without the loss of generality – consider $|B| \geq 2$, choose $b \in B$ and put $B' = B \setminus \{b\}$. By the induction assumption, (1.10) implies both $A \perp\!\!\!\perp B' \,|\, C \; [\mathcal{M}]$ and $A \perp\!\!\!\perp b \,|\, B'C \; [\mathcal{M}]$. Thus, the contraction property gives $A \perp\!\!\!\perp B \,|\, C \; [\mathcal{M}]$. □

One can also easily show that $\mathcal{N} \subseteq \mathcal{T}_\epsilon(N)$ is a trace of a semi-graphoid iff the *symmetry* condition

$$i \perp\!\!\!\perp j \,|\, K \; [\mathcal{N}] \;\Leftrightarrow\; j \perp\!\!\!\perp i \,|\, K \; [\mathcal{N}]$$

and the *exchange* property

$$i \perp\!\!\!\perp j \,|\, kL \; [\mathcal{N}] \;\&\; i \perp\!\!\!\perp k \,|\, L \; [\mathcal{N}] \quad\Leftrightarrow\quad i \perp\!\!\!\perp k \,|\, jL \; [\mathcal{N}] \;\&\; i \perp\!\!\!\perp j \,|\, L \; [\mathcal{N}]$$

hold. Thus, the semi-graphoid closure can be described in terms of elementary triplets. Since $|\mathcal{T}_\epsilon(N)| = |N| \cdot (|N| - 1) \cdot 2^{|N|-2}$ it is enough to have $\binom{|N|}{2} \cdot 2^{|N|-2}$ bits to represent a semi-graphoid over N.

Matúš [35] was interested in the intricacy of the semi-graphoid inference between elementary CI statements and showed that the length of the derivation sequence can be exponential in $|N|$. Nonetheless, there is an alternative way to represent semi-graphoids in the memory of a computer.

Definition 1.5.4. We say that $\langle A, B|C\rangle \in \mathcal{T}(N)$ *dominates* $\langle A', B'|C'\rangle \in \mathcal{T}(N)$ if $A' \subseteq A$, $B' \subseteq B$ and $C \subseteq C' \subseteq ABC$. The triplets in a semi-graphoid which are maximal with respect to this partial order on $\mathcal{T}(N)$ are called *dominant.*

If one restricts oneself to non-trivial triplets then elementary triplets in a (fixed) semi-graphoid $\mathcal{M}$ are minimal with respect to the dominance ordering; thus, the dominant and elementary triplets are somehow opposite to each other. An alternative way to represent a semi-graphoid in the memory of a computer is by a list of its non-trivial (symmetrized) dominant triplets.

One can also implement the semi-graphoid and graphoid closures in these terms, as shown by Baioletti, Busanello and Vantaggi [2]. Dominant triplets were also employed as an useful tool in [55] to show that the semi-graphoid closure of two disjoint triplets over N is always a probabilistic CI structure. This fact can be interpreted as a result on *relative completeness* of semi-graphoid implications for probabilistic CI inference if the input list has at most 2 items (see § 1.11). Semi/graphoids over a fixed set N can also be classified according to their *semi/graphoid complexity,* by which we mean the minimal cardinality of a semi/graphoid generator [56].

For readers familiar with (advanced) polyhedral geometry, we mention two interesting equivalent geometric definitions/interpretations of the concept of a semi-graphoid, which were offered by Morton and his co-authors [37, 38]. Both equivalent geometric definitions come from the semi-graphoid description in terms of elementary triplets.

The first equivalent definition is related to a special polytope, called a *permutohedron,* which was previously introduced by Shouté in 1911 [46]. The idea is that all permutations over a set $N = \{1, 2, \ldots, n\} \equiv [n]$ are interpreted as vectors in $\mathbb{R}^N$ and their convex hull is taken. There is a certain standard way to label one-dimensional faces (= geometric edges) of this polytope by elementary triplets over N. Thus, $\mathcal{N} \subseteq \mathcal{T}_\epsilon(N)$ is identified with a set of geometric edges of the permutohedron. The two above-mentioned conditions on $\mathcal{N}$ characterizing a semi-graphoid then have an elegant geometric interpretation. Every two-dimensional face of the permutohedron is either a square or a regular hexagon. The

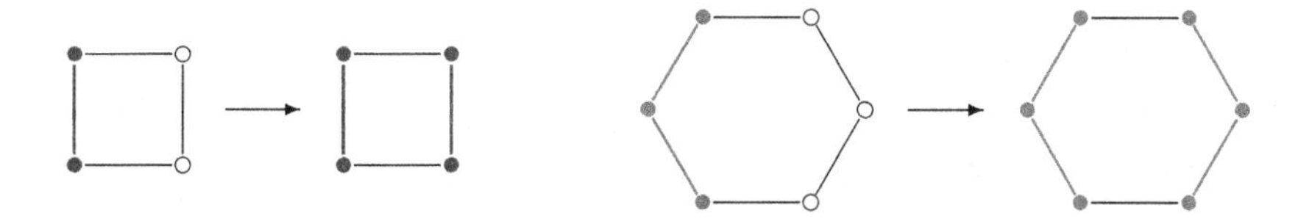

FIGURE 1.2: Illustration of the square and hexagon axioms.

symmetry condition can be interpreted as a *square axiom* requiring that if a geometric edge of a square belongs to $\mathcal{N}$ then so does the opposite edge. The exchange property corresponds to a *hexagon axiom*, which says that if a pair of touching edges of a hexagon belongs to $\mathcal{N}$ then the same holds for the pair of edges opposite to them in the hexagon; see Figure 1.2.

The second equivalent definition is in terms of (complete) *polyhedral fans*, which are certain collections of polyhedral cones covering $\mathbb{R}^N$. There is a prominent polyhedral fan induced by a special equivalence of vectors in $\mathbb{R}^N$, where $u, v \in \mathbb{R}^N$ are equivalent if for all $i, j \in N$ one has $u_i \leq u_j \Leftrightarrow v_i \leq v_j$. That fan is called the S_n*-fan* (for $n = |N|$) by Morton [38] or *braid arrangement* by other authors. Semi-graphoids are then in a one-to-one correspondence with polyhedral fans which coarsen the prominent S_n-fan.

1.6 Elementary Graphical Concepts

In this section we introduce basic graphical concepts to be used in the following three sections. Recall from § 1.2 that N is a generic symbol for a finite non-empty index set whose elements correspond to random variables and occur as nodes of graphs in a graphical context.

By a graph *over* N we will understand a graph which has N as the set of *nodes*. Graphs considered in this chapter have no multiple edges; and there are two possible types of their edges.

Undirected edges are unordered pairs of distinct nodes, that is, two-element subsets of N. We will write $i - j$ to denote an undirected edge between nodes i and j from N; a pictorial representation is analogous. An *undirected graph* (UG) is a graph whose edges are all undirected; if $i - j$ in an undirected graph G then we say that i and j are *neighbors* in G. The symbol $\mathrm{ne}_G(i) := \{j \in N : i - j \text{ in } G\}$ will denote the set of all neighbors of $i \in N$ in G. A set of nodes $A \subseteq N$ is *complete* in an UG G if $i - j$ in G is true for all distinct $i, j \in A$. Maximal complete sets in G with respect to the set-inclusion ordering are called *cliques* of G. An UG G over N is *complete* if N is complete in G.

Directed edges, also called *arrows*, are ordered pairs of distinct nodes. We will write $i \rightarrow j$ to denote an arrow from node i to node j in N; similarly in figures. A *directed graph* is a graph whose edges are all arrows. If $i \rightarrow j$ in a directed graph G then we say that i is a *parent* of j in G or, dually, that j is a *child* of i. The symbol $\mathrm{pa}_G(j) := \{i \in N : i \rightarrow j \text{ in } G\}$ will denote the set of all parents of $j \in N$ in G.

Given a graph G over N (either directed or undirected) and a non-empty set of nodes $T \subseteq N$, the *induced subgraph* of G for T, denoted by G_T, is a graph over T with just those edges in G which run between nodes of T.

A *walk* in a graph G over N (either directed or undirected) is a sequence of nodes $i_1, \ldots, i_k$, $k \geq 1$, such that each consecutive pair of nodes in the sequence is adjacent by an

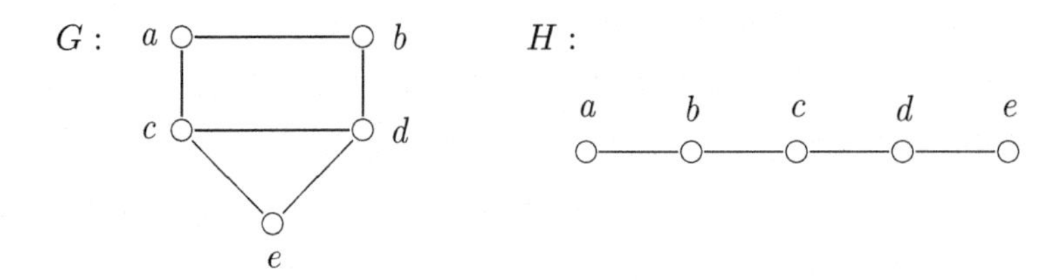

FIGURE 1.3: Two examples of undirected graphs.

edge in the graph G. The end-nodes of the walk are i_1 and i_k; if $k \geq 3$ then the remaining nodes i_ℓ, $1 < \ell < k$, are *internal nodes*. The number of edges in the walk, that is, $k-1$, is called the *length* of the walk. A walk in G is called a *path* if $i_1, \ldots, i_k$ are distinct; it is called a *cycle* if $k \geq 4$, $i_1 = i_k$ and $i_1, \ldots, i_{k-1}$ are distinct. In the case of a directed graph G, a path or a cycle is called *directed* if $i_\ell \to i_{\ell+1}$ for $\ell = 1, \ldots, k-1$.

A directed graph G is called *acyclic* if it contains no directed cycle. Directed graphs that are acyclic are conventionally called *directed acyclic graphs* (DAGs). A well-known equivalent characterization of a DAG is that it is such a directed graph G which admits an enumeration of nodes $i_1, \ldots, i_{|N|}$ which is *consonant* with the direction of arrows: that is, if $i_\ell \to i_k$ in G then $\ell < k$.

An important concept is that of a *chordal* (undirected) graph. It is an UG G such that each cycle in G of the length at least 4 has a *chord*, that is, an edge between nodes in the cycle which is not an edge forming the cycle. A well-known equivalent definition of a chordal graph G is that the cliques of G can be ordered into a sequence $C_1, \ldots, C_m$, $m \geq 1$, satisfying the *running intersection property*: $\forall\, k \geq 2$ exists $\ell < k$ such that $C_k \cap (\bigcup_{r<k} C_r) \subseteq C_\ell$.

1.7 Markov Properties for Undirected Graphs

This section contains some theoretical results concerning undirected graphical models, called *Markov networks* in the context of probabilistic reasoning [42].

1.7.1 Global Markov property for an UG

Given an undirected graph G over N and a disjoint triplet $\langle A, B|C\rangle \in \mathcal{T}(N)$, we say that A and B are *separated* by C in G and write $A \perp\!\!\!\perp B \,|\, C\ [G]$ if every walk in G from a node in A to a node in B contains a node in C. Of course, this is equivalent to an identical condition with paths in place of walks. Another formulation is that after the removal of the set of nodes in C (including the edges leading to those nodes) there is no path between A and B; that is, no connected component of the induced graph $G_{N\setminus C}$ meets both A and B.

To illustrate the concept of (undirected graphical) separation, consider the graphs G and H in Figure 1.3. Clearly $A = \{a\}$ and $B = \{e\}$ are not separated by $C = \{c\}$ in G because of the path $a - b - d - e$, which avoids $C = \{c\}$. But they are separated by $C = \{c, d\}$, which means that $a \perp\!\!\!\perp e \,|\, cd\ [G]$. One can also easily observe that $a \perp\!\!\!\perp e \,|\, cd\ [H]$.

Every undirected graph G over N induces a formal independence model over N by means of the *undirected separation* criterion

$$\mathcal{M}_G = \{\, \langle A, B|C\rangle \in \mathcal{T}(N) \,:\, A \perp\!\!\!\perp B \,|\, C\ [G]\,\},$$

which appears to be a (disjoint) graphoid. A probability measure P over N with $\mathcal{M}_G \subseteq \mathcal{M}_P$ is then called *Markovian* with respect to G; in an alternative terminology, P satisfies the *global Markov property* relative to G:

(G) if A and B are separated by C in G then $A \perp\!\!\!\perp B \mid C\ [P]$.

The (statistical) *undirected graphical model* $\mathbb{M}_G$ then consists of Markovian distributions with respect to G. As explained in § 1.4.2, the class $\mathbb{M}_G$ can be interpreted as the statistical model of the CI structure given by $\mathcal{M}_G$.

A probability measure P over N is called *perfectly Markovian* with respect to G if $\mathcal{M}_G = \mathcal{M}_P$. The existence of a discrete perfectly Markovian measure with respect to any given UG G was shown by Geiger and Pearl in [19, Theorem 11]. In particular, $\mathcal{M}_G$ is indeed a probabilistic CI structure for any UG G and the statistical model $\mathbb{M}_G$ is non-empty (in the case of non-degenerate sample spaces X_i, $i \in N$). Another related result says that formal independence models induced by UGs can be described in an axiomatic way, that is, they are characterized in terms of finitely many implications [43].

1.7.2 Local and pairwise Markov properties for an UG

Verification whether a probability measure over N is Markovian with respect to an UG over N can be difficult because of the number of CI statements to be tested, which may be very high. Nevertheless, in the case of a measure with a (strictly) positive density certain reasonable sufficient conditions exist.

We say that a probability measure P over N satisfies the *local/pairwise Markov property* relative to G if

(L) for all $i \in N$ $\qquad i \perp\!\!\!\perp N \setminus (i \cup \mathrm{ne}_G(i)) \mid \mathrm{ne}_G(i)\ [P]$,

(P) for all distinct $i, j \in N$ with $\neg(i - j \text{ in } G)$ $\qquad i \perp\!\!\!\perp j \mid N \setminus \{i, j\}\ [P]$.

It is easy to verify, using Observation 1.4.1(iii), that (G)$\Rightarrow$(L)$\Rightarrow$(P); however, examples are available in which (P)$\not\Rightarrow$(L)$\not\Rightarrow$(G) for discrete distributions [24].

Observation 1.7.1. Assume that a probability measure P over N has a strictly positive density. Then one has (G)$\Leftrightarrow$(L)$\Leftrightarrow$(P) for P.

Proof. The key fact is the property specified in Observation 1.4.1(iv), which implies that the CI structure induced by G is a graphoid. Thus, it is enough to show that the graphoid closure of the set of triplets of the form $\langle i, j | N \setminus \{i, j\}\rangle$ for non-edges $i, j \in N$, $\neg(i - j \text{ in } G)$, contains the whole formal independence model $\mathcal{M}_G$. This observation is left to the reader as an exercise. □

Of course, it is clear from the presented proof that P need not necessarily have a strictly positive density; it is sufficient for the CI structure induced by P to be a graphoid. There are weaker conditions [45] which ensure the validity of that assertion. To illustrate the above-mentioned concepts, let us again consider the UGs in Figure 1.3. The reader can easily check that the following lists of independencies are the respective requirements.

(L) for G:	(P) for G:	(L) for H:	(P) for H:
$a \perp\!\!\!\perp de \mid bc$	$a \perp\!\!\!\perp d \mid bce$	$a \perp\!\!\!\perp cde \mid b$	$a \perp\!\!\!\perp c \mid bde$
$b \perp\!\!\!\perp ce \mid ad$	$a \perp\!\!\!\perp e \mid bcd$	$b \perp\!\!\!\perp de \mid ac$	$a \perp\!\!\!\perp d \mid bce$
$c \perp\!\!\!\perp b \mid ade$	$b \perp\!\!\!\perp c \mid ade$	$c \perp\!\!\!\perp ae \mid bd$	$a \perp\!\!\!\perp e \mid bcd$
$d \perp\!\!\!\perp a \mid bce$	$b \perp\!\!\!\perp e \mid acd$	$d \perp\!\!\!\perp ab \mid ce$	$b \perp\!\!\!\perp d \mid ace$
$e \perp\!\!\!\perp ab \mid cd$		$e \perp\!\!\!\perp abc \mid d$	$b \perp\!\!\!\perp e \mid acd$
			$c \perp\!\!\!\perp e \mid abd$

Thus, by Observation 1.7.1, to check whether a probability measure P with a strictly positive density satisfies the global Markov property relative to G, it is enough to verify that four CI statements (P) for G are valid with respect to P. Analogously, as concerns H, five CI statements in (L) for H are enough to verify the global Markov property relative to H.

Note that the undirected *separation criterion* from § 1.7.1 was a result of a certain development in the theory of Markov fields, which stemmed from statistical physics. The authors, who had developed this theory in the 1970s, restricted their attention to strictly positive discrete probability distributions. Several types of Markov conditions were proposed in [40]: the original pairwise Markov property was strengthened to the local and global versions. The reader can ask whether one can possibly even strengthen the global Markov property. Note that it follows from the result on the existence of a perfectly Markovian positive discrete measure [19] that the global Markov property cannot be strengthened. Moreover, it also occurs to be the strongest possible Markov property within the framework of regular Gaussian measures.

1.7.3 Factorization property for an UG

There is another sufficient condition for the global Markov property, which does not demand for the distribution to have a positive density. Specifically, we say that a marginally continuous measure P over N *is factorized* according to an UG G over N if a dominating system of σ-finite measures μ^i, $i \in N$, exists such that, for the respective joint density f, one has

(F) there exists potentials $\psi_C : \mathsf{X}_C \to [0, \infty)$, $C \in \mathcal{C}_G$, with

$$f(x) = \prod_{C \in \mathcal{C}_G} \psi_C(x_C) \qquad \text{for } \mu\text{-a.e. } x \in \mathsf{X}_N\,,$$

where $\mathcal{C}_G$ denotes the collection of cliques of G.

Note that one always has (F)$\Rightarrow$(G); this observation can be derived from repeated application of the fact that the factorization condition (1.5) is an equivalent definition of CI; see [25, Proposition 1]. On the other hand, examples of discrete measures showing (G)$\not\Rightarrow$(F) exist [31]. Nevertheless, the conditions are quite often equivalent. The following result, whose proof is omitted, is known as the *Hammersley-Clifford theorem*, see [24, Theorem 3.9]. It is a very useful observation as discussed in Chapter 3 of this handbook.

Observation 1.7.2. Assume that a probability measure P over N has a strictly positive density. Then one has (F)$\Leftrightarrow$(G) for P.

1.8 Markov Properties for Directed Graphs

This section deals with directed acyclic graphical models, called *Bayesian networks* in the context of probabilistic reasoning [42].

1.8.1 Directional separation criteria

In the directed case, several separation criteria are available to decide whether a disjoint triplet is represented in a graph; however, these apparently different criteria are equivalent

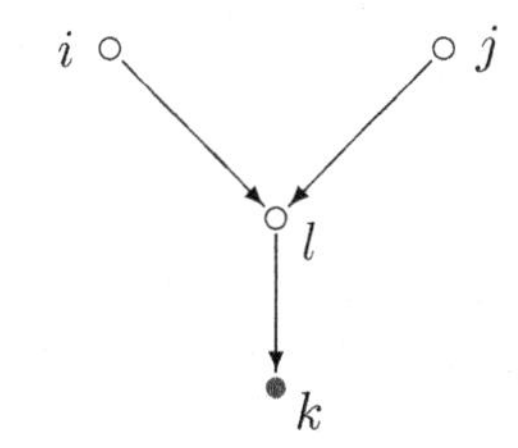

FIGURE 1.4: A simple example of a DAG.

with each other. They are described in this subsection, throughout which we assume that G is a directed graph over N; indeed, to introduce the criteria it is not substantial whether G is acyclic or not.

Straightforward criterion in terms of walks

Let us start with a straightforward separation criterion for walks, which is the simplest one. Let $\rho : i_1, \ldots, i_k$, $k \geq 1$, be a walk in G. We say that a node i_ℓ in ρ occurs as a *collider* in ρ if it is an internal node in ρ and $i_{\ell-1} \rightarrow i_\ell \leftarrow i_{\ell+1}$ in G. Other occurrences of nodes in ρ, including its end-nodes, are called *non-colliders.* We say that ρ is *interrupted* by a set of nodes $C \subseteq N$ if

either a node exists which occurs as a *non-collider* in ρ and *belongs to* C,

or a node exists which occurs as a *collider* in ρ and *is outside of* C.

A walk in G which is not interrupted by a set $C \subseteq N$ will be called *free* for C, or just briefly *C-free.* Thus, $\rho : i_1, \ldots, i_k$, $k \geq 1$, is C-free provided one has, for all i_ℓ, $1 \leq \ell \leq k$:

- if i_ℓ is a non-collider node occurrence in ρ then $i_\ell \notin C$,
- if i_ℓ is a collider node occurrence in ρ then $i_\ell \in C$.

Given $\langle A, B|C\rangle \in \mathcal{T}(N)$, we say that A and B are *directionally separated* by C in G if every *walk* in G from a node in A to a node in B is interrupted by C and write $A \perp\!\!\!\perp B \,|\, C\ [G]$ then.

Thus, the interrupting condition for non-colliders is the same as in the undirected case (see § 1.7.1), while the condition for colliders is completely converse. It also follows from the definition that if a walk has a node with occurrences of both a collider and a non-collider then it must be interrupted by any $C \subseteq N$.

Note that, when testing $A \perp\!\!\!\perp B \,|\, C\ [G]$, one has to consider all walks from A to B, not just paths. For example, the only path from i to j in the graph in Figure 1.4 is $i \rightarrow l \leftarrow j$ and this path is interrupted by the set $C = \{k\}$. Nevertheless, a walk $i \rightarrow l \rightarrow k \leftarrow l \leftarrow j$ exists in the graph which is C-free.

A natural question arises whether the walk-based criterion is decidable. Below we describe a propagation algorithm which, for given disjoint sets of nodes A and C, finds the set $\bar{A}$ of nodes to which a C-free walk exists from a node in A. Thus, if B is disjoint with $\bar{A} \cup C$, then directional separation $A \perp\!\!\!\perp B \,|\, C\ [G]$ holds, otherwise it does not. The algorithm can be viewed as a kind of modification of the *Bayes-ball* algorithm [47] by Shachter.

Input: Directed graph G over N; $A, C \subseteq N$ disjoint sets of nodes.
Auxiliary sets of nodes: $U, V, W \subseteq N$.
Put $U := A$, $V := \emptyset$, $W := \emptyset$.
Apply exhaustively the following three propagation rules:

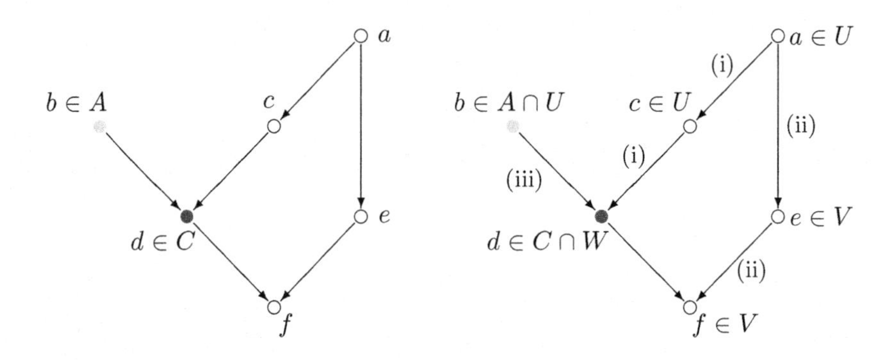

FIGURE 1.5: Illustration of the propagation algorithm for directed graphs.

(i) $w \in U \cup W,\, w \leftarrow u,\, u \notin C \quad \Rightarrow \quad u \in U,$

(ii) $u \in U \cup V,\, u \rightarrow v,\, v \notin C \quad \Rightarrow \quad v \in V,$

(iii) $u \in U \cup V,\, u \rightarrow w,\, w \in C \quad \Rightarrow \quad w \in W.$

Output: Put $\bar{A} := U \cup V$ when the algorithm terminates.

We leave to the reader to verify that the output of the algorithm is indeed the set $\bar{A}$. This follows from the interpretation of the auxiliary sets of nodes:

- U is the set of nodes u in N such that either $u \in A$ or there exists a C-free walk from A to u which ends by an arrow pointing *out of* u,
- V is the set of nodes v in N such that there exists a C-free walk from A to v which ends by an arrow pointing *into* v,
- W is the set of nodes w in C such that there exists a C-free walk from A to some $u \in \mathrm{pa}_G(w)$.

An example of application of the algorithm is in Figure 1.5: here we start with $A = \{b\}$ and $C = \{d\}$ and by consecutive application of (iii), (i) and (ii) get $U = \{a, b, c\}$, $V = \{e, f\}$ and $W = \{d\}$. Thus, $\bar{A} = U \cup V = \{a, b, c, e, f\}$.

D-separation criterion

Another option to solve the verification problem is to modify the criterion so that only paths are considered. This leads to a traditional directional criterion, often abbreviated as *d-separation criterion*, which was promoted by Pearl and his coauthors [42]. To formulate this criterion, one needs an additional graphical concept: if there exists a directed path in G from node $i \in N$ to node $j \in N$ then we say that i is an *ancestor* of j in G, or, dually, that j is a *descendant* of i in G. Note that any node is its own descendant.

Given a path $\rho : i_1, \ldots, i_k$, $k \geq 1$, in G and $C \subseteq N$ we say that ρ is *active* for C, briefly *C-active* if, for all i_ℓ, $1 \leq \ell \leq k$:

- if i_ℓ is a non-collider in ρ then $i_\ell \notin C$,
- if i_ℓ is a collider in ρ then i_ℓ has a descendant in C.

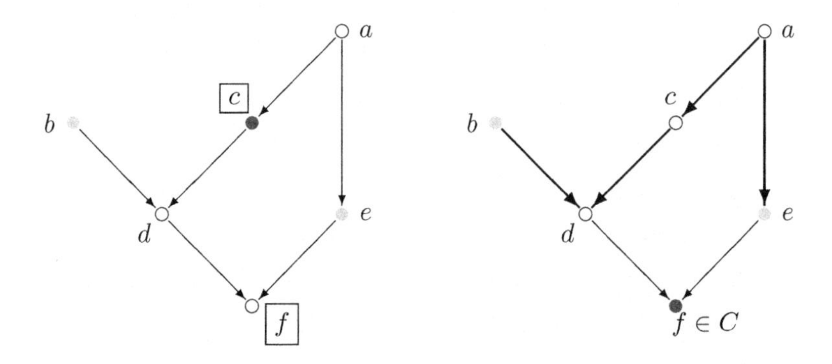

FIGURE 1.6: Illustration of the d-separation criterion for directed graphs.

A path which is not active with respect to C is *blocked by* C. Finally, $\langle A, B|C\rangle \in \mathcal{T}(N)$ is represented in G according to the *d-separation criterion*, if every *path* in G from A to B is blocked by C.

Note that, in the case of a path, any node occurs at most once in ρ and must be either a collider or a non-collider. The concept of a C-active path only slightly differs from that of a C-free walk: there is a weaker requirement in the case of collider nodes. To illustrate the application of d-separation criterion, consider the graph in Figure 1.6. If we test $\langle b, e|c\rangle$ then we can observe that any path from b to e either goes through a non-collider $c \in C$ and is blocked there or it goes through a collider f and is blocked there (because f is *not* an ancestor of c). In order to see that $\langle b, e|f\rangle$ is not represented, let us consider the path $b \to d \leftarrow c \leftarrow a \to e$, which is C-active because the only collider d is an ancestor of $f \in C$.

Moralization criterion

The *moralization criterion*, promoted by Lauritzen and his co-authors [24, 8], is not straightforward in the sense that the graph is modified during the test. This criterion is based on transformation of the directed graph into a certain UG and then using the undirected separation criterion. It has three steps:

1. one removes some nodes and gets an induced subgraph of the original graph, which is relevant for the tested triplet,

2. this induced subgraph is transformed to a certain UG over the same set of nodes, which is – for certain reasons – called the *moral graph*,

3. finally the undirected separation criterion from § 1.7.1 is applied to the tested triplet and the moral graph.

Because of the first step, the moral graphs assigned to different tested triplets may be different. To formulate the criterion, one also needs additional graphical concepts. Specifically, an *immorality* in G is an induced subgraph of G of the form $i \to k \leftarrow j$. The *moral graph* of a directed graph G over N is an UG G^{mor} over the same set of nodes N such that $i — j$ in G^{mor} if

either $[i, j]$ is an edge in the original graph G,

or there exists an immorality in G of the form $i \to k \leftarrow j$.

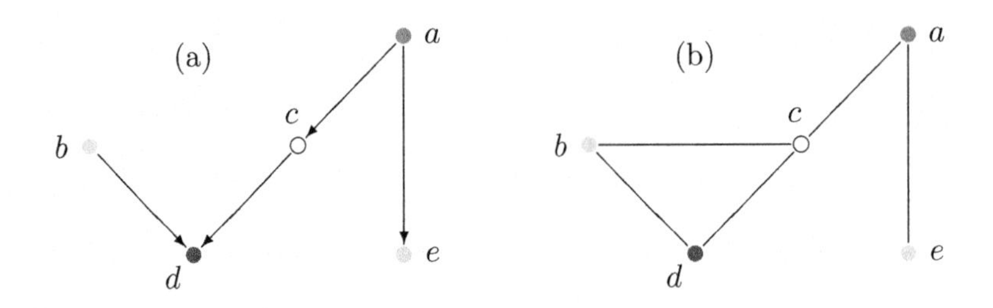

FIGURE 1.7: Illustration of the moralization criterion.

We say that a triplet $\langle A, B|C\rangle \in \mathcal{T}(N)$ is represented in G according to the *moralization criterion* if A and B are separated by C in the undirected graph $H = (G_{\mathrm{an}_G(ABC)})^{mor}$, where the symbol $\mathrm{an}_G(ABC)$ denotes the set of *ancestors* of nodes in ABC (see the text about d-separation).

To illustrate the application of the moralization criterion, consider the graph in Figure 1.6. To test $\langle b, e|ad\rangle$, we first transform the graph into the induced subgraph for the set of ancestors of nodes in $\{a, b, d, e\}$ in Figure 1.7(a) and then to the moral graph in Figure 1.7(b). We observe that every path from b to e in the moral graph goes through a. Thus, $\langle b, e|ad\rangle$ is represented in the graph according to the moralization criterion. Testing $\langle b, e|d\rangle$ leads to the same moral graph in Figure 1.7(b); but, this time, a path $b — c — a — e$ exists which avoids the set $C = \{d\}$. This implies that $\langle b, e|d\rangle$ is not represented.

The third option

There is another criterion based on transformation of the graph, in which some edges are removed instead of added. This criterion was suggested by Massey [29] and also independently by Darwiche in his book [11]. To test a triplet $\langle A, B|C\rangle \in \mathcal{T}(N)$ the following steps are made:

1. the induced subgraph for the ancestors of nodes in ABC is constructed (this is identical with the first step of the moralization criterion),
2. the subgraph is pruned by the removal of arrows outgoing from C,
3. if there is no path between A and B in the resulting directed graph then the triplet is represented according to this criterion.

To illustrate the criterion, let us again consider the graph in Figure 1.6 and test $\langle b, e|d\rangle$. The first step leads to the graph in Figure 1.7(a), and no arrow is removed from that graph in the second step. Since there is a path between b and e in it, the triplet is not represented. On the other hand, when $\langle b, e|ad\rangle$ is tested, the graph in Figure 1.7(a) is pruned in the second step by the removal of arrows $a \to c$ and $a \to e$. Thus, e is an isolated node in the resulting graph and the triplet is represented according to this special criterion.

Equivalence of directional criteria

The equivalence of d-separation and moralization criteria (in the case of a DAG) was shown in [24, Proposition 3.25]. The equivalence of the last criterion with d-separation was proved in [11, Theorem 4.1].

Nonetheless, all the criteria are mutually equivalent even in the case of a general directed graph. Thus, the following fact is left to the reader as an exercise. A hint is that one shows, for any of the three criteria, a directed graph G over N and $\langle A,B|C\rangle \in \mathcal{T}(N)$, that the triplet is *not represented* in G according to the respective criterion iff there exists a C-free walk between A and B, that is, $A \not\perp\!\!\!\perp B \,|\, C$ $[G]$.

Observation 1.8.1. Let G be a directed graph over N, and let $\langle A,B|C\rangle \in \mathcal{T}(N)$. Then $A \perp\!\!\!\perp B \,|\, C$ $[G]$ iff $\langle A,B|C\rangle \in \mathcal{T}(N)$ is represented in G according to any of the three above-mentioned path-based criteria.

1.8.2 Global Markov property for a DAG

Every directed acyclic graph G over N induces a formal independence model over N through the *directional separation* criterion

$$\mathcal{M}_G = \{\, \langle A,B|C\rangle \in \mathcal{T}(N) \,:\, A \perp\!\!\!\perp B \,|\, C \; [G] \,\},$$

which is a disjoint graphoid. A probability measure P over N with $\mathcal{M}_G \subseteq \mathcal{M}_P$ is called *Markovian* with respect to G and we also say that P satisfies the *directed global Markov property* relative to G:

(DG) if A and B are directionally separated by C in G then $A \perp\!\!\!\perp B \,|\, C$ $[P]$.

The statistical *directed graphical model* $\mathbb{M}_G$ consists of all Markovian measures with respect to G. The class $\mathbb{M}_G$ can be interpreted as the statistical model of the CI structure given by $\mathcal{M}_G$ (see § 1.4.2).

A probability measure P over N is called *perfectly Markovian* with respect to a DAG G if $\mathcal{M}_G = \mathcal{M}_P$. The existence of a perfectly Markovian measure with respect to any given DAG was shown by Geiger and Pearl [18].

Note that formal independence models induced by DAGs cannot be described completely in an axiomatic way. The reason is that these models are not closed under marginalization operation; see [57, Remark 3.5].

1.8.3 Local Markov property for a DAG

In the directed case, several variations of both local and pairwise Markov properties exist. One can distinguish between ordered versions, when an enumeration of nodes consonant with the direction of arrows is given, and the Markov property is relative to it on the one hand, and unordered versions on the other hand; see [8, § 5.3]. In this section, a basic unordered version of the local Markov property is presented.

An auxiliary graphical concept is needed to formulate this property. Recall from § 1.8.1 that a node j is a *descendant* of a node i in G if a directed path exists in G from i to j; denote the set of all descendants of node $i \in N$ in G by $\mathrm{ds}_G(i)$. Note that $i \in \mathrm{ds}_G(i)$.

A probability measure P over N satisfies a *directed local Markov property* relative to a DAG G over G if

(DL) for all $i \in N$ $\quad i \perp\!\!\!\perp N \setminus (\mathrm{ds}_G(i) \cup \mathrm{pa}_G(i)) \,|\, \mathrm{pa}_G(i)$ $[P]$.

Observation 1.8.2. For any probability measure P over N, (DG)$\Leftrightarrow$(DL).

Proof. Given any enumeration $i_1, \ldots, i_{|N|}$ of nodes which is consonant with the direction of arrows G, it was shown in [64] that $\mathcal{M}_G$ is the semi-graphoid closure of the list of triplets of the form $\langle i_\ell, \{i_1, \ldots, i_{\ell-1}\} \setminus \mathrm{pa}_G(i_\ell) \,|\, \mathrm{pa}_G(i_\ell)\rangle$, $\ell = 2, \ldots, |N|$. Hence, $\mathcal{M}_G$ can be shown to be the semi-graphoid closure of the set of triplets of the form $\langle i, N \setminus (\mathrm{ds}_G \cup \mathrm{pa}_G(i)) \,|\, \mathrm{pa}_G(i)\rangle$; use Observation 1.4.1. $\square$

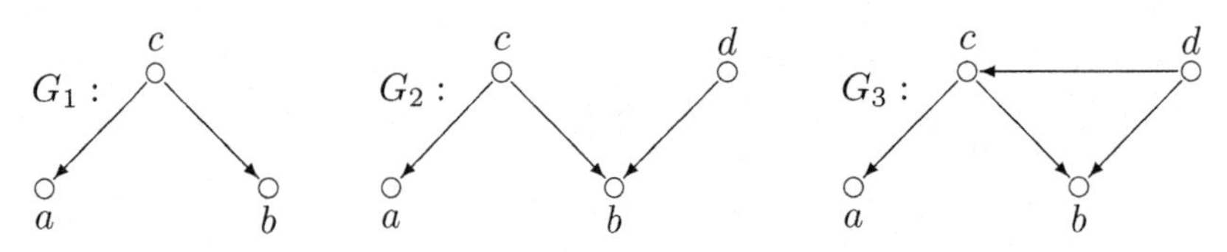

FIGURE 1.8: Illustration of the concept of a legally reversible arrow.

1.8.4 Factorization property for a DAG

Recursive factorization is a necessary and sufficient condition for a marginally continuous measure to be Markovian with respect to a *directed acyclic graph*. In the case of a discrete measure P over N it has the form

$$\textbf{(DF)} \qquad p(x) = \prod_{i \in N} p_{i|\,\mathrm{pa}_G(i)}(x_i | x_{\mathrm{pa}_G(i)}) \qquad \text{for every } x \in \mathsf{X}_N,$$

where a convention is accepted that $p_{A|C}(a|c) = 0$ whenever $p_C(c) = 0$ for $a \in \mathsf{X}_A$, $c \in \mathsf{X}_C$, $A, C \subseteq N$ disjoint.

The definition in the case of a marginally continuous measure is analogous, but one has to correctly introduce the conditional densities and the equation in (DF) is meant in the μ-a.e. sense, where μ is a dominating joint product measure. One can show that (DF)$\Leftrightarrow$(DG) then; see [25, Theorem 1].

Since the statistical model $\mathbb{M}_G$ for a DAG G coincides with the class of recursively factorizable distributions, there is a natural *parameterization* of this class in the discrete case; the elementary parameters are interpreted as (the values of) conditional probabilities [57, Lemma 8.1].

1.8.5 Markov equivalence for DAGs

We say that two DAGs G and H over N are *Markov equivalent* if they define the same statistical model, that is, $\mathbb{M}_G = \mathbb{M}_H$ (see § 1.8.2); note that this concept depends on the considered distribution framework Ψ (see § 1.4.2).

Analogously, two DAGs G and H over N are *independence equivalent* if they induce the same formal independence model: $\mathcal{M}_G = \mathcal{M}_H$; this notion, however, does not depend on the considered distribution framework. Clearly, independence equivalence implies Markov equivalence and the converse is also true provided that the distribution framework Ψ is non-degenerate [57, § 6.1]. For example, in the discrete case, non-degeneracy means that, for any $i \in N$, the individual sample space X_i has at least two elements. Thus, these two concepts of equivalence for DAGs typically coincide.

There could be different DAGs which are independence equivalent and a natural task is to characterize them graphically. A classic characterization of this kind was mentioned by Verma and Pearl [65]. One crucial concept here is that of *underlying undirected graph* of a DAG G, called alternatively a *skeleton* by some authors [1]: it is an UG over N in which an edge between a and b exists if either $a \to b$ in G or $a \leftarrow b$ in G. The second substantial concept is that of *immorality*: recall from § 1.8.1 that it is an induced subgraph of G of the form $i \to k \leftarrow j$. The classic graphical characterization says that two DAGs are independence equivalent iff they have the same skeleton and immoralities.

Nevertheless, there is also an indirect *transformational characterization* of equivalent DAGs proposed by Chickering [6], which often appears to be useful. It reveals an elementary graphical operation preserving independence equivalence of graphs. More specifically, given a DAG G over N, we say that an arrow $a \to b$ in G is *legally reversible* if the graph H obtained from G by replacing $a \to b$ by $a \leftarrow b$ in H (and keeping remaining arrows untouched) is also acyclic, and, moreover, independence equivalent to G. We say then that H is obtained from G by *legal arrow reversal.* The following fact is true.

Observation 1.8.3. Given a DAG G over N with an arrow $a \to b$, it is legally reversible iff $\mathrm{pa}_G(b) = \mathrm{pa}_G(a) \cup \{a\}$.

The proof can be found in [6, Lemma 1]. Note, however, that Chickering's terminology is different: he talks about a *covered edge* if the condition from Observation 1.8.3 holds. To illustrate the concept of a legally reversible arrow consider the DAGs in Figure 1.8. Both arrows in the graph G_1 are legally reversible because the parent set for c is empty and the other two nodes have c as the only parent node. If one modifies the graph by adding an arrow $d \to b$ (and the node d) then one gets the graph G_2 in which the arrow $c \to b$ will *not* be legally reversible. This can again be changed by adding a further arrow $d \to c$: then the arrow $c \to b$ will become legally reversible (in G_3).

The next observation, shown in [23, Lemma 4.2], gathers both graphical characterizations. Condition (B) is the classic direct characterization, while condition (C) is a transformational characterization.

Observation 1.8.4. The following three conditions are equivalent for any two given DAGs G and H over N:

(A) G and H are independence equivalent (that is, $\mathcal{M}_G = \mathcal{M}_H$),

(B) G and H have the same skeleton and immoralities,

(C) there exists a sequence $G = G_1, \ldots, G_m = H$, $m \geq 1$, of graphs over N, such that G_{i+1} is obtained from G_i by a legal arrow reversal for all $i = 1, \ldots, m-1$. (The graphs must be DAGs then.)

1.9 Remarks on Chordal Graphs

The class of *chordal undirected graphs* (see § 1.6 for the definition) plays a central role in graphical models. These graphs have widely been studied in graph-theoretical literature and a plenty of equivalent definitions/characterizations have been introduced; see, for example, [24, § 2.1].

A common alternative name is a decomposable graph, which is related to an equivalent definition of a chordal graph in terms of decompositions. Specifically, a non-trivial *decomposition* of an UG G over N is defined by a pair of sets $S, T \subseteq N$ such that

- $S \cup T = N$, $S \setminus T \neq \emptyset \neq T \setminus S$,
- $S \cap T$ is a complete set in G (see § 1.6),
- $S \setminus T \perp\!\!\!\perp T \setminus S \,|\, S \cap T\ [G]$ (see § 1.7.1).

Then G is decomposed into its induced subgraphs G_S and G_T. An UG G is chordal iff it is *decomposable*, which means that it is either complete or can be non-trivially decomposed into decomposable graphs (over smaller sets of nodes); see [24, Proposition 2.5].

The statistical models ascribed to decomposable graphs exhibit elegant properties. For example, an explicit closed form expression for the maximum likelihood estimate exists; see

[24, § 4.4.2]. There is an analogous formula for the joint density of (any globally) Markovian measure with respect to a chordal UG in terms of the marginal densities for its cliques; see [57, § 3.4.1].

Another related equivalent definition is the existence of a *junction tree* of its cliques; see [8, Theorem 4.6]. Junction trees then form a mathematical basis for miscellaneous effective computational methods which originate from the *local computation method* [26]. Some of the chapters in Part II of the handbook discuss the computation methods.

An interesting fact illustrating the mathematical beauty of these graphs is as follows: a formal independence model $\mathcal{M}$ is induced by a chordal graph (by undirected separation) iff it is a model induced by a certain UG (by undirected separation) and by a certain DAG (by directional separation). There is a finite axiomatization of such formal independence models found by de Campos [14].

1.10 Imsets and Geometric Views

In this section we mention the method of structural imsets, which offers a geometric point of view on the (description of) CI structures.

1.10.1 The concept of a structural imset

Although graphs offer an elegant and intuitive interpretation of some CI structures, they are not able to describe all possible probabilistic CI structures. This motivates a proposal for a non-graphical method of their description by means of vectors, whose components are integers indexed by subsets of N; such vectors are called *imsets*.

A starting point is the concept of an *elementary imset* from [57, § 4.2.1], which is a vector in $\mathbb{R}^{\mathcal{P}(N)}$ encoding the elementary CI statement $i \perp\!\!\!\perp j \,|\, K$ corresponding to $\langle i,j|K\rangle \in \mathcal{T}_\epsilon(N)$ (see § 1.5.1). Specifically, we put

$$u_{\langle i,j|K\rangle} \; := \; \delta_{ijK} + \delta_K - \delta_{iK} - \delta_{jK},$$

where $\delta_A \in \mathbb{R}^{\mathcal{P}(N)}$ denotes the zero-one vector identifier of a set $A \subseteq N$.

One can consider the cone $\mathcal{S}(N)$ in $\mathbb{R}^{\mathcal{P}(N)}$ of non-negative linear combinations of elementary imsets over N. *Structural imsets*, used to describe CI structures, can equivalently be introduced as vectors in $\mathcal{S}(N) \cap \mathbb{Z}^{\mathcal{P}(N)}$ [22]. There was an open problem whether every structural imset is also a *combinatorial imset*, that is, a combination of elementary imsets with non-negative integer coefficients. This is indeed true if $|N| \leq 4$ but Hemmecke et al. [21] gave an example of a structural imset over N with $|N| = 5$ which is not a combinatorial imset.

The next step is to ascribe a formal independence model over N to any structural imset u over N. There is a certain linear-algebraic criterion to decide, for each $\langle A,B|C\rangle \in \mathcal{T}(N)$, whether $A \perp\!\!\!\perp B \,|\, C \; [u]$ holds; this criterion is omitted in this chapter and can be found in [57, § 4.4.1]. The criterion can be viewed as an analogue of separation criteria used in graphical description of CI structures. The formal independence models

$$\mathcal{M}_u \; = \; \{\, \langle A,B|C\rangle \in \mathcal{T}(N) \,:\, A \perp\!\!\!\perp B \,|\, C \; [u] \,\} \qquad \text{for } u \in \mathcal{S}(N) \cap \mathbb{Z}^{\mathcal{P}(N)}$$

appear to be semi-graphoids, called *structural semi-graphoids*. Every such semi-graphoid is, in fact, induced by a combinatorial imset, which means that one can limit oneself to combinatorial imsets. Following the analogy with graphical models, one can introduce, for any structural imset u, the corresponding statistical model $\mathbb{M}_u$ of *Markovian distributions* P with respect to u satisfying $\mathcal{M}_u \subseteq \mathcal{M}_P$. Moreover, it was shown [57, Theorem 4.1] that, for marginally continuous measure P over N the Markov property with respect to a structural imset u is equivalent to a certain factorization property, which generalizes the recursive factorization for DAGs mentioned in § 1.8.4.

The crucial result concerning structural imsets is that, for any probability measure P over N with *finite multiinformation*, that is, with finite relative entropy of P with respect to $\bigotimes_{i\in N} P_i$, the CI structure induced by P is a structural semi-graphoid [57, Theorem 5.2]. In other words, any such distribution is *perfectly Markovian* with respect to some combinatorial imset u, which means that $\mathcal{M}_u = \mathcal{M}_P$. Note that all discrete measures and all regular Gaussian measures over N have finite multiinformation values.

Structural semi-graphoids also coincide with semi-graphoids ascribed to supermodular functions mentioned in § 1.5. A remark, which may interest a reader familiar with advanced polyhedral geometry, is that one can extend the observation according to which semi-graphoids correspond to polyhedral fans coarsening the S_n-fan (see § 1.5.1). Morton [37] also mentioned that a semi-graphoid is structural iff the corresponding polyhedral fan is a normal fan of a polytope.

1.10.2 Imsets for statistical learning

Imsets can also be applied in the context of learning Bayesian network (BN) structure. There is a certain standard translation of a DAG G over N into a combinatorial imset u_G, called the *standard imset* (for G), which has the property that the usual criteria for learning BN structure become affine functions (= sums of linear functions with constants) of the standard imset [61]. Thus, the learning task can be transformed into a *linear programming* problem; a mathematical task is then to characterize the domain in the form of finitely many linear inequalities.

It is sometimes advantageous in combinatorial optimization to work with zero-one vectors. Therefore, standard imsets were transformed by an affine invertible self-transformation of $\mathbb{Z}^{\mathcal{P}(N)}$ into *characteristic imsets*, which are zero-one vectors with an elegant graphical interpretation [20], and these vectors were applied to learning the BN structure by tools of integer linear programming [60]. This approach seems to be particularly suitable for learning decomposable models [59], in which case there is hope that the corresponding polytope will be completely characterized by linear inequalities.

1.11 CI Inference

This section is concerned with the following task: given an input list $\mathcal{L}$ of CI statements over N, characterize its probabilistic *CI closure*, which is the smallest CI structure containing $\mathcal{L}$. A traditional aim is to obtain the CI closure by applying interpretable formal CI implications, analogous to the semi-graphoid inference rules from Definition 1.5.1. Although there is no finite set of inference rules characterizing probabilistic CI inference [53], one can find such an axiomatic characterization in some special instances. The semi-graphoid implications are sufficient in the case of $|\mathcal{L}| = 2$ [55] or if $\mathcal{L}$ consists of special CI statements, such as the marginal CI statements $A \perp\!\!\!\perp B \,|\, \emptyset$ [17, 32] or saturated CI statements $A \perp\!\!\!\perp B \,|\, C$ with $ABC = N$ [28, 19].

Matúš [34] characterized the CI closure for discrete measures for $|N| = 4$; in this case 24 formal properties are enough [58]. Several methods to derive implications among CI statements can be used. The method of structural imsets [57, § 6.2] provides a sufficient condition for probabilistic CI implication; the respective linear-algebraic criterion can be tested using a computer [5]. The most efficient methods for computer testing of that linear-algebraic condition seem to be linear programming ones [4, 41]. On the other hand, there are linear-algebraic tools to derive CI implications based on different principles [62]. On top of that, advanced methods of modern algebra can be used to derive CI implications; Chapter 3 of this handbook gives more details on this topic.

Acknowledgments

I am indebted to Fero Matúš for his cooperation on the topic of CI. Our work has been supported from Grant Project GAČR n. 16-12010S. I would also like to express my thanks to the reviewers, whose comments helped me improve the quality of presentation. I am also grateful to Antonín Otáhal and Cheri Dohnal for correcting my English.

Bibliography

[1] S. A. Andersson, D. Madigan, and M. D. Perlman. A characterization of Markov equivalence classes for acyclic digraphs. *Ann. Statist.*, 25(2):505–541, 1997.

[2] M. Baioletti, G. Busanello, and B. Vantaggi. Conditional independence structure and its closure: inferential rules and algorithms. *Internat. J. Approx. Reason.*, 50(7):1097–1114, 2009.

[3] C. Beeri, R. Fagin, and J. H. Howard. A complete axiomatization for functional and multivalued dependencies in database relations. In *Proceedings of the 1977 ACM SIGMOD International Conference on Management of Data*, pages 47–61. ACM, 1977.

[4] R. Bouckaert, R. Hemmecke, S. Lindner, and M. Studený. Efficient algorithms for conditional independence inference. *J. Mach. Learn. Res.*, 11:3453–3479, 2010.

[5] R. R. Bouckaert and M. Studený. Racing algorithms for conditional independence inference. *Internat. J. Approx. Reason.*, 45(2):386–401, 2007.

[6] D. M. Chickering. A transformational characterization of equivalent Bayesian network structures. In *Uncertainty in Artificial Intelligence 11*, pages 87–98. Morgan Kaufmann, San Francisco, 1995.

[7] E. F. Codd. A relational model of data for large shared data banks. *Communications of the ACM*, 13:377–387, 1970.

[8] R. G. Cowell, A. P. Dawid, S. L. Lauritzen, and D. J. Spiegelhalter. *Probabilistic Networks and Expert Systems.* Springer, New York, 1999.

[9] F. G. Cozman and P. Walley. Graphoid properties of epistemic irrelevance and independence. *Ann. Math. Artif. Intell.*, 45(1/2):173–195, 2005.

[10] J. N. Darroch, S. L. Lauritzen, and T. P. Speed. Markov fields and log-linear interaction models for contingency tables. *Ann. Statist.*, 8(3):522–539, 1980.

[11] A. Darwiche. *Modeling and Reasoning with Bayesian Networks.* Cambridge University Press, New York, 2009.

[12] A. P. Dawid. Conditional independence in statistical theory. *J. R. Stat. Soc. Ser. B. Stat. Methodol.*, 41:1–31, 1979.

[13] A. P. Dawid. Separoids: a mathematical framework for conditional independence and irrelevance. *Ann. Math. Artif. Intell.*, 31(1/4):335–372, 2001.

[14] L. M. de Campos. Characterization of decomposable dependency models. *J. Artif. Intell. Res.*, 5:289–300, 1996.

[15] A. P. Dempster. Covariance selection. *Biometrics*, 28:157–175, 1972.

[16] M. Drton, B. Sturmfels, and S. Sullivant. *Lectures on Algebraic Statistics.* Birkhäuser, 2009.

[17] D. Geiger, A. Paz, and J. Pearl. Axioms and algorithms for inferences involving probabilistic independence. *Inform. and Comput.*, 91(1):128–141, 1991.

[18] D. Geiger and J. Pearl. On the logic of causal models. In *Uncertainty in Artificial Intelligence 4*, pages 3–14. North-Holland, Amsterdam, 1990.

[19] D. Geiger and J. Pearl. Logical and algorithmic properties of conditional independence and graphical models. *Ann. Statist.*, 21(4):2001–2021, 1993.

[20] R. Hemmecke, S. Lindner, and M. Studený. Characteristic imsets for learning Bayesian network structure. *Internat. J. Approx. Reason.*, 53:1336–1349, 2012.

[21] R. Hemmecke, J. Morton, A. Shiu, B. Sturmfels, and O. Wienand. Three counter-examples on semi-graphoids. *Combin. Probab. Comput.*, 17:239–257, 2008.

[22] T. Kashimura, T. Sei, A. Takemura, and K. Tanaka. Cones of elementary imsets and supermodular functions: a review and some new results. In *Proceedings of the 2nd CREST-SBM International Conference*, pages 357–363. World Scientific, 2012.

[23] T. Kočka, R. R. Bouckaert, and M. Studený. On characterizing inclusion of Bayesian networks. In *Uncertainty in Artificial Intelligence 17*, pages 261–268. Morgan Kaufmann, San Francisco, 2001.

[24] S. L. Lauritzen. *Graphical Models.* Clarendon Press, Oxford, 1996.

[25] S. L. Lauritzen, A. P. Dawid, B. N. Larsen, and H.-G. Leimer. Independence properties of directed Markov fields. *Networks*, 20(5):491–505, 1990.

[26] S. L. Lauritzen and D. J. Spiegelhalter. Local computation with probabilities on graphical structures and their application to expert systems. *J. R. Stat. Soc. Ser. B. Stat. Methodol.*, 50(2):157–224, 1988.

[27] M. Loève. *Probability Theory, Foundations, Random Processes.* Van Nostrand, Toronto, 1955.

[28] F. M. Malvestuto. A unique formal system for binary decomposition of database relations, probability distributions and graphs. *Inform. Sci.*, 59:21–52, 1992.

[29] J. L. Massey. Causal interpretation of random variables (in Russian). *Problemy Peredachi Informatsii*, 32(1):112–116, 1996.

[30] F. Matúš. Abstact functional dependency structures. *Theoret. Comput. Sci.*, 81:117–126, 1991.

[31] F. Matúš. On equivalence of Markov properties over undirected graphs. *J. Appl. Probab.*, 29(3):745–749, 1992.

[32] F. Matúš. Stochastic independence, algebraic independence and abstract connectedness. *Theoret. Comput. Sci. A*, 134(2):445–471, 1994.

[33] F. Matúš. Conditional independences among four random variables II. *Combin. Probab. Comput.*, 4(4):407–417, 1995.

[34] F. Matúš. Conditional independences among four random variables III., final conclusion. *Combin. Probab. Comput.*, 8(3):269–276, 1999.

[35] F. Matúš. Lengths of semigraphoid inferences. *Ann. Math. Artif. Intell.*, 35:287–294, 2002.

[36] F. Matúš and M. Studený. Conditional independences among four random variables I. *Combin. Probab. Comput.*, 4(4):269–278, 1995.

[37] J. Morton. *Geometry of Conditional Independence.* PhD thesis, University of California Berkeley, 2007.

[38] J. Morton, L. Pachter, A. Shiu, B. Sturmfels, and O. Wienand. Convex rank tests and semigraphoids. *SIAM J. Discrete Math.*, 23(3):1117–1134, 2009.

[39] M. Mouchart and J.-M. Rolin. A note on conditional independence with statistical applications. *Statistica*, 44(4):557–584, 1984.

[40] J. Moussouris. Gibbs and Markov properties over undirected graphs. *J. Stat. Phys.*, 10(1):11–31, 1974.

[41] M. Niepert, M. Gyssens, B. Sayrafi, and D. van Gucht. On the conditional independence implication problem: a lattice-theoretic approach. *Artificial Intelligence*, 202:29–51, 2013.

[42] J. Pearl. *Probabilistic Reasoning in Intelligent Systems: Networks of Plausible Inference.* Morgan Kaufmann, San Mateo, 1988.

[43] J. Pearl and A. Paz. Graphoids, graph-based logic for reasoning about relevance relations. In *Advances in Artificial Intelligence II*, pages 357–363. North-Holland, Amsterdam, 1987.

[44] Y. Sagiv and S. F. Walecka. Subset dependencies and completeness result for a subclass of embedded multivalued dependencies. *J. ACM*, 29(1):103–117, 1982.

[45] E. San Martín, M. Mouchart, and J.-M. Rolin. Ignorable common information, null sets and Basu's first theorem. *Sankhyā*, 67:674–697, 2005.

[46] P. H. Schouté. Analytic treatment of the polytopes regularly derived from regular polytopes. *Verhandelingen der Koninklijke Akademie van Wetenschappen te Amsterdam*, 11(3):370–381, 1911.

[47] R. D. Shachter. Bayes-ball, the rational pastime (for determining irrelevance and requisite information in belief networks and influence diagrams). In *Uncertainty in Artificial Intelligence 14*, pages 480–487. Morgan Kaufmann, San Francisco, 1998.

[48] P. P. Shenoy. Conditional independence in valuation-based systems. *Internat. J. Approx. Reason.*, 10(3):203–234, 1994.

[49] J. Q. Smith. Influence diagrams for statistical modelling. *Ann. Statist.*, 17(2):654–672, 1989.

[50] W. Spohn. Stochastic independence, causal independence and shieldability. *J. Philos. Logic*, 9(1):73–99, 1980.

[51] W. Spohn. Ordinal conditional functions: a dynamic theory of epistemic states. In *Causation in Decision, Belief Change, and Statistics II.*, pages 105–134. Kluwer, Dordrecht, 1988.

[52] M. Studený. Multiinformation and the problem of characterization of conditional independence relations. *Probl. Control Inform.*, 18:3–16, 1989.

[53] M. Studený. Conditional independence relations have no finite complete characterization. In *Information Theory, Statistical Decision Functions and Random Processes, Transactions of 11th Prague Conference, Vol. B*, pages 377–396. Kluwer, Dordrecht, 1992.

[54] M. Studený. Conditional independence and natural conditional functions. *Internat. J. Approx. Reason.*, 12(1):43–68, 1995.

[55] M. Studený. Semigraphoids and structures of probabilistic conditional independence. *Ann. Math. Artif. Intell.*, 21(1):71–98, 1997.

[56] M. Studený. Complexity of structural models. In *Proceedings of the joint session of 6th Prague Symposium on Asymptotic Statistics and 13th Prague Conference*, pages 523–528. Union of Czech Mathematicians and Physicists, 1998.

[57] M. Studený. *Probabilistic Conditional Independence Structures.* Springer, London, 2005.

[58] M. Studený and P. Boček. CI-models arising among 4 random variables. In *Proceedings of WUPES'94, September 11-15, 1994, Czech Republic*, pages 268–282, 1994.

[59] M. Studený and J. Cussens. Towards using the chordal graph polytope in learning decomposable models. *Internat. J. Approx. Reason.*, 88:259–281, 2017.

[60] M. Studený and D. Haws. Learning Bayesian network structure: towards the essential graph by integer linear programming tools. *Internat. J. Approx. Reason.*, 55:1043–1071, 2014.

[61] M. Studený, J. Vomlel, and R. Hemmecke. A geometric view on learning Bayesian network structures. *Internat. J. Approx. Reason.*, 51:578–586, 2010.

[62] K. Tanaka, M. Studený, A. Takemura, and T. Sei. A linear-algebraic tool for conditional independence inference. *J. Algebr. Stat.*, 6(2):150–167, 2015.

[63] J. Vejnarová. Conditional independence in possibility theory. *Internat. J. Uncertain. Fuzziness Knowledge-Based Systems*, 12:253–269, 2000.

[64] T. Verma and J. Pearl. Causal networks, semantics and expressiveness. In *Uncertainty in Artificial Intelligence 4*, pages 69–76. North-Holland, Amsterdam, 1990.

[65] T. Verma and J. Pearl. Equivalence and synthesis of causal models. In *Uncertainty in Artificial Intelligence 6*, pages 220–227. Elsevier, Amsterdam, 1991.

[66] P. Šimeček. *Independence Models (in Czech).* PhD thesis, Charles University, 2007.

[67] N. Wermuth. Analogies between multiplicative models for contingency tables and covariance selection. *Biometrics*, 32:95–108, 1976.

[68] J. Whittaker. *Graphical Models in Applied Multivariate Statistics.* John Wiley, Chichester, 1990.

2

Markov Properties for Mixed Graphical Models

Robin Evans

Department of Statistics, University of Oxford

CONTENTS

2.1 Introduction

In the previous chapter we were introduced to two different types of graph, each having its own specific edge type: directed and undirected. *Mixed graphs* have more than one type of edge, which may be undirected, directed, bidirected ($\leftrightarrow$) or take some other form. In this chapter we consider models based on some of the more common varieties of these

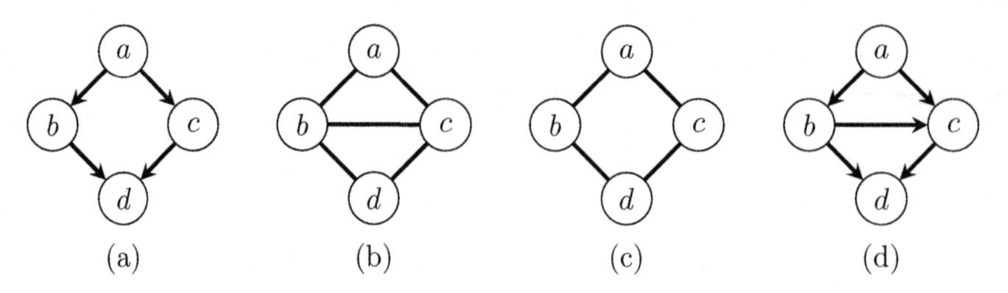

FIGURE 2.1: (a) Directed acyclic graph and (b) its moral graph. (c) A non-decomposable undirected graph and (d) a directed model containing it. The models corresponding to (b) and (d) are equivalent.

mixed graphs, specifically those introduced for two purposes: the unification of directed and undirected graphs; and models that are closed under marginalization.

Mixed graphs are a rich subject and, as such, the chapter is necessarily rather sparse on details. The reader is encouraged to work through some of the examples on their own to help solidify the ideas, and also to consult some of the references for further information.

2.1.1 Decomposable graphs

Given the two classes of graphical model defined in Chapter 1, one might reasonably ask what—if any—the connection between them is. We do not have to look too hard to see that there are models that may be represented by either a directed or an undirected graph. For example, the model defined by the independence $X_a \perp\!\!\!\perp X_d \mid X_b, X_c$ corresponds to both of the graphs in Figures 2.1(b) and (d) under their respective global Markov properties. In this short section, we ask which directed graphical models can be represented by an undirected graph, and vice versa.

Given a DAG $\mathcal{H}$ and a density p that obeys the Markov property for $\mathcal{H}$, we can write

$$p(x_V) = \prod_{v \in V} p(x_v \mid x_{\mathrm{pa}_{\mathcal{H}}(v)}) = \prod_{v \in V} \psi_v(x_v, x_{\mathrm{pa}_{\mathcal{H}}(v)}).$$

One can see that the distribution is contained in an undirected graph model with an edge between every pair of vertices (or nodes) contained in any set of the form $\{v\} \cup \mathrm{pa}_{\mathcal{G}}(v)$; this graph is called the *moral graph*, and we denote it $\mathcal{H}^m$. The moral graph can be obtained from a DAG by adding an edge between any pair of vertices with a common child (if they are not already adjacent), and then dropping the orientation of all edges. In other words it joins the two parents in any *v-structure* $a \to c \leftarrow b$ with a and b not adjacent.

The moral graph is the smallest undirected graph such that, when interpreted using the global Markov property, does not imply any independences not already implied by the original graph. The models for $\mathcal{H}$ and $\mathcal{H}^m$ are equivalent if and only if $\mathcal{H}$ contains no v-structures. If no v-structures are present then the moral graph is just the undirected 'skeleton' obtained by dropping the orientations. Otherwise, the model implied by the moral graph is strictly larger than the original model, and it follows that no undirected graphical model can represent the same independences as the directed one. For example, the model associated with the graph in Figure 2.1(a) is contained in the model associated with its moral graph in Figure 2.1(b); however the first graph implies the independence $X_b \perp\!\!\!\perp X_c \mid X_a$, while the second does not.

Correspondingly, not every undirected graph model can be represented using a directed acyclic graph. The model in Figure 2.1(c), for example, induces the independences $X_a \perp\!\!\!\perp$

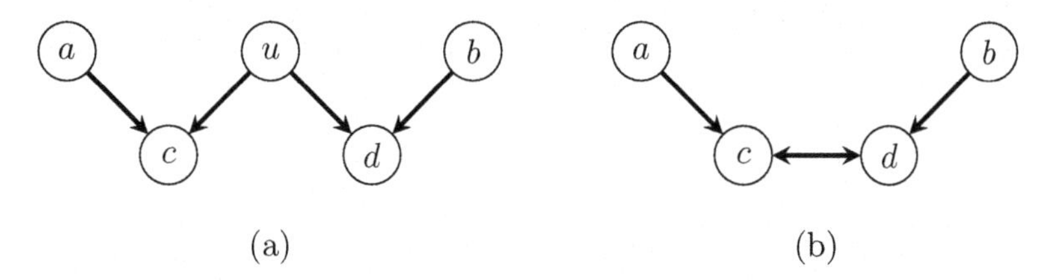

FIGURE 2.2: (a) Directed acyclic graph whose margin over $\{a, b, c, d\}$ is represented by the mixed graph (b), its latent projection to an ADMG.

$X_d \mid X_b, X_c$ and $X_b \perp\!\!\!\perp X_c \mid X_a, X_d$ (under any of the Markov properties); we can choose to represent one of these independences using a directed graph, such as the one in Figure 2.1(d), but not both.

The classes of models defined by undirected and directed graphs are distinct, but they have a non-empty intersection. The set of models in this intersection are the decomposable models, and they are represented by decomposable graphs.

Definition 2.1.1. *A* decomposition *of an undirected graph $\mathcal{G}$ is a triple (A, S, B) of disjoint sets that partition the vertices $V = A \cup B \cup S$, and such that (i) the induced subgraph $\mathcal{G}_S$ is complete, and (ii) A is separated from B by S. The decomposition is non-trivial if both A and B are non-empty.*

We say that a graph is decomposable *if it is either complete, or there is a non-trivial decomposition such that $\mathcal{G}_{A\cup S}$ and $\mathcal{G}_{B\cup S}$ are also decomposable.*

As an example, the graph in Figure 2.1(b) has a non-trivial decomposition into $A = \{a\}$, $B = \{d\}$, $S = \{b, c\}$. The two subgraphs obtained from the decomposition above are complete, so the graph is decomposable.

Proposition 2.1.2. *Let $\mathcal{G}$ be undirected graphical model and $\mathcal{H}$ be a DAG. Then $\mathcal{G}$ and $\mathcal{H}$ induce the same conditional independences under their respective global Markov properties if and only if they have the same adjacencies and $\mathcal{H}$ has no v-structures.*

This occurs for some $\mathcal{H}$ if and only if $\mathcal{G}$ is decomposable.

2.1.2 Unification

The previous section characterizes the intersection of these two classes of graphical models. A natural question to ask is whether there is some larger and more flexible class of graphical models which unifies undirected and directed graphical models.

> **Question 1.** Is there a class of graphical models of which *both* undirected and directed graphical models are special cases?

The answer is indeed yes, and in fact there are several such classes, each of which combine directed and undirected edges. We will study one of these classes, known as *chain graphs*, in Section 2.2.

2.1.3 Marginalizing and conditioning

A separate question is motivated specifically by directed graphs. Consider the directed graph in Figure 2.2(a): using the global Markov property we can see that the independences

$$X_a \perp\!\!\!\perp X_u, X_b, X_d \qquad X_a, X_c \perp\!\!\!\perp X_b, X_d \mid X_u \qquad X_b \perp\!\!\!\perp X_u, X_a, X_c$$

hold, and the reader is invited to verify that all other independences can be deduced from these using graphoid properties. Now suppose that we do not observe X_u, and so we are interested only in conditional independence relationships observable over X_a, X_b, X_c, X_d; in this case these are

$$X_a \perp\!\!\!\perp X_b, X_d, \qquad\qquad X_b \perp\!\!\!\perp X_a, X_c.$$

There is no directed graphical model (nor indeed an undirected graphical model), that represents precisely these independence relationships. In this sense we say that directed graphical models are not 'closed under marginalization'. That is, given an independence model corresponding to a directed graph, the independence model over a subset of variables is not itself a directed graphical model.

This problem was noted by Verma and Pearl [36] in the context of searching for causal models represented by directed graphs. In practice it is often implausible to assume that all relevant variables have been measured, and hence it is necessary to allow for the possibility that one is observing some margin of a directed graphical model. This begs our second question:

> **Question 2.** Is there a class of graphical models that includes all directed acyclic graph models, *and* is closed under marginalization?

As with Question 1, the answer is yes: in fact there are several such classes. We introduce one class of graphs, known as ADMGs, in Section 2.3. These resolve the problem by allowing *bidirected* edges ($\leftrightarrow$) to represent the existence of an unobserved common parent. See Figure 2.2(b) for the graph specific to the example above.

2.1.4 Outline of the chapter

The classes of graphs we will use to answer Questions 1 and 2 are quite distinct; however, the methods share some important features and can be attacked via similar methods. In particular, we will define models associated with each class of graphs by factorizing the distribution according to special components of the graph, and then imposing a particular structure on each factor. Classes of graphs that answer Questions 1 and 2 are the subject of Sections 2.2 and 2.3 respectively. The idea of being 'closed' that we consider in Section 2.3 is in terms of conditional independence relations. We follow this in Sections 2.4 and 2.5 by considering interpretations of graphs that do not correspond purely to conditional independence models.

The list of models discussed in this chapter is far from exhaustive, and the zoo of graphical models has expanded greatly in the past 20 years. More detailed overviews of varieties of mixed graphs can be found in [31, 20], for example.

Throughout this chapter we will assume that we have a finite collection of random variables $(X_v : v \in V)$ defined on a product space $\mathcal{X}_V$ and with joint density function $p(x_V)$. We will assume, for simplicity, that $p(x_V) > 0$; the Hammersley-Clifford Theorem (Observation 1.7.2 in Chapter 1) therefore applies and the various Markov properties for undirected graphs are equivalent. Our graphs do not contain self-loops: i.e. there is never an edge from a vertex to itself.

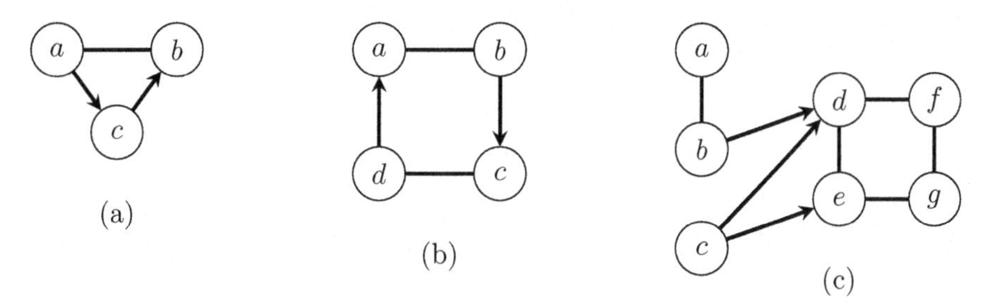

FIGURE 2.3: (a) & (b) Examples of mixed graphs which are not chain graphs. (c) A chain graph.

2.2 Chain Graphs

To answer Question 1, we will introduce chain graphs. These unify directed and undirected graphs by imposing an acyclic directed structure on disjoint groups of vertices called the *chain components*. Variables in a particular component are considered to be 'on an equal footing'—i.e. unordered—and an undirected graphical structure is used to model relationships between them.

In this section we consider *mixed graphs* of the form (V, E, D), where V is a set of vertices, and E, D respectively are sets of undirected and directed edges between those vertices; in other words, (V, E) is an undirected graph and (V, D) a directed one. We use $a \sim b$ to denote that two vertices a and b are adjacent, but without specifying whether they are joined by a directed or an undirected edge. Throughout this section we will also assume that graphs are *simple*, so that there is at most one edge between any pair of vertices. For a mixed graph $\mathcal{G}$, let $\mathcal{G}_-$ denote the subgraph (V, E) consisting of only the undirected edges.

A *path* is a sequence of distinct vertices that are adjacent in the graph whenever they are adjacent in the sequence; for example, $b \to d - e - g$ is a path in Figure 2.3(c). The *length* of a path is the number of edges contained in it; i.e. it is one less than the length of the sequence of vertices. A *cycle* consists of a path from a to b of length at least two together with an edge $a \sim b$; the sequence $d - e - g - f - d$ is a cycle in Figure 2.3(c). A cycle or path is said to be *semi-directed* if at least one edge is directed, and all directed edges are oriented in the same direction; $a - b \to c - d \to a$ in Figure 2.3(b) is such a cycle.

Definition 2.2.1. *A* chain graph *is a simple mixed graph with directed and undirected edges, such that there are no semi-directed cycles.*

Of the graphs in Figure 2.3, (a) and (b) are not chain graphs, since they each contain a semi-directed cycle; (c) is a chain graph. As a consequence of the absence of semi-directed cycles, a chain graph can be decomposed into ordered *chain components* $\mathcal{T}$, which are the connected components of $\mathcal{G}_-$.

Proposition 2.2.2. *Let $\mathcal{G}$ be a chain graph; then there exists an ordering $T_1, \ldots, T_k$ of the chain components such that if $c \in T_i$, $d \in T_j$ for $i < j$, then $c \sim d$ if and only if $c \to d$.*

In other words, we can always totally order the vertices in a way that respects the directed edges, and also keeps vertices in the same component together. For example, the chain components from Figure 2.3(c) can be ordered as $\{a, b\}$, $\{c\}$, $\{d, e, f, g\}$.

Consider a directed graph $\mathcal{G}^c$ with vertices $\mathcal{T}$, and edges $T_i \to T_j$ precisely when there is an edge $c \to d$ in $\mathcal{G}$ from any $c \in T_i$ to any $d \in T_j$. The proposition shows that this graph will be acyclic; an example of this component graph is shown in Figure 2.4(b).

The definitions of neighbors $\text{ne}_\mathcal{G}(v)$ and parents $\text{pa}_\mathcal{G}(v)$ from Section 1.6 in Chapter 1 are applied to mixed graphs by considering the undirected and directed parts of the graph respectively. That is:

$$\text{pa}_\mathcal{G}(v) = \{w \in V : w \to v\}, \qquad \text{ne}_\mathcal{G}(v) = \{w \in V : w - v\}.$$

The *non-descendants* of v, denoted $\text{nd}_\mathcal{G}(v)$, are the vertices that cannot be reached from v by an undirected path nor by a semi-directed path. This means that if v is in a chain component T, then the non-descendants of v in $\mathcal{G}$ is the set of all the vertices contained within the non-descendants of T in $\mathcal{G}^c$.

The definition of parents is applied disjunctively to sets: that is if $A \subseteq V$ we have

$$\text{pa}_\mathcal{G}(A) = \bigcup_{v \in A} \text{pa}_\mathcal{G}(v).$$

Note that this set may include elements of A, which is not the convention used by all authors.

2.2.1 Factorization

The chain graph model is most easily defined via its factorization property, which combines those properties from the directed acyclic and undirected graph cases. The factorizations are at two different levels: the first level is over entire chain components, induced by the DAG structure of those components. If X_V has a joint density p,

$$p(x_V) = \prod_{T \in \mathcal{T}} p(x_T \mid x_{\text{pa}(T)}). \tag{2.1}$$

In other words, if we think of each chain component X_T as a single variable, these variables would factorize according to the directed acyclic graph model on $\mathcal{G}^c$. Here $\text{pa}(T) \equiv \text{pa}_{\mathcal{G}^c}(T)$ refers to the parent components of T in $\mathcal{G}^c$, rather than the parents of vertices in T in the original graph[1] $\mathcal{G}$. However, it follows from the further factorization restriction below that the two are equivalent under our model.

The second level is a factorization determined by the structure of each component and its parent components. Let $\mathcal{G}$ be a chain graph with components $\mathcal{T}$. Given $T \in \mathcal{T}$, let $\mathcal{G}[T]$ be the graph with vertices $T \cup \text{pa}_\mathcal{G}(T)$, and edges from $\mathcal{G}$ as well as undirected edges between every $s, t \in \text{pa}_\mathcal{G}(T)$. This object is sometimes known as the *closure graph* [35].

For example, consider the chain graph $\mathcal{G}$ in Figure 2.4(a), which has chain components $\{a\}$, $\{b, e\}$ and $\{c, d\}$. The graph $\mathcal{G}[\{c, d\}]$ is shown in Figure 2.4(c). We draw the vertices in $\text{pa}_\mathcal{G}(T)$ with square nodes, and the edges between them with dashed edges, so that it is clear which are which. Let

$$\mathcal{C}_T \equiv \{S \subseteq T \cup \text{pa}_\mathcal{G}(T) \mid S \text{ a clique in } \mathcal{G}[T]\},$$

that is the collection of maximal complete sets in $\mathcal{G}[T]$. Then each of the factors in (2.1) should factorize as:

$$p(x_T \mid x_{\text{pa}(T)}) = \prod_{C \in \mathcal{C}_T} \psi_C(x_C). \tag{2.2}$$

[1]To see that these are different, note that in Figure 2.4(a) we have $\text{pa}_\mathcal{G}(\{c, d\}) = \{a, b\}$, but $\text{pa}_{\mathcal{G}^c}(\{c, d\}) = \{a, b, e\}$.

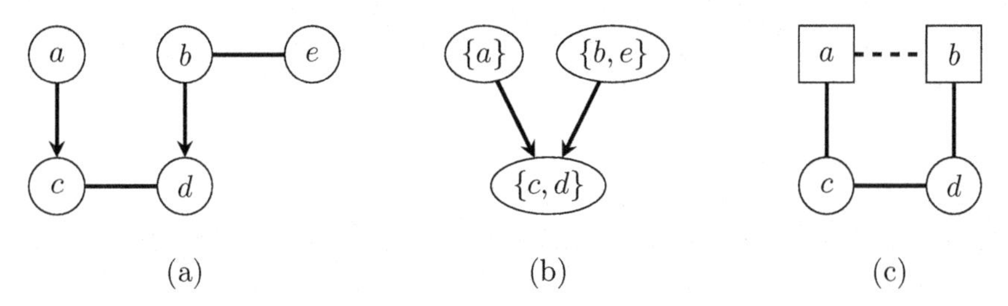

FIGURE 2.4: (a) A chain graph $\mathcal{G}$; (b) directed graph of chain components $\mathcal{G}^c$; (c) undirected graph induced by $\{c, d\}$ component $\mathcal{G}[\{c, d\}]$.

So each factor itself obeys the factorization property[2] of the undirected graph $\mathcal{G}[T]$.

The model corresponding to (2.1) and (2.2) is generally known as the LWF chain graph model, after three of its pioneers: Lauritzen, Wermuth and Frydenberg [21, 14, 17].

2.2.2 Local Markov property

The factorization in (2.1) can easily be turned into a collection of independences using the local Markov property for the DAG $\mathcal{G}^c$; this gives us that for any chain component T,

$$T \perp\!\!\!\perp \mathrm{nd}_{\mathcal{G}^c}(T) \setminus \mathrm{pa}_{\mathcal{G}^c}(T) \mid \mathrm{pa}_{\mathcal{G}^c}(T). \tag{2.3}$$

Here we abuse notation slightly and write, for example, T instead of X_T to reduce notational overhead. Here $\mathrm{nd}_{\mathcal{G}^c}(T)$, the *non-descendants* of T are those chain components from $\mathcal{G}$ that cannot be reached by a directed path in $\mathcal{G}^c$; equivalently it is the set of vertices that cannot be reached from any $t \in T$ by an undirected or semi-directed path in $\mathcal{G}$.

Applied to the graph in Figure 2.4(a) for example, we have $X_a \perp\!\!\!\perp X_b, X_e$, but no other non-trivial independences.

The second factorization gives us independences that can be shown to hold within a chain component, T. If we take the local Markov property for the undirected graph $\mathcal{G}[T]$, we obtain for any $t \in T$:

$$t \perp\!\!\!\perp T \cup \mathrm{pa}_{\mathcal{G}}(T) \setminus (\{t\} \cup \mathrm{ne}_{\mathcal{G}}(t) \cup \mathrm{pa}_{\mathcal{G}}(t)) \mid \mathrm{ne}_{\mathcal{G}}(t), \mathrm{pa}_{\mathcal{G}}(t). \tag{2.4}$$

Combining this with (2.3) gives

$$t \perp\!\!\!\perp (\mathrm{nd}_{\mathcal{G}}(t) \cup T) \setminus (\{t\} \cup \mathrm{ne}_{\mathcal{G}}(t) \cup \mathrm{pa}_{\mathcal{G}}(t)) \mid \mathrm{ne}_{\mathcal{G}}(t), \mathrm{pa}_{\mathcal{G}}(t). \tag{2.5}$$

The independences (2.5) define a *local Markov property* for chain graphs [14]. If p is positive then we can reverse this derivation and show that the local Markov property implies our original factorization.

Example 2.2.3. Consider the chain graph in Figure 2.4(a). The factorization (2.1) gives that

$$p(x_a, x_b, x_c, x_d, x_e) = p(x_a) \cdot p(x_b, x_e) \cdot p(x_c, x_d \mid x_a, x_b, x_e),$$

and implies that $X_a \perp\!\!\!\perp X_b, X_e$. Looking only at the block $\{c, d\}$, we consider the undirected graph in Figure 2.4(c); applying (2.2) gives

$$p(x_c, x_d \mid x_a, x_b, x_e) = \psi_{ac}(x_a, x_c)\psi_{bd}(x_b, x_d)\psi_{cd}(x_c, x_d)\psi_{ab}(x_a, x_b),$$

[2]Note that, similarly to the previous footnote we could take $\mathcal{G}[T]$ to include all of $\mathrm{pa}_{\mathcal{G}^c}(T)$ rather than only vertices that are actually parents of some vertex $t \in T$. Again, it does not change the model induced by the factorization.

which implies the independences $X_c \perp\!\!\!\perp X_b, X_e \mid X_a, X_d$ and $X_d \perp\!\!\!\perp X_a, X_e \mid X_b, X_c$ (note that in fact x_e does not appear at all in the expression given). If $p > 0$ then these independences imply the factorizations.

If we apply the local Markov property (2.5) directly we also obtain that

$$X_a \perp\!\!\!\perp X_b, X_e, \qquad X_c \perp\!\!\!\perp X_b, X_e \mid X_a, X_d, \qquad X_d \perp\!\!\!\perp X_a, X_e \mid X_b, X_c.$$

A pairwise and a global Markov property for chain graphs were proven to be equivalent to the local Markov property and factorization property by Frydenberg [14], under the assumption of positivity. A separation criterion known as c-separation was introduced in [34]. The latter is significantly more involved than the directional separation criterion (d-separation) introduced in Section 1.8.1 of Chapter 1, both conceptually and computationally; it is not enough to look at all paths, in general one may have to consider 'walks' in which vertices are repeated.

2.2.3 General remarks

It is clear that LWF chain graph models fulfill the requirement of generalizing directed and undirected graphical models: if all the chain components are of size one, then we obtain an arbitrary DAG; if there are no directed edges then we obtain an arbitrary undirected graph.

Chain graphs are a parsimonious statistical model, and rather computationally easy to work with because of the factorization property. This factorization means that chain graphs can—like directed and undirected graphs—be treated using the junction tree and message passing techniques introduced in Part II of this handbook.

As for statistical modeling, optimal model selection in a chain graph without knowing an ordering in advance is at least as hard as learning a directed acyclic graph, and therefore is an NP-hard problem [4]. If we fix chain components $\mathcal{T}$ and a total ordering upon them in advance, then selecting a graph becomes more feasible. It essentially amounts to sequentially fitting undirected graphs on $T_1 \cup \cdots \cup T_j$ for each $j = 1, \ldots, k$, and then compiling the edges. See [22] for methods for learning.

2.2.4 Other Markov properties for chain graphs

We can split the independence (2.4) into two parts using the graphoid properties. The two parts describe the independence of each variable from other variables in the same chain component, and from variables in the parent chain components respectively:

$$\begin{aligned} t &\perp\!\!\!\perp T \setminus (t \cup \mathrm{ne}_{\mathcal{G}}(t)) \mid \mathrm{ne}_{\mathcal{G}}(t), \mathrm{pa}_{\mathcal{G}}(T) \\ t &\perp\!\!\!\perp \mathrm{pa}_{\mathcal{G}}(T) \setminus \mathrm{pa}_{\mathcal{G}}(t) \mid \mathrm{ne}_{\mathcal{G}}(t), \mathrm{pa}_{\mathcal{G}}(t). \end{aligned} \tag{2.6}$$

If chain graphs are to generalize undirected graphs, then it is clear that the first independence requirement in (2.6) is necessary: one should need to condition on neighbors of a vertex in order to obtain a conditional independence from other parts of the same chain component. However, there is room for variation in how missing edges *between* a block and its parent blocks are interpreted.

The *alternative Markov property* (AMP) for chain graphs [2] replaces the second requirement in (2.6) with

$$t \perp\!\!\!\perp \mathrm{pa}_{\mathcal{G}}(T) \setminus \mathrm{pa}_{\mathcal{G}}(t) \mid \mathrm{pa}_{\mathcal{G}}(t),$$

whilst keeping independences within chain components unchanged. In other words, each variable is independent of variables in any parent components conditional on one's own

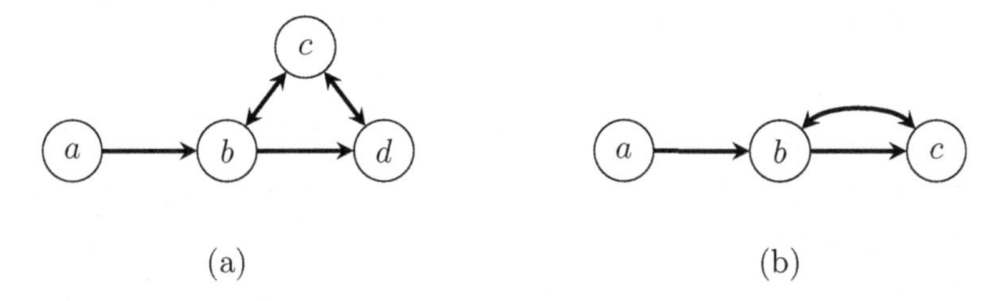

FIGURE 2.5: Two examples of acyclic directed mixed graphs.

parents, but *marginally* on the rest of one's own chain component. For a graph like the one in Figure 2.4(a), the independence $X_d \perp\!\!\!\perp X_a, X_e \mid X_b, X_c$ is replaced by $X_d \perp\!\!\!\perp X_a, X_e \mid X_b$.

Drton [6] shows that these models are generally non-smooth, so that standard properties of statistical models such as the asymptotic distribution of a likelihood ratio statistic do not apply everywhere in the parameter space. Drton also gives two other Markov properties that can be applied to chain graphs. The *multivariate regression* chain graph property is a special case of the Markov property for ADMGs that we discuss in Section 2.3; see also Chapter 8 in Part III. The final, '*Type III*' property, gives models that, as for the AMP, are non-smooth; it has not been used in practice.

2.3 Closed Independence Models

2.3.1 ADMGs

Consider again the directed model from Figure 2.2, where we supposed that the variable X_u is unobserved. Clearly this will induce dependence between its children X_c, X_d, and therefore a graph that represents this situation should include an edge between c and d. A suitable type of graph is an *acyclic directed mixed graph* (ADMG).

Definition 2.3.1. *An* acyclic directed mixed graph *(ADMG) is a triple* (V, D, B)*, where* (V, D) *is a DAG, and* B *is a collection of unordered pairs of vertices, known as* bidirected edges. *If* $\{v, w\} \in B$ *we write* $v \leftrightarrow w$.

Some examples of ADMGs are shown in Figure 2.5. We can informally think of a bidirected edge as representing a latent common parent. Unlike chain graphs, ADMGs are not simple: it is possible to have both a directed and a bidirected edge between each pair of vertices; see Figure 2.5(b).

To obtain an ADMG from a DAG, we use a *latent projection* operation for removing vertices corresponding to unobserved variables, and adding in edges where dependence between the remaining variables has been induced.

Definition 2.3.2 (Latent Projection). *Let* $\mathcal{G}$ *be an ADMG and* $u \in V$*. Define a new graph* $\mathcal{G}'$ *starting with the induced subgraph* $\mathcal{G}' = \mathcal{G}_{V\setminus\{u\}}$ *and adding edges as follows:*

$$\text{whenever} \left\{ \begin{array}{c} a \leftarrow u \rightarrow b \\ a \leftrightarrow u \rightarrow b \\ a \rightarrow u \rightarrow b \end{array} \right\} \text{ in } \mathcal{G}, \text{ add } \left\{ \begin{array}{c} a \leftrightarrow b \\ a \leftrightarrow b \\ a \rightarrow b \end{array} \right\} \text{ to } \mathcal{G}'.$$

$\mathcal{G}'$ *is the* latent projection *of* $\mathcal{G}$ *over* $V \setminus \{u\}$.

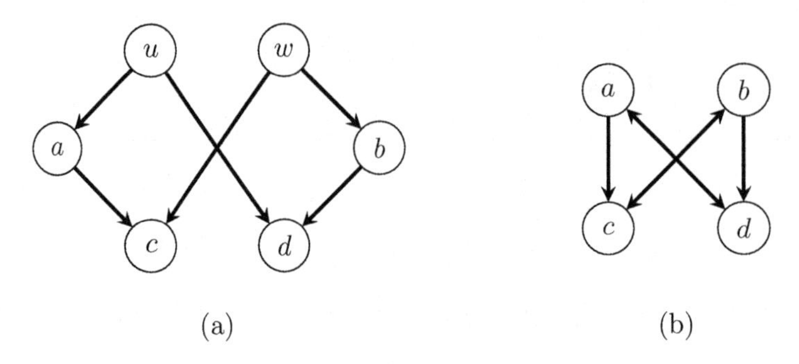

FIGURE 2.6: (a) A DAG and (b) its latent projection over $\{a, b, c, d\}$.

Other combinations of edges meeting at u (for example, $a \to u \leftrightarrow b$) do not result in edges being added in the latent projection. Typically the projection starts from a DAG, and some simple examples of latent projection are given in Figures 2.2, 2.6 and 2.8. The definition can be applied inductively to sets of vertices, and the result is independent of the order in which the vertices are projected out. We can then unambiguously talk about the latent projection over any subset of the vertices.

The key fact about latent projection is that it preserves the directional separation criterion seen in Section 1.8.1. We summarize this in the following result.

Proposition 2.3.3. *Suppose $\mathcal{G}$ and $\tilde{\mathcal{G}}$ are DAGs whose latent projections over a set V are both equal to the ADMG $\mathcal{H}$. For any $A, B, C \subseteq V$, if $X_A \perp\!\!\!\perp X_B \mid X_C$ is implied by the global Markov property for $\mathcal{G}$, then it is also implied by that for $\tilde{\mathcal{G}}$.*

In other words, the structure of the conditional independence model over X_V is preserved in the latent projection over V. This means it will be possible to define a conditional independence model over ADMGs that describes precisely those over the observed variables from the original DAG.

This result also motivates the idea of the *canonical DAG* $\bar{\mathcal{G}}$ associated with an ADMG $\mathcal{G}$; this replaces the bidirected edges with latent variables, to partly 'undo' the latent projection. Define $\bar{\mathcal{G}}$ by starting with only the directed edges of $\mathcal{G}$, and adding new vertices and edges as follows: whenever $a \leftrightarrow b$ in $\mathcal{G}$, add $a \leftarrow u_{ab} \to b$ to $\bar{\mathcal{G}}$. In other words, each bidirected edge is replaced with exactly one new variable with two observed children. For example, the canonical DAG associated with Figure 2.6(b) is the graph in Figure 2.6(a).

Latent projection also preserves causal structure, in the sense that the effects of interventions on the observed variables commute with the projection operation: hence we can see the effect of an intervention on the conditional independence structure of a model directly from the ADMG, without knowing the exact structure of the original DAG. See [8] for more details.

The name ADMG was coined by Richardson [26], but the same structures were used by Verma and Pearl [36], and can be traced back to Sewall Wright's path diagrams [37]. Latent projection was introduced by Pearl and Verma [25], and more recently generalized in [8].

2.3.2 Ancestral sets

Whereas chain components play the key role in the factorization of chain graphs models, in ADMG models the most important subsets are *districts* and *ancestral sets*. Given an ADMG $\mathcal{G}$, we say that a set A is *ancestral* if it contains all its own parents. Ancestrality plays a key role because ancestral structures are preserved under marginalization.

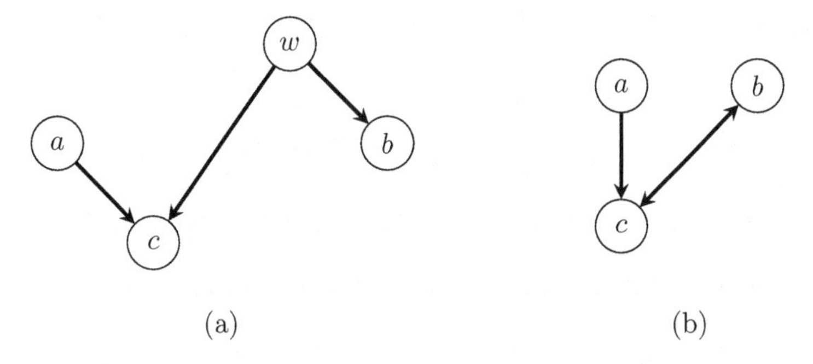

FIGURE 2.7: The respective ancestral subgraphs for Figures 2.6 (a) and (b) after removing d. Note that (b) is also the latent projection of (a).

Proposition 2.3.4. *Let $\mathcal{G}$ be a DAG with vertices V, and suppose that $A \subseteq V$ is an ancestral set. Then if $p(x_V)$ is Markov with respect to $\mathcal{G}$, the margin $p(x_A)$ is Markov with respect to the induced subgraph $\mathcal{G}_A$.*

Proof. If $V = A$ we are done. Otherwise, since A is ancestral, there is at least one vertex $w \in V \setminus A$ that has no children in $\mathcal{G}$. Hence

$$\begin{aligned} p(x_{V\setminus\{w\}}) &= \int_{\mathcal{X}_w} \prod_{v\in V} p(x_v \mid x_{\mathrm{pa}(v)})\, dx_w \\ &= \prod_{v\in V\setminus\{w\}} p(x_v \mid x_{\mathrm{pa}(v)}) \int_{\mathcal{X}_w} p(x_w \mid x_{\mathrm{pa}(w)})\, dx_w \\ &= \prod_{v\in V\setminus\{w\}} p(x_v \mid x_{\mathrm{pa}(v)}) \end{aligned}$$

since w is not in any parent sets. Now we see that this is just the expression for the margin of the induced subgraph without w. By repeating over vertices in $V \setminus A$, we obtain the result. □

In other words, an ancestral margin of a DAG is described by the associated induced subgraph. This result carries over to ADMGs in the following manner. Let $\mathcal{G}$ be an ADMG with vertices V and $\bar{\mathcal{G}}$ its canonical DAG with vertices $V \cup U$; if $A \subseteq V$ is an ancestral set of $\mathcal{G}$, then $A \cup U$ is an ancestral set in $\bar{\mathcal{G}}$, and further the projection of $\bar{\mathcal{G}}_{A\cup U}$ over A is precisely $\mathcal{G}_A$. Now, since $\bar{\mathcal{G}}_{A\cup U}$ represents the conditional independence structure over $\{X_A, X_U\}$ by Proposition 2.3.4, and this is preserved by latent projection by Proposition 2.3.3, it follows that $\mathcal{G}_A$ represents the conditional independence structure of the variables X_A.

For example, the set $\{a, b, c\}$ is ancestral in the ADMG in Figure 2.6(b). The distribution over X_a, X_b, X_c is therefore Markov with respect to the induced subgraph over $\{a, b, c\}$, shown in Figure 2.7(b). This graph suggests that $X_a \perp\!\!\!\perp X_b$, which indeed we can verify from the original DAG in Figure 2.6(a), and in the canonical DAG corresponding to the ancestral subgraph, shown in Figure 2.7(a).

2.3.3 Districts

The districts of a graph are the connected components of $\mathcal{G}_{\leftrightarrow}$, the subgraph consisting of only the bidirected edges in $\mathcal{G}$. Districts are similar to chain components, but there is no ordering restriction analogous to the requirement for no semi-directed cycles. The ADMG in Figure 2.6(b) illustrates this: there is no way to order the two districts $\{a, d\}$ and $\{b, c\}$ such that no vertex in the second district is a parent of a vertex in the first. Compare this with the graph in Figure 2.3(b), which is not a chain graph for precisely this reason.

What makes districts so important? Well, suppose we start with a DAG $\mathcal{H}$ with variables $V \cup U$, and then project onto V to obtain an ADMG $\mathcal{G}$. The districts of $\mathcal{G}$ are precisely the groups of observed vertices that are connected by latent variables in $\mathcal{H}$. To see this, assume for simplicity the DAG is canonical. We can partition U into sets $U_D \equiv U \cap \mathrm{pa}_{\mathcal{H}}(D)$ for each district D. Let $\mathcal{D}$ be the collection of districts in $\mathcal{G}$, and let $D^* = D \cup U_D$ for each $D \in \mathcal{D}$; one can check that the sets D^* partition $V \cup U$. Then

$$p(x_V) = \int \prod_{v \in V \cup U} p(x_v \,|\, x_{\mathrm{pa}_{\mathcal{H}}(v)})\, dx_U \tag{2.7}$$

$$= \prod_{D \in \mathcal{D}} \int \prod_{v \in D^*} p(x_v \,|\, x_{\mathrm{pa}_{\mathcal{H}}(v)})\, dx_{U_D}$$

$$= \prod_{D \in \mathcal{D}} f(x_D, x_{\mathrm{pa}(D)}). \tag{2.8}$$

For example, in Figure 2.6 we see that the districts in (b), $\{a, d\}$ and $\{b, c\}$, are respectively connected in (a) by the latent vertices u and w. We therefore have

$$D_1 = \{a, d\}, \qquad D_1^* = \{a, d, u\} \qquad \text{and} \qquad D_2 = \{b, c\}, \qquad D_2^* = \{b, c, w\}.$$

A joint distribution Markov to Figure 2.6(a) can be written as

$$p(x_V) = p(x_u)p(x_a|x_u)p(x_d|x_u, x_b) \cdot p(x_w)p(x_b|x_w)p(x_c|x_w, x_a),$$

so integrating out x_u, x_w we end up with

$$p(x_V) = \int p(x_u)p(x_a|x_u)p(x_d|x_u, x_b)\, dx_u \cdot \int p(x_w)p(x_b|x_w)p(x_c|x_w, x_a)\, dx_w$$
$$= f(x_a, x_b, x_d) \cdot g(x_a, x_b, x_c);$$

from this we deduce that $X_c \perp\!\!\!\perp X_d \mid X_a, X_b$.

2.3.4 A conditional independence model

The two properties given in Sections 2.3.2 and 2.3.3 can be turned into a recursive, axiomatic characterization of marginal independence models.

Definition 2.3.5. *We say that* $p(x_V)$ *is* Markov with respect to an ADMG $\mathcal{G}$ *if:*

1. $p(x_V) = \prod_{D \in \mathcal{D}(\mathcal{G})} f(x_D, x_{\mathrm{pa}(D)})$ *for some non-negative functions* f*;*

2. *for every ancestral set* $A \subset V$, $p(x_A)$ *is Markov with respect to* $\mathcal{G}_A$.

This is a conditional independence model, since it imposes factorizations on (ancestral) margins of the joint distribution. If all districts are of size one then we have a DAG, and the conditions just reduce to the usual DAG model. Consider the ADMG in Figure 2.5(a). In this case the ancestral subgraphs $\{a, b, d\}$ and $\{a, c\}$ respectively give us the factorizations:

$$p(x_a, x_b, x_d) = f_a(x_a) \cdot f_b(x_a, x_b) \cdot f_d(x_b, x_d)$$
$$p(x_a, x_c) = f_a(x_a) \cdot f_c(x_c),$$

and hence the independences $X_a \perp\!\!\!\perp X_c$ and $X_a \perp\!\!\!\perp X_d \mid X_b$. Note that there is no DAG-like factorization that could give us these two independences.

Unlike undirected, directed and chain graph models, a missing edge in an ADMG does not necessarily induce a conditional independence. For example, any distribution is Markov with respect to the graph in Figure 2.5(b), as the factorizations of each of its ancestral margins are trivial. Applying the definition to the graph in Figure 2.8(b) gives us $X_a \perp\!\!\!\perp X_c \mid X_b$, but there is no independence corresponding to the missing $a - d$ edge.

The model in Definition 2.3.5 defines all the observable conditional independence constraints induced by a DAG with hidden variables; there are several other equivalent Markov properties that describe the same model, including local and global properties [26, 11]. These are all equivalent, regardless of the positivity or otherwise of the joint distribution.

2.3.5 Connection to chain graphs

If we restrict the class of ADMGs not to have semi-directed cycles[3] and apply the Markov property above, then we obtain the class of models known as the *multivariate regression chain graph models*, or 'Type IV' chain graph models [5, 6]. These are sometimes represented with dashed undirected edges in place of the solid bidirected ones we use. Shpitser [32] gives a class of graphs called 'segregated' that is able to model conditional independences induced by marginals of LWF chain graphs; these graphs, which use undirected, directed and bidirected edges, include both chain components and districts.

2.4 Non-Independence Constraints

So far we have considered the idea of a class of models being 'closed' only in the context of conditional independence restrictions. However, there is no reason to suppose that conditional independences are the *only* sort of constraint on the margin of a conditional independence model; indeed the picture is rather more complicated.

To describe such models in more detail, we need to extend the notion of an ADMG very slightly, and introduce the *conditional acyclic directed mixed graph*, or CADMG [28]. A CADMG $\mathcal{G}$ is an ADMG whose vertices are split into two distinct sets V and W; they are subject to the restriction that the only edges incident to vertices in W are directed away from them (and into a vertex in V). The vertices in V are called *random*, and those in W are *fixed*. The terminology and the restriction reflect the fact that we will view fixed vertices not as random variables, but rather as conditioned upon variables that index the distribution.

All our previous definitions for ADMGs carry over to CADMGs in the obvious way, although note that we ignore fixed vertices for the purposes of determining the districts of a CADMG.

Given a district D in a (C)ADMG $\mathcal{G}$, we define the CADMG $\mathcal{G}[D]$ as the graph with random vertices D, fixed vertices $\mathrm{pa}_{\mathcal{G}}(D) \setminus D$, and all edges inherited from $\mathcal{G}$ that are compatible with the constraint above: i.e. those edges that are within D, or are directed from $\mathrm{pa}_{\mathcal{G}}(D) \setminus D$ to D. For example, in Figure 2.8(b) there is a district $\{b, d\}$, and the associated graph $\mathcal{G}[\{b, d\}]$ is shown in Figure 2.8(d). Note that the edge $b \to c$ is dropped

[3] Here we define 'semi-directed cycle' in the same manner as for a chain graph, but with bidirected edges replacing undirected edges.

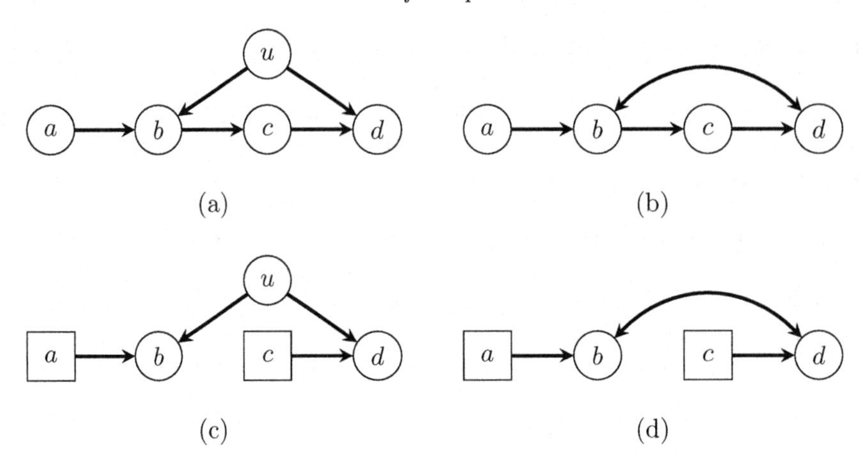

FIGURE 2.8: (a) Directed acyclic graph whose margin over $\{a, b, c, d\}$ is represented by the latent projection in (b). In this case, the induced model is not a conditional independence model. (c) Graph corresponding to intervening on a and c, and its latent projection (d).

because it does not satisfy the restriction of being directed from $\text{pa}_{\mathcal{G}}(D) \setminus D = \{a, c\}$ to $D = \{b, d\}$.

We depict the fixed vertices with square nodes, and the random ones with circles. For consistency with the notation use in chain graphs, it is possible to add edges in $\mathcal{G}[D]$ between every pair of vertices in $\text{pa}_{\mathcal{G}}(D) \setminus D$, and this does not change any of the following discussion.

2.4.1 Verma constraints

In our discussion on districts in Section 2.3.3, we did not stop to consider whether the factors $f(x_D, x_{\text{pa}(D)})$ in (2.8) had any particular structure of their own. In fact, from that expression we have

$$f(x_D, x_{\text{pa}(D)}) = \int \prod_{v \in D^*} p(x_v \,|\, x_{\text{pa}_{\mathcal{H}}(v)})\, dx_{U_D},$$

so the integrand is a product of univariate densities ordered according to the original DAG $\mathcal{H}$; it is therefore a probability density over X_{D^*} (conditional on $x_{\text{pa}(D) \setminus D}$). Integrating out x_{U_D} we obtain a density over X_D conditional on $x_{\text{pa}(D) \setminus D}$; call this q_D. More than this, it is clear that the conditional distribution is a margin of a distribution that is Markov with respect to the DAG over just those variables in D^* (conditional on $x_{\text{pa}(D) \setminus D}$). This means that we should expect q_D to obey any structure induced by this marginal model. For example, the factor $q_{\{b,d\}}$ associated with Figure 2.8(b) should follow the model associated with the CADMG in Figure 2.8(d).

We can therefore refine (2.8) and note that

$$p(x_V) = \prod_{D \in \mathcal{D}} q_D(x_D \,|\, x_{\text{pa}(D) \setminus D}),$$

where each q_D is Markov with respect to the CADMG $\mathcal{G}[D]$.

Example 2.4.1. Consider the graphs in Figure 2.8. We have

$$
\begin{aligned}
&p(x_a, x_b, x_c, x_d) \\
&= \int p(x_u) \cdot p(x_a) \cdot p(x_b \mid x_a, x_u) \cdot p(x_c \mid x_b) \cdot p(x_d \mid x_c, x_u)\, dx_u \\
&= p(x_a) \cdot p(x_c \mid x_b) \cdot \int p(x_u) \cdot p(x_b \mid x_a, x_u) \cdot p(x_d \mid x_c, x_u)\, dx_u \qquad (2.9) \\
&\equiv q_a(x_a) \cdot q_c(x_c \mid x_b) \cdot q_{\{b,d\}}(x_b, x_d \mid x_a, x_c).
\end{aligned}
$$

Note that for the districts $\{a\}$ and $\{c\}$, this distribution $q_D(x_D \mid x_{\mathrm{pa}(D)\setminus D})$ is the same as $p(x_D \mid x_{\mathrm{pa}(D)\setminus D})$; however, by rearranging the previous equation and using the fact that $p(x_c \mid x_b) = p(x_c \mid x_a, x_b)$, we have

$$
\begin{aligned}
q_{\{b,d\}}(x_b, x_d \mid x_a, x_c) &= \frac{p(x_a, x_b, x_c, x_d)}{p(x_a) \cdot p(x_c \mid x_a, x_b)} \qquad (2.10) \\
&= p(x_b \mid x_a) \cdot p(x_d \mid x_a, x_b, x_c),
\end{aligned}
$$

which is not the same as $p(x_b, x_d \mid x_a, x_c)$. In fact, if the graph is interpreted causally, this object is the *interventional* distribution, denoted $p(x_b, x_d \mid do(x_a, x_c))$ by Pearl [24]. That is, it corresponds to the distribution one would obtain if X_a, X_c were intervened on experimentally, rather than simply conditioned upon statistically.

In any case, it is clear from the form of the integral in (2.9) that $q_{\{b,d\}}$ is the margin of a distribution that is Markov with respect to the DAG in Figure 2.8(c). Under the usual Markov property for this graph we can see that $X_a \perp\!\!\!\perp X_d \mid X_c$, and this is reflected in $q_{\{b,d\}}$ by the fact that its margin over X_d does not depend upon x_a:

$$
\begin{aligned}
q_{\{b,d\}}(x_d \mid x_a, x_c) &= \int q_{\{b,d\}}(x_b, x_d \mid x_a, x_c)\, dx_b \\
&= \int\!\!\int p(x_u) \cdot p(x_b \mid x_a, x_u) \cdot p(x_d \mid x_c, x_u)\, dx_u\, dx_b \\
&= \int p(x_u) \left(\int p(x_b \mid x_a, x_u)\, dx_b \right) p(x_d \mid x_c, x_u)\, dx_u \\
&= \int p(x_u) \cdot p(x_d \mid x_c, x_u)\, dx_u.
\end{aligned}
$$

Furthermore, it is not hard to see that this independence holds precisely because there is no edge $a \to d$, so this fact places a non-trivial constraint on p. Importantly, however, the conditional independence $X_a \perp\!\!\!\perp X_d \mid X_c$ does *not* hold in p; it is only after performing the division (2.10) to obtain the interventional distribution that this independence 'appears'.

The 'independence' in this example is often referred to as the *Verma Constraint*, after [36]. It was first made use of by Robins [29], in the context of longitudinal treatments. In order to define a Markov property that incorporates these extra constraints, we modify our earlier axioms slightly to define what it means for a distribution to be *nested Markov* with respect to a CADMG.

Definition 2.4.2. *We say a conditional distribution $p(x_V \mid x_W)$ is* nested Markov *with respect to a CADMG $\mathcal{G}$ if either $|V| = 1$ and $\mathrm{pa}_{\mathcal{G}}(V) = W$, or:*

1'. if $\mathcal{G}$ has more than one district, then $p(x_V) = \prod_{D \in \mathcal{D}(\mathcal{G})} q_D(x_D \mid x_{\mathrm{pa}(D)\setminus D})$, and each q_D is nested Markov with respect to $\mathcal{G}[D]$;

2. for every ancestral set $A \subset V$, $p(x_A)$ is nested Markov with respect to $\mathcal{G}_A$.

The second axiom is essentially unchanged from Definition 2.3.5, but the first axiom now leads to a recursion; this terminates only when there is a single random node with all the fixed nodes as its parents, since such a graph does not induce any constraints. The model itself is defined by the constraints in the factorizations (1').

Applying the definition to the graph in Figure 2.8(b) gives precisely the independence $X_a \perp\!\!\!\perp X_c \mid X_b$ noted in Section 2.3, and the Verma constraint from Example 2.4.1.

There is much more to say about the nested Markov property, and the *nested Markov model* it induces. For a full description, including local and global Markov properties, see [28]; a discrete parameterization is given in [12].

2.4.2 Inequalities

At this point it is reasonable to ask whether further non-independence constraints exist, and therefore more refined axioms might be required to describe them. In terms of dimension reducing constraints—that is, equality constraints on probabilities—the answer is that there are none, at least for discrete variables [9]. However, there are *inequality* constraints that restrict the joint distributions possible to observe.

The first known inequality constraint of this kind predates the DAG model itself. Consider the graph in Figure 2.8(c) and its latent projection (d): according to the nested Markov property, the only constraints correspond to $X_a \perp\!\!\!\perp X_c, X_d$ and $X_c \perp\!\!\!\perp X_a, X_b$. However, John Stewart Bell noted in his seminal paper [3] that the correlations of X_b, X_d conditional on X_a, X_c are further restricted by what has become known as *Bell's Inequality*. This has important implications in quantum mechanics: see Section 2.5.3.

Pearl [23] was the first to consider inequalities directly on ADMG models, finding that distributions derived from the latent variable model corresponding to Figure 2.5(b) should satisfy (for binary observed variables)

$$p(x_b, x_c \mid x_a) + p(x_b, x_c' \mid x_a') \leq 1, \qquad x_c \neq x_c'.$$

This is known as the *instrumental inequality*. More recently a general approach to finding inequality constraints has been developed [7], which leads to a possible additional axiom:

3. For every $D \subseteq V$ and $z_D \in \mathcal{X}_D$, there must exist a distribution p^* such that

$$p(x_{V\setminus D}, z_D) = p^*(x_{V\setminus D}, z_D) \qquad \text{for all } x_{V\setminus D} \in \mathcal{X}_{V\setminus D},$$

and p^* is Markov with respect to $\mathcal{G}_{V\setminus D}$.

This axiom yields Pearl's instrumental inequality, though not Bell's inequality. There are further papers [16, 13] that give inequalities on marginal DAG models, and stronger results are possible if we make parametric assumptions about the hidden variables [1, 38]. Recent results [30] show that the model obtained from marginalizing a Bayesian network with discrete observed variables and arbitrary latent variables is always *semi-algebraic*, which means that it can be described with a finite number of polynomial equalities and inequalities; see also Section 3.2.4 in Chapter 3. The equalities are the conditional independences and Verma constraints [9], but a complete characterization of inequalities is not known.

2.4.3 mDAGs

As a class of graphs, ADMGs are not quite rich enough to faithfully represent the range of possible models that can be induced by marginalization; they fail to capture certain constraints that arise because of weak interactions between several variables. This happens

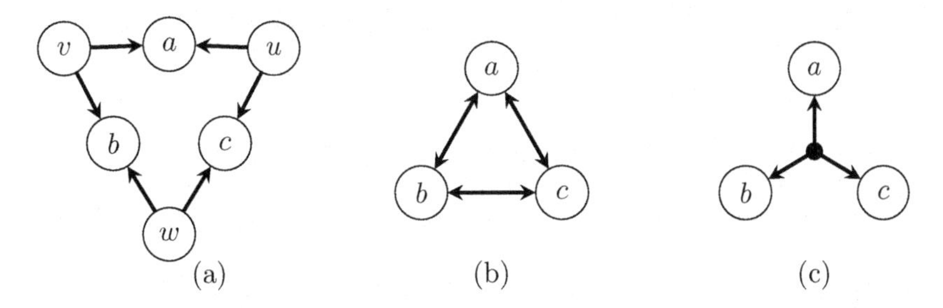

FIGURE 2.9: (a) A DAG over six variables, and (b) its latent projection over $\{a, b, c\}$. (c) An mDAG hypergraph representing a single latent variable. When (b) is interpreted as an mDAG model, it satisfies an inequality constraint.

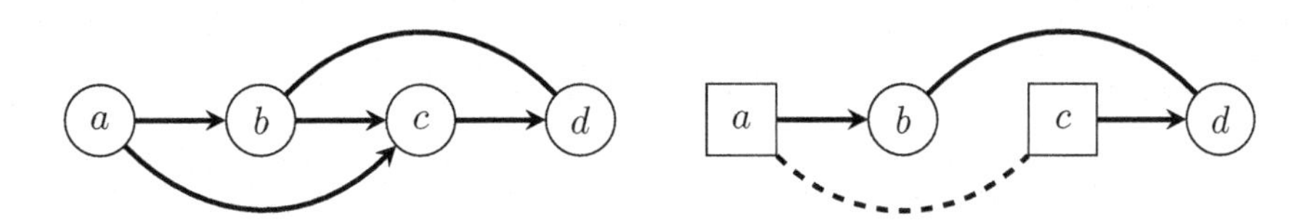

FIGURE 2.10: A mixed graph which is not a chain graph.

in the case illustrated in Figure 2.9(a); in the margin of this DAG over $\{a, b, c\}$ it is not possible for all three variables to be perfectly correlated [13]. However, if a single latent variable is a parent of all of $\{a, b, c\}$ then any distribution can be obtained as a margin, provided the latent variable is allowed a sufficiently large state-space. Despite these different models, the two scenarios lead to the same ADMG under latent projection, which is given in Figure 2.9(b). To resolve this Evans [8] introduced hypergraphs called mDAGs, which can distinguish between these two types of joint correlation, and would result in the graphs in Figures 2.9(b) and (c), depending on the latent set up.

2.5 Other Graphs and Models

2.5.1 Other models

One might be tempted to ask whether we can extend chain graphs to allow the chain components not to be totally ordered, making them more like ADMGs. Suppose we consider the mixed graph in Figure 2.10, in which the components are not totally ordered, and attempt to impose a factorization model.

Presuming we had a factorization into components conditional on parent components as before, we would obtain

$$p(x_a, x_b, x_c, x_d) = q_a(x_a) \cdot q_c(x_c \mid x_a, x_b) \cdot q_{\{b,d\}}(x_b, x_d \mid x_a, x_c).$$

Note that

$$\begin{aligned} p(x_a, x_b, x_c) &= q_a(x_a) \cdot q_c(x_c \mid x_a, x_b) \cdot \int q_{\{b,d\}}(x_b, x_d \mid x_a, x_c)\, dx_d \\ &= q_a(x_a) \cdot q_c(x_c \mid x_a, x_b) \cdot q_{\{b,d\}}(x_b \mid x_a, x_c). \end{aligned}$$

The construction of this product implies that the final factor satisfies the independence $X_b \perp\!\!\!\perp X_c \mid X_a$, since otherwise it is not clear how it could be identifiably separated from $q(x_c \mid x_a, x_b)$.

Now, supposing that we want this final factor to be Markov with respect to the graph in Figure 2.10(b), we also need $X_b \perp\!\!\!\perp X_c \mid X_a, X_d$; combining this with $X_b \perp\!\!\!\perp X_c \mid X_a$ gives a model that is non-smooth. In fact, one can show that if the variables are binary, this becomes a union of four models: for each level x of X_a we must have

$$\text{either} \quad X_b \perp\!\!\!\perp X_c, X_d \mid X_a = x \quad \text{or} \quad X_c \perp\!\!\!\perp X_b, X_d \mid X_a = x.$$

Combining these gives an algebraically *reducible* model, which implies that there can be no regular parameterization.

One can choose to interpret these mixed graphs as pure conditional independence models using separation criteria [31]. This too leads to non-smooth models, but in principle does not prevent their use as statistical models. More recently, Lauritzen and Sadeghi [20] give a generalization of ADMGs and chain graphs using a separation criterion.

2.5.2 Ancestral graphs

So far we have only discussed graphical models under marginalization; i.e. what happens if a variable is unobserved? Another natural question that arises in statistics is what happens if a variable is conditioned upon: for example, in a case-control study we condition on presence or absence of a disease.

Ancestral graphs models [27] are a class of graphical models that are closed with respect to conditional independence under both marginalization *and* conditioning; they include directed and undirected graphs (but not chain graphs) as special cases, and are obtained by using an operation similar to latent projection. To get a sense of this, consider the DAG in Figure 2.1(a), and suppose we ask what happens if we observe X_a, X_c, X_d conditional on some (arbitrary) value of X_b. We know that $X_a \perp\!\!\!\perp X_d \mid X_b, X_c$, so in data that are obtained conditional on a particular value of X_b we will observe $X_a \perp\!\!\!\perp X_d \mid X_c$. The relevant ancestral graph is the induced subgraph over $\{a, c, d\}$, which displays this new independence. Conversely, suppose we observe data conditionally on X_d; although $X_b \perp\!\!\!\perp X_c \mid X_a$ is implied by the original graph, we have $X_b \not\perp\!\!\!\perp X_c \mid X_a, X_d$, and therefore under selection there are no conditional independences. The relevant ancestral graph is then the complete undirected graph on $\{a, b, c\}$.

Ancestral graphs are used for causal learning, and there are well developed algorithms that are consistent for the correct causal model under selection and marginalization [33]. However, since ancestral graph models are pure conditional independence models they ignore the other constraints discussed in Section 2.4. In principle, using these as well would lead to more powerful learning methods, but it remains an open problem to find a way to do this efficiently.

Indeed, conditioning induces additional constraints that are not fully represented by ancestral graphs. To see this, consider the DAG in Figure 2.9(a) again, but suppose that we are interested in the conditional distribution of $p(x_u, x_v, x_w \mid x_a, x_b, x_c)$. This can be seen as a dual problem to finding the marginal distribution. In this case, we can see that (for a fixed x_a, x_b, x_c)

$$\begin{aligned}
&p(x_u, x_v, x_w \mid x_a, x_b, x_c) \\
&\quad = \frac{p(x_u)p(x_v)p(x_w)p(x_a \mid x_u, x_v)p(x_b \mid x_v, x_w)p(x_c \mid x_u, x_w)}{p(x_a, x_b, x_c)} \\
&\quad = f(x_u, x_v) \cdot g(x_v, x_w) \cdot h(x_w, x_u).
\end{aligned}$$

In fact, it is not hard to prove that any distribution that factorizes in this way can be obtained from the selection model. Hence, we have a distribution in which there are no independence constraints, but there is no *three-way interaction* term between X_u, X_v, X_w: that is, there is no factor which varies with all three variables. This example is due to Lauritzen [18], who shows that any hierarchical model can be generated in this way. Such models can be represented using factor graphs.

More recently Evans and Didelez [10] consider constraints on 'full conditionals': that is, on distributions of the form $p(x_V \mid x_W)$ for all x_W. These induce complicated rank constraints in the discrete case, and lead to statistical models with singularities, similar to latent variable models. However, the constraints can in principle be used to learn the structure of a model under selection bias, and even to recover the original joint distribution.

2.5.3 Quantum states

Bell's inequality was first derived as a property which holds if the hidden information is a classical random variable, but does not hold if it instead represents a quantum mechanical state. The quantum interpretation induces a weaker but non-trivial inequality of its own, but one that has not been explicitly described; for a statistical overview see [15].

In fact there is a hierarchy of models that can be used to interpret the graph in Figure 2.8(d), obtained by using more or fewer of the following constraints:

(1) $X_a \perp\!\!\!\perp X_c, X_d$ and $X_c \perp\!\!\!\perp X_a, X_b$;

(2) Quantum inequality;

(3) Bell's inequality;

(4) latent variable constraints.

All models agree on (1); adding (2) gives the quantum mechanical model, and taking the stronger inequality (3) gives what is called the *marginal model* in [8]: that is, the set of distributions that can be obtained as the margin of a DAG with arbitrary latent variables. If we make additional restrictions about the state-space of the hidden variable, as in (4), we may obtain additional restrictions. For example, if this hidden variable has only d_u states, then this implies that the matrix with entries

$$M(x_a, x_b; x_c, x_d) = P(X_b = x_b, X_d = x_d \mid X_a = x_a, X_c = x_c)$$

(with rows indexed by x_a, x_b, and columns by x_c, x_d) has rank at most d_u. These more exotic constraints are explored, for example, by Allman et al. [1].

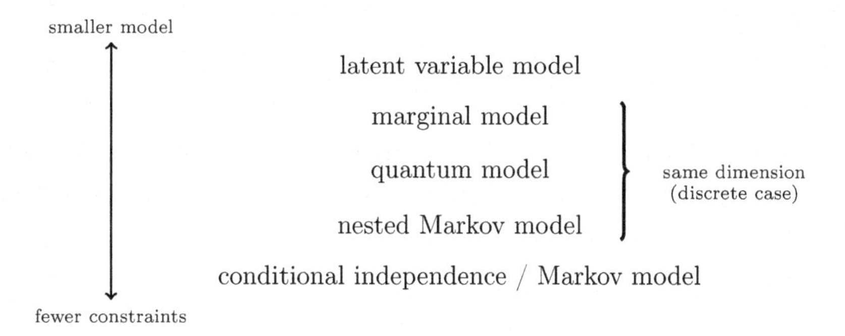

2.6 Summary

The term 'mixed graphs' applies to a highly varied collection of objects, each of which can be interpreted in a variety of different ways to obtain a huge range of possible models. More recent efforts at unification have been made [31, 20], but the resulting graphs are often complex. Lauritzen and Sadeghi [20] use four different edge types in order to unify LWF chain graph models and ancestral graphs, and the authors note that "we have chosen to avoid introducing a fifth type of edge to accommodate [the alternative Markov property]".

Different subclasses have different properties, and practitioners are advised to select a class of models with these attributes in mind. LWF chain graphs are parsimonious but not easily causally interpretable (though see [19]); ancestral graphs are causally interpretable but are restricted to conditional independence models; ADMGs and mDAGs under the nested Markov property are true to causal models with latent variables but may be computationally infeasible to use if districts are large.

Bibliography

[1] E. S. Allman, J. A. Rhodes, B. Sturmfels, and P. Zwiernik. Tensors of nonnegative rank two. *Linear Algebra and Its Applications*, 473:37–53, 2015.

[2] S. A. Andersson, D. Madigan, and M. D. Perlman. Alternative Markov properties for chain graphs. *Scandinavian Journal of Statistics*, 28:33–85, 2001.

[3] J. S. Bell. On the Einstein-Podolsky-Rosen paradox. *Physics*, 1:195–200, 1964.

[4] D. M. Chickering, D. Heckerman, and C. Meek. Large-sample learning of Bayesian networks is NP-hard. *Journal of Machine Learning Research*, 5:1287–1330, 2004.

[5] D. R. Cox and N. Wermuth. *Multivariate Dependencies: Models, Analysis and Interpretation.* CRC Press, 1996.

[6] M. Drton. Discrete chain graph models. *Bernoulli*, 15:736–753, 2009.

[7] R. J. Evans. Graphical methods for inequality constraints in marginalized DAGs. In *Machine Learning for Signal Processing (MLSP)*, 2012.

[8] R. J. Evans. Graphs for margins of Bayesian networks. *Scandinavian Journal of Statistics*, 43:625–648, 2016.

[9] R. J. Evans. Margins of discrete Bayesian networks. *Annals of Statistics*, 2018. To appear.

[10] R. J. Evans and V. Didelez. Recovering from selection bias using marginal structure in discrete models. In *Uncertainty in Artificial Intelligence, Advances in Causal Inference Workshop*, 2015.

[11] R. J. Evans and T. S. Richardson. Markovian acyclic directed mixed graphs for discrete data. *Annals of Statistics*, 42:1452–1482, 2014.

[12] R. J. Evans and T. S. Richardson. Smooth, identifiable supermodels of discrete DAG models with latent variables. *Bernoulli*, 2018. To appear.

[13] T. Fritz. Beyond Bell's theorem: correlation scenarios. *New Journal of Physics*, 14: 103001, 2012.

[14] M. Frydenberg. The chain graph Markov property. *Scandinavian Journal of Statistics*, 17:333–353, 1990.

[15] R. D. Gill. Statistics, causality and Bell's theorem. *Statistical Science*, 29:512–528, 2014.

[16] C. Kang and J. Tian. Inequality constraints in causal models with hidden variables. In *Proceedings of the 22nd Conference on Uncertainty in Artificial Intelligence (UAI-06)*, pages 411–420, 2006.

[17] S. L. Lauritzen. *Graphical Models.* Clarendon Press, Oxford, UK, 1996.

[18] S. L. Lauritzen. Generating mixed hierarchical interaction models by selection. Technical report, Aalborg University, Denmark, 1999.

[19] S. L. Lauritzen and T. S. Richardson. Chain graph models and their causal interpretations. *Journal of the Royal Statistical Society, Series B*, 64:321–348, 2002.

[20] S. L. Lauritzen and K. Sadeghi. Unifying Markov properties for graphical models. *Annals of Statistics*, 2018. To appear.

[21] S. L. Lauritzen and N. Wermuth. Graphical models for associations between variables, some of which are qualitative and some quantitative. *Annals of Statistics*, 17:31–57, 1989.

[22] Z. Ma, X. Xie, and Z. Geng. Structural learning of chain graphs via decomposition. *Journal of Machine Learning Research*, 9:2847–2880, 2008.

[23] J. Pearl. On the testability of causal models with latent and instrumental variables. In *Proceedings of the 11th Conference on Uncertainty in Artificial Intelligence (UAI)*, pages 435–443, 1995.

[24] J. Pearl. *Causality: Models, Reasoning, and Inference.* Cambridge University Press, second edition, 2009.

[25] J. Pearl and T. S. Verma. A statistical semantics for causation. *Statistics and Computing*, 2:91–95, 1992.

[26] T. S. Richardson. Markov properties for acyclic directed mixed graphs. *Scandinavian Journal of Statistics*, 30:145–157, 2003.

[27] T. S. Richardson and P. Spirtes. Ancestral graph Markov models. *Annals of Statistics*, 30:962–1030, 2002.

[28] T. S. Richardson, R.J. Evans, J. M. Robins, and I. Shpitser. Nested Markov properties for acyclic directed mixed graphs. arXiv:1701.06686, 2017.

[29] J. M. Robins. A new approach to causal inference in mortality studies with a sustained exposure period—application to control of the healthy worker survivor effect. *Mathematical Modelling*, 7:1393–1512, 1986.

[30] D. Rosset, N. Gisin, and E. Wolfe. Universal bound on the cardinality of local hidden variables in networks. arXiv:1709.00707, 2017.

[31] K. Sadeghi and S. L. Lauritzen. Markov properties for mixed graphs. *Bernoulli*, 20: 676–696, 2014.

[32] I. Shpitser. Segregated graphs and marginals of chain graph models. In *Advances in Neural Information Processing Systems*, pages 1720–1728, 2015.

[33] P. Spirtes, C. Glymour, and R. Scheines. *Causation, Prediction and Search.* MIT Press, 2000.

[34] M. Studeny and R. R. Bouckaert. On chain graph models for description of conditional independence structures. *Annals of Statistics*, 26:1434–1495, 1998.

[35] M. Studený, A. Roverato, and Š. Štěpánová. Two operations of merging and splitting components in a chain graph. *Kybernetika*, 45:208–248, 2009.

[36] T. S. Verma and J. Pearl. Equivalence and synthesis of causal models. In *Proceedings of the 6th Conference on Uncertainty in Artificial Intelligence (UAI-90)*, pages 255–270, 1990.

[37] S. Wright. Correlation and causation. *Journal of Agricultural Research*, 20:557–585, 1921.

[38] P. Zwiernik. *Semialgebraic Statistics and Latent Tree Models.* CRC Press, 2015.

3

Algebraic Aspects of Conditional Independence and Graphical Models

Thomas Kahle

Faculty of Mathematics, Otto von Guericke University of Magdeburg

Johannes Rauh

Max Planck Institute for Mathematics in the Sciences

Seth Sullivant

Department of Mathematics, North Carolina State University

CONTENTS

3.1 Introduction

Consider a finite set of random variables X_v, $v \in V$. Section 1.7 in Chapter 1 describes how to use a simple undirected graph $G = (V, E)$ to encode conditional independence (CI) statements among the random variables. One can also naturally associate a parametrized family of joint probability distributions of the X_v to a graph. For undirected graphs, the Hammersley-Clifford theorem (Observation 1.7.2) shows that both the implicit method and the parametric method lead to the same families of probability distributions (called graphical models), as long as all distributions are assumed strictly positive.

When probabilities are allowed to go to zero, the models defined by the collections of

CI statements contain probability distributions that do not lie in the parametric graphical model, which, by definition, consists of strictly positive probability distributions. In fact, these additional distributions do not even lie in the closure of the parametric graphical model, so they cannot be approximated by distributions from the parametric graphical model. Moreover, models defined by collections of CI statements (pairwise Markov properties, local Markov properties, global Markov properties) differ from one another. As an example, consider the four-cycle C_4.

Proposition 3.1.1. *The binary random variables* $X = (X_1, X_2, X_3, X_4)$ *satisfy the global Markov statements of* C_4, $1 \perp\!\!\!\perp 3 \,|\, \{2,4\}$ *and* $2 \perp\!\!\!\perp 4 \,|\, \{1,3\}$, *if and only if one (or more) of the following statements is true:*

1. *The joint distribution lies in the closure of the graphical model.*
2. *There is a pair* (X_i, X_{i+1}) *of neighboring nodes such that* $X_i = X_{i+1}$ *a.s.*
3. *There is a pair* (X_i, X_{i+1}) *of neighboring nodes such that* $X_i \neq X_{i+1}$ *a.s.*

This chapter shows how to prove results such as Proposition 3.1.1 using algebraic tools. The algebraic method can also be used to study implications between conditional independence statements. Here is an example:

Proposition 3.1.2. *Suppose that* X, Y, Z *are binary random variables or jointly normal random variables. If* $X \perp\!\!\!\perp Y$ *and* $X \perp\!\!\!\perp Y \,|\, Z$ *then either* $X \perp\!\!\!\perp (Y, Z)$ *or* $(X, Z) \perp\!\!\!\perp Y$.

The CI implication in Proposition 3.1.2 is a special case of the *gaussoid axiom* [26]. One may wonder what is special about jointly normal or binary random variables. For instance, is there a variant of this implication when X, Y, Z are discrete but not binary? How can one systematically find and study implications like this?

CI implications can also be interpreted as intersections of graphical models. For example, the two CI statements $X \perp\!\!\!\perp Y$ and $X \perp\!\!\!\perp Y \,|\, Z$ in Proposition 3.1.2 correspond to the two graphical models $X \to Z \leftarrow Y$ and $X - Z - Y$, respectively. Thus, Proposition 3.1.2 says that the intersection of these two graphical models equals the union of the two graphical models $X \;\; Z - Y$ and $X - Z \;\; Y$, provided the random variables are either binary or jointly normal. As the example shows, the intersection of two graphical models need not be a graphical model. How can one compute this intersection?

The goal of this chapter is to explore these questions and introduce tools from computational algebra for studying them. Our perspective is that, for a fixed type of random variable, the set of distributions that satisfy a collection of independence constraints is the zero set of a collection of polynomial equations. Solutions of systems of polynomial equations are the objects of study of algebraic geometry, and so tools from algebra can be brought to bear on the problem. The next section contains an overview of basic ideas in algebraic geometry which are useful for the study of conditional independence structures and graphical models. In particular, it introduces algebraic varieties, polynomial ideals, and primary decomposition. Section 3.3 introduces the ideals associated to families of conditional independence statements, and explains how to apply the basic techniques to deduce conditional independence implications. Section 3.4 illustrates the main ideas with some deeper examples coming from the literature. Section 3.5 concerns the vanishing ideal of a graphical model, which is a complete set of implicit restrictions for that model. This set of restrictions is usually much larger than the set of conditional independence constraints that come from the graph, but it can illuminate the structure of the model especially with more complex families of models involving mixed graphs or hidden random variables. Section 3.6 highlights some key references in this area.

3.2 Notions of Algebraic Geometry and Commutative Algebra

Commutative algebra is the study of systems of polynomial equations, and algebraic geometry is the study of geometric properties of their solutions. Both are rich fields with many deep results. This section only gives a very coarse introduction to the basic facts that hopefully makes it possible for the reader to understand the phenomena and algorithms discussed in later parts of this chapter. For a more detailed introduction, the reader is referred to the standard textbook [7].

3.2.1 Polynomials, ideals and varieties

Let $\mathbb{k}$ be a field, for example the rational numbers $\mathbb{Q}$, the real numbers $\mathbb{R}$, or the complex numbers $\mathbb{C}$. Let $\mathbb{N} = \{0, 1, \ldots\}$ denote the natural numbers. Let $p_1, p_2, \ldots, p_r$ be a collection of *indeterminates* or *variables*. A *monomial* in the indeterminates $p_1, p_2, \ldots, p_r$ is an expression of the form $p_1^{u_1} p_2^{u_2} \cdots p_r^{u_r}$ where $u_1, \ldots, u_r$ are nonnegative integers. Writing $u = (u_1, \ldots, u_r)$, let

$$p^u := p_1^{u_1} p_2^{u_2} \cdots p_r^{u_r}.$$

A *polynomial* is a finite linear combination of monomials, i.e.

$$f = \sum_{u \in U} c_u p^u$$

where $U \subset \mathbb{N}^r$ is a finite set and $c_u \in \mathbb{k}$. Of course, one is used to thinking of a polynomial as a function, $f : \mathbb{k}^r \to \mathbb{k}$, which can be evaluated in a point $a \in \mathbb{k}^r$ for p. In the following, this function will usually have the role of a constraint; i.e., the object of interest is the zero set $\{a \in \mathbb{k}^r : f(a) = 0\}$. In algebra, it is also useful to think of a polynomial as a formal object, i.e. the indeterminates are simply symbols that are used in manipulations, with no need for them to be evaluated.

The set of all polynomials in indeterminates $p_1, \ldots, p_r$ with coefficients in $\mathbb{k}$ is called the *polynomial ring* and denoted $\mathbb{k}[p_1, \ldots, p_r]$. The word *ring* means that $\mathbb{k}[p_1, \ldots, p_r]$ has two operations, namely addition of polynomials and multiplication of polynomials, and that these operations satisfy all the usual properties of addition and multiplication (associativity, commutativity, distributivity). However, multiplicative inverses need not exist: The result of dividing one polynomial by another non-constant polynomial is in general not a polynomial, but a rational function.

Definition 3.2.1. *Let $\mathcal{F} \subseteq \mathbb{k}[p_1, \ldots, p_r]$. The* variety *defined by $\mathcal{F}$ is the vanishing set of the polynomials in $\mathcal{F}$, that is,*

$$V(\mathcal{F}) = \{a \in \mathbb{k}^r : f(a) = 0 \text{ for all } f \in \mathcal{F}\}.$$

Example 3.2.2. Let $r = 2$ and consider $\mathcal{F} = \{x^2 - y\} \subseteq \mathbb{k}[x, y]$. The variety $V(\{x^2 - y\}) \subseteq \mathbb{k}^2$ is the familiar parabola "$y = x^2$" in the plane. For $r = 4$ and $\mathcal{F} = \{p_{11}p_{22} - p_{12}p_{21}\} \subseteq \mathbb{k}[p_{11}, p_{12}, p_{21}, p_{22}]$, the variety $V(\{p_{11}p_{22} - p_{12}p_{21}\}) \subseteq \mathbb{k}^4$ is the set of all singular 2×2 matrices $\left(\begin{smallmatrix} p_{11} & p_{12} \\ p_{21} & p_{22} \end{smallmatrix}\right)$ with entries in $\mathbb{k}$.

Example 3.2.3. Let $\mathcal{F} = \{x^2 - y, x^2 + y^2 - 1\}$. The variety $V(\mathcal{F})$ is the set of points

$$\{(x, y) \in \mathbb{k}^2 : y = x^2 \text{ and } x^2 + y^2 = 1\},$$

in other words, the intersection of a parabola and a circle. The number of points in this intersection varies depending on whether the underlying field is $\mathbb{Q}$, $\mathbb{R}$, or $\mathbb{C}$ (or some other field). If $\Bbbk = \mathbb{Q}$ the variety is empty, if $\Bbbk = \mathbb{R}$ the variety has two points, and if $\Bbbk = \mathbb{C}$, the variety has four points. In statistical applications one is usually interested in solutions over $\mathbb{R}$. However, it is often easier to first perform computations in algebraically closed fields like $\mathbb{C}$ before restricting to the real numbers, at least in theory. On the other hand, when using a computer algebra system, it may be advantageous to work with $\mathbb{Q}$, if possible, because rational numbers can be represented exactly on a computer.

The examples so far have always used finite sets $\mathcal{F}$. This is not necessary for the definition of a variety, and it is often worthwhile to consider the variety $V(\mathcal{F})$ where $\mathcal{F}$ is an infinite set of polynomials. In fact, it is often convenient to replace the original set of polynomials $\mathcal{F}$ by an infinite set, the *ideal* generated by $\mathcal{F}$, which is equivalent to $\mathcal{F}$ in some sense but has more structure.

One reason is that different families of polynomials may have the same varieties. For example, if $f, g \in \mathcal{F}$, then the variety of $\mathcal{F} \cup \{f + g\}$ equals $V(\mathcal{F})$. Similarly, for $f \in \mathcal{F}$ and $\lambda \in k$, the variety of $\mathcal{F} \cup \{\lambda f\}$ equals $V(\mathcal{F})$.

Definition 3.2.4. *A set $I \subseteq \Bbbk[p_1, \dots, p_r]$ is an* ideal *if for all $f, g \in I$, $f + g \in I$ and for all $f \in I$ and $h \in \Bbbk[p_1, \dots, p_r]$, $hf \in I$.*

Definition 3.2.5. *Let $\mathcal{F} \subseteq \Bbbk[p_1, \dots, p_r]$ be a set of polynomials. The* ideal generated by $\mathcal{F}$ *is the smallest ideal in $\Bbbk[p_1, \dots, p_r]$ that contains $\mathcal{F}$. Equivalently, the ideal generated by $\mathcal{F}$ consists of all polynomials $h_1 f_1 + \dots + h_k f_k$ for $h_1, \dots, h_k \in \Bbbk[p_1, \dots, p_r]$, $f_1, \dots, f_k \in \mathcal{F}$, and any k. The ideal generated by $\mathcal{F}$ is denoted $\langle \mathcal{F} \rangle$.*

Proposition 3.2.6. *Let $\mathcal{F} \subseteq \Bbbk[p_1, \dots, p_r]$ be a set of polynomials. Then $V(\mathcal{F}) = V(\langle \mathcal{F} \rangle)$.*

Example 3.2.7. The ideal $\langle x^2 - y, x^2 + y^2 - 1 \rangle$ generated by the set $\mathcal{F}$ from Example 3.2.3 has many different possible generating sets. For example, an alternate generating set is $\{x^2 - y, x^4 + x^2 - 1\}$. This allows to easily find all the solutions of the polynomial system because all roots of the univariate polynomial $x^4 + x^2 - 1 = 0$ can be plugged into the second polynomial $x^2 - y = 0$, which can then be solved for y.

Hilbert's basis theorem implies that for any ideal $I \subseteq \Bbbk[p_1, \dots, p_r]$ there exists a finite set of polynomials $\mathcal{F} \subseteq \Bbbk[p_1, \dots, p_r]$ such that $I = \langle \mathcal{F} \rangle$.

Even though it is, for theoretical considerations, often easier to think about systems of polynomial equations in terms of ideals, in practice (i.e. when working with computer algebra systems), the ideal is almost always specified in terms of a finite set of generators (or such a finite set of generators has to be computed on the way). On the other hand, during a computation it is often necessary to replace this set of generators by a more convenient set of generators (e.g. a Gröbner basis), so the generators may change even though the ideal stays the same along a computation.

Definition 3.2.8. *Let $S \subseteq \Bbbk^r$. The* vanishing ideal *of S is the set*

$$I(S) := \{f \in \Bbbk[p_1, \dots, p_r] : f(a) = 0 \text{ for all } a \in S\}.$$

It is easy to check that a vanishing ideal is indeed an ideal. Clearly, any ideal J satisfies $J \subseteq I(V(J))$. However, the converse inclusion does not hold in general. For instance, $I(V(\langle x^2 \rangle)) = \langle x \rangle$, and $I(V(\langle x^2 y, xy^2 \rangle)) = \langle xy \rangle$ (over any field $\Bbbk$).

Definition 3.2.9. *The ideal $I(V(J))$ is the* $\Bbbk$-radical of J. *An ideal J such that $I(V(J)) = J$ is a* $\Bbbk$-radical ideal. *If $\Bbbk$ is algebraically closed (e.g. if $\Bbbk = \mathbb{C}$), such an ideal J is simply called a* radical ideal.

Radical ideals can also be characterized algebraically, and there are algorithms to compute radicals. The radical is usually a simpler ideal, and if the radical of an ideal can be computed, it is advantageous to do this in a first step in each calculation (as long as one is only interested in properties of $V(J)$, and not in algebraic properties of J).

The following proposition illustrates the close relation between ideals and varieties.

Proposition 3.2.10.

- *Let* $S_1, S_2 \subseteq \Bbbk^r$. *Then* $I(S_1 \cup S_2) = I(S_1) \cap I(S_2)$ *and* $I(S_1 \cap S_2) \supseteq I(S_1) + I(S_2) := \{f + g : f \in I(S_1), g \in I(S_2)\}$.
- *Let* $I, J \subseteq \Bbbk[p_1, \ldots, p_r]$. *Then* $V(I \cup J) = V(I + J) = V(I) \cap V(J)$ *and* $V(I \cap J) = V(I) \cup V(J)$.

3.2.2 Irreducible and primary decomposition

Proposition 3.2.10 shows that the union of two varieties is again a variety. Interestingly, not every variety can be written as a non-trivial finite union.

Definition 3.2.11. *A variety V is* reducible *if there are two varieties $V_1, V_2 \neq V$ such that $V_1 \cup V_2 = V$. Otherwise V is* irreducible.

Theorem 3.2.12. *Any variety $V \subseteq \Bbbk^r$ has a unique decomposition into finitely many irreducible varieties $V = V_1 \cup \cdots \cup V_k$ (with $V_i \not\subseteq V_j$ for $i \neq j$).*

The varieties $V_1, \ldots, V_k$ are called the *irreducible components* of V.

Theorem 3.2.13.

1. *Let V be an irreducible variety, and let $\phi : \Bbbk^r \to \Bbbk^s$ be a rational map. Then $V(I(\phi(V)))$ is irreducible.*
2. *Let V be a variety that has a rational parametrization $\phi : \Bbbk^r \to V$ such that the image of ϕ is dense in V. Then V is irreducible.*

According to Proposition 3.2.10, the corresponding decomposition operation for ideals is to write ideals as the intersection of other ideals. However, for general ideals, the situation is much more complicated than for varieties. The situation simplifies for radical ideals (which are in a one-to-one correspondence with varieties). This case is discussed next. The general case is summarized afterwards.

Definition 3.2.14. *An ideal $I \subseteq \Bbbk[p_1, \ldots, p_r]$ is* prime *if for all $f, g \in \Bbbk[p_1, \ldots, p_r]$ with $f \cdot g \in I$, one of the factors f, g belongs to I.*

For example, $I := \langle xy \rangle$ is not prime, because $xy \in I$, but neither $x \in I$ nor $y \in I$.

Theorem 3.2.15. *A variety V is irreducible if and only if $I(V)$ is prime.*

Definition 3.2.16. *A prime ideal $P \subseteq \Bbbk[p_1, \ldots, p_r]$ is a* minimal prime *of an ideal $I \subseteq \Bbbk[p_1, \ldots, p_r]$ if and only if $V(P)$ is an irreducible component of $V(I)$.*

There is also an algebraic definition of the minimal primes, and there are algorithms to compute the minimal primes. By definition, the minimal primes of an ideal encode the irreducible decomposition of the corresponding variety:

Theorem 3.2.17.

1. *Any ideal $I \subseteq \Bbbk[p_1, \ldots, p_r]$ has finitely many minimal primes $P_1, \ldots, P_k$.*

2. *The ideal $P_1 \cap \cdots \cap P_k$ equals the radical of I.*

3. *The irreducible components of $V(I)$ are $V(P_1), \ldots, V(P_k)$.*

If I is not radical, then $P_1 \cap \cdots \cap P_k \subseteq I$. In this case, it is still possible to write I as an intersection of special ideals (called *primary ideals*) in a way that is algebraically and geometrically meaningful. This intersection is called a *primary decomposition.* The precise definitions are omitted, since a primary decomposition often adds little to the statistical understanding. However, some works in algebraic statistics written by algebraists who do care about the differences between ideals and their radicals use this notation. The following result explains how a primary decomposition is related to the minimal primes.

Theorem 3.2.18. *Let $I = I_1 \cap \cdots \cap I_l$ be a primary decomposition of $I \subseteq \Bbbk[p_1, \ldots, p_r]$, and let P_i be the radical of I_i.*

1. $V(I) = V(I_1) \cup V(I_2) \cup \cdots \cup V(I_l) = V(P_1) \cup V(P_2) \cup \cdots \cup V(P_l)$.

2. *Each P_i is prime.*

3. *Each minimal prime of I is among the P_i.*

4. *If P_i is not a minimal prime of I, then there is a minimal prime P_j of I with $P_j \subset P_i$ (and so $V(P_i) \subset V(P_j)$).*

Example 3.2.19. Let $I = \langle xy, xz \rangle \in \Bbbk[x, y, z]$. The variety $V(I)$ consists of the union of the plane where $x = 0$, and the line where $y = 0, z = 0$. Hence $V(\langle xy, xz \rangle) = V(\langle x \rangle) \cup V(\langle y, z \rangle)$ is a decomposition into irreducibles. This corresponds to the ideal decomposition $\langle xy, xz \rangle = \langle x \rangle \cap \langle y, z \rangle$.

The primary decomposition need not be unique.

Example 3.2.20. The ideal $\langle x^2, xy \rangle$ has several different primary decompositions, e.g.

$$\langle x^2, xy \rangle = \langle x \rangle \cap \langle x^2, y \rangle = \langle x \rangle \cap \langle x^2, x + y \rangle.$$

The variety $V(\langle x^2, xy \rangle)$ equals the line where $x = 0$, corresponding to the unique minimal prime $\langle x \rangle$. The non-uniqueness of the primary decomposition is related to the fact that the variety of the "extra" component is a subset of one of the other components. This variety (which is superfluous in the irreducible decomposition) is called an *embedded component.*

This example can be analyzed as follows using the computer algebra system MACAULAY2 [19]. First set up a polynomial ring in the indeterminates x, y with the rational numbers $\mathbb{Q}$ as the coefficient field. In MACAULAY2 it is advisable to work with $\mathbb{Q}$ rather than $\mathbb{R}$ or $\mathbb{C}$ since the arithmetic in $\mathbb{Q}$ can be carried out exactly on a computer.

```
i1 : R = QQ[x,y]
o1 = R
o1 : PolynomialRing
```

The system reports that it understands `R` as a polynomial ring. The following input makes MACAULAY2 decompose the ideal. The decomposition is computed over $\mathbb{Q}$, but in this case it happens to be valid over any field $\Bbbk$.

```
i2 : primaryDecomposition ideal (x^2, x*y)
                              2
o2 = {ideal x, ideal (x , y)}
```

If one is only interested in the irreducible decomposition, the command `decompose` returns the minimal primes corresponding to the irreducible components, discarding all embedded components:

```
i3 : decompose ideal (x^2, x*y)
o3 = {ideal x}
```

3.2.3 Binomial ideals

This section ends with a short discussion of binomial ideals and toric ideals, which make frequent appearance in applications.

Definition 3.2.21. *A* binomial *is a polynomial $p^u - \lambda p^v$, $\lambda \in \mathbb{k}$ with at most two terms. An ideal I is a* binomial ideal *if it has a generating set of binomials. A binomial ideal that is prime and does not contain any variable is a* toric ideal.

The main reason why it is important whether an ideal is binomial is that there are dedicated algorithms for binomial ideals that are much faster than the generic algorithms that work for any ideal [13, 10, 22, 23]. Note that there are some instances of ideals that arise in algebraic statistics that are not binomial in their natural coordinate systems but become binomial ideals after a linear change of coordinates [31].

Let $A \in \mathbb{Z}^{h \times r}$ be an integer matrix, and consider the ideal

$$I_A := \langle p^{u_+} - p^{u_-} : u = u_+ - u_- \in \ker_{\mathbb{Z}} A \rangle$$

in the polynomial ring $\mathbb{k}[p_1, \dots, p_r]$, where $u = u_+ - u_-$ is the decomposition of u into its positive and negative part $u_+, u_- \in \mathbb{N}^r$. Clearly, I_A is binomial and does not contain any of the p_i. One can also show that I_A is prime, and thus it is an example of a toric ideal. In fact, any toric ideal is of this form up to a scaling of coordinates [13, Corollary 2.6]. The generating set above is infinite, but Theorem 3.1 in [9] shows that finite generating sets of toric ideals are related to Markov bases, which can be computed using the software `4ti2` [1].

3.2.4 Real algebraic geometry

In addition to polynomial equations, in many situations in statistics it is useful to consider solutions to polynomial inequalities as well. This is the subject of the field *real algebraic geometry*. Inequalities only make sense over an ordered field like $\mathbb{R}$ (but not over $\mathbb{C}$). For simplicity, the following definitions and results are formulated with $\mathbb{R}$. Again, this text only contains the basic definitions. For more details the reader is referred to [4, 5].

Definition 3.2.22. *Let $\mathcal{F}, \mathcal{G} \subseteq \mathbb{R}[p_1, \dots, p_r]$ be sets of polynomials with $\mathcal{G}$ finite. The* basic semialgebraic set *defined by $\mathcal{F}$ and $\mathcal{G}$ is*

$$\{a \in \mathbb{R}^r : f(a) = 0 \text{ for all } f \in \mathcal{F} \text{ and } g(a) > 0 \text{ for all } g \in \mathcal{G}\}.$$

A semialgebraic set *is a finite union of basic semialgebraic sets.*

Here are some common examples of semialgebraic sets arising in statistics.

Example 3.2.23. The open probability simplex

$$\text{int}(\Delta_{r-1}) := \{p \in \mathbb{R}^r : \sum_{i=1}^{r} p_i = 1, p_i > 0, i = 1, \dots, r\}$$

consists of all probability distributions for a categorical random variable with r states. It is a basic semialgebraic set: In the above definition, one may take $\mathcal{F} = \{ \sum_{i=1}^r p_i - 1\}$ and $\mathcal{G} = \{p_1, \ldots, p_r\}$. The probability simplex

$$\Delta_{r-1} := \{p \in \mathbb{R}^r : \sum_{i=1}^{r} p_i = 1, p_i \geq 0, i = 1, \ldots, r\}$$

is a semialgebraic set. It can be written as the union of $2^r - 1$ basic semialgebraic sets.

Example 3.2.24. The cone PD_m of $m \times m$ positive definite symmetric matrices is an example of a basic semialgebraic set in $\mathbb{R}^{\binom{m+1}{2}}$, where $\mathcal{F} = \emptyset$ and where $\mathcal{G}$ consists of the principal subdeterminants of an $m \times m$ symmetric matrix of indeterminates. For instance, if $m = 3$ consider the polynomial ring $\mathbb{R}[\sigma_{11}, \sigma_{12}, \sigma_{13}, \sigma_{22}, \sigma_{23}, \sigma_{33}]$ and the symmetric matrix of indeterminates

$$\Sigma = \begin{pmatrix} \sigma_{11} & \sigma_{12} & \sigma_{13} \\ \sigma_{12} & \sigma_{22} & \sigma_{23} \\ \sigma_{13} & \sigma_{23} & \sigma_{33} \end{pmatrix}.$$

The symmetry has been enforced by making certain entries in the matrix equal. The set of polynomials defining PD_3 can be chosen to be

$$\mathcal{G} = \{\sigma_{11}, \sigma_{11}\sigma_{22} - \sigma_{12}^2, \det \Sigma\},$$

the set of leading principal minors of Σ. The cone of positive semidefinite symmetric matrices is a semialgebraic set, which can be realized by using non-strict inequalities with the much larger set of all principal minors of Σ.

3.3 Conditional Independence Ideals

This section shows how the algebraic tools introduced in Section 3.2 can be used to analyze conditional independence structures. The tools can be applied in the settings of discrete random variables and jointly normal variables, but in different ways.

3.3.1 Discrete random variables

Let $X_1, X_2, \ldots, X_m$ be finite discrete random variables. Suppose that the state space of X_i is $[r_i] := \{1, 2, \ldots, r_i\}$. There is an algebraic description of the set of all distributions that satisfy a given conditional independence statement. The first example comes from the simplest CI statement: $1 \perp\!\!\!\perp 2$.

Proposition 3.3.1. *Let X_1, X_2 be discrete random variables where the state space of X_i is $[r_i]$. Let $p_{i_1 i_2} = P(X_1 = i_1, X_2 = i_2)$ and let $p = (p_{i_1 i_2})_{i_1 \in [r_1], i_2 \in [r_2]}$ be the joint probability mass function of X_1 and X_2. Then $1 \perp\!\!\!\perp 2$ if and only if p is a rank one matrix.*

Proof. If $1 \perp\!\!\!\perp 2$ then $P(X_1 = i_1, X_2 = i_2) = P(X_1 = i_1)P(X_2 = i_2)$. This expresses the joint probability mass function as an outer product of two nonzero vectors, hence p has rank one.

Conversely, if p has rank one, it is expressed as the outer product of two vectors $p = \alpha^T \beta$. Since p is a matrix of nonnegative real numbers, one can assume that α and β are also nonnegative. Let $\|.\|_1$ denote the l_1-norm. Replacing α by $\alpha/\|\alpha\|_1$ and β by $\beta/\|\beta\|_1$, yields a rank one factorization for p where the two factors are necessarily the marginal distributions of X_1 and X_2 respectively. Hence $1 \perp\!\!\!\perp 2$. □

A nonzero matrix having rank one is characterized by the vanishing of all its 2×2 subdeterminants. Hence, one can associate an ideal to the independence statement $1 \perp\!\!\!\perp 2$.

Definition 3.3.2. *The* conditional independence ideal *for the statement* $1 \perp\!\!\!\perp 2$ *is*

$$\begin{aligned} I_{1 \perp\!\!\!\perp 2} &= \langle p_{i_1 i_2} p_{j_1 j_2} - p_{i_1 j_2} p_{j_1 i_2} : i_1, j_1 \in [r_1], i_2, j_2 \in [r_2] \rangle \\ &= \langle 2 \times 2 \text{ subdeterminants of } p \rangle \subseteq \mathbb{R}[p_{i_1, i_2} : i_1 \in [r_1], i_2 \in [r_2]]. \end{aligned}$$

Example 3.3.3. Let $r_1 = 2$ and $r_2 = 3$. Then

$$I_{1 \perp\!\!\!\perp 2} = \langle p_{11} p_{22} - p_{12} p_{21}, p_{11} p_{23} - p_{13} p_{21}, p_{12} p_{23} - p_{13} p_{22} \rangle.$$

The conditional independence ideal $I_{1 \perp\!\!\!\perp 2}$ captures the algebraic structure of the independence condition. Although all probability distributions would satisfy the additional constraint that $\sum_{i_1 \in [r_1], i_2 \in [r_2]} p_{i_1 i_2} - 1 = 0$, this trivial constraint is not included in the conditional independence ideal because leaving it out tends to simplify certain algebraic calculations. For example, without this constraint $I_{1 \perp\!\!\!\perp 2}$ is a binomial ideal.

More generally, any conditional independence condition for discrete random variables can be expressed by similar determinantal constraints. This requires a bit of notation. The determinantal constraints are written in terms of the entries of the joint distribution of $X_1, \ldots, X_m$. This is a *tensor* $p = (p_{i_1, \ldots, i_m})_{i_j \in [r_j]}$.

Let $A, B, C \subset [m]$ be disjoint subsets of indices of the random variables $X_1, \ldots, X_m$, and $D = [m] \setminus (A \cup B \cup C)$ the set of indices appearing in none of A, B, C. Any such assignment yields a grouping of indices and random variables. The random vector $X_A = (X_j)_{j \in A}$ takes values in $\mathcal{R}_A = \prod_{j \in A} [r_j]$. Let $\mathcal{R}_B, \mathcal{R}_C$ and $\mathcal{R}_D$ be defined analogously. The grouping allows one to write $p = (p_{i_A, i_B, i_C, i_D})$ where now $i_A \in \mathcal{R}_A$ and similarly for i_B, i_C, and i_D. The final notational gadget is the marginalization of p over D. The entries of this marginal distribution are indexed by $\mathcal{R}_A, \mathcal{R}_B, \mathcal{R}_C$ and have entries

$$p_{i_A, i_B, i_C, +} = \sum_{i_D \in \mathcal{R}_D} p_{i_A, i_B, i_C, i_D}.$$

The $+$ indicates the summation.

Definition 3.3.4. *The conditional independence ideal for the conditional independence statement* $A \perp\!\!\!\perp B \,|\, C$ *is*

$$I_{A \perp\!\!\!\perp B \,|\, C} = \Big\langle p_{i_A, i_B, i_C, +} \cdot p_{j_A, j_B, i_C, +} - p_{i_A, j_B, i_C, +} \cdot p_{j_A, i_B, i_C, +}, \text{ for all } \\ i_A, j_A \in \mathcal{R}_A, i_B, j_B \in \mathcal{R}_B, i_C \in \mathcal{R}_C \Big\rangle.$$

The notation simplifies for *saturated conditional independence statements*, for which $A \cup B \cup C = [m]$. With this condition there is no marginalization, and the defining polynomials of $I_{A \perp\!\!\!\perp B \,|\, C}$ are binomials.

Example 3.3.5. Consider three binary random variables X_1, X_2, X_3. Let $p_{111}, \ldots, p_{222}$ denote the indeterminates standing for the elementary probabilities in the joint distribution. The conditional independence ideal of the statement $1 \perp\!\!\!\perp 3 \,|\, 2$ is

$$I_{1 \perp\!\!\!\perp 3 \,|\, 2} = \langle p_{111} p_{212} - p_{211} p_{112}, p_{121} p_{222} - p_{221} p_{122} \rangle.$$

The conditional independence ideal of the statement $1 \perp\!\!\!\perp 3$ is

$$I_{1 \perp\!\!\!\perp 3} = \langle (p_{111} + p_{121})(p_{212} + p_{222}) - (p_{112} + p_{122})(p_{211} + p_{221}) \rangle.$$

Proposition 3.3.6. *For any conditional independence statement $A \perp\!\!\!\perp B \,|\, C$, the conditional independence ideal $I_{A \perp\!\!\!\perp B \,|\, C}$ is a prime ideal and hence $V(I_{A \perp\!\!\!\perp B \,|\, C})$ is an irreducible variety.*

Proposition 3.3.6 is a consequence of the fact that general determinantal ideals are prime (see [6]). Irreducibility of the variety $V(I_{A \perp\!\!\!\perp B \,|\, C})$ can also be deduced from the fact that this variety can be parametrized, for instance, the set of all probability distributions in $V(I_{A \perp\!\!\!\perp B \,|\, C})$ can be realized as the set of probability distributions in a graphical model.

Example 3.3.7. A strictly positive joint distribution p of binary random variables X_1, X_2, X_3 satisfies $1 \perp\!\!\!\perp 3 \,|\, 2$ if and only if

$$p_{i_1,i_2,i_3} = s_{i_1,i_2} t_{i_2,i_3} \tag{3.1}$$

for some vectors $(s_{i_1,i_2})_{i_1 \in [r_1], i_2 \in [r_2]}$, $(t_{i_2,i_3})_{i_2 \in [r_2], i_3 \in [r_3]}$; see Section 1.3. That is, it lies in the undirected graphical model

$$X_1 - X_2 - X_3.$$

Since $V(I_{A \perp\!\!\!\perp B \,|\, C})$ is irreducible, any joint distribution (possibly with zeros) that satisfies $1 \perp\!\!\!\perp 3 \,|\, 2$ lies in the closure of the undirected graphical model. In fact, any such joint distribution has a parametrization of the form (3.1), where s or t also may have zeros.

More interesting than just single statements are combinations of two or more conditional independence statements. To determine the classes of distributions satisfying a collection of independence statements leads to interesting problems in computational algebra. Such sets are typically not irreducible varieties and cannot be parametrized with a single parametrization. The first task is to break such a set into components, and to see if those components have natural interpretations in terms of conditional independence and can be parametrized.

Definition 3.3.8. *Let $\mathcal{C} = \{A_1 \perp\!\!\!\perp B_1 \,|\, C_1, A_2 \perp\!\!\!\perp B_2 \,|\, C_2, \ldots\}$ be a set of conditional independence statements for the random variables $X_1, X_2, \ldots, X_m$. The* conditional independence ideal of $\mathcal{C}$ *is the sum of the conditional independence ideals of the elements of $\mathcal{C}$:*

$$I_{\mathcal{C}} = I_{A_1 \perp\!\!\!\perp B_1 \,|\, C_1} + I_{A_2 \perp\!\!\!\perp B_2 \,|\, C_2} + \cdots.$$

Understanding the probability distributions that satisfy $\mathcal{C}$ can be accomplished by analyzing an irreducible decomposition of $V(I_{\mathcal{C}})$, which can be obtained from a primary decomposition of $I_{\mathcal{C}}$.

Example 3.3.9. Let X_1, X_2, X_3 be binary random variables, and consider $\mathcal{C} = \{1 \perp\!\!\!\perp 3 \,|\, 2, 1 \perp\!\!\!\perp 3\}$. The conditional independence ideal $I_{\mathcal{C}}$ is generated by three polynomials of degree 2:

$$\begin{aligned} I_{\mathcal{C}} = I_{1 \perp\!\!\!\perp 3 \,|\, 2} + I_{1 \perp\!\!\!\perp 3} = \langle & p_{111}p_{212} - p_{112}p_{211}, p_{121}p_{222} - p_{122}p_{221}, \\ & (p_{111} + p_{121})(p_{212} + p_{222}) - (p_{112} + p_{122})(p_{211} + p_{221}) \rangle. \end{aligned}$$

The following MACAULAY2 code asks for the primary decomposition of this ideal over $\mathbb{Q}$. It can be shown that the decomposition is the same over $\mathbb{R}$ and $\mathbb{C}$.

```
loadPackage "GraphicalModels"
S = markovRing (2,2,2)
L = {{{1},{3},{2}}, {{1},{3},{}}}
I = conditionalIndependenceIdeal(S,L)
primaryDecomposition I
```

This code uses the GRAPHICALMODELS package of MACAULAY2 which implements many convenient functions to work with graphical and other conditional independence models. In particular, it allows to easily set up the polynomial ring with eight variables $p_{111}, \ldots, p_{222}$ with `markovRing` and write out the equations for $I_{\mathcal{C}}$ with `conditionalIndependenceIdeal`. The command `primaryDecomposition` is a generic MACAULAY2 command. The output of this code consists of two ideals which upon inspection can be recognized as binomial conditional independence ideals themselves. The result is

$$I_{\mathcal{C}} = I_{\{1,2\} \perp\!\!\!\perp 3} \cap I_{1 \perp\!\!\!\perp \{2,3\}}.$$

According to Section 3.2 this implies a decomposition of varieties

$$V(I_{\mathcal{C}}) = V(I_{\{1,2\} \perp\!\!\!\perp 3}) \cup V(I_{1 \perp\!\!\!\perp \{2,3\}}).$$

On the level of probability distributions, this proves the binary case of Proposition 3.1.2.

The general situation may be less favorable than that in Example 3.3.9. In particular, the components that appear need not have interpretations in terms of conditional independence. The appearing ideals also need not be prime ideals (in general they are only primary) and it is unclear what this algebraic extra information may reveal about conditional independence. For examples on how to extract information from primary decompositions see [20, 24].

3.3.2 Gaussian random variables

Algebraic approaches to conditional independence can also be applied to Gaussian random variables. Let $X \in \mathbb{R}^m$ be a nonsingular multivariate Gaussian random vector with mean $\mu \in \mathbb{R}^m$ and covariance matrix $\Sigma \in PD_m$, the cone of $m \times m$ symmetric positive definite matrices. One writes $X \sim \mathcal{N}(\mu, \Sigma)$. For subsets $A, B \subseteq [m]$ let $\Sigma_{A,B}$ be the submatrix of Σ obtained by extracting rows indexed by A and columns indexed by B, that is $\Sigma_{A,B} = (\sigma_{a,b})_{a \in A, b \in B}$.

Proposition 3.3.10. *Let $X \sim \mathcal{N}(\mu, \Sigma)$ with $\Sigma \in PD_m$. Let $A, B, C \subseteq [m]$ be disjoint subsets. Then the conditional independence statement $A \perp\!\!\!\perp B \,|\, C$ holds if and only if the matrix $\Sigma_{A \cup C, B \cup C}$ has rank $\leq \#C$.*

A proof of this proposition recognizes $\Sigma_{A \cup C, B \cup C}$ as a Schur complement of a submatrix of Σ. The details can be found in [11, Proposition 3.1.13]; see also Section 9.1 in the present handbook.

Just as the rank one condition on a matrix was characterized by the vanishing of 2×2 subdeterminants, higher rank conditions on matrices can also be characterized by the vanishing of subdeterminants. Indeed, a basic fact of linear algebra is that a matrix has rank $\leq r$ if and only if the determinant of every $(r + 1) \times (r + 1)$ submatrix is zero. This leads to the conditional independence ideals for multivariate Gaussian random variables.

Let $\mathbb{R}[\Sigma] := \mathbb{R}[\sigma_{ij} : 1 \leq i \leq j \leq m]$ be the polynomial ring with real coefficients in the entries of the symmetric matrix Σ.

Definition 3.3.11. *The* Gaussian conditional independence ideal *for the conditional independence statement $A \perp\!\!\!\perp B \,|\, C$ is the ideal*

$$J_{A \perp\!\!\!\perp B \,|\, C} := \langle (\#C + 1)\text{-minors of } \Sigma_{A \cup C, B \cup C} \rangle.$$

If $\mathcal{C} = \{A_1 \perp\!\!\!\perp B_1 \,|\, C_1\,,\, A_2 \perp\!\!\!\perp B_2 \,|\, C_2\,, \ldots, \}$ is a collection of conditional independence statements, the Gaussian conditional independence ideal is

$$J_{\mathcal{C}} = J_{A_1 \perp\!\!\!\perp B_1 \,|\, C_1} + J_{A_2 \perp\!\!\!\perp B_2 \,|\, C_2} + \cdots.$$

Remark 3.3.12. A common criterion in statistics says that, in fact, $A \perp\!\!\!\perp B \,|\, C$ holds if and only if $\det(\Sigma_{\{a\}\cup C,\{b\}\cup C})$ vanishes for all $a \in A$, $b \in B$. Since, by assumption, C is non-singular, it is easy to see that this condition is, in fact, equivalent to the vanishing of all $(\#C+1)$-minors of $\Sigma_{A\cup C, B\cup C}$.

Example 3.3.13. Consider the conditional independence statement $2 \perp\!\!\!\perp \{1,3\} \,|4$. The ideal $J_{2\perp\!\!\!\perp\{1,3\}\,|4}$ is generated by the 2×2 minors of the matrix

$$\Sigma_{\{2,4\},\{1,3,4\}} = \begin{pmatrix} \sigma_{12} & \sigma_{23} & \sigma_{24} \\ \sigma_{14} & \sigma_{34} & \sigma_{44} \end{pmatrix}.$$

Since Σ is a symmetric matrix, $\sigma_{ij} = \sigma_{ji}$ and one always writes σ_{ij} with $i \le j$. Then

$$J_{2\perp\!\!\!\perp\{1,3\}\,|4} = \langle \sigma_{12}\sigma_{34} - \sigma_{14}\sigma_{23}, \sigma_{12}\sigma_{44} - \sigma_{14}\sigma_{24}, \sigma_{23}\sigma_{44} - \sigma_{34}\sigma_{24} \rangle.$$

Example 3.3.14. Let X_1, X_2, X_3 be jointly Gaussian random variables. The conditional independence ideal of $\mathcal{C} = \{1 \perp\!\!\!\perp 3\,|2 \,,\, 1 \perp\!\!\!\perp 3\}$ is

$$J_{\mathcal{C}} = J_{1\perp\!\!\!\perp 3\,|2} + J_{1\perp\!\!\!\perp 3} = \langle \sigma_{13}\sigma_{22} - \sigma_{12}\sigma_{23}, \sigma_{13} \rangle.$$

Straightforward manipulations of these ideals show

$$\begin{aligned} J_{\mathcal{C}} = \langle \sigma_{13}\sigma_{22} - \sigma_{12}\sigma_{23}, \sigma_{13} \rangle &= \langle \sigma_{12}\sigma_{23}, \sigma_{13} \rangle = \langle \sigma_{12}, \sigma_{13} \rangle \cap \langle \sigma_{23}, \sigma_{13} \rangle \\ &= J_{\{1,2\}\perp\!\!\!\perp 3} \cap J_{1\perp\!\!\!\perp\{2,3\}}\,. \end{aligned}$$

This last primary decomposition proves the Gaussian case of Proposition 3.1.2.

3.3.3 The contraction axiom

When computing the decompositions of conditional independence ideals, there might be components that are "uninteresting" from the statistical standpoint. These components might not intersect the region of interest in probabilistic applications (e.g. they might miss the probability simplex or the cone of positive definite matrices) or they might have non-trivial intersections but that intersection is contained in some other component.

Example 3.3.15. Let $X = (X_1, X_2, X_3)$ be a multivariate Gaussian random vector. The conditional independence ideal of $\mathcal{C} = \{1 \perp\!\!\!\perp 2\,|3 \,,\, 2 \perp\!\!\!\perp 3\}$ is

$$J_{\mathcal{C}} = \langle \sigma_{12}\sigma_{33} - \sigma_{13}\sigma_{23}, \sigma_{23} \rangle$$

which has primary decomposition

$$J_{\mathcal{C}} = \langle \sigma_{12}\sigma_{33}, \sigma_{23} \rangle = \langle \sigma_{12}, \sigma_{23} \rangle \cap \langle \sigma_{33}, \sigma_{23} \rangle.$$

This decomposition shows that

$$V(J_{\mathcal{C}}) = V(\langle \sigma_{12}, \sigma_{23} \rangle) \cup V(\langle \sigma_{33}, \sigma_{23} \rangle).$$

However, the second component does not intersect the positive definite cone, because $\sigma_{33} > 0$ for all $\Sigma \in PD_3$. The first component is the conditional independence ideal $J_{1,3\perp\!\!\!\perp 2}$. From this decomposition it is visible that $1 \perp\!\!\!\perp 2\,|3$ and $2 \perp\!\!\!\perp 3$ imply that $1,3 \perp\!\!\!\perp 2$. This implication is called the *contraction axiom*. See Section 1.5 for other CI axioms.

The contraction axiom also holds for non-Gaussian random variables. For discrete random variables, it can again be checked algebraically. The primary decomposition associated to the discrete contraction axiom is worked out in detail in [17]. The next example discusses the binary case as an illustration:

Example 3.3.16. Let X_1, X_2, X_3 be binary random variables. The conditional independence ideal of $\mathcal{C} = \{1 \perp\!\!\!\perp 2\,|3\,,\,2 \perp\!\!\!\perp 3\}$ is

$$\begin{aligned} I_{\mathcal{C}} = \big\langle & p_{111}p_{221} - p_{121}p_{211}, p_{112}p_{222} - p_{122}p_{212}, \\ & (p_{111} + p_{211})(p_{122} + p_{222}) - (p_{112} + p_{212})(p_{121} + p_{221})\big\rangle \end{aligned}$$

which has primary decomposition

$$\begin{aligned} I_{\mathcal{C}} = I_{1,3 \perp\!\!\!\perp 2} & \cap \langle p_{122} + p_{222}, p_{112} + p_{212}, p_{121}p_{211} - p_{111}p_{221}\rangle \\ & \cap \langle p_{121} + p_{221}, p_{111} + p_{211}, p_{122}p_{212} - p_{112}p_{222}\rangle. \end{aligned}$$

The intersection of the second component with the probability simplex forces that $p_{122} = p_{222} = p_{112} = p_{212} = 0$. This in turn implies that

$$V(\langle p_{122} + p_{222}, p_{112} + p_{212}, p_{121}p_{211} - p_{111}p_{221}\rangle) \cap \Delta_7 \subseteq V(I_{1,3 \perp\!\!\!\perp 2})$$

A similar argument holds for the third component. So although the variety $V(I_{\mathcal{C}})$ has three components, only one of them is statistically meaningful:

$$V(I_{\mathcal{C}}) \cap \Delta_7 = V(I_{1,3 \perp\!\!\!\perp 2}) \cap \Delta_7.$$

3.4 Examples of Decompositions of Conditional Independence Ideals

This section studies some examples of families of conditional independence statements and how algebraic tools can be used to understand them. The first example is a detailed study of the intersection axiom, and the second example concerns the conditional independence statements associated to the 4-cycle graph.

3.4.1 The intersection axiom

The *intersection axiom* from Section 1.5 is the following implication of conditional independence statements:

$$A \perp\!\!\!\perp B\,|C \cup D \text{ and } A \perp\!\!\!\perp C\,|B \cup D \Rightarrow A \perp\!\!\!\perp B \cup C\,|D\,. \tag{3.2}$$

This implication is valid for strictly positive probability distributions. Algebraic techniques can be used to study how its validity extends beyond this.

The question about the primary decomposition of the ideal(s) $I_{\{A \perp\!\!\!\perp B\,|C \cup D\,,\,A \perp\!\!\!\perp C\,|B \cup D\}}$ was first asked in [11, Chapter 6.6], and the answer is due to [15]. Grouping variables if necessary, one can assume $A = \{1\}$, $B = \{2\}$, $C = \{3\}$. Moreover, let $D = \emptyset$. From this one can always recover the general case by adding conditioning constraints.

Proposition 3.4.1 (Proposition 1 in [15]). *The ideal $I_{\{X_1 \perp\!\!\!\perp X_2\,|X_3\,,\,X_1 \perp\!\!\!\perp X_3\,|X_2\}}$ is radical, that is, its irredundant primary decompositions consists only of prime ideals. These minimal primes correspond to pairs of partitions $[r_2] = A_1 \cup \cdots \cup A_s$, $[r_3] = B_1 \cup \cdots \cup B_s$ of the same size. The minimal prime P corresponding to two partitions is*

$$\begin{aligned} P = & \langle p_{i_1 i_2 i_3} : i_1 \in [r_1], i_2 \in A_j, i_3 \in B_k \text{ for some } j \neq k\rangle \\ & + \langle p_{i_1 i_2 i_3} p_{i_1' i_2' i_3'} - p_{i_1 i_2' i_3'} p_{i_1' i_2 i_3} : i_1, i_1' \in [r_1], i_2, i_2' \in A_j, i_3, i_3' \in B_j \text{ for some } j\rangle. \end{aligned}$$

The paper [15] uses a different formulation in terms of complete bipartite graphs. It can be seen that our formulation is equivalent.

To give a statistical interpretation to Proposition 3.4.1, whenever the joint distribution of X_1, X_2, X_3 lies in the prime P corresponding to the two partitions $[r_2] = A_1 \cup \cdots \cup A_s$, $[r_3] = B_1 \cup \cdots \cup B_s$, construct a random variable B as follows: put $B := j$ whenever $X_2 \in A_j$ and $X_3 \in B_j$. Thus, B is uniquely defined except on a set of measure zero, since $P(X_2 \in A_j, X_3 \in B_k) = 0$ for $j \neq k$, which follows from the containment of monomials in P. The variable B specifies in which blocks of the two partitions the random variables X_2 and X_3 lie. Now the binomials in P imply that $X_1 \perp\!\!\!\perp \{X_2, X_3\} \,|B$.

Corollary 3.4.2. *Suppose that X_1, X_2, X_3 satisfy $X_1 \perp\!\!\!\perp X_2 \,|X_3$ and $X_1 \perp\!\!\!\perp X_3 \,|X_2$. Then there is a random variable B that satisfies:*

1. *B is a (deterministic) function of X_2;*
2. *B is a (deterministic) function of X_3;*
3. *$X_1 \perp\!\!\!\perp \{X_2, X_3\} \,|B$.*

Conversely, whenever there exists a random variable B with properties 1. to 3., the random variables X_1, X_2, X_3 satisfy $X_1 \perp\!\!\!\perp X_2 \,|X_3$ and $X_1 \perp\!\!\!\perp X_3 \,|X_2$.

The case where B is a constant corresponds to the CI statement $X_1 \perp\!\!\!\perp \{X_2, X_3\}$. The intersection axiom can be recovered by noting that a function B that is a function of only X_2 as well as a function of only X_3 is necessarily constant, if the joint distribution of X_2 and X_3 is strictly positive.

Similar results hold for all families of CI statements of the form $A \perp\!\!\!\perp B_i \,|C_i$, where $A \cup B_i \cup C_i = V$ and where A is fixed for all statements, see [28, 29] (the case where X_A is binary was already described in [20]). The corresponding CI ideal is still radical, and the minimal primes have a similar interpretation. However, finding the minimal primes is more difficult and involves solving a combinatorial problem. Finally, Corollary 3.4.2 can be generalized to continuous random variables [27].

3.4.2 The four-cycle

Consider four discrete random variables X_1, X_2, X_3, X_4 and the (undirected) graphical model of the four cycle $C_4 = (V, E)$ with edge set $E = \{(1,2), (2,3), (3,4), (1,4)\}$. The global Markov CI statements of this graph are

$$\text{global}(C_4) = \{1 \perp\!\!\!\perp 3 \,|\{2,4\} \,,\; 2 \perp\!\!\!\perp 4 \,|\{1,3\} \,\}.$$

The primary decomposition of the corresponding CI ideal was studied in [24] in the case where X_1 and X_3 are binary. The case that all variables are binary is as follows:

Proposition 3.4.3 (Theorem 5.6 in [24]). *The minimal primes of the CI ideal $I_{\text{global}(C_4)}$ of the binary four cycle are the toric ideal I_{C_4} and the monomial ideals*

$$P_i = \langle p_x : x_i = x_{i+1} \rangle, \qquad P_i' = \langle p_x : x_i \neq x_{i+1} \rangle \quad \textit{for } 1 \leq i < 4.$$

The ideal I_{C_4} equals the vanishing ideal of the graphical model, to be discussed in Section 3.5. Interestingly, in this primary decomposition, all ideals except I_{C_4} are monomial ideals; that is, they only give support restrictions on the probability distribution.

The primary decomposition of the CI ideal gives an irreducible decomposition of the corresponding set of probability distributions. This leads to the statement of Proposition 3.1.1 in the introduction.

When X_2 and X_4 are not binary, the decomposition of $I_{\text{global}(C_4)}$ involves prime ideals parametrized by $i \in \{2,4\}$ and two sets $\emptyset \neq C, D \subsetneq [r_i]$. For such a choice of i, C, D, let j denote the element of $\{2,4\} \setminus \{i\}$, and let

$$\begin{aligned} P_{i,C,D} = &\langle p_{1x_21x_4} : x_i \in C, x_j \in [r_j]\rangle + \langle p_{1x_22x_4} : x_i \in D, x_j \in [r_j]\rangle \\ &+ \langle p_{2x_21x_4} : x_i \notin D, x_j \in [r_j]\rangle + \langle p_{2x_22x_4} : x_i \notin C, x_j \in [r_j]\rangle + I_{\text{global}(C_4)} \end{aligned}$$

the result is the following:

Proposition 3.4.4 (Theorem 6.5 in [24]). *Let X_1 and X_3 be binary random variables. The minimal primes of the CI ideal $I_{\text{global}(C_4)}$ of the four cycle are the toric ideal I_{C_4} and the ideals $P_{i,C,D}$ for $i \in \{2,4\}$ and $\emptyset \neq C, D \subsetneq [r_i]$. Furthermore the ideal is radical and thus equals the intersection of its minimal primes.*

In this case, the non-toric primes are not monomial, but consist of monomials and binomials. This fact is independent of the field $\mathbb{k}$. The following is one example of the kind of information that can be extracted from knowing the minimal primes of a conditional independence ideal.

Corollary 3.4.5. *Let X_1, X_2, X_3, X_4 be finite random variables that satisfy* global(C_4)*, and suppose that X_1 and X_3 are binary. Then one (or more) of the following statements is true:*

1. *The joint distribution lies in the closure of the graphical model.*
2. *There is $i \in \{2,4\}$ and there are sets $E, F \subseteq [r_i]$ such that the following holds:*

$$\text{If } (X_1, X_3) = \begin{Bmatrix} (1,1) \\ (1,2) \\ (2,1) \\ (2,2) \end{Bmatrix}, \text{ then } \begin{Bmatrix} X_i \in E \\ X_i \in F \\ X_i \notin F \\ X_i \notin E \end{Bmatrix}.$$

Conversely, any probability distribution that satisfies one of these statements and that satisfies $2 \perp\!\!\!\perp 4 \,|\, \{1,3\}$ also satisfies global(C_4).

3.5 The Vanishing Ideal of a Graphical Model

Graphical models can be represented via either parametric descriptions (e.g. factorizations of the density function) or implicit descriptions (e.g. Markov properties and conditional independence constraints). One use of the algebraic perspective on graphical models is to find the complete implicit description of the model, in particular, to find the vanishing ideal of the model. As described in Definition 3.2.8, the vanishing ideal of a set S is the set of all polynomial functions that evaluate to zero at every point in S. Although some graphical models have complete descriptions only in terms of conditional independence constraints, understanding the vanishing ideal can be useful for more complex models or hidden variable models where conditional independence is not sufficient to describe the model, for instance, the mixed graph models studied in Chapter 2.

Example 3.5.1. Consider the four cycle C_4 and let X_1, X_2, X_3, X_4 be binary random variables. The vanishing ideal $I_{C_4} \subseteq \mathbb{R}[p_{i_1i_2i_3i_4} : i_1, i_2, i_3, i_4 \in \{1,2\}]$ is generated by 16 binomials, 8 of which have degree 2 and 8 of which have degree 4. The degree 2 binomials are all implied by the two conditional independence statements $1 \perp\!\!\!\perp 3 \mid \{2,4\}$ and $2 \perp\!\!\!\perp 4 \mid \{1,3\}$. On the other hand, the degree 4 binomials are not implied by the conditional independence constraints, even when we restrict to probability distributions. One example degree four polynomial is

$$p_{1111}p_{1222}p_{2122}p_{2211} - p_{1122}p_{1211}p_{2111}p_{2222}$$

and the others are obtained by applying the symmetry group of the four cycle and permuting levels of the random variables.

Even in the simple example of the four cycle, there are generators of the vanishing ideal that do not correspond to conditional independence statements. It seems an important problem to try to understand what other types of equations can arise. Theorem 3.2 in [18] shows that the vanishing ideal of a graphical model of discrete random variables is the toric ideal I_A, where A is the design matrix of the graphical model, defined in the end of Section 3.2. A classification for discrete random variables and undirected graphs of when no further polynomials are needed beyond conditional independence constraints is obtained in the following theorem:

Theorem 3.5.2 (Theorem 4.4 in [18]). *Let G be an undirected graph and let $\mathcal{M}_G$ be its graphical model for discrete random variables. Then the vanishing ideal $I(\mathcal{M}_G)$ is equal to the conditional independence ideal $I_{\text{global}(G)}$ if and only if G is a chordal graph.*

It is unknown what the appropriate analogue of Theorem 3.5.2 is for other families of graphical models, either with different classes of graphs (e.g. DAGs) or with other types of random variables (e.g. Gaussian). Computational studies of the vanishing ideals appear in many different papers: for Bayesian networks with discrete random variables [17], for Bayesian networks with Gaussian random variables [33], for undirected graphical models with Gaussian random variables [32]. A characterization for which graph families the vanishing ideal is equal to the conditional independence ideal of global Markov statements is lacking in all these cases.

One natural question is to determine the other generators of the vanishing ideal that do not come from conditional independence, and to give combinatorial structures in the underlying graphs that imply that these more general constraints hold. For instance, for mixed Gaussian graphical models, conditional independence constraints are determinantal, but not every determinantal constraint comes from a conditional independence statement, and there is a characterization of which determinantal constraints come from conditional independence:

A mixed graph $G = ([m], B, D)$ is a triple where $[m] = \{1, 2, \ldots, m\}$ is the vertex set, B is a set of unordered pairs of $[m]$ representing bidirected edges in G, and D is a set of ordered pairs of $[m]$ representing directed edges in G. There might also be both directed and bidirected edges between a pair of vertices. To the set of B of bidirected edges one associates the set of symmetric positive definite matrices

$$PD(B) = \{\Omega \in PD_m : \omega_{ij} = 0 \text{ if } i \neq j \text{ and } i \leftrightarrow j \notin B\}.$$

To the set of directed edges one associates the set of $m \times m$ matrices

$$\mathbb{R}^D = \{\Lambda \in \mathbb{R}^{m \times m} : \lambda_{ij} = 0 \text{ if } i \to j \notin D\}.$$

Let $\epsilon \sim \mathcal{N}(0, \Omega)$ and let X be a jointly normal random vector satisfying the structural equation system

$$X = \Lambda^T X + \epsilon.$$

This is an example of a linear structural equation model, and contains as special cases

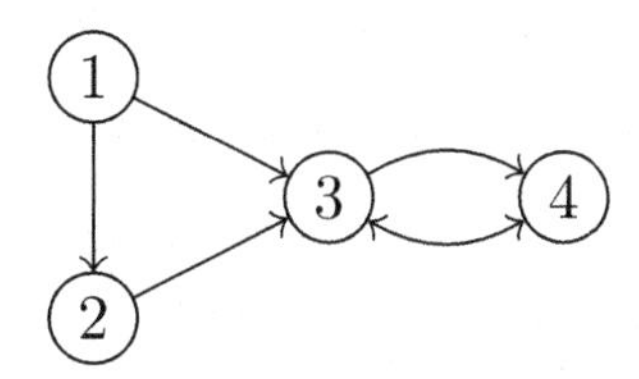

FIGURE 3.1: The mixed graph from Section 3.5.

various families of graphical models. Let Id denote the $m \times m$ identity matrix. With these assumptions, if $(\mathrm{Id} - \Lambda)$ is invertible,

$$X \sim \mathcal{N}(0, \Sigma) \text{ where } \Sigma = (\mathrm{Id} - \Lambda)^{-T}\Omega(\mathrm{Id} - \Lambda)^{-1}.$$

Example 3.5.3. Consider the mixed graph G from Figure 3.1. In this case, $PD(B)$ is the set of positive definite matrices of the form

$$\Omega = \begin{pmatrix} \omega_{11} & 0 & 0 & 0 \\ 0 & \omega_{22} & 0 & 0 \\ 0 & 0 & \omega_{33} & \omega_{34} \\ 0 & 0 & \omega_{34} & \omega_{44} \end{pmatrix}$$

and $\mathbb{R}^D$ is the set of real matrices of the form

$$\Lambda = \begin{pmatrix} 0 & \lambda_{12} & \lambda_{13} & 0 \\ 0 & 0 & \lambda_{23} & 0 \\ 0 & 0 & 0 & \lambda_{34} \\ 0 & 0 & 0 & 0 \end{pmatrix}.$$

A positive definite matrix Σ belongs to the graphical model associated to this mixed graph, if and only if there are $\Omega \in PD(B)$ and $\Lambda \in \mathbb{R}^D$ such that $\Sigma = (\mathrm{Id} - \Lambda)^{-T}\Omega(\mathrm{Id} - \Lambda)^{-1}$.

Definition 3.5.4. *Let $G = ([m], B, D)$ be a mixed graph. A* trek *between vertices i and j in G consists of either*

- *a pair (P_L, P_R) where P_L is a directed path ending in i and P_R is a directed path ending in j where both P_L and P_R have the same source, or*
- *a pair (P_L, P_R) where P_L is a directed path ending in i and P_R is a directed path ending in j such that the source of P_L and the source of P_R are connected by a bidirected edge.*

Let $\mathcal{T}(i, j)$ denote the set of all treks in G between i and j.

To each trek $T = (P_L, P_R)$ one associates the trek monomial m_T which is the product with multiplicities of all λ_{st} over all directed edges appearing in T times ω_{st} where s and t are the sources of P_L and P_R. One reason for the interest in treks is the *trek rule*, which says for the Gaussian graphical model associated to G

$$\sigma_{ij} = \sum_{T \in \mathcal{T}(i,j)} m_T.$$

For instance, for the mixed graph in Figure 3.1, the pair $(\{1 \to 2, 2 \to 3\}, \{1 \to 3\})$ is a trek from 3 to 3. The corresponding trek monomial m_T is $\omega_{11}\lambda_{12}\lambda_{23}\lambda_{13}$.

Definition 3.5.5. *Let A, B, C_A, C_B be four sets of vertices of G, not necessarily disjoint. The pair of sets (C_A, C_B)* t-separates *(short for* trek separates*) A from B if for every $a \in A$ and $b \in B$ and every trek $(P_L, P_R) \in \mathcal{T}(a,b)$, either P_L has a vertex in C_A or P_R has a vertex in C_B, or both.*

Theorem 3.5.6. *[35] Let $G = ([m], B, D)$ be a mixed graph and A and B two subsets of $[m]$ with $|A| = |B| = k$. Then the minor* $\det \Sigma_{A,B}$ *belongs to the vanishing ideal I_G if and only if there is a pair of sets C_A, C_B such that (C_A, C_B) t-separates A and B and such that $|C_A| + |C_B| < k$.*

The t-separation criterion can produce implicit constraints for structural equation models in situations where there are no conditional independence constraints.

Example 3.5.7. Consider the mixed graph G from Figure 3.1. The vanishing ideal of the model is $I_G = \langle \det \Sigma_{\{1,2\},\{3,4\}} \rangle$. This determinantal constraint is not a conditional independence constraint. It is implied by the t-separation criterion because the pair $(\emptyset, \{3\})$ t-separates $\{1,2\}$ and $\{3,4\}$.

Remark 3.5.8. In the case where there are hidden random variables, the vanishing ideal is typically not sufficient to completely describe the set of probability distributions that come from the graphical model. Usually one also needs to consider inequality constraints and other semialgebraic conditions. This problem is discussed in more detail in [3, 2, 37], among others.

3.6 Further Reading

Diaconis, Eisenbud and Sturmfels [8] were the first to consider primary decompositions for statistical applications, in particular the analysis of the connectivity of certain random walks. This perspective was also picked up in [24] using conditional independence ideals. Primary decomposition of conditional independence ideals also makes an appearance in the following papers not already mentioned [12, 16, 25, 34, 36].

The algebraic view on undirected graphical models was presented in [18], which began extensive study of the vanishing ideals of undirected graphical models for discrete random variables. Focus has been on developing techniques for constructing generating sets of the vanishing ideals with [14, 21, 30] being representative papers in this area. Vanishing ideals of undirected models with Gaussian random variables and models for DAGs have not been much studied. Some papers that initiated their study include [17, 32, 33].

Bibliography

[1] 4ti2 team. 4ti2—a software package for algebraic, geometric and combinatorial problems on linear spaces. available at `http://www.4ti2.de`.

[2] Elizabeth S. Allman, John A. Rhodes, Bernd Sturmfels, and Piotr Zwiernik. Tensors of nonnegative rank two. *Linear Algebra Appl.*, 473:37–53, 2015.

[3] Elizabeth S. Allman, John A. Rhodes, and Amelia Taylor. A semialgebraic description

of the general Markov model on phylogenetic trees. *SIAM J. Discrete Math.*, 28(2):736–755, 2014.

[4] Saugata Basu, Richard Pollack, and Marie-Françoise Roy. *Algorithms in Real Algebraic Geometry*, volume 10 of *Algorithms and Computation in Mathematics*. Springer-Verlag, Berlin, second edition, 2006.

[5] Jacek Bochnak, Michel Coste, and Marie-Françoise Roy. *Real Algebraic Geometry*, volume 36 of *Ergebnisse der Mathematik und ihrer Grenzgebiete (3) [Results in Mathematics and Related Areas (3)]*. Springer-Verlag, Berlin, 1998. Translated from the 1987 French original, Revised by the authors.

[6] Winfried Bruns and Udo Vetter. *Determinantal rings*, volume 45 of *Monografías de Matemática [Mathematical Monographs]*. Instituto de Matemática Pura e Aplicada (IMPA), Rio de Janeiro, 1988.

[7] David A. Cox, John Little, and Donal O'Shea. *Ideals, Varieties, and Algorithms: An Introduction to Computational Agebraic Geometry and Commutative Algebra*. Springer, fourth edition, 2015.

[8] Persi Diaconis, David Eisenbud, and Bernd Sturmfels. Lattice walks and primary decomposition. In *Mathematical essays in honor of Gian-Carlo Rota (Cambridge, MA, 1996)*, volume 161 of *Progr. Math.*, pages 173–193. Birkhäuser Boston, Boston, MA, 1998.

[9] Persi Diaconis and Bernd Sturmfels. Algebraic algorithms for sampling from conditional distributions. *Ann. Statist.*, 26:363–397, 1998.

[10] Alicia Dickenstein, Laura Felicia Matusevich, and Ezra Miller. Combinatorics of binomial primary decomposition. *Math. Z.*, 264(4):745–763, 2010.

[11] Mathias Drton, Bernd Sturmfels, and Seth Sullivant. *Lectures on Algebraic Statistics*, volume 39 of *Oberwolfach Seminars*. Springer, Berlin, 2009. A Birkhäuser book.

[12] Mathias Drton and Han Xiao. Smoothness of Gaussian conditional independence models. In *Algebraic methods in statistics and probability II*, volume 516 of *Contemp. Math.*, pages 155–177. Amer. Math. Soc., Providence, RI, 2010.

[13] David Eisenbud and Bernd Sturmfels. Binomial ideals. *Duke Math. J.*, 84(1):1–45, 1996.

[14] Alexander Engström, Thomas Kahle, and Seth Sullivant. Multigraded commutative algebra of graph decompositions. *J. Algebraic Combin.*, 39(2):335–372, 2014.

[15] Alex Fink. The binomial ideal of the intersection axiom for conditional probabilities. *J. Algebraic Combin.*, 33(3):455–463, 2011.

[16] Alex Fink, Jenna Rajchgot, and Seth Sullivant. Matrix Schubert varieties and Gaussian conditional independence models. srXiv:1510.04124, 2015.

[17] Luis David Garcia, Michael Stillman, and Bernd Sturmfels. Algebraic geometry of Bayesian networks. *J. Symbolic Comput.*, 39(3-4):331–355, 2005.

[18] Dan Geiger, Christopher Meek, and Bernd Sturmfels. On the toric algebra of graphical models. *Ann. Statist.*, 34(3):1463–1492, 2006.

[19] Daniel R. Grayson and Michael E. Stillman. Macaulay2, a software system for research in algebraic geometry. Available at `http://www.math.uiuc.edu/Macaulay2/`.

[20] Jürgen Herzog, Takayuki Hibi, Freyja Hreinsdóttir, Thomas Kahle, and Johannes Rauh. Binomial edge ideals and conditional independence statements. *Adv. in Appl. Math.*, 45(3):317–333, 2010.

[21] Serkan Hoşten and Seth Sullivant. Gröbner bases and polyhedral geometry of reducible and cyclic models. *J. Combin. Theory Ser. A*, 100(2):277–301, 2002.

[22] Thomas Kahle. Decompositions of binomial ideals. *J. Softw. Algebra Geom.*, 4:1–5, 2012.

[23] Thomas Kahle, Ezra Miller, and Christopher O'Neill. Irreducible decomposition of binomial ideals. *Compos. Math.*, 152(6):1319–1332, 2016.

[24] Thomas Kahle, Johannes Rauh, and Seth Sullivant. Positive margins and primary decomposition. *J. Commut. Algebra*, 6(2):173–208, 2014.

[25] George A. Kirkup. Random variables with completely independent subcollections. *J. Algebra*, 309(2):427–454, 2007.

[26] Radim Lněnička and František Matúš. On Gaussian conditional independent structures. *Kybernetika (Prague)*, 43(3):327–342, 2007.

[27] Jonas Peters. On the intersection property of conditional independence and its application to causal discovery. *Journal of Causal Inference*, 3(1):97–108, 2015.

[28] Johannes Rauh. Generalized binomial edge ideals. *Adv. in Appl. Math.*, 50(3):409–414, 2013.

[29] Johannes Rauh and Nihat Ay. Robustness, canalyzing functions and systems design. *Theory Biosci.*, 133(2):63–78, 2014.

[30] Johannes Rauh and Seth Sullivant. Lifting Markov bases and higher codimension toric fiber products. *J. Symbolic Comput.*, 74:276–307, 2016.

[31] Bernd Sturmfels and Seth Sullivant. Toric ideals of phylogenetic invariants. *J. Comp. Biol.*, 12:204–228, 2005.

[32] Bernd Sturmfels and Caroline Uhler. Multivariate Gaussian, semidefinite matrix completion, and convex algebraic geometry. *Ann. Inst. Statist. Math.*, 62(4):603–638, 2010.

[33] Seth Sullivant. Algebraic geometry of Gaussian Bayesian networks. *Adv. in Appl. Math.*, 40(4):482–513, 2008.

[34] Seth Sullivant. Gaussian conditional independence relations have no finite complete characterization. *J. Pure Appl. Algebra*, 213(8):1502–1506, 2009.

[35] Seth Sullivant, Kelli Talaska, and Jan Draisma. Trek separation for Gaussian graphical models. *Ann. Statist.*, 38(3):1665–1685, 2010.

[36] Irena Swanson and Amelia Taylor. Minimal primes of ideals arising from conditional independence statements. *J. Algebra*, 392:299–314, 2013.

[37] Piotr Zwiernik. *Semialgebraic statistics and latent tree models*, volume 146 of *Monographs on Statistics and Applied Probability*. Chapman & Hall/CRC, Boca Raton, FL, 2016.

Part II

Computing with factorizing distributions

4

Algorithms and Data Structures for Exact Computation of Marginals

Jeffrey A. Bilmes

Department of Electrical Engineering, University of Washington

CONTENTS

As seen in Chapters 1-3 in Part I of this handbook, a graphical model represents a potentially infinite-size family of joint probability distributions over a set of random variables, and each member of the family must obey certain Markov properties (or factorization requirements) that are mathematically expressed by the graph [53, 36, 40, 38].

In this chapter, we will develop computational strategies to compute exact marginal distributions for any distribution within the family of distributions of a graphical model, all based only on properties of the graph itself. Marginal distributions are useful in a number of applications. For example, most machine learning and artificial intelligence (AI) systems require marginals. During an iterative process of machine learning, marginal distributions are needed at each iteration for updating parameters based on available data. In an AI based system, marginals can be used to make decisions, where we might make a choice based on the variable assignment that maximizes the probability of a marginal. Marginals can also be useful to understand a physical system modeled via a probability distribution; see for instance the applications discussed in Chapter 19. Being able to efficiently compute marginals of a distribution is therefore an important component in data science, machine learning, and AI.

The computational strategies we discuss in this chapter are actually meta algorithms—we will develop strategies that will produce an algorithm valid for all distributions in a graphical model, rather than being valid only for one particular distribution. This will also involve characterizing the computational complexity of computing a set of marginals, where the complexity will depend both on the graph and the desired marginal queries. We will see that for a given graph and query, the problem of finding the optimal way of computing the marginal, a problem of combinatorial optimization, can itself be hard in a complexity theoretic sense. Once we have determined a strategy to compute an exact marginal, we may find it too computationally costly (either because it is suboptimal and the process of finding the optimal strategy is too hard, or because it is optimal and the optimal strategy is itself too complex [16]) and thus may have to take recourse in inference approximation schemes outlined in the remaining chapters of Part II of this book. Nevertheless, it is important to be well grounded in exact marginal methods, which this chapter strives to make accessible for the reader.

4.1 Introduction

All exact inference methods [53, 18, 36, 41], in one way or another, perform inference on trees. Either the given graph is itself a tree on which one may perform inference directly, or the graph is transformed into a form of tree (specifically, a hypertree [48, 30]) on which inference is performed. The local computational steps performed on each node of the tree have a large degree of flexibility. These steps can be seen either as summation or can be viewed as a form of search (which itself can be viewed as a tree search) — the goal of this chapter is to explore the former summation case.

In order to limit scope, we will concentrate on discrete-valued probability distributions. If the distribution is jointly Gaussian, much of this chapter is still relevant, but many of the marginalization steps become much more efficient since they can be performed by efficient matrix operations. Distributions that involve continuous random variables that are not linear-Gaussian, or that involve mixed discrete/continuous variables would follow the same underlying steps but the operations (numerical integration) entail out-of-scope details and subtleties.

4.1.1 Graphical models

As in Section 1.7 in Chapter 1, assume that we have an undirected graph $G = (V, E)$ with a size $n = |V|$ node set V and a size $m = |E|$ edge set E. A subset of nodes that are fully connected (equivalently, that are *complete*) is called a *clique* — any two nodes in a clique of G are connected by an edge. We use $\mathcal{C}(G)$ to denote the set of all cliques of G, hence $\mathcal{C}(G)$ is a set of sets of graph nodes. We are given a distribution p that lies within the family of distributions that obey the Markov properties [40] associated with G. Hence, p is a distribution over a set of n discrete random variables $x_1, x_2, \ldots, x_n$, that (for simplicity) all have domain $\mathcal{D}$. For all assignments of the variables $x = (x_1, x_2, \ldots, x_n) \in \mathcal{D}^n$, the distribution p provides the value $p(x_1, x_2, \ldots, x_n)$. Since p obeys the Markov properties of G, it means that p lies within the family $\mathcal{F}(G)$ of distributions represented by G, and we notate this as $p \in \mathcal{F}(G)$. This essentially means that p may be written as:

$$p(x) = \prod_{C \in \mathcal{C}(G)} \phi_C(x_C) \tag{4.1}$$

where x_C is a subvector of X involving only variables with indices in C, and where $\phi_C(x_C)$ is a potential function over those variables. Ordinarily, computing the list of cliques in a graph is NP-hard [27] but for the instances discussed in this chapter, we assume that the cliques are given and correspond to factors of distributions (i.e., each factor corresponds to a cluster of variables that have the ability to directly interact in the distribution, and these clusters are determined either by a domain-specific scientist or via data-driven methods [14, 32, 50]).

4.1.2 Probabilistic inference

Probabilistic inference entails computing marginals of this distribution. Given a set $A = \{a_1, a_2, \ldots, a_{|A|}\} \subset V$ of random variable indices, the corresponding set of random variables, denoted X_A, might take value $x_A = (x_{a_1}, x_{a_2}, \ldots, x_{a_{|A|}}) \in \mathcal{D}^A$. We might come to know (say, as evidence) these values, or they could be used to condition on. For example, we may wish to compute the marginal $p(x_A)$, or the posterior probability of some non-evidence variables $p(x_B|x_A) = p(x_B, x_A)/p(x_A)$ where $B \subseteq V \setminus A$, or the most probable assignment $\text{argmax}_{x_B \in \mathcal{D}_{X_B}} p(x_B, x_A)$. In this last case, there can be a big computational difference between when $B = V \setminus A$ and when $B \subset V \setminus A$, the latter being considerably harder. The computation can be parameterized either by the subset $A \subseteq V$ (the desired marginal) or the pair of subsets $(A \subseteq V, B \subseteq V \setminus A)$ giving the denominator and numerator of the posterior.

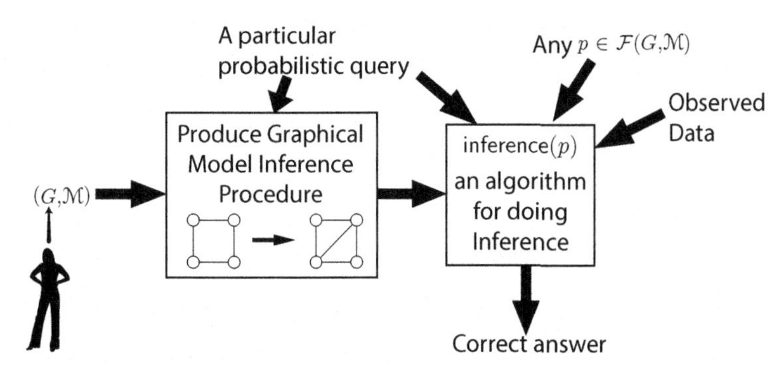

FIGURE 4.1: Graphical model inference: take a graph $G = (V, E)$ and its associated Markov properties produce an inference procedure so that for any $p \in \mathcal{F}(G)$, it will produce the correct quantity. The cost of producing the procedure is amortized over the number of times it is run for different $p \in \mathcal{F}(G)$.

A goal of graphical model inference is to produce a *meta algorithm* that involves the production of a specific method that can perform exact inference on any $p \in \mathcal{F}(G)$, not just on one particular p, where $\mathcal{F}(G)$ denotes all probability distributions that obey the Markov properties of G. This method is developed using only the graph G and its associated Markov properties, without knowledge of any particular p. This idea is shown in Figure 4.1. The cost of producing the method is amortized over the many members of the family $\mathcal{F}(G)$. In many cases, an inference algorithm for $G = (V, E)$ also is an exact algorithm for a different graph $\tilde{G}$ — as an example, if $\tilde{G} = (V, \tilde{E})$ has a subset of the edges of 6 (so $\tilde{E} \subseteq E$), then any method for G will also work for $\tilde{G}$, although there might be a more efficient and more specific algorithm for $\tilde{G}$. In general we have the following:

Proposition 4.1.1 (Sub-graphs mean sub-families). *Let $G = (V, E)$ be a graph and let $\tilde{E} \subseteq E$ with $\tilde{G} = (V, \tilde{E})$ being the corresponding graph. Then $\mathcal{F}(\tilde{G}) \subseteq \mathcal{F}(G)$.*

4.1.3 Benefits of graphical models for inference

Attractive features of graphical model inference include the following: (1) we can map from G to an inference procedure automatically, without human intervention; (2) the meta algorithm uses only graph-theoretic properties of G; (3) it is sometimes possible to find an inference procedure that is as computationally efficient as possible for certain G; it is also possible to know when you have found such an algorithm. On the other hand, there are some limitations to this process. For example, it is typical to encounter a G for which it is an NP-hard problem to determine the most efficient inference procedure for G. Moreover, there are some instances for which an inference algorithm developed for any $p \in \mathcal{F}(G)$ is not as efficient as an algorithm developed for a specific single $p \in \mathcal{F}(G)$. One example is mentioned above, where if $\tilde{G}$ has a subset of the edges of G, and if $p \in \mathcal{F}(\tilde{G})$, it may be more efficient to develop an algorithm for all members of $\mathcal{F}(\tilde{G})$. Another example is where a particular $p \in \mathcal{F}(G)$ has an exploitable pattern of zeros, where $p(x) = 0$ for certain x. Computing marginals for sparse distributions can sometimes be faster than methods developed for any $p \in \mathcal{F}(G)$ [24, 22, 4, 29]. Alternatively, approximate inference methods can be more accurate for certain specific distributions.

Chapter 2 describes different types of graphical models; see also [53, 35, 40, 55, 39]. We will not consider these different models here, with the one exception that we briefly address the distinction between Bayesian networks [35, 53] and the Markov random fields utilized below. A Bayesian network is expressed via a directed acyclic graph (DAG), while a Markov random field is expressed via an undirected graph. The inference procedures below are developed for Markov random fields, so how can we ensure they are still exact for Bayesian networks? Assume G_{BN} is a DAG and that $\mathcal{F}(G_{\mathrm{BN}})$ is the family of distributions that obey the Bayesian network's Markov properties (e.g., d-separation) on that graph. To ensure our procedures work, we must find a G such that $\mathcal{F}(G_{\mathrm{BN}}) \subseteq \mathcal{F}(G)$. One strategy is the "moralization" procedure, one that connects together into a clique all parents of any child, and then removes all edge directions. Let $m(G_{\mathrm{BN}})$ be the undirected graph obtained by moralizing the directed graph G_{BN}; then we have that $\mathcal{F}(G_{\mathrm{BN}}) \subseteq \mathcal{F}(m(G_{\mathrm{BN}}))$. Therefore, if we develop an inference procedure for any $p \in \mathcal{F}(m(G_{\mathrm{BN}}))$, it will assuredly work for any $p \in \mathcal{F}(G_{\mathrm{BN}})$. There are still other ways of producing graph transformations prior to performing inference that ensures the family of distributions covered by inference can only grow (see Chapter 19 in the present book).

In what follows, we assume we have an undirected graph G — we will develop an associated exact inference procedure to produce desired marginals. We start with trees, however, since they are fundamental.

4.2 Inference on Trees

Inference on trees means inference for all $p \in \mathcal{F}(T)$ where $T = (V, E)$ is a graph that is a tree, or more generally a forest.

4.2.1 Trees and tree structured factorization

Trees and forests have the following definition:

Definition 4.2.1. *A graph $G = (V, E)$ is a* forest *if for all $u, v \in V$, there is no more than one path that connects u to v in G. Given a forest G, if for all $u, v \in G$ there is a unique path connecting u and v, then it is called a* tree.

We can equivalently define a forest as a graph that has no cycles, and a tree is a connected graph without cycles. Figure 4.2-A shows an example of a tree. Two other identical and relevant characterizations of trees follow:

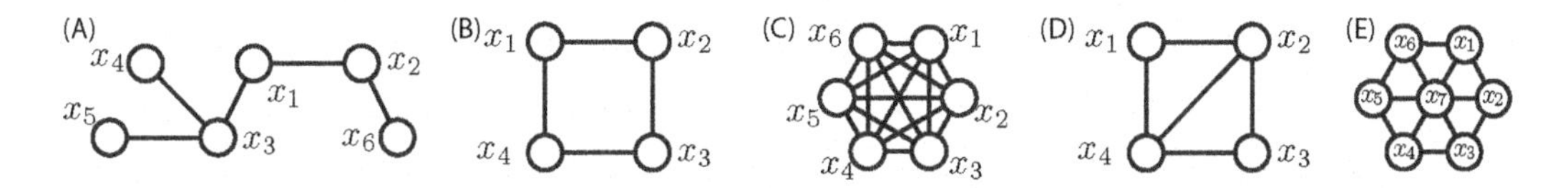

FIGURE 4.2: (A) A tree with six nodes. (B) A 4-cycle. (C) A 6-clique. (D) A chorded 4-cycle. (E) A graph comprising six joined triangles.

Theorem 4.2.2 (Trees). *Let $G = (V, E)$ be an undirected graph with $|V| = n > 2$. The following properties each define a tree.*

(a) G has $n - 1$ edges and has no cycles,

(b) G can be generated as follows: Start with a node v, and repeatedly choose a new node and connect it with an edge to any one previously chosen node.

The maximum clique size in any forest is two since there can be no cycles, and any set of nodes larger than two will constitute a sub-forest of the tree. From the perspective of a $p \in \mathcal{F}(T)$ this means that all factors of p involve at most two variables connected via an edge, i.e., we have:

$$p(x_1, \dots, x_n) = \prod_{e \in E(T)} \psi_e(x_e) = \prod_{(i,j) \in E(T)} \psi_{i,j}(x_i, x_j) \tag{4.2}$$

where $E(T)$ are the edges of the tree, and $\psi_e(x_e) = \psi_{i,j}(x_i, x_j)$, for all e are edge factors. Any such distribution $p \in \mathcal{F}(T)$ is called a *tree structured distribution.* Any tree structured distribution p may be expressed in the following way:

$$p(x_1, \dots, x_N) = \frac{\prod_{(i,j) \in E(T)} p_{i,j}(x_i, x_j)}{\prod_{v \in V(T)} p_v(x_v)^{d(v)-1}} = \prod_{v \in V(T)} p_v(x_v) \prod_{(i,j) \in E(T)} \frac{p_{i,j}(x_i, x_j)}{p_i(x_i) p_j(x_j)}, \tag{4.3}$$

where $p_{i,j}$ are pairwise and p_v unary marginals (i.e., $p_{i,j}(x_i, x_j) = \sum_{x_{V \setminus \{i,j\}}} p(x_1, \dots, x_N)$ and $p_v(x_v) = \sum_{x_{V \setminus \{v\}}} p(x_1, \dots, x_N)$) and $d(v)$ is the degree of (number of edges incident to) v in the tree. To show that this is true, we can use induction. It is clearly true for a tree with either one or two nodes. We next assume it is true for any tree with $i - 1$ nodes and show it is true for i nodes. We use the tree generation procedure of Theorem 4.2.2-(b) where node i connects to some node $j < i$, and also use the property of conditional independence for any such tree-structured distribution. With such a tree, given X_j, X_i is conditionally independent of all other previous nodes. Thus we can write $p(x_1, \dots, x_i) = p(x_i | x_j) p(x_1, \dots, x_{i-1}) =$

$p(x_1,\ldots,x_{i-1})p(x_i,x_j)/p(x_j) = p(x_1,\ldots,x_{i-1})p(x_i)p(x_i,x_j)/(p(x_i)p(x_j))$. Equation (4.3) is notable since it has an equivalent form in terms of junction trees, as we will see in Equation (4.13). For a given node $v \in V(T)$, let $\delta(v)$ be the set of neighbors of v in T. A tree is simpler than an arbitrary graph but it nonetheless captures the essence of exact inference on even non-tree graphs. For this reason, we first describe inference on trees.

4.2.2 Eliminating variables via marginalization

Consider the tree $T = (\{1,2,3,4,5,6\}, E)$ in Figure 4.2-A, and suppose we have a $p \in \mathcal{F}(T)$, which means p may be written as:

$$p(x) = \psi_{1,2}(x_1,x_2)\psi_{2,6}(x_2,x_6)\psi_{1,3}(x_1,x_3)\psi_{3,4}(x_3,x_4)\psi_{3,5}(x_3,x_5) \tag{4.4}$$

for some set of functions $\{\psi_{u,v}\}_{\{u,v\}\in E}$ corresponding to the edges of the tree. Naively computing the marginal quantity $p(x_1,x_2)$ might proceed as:

$$p(x_1,x_2) = \sum_{x_3}\sum_{x_4}\cdots\sum_{x_6} p(x_1,x_2,\ldots,x_6), \tag{4.5}$$

requiring $O(r^6)$ operations as Algorithm 4.1 suggests, where $r = |\mathcal{D}|$ is the domain size. Line one of Algorithm 4.1 is performed r^2 times, and line two requires $O(r^{n-2})$ operations, leading to an $O(r^n)$ complexity in general.

Algorithm 4.1 Naive marginalization.

1: **for all** $(x_1,x_2) \in \mathcal{D}_{X_1} \times \mathcal{D}_{X_2}$ **do**
2: compute $\sum_{x_3}\sum_{x_4}\cdots,\sum_{x_6} p(x_1,x_2,\ldots,x_8)$
3: **end for**

The above computation can be significantly improved by taking advantage of structure in the graph, the assumption that any distribution obeys the graph's Markov properties, and the distributive law over the reals (and which can be generalized [1, 51]), which states that $ab+ac = a(b+c)$ for real numbers a,b, and c. Specifically, when wishing to marginalize out variable x_i, the distributive law states:

$$\begin{aligned}
&\sum_{x_1,x_2,\ldots,x_N}\left(\prod_{c\in\text{factors not involving }x_i}\psi_c\right)\left(\prod_{c\in\text{factors involving }x_i}\psi_c\right)\\
&= \sum_{x_1,\ldots,x_{i-1},x_{i+1},\ldots,x_N}\left(\prod_{c\in\text{factors not involving }x_i}\psi_c\right)\sum_{x_i}\left(\prod_{c\in\text{factors involving }x_i}\psi_c\right)
\end{aligned} \tag{4.6}$$

Hence, we should first distribute the sum over x_6 into the product. While there is some choice where it should go, Equation (4.6) suggests that it should involve no more than the factors that involve x_6. Equation (4.5) becomes:

$$\begin{aligned}
p(x_1,x_2) &= \sum_{x_3}\sum_{x_4}\sum_{x_5}\psi(x_1,x_2)\psi(x_1,x_3)\psi(x_3,x_4)\psi(x_3,x_5)\underbrace{\sum_{x_6}\psi(x_2,x_6)}_{\phi_{\not{6},2}(x_2)}\\
&= \sum_{x_3}\sum_{x_4}\sum_{x_5}\psi(x_1,x_2)\psi(x_1,x_3)\psi(x_3,x_4)\psi(x_3,x_5)\phi_{\not{6},2}(x_2)
\end{aligned} \tag{4.7}$$

where $\phi_{\not{6},2}(x_2)$ is a function of x_2 only. The notation $\not{6}$ indicates that x_6 has been summed away. Constructing $\phi_{\not{6},2}(x_2)$ costs only $O(r^2)$. Also, for convenience, we write $\psi_{u,v}(x_u, x_v)$ as $\psi(x_u, x_v)$. We say that the variable x_6 has been "eliminated" because once it is marginalized away, it is no longer part of the computation. As shown below, the operation of marginalizing or summing away variables corresponds to a purely graphical operation of eliminating nodes (defined below) and is also the source of the name of the variable elimination algorithm.

We now, in Equation (4.7), have an expression that does not involve x_6. We can further sum out the other variables. If we were next to choose x_3, this results in:

$$p(x_1, x_2) = \sum_{x_4}\sum_{x_5} \psi(x_1, x_2)\phi_{\not{6},2}(x_2) \sum_{x_3} \psi(x_1, x_3)\psi(x_3, x_4)\psi(x_3, x_5) \tag{4.8}$$

$$= \sum_{x_4}\sum_{x_5} \psi(x_1, x_2)\phi_{\not{6},2}(x_2)\phi_{\not{3},1,4,5}(x_1, x_4, x_5). \tag{4.9}$$

The expression $\phi_{\not{3},1,4,5}(x_1, x_4, x_5)$ indicates we have marginalized away x_3 but now have a function of the three coupled variables x_1, x_4, x_5 since a sum of factors, in general, does not itself factor. Hence, computing $\phi_{\not{3},1,4,5}(x_1, x_4, x_5)$ alone already involves $O(r^4)$ operations (for each of the $O(r^3)$ possible values of the triple (x_1, x_4, x_5), there are $O(r)$ terms in the sum over x_3). If rather than summing out x_3, we instead chose x_5, this yields:

$$p(x_1, x_2) = \sum_{x_4}\sum_{x_3} \psi(x_1, x_2)\phi_{\not{6},2}(x_2)\psi(x_1, x_3)\psi(x_3, x_4) \sum_{x_5} \psi(x_3, x_5) \tag{4.10}$$

$$= \sum_{x_4}\sum_{x_3} \psi(x_1, x_2)\phi_{\not{6},2}(x_2)\psi(x_1, x_3)\psi(x_3, x_4)\phi_{\not{5},3}(x_3) \tag{4.11}$$

where $\phi_{\not{5},3}(x_3)$ involves $O(r^2)$ computation. The order of the variables we eliminate therefore makes a big difference to the resulting computation.

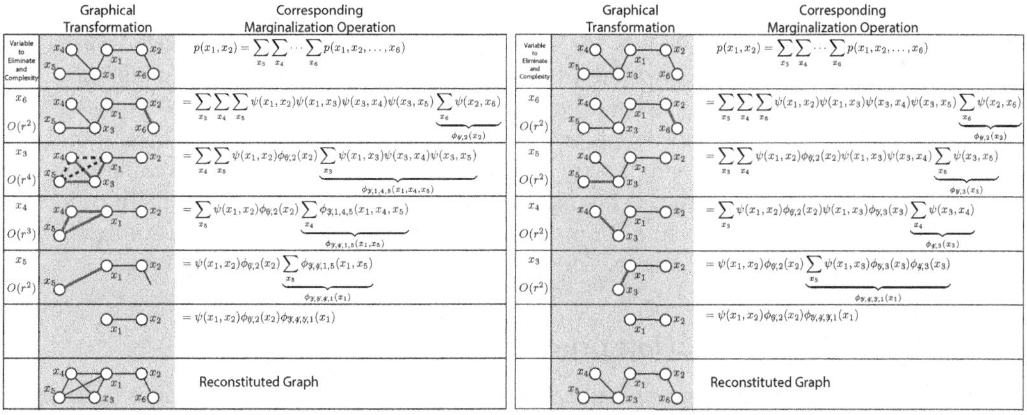

Variable to Eliminate and Complexity	Graphical Transformation	Corresponding Marginalization Operation
		$p(x_1,x_2)=\sum_{x_3}\sum_{x_4}\cdots\sum_{x_6} p(x_1,x_2,\ldots,x_6)$
x_6 $O(r^2)$		$=\sum_{x_3}\sum_{x_4}\sum_{x_5}\psi(x_1,x_2)\psi(x_1,x_3)\psi(x_3,x_4)\psi(x_3,x_5)\underbrace{\sum_{x_6}\psi(x_2,x_6)}_{\phi_{\not{6},2}(x_2)}$
x_3 $O(r^4)$		$=\sum_{x_4}\sum_{x_5}\psi(x_1,x_2)\phi_{\not{6},2}(x_2)\underbrace{\sum_{x_3}\psi(x_1,x_3)\psi(x_3,x_4)\psi(x_3,x_5)}_{\phi_{\not{3},1,4,5}(x_1,x_4,x_5)}$
x_4 $O(r^3)$		$=\sum_{x_5}\psi(x_1,x_2)\phi_{\not{6},2}(x_2)\underbrace{\sum_{x_4}\phi_{\not{3},1,4,5}(x_1,x_4,x_5)}_{\phi_{\not{3},\not{4},1,5}(x_1,x_5)}$
x_5 $O(r^2)$		$=\psi(x_1,x_2)\phi_{\not{6},2}(x_2)\underbrace{\sum_{x_5}\phi_{\not{3},\not{4},1,5}(x_1,x_5)}_{\phi_{\not{3},\not{5},\not{4},1}(x_1)}$
		$=\psi(x_1,x_2)\phi_{\not{6},2}(x_2)\phi_{\not{3},\not{4},\not{5},1}(x_1)$
		Reconstituted Graph

Variable to Eliminate and Complexity	Graphical Transformation	Corresponding Marginalization Operation
		$p(x_1,x_2)=\sum_{x_3}\sum_{x_4}\cdots\sum_{x_6} p(x_1,x_2,\ldots,x_6)$
x_6 $O(r^2)$		$=\sum_{x_3}\sum_{x_4}\sum_{x_5}\psi(x_1,x_2)\psi(x_1,x_3)\psi(x_3,x_4)\psi(x_3,x_5)\underbrace{\sum_{x_6}\psi(x_2,x_6)}_{\phi_{\not{6},2}(x_2)}$
x_5 $O(r^2)$		$=\sum_{x_3}\sum_{x_4}\psi(x_1,x_2)\phi_{\not{6},2}(x_2)\psi(x_1,x_3)\psi(x_3,x_4)\underbrace{\sum_{x_5}\psi(x_3,x_5)}_{\phi_{\not{5},3}(x_3)}$
x_4 $O(r^2)$		$=\sum_{x_3}\psi(x_1,x_2)\phi_{\not{6},2}(x_2)\psi(x_1,x_3)\phi_{\not{5},3}(x_3)\underbrace{\sum_{x_4}\psi(x_3,x_4)}_{\phi_{\not{4},3}(x_3)}$
x_3 $O(r^2)$		$=\psi(x_1,x_2)\phi_{\not{6},2}(x_2)\underbrace{\sum_{x_3}\psi(x_1,x_3)\phi_{\not{5},3}(x_3)\phi_{\not{4},3}(x_3)}_{\phi_{\not{5},\not{4},\not{3},1}(x_1)}$
		$=\psi(x_1,x_2)\phi_{\not{6},2}(x_2)\phi_{\not{5},\not{4},\not{3},1}(x_1)$
		Reconstituted Graph

FIGURE 4.3: Node elimination on the tree in Figure 4.2 and corresponding marginalization. Left: order $(6, 3, 4, 5)$, Right: order $(6, 5, 4, 3)$. Bold edges show elimination cliques.

Figure 4.3 shows two such orders, in each case showing the graphical transformation and the corresponding mathematical operations. We first discuss the left case, a variable marginalization using order $(6, 3, 4, 5)$. We see a sequence of marginalization operation equations according to that order. For each of these equations, we display to the left the corresponding graphical analogy. For example, when we eliminate variable x_6 only the factor $\psi(x_2, x_6)$ is involved, leading to an $O(r^2)$ computation. Graphically, the edge between x_2 and x_6 is bold, indicating a size-two clique. Notice that the clique size (two) is the exponent in the computation. Next, when we eliminate x_3 there are three other variables involved in factors, namely x_1, x_4, and x_5, leading to computation of $O(r^4)$. On the left, we show in bold the edges between the four variables involved in this computation. Again, the clique size is identical to the computational cost's exponent. Some of the bold edges are dashed, indicating that they are not edges in the original graph, rather they correspond to the coupling that happens by marginalizing x_3. The next variable being marginalized, x_4, has an $O(r^3)$ computation and three is the corresponding clique size. All the remaining marginalizations have $O(r^2)$ computation. The total computation is therefore $O(r^4)$.

We next discuss the right case, using order $(6, 5, 4, 3)$. We start, again, with x_6, but things differ in the second variable, x_5, which produces a computation of only $O(r^2)$ and a corresponding clique size of two. In fact, the figure shows that each successive variable has a computation of $O(r^2)$ and a clique size corresponding to the exponent, leading to a total complexity of $O(r^2)$. Hence, by just changing the order, we've achieved a speedup of $O(r^4)/O(r^2) = O(r^2)$ which grows unboundedly with r.

With the first order, the three added edges (called "fill-in" edges) are caused by the marginalization of x_3 while with the second order, no new edges are added. Some of the possible node elimination orders, therefore, are much worse than others. Specifically, under some orders, we inextricably couple together factors so that they never again can be decoupled. Given a set of distinct variable indices $\{a_1, a_2, \ldots, a_k, b\}$ the function where x_b is marginalized out (or eliminated),

$$f(x_{a_1}, x_{a_2}, \ldots, x_{a_k}) = \sum_{x_b} \prod_{i=1}^{k} f_i(x_{a_i}, x_b), \tag{4.12}$$

does **not** in general factor, meaning there are no functions $h_1, h_2, \ldots, h_k$ so that $f(x_{a_1}, x_{a_2}, \ldots, x_{a_k}) = \prod_i h_i(x_{a_i})$. Marginalizing out x_b can be graphically represented by adding an edge between any pair of nodes with index set $\{a_1, a_2, \ldots, a_k\}$, creating at most $k(k-1)/2$ fill-in edges, turning the neighbors of x_b into a clique. In Figure 4.3-left when we eliminate x_3, $k = 3$ with $a_1 = 1$, $a_2 = 4$, $a_3 = 5$, and $b = 3$. The extra fill-in edges added to a graph represent the coupling of variables together when performing a marginalization. The original graph, along with any extra fill-in edges added during the elimination process, is called the *reconstituted* graph. The clique sizes in the reconstituted graph then correspond to the exponents of the resulting computation. With computational impunity, we could have started with the reconstituted graph, if we insist on the ordering that produced it. In order to reduce computation, we try to find an ordering that results in no fill-in, as we saw in Figure 4.3 on the right.

4.2.3 The variable elimination process on graphs

One of the key benefits of using a graphical model is that the order can be determined based on only the graph itself before doing the actual (and potentially very costly) marginalization computation. The reason is that every marginalization corresponds to the purely graphical operation of **elimination**, which means to connect all unconnected neighbors of a variable (with fill-in edges) and then remove the variable and its incident edges. We consider the following purely graph-theoretic operation for elimination:

Definition 4.2.3. Elimination: *To* eliminate *a node $v \in V$ in an undirected graph G, we first connect all neighbors of v to each other and then remove from the graph v and all v's incident edges.*

We eliminate variables repeatedly until the resulting marginal we desire is all that remains. In Figure 4.3, the right case chose an order that ensured no fill-in, leading to the $O(r^2)$ complexity, while on the left choosing x_3 resulted in fill-in.

Graphs that have at least one variable order that does not cause fill-in edges during the elimination process are special. Trees and forests are examples, as we will soon see. Also, if we run an elimination process on a graph to produce a reconstituted graph, and then run an elimination process again on that reconstituted graph using that same order, then there will be no additional fill-in edges added (the elimination process has, for that order, reached a fixed point). Consider, for example, the ordering $(6, 3, 4, 5)$ on the reconstituted graph shown on the bottom of Figure 4.3-left, once we get to x_3 all of its neighbors are already coupled via edges connecting them.

Many graphs do not have the property that there exists an order that yields no fill-in, and some graphs do have elimination orders that yield no fill-in but still have high complexity. Before we discuss this, let us further consider elimination in a non-tree graph, namely one that has cycles. Consider the four-cycle in Figure 4.2-(B), corresponding to any distribution with $p(x_1, x_2, x_3, x_4) = \psi(x_1, x_2)\psi(x_2, x_3)\psi(x_3, x_4)\psi(x_4, x_1)$. If we were to marginalize out any variable, it would involve coupling together two of the other variables. Hence, there is no ordering that yields no fill-in, and the complexity of this graph is $O(r^3)$ despite the fact that the original graph involves direct interactions between only pairs of variables. Another example is the six-clique in Figure 4.2-(C). All variables are already coupled and the distribution could be $p(x_1, x_2, \ldots, x_6) = \phi(x_1, x_2, \ldots, x_6)$, without any factorization properties at all where marginalization of any variable results in an $O(r^6)$ computation, despite the fact that there is no fill-in.

What we need is a strategy that helps us to usefully navigate through these issues. Can we characterize the graphs for which there is an elimination order that results in no fill-in, and in such case can we find it efficiently, and characterize the computational cost beforehand, all using purely graphical operations? Can we identify those graphs for which there is no fill-in free order, and for those graphs find an elimination order that either leads to the minimal fill-in, or the minimal computation, again using only graph operations? The answer to some of these questions is yes, but others (as we will soon see) are NP-complete to determine.

4.3 Triangulated Graphs and Fill-in Free Elimination Orders

With a four-cycle, we saw that running the elimination process according to any order produces a fill-in edge. On the other hand, for the tree, the right of Figure 4.3 used an order that did not cause fill-in edges. In fact, any tree has the property that there exists a fill-in free order (although not all orders are fill-in free, as we saw on the left in Figure 4.3). Consider Theorem 4.2.2-(b) that explains a generative process that can produce any tree. If we start with a tree, and choose its generating order but in reverse, and then run the elimination process according to this reversed order, then no fill-in is produced. The reason is that we are always eliminating a leaf in the tree. In any tree, a fill-in free order can be produced by choosing a leaf node to eliminate at each step, one that is guaranteed to exist thanks to the following:

Lemma 4.3.1. *A tree with more than one node always has at least two leaf nodes.*

Once we eliminate a leaf node, we still have a tree so leaf nodes still exist.

4.3.1 Graphs with fill-in free orders

Are there other classes of graphs that have fill-in free elimination orders? Figure 4.2-(C) shows the 6-clique in which all orders are elimination free. The reconstituted graph on the bottom left of Figure 4.3 also, as we saw, has a fill-in free elimination order.

As another example, Figure 4.2-(D), shows a four cycle with an additional chord. A chord [26, 28] is an edge between two non-adjacent nodes along the cycle. A chord is defined relative to but is not part of a cycle — the existence of a chord does not make its incident nodes adjacent in the cycle as only the cycle's edges can do that. For example, Figure 4.2-(D) shows a cycle (x_1, x_2, x_3, x_4) with a chord $\{x_2, x_4\}$ between nodes x_2 and x_4 that are not adjacent in the cycle although they are adjacent in the graph. Here, if we eliminate x_3 first, no fill-in is produced, leaving a three cycle as a result. Hence, this graph, like a tree albeit not a tree, also has a fill-in free elimination order. Not all elimination orders are fill-in free, however, as if we eliminate x_4 first, it will add a fill-in edge between x_1 and x_3.

As we saw above for the four-cycle, the elimination process (both the graphical operation and the marginalization process) is defined on not just trees but on any graph. The graphical operation (see Definition 4.2.1) and the marginalization operation (see Equation (4.12)) are therefore strongly related, as suggested by Figure 4.3. Therefore, given a graph G that has a fill-in free elimination order, for any $p \in \mathcal{F}(G)$ the corresponding marginalization equations, done in the same order, results in no additional variable coupling beyond what already exists. For such graphs, the computation will stay bounded in an immediate property of the graph, namely the sizes of the cliques in the graph.

The above suggests we should choose nodes to eliminate that either: (1) have only zero or one neighbor (so that no new edges are added), or (2) have neighbors that are already connected so that elimination results in no new edges. When there are zero neighbors, the elimination step costs $O(r)$ and when there is one neighbor, the cost is $O(r^2)$, the only two cases we encounter with a forest. In general, the cost of an elimination step is exponential in one more than the neighbor set size. Graphically, v along with its fully connected neighbors constitutes a clique in the reconstituted G and the appearance of this clique means that any $p \in \mathcal{F}(G)$ is allowed to have a factor of the form $f(x_v, x_{\delta(v)})$, where $\delta(v)$ are the neighbors of v at the time it is eliminated. In fact, the set $\{v \cup \delta(v)\}$ is called an "elimination clique", and ends up being a subclique of some clique in the final reconstituted graph. Since each elimination step has cost exponential in $|\{v\} \cup \delta(v)|$, we wish for an order that has the smallest maximum elimination clique size.

Lemma 4.3.2. *Given an elimination order, the computational complexity of the variable elimination process is $O(nr^{k+1})$ where k is the largest set of neighbors encountered during elimination.*

The exponent, $k+1$, is the largest clique size in the reconstituted graph.

4.3.2 Triangulated graphs

Any graph for which there exists at least one variable order that results in no fill-in edges when eliminating variables is known as a triangulated graph. Triangulated graphs have a more standard definition which is as follows.

Definition 4.3.3 (Triangulated graph [47, 15, 37, 10])**.** *A graph G is triangulated if all cycles greater than length three have a chord.*

All trees are triangulated since there are no cycles. A Markov chain is a sequence of successively connected nodes, so it is a tree, and hence is also triangulated. A hidden Markov model (HMM) [8] is another example of a tree. A trivial example is a set of disconnected nodes. Figure 4.2-(B) is not triangulated, since the four-cycle has no chord. Figure 4.2-(C) is triangulated, as is any clique, since all cycles are chorded. Figure 4.2-(D) is also triangulated since the four-cycle has a chord. Figure 4.2-(E) is *not* triangulated despite being composed of six triangles. Any node-induced subgraph (i.e., a subset of the nodes and corresponding incident edges) of a triangulated graph is still triangulated. We care about triangulated graphs since they have fill-in free elimination orders. In fact:

Theorem 4.3.4. *A graph G is triangulated iff there exists a fill-in free elimination order over the nodes in G.*

Since a reconstituted graph of the elimination process is one such that using the same order for elimination results in no additional fill-in edges, any reconstituted graph is triangulated.

Corollary 4.3.5. *Take any graph G and an elimination order σ, then the reconstituted graph $G' = (V, E \cup F_\sigma)$ is triangulated, where F_σ is the set of fill-in edges added during elimination.*

For example, the two reconstituted graphs on the bottom row of Figure 4.3 are triangulated. Thus, when we compute the marginal distribution, the computational cost is the same as if the original graph was the resulting reconstituted, and hence triangulated, graph. That is, from the perspective of computation, we could have had those additional fill-in edges (if any, for a given order) in the graph to begin with. Any set of neighbors $w, u \in \delta(v)$ that are not connected by an edge at the time we eliminate v could have been originally connected at the start since they are inevitable.

It is thus inevitable that, for exact inference, and from a computational perspective, we are always indirectly utilizing a triangulated graph corresponding to the reconstituted graph (under some order) after eliminating the nodes. Since the reconstituted graph has no fewer edges than the original graph (e.g., compare the bottom row of Figure 4.3 to the two original graphs), the family of distributions is enlarged (c.f. Proposition 4.1.1). That is, let G be an original graph and $G' = (V, E \cup F_\sigma)$ be a reconstituted graph resulting from eliminating with order σ, which caused additional fill in F_σ. Then the inference procedure under σ is still exact for any $p \in \mathcal{F}(G')$, even if we started with a $p \in \mathcal{F}(G) \subset \mathcal{F}(G')$. We say that $G = (V, E)$ can be embedded [37, 3] into another graph $G' = (V, E')$ if $E \subseteq E'$. Hence, if $p \in \mathcal{F}(G)$ then $p \in \mathcal{F}(G')$ for any G' into which G may be embedded.

The key difference between the two orderings in Figure 4.3 is that fill-in edges are produced in the right case, but no fill-in edges are produced in the left case. The largest clique in the original graph has size two, but the left order's reconstituted graph has a largest clique size of four. The number of nodes in the largest clique is encountered as an exponent of the computational cost during the elimination of the first variable in that clique. Since the cost of eliminating on the reconstituted graph is the same as that on the original graph (under a given order), the resulting triangulated graph's largest clique gives us the exponent of the computational cost of computing the desired marginal. Hence, to minimize computation of exact inference, we need to find the triangulated graph having the smallest maximum clique size into which the original graph may be embedded. One way to do this is to consider all possible elimination orders, and choose the one that results in the smallest maximum clique. Of course, this is not feasible since there are $n!$ possible orders.

Unfortunately, finding a triangulated graph into which G can be embedded that has the smallest maximum clique size is an NP-hard optimization problem.

Theorem 4.3.6. *For an arbitrary graph $G = (V, E)$, finding the smallest k such that G can be embedded into a triangulated graph with maximum clique size $k + 1$ is an NP-complete optimization problem [3].*

The good news is that it is easy to determine if a graph is triangulated. One way, for example, would be to find a fill-in free elimination order, something that can be done by successfully finding and then eliminating a node that results in no fill-in. This is something guaranteed to be easy whenever one has a triangulated graph as we next see.

We first discuss a generalization of a leaf node in a tree to a corresponding type of node in a triangulated graph, namely those nodes that produce no fill-in when eliminated. These are called *simplicial* nodes. A simplicial node, equivalently, is one whose neighbors form a clique. Leaf nodes are in fact simplicial nodes, but a simplicial node can exist in non-trees. For example, in Figure 4.2-(C), all nodes are simplicial. Triangulated graphs always have a simplicial node and in fact:

Lemma 4.3.7. *A triangulated graph on $n \geq 2$ nodes is either a clique, or there are at least two non-adjacent nodes that are simplicial.*

This is directly analogous to how (two or more node) trees always have at least two leaf nodes. And since any node-induced subgraph of a triangulated graph is still triangulated, once we eliminate a simplicial node, the residual is also triangulated and, if it has at least two nodes, also at least two simplicial nodes. Therefore, in addition to Theorem 4.3.4, we can produce a sequence of elimination steps without causing fill in.

Being triangulated does not mean that all nodes are simplicial. For example, consider the chorded 4-cycle in Figure 4.2-(D). It is triangulated (and has two simplicial nodes, namely x_1 and x_3) but the two other nodes are not simplicial.

Trees and triangulated graphs have more in common than the simplicial node generalization of leafs. In fact, we can generalize Theorem 4.2.2-(b) as follows.

Corollary 4.3.8. *All triangulated graphs can be generated by starting with a clique. We then repeatedly chose a new node and connect it to a set of previously chosen nodes that constitute a clique or sub-clique.*

Of course, a tree can be generated this way, where we connect a new node to a single previous node in the tree (Theorem 4.2.2-(b)). The above theorem states that any triangulated graph can be obtained in a generative fashion using an elimination order in reverse. A clique, for example, can be generated by adding an edge between any new node and all previously added nodes. In the case of Figure 4.2-(D), we can generate it as follows: emit x_1, emit x_2 and connect it to x_1, then emit x_4 and connect it to both x_1 and x_2, and last emit x_3 and connect it to both x_4 and x_2. Note that the emission order is a reverse fill-in free elimination order.

While a reverse elimination order can generate any triangulated graph, this does not mean that all triangulations of a particular graph are achievable by the elimination process, via some order, on that graph. For example, the four clique over x_1, x_2, x_3, x_4 is a triangulation of Figure 4.2-(D), but no elimination order will generate it — once we eliminate the first node, w.l.o.g., x_1, it will never be connected to its diagonally opposite node, x_3. The 4-clique, however, is a special type of triangulation over the 4-cycle since there are edges that can be removed without destroying the triangulation property. In fact, we have:

Theorem 4.3.9. *Let $G = (V, E)$ be a graph and let $G' = (V, E \cup F)$ be a triangulation of G with F the required edge fill-in. If the triangulated graph is* minimal, *meaning for any $F'' \subset F$ the graph $G'' = (V, E \cup F'')$ is no longer triangulated, then F can be obtained by the result of an elimination order.*

This means that the elimination algorithm and all the various variable orderings do include all minimal triangulations [31] of a graph G. Minimal triangulations are of interest since to find a triangulation with minimal maximum clique size, it is sufficient to consider a minimal triangulation (any non-minimal triangulation can only possibly have larger cliques) whenever there is no sparsity in the distribution [6]. The main point of the above discussion is that, if a graph is triangulated to begin with, then it is easy to detect; we just find a fill-in free elimination order, something we can do by repeatedly eliminating simplicial nodes.

When running the elimination process, one encounters all of the maximum cliques of the resulting reconstituted triangulated graph, even if the original graph is not triangulated. The reason is that each time we eliminate a node, that node, and its neighbors, become a clique. That clique must be a sub-clique of some maximal clique in the reconstituted graph, where a *maximal clique* is a clique that, when any additional node is added, is no longer complete. Once we have eliminated a node, the clique consisting of that node and its neighbors can never again be encountered in any future elimination step (since the node is eliminated). Hence, the clique is either itself a max-clique or is a subclique of some previous maximum clique. Thus, to find the set of maximal cliques in a triangulated graph, we find an elimination order, keeping track of the set of maximum cliques — if an elimination clique is not a subset of some previously stored maximum clique (which, therefore, includes the very first elimination clique), then it is a maximum clique and is added to the stored set.

Lemma 4.3.10. *When running the elimination algorithm, all maximal cliques in the resulting reconstituted graph are encountered as elimination cliques.*

As an immediate consequence:

Corollary 4.3.11. *The first node eliminated in a graph, along with its neighbors, forms a maximal clique.*

4.3.3 Good heuristics for choosing an elimination order

There are a number of heuristics that attempt to find the best elimination ordering that work quite well in practice, although they provide no guarantees of approximate optimality. They are, in roughly increasing order of complexity, as follows:

1. **min fill-in heuristic:**. Eliminate next the node n that would result in the smallest number of fill-in edges at that step. Break ties arbitrarily.

2. **min size heuristic:** Eliminate next the node that would result in the smallest clique when eliminated (i.e., choose the node that, at the time of elimination, has the smallest edge degree). Break ties arbitrarily.

3. **min weight heuristic:** If the nodes have non-uniform domain sizes (meaning each random variable X_i can have its own $r_i = |\mathcal{D}_{X_i}|$), then we choose next the node that would result in the clique with the smallest state space, which is defined as the product of the domain sizes. Break ties arbitrarily.

4. **tie-breaking:** When one heuristic has a tie on a node elimination, then choose one of the other heuristics to break that tie.

5. **non-greedy:** Rather than choosing the node that greedily looks best at the moment, take the m-best nodes (e.g., the $m < n$ nodes that would result in, say, the smallest fill-in) and eliminate one of them.

6. **random next step:** Create a distribution over the m-best nodes at any given step, where the probability of eliminating that node is either 1) uniform, or 2) inversely proportional to the greedy score (e.g., say the inverse of the required fill-in), and draw from that distribution to choose what node to eliminate next.

7. **random repeats:** Run any of the above complete heuristics multiple times, each time producing a different elimination order. As the final order, choose the one that results in the smallest maximum clique size.

Other than these heuristics, there are other methods for attempting to achieve a better triangulated embedding [11, 7, 3, 52], some of which, however, can run in exponential time in the worst case.

Identifying if a given graph is triangulated or not is easy. As suggested above, one simple way keeps eliminating simplicial nodes as long as possible, and outputs "not-triangulated" only if a point is reached where simplicial nodes no longer exist. A naive implementation would find the fill-in of each node, and eliminate one (if any) with no fill-in, yielding an $O(n^3)$ algorithm.

There are much better methods than this, however. For example, the *maximum cardinality search* procedure [57, 59] does the following: all nodes of the graph are given a binary flag indicating if they are labeled or not. We define an ordering of the nodes by the order in which they are chosen by the algorithm. Initially all nodes are unlabeled. We repeatedly choose as a next node the one that has the greatest number of previously labeled neighbors (i.e., we choose the node whose set of previously labeled neighbors has the maximum cardinality). Once a variable is chosen it is considered labeled. If at any time these neighbors are not a clique, the graph is not triangulated. Otherwise, the algorithm generates a perfect elimination order in reverse, since considering the order in reverse, each node so chosen will be simplicial. The complexity of the algorithm can be made to run in $O(|V| + |E|)$ (via amortized analysis [17]) using a Fibonacci heap to determine the next label (the one that has maximum cardinality of the set of previously labeled neighbors). This procedure has the following useful property:

Corollary 4.3.12. *Every maximum cardinality search of a triangulated graph G corresponds to a reverse fill-in free eliminating order of G.*

It is also possible to use this procedure to find a lower bound on the maximum clique size of a graph [44].

4.3.4 The running intersection property and junction trees

Triangulated graphs have still more in common with trees. One can in fact cluster the nodes of G together and then form a tree where the node clusters are vertices[1] of this new tree. In a triangulated graph, the clusters could correspond to the set of cliques or the set of maximal cliques of the triangulated graph. From such clusters, we can form a cluster tree:

Definition 4.3.13 (Cluster Tree). *Let $\mathcal{C} = \{C_1, C_2, \ldots, C_{|I|}\}$ be a set of node clusters ($C_i \subseteq V, \forall i$) of a graph $G = (V, E)$. A cluster tree is a tree $\mathcal{T} = (I, \mathcal{E}_T)$ with vertices corresponding to clusters in $\mathcal{C}$ and edges corresponding to pairs of clusters $C_1, C_2 \in \mathcal{C}$. We can label each vertex in $i \in I$ by the set of graph nodes in the corresponding cluster C_i, and we label each edge $(i, j) \in \mathcal{E}_T$ by the set of graph nodes in the intersection $S_{ij} = C_i \cap C_j$.*

Examples of cluster trees for a graph is shown in Figure 4.4. A particularly important type of cluster tree is called a junction tree. As we will see below, a *junction tree* is a cluster tree that (if it exists for a graph) allows a message-passing algorithm on trees to produce mathematically exact probabilistic inference for any distribution that is in the original graph family. Before defining junction trees, we define a property of sets of clusters in a tree.

[1]In this paper, we use the term "nodes" for nodes of the original graph G, and the term "vertices" for vertices (depicted as clouds, or clusters of nodes, in Figure 4.4) of the junction tree.

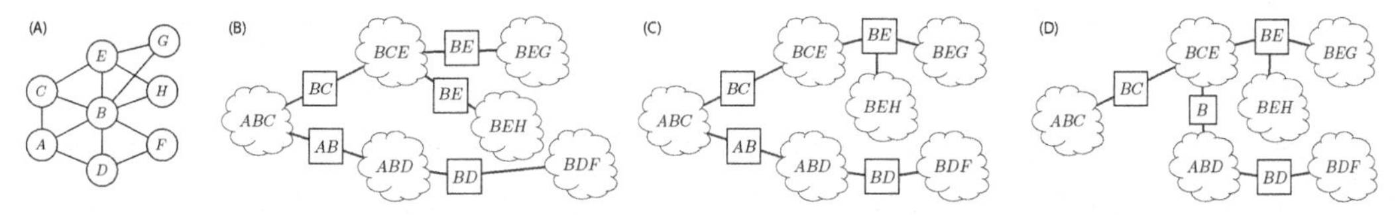

FIGURE 4.4: An example of a graph (A) and trees of clusters (shown as clouds) of nodes of the graph (B-D), where the clusters are $\{A, B, C\}$, $\{B, C, E\}$, $\{B, E, H\}$, $\{B, E, G\}$, $\{A, B, D\}$, and $\{B, D, F\}$. Tree edges are labeled with squares indicating the intersection between the incident clusters. In (B) and (C), the clusters are arranged to satisfy c.i.p. and also r.i.p. (consider the connected path containing $\{B, E, H\} \cap \{B, D, F\} = \{B\}$ running between $\{B, E, H\}$ and $\{B, D, F\}$), properties that are not true for (D). In (C), the separator $\{B, E\}$ has three incident edges so $d(\{B, E\}) = 3$, while (B) has two distinct identically $\{B, E\}$ labeled edges. Both (B) and (C) satisfy c.i.p./r.i.p.

Definition 4.3.14 (Cluster Intersection Property (c.i.p.)). *We are given a cluster tree $\mathcal{T} = (I, \mathcal{E}_\mathcal{T})$. The cluster intersection property states that for any two clusters C_1, C_2 in the tree, $C_1 \cap C_2 \subseteq C_i$ for all nodes, $\{C_i\}_{i \in P}$, on the path P between C_1 and C_2 in the tree $\mathcal{T}$.*

A given cluster tree might or might not have this property. The cluster tree in Figure 4.4-(C) satisfies c.i.p., while the one in Figure 4.4-(D) does not. Cluster trees that do satisfy the property are important and have a particular name.

Definition 4.3.15 (Junction Tree). *A* junction tree *is a cluster tree that satisfies c.i.p.*

A junction tree could have vertices that consist of cliques $\mathcal{C}(G)$, maxcliques, or even other forms of original graph node clusters. In fact, an extreme example would take a single cluster composed of all the original graph nodes — this is a tree with one (large cluster) node and that satisfies c.i.p., although it is not useful computationally.

The c.i.p., i.e., that the intersection of any two nodes in the tree is contained in the path between those two nodes, is identical to the *running intersection property*, or just r.i.p. Given a set of node clusters, r.i.p. means that the clusters can be ordered in a particular way, as defined next.

Definition 4.3.16 (Running Intersection Property (r.i.p.)). *Let $C_1, C_2, \ldots, C_\ell$ be a sequence of subsets of $V(G)$. Then the sequence order obeys the running intersection property (r.i.p.) if for all $i > 1$, there exists $j < i$ such that $C_i \cap (\cup_{k<i} C_k) = C_i \cap C_j$.*

Like the vertices of a junction tree, the running intersection property is defined not necessarily using cliques or even maximal cliques in the graph, but rather can be based on any clusters of the nodes of G.

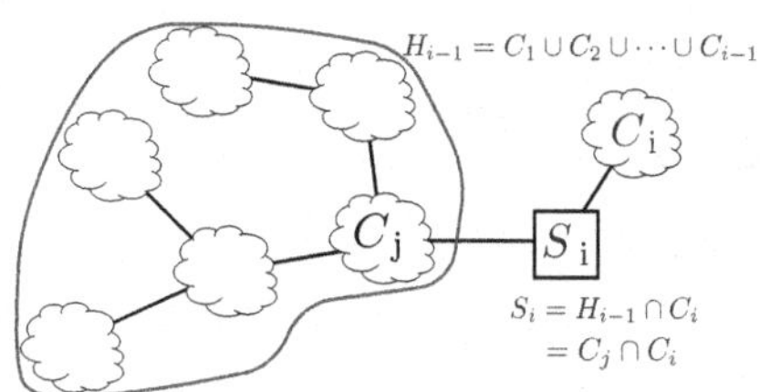

FIGURE 4.5: Example of an r.i.p. order.

It is easier to understand r.i.p. by defining and then explaining various additional terms. Suppose we are given a sequence of clusters $C_1, C_2, \ldots, C_\ell$. First, we define the accumulated history of the sequence at position i as $H_i = C_1 \cup C_2 \cup \cdots \cup C_i$. Next, we define the residual (or new) elements of C_i not encountered in the history as $R_i = C_i \setminus H_{i-1}$. Lastly, we define the commonality or separation elements between the new i^{th} clique and the history $S_i = C_i \cap H_{i-1}$. Since $C_i = R_i \cup S_i$, the ith clique consists of the innovation R_i and the commonality S_i. The cliques are in r.i.p. order if the commonality S_i between the new clique and the entire history is fully contained in one element, namely j, of that history; i.e., there exists an $j < i$ such that $S_i \subseteq C_j$. Adding further intuition to the r.i.p., we have the following:

Lemma 4.3.17. *Given a cluster tree, c.i.p. holds iff the clusters can be ordered to satisfy r.i.p.*

Proof. Start with the clusters that satisfy r.i.p. and that are in r.i.p. order. Using this order, we construct a tree with clusters as nodes by starting with C_1, and then for each position $i = 2 \dots \ell$, connecting an edge between C_i and the corresponding C_j where we know $j < i$. Clearly, this will produce a tree (consider the generative tree property of Theorem 4.2.2).

Take any C_i and C_k and suppose w.l.o.g. that C_i is later in the order than C_k (i.e., $k < i$). By the r.i.p., we know that S_i contains everything common between H_{i-1} and C_i, and so $C_i \cap C_k \subseteq C_i \cap H_{i-1} = S_i \subseteq C_j$ where $j < i$ is the earlier neighbor of i in the tree according to r.i.p. Thus, $C_k \cap C_i \subseteq C_k \cap C_j$. Apply this argument again with j taking the place of i (possibly swapping k and j so that $k < j$), and then repeat this argument until k and j are neighbors. Hence, $C_k \cap C_j$ is contained on the path between them in this tree and therefore the tree satisfies c.i.p.

Conversely, consider a cluster tree satisfying c.i.p. We perform a breadth first, or a preorder depth first (where we visit and number the parent node before any of the children), traversal of this cluster tree to produce an ordered sequence of cliques, where we define H_i and S_i as above according to this order. By the c.i.p., for every C_i, and C_k with $k < i$, $C_i \cap C_k$ is a subset of every node on the path from C_i to C_k. In particular, $C_i \cap C_k \subseteq C_j$ for $j < i$ being i's immediate earlier neighbor in the tree. Hence, $\bigcup_{k<i}(C_i \cap C_k) \subseteq C_j$ implying $C_i \cap \bigcup_{k<i} C_k = C_i \cap H_{i-1} \subseteq C_j$, and hence $C_i \cap H_{i-1} \subseteq C_i \cap C_j$. On the other hand, it is always true that $C_i \cap H_{i-1} \supseteq C_i \cap C_j$. Hence, $C_i \cap H_{i-1} = C_i \cap C_j = S_i$ which is r.i.p. □

Figure 4.4-(C) shows a cluster tree that satisfies both r.i.p. and c.i.p. As can be seen, the intersection of any pair of clusters is a subset of all of the clusters on the path between those two clusters. For example, $\{B, E, H\} \cap \{B, D, F\} = \{B\}$ is within every cluster in the tree. The example also shows that not all trees of maximal cliques satisfy the r.i.p. and therefore not all trees of maximal cliques are junction trees. The tree of maximal cliques on the right (Figure 4.4-(D)) is not a junction tree since a traversal of the tree does not produce a r.i.p. order. For example, $\{A, B\} = \{A, B, D\} \cap (\{A, B, C\} \cup \{B, C, E\}) \neq \{A, B, D\} \cap \{B, C, E\} = \{B\}$. This shows r.i.p. is not satisfied since $\{B, C, E\}$, which is what is connected to $\{A, B, D\}$ in the tree, is not representative of everything in the history in this traversal (which is $\{A, B, C\} \cup \{B, C, E\}$). Equivalently, we see that $\{A, B\} = \{A, B, C\} \cap \{A, B, D\} \not\subseteq \{B, C, E\}$ meaning that the intersection is not a subset of all cliques on the path between the two clusters.

Another equivalent way of stating the junction tree property is called the induced sub-tree property.

Definition 4.3.18 (Induced Sub-tree Property). *Given a cluster tree $\mathcal{T}$ for graph G, the induced sub-tree property is true if for all $v \in V$, the set of cliques $C \in \mathcal{C}$ such that $v \in C$ induces a (necessarily connected) sub-tree $\mathcal{T}(v)$ of $\mathcal{T}$.*

This property means that it is never possible to "forget" about any node with respect to constructing marginals using only neighboring clusters (this is critical for message passing on junction trees as we discuss below). The induced sub-tree property is useful since it offers us another way to view r.i.p. In fact, we have the following:

Lemma 4.3.19. *Given a cluster tree, c.i.p. holds iff the induced sub-tree property holds.*

We leave the proof as an exercise for the reader.

The relationship between triangulated graphs and junction trees is a strong one. In fact:

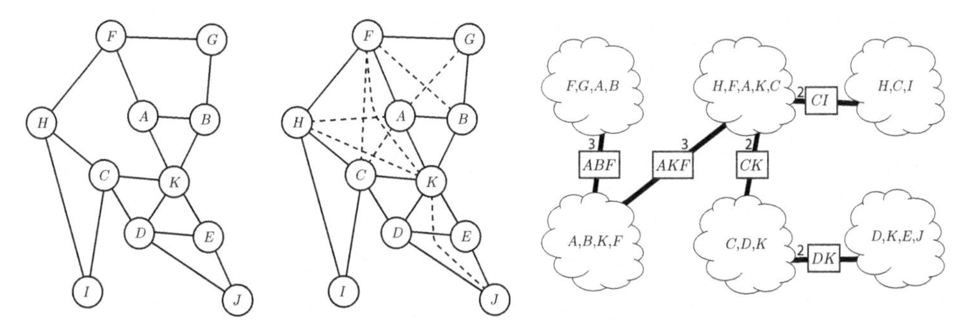

FIGURE 4.6: Given the graph (left), and a version with additional fill-in edges (middle), we can form a junction tree (right) where the clusters are maximal cliques of the triangulated graph, but not maximal cliques of the original graph. The triangulated graph can be constructed by starting with a graph with no edges, and then completing the nodes corresponding to each cluster (i.e., vertex) in the junction tree.

Theorem 4.3.20. *A graph G is triangulated iff one can form a junction tree with tree nodes corresponding to all of the maximal cliques of G.*

If we form the clusters of all maximal cliques of a non-triangulated graph, we cannot construct a junction tree, as Figure 4.6 shows. We could also form a junction tree of maximal cliques and also some, or all, of the non-maximal cliques of the graph. For example, given a junction tree of maximal cliques, any subset of the maximal clique is still a clique. Any clique A could be added as a node in a junction tree as a pendant neighbor of another existing clique C in the junction tree where $A \subseteq C$, and this would preserve r.i.p. A reason for doing this is the following. Given a general software toolkit to transform all junction tree cliques into marginals $p(x_C)$ (the methodology of which we describe in Section 4.4), adding a pendant subclique as above would be an easy way to use the existing toolkit's capabilities to automatically produce a desired subclique marginal. In other words, no additional marginalization code to convert from the clique marginal $p(x_C)$ to the desired marginal $p(x_A)$ is needed.

Given a junction tree, it is possible to recover the original triangulated graph. To do this, we start with a set of nodes $V = \cup_{C \in \mathcal{C}} C$. Then, we add an edge between $u, v \in V$ if there exists a $C \in \mathcal{C}$ such that $u, v \in C$. The maximal cliques of the junction tree, therefore, form a type of cover for the graph — every edge in the graph must lie within one of the cliques $C \in \mathcal{C}$. In fact, a triangulated graph can just as well be expressed as a set of nodes and a set of cliques.

Proposition 4.3.21. *Start with a triangulated graph $G = (V, E)$ and form all the cliques of the graph $\mathcal{C}(G)$. Form a new graph $G' = (V, E')$ where $\{u, v\} = e \in E'$ if and only if $\exists C \in \mathcal{C}(G)$ with $u, v \in C$. Then $G' = G$.*

In general, there are two ways to create a junction tree from a triangulated graph. The first way is to use the maximum cardinality search algorithm.

Theorem 4.3.22. *If the input to the maximum cardinality search (MCS) algorithm is already triangulated, then it produces a list of maximal cliques in a r.i.p.-respecting order.*

Given maximal cliques in r.i.p. order, we can form a junction tree by using the tree-generative procedure mentioned above where we connect an edge between C_i and C_j with j being the appropriate $j < i$ for i (see Figure 4.5). This j is obtained as $j \in \text{argmax}_{j' < i} |C_{j'} \cap C_i|$.

Alternatively, given the set of maximal cliques of a triangulated graph (obtained, say, through an elimination order computed via one of the heuristics mentioned above), we can produce a junction tree by first constructing a weighted graph of maximal cliques, where nodes in the graph correspond to maximal cliques, and where edges exist between two nodes only if the corresponding cliques have a nonempty intersection, and where edges are weighted by the size of this intersection (meaning all weights are integers greater than 0). As an example, consider the edge weights on the right of Figure 4.6. We have the following.

Theorem 4.3.23. *A tree of maximal cliques $\mathcal{T}$ is a junction tree of maximal cliques iff it is a maximum spanning tree*[2] *on the maximal cliques graph, with edge weights equal to the cardinality of the separator between the two maximal cliques.*

The graph must be triangulated for this to be true — a maximum spanning tree of a cluster graph where the clusters are maximal cliques but the graph is not triangulated will not produce a junction tree. This is a very standard and practical way to get a junction tree in inference code since the elimination algorithm, as mentioned above, produces the set of maximal cliques, and there are a number of simple but fast greedy strategies to form a maximal spanning tree, including Kruskal's or Prim's MST algorithm [17]. Prim's algorithm in fact can be made to run in $O(|m'| + |n'| \log |n'|)$ using a Fibonacci heap, where n' (resp. m') is the number of nodes (resp. edges) in the graph of cliques.

A junction tree of maximal cliques is unique in that any junction tree of all maximal cliques will always contain the same maximal cliques. There can, however, be multiple junction trees of maximal cliques since there might be several ways of connecting together the maximal cliques into a tree such that r.i.p. still holds. This can be seen by considering that there could be multiple equivalent weight maximum spanning trees of the graph of cliques, and since they each have maximum weight, they each constitute a junction tree.

4.3.5 Entanglement

There is one last topic to cover before we move on to how to perform inference on junction trees. At the beginning of this chapter, we discussed how the elimination algorithm can be used to produce a marginal $p(x_A)$ on a set A of variables x_A. In Equation (4.5), we had $A = \{1, 2\}$. In that example, the set $\{x_1, x_2\}$ was already a clique in the graph in Figure 4.2-A, and we eliminated, in a good order, all of the variables other than x_1 and x_2 to get the desired marginal. In some cases, however, the marginal we desire might not already be a clique. For example, suppose we wished, in Figure 4.2-A, the marginal $p(x_4, x_6)$ so we would eliminate all variables except for x_4 and x_6. We see that there is no order that results in a computation less than $O(r^3)$, even though the original graph is a tree.

We can see this issue even more directly when considering a chain with graph x_1—x_2—x_3, meaning that $p(x_1, x_2, x_3) = \phi(x_1, x_2)\phi(x_2, x_3)$. If we wish the marginal $p(x_2, x_3)$, which is already a clique in the graph, we marginalize x_1 leading to the $O(r^2)$ computation we expect from a tree. If we wish the marginal $p(x_1, x_3)$, however, we must compute $p(x_1, x_3) = \sum_{x_2} \phi(x_1, x_2)\phi(x_2, x_3)$ coupling together the two factors, leading to an $O(r^3)$ computation. It is as if the graph we started from had an additional edge between x_1 and x_3 that is caused by the desired marginal. This is made formal in the following

Theorem 4.3.24 (Rose's Entanglement Theorem [56])**.** *Let $G = (V, E)$ be an undirected graph with a given elimination ordering σ that maps G to $G^{\triangle} = (V, E') = (V, E \cup F_\sigma)$ where $E' = E \cup F_\sigma$, and where F_σ are the fill-in edges added during elimination. Then $\{v, w\} \in E'$*

[2]A maximum spanning tree is a tree over the same node set with the maximum possible value sum of its edge weights.

is an edge in $G^{\triangle}$ **iff** *there is a path in* G *with endpoints* v *and* w*, and where any nodes on this path other than* v *and* w *are eliminated before* v *and* w *in order* σ*.*

In other words, if there is a path $(v = v_1, v_2, \ldots, v_{k+1} = w)$ in G such that nodes $v_2, v_3, \ldots, v_k$ are eliminated before v and w, then there will be an edge between v and w as a result. If there are no nodes on the path other than v and w, this means $\{v, w\} \in E$, an edge that is retained in E'. In Figure 4.2-A, there indeed is a path between x_4 and x_6 causing a resulting edge.

One consequence of this theorem is that if A is a set of variable indices we wish to compute the marginal over, and if for all pairs of variables in A there is a path outside of A between those two variables, then A will become a clique. This means that A could have been a clique to begin with, and $|A|$ will be at least the exponent in the computational cost. Similar to how elimination might as well have started from a triangulated graph, where we are solving inference in the enlarged triangulated graph family into which the original graph is embedded, elimination could also have started with the enlarged graph where the desired marginal is a clique (when the conditions of Theorem 4.3.24 apply). Hence, when designing an exact marginal procedure, one should keep in mind the computational consequences not only of the natural relationships between variables but also the desired marginal.

4.4 Inference on Junction Trees

In this section, we describe a message-passing algorithm that can produce exact marginal distributions for the maximal cliques of a graph that is triangulated and has been converted into a junction tree.

We start with a triangulated graph $G = (V, E)$ that has been obtained using the methods described in earlier sections. Thus, we are solving inference for all distributions $p \in \mathcal{F}(G)$ and also for any distribution $p \in \mathcal{F}(G')$ where $G' = (V, E')$ with $E' \subseteq E$. The junction tree associated with G is designated as $\mathcal{T} = (\mathcal{C}, \mathcal{S})$ where $\mathcal{C}$ is the set of clusters of nodes of G (vertices in $\mathcal{T}$) and $\mathcal{S}$ is a set of G's separators (edges in $\mathcal{T}$). When we wish to designate the particular graph that the junction tree is derived from, we'll say $\mathcal{T}(G) = (\mathcal{C}(G), \mathcal{S}(G))$ where $\mathcal{C}(G)$ and $\mathcal{S}(G)$ are the corresponding vertices and edges in the junction tree $\mathcal{T}(G)$. In most cases, these clusters will correspond to maximal cliques (or just cliques) of G where $\mathcal{S}(G)$ will be separators within G, but to be mathematically correct, we only need that the r.i.p. be satisfied.

4.4.1 Benefits of junction trees

A junction tree allows us to treat any graph, and itself can be treated, as a standard tree. Theorem 4.3.20 is a bijection, so while any graph may have its nodes clustered to form a tree, only with a triangulated graph can we construct a junction tree where a message-passing algorithm is a mathematically correct form of inference, as we elaborate upon below. Therefore, the methods we earlier discussed for standard trees can be applied to any graph.

An additional benefit of a junction tree, and the elimination process, is that it can be used for all marginals. In earlier sections, we discussed the case where we wish only for one marginal $p(x_A)$ but the algorithm presented in this section will optionally produce a marginal for all $C \in \mathcal{C}$. The desired marginal $p(x_A)$ will be a subset of one of the junction tree clusters (so $A \subseteq C$ for some $C \in \mathcal{C}$), but we know it will not be a superset since, from the discussion at the end of Section 4.3, the variables comprising the desired marginal (those having index A) have been completed in the graph before any triangulation process begins. Therefore, A is a subset of some maximal clique in the triangulated graph. The cost of producing all of these marginals is $O(r^{\max_{C \in \mathcal{C}} |C|})$, i.e., it is exponential in the largest maximal clique size in the triangulated graph used to form the junction tree. This is the same cost, ignoring constants w.r.t. the clique sizes, as what it would take to produce only one marginal $p(x_A)$ (assuming the same triangulated graph).

We also discussed triangulating the graph from the perspective of only a single marginal. Since the algorithm described below produces marginals for all $C \in \mathcal{C}$, a natural worry is whether forming the triangulation based on knowing only A is detrimental relative to one formed knowing that we wish to compute all marginals. Luckily, this is not the case. Firstly, the set A is completed in any event, as a consequence of Theorem 4.3.24. Secondly, Theorem 4.3.9 states that all minimal triangulations are achievable via some elimination order. Hence, the process of forming a triangulated graph involves two steps: 1) complete any edge set (such as A) we wish to ensure is available as a marginal contained within some maximal clique in the resulting triangulation; 2) construct a triangulation using a good elimination order (using, say, the heuristics described above). Since we attempt to minimize the maximal clique size in the triangulation, the computation costs for computing all marginals will benefit simultaneously. Therefore, we do not need to spend time triangulating the graph multiple times and can operate on one junction tree for all desired marginals. Hence, only one effort (i.e., using the aforementioned heuristics) to find one good order is sufficient to produce a triangulated graph that is optimal relative to doing multiple efforts of elimination for each desired marginal.

A junction tree is not just an interesting graph-theoretic alternative representation of a triangulated graph, but it also yields a data structure useful to compute exact marginals. The junction tree, treated as a data structure, allocates memory for the potential function $\psi_C(x_C)$ associated with every maximal clique $C \in \mathcal{C}$, and memory also for the potential function $\phi_S(x_S)$ associated with every separator $S \in \mathcal{S}$ in the junction tree. When x is a discrete vector, we can view $\psi_C(x_C)$ (respectively $\phi_S(x_S)$) as a $|C|$-dimensional (respectively $|S|$-dimensional) table, indexed by vector arguments. We initialize these potential functions as described below, and then perform the $O(r^{\max_{C \in \mathcal{C}} |C|})$ computation to achieve the desired marginals.

Given an $S \in \mathcal{S}$, and since S is a separator, we know that S will break the graph into at least two connected components. For example, in Figure 4.6, all separators break the junction tree into two connected (subtree) components, namely those components separated by the separator, while in Figure 4.4-(C), separator $\{B, E\}$ breaks the tree into three components (indeed, there is some choice when forming a junction tree in some cases [33]). Hence, a separator might break a graph into more than two connected components. We use the notation $d(S)$ to refer to the number of connected components S breaks G into, where $d(S) \geq 2$. We use the symbol "d" since this also corresponds to the degree of a separator S in a junction tree when multiple identical separators are not redundantly represented (e.g., Figure 4.4-(C) rather than Figure 4.4-(B)).

4.4.2 Factorization

A junction tree expresses a form of factorization-based decomposition, where the separators in a junction tree correspond to the sets of random variables that, when conditioned on, render the remaining variables conditionally independent. For example, given any distribution p in the family corresponding to the graph in Figure 4.6, $\{C, K\}$ renders $\{D, E, J\}$ independent of $\{A, B, F, G, H, I\}$. Conditional independence between sets of random variables is expressed as a factorization property, and factorization is what enables the speedups relative to the naive marginalization in Algorithm 4.1. Accordingly, analogous to Equation (4.3) for trees, it is possible to write p as follows:

$$p(x) = \frac{\prod_{C \in \mathcal{C}(G)} p(x_C)}{\prod_{S \in \mathcal{S}(G)} p(x_S)^{d(S)-1}} = \frac{\prod_{C \in \mathcal{C}(G)} p(x_C)}{\prod_{S \in \mathcal{S}(G)} p(x_S)} \tag{4.13}$$

where the right-hand side is true only if $d(S) = 2$ for all S. It is mathematically possible to write the distribution as above in terms of the potentials corresponding to all maximal cliques and all separators of the graph (i.e., $\mathcal{C}(G)$ is the set of G's maximal cliques, and $\mathcal{S}(G)$ is the set of G's separators). This is not true in the case of a non-triangulated graph (e.g., consider the four-cycle in Figure 4.2-(B)).

4.4.3 Potentials as true marginals

Our goal is to obtain true marginals for all clusters of variables corresponding to junction tree nodes. A *true marginal* over any set of variables with index set A, namely $p(x_A)$ has the property that $p(x_A) = \sum_{x_{V \setminus A}} p(x_V)$, meaning we are marginalizing away all of the variables except for x_A. While mathematically we can form any true marginal in this fashion for any desired A, doing so is horribly inefficient.

The potentials in Equation (4.13) may already be true marginals, meaning:

$$\psi_C(x_C) = p(x_C), \forall C \in \mathcal{C}(G), \quad \text{and} \quad \phi_S(x_S) = p(x_S), \forall S \in \mathcal{S}(G), \tag{4.14}$$

in which case we are done. Typically, we start with a configuration where the maximal cluster and separator potentials are not true marginals, but an equation similar to Equation (4.13) is true. That is, we initialize the clique and separator potentials to ensure the following holds:

$$p(x) = \frac{\prod_{C \in \mathcal{C}} \psi_C(x_C)}{\prod_{S \in \mathcal{S}(G)} \phi_S(x_S)^{d(S)-1}} = \frac{\prod_{C \in \mathcal{C}} \psi_C(x_C)}{\prod_{S \in \mathcal{S}(G)} \phi_S(x_S)}, \tag{4.15}$$

where, again, the r.h.s. holds only when $d(S) = 2$ for all S.

The algorithm we describe performs computation that can be viewed as a *message* along a junction tree edge. The cost of the algorithm is $O(r^{\max_{C \in \mathcal{C}} |C|})$. While the algorithm runs, Equation (4.15) remains true after each step. Eventually, the potential functions are transformed, and the distribution is reparameterized, so that Equation (4.13) becomes true (i.e., the potentials become true marginals). At that point, the desired marginals may be computed, again at a cost of $O(r^{\max_{C \in \mathcal{C}} |C|})$, directly from the potential functions.

There are a variety of forms of messages [41, 58, 34, 42, 45, 49, 54] but they all have the same asymptotic computational cost (our exposition concentrates on what is known as the "Hugin" style [34] of messages).

4.4.4 Message initialization

We next describe how we can initialize the clique and separator potentials achieving Equation (4.15). We start from an undirected graph $G' = (V, E')$ that is not necessarily triangulated. Given a $p \in \mathcal{F}(G')$, we have a set of potential functions associated with p, and let's say that $\mathcal{C}'$ is the set of cliques of G' (recall from earlier in the chapter that these are given to us), so we may write $p \in \mathcal{F}(G')$ as $p(x) = \prod_{C' \in \mathcal{C}'} \psi_{C'}(x_{C'})$ for any x. Once we triangulate G' into $G = (V, E \cup F)$, any clique in G' is a subset of a maximal clique in G. Correspondingly, for any factor $\phi_{C'}$ of $p \in \mathcal{F}(G')$ over a set of variables, there is a maximal clique C of G that is a superset, meaning $\exists C \in \mathcal{C}$ with $C' \subseteq C$. Thus, any $p \in \mathcal{F}(G')$ can be represented as Equation (4.1), where G is a triangulation of G' and $\mathcal{C}(G)$ are the maximal cliques of G. We can use this factorization in the triangulated graph as an initial assignment to the junction tree maximal clique and separator potentials. Our goal is to initialize $\{\psi_C(x_C) : C \in \mathcal{C}\}$ and $\{\phi_S(x_S) : S \in \mathcal{S}\}$ so that

$$p(x) = \prod_{C' \in \mathcal{C}'(G')} \psi_{C'}(x_{C'}) = \frac{\prod_{C \in \mathcal{C}} \psi_C(x_C)}{\prod_{S \in \mathcal{S}(G)} \phi_S(x_S)}. \tag{4.16}$$

We do this as follows. First, all of the separator potentials are set identically to unity, meaning for all $S \in \mathcal{S}$, and for all x_S, we set $\phi_S(x_S) = 1$, thereby rendering the denominator on the r.h.s. of Equation (4.16) inconsequential (for the moment).

Next, to initialize the maximal clique potentials, we do the following. For each clique C' of the non-triangulated graph G', we find one and only one maximal clique $C \in \mathcal{C}$ of the corresponding junction tree of a triangulation G of G' such that $C' \subseteq C$. We are finding any maximal clique C of G that covers the clique C' of G'. Any such maximal clique will work (since multiplication is commutative) as long as this is not done more than once (which would be mathematically incorrect). If there is more than one choice, we select only one from amongst those choices arbitrarily. We then assign that clique C' of G' to C of G, and $\phi_{C'}(x_{C'})$ gets correspondingly assigned multiplicatively to the potential $\phi_C(x_C)$. When we are done with this process, each maximal clique C has a set of cliques, say $\mathcal{A}(C)$, of G' assigned to it. Given $\mathcal{A}(C)$, and the corresponding potential functions $\phi_A(x_A)$ for each $A \in \mathcal{A}(C)$ (which are factors of p), the resulting potential function over just C takes the following form:

$$\phi_C(x_C) = \prod_{A \in \mathcal{A}(C)} \psi_A(x_A). \tag{4.17}$$

Hence, $p(x) = \prod_{C \in \mathcal{C}} \prod_{A \in \mathcal{A}(C)} \psi_A(x_A)$.

If there is a maximal clique C in the junction tree containing a $v \in C$ where no potential has been assigned that involves v, then there will be some other potential in $\mathcal{C}$ that involves v and that has a factor involving v assigned to it. It might be that $\bigcup_{A \in \mathcal{A}(C)} = C$, meaning that all variables in the resulting potential function $\phi_C(x_C)$ have a corresponding factor (i.e., $\mathcal{A}(C)$ is a cover of C). On the other hand, it might also be that $\cup_{A \in \mathcal{A}(C)} A \subset C$ which can happen, for example, if some factor whose variables are fully contained within C are assigned to some other maximal clique's potential function. In this case, for any $v \in C \setminus (\cup_{A \in \mathcal{A}(C)} A)$, changing the value of x_v in $\phi_C(x_C)$ has no effect on $\phi_C(x_C)$'s value (i.e., $\phi_C(x_{C \setminus \{v\}}, x_v) = \phi_C(x_{C \setminus \{v\}}, x'_v)$ for any $x_v \neq x'_v$). In either case, the assignment is mathematically valid.

Algorithm 4.2 gives pseudo-code showing the assignment process. With this assignment to $\phi_C(x_C)$ for each C, and the corresponding unity assignment to the separator potentials, we have reached a state where Equation (4.16) is true.

We next address the question of why this initialization alone does not already achieve the junction tree potentials as marginals as in Equation (4.14). The answer depends on what the original factors $\phi_{C'}(x_{C'})$ are to begin with. When starting from a Bayesian network, the factors will be local conditional distributions of the form $p(x_a|x_{\pi_a})$, and hence are not required to constitute a marginal over $x_{C'}$ where $C' = \{a\} \cup \pi_a$. When starting from an undirected model, the distribution has the form $p(x) = \frac{1}{Z} \prod_{C'} \phi_{C'}(x_{C'})$ where the $\phi_{C'}(x_{C'})$s could be any arbitrary finite-valued non-negative function, and so need not be marginals. The constant Z is the partition function which ensures $p(x)$ is a valid distribution. This constant might be absorbed multiplicatively by the factors when the distribution is expressed without an explicit mention of the $1/Z$ factor in front. Another reason involves evidence variables, i.e., a subset of variables $I \subseteq V$ that are known and constant, i.e., the values of the vector x_I do not get marginalized since they are fixed. We denote such variables as $\bar{x}_I$

Algorithm 4.2 Initialize junction tree potentials for Equation (4.16).

Input: An undirected graph $G' = (V, E)$ and corresponding $p \in \mathcal{F}(G)$ with potential functions $p_{C'}(x_{C'})$ for each $C' \in \mathcal{C}'(G')$. A triangulation $G = (V, E \cup F)$ of G' and corresponding junction tree $\mathcal{T} = (\mathcal{C}, \mathcal{S})$ for G.

Output: An assignment to junction tree (clique and separator) potentials satisfying Equation (4.16).

1: **for all** $C \in \mathcal{C}$ **do**
2: $\quad \psi_C(x_C) \leftarrow 1$
3: **end for**
4: **for all** $S \in \mathcal{S}$ **do**
5: $\quad \phi_S(x_S) \leftarrow 1$
6: **end for**
7: **for all** $C' \in \mathcal{C}'(G')$ **do**
8: $\quad$ Find one $C \in \mathcal{C}$ such that $C' \subseteq C$, if more than one, choose one arbitrarily $\psi_C(x_C) \leftarrow \psi_C(x_C)\psi_{C'}(x_{C'})$
9: **end for**
10: return the finished junction tree potentials

---

to indicate they are fixed. The marginal we wish to compute, at each junction tree node C, becomes $p(x_{C\setminus I}, \bar{x}_I)$ from which we can compute conditionals such as $p(x_{A\setminus I}|\bar{x}_I)$ where $A \subseteq C$. When evidence is introduced, each potential function over a clique C becomes $\psi_C(x_{C\setminus I}, \bar{x}_{C\cap I})$, something that, for the reasons mentioned above, need not be the same as the desired $p_C(x_{C\setminus I}, \bar{x}_I)$. In other words, evidence I initially affects the potential via only the variables $C \cap I$, but after the potentials are converted to marginals, all of the evidence I affects every marginal which has the form $p_C(x_{C\setminus I}, \bar{x}_I)$ rather than $p_C(x_{C\setminus I}, \bar{x}_{I\cap C})$. This is useful since we can see how the evidence influences random variables regardless of where they are in the junction tree.

4.4.5 Necessary condition for true marginals

We next identify a necessary (but not a sufficient) condition for the potential functions to be true marginals. If our final configuration of the potentials has this property, these potentials must agree on at least the nodes that they mutually have in common. To agree means to be consistent, or that they give the same sub-marginals over the variables that lie in their common intersection. Specifically, given any two maximal cliques C_1 and C_2 of $\mathcal{C}$, with $S_{12} = C_1 \cap C_2$, we must have that

$$\sum_{x_{C_1 \setminus S_{12}}} \psi_{C_1}(x_{C_1}) = \sum_{x_{C_2 \setminus S_{12}}} \psi_{C_2}(x_{C_2}). \tag{4.18}$$

If Equation (4.18) is true, we say that the two potentials are *consistent.* If junction tree potentials are already equal to the marginals, it would follow that

$$\sum_{x_{C_1 \setminus S_{12}}} \psi_{C_1}(x_{C_1}) = \sum_{x_{C_1 \setminus S_{12}}} p(x_{C_1}) = \sum_{x_{C_2 \setminus S_{12}}} p(x_{C_2}) = \sum_{x_{C_2 \setminus S_{12}}} \psi_{C_2}(x_{C_2}), \tag{4.19}$$

even if C_1 and C_2 are not immediate neighbors in the junction tree, thanks to the properties of true marginal distributions.

4.4.6 Achieving true marginals via message passing

The discussion below will present an algorithm that ensure that pairs of potentials only between neighbors in a junction tree become consistent in the aforementioned fashion, a property that is a necessary condition of the potentials being true marginals (which is our goal). This property is called *local consistency*, since the potentials are consistent locally (immediate neighbors). The properties of a junction tree itself will then ensure that, along with local consistency, all potentials that have any variable overlap will also be consistent (known as *global consistency*), and as a further consequence, the potentials become true marginals. In other words, local consistency along with the junction tree property, together, ensure that global consistency is achieved.

Consider two neighboring maximal cliques U and W with separator $S = U \cap W$, and potential functions ψ_U, ψ_W, and ϕ_S, arranged along a single-edge (two node) junction tree as shown on the bottom of Figures 4.7 and 4.8.

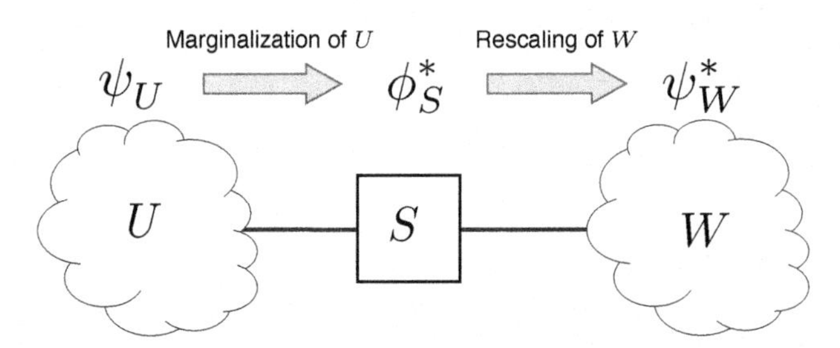

FIGURE 4.7: A simple two-node, U and V, junction tree and separator, S. We also see a message from U to W via S consisting of two steps: (1) marginalization of the potential $\psi_U(x_U)$ into a new potential $\phi_S^*(x_S)$, and (2) rescaling of the potential $\psi_W(x_S)$ by $\phi_S^*(x_S)$, and dividing out any previous value of the potential at S. This also corresponds to the lower region in Figure 4.8.

Suppose the junction tree potentials start out being inconsistent, meaning:

$$\sum_{x_{U \setminus S}} \psi_U(x_U) \neq \sum_{x_{W \setminus S}} \psi_W(x_W) \quad \text{and} \quad \phi_S(x_S) = 1. \tag{4.20}$$

Before proceeding, it should be clear that $\psi_U(x_U) = \psi_U(x_{U \setminus S}, x_S)$, and that $\sum_{x_{U \setminus S}} \psi_U(x_U)$ is a function only of x_S as is $\sum_{x_{W \setminus S}} \psi_W(x_W)$. Since we initialized the potentials appropriately, we have $p(x_{V \setminus I}, \bar{x}_I) = \psi_U(x_{U \setminus I}, \bar{x}_{U \cap I})\psi_W(x_{W \setminus I}, \bar{x}_{W \cap I})/\phi_S(x_{S \setminus I}, \bar{x}_{S \cap I})$ (so $U \cup W = V$) which follows from Equation (4.15).

We describe a message operation passed between the two maximal cliques, via the separator, in the junction tree. We start off with a message from U to W, the first part of which involves marginalizing U as follows:

$$\forall x_S, \text{ compute } \phi_S^*(x_S) = \sum_{x_{U \setminus S}} \psi_U(x_U). \tag{4.21}$$

This leads to a new separator potential $\phi_S^*(x_S)$ and can be seen as a partial message, as shown by the left arrow in Figure 4.7. The next operation involves a re-scaling of W, as follows:

$$\forall x_W, \text{ compute } \psi_W^*(x_W) = \phi_W(x_W)\frac{\phi_S^*(x_S)}{\phi_S(x_S)} \tag{4.22}$$

This produces a new potential on W based on the updated separator potential at S. This can also be seen as a partial message, as shown as the right arrow in Figure 4.7. After these operations, the new joint $p(x_{V\setminus I}, \bar{x}_I)$ has not changed. To see this, define an alias $\psi^*_U(x_U) = \psi_U(x_U)$ for convenience, and consider the following:

$$\frac{\psi^*_U(x_U)\psi^*_W(x_W)}{\phi^*_S(x_S)} = \frac{\psi_U(x_U)\psi_W(x_W)\phi^*_S(x_S)}{\phi_S(x_S)\phi^*_S(x_S)} = \frac{\psi_U(x_U)\psi_W(x_W)}{\phi_S(x_S)} \tag{4.23}$$

While the joint distribution over these potentials has not changed, we have not yet achieved consistency between the potentials, since

$$\sum_{x_{U\setminus S}} \psi^*_U(x_U) = \sum_{x_{U\setminus S}} \psi_U(x_U) = \phi^*_S(x_S) \neq \sum_{x_{W\setminus S}} \psi^*_W(x_W) = \frac{\phi^*_S(x_S)}{\phi_S(x_S)} \sum_{x_{W\setminus S}} \psi_W(x_W) \tag{4.24}$$

which follows because

$$\phi_S(x_S) \neq \sum_{x_{W\setminus S}} \psi_W(x_S). \tag{4.25}$$

We have, however, achieved at least one marginal in $\psi^*_W(x_W)$, which follows because we started with: $p(x_{V\setminus I}, \bar{x}_I) = \psi_U(x_U)\psi_W(x_S)/\phi_S(x_S)$ and

$$\psi^*_W(x_W) = \frac{\phi^*_S(x_S)}{\phi_S(x_S)}\psi_W(x_W) = \psi_W(x_W)\sum_{x_{U\setminus S}} \psi_U(x_U) = \sum_{x_{U\setminus S}} p(x_{V\setminus I}, \bar{x}_I) = p(x_{W\setminus I}, \bar{x}_I) \tag{4.26}$$

is one of the marginals we desire. As mentioned above, we may view this as a message-passing procedure, passing a message from maximal clique U through S and to W, shown in Figure 4.7.

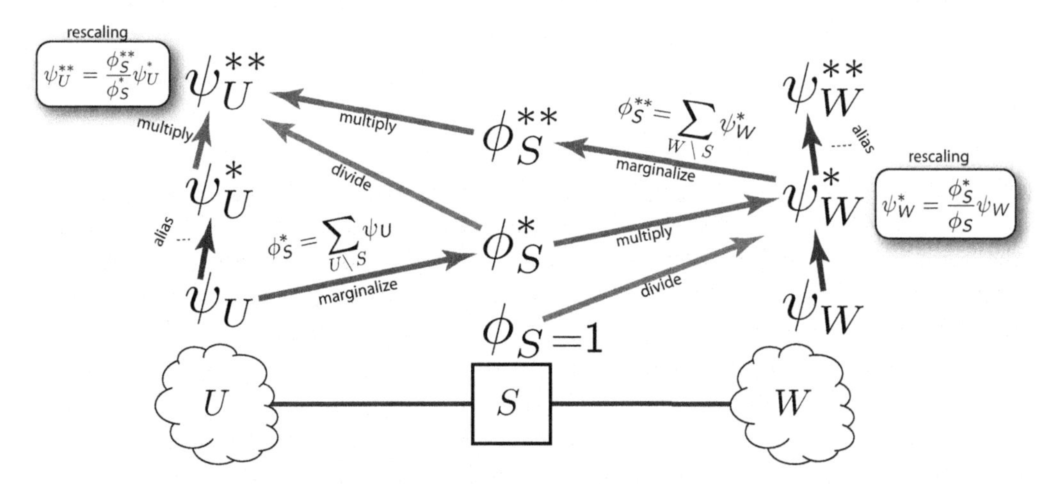

FIGURE 4.8: Message from U to W via S and back from W to U along a single edge (two node) junction tree corresponding to a triangulated graph with two maximal cliques, U and W. In the figure, we use a short-hand notation, where, for example, $\phi^{**}_S = \sum_{W\setminus S} \psi^*_W$ means to compute $\phi^{**}_S(x_S) = \sum_{x_{W\setminus S}} \psi^*_W(x_W) = \sum_{x_{W\setminus S}} \psi^*_W(x_{W\setminus S}, x_S)$ for all $x_S \in \mathcal{D}^S$.

We next perform the same set of operations in reverse, i.e., send a message from W back to U using the new state of the potential functions. I.e., we first marginalize W, as in the following:

$$\forall x_S, \text{ compute } \phi_S^{**}(x_S) = \sum_{x_{W \setminus S}} \psi_W^*(x_W), \tag{4.27}$$

resulting in yet another new separator potential $\phi_S^{**}(x_S)$, and then rescale U as in:

$$\forall x_U, \text{ compute } \psi_U^{**}(x_U) = \frac{\phi_S^{**}(x_S)}{\phi_S^*(x_S)} \psi_U^*(x_S) \tag{4.28}$$

resulting in a new potential on U. After these operations, the new joint $p(x_H, \bar{x}_E)$ has again not changed. To see this, define an alias $\psi_W^{**} = \psi_W^*$ for convenience, and consider

$$\frac{\psi_U^{**}(x_U)\psi_W^{**}(x_W)}{\phi_S^{**}(x_S)} = \frac{\psi_U(x_U)\phi_S^{**}(x_S)\psi_W(x_W)\phi_S^*(x_S)}{\phi_S^{**}(x_S)\phi_S(x_S)\phi_S^*(x_S)} = \frac{\psi_U(x_U)\psi_W(x_W)}{\phi_S(x_S)}. \tag{4.29}$$

Most importantly, after this second "backwards" message pass, consistency is now achieved. In particular, $\psi_U^{**}(x_U)$ and $\psi_W^{**}(x_W)$ are now consistent since:

$$\sum_{x_{U \setminus S}} \psi_U^{**}(x_U) = \sum_{x_{U \setminus S}} \frac{\phi_S^{**}(x_S)}{\phi_S^*(x_S)} \psi_U^*(x_U) = \frac{\phi_S^{**}(x_S)}{\phi_S^*(x_S)} \sum_{x_{U \setminus S}} \psi_U^*(x_U) \tag{4.30}$$

$$= \frac{\phi_S^{**}(x_S)}{\phi_S^*(x_S)} \phi_S^*(x_S) = \phi_S^{**}(x_S) = \sum_{x_{W \setminus S}} \psi_W^{**}(x_S) \tag{4.31}$$

We also now have the other marginal we wish at $\psi_U^{**}(x_U)$ since:

$$\psi_U^{**}(x_U) = \frac{\phi_S^{**}(x_S)}{\phi_S^*(x_S)} \psi_U(x_U) = \psi_U(x_U) \frac{\sum_{x_{W \setminus S}} \psi_W^*(x_W)}{\sum_{x_{U \setminus S}} \psi_U(x_U)} = \psi_U(x_U) \frac{\sum_{x_{W \setminus S}} \frac{\phi_S^*(x_S)}{\phi_S(x_S)} \psi_W(x_W)}{\sum_{x_{U \setminus S}} \psi_U(x_U)}$$

$$= \psi_U(x_U) \frac{\sum_{x_{W \setminus S}} \psi_W(x_W) \sum_{x_{U \setminus S}} \psi_U(x_U)}{\sum_{x_{U \setminus S}} \psi_U(x_U)} = \psi_U(x_U) \sum_{x_{W \setminus S}} \psi_W(x_W) = \sum_{x_{W \setminus S}} p(x_H, \bar{x}_E)$$

$$= p(x_{H \cap U}, \bar{x}_E)$$

We have therefore performed a forward and then a backward message on the junction tree. A diagrammatic summary of these messages is shown in Figure 4.8, where arrows show the dependency graph based on the definitions of the messages. The procedure offers a strategy to produce the two desired marginals and thus achieves local consistency, at least when we have only two cliques.

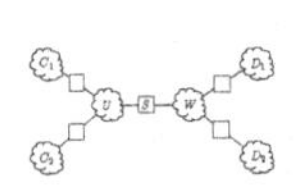

a A slightly larger tree into which the sub-tree edge U, W has been embedded.

b Valid and invalid messages according to the message passing protocol.

FIGURE 4.9: (a) A tree of maximal cliques. (b) The message passing protocol.

We next generalize this procedure to more than two cliques arranged in a junction tree.

Consider Figure 4.9a which shows the cliques U and W from Figure 4.7 but in a broader context of other cliques, C_1, C_2, D_1, and D_2. When a message is sent from U to W and back, we know W and U are consistent with each other. If, next, a message is sent from W to D_1 and back, then W and D_1 will be consistent, but U and W might no longer be consistent. The essential worry is that future messages might destroy the previously achieved local marginal consistency, thereby removing the aforementioned necessary condition for the potentials being true marginals. In general, subsequent messages will indeed cause previously consistent pairs of cliques to be no longer consistent.

If the messages are sent in a particular order, however, then consistency is assured. One possible set of orders ensuring consistency are those that abide by the *message passing protocol*, or MPP.

Definition 4.4.1 (Message passing protocol). *A node may send a message to a neighboring node in a junction tree* **only** *after it has received messages from all of its* other *neighbors.*

Figure 4.9b depicts valid and invalid messages according to MPP. MPP states that a message may be sent out from a clique to a neighbor only at certain points (namely when it has received enough incoming messages from other edges). Leaf nodes in the tree can send messages immediately since they have no other adjacent edges. Once immediate parents of leaf nodes have received messages from their children, then the parents can send messages to grandparents, and so on. This may proceed up until the root of the tree. The root, in fact, can be chosen arbitrarily. The protocol does not specify one set of messages, but rather a condition on a set of message orders between cliques. There can be many message schedules that abide by MPP. Examples include different selections of roots, and different selections of orders among the leaf nodes. With MPP, we have the following:

Theorem 4.4.2. *Suppose enough messages have been sent so that every edge in the tree of cliques has had a message sent in both directions along that edge. If the messages followed the message passing protocol (Definition 4.4.1), then the cliques are locally consistent between all pairs of connected cliques in the tree.*

This theorem means that if MPP is obeyed, the message operations described in Equations (4.21), (4.22), (4.27), and (4.28) (and shown in Figure 4.8) will render the potentials locally consistent. This theorem is a consequence of the commutativity of multiplication. One can also see, looking at the equations, that once such consistency is achieved, any further messages do not change the potentials (i.e., they are identity transforms) and so no additional messages would have an effect (and hence would only entail redundant computation). Hence, this also gives a stopping condition. Another interesting point is that the above holds whenever we have a tree of clusters, meaning that local consistency is assured even if the r.i.p. property of the tree is not satisfied.

While local consistency is a necessary condition of the potentials being true marginals, it is not a sufficient condition in general. A stronger property of true marginals is that potentials over any two clusters having common variables, not just neighbors in the tree, marginally agree on their common variables. This is a property called *global consistency.* It is an interesting fact that if the tree has reached a state of local consistency, and if the tree is a junction tree, global consistency is also assured. That is, we have:

Theorem 4.4.3. *In any* junction tree *of clusters, any configuration of cluster functions that are locally (neighbor) consistent will be globally consistent. I.e., for* **any** *clusters pair* C_1, C_2*, even non-neighbors, with* $C_1 \cap C_2 \neq \emptyset$*, we have:*

$$\sum_{x_{C_1 \setminus C_2}} \psi_{C_1}(x_{C_1}) = \psi_{C_1}(x_{C_1 \cap C_2}) = \psi_{C_2}(x_{C_1 \cap C_2}) = \sum_{x_{C_2 \setminus C_1}} \psi_{C_2}(x_{C_2}) \tag{4.32}$$

for all values $x_{C_1 \cap C_2}$.

We can get an intuitive sense for why this theorem is true by considering the cluster intersection property (c.i.p., Definition 4.3.14), the induced sub-tree property (Definition 4.3.18), and transitivity along junction tree edges. That is, any marginals that two clusters have in common with each other are shared along the path in the tree (thanks to c.i.p.), and at each edge step there is local consistency, so there must be global consistency as well. Each variable induces a connected subtree in the junction tree. Hence, as long as information is passed around in the tree in the appropriate order, information regarding that variable is not lost and that variable will not become inconsistent. If the induced sub-tree property was not true, then a variable might become inconsistent after message passing.

Global consistency does not necessarily guarantee true marginals in general, but in the present case it does. I.e., we also have the following:

Theorem 4.4.4. *Given a junction tree of clusters $\mathcal{C}$ and separators $\mathcal{S}$, and corresponding potential functions that satisfy Equation (4.15), if the potential functions are in a state of global consistency, then the potentials will be true marginals, i.e.,:*

$$\psi_C(x_C) = p(x_C) \text{ and } \phi_S(x_S) = p(x_S) \tag{4.33}$$

The theorem can be proven inductively starting with a base case of two cliques (which we showed in Equations (4.21), (4.22), (4.27), and (4.28)).

Algorithm 4.3 PostOrder($U \to W$)

1: **for all** $C \in \text{child}(U)$ **do**
2: Call PostOrder ($C \to U$)
3: **end for**
4: Send message from U to W according to Equations (4.21) and (4.22)

Algorithm 4.4 PreOrder($W \to U$)

1: Send message from W to U according to Equations (4.27), and (4.28)
2: **for all** $C \in \text{child}(U)$ **do**
3: Call PreOrder($U \to C$)
4: **end for**

Algorithm 4.5 MPP Messages

1: Designate arbitrary root as R.
2: **for all** $C \in \text{child}(R)$ **do**
3: Call PostOrder($C \to R$)
4: **end for**
5: **for all** $C \in \text{child}(R)$ **do**
6: Call PreOrder($R \to C$)
7: **end for**

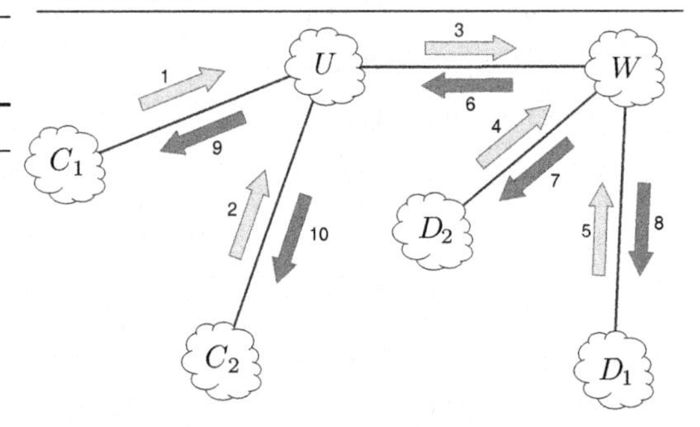

FIGURE 4.10: Messages in MPP order.

4.4.7 Message schedules

One last important issue to address is how to generate a schedule of messages that obey the MPP. This turns out to be fairly simple. First, we choose an arbitrary junction tree node to be the root of the tree. This immediately turns the edges into directed edges, and also allows one to refer to a parent of a node (the adjacent node one step closer to the root, if any) and children of a node (the set of adjacent nodes one step farther away from the root). Next, we send all messages starting from the leaves up to this root. Sending messages that way guarantees messages may be sent obeying the message passing protocol. This can be done simply by performing a post-order traversal [17] of the tree (as shown in Algorithm 4.3, a process that is sometimes called *Collect Evidence*). Once all messages have been sent to the root, the root may start sending messages down to the children, and so on, until they reach the leaves. This can be done using a pre-order traversal [17] of the tree, as shown in Algorithm 4.4 (a process sometimes called *Distribute Evidence*). When done, all nodes will have received messages from all neighbors, and we may thus form the desired marginals. To fully traverse the tree, we must call the above algorithms appropriately, by first doing a post-order and then a pre-order tree traversal, as shown in Algorithm 4.5.

It is clear that this procedure produces messages that abide by the message passing protocol. At the collect evidence stage, a message is not sent to a node's (single) parent until it has received messages from all its children, so there is only one node it has not yet received a message from, namely the parent. At the distribute evidence stage, once a node has received a message from its parent, it has received a message from all of its neighbors (since it received a message from all its children earlier, during the collect evidence phase) so it is free to send a message to any child that it likes. Different message schedules can be achieved by choosing different orders of the children at each step in line 1 of Algorithm 4.3 and line 2 of Algorithm 4.4. Figure 4.10 shows an example of the ordering of the messages for the junction tree of Figure 4.9a, where W has been designated as the root.

We reflect for a moment on why the algorithm has historically been called collect and distribute *evidence* which may seem curious since evidence is not explicitly mentioned. Recall from above, each of the factors that involve evidence variables I treat those variables as constants, so they are not marginalized over and they cause no computation. Hence, each marginal is joint with the evidence variables, and the resulting marginal configurations are of the form $p(x_{C\setminus I}, \bar{x}_I)$. The evidence variables might not initially have direct intersection with a cluster (i.e., we can have $C \cap I = \emptyset$), so what the algorithms above are doing, in addition to achieving global consistency and marginals, is propagating the evidence around the junction tree, hence the name.

4.5 Discussion

In this chapter, we have offered the basics of the exact computation of marginal distributions in graphical models, and have focused on the undirected graphical model perspective, where any other type of graphical model (such as a Bayesian network) is embedded in a family large enough to ensure it is mathematically valid for the original distribution.

This chapter is only a brief introduction, and there are a number of additional aspects of exact inference that space limitations have prevented us from addressing. We conclude the chapter by mentioning some of them.

As mentioned above, the computational cost needed for the above is $O(r^{\max_{C\in\mathcal{C}}|C|})$, and this is both in terms of time and memory. Very similar operations (e.g., the most probable random variable configuration, and sampling) can be achieved with very similar patterns of computation [19, 1]. There are a number of other interesting points that are achievable on what could be called the time-space frontier of achievable exact inference. For example, methods such as cutset and recursive conditioning [21, 20, 2] can perform the above computation with only $O(n)$ space-complexity, but with a time cost of $O(r^{\log n \max_{C\in\mathcal{C}}|C|})$, a tradeoff that might be worth making in some large memory instances.

Another set of methods to pursue has to do with the messages themselves. Equations (4.21), (4.22), (4.27), and (4.28) involve exact summation but one may view these as a form of clique-bounded AI-style tree search, where expanding each clique potential involves a search through the space of random variable assignments, and one reaches a valid table entry for each leaf in this search tree, an approach useful for dynamic graphical models [9] as well. This allows one to use a myriad of different search pruning, caching, and re-compilation strategies [12, 13, 25, 46, 43] and these can be applied for each message. These strategies commonly are used to expand larger cliques, and can be used to expand to non-minimal cliques, or even (as is typical) an implicit clique consisting of the entire set of random variables (treating the graph as a giant clique, so no explicit junction tree is ever created). As long as a proper caching and tree search strategies are used, this can be as efficient as the methods above. This procedure becomes more interesting and effective as the amount of sparsity (and other forms of redundancy [4]) in the probability distribution increases. It is interesting to note that the caching [24, 22, 4] strategies (i.e., the trigger to cache a portion of the sub-tree) can be made based on the separators of a junction tree. This begins to address strategies that have been used in the constraint satisfaction [5, 23, 60] and SAT-solver communities, whereby large portions of the search tree are either cached or pruned, with no loss of accuracy. Still other strategies look at the most probable assignment problem as linear programming relaxations of integer programming problems [61]. In general, the above strategies have lead to methods that have allowed exact inference procedures to occur on unprecedentedly large model sizes.

Of course, even with the above strategies, exact inference in general will take us only so far. For many real-world distributions, even the best engineered implementations are unable to tame the inherent $O(r^{\max_{C \in \mathcal{C}} |C|})$ computational and memory complexity mentioned above. At that point, one must take recourse in approximate procedures, as expounded upon in the next chapter in this book.

Bibliography

[1] S. M. Aji and R. J. McEliece. The generalized distributive law. *IEEE transactions on Information Theory*, 46(2):325–343, 2000.

[2] D. Allen and A. Darwiche. New advances in inference by recursive conditioning. In *Proceedings of the Nineteenth Conference on Uncertainty in Artificial Intelligence*, pages 2–10. Morgan Kaufmann Publishers Inc., 2002.

[3] S. Arnborg, D. G. Corneil, and A. Proskurowski. Complexity of finding embeddings in a k-tree. *SIAM Journal on Algebraic Discrete Methods*, 8(2):277–284, 1987.

[4] F. Bacchus, S. Dalmao, and T. Pitassi. Value elimination: Bayesian inference via backtracking search. In *Proceedings of the Nineteenth Conference on Uncertainty in Artificial Intelligence*, pages 20–28. Morgan Kaufmann Publishers Inc., 2002.

[5] R. Barták. Theory and practice of constraint propagation. In *Proceedings of the 3rd Workshop on Constraint Programming in Decision and Control*, 2001.

[6] C. Bartels and J. A. Bilmes. Creating non-minimal triangulations for use in inference in mixed stochastic / deterministic graphical models. *Machine Learning Journal*, 84(3):249–289, 2011.

[7] A. Becker and D. Geiger. A sufficiently fast algorithm for finding close to optimal junction trees. In *Proceedings of the Twelfth Conference on Uncertainty in Artificial Intelligence*, pages 81–89. Morgan Kaufmann Publishers Inc., 1996.

[8] J. A. Bilmes. What HMMS can do. *IEICE Transactions on Information and Systems*, 89(3):869–891, 2006.

[9] J. A. Bilmes and C. Bartels. On triangulating dynamic graphical models. In *Proceedings of the Nineteenth Conference on Uncertainty in Artificial Intelligence*, pages 47–56. Morgan Kaufmann Publishers Inc., 2002.

[10] J. R. S. Blair and B. Peyton. An introduction to chordal graphs and clique trees. In *Graph Theory And Sparse Matrix Computation*, pages 1–29. Springer, 1993.

[11] H. L. Bodlaender, A. M. C. A. Koster, F. van den Eijkhof, and L. C. van der Gaag. Preprocessing for triangulation of probabilistic networks. In *Proceedings of the Seventeenth Conference on Uncertainty in Artificial Intelligence*, pages 32–39. Morgan Kaufmann Publishers Inc., 2001.

[12] M. Chavira and A. Darwiche. Compiling Bayesian networks with local structure. In *International Joint Conferences on Artificial Intelligence*, volume 5, pages 1306–1312, 2005.

[13] M. Chavira, A. Darwiche, and M. Jaeger. Compiling relational Bayesian networks for exact inference. *International Journal of Approximate Reasoning*, 42(1):4–20, 2006.

[14] C. K. Chow and C. N. Liu. Approximating discrete probability distributions with dependence trees. *IEEE Transactions on Information Theory*, 14, 1968.

[15] V. Chvátal and C. Berge. *Topics on Perfect Graphs*, Elsevier, 1984.

[16] G. F. Cooper. The computational complexity of probabilistic inference using Bayesian belief networks. *Artificial Intelligence*, 42(2-3):393–405, 1990.

[17] T. H. Cormen, C. E. Leiserson, R. L. Rivest, and C. Stein. *Introduction to Algorithms*, volume 6. MIT Press, 2001.

[18] R. G. Cowell, A. P. Dawid, S. L. Lauritzen, and D. J. Spiegelhalter. *Probabilistic Networks and Expert Systems.* Springer, 1999.

[19] R. Cowell. Advanced inference in bayesian networks. In *Learning in Graphical Models*, pages 27–49. Springer, 1998.

[20] A. Darwiche. Recursive Conditioning. *Artificial Intelligence*, 126(1):5–41, 2001.

[21] R. Dechter. Enhancement schemes for constraint processing: Backjumping, learning, and cutset decomposition. *Artificial Intelligence*, 41(3):273–312, 1990.

[22] R. Dechter. Bucket elimination: A unifying framework for probabilistic inference. In *Learning in Graphical Models*, pages 75–104. Springer, 1998.

[23] R. Dechter. *Constraint processing.* Morgan Kaufmann Publishers Inc., 2003.

[24] R. Dechter and R. Mateescu. Mixtures of deterministic-probabilistic networks and their and/or search space. In *Proceedings of the 20th Conference on Uncertainty in Artificial Intelligence*, pages 120–129. AUAI Press, 2004.

[25] R. Dechter and R. Mateescu. And/or search spaces for graphical models. *Artificial Intelligence*, 171(2):73–106, 2007.

[26] G. A. Dirac. On rigid circuit graphs. In *Abhandlungen aus dem Mathematischen Seminar der Universität Hamburg*, volume 25, pages 71–76. Springer, 1961.

[27] M. R. Garey and D. S. Johnson. *Computers and Intractability: A Guide to the Theory of NP-Completeness*, W. H. Freeman & Co., 1979.

[28] F. Gavril. Algorithms for minimum coloring, maximum clique, minimum covering by cliques, and maximum independent set of a chordal graph. *SIAM Journal on Computing*, 1(2):180–187, 1972.

[29] R. Gens and D. Pedro. Learning the structure of sum-product networks. In *Proceedings of the 30th International Conference on Machine Learning*, pages 873–880, 2013.

[30] M. C. Golumbic. *Algorithmic Graph Theory and Perfect Graphs*. Elsevier, 2004.

[31] P. Heggernes. Minimal triangulations of graphs: A survey. *Discrete Mathematics*, 306(3):297–317, 2006.

[32] K.-U. Höffgen. Learning and robust learning of product distributions. In *Proceedings of the Sixth Annual Conference on Computational Learning Theory*, pages 77–83. ACM Press, 1993.

[33] F. V. Jensen and F. Jensen. Optimal junction trees. In *Proceedings of the Tenth Conference on Uncertainty in Artificial Intelligence*, pages 360–366. Morgan Kaufmann Publishers Inc., 1994.

[34] F. V. Jensen, K. G. Olesen, and S. K. Andersen. An algebra of Bayesian belief universes for knowledge-based systems. *Networks*, 20(5):637–659, 1990.

[35] F. V. Jensen. *An Introduction to Bayesian Networks*. Springer, 1996.

[36] F. V. Jensen. *Bayesian Networks and Decision Graphs*. Springer, 2001.

[37] T. Kloks. *Treewidth*. Springer, 1994.

[38] D. Koller and N. Friedman. *Probabilistic Graphical Models: Principles and Techniques*. MIT Press, 2009.

[39] F. R. Kschischang, B. Frey, and H.-A. Loeliger. Factor graphs and the sum-product algorithm. *IEEE Transactions on Information Theory*, 47(2):498–519, 2001.

[40] S. L. Lauritzen. *Graphical Models*. Oxford University Press, 1996.

[41] S. L. Lauritzen and D. J. Spiegelhalter. Local computations with probabilities on graphical structures and their application to expert systems. *Journal of the Royal Statistical Society. Series B (Methodological)*, pages 157–224, 1988.

[42] V. Lepar and P. P. Shenoy. A comparison of Lauritzen-Spiegelhalter, Hugin, and Shenoy-Shafer architectures for computing marginals of probability distributions. In *Proceedings of the Fourteenth Conference on Uncertainty in Artificial Intelligence*, 328–337. Morgan Kaufmann Publishers Inc., 1998.

[43] Z. Li and B. d'Ambrosio. Efficient inference in Bayes networks as a combinatorial optimization problem. *International Journal of Approximate Reasoning*, 11(1):55–81, 1994.

[44] B. Lucena. A new lower bound for tree-width using maximum cardinality search. *SIAM Journal on Discrete Mathematics*, 16(3):345–353, 2003.

[45] A. L. Madsen and D. Nilsson. Solving influence diagrams using Hugin, Shafer-Shenoy and lazy propagation. In *Proceedings of the Seventeenth Conference on Uncertainty in Artificial Intelligence*, pages 337–345. Morgan Kaufmann Publishers Inc., 2001.

[46] R. Marinescu and R. Dechter. And/or branch-and-bound search for combinatorial optimization in graphical models. *Artificial Intelligence*, 173(16):1457–1491, 2009.

[47] T. A. McKee. How chordal graphs work. *Bulletin of the Institute of Combinatorics and Its Applications*, 9:27–39, 1993.

[48] T. A. McKee and F. R. McMorris. *Topics in Intersection Graph Theory.* SIAM Monographs on Discrete Mathematics and Applications, 1999.

[49] J. M. Mooij. *Understanding and Improving Belief Propagation.* PhD thesis, Radboud University, Nijmegen. 2008.

[50] M. Narasimhan and J. Bilmes. PAC-learning bounded tree-width graphical models. In *Proceedings of the Twentieth Conference on Uncertainty in Artificial Intelligence.* Morgan Kaufmann Publishers Inc., 2004.

[51] P. Pakzad and V. Anantharam. A new look at the generalized distributive law. *IEEE Transactions on Information Theory*, 50(6):1132–1155, 2004.

[52] A. Parra and P. Scheffler. How to use the minimal separators of a graph for its chordal triangulation. In *International Colloquium on Automata, Languages, and Programming*, pages 123–134. Springer, 1995.

[53] J. Pearl. *Probabilistic Reasoning in Intelligent Systems: Networks of Plausible Inference.* Morgan Kaufmann Publishers Inc., 2nd edition, 1988.

[54] M. A. Peot and R. D. Shachter. Fusion and propagation with multiple observations in belief networks. *Artificial Intelligence*, 48(3):299–318, 1991.

[55] T. S. Richardson. Chain graphs and symmetric associations. In: *Learning in Graphical Models.* Kluwer Academic Publishers, 1998.

[56] D. J. Rose, R. E. Tarjan, and G. S. Lueker. Algorithmic aspects of vertex elimination on graphs. *SIAM Journal Computing*, 5(2):266–282, 1976.

[57] D. J. Rose, R. E. Tarjan, and G. S. Lueker. Algorithmic aspects of vertex elimination on graphs. *SIAM Journal on Computing*, 5(2):266–283, 1976.

[58] P. P. Shenoy and G. Shafer. Axioms for probability and belief-function propagation. In *Proceedings of the Fourteenth conference on Uncertainty in Artificial Intelligence*, pages 169–198. Morgan Kaufmann Publishers Inc., 1990.

[59] R. E. Tarjan and M. Yannakakis. Simple linear-time algorithms to test chordality of graphs, test acyclicity of hypergraphs, and selectively reduce acyclic hypergraphs. *SIAM Journal on Computing*, 13(3):566–579, 1984.

[60] E. Tsang. *Foundations of Constraint Satisfaction: The Classic Text.* BoD–Books on Demand, 2014.

[61] T. Werner. A linear programming approach to max-sum problem: A review. *IEEE Transactions on Pattern Analysis and Machine Intelligence*, 29(7):1165–1179, 2007.

5

Approximate Methods for Calculating Marginals and Likelihoods

Nicholas Ruozzi

Erik Jonsson School of Engineering and Computer Science, University of Texas at Dallas

CONTENTS

Exact marginal inference, as discussed in the preceding Chapter 4, is tractable only in the simplest of models (trees, cycles, low tree width graphs, etc.). Further, marginal inference is often a basic subproblem that needs to be solved repeatedly in order to learn graphical models, that is, to fit a graphical model to a set of data observations. As a result fast, approximate inference algorithms are often necessary in practice. This chapter reviews the basic theory of variational approximations and how they can be used to design algorithms for approximate marginal inference and learning in graphical models. Closely related approximations for MAP inference are discussed in Chapter 6.

Our starting point are factorizations of probability distributions as in (4.1). Instead of thinking of factors coming from cliques of a graph, we generalize to the setting of a hypergraph $G = (V, \mathcal{A})$ determining the factorization. In contrast to the setting of graphs, a (hyper-)edge in $\mathcal{A}$ may be any non-empty set of nodes in V. Consider then probability distributions that factor over the hypergraph $G = (V, \mathcal{A})$ as

$$p(x_V) = \frac{1}{Z} \prod_{i \in V} \phi_i(x_i) \prod_{\alpha \in \mathcal{A}} \psi_\alpha(x_\alpha), \tag{5.1}$$

where the potential functions $\phi_{i\in V} : \mathcal{X}_i \to \mathbb{R}_{\geq 0}$ and $\psi_{\alpha \in \mathcal{A}} : \mathcal{X}_\alpha \to \mathbb{R}_{\geq 0}$ are nonnegative functions, $\mathcal{X}_i$ is discrete set for each $i \in V$ and $\mathcal{X}_\alpha = \prod_{i\in\alpha} \mathcal{X}_i$ for each $\alpha \subseteq V$. For vectors $x \in \mathcal{X}_V$ and $\alpha \subset V$, x_α is used to denote the vector formed by taking only those indices of the vector x that appear in the set α. The normalizing constant Z, usually called the partition function, is chosen to ensure that $p(x)$ is a probability distribution.

$$Z = \sum_{x \in \mathcal{X}_V} \left[\prod_{i \in V} \phi_i(x_i) \prod_{\alpha \in \mathcal{A}} \psi_\alpha(x_\alpha) \right]$$

The partition function, Z, is often not provided in advance and must be computed from the product of the potential functions. In this chapter, we will explore methods for approximate inference, e.g., computing Z and the closely related problem of computing marginal distributions of p, and learning. The theory and algorithms for approximate inference will turn out to form the backbone of methods for approximate learning, as such, the chapter begins with a discussion of inference and concludes with its application in learning.

5.1 Inference as Optimization

The fundamental observation of variational methods for approximate marginal inference is that the computation of the partition function can be formulated as a concave optimization problem. Despite being concave this optimization problem remains intractable, but the form of the optimization problem suggests a variety of different approximations that turn out to be useful in both theory and practice.

5.1.1 The Kullback-Leibler divergence

The principal tool in variational methods is a notion of closeness between probability distributions known as the Kullback-Leibler (KL) divergence. The KL-divergence between a pair of probability distributions p and q over a space $\mathcal{X}$ is given by[1]

$$d(p||q) = \sum_{x \in \mathcal{X}} p(x) \log \frac{p(x)}{q(x)}.$$

The KL-divergence acts almost like a distance measure between two discrete probability distributions: For all distributions p and q, $d(p||q) \geq 0$ with equality if and only if p is equal

[1]All logarithms in this chapter are natural logarithms unless otherwise specified.

to q. Indeed,

$$\begin{aligned} d(p||q) &= \sum_{x\in\mathcal{X}} p(x)\log\frac{p(x)}{q(x)} \\ &= -\sum_{x\in\mathcal{X}} p(x)\log\frac{q(x)}{p(x)} \\ &\geq -\log\sum_{x\in\mathcal{X}}\left(p(x)\frac{q(x)}{p(x)}\right) \\ &= \log 1 \\ &= 0. \end{aligned}$$

However, the KL-divergence is not symmetric. That is, $d(p||q) \neq d(q||p)$ for all probability distributions p and q. As a result, the KL-divergence is not considered a proper distance measure. The KL-divergence can be challenging to compute in general as it requires summing over all $|\mathcal{X}_V|$ different possible assignments to the variables in the model.

5.1.2 The Gibbs free energy

Consider a probability distribution, p, that factors over the hypergraph $G = (V, \mathcal{A})$ as in (5.1). The KL-divergence between the model distribution, p, and an arbitrary distribution, q, can be used to estimate Z. That is, consider

$$\begin{aligned} d(q||p) &= \sum_{x\in\mathcal{X}} q(x)\log\frac{q(x)}{p(x)} \\ &= \sum_{x\in\mathcal{X}_V} q(x)\log q(x) - \sum_{x\in\mathcal{X}_V} q(x)\log p(x) \\ &= \sum_{x\in\mathcal{X}_V} q(x)\log q(x) - \sum_{x\in\mathcal{X}_V} q(x)\log\left(\frac{1}{Z}\prod_{i\in V}\phi_i(x_i)\prod_{\alpha\in\mathcal{A}}\psi_\alpha(x_\alpha)\right) \\ &= \sum_{x\in\mathcal{X}_V} q(x)\log q(x) + \log Z - \sum_{x\in\mathcal{X}_V}\sum_{i\in V} q(x)\log\phi_i(x_i) - \sum_{x\in\mathcal{X}_V}\sum_{\alpha\in\mathcal{A}} q(x)\log\psi_\alpha(x_\alpha) \\ &\geq 0. \end{aligned}$$

As the KL-divergence is nonnegative, this can be turned into a lower bound on $\log Z$.

$$\log Z \geq H(q) + \sum_{x\in\mathcal{X}_V}\sum_{i\in V} q(x)\log\phi_i(x_i) + \sum_{x\in\mathcal{X}_V}\sum_{\alpha\in\mathcal{A}} q(x)\log\psi_\alpha(x_\alpha),$$

where $H(q) \triangleq -\sum_{x\in\mathcal{X}_V} q(x)\log q(x)$ is known as the entropy of the distribution q. This lower bound holds for any distribution q and is satisfied with equality whenever $q(x) = p(x)$ for all $x \in \mathcal{X}_V$. Further, the lower bound is a concave function of $q(x)$ for each $x \in \mathcal{X}_V$. As a result, $\log Z$ can be expressed as a solution to the following concave optimization problem.

$$\log Z = \sup_{q\in\mathcal{P}}\left[H(q) + \sum_{x\in\mathcal{X}_V}\sum_{i\in V} q(x)\log\phi_i(x_i) + \sum_{x\in\mathcal{X}_V}\sum_{\alpha\in\mathcal{A}} q(x)\log\psi_\alpha(x_\alpha)\right] \tag{5.2}$$

Here, $\mathcal{P}$ is the set of all probability distributions over the set $\mathcal{X}_V$.

Despite the concavity of the lower bound, the partition function of a graphical model is $\#P$-hard to compute in general. This follows from the observation that counting the number of satisfying assignments of a 3-SAT instance can be formulated as a graphical model in which there is a hyperedge for each clause whose corresponding potential function simply indicates whether or not the clause is satisfied. However, the partition function can be computed exactly in polynomial time for tree-structured models and models with low tree width (see Chapter 4).

Example 5.1.1 (Tree-structured Models). Suppose that the hypergraph $G = (V, \mathcal{A})$ is a tree and that p factorizes over G as in (5.1). Distributions that factorize over a tree have the special property that they admit a factorization in terms of their marginal distributions. That is, for any $\beta \subseteq V$, let p_β denote the marginal distribution obtained from p by fixing the variables x_β and summing over the remaining variables. With this notation, it can be shown that

$$p(x) = \prod_{i \in V} p_i(x_i) \prod_{\alpha \in \mathcal{A}} \frac{p_\alpha(x_\alpha)}{\prod_{j \in \alpha} p_j(x_j)}. \tag{5.3}$$

This type of rewriting is called a **reparameterization** of p. In general, there are many equivalent factorizations of p over the graph G using different potential functions. The proof that every tree-structured distribution admits a factorization as in (5.3) can be demonstrated by induction on the number of nodes in the hypergraph and can be found in the appendix. Given the reparameterization (5.3), the entropy of a tree structured distribution can also be expressed in terms of its marginal distributions.

$$\begin{aligned}
H(p) &= -\sum_{x \in \mathcal{X}_V} p(x) \log p(x) \\
&= -\sum_{x \in \mathcal{X}_V} p(x) \log \left(\prod_{i \in V} p_i(x_i) \prod_{\alpha \in \mathcal{A}} \frac{p_\alpha(x_\alpha)}{\prod_{j \in \alpha} p_j(x_j)} \right) \\
&= -\sum_{i \in V} \sum_{x \in \mathcal{X}_V} p(x) \log p_i(x_i) - \sum_{\alpha \in \mathcal{A}} \sum_{x \in \mathcal{X}_V} p(x) \log \frac{p_\alpha(x_\alpha)}{\prod_{j \in \alpha} p_j(x_j)} \\
&= -\sum_{i \in V} \sum_{x_i \in \mathcal{X}_i} p_i(x_i) \log p_i(x_i) - \sum_{\alpha \in \mathcal{A}} \sum_{x_\alpha \in \mathcal{X}_\alpha} p_\alpha(x_\alpha) \log \frac{p_\alpha(x_\alpha)}{\prod_{j \in \alpha} p_j(x_j)}
\end{aligned}$$

In terms of the partition function, this yields the following optimization problem for tree-structured distributions.

$$\begin{aligned}
\log Z = \max_{q \in \mathcal{P}} \Bigg[&\sum_{i \in V} \sum_{x_i \in \mathcal{X}_i} q_i(x_i) \log \phi_i(x_i) + \sum_{\alpha \in \mathcal{A}} \sum_{x_\alpha \in \mathcal{X}_\alpha} q_\alpha(x_\alpha) \log \psi_\alpha(x_\alpha) + \\
&- \sum_{i \in V} \sum_{x_i \in \mathcal{X}_i} q_i(x_i) \log q_i(x_i) - \sum_{\alpha \in \mathcal{A}} \sum_{x_\alpha \in \mathcal{X}_\alpha} q_\alpha(x_\alpha) \log \frac{q_\alpha(x_\alpha)}{\prod_{i \in \alpha} q_i(x_i)} \Bigg]
\end{aligned} \tag{5.4}$$

□

5.2 The Bethe Free Energy

To overcome the difficulties inherent in the computation of the Gibbs free energy a number of approximations have been suggested. One of the most popular approximations, known as the Bethe free energy approximation, originated in the statistical physics community. This approximation simplifies the optimization problem (5.2) in two ways. First, it replaces the optimization over the set of all probability distributions, $\mathcal{P}$, with an optimization over locally consistent marginal distributions. Second, it approximates the entropy as if the model were tree-structured, see Example 5.1.1. Specifically, the *local marginal polytope*, $\mathcal{T}$, consists of vectors of probability distributions. There is exactly one entry in the vector $\tau \in \mathcal{T}$ for each $i \in V, x_i \in \mathcal{X}_i$ and one entry for each $\alpha \in \mathcal{A}, x_\alpha \in \mathcal{X}_\alpha$. The marginals in any given vector must agree on single variable overlaps. More formally,

$$\mathcal{T} \triangleq \{\tau \geq 0 \mid \forall \alpha \in \mathcal{A}, x_i \in \mathcal{X}, \sum_{x_{\alpha \setminus i}} \tau_\alpha(x_i, x_{\alpha \setminus i}) = \tau_i(x_i) \text{ and } \forall i \in V, \sum_{x_i} \tau_i(x_i) = 1\}. \quad (5.5)$$

The Bethe free energy approximation is given by

$$\log F_{\mathrm{B}}(\tau; G, \phi, \psi) \triangleq U(\tau; G, \phi, \psi) + \tilde{H}(\tau; G) \quad (5.6)$$

where U is the energy,

$$U(\tau; G, \phi, \psi) \triangleq \sum_{i \in V} \sum_{x_i} \tau_i(x_i) \log \phi_i(x_i) + \sum_{\alpha \in \mathcal{A}} \sum_{x_\alpha} \tau_\alpha(x_\alpha) \log \psi_\alpha(x_\alpha),$$

and $\tilde{H}$ is the tree-structured entropy approximation,

$$\tilde{H}(\tau; G) \triangleq -\sum_{i \in V} \sum_{x_i} \tau_i(x_i) \log \tau_i(x_i) - \sum_{\alpha \in \mathcal{A}} \sum_{x_\alpha} \tau_\alpha(x_\alpha) \log \frac{\tau_\alpha(x_\alpha)}{\prod_{i \in \alpha} \tau_i(x_i)}.$$

The Bethe partition function, Z_{B}, is then expressed in terms of the maximum value achieved by this approximation over $\mathcal{T}$.

$$\log Z_{\mathrm{B}}(G, \phi, \psi) \triangleq \max_{\tau \in \mathcal{T}} F_{\mathrm{B}}(\tau; G, \phi, \psi) \quad (5.7)$$

In general, the Bethe free energy optimization problem (5.7) is not concave. As a result, most approximation algorithms settle for a local optimum of this objective.

Example 5.2.1 (Partition Function of a Tree-structured Model). The Bethe partition function, Z_{B}, is equal to the true partition function Z when the graph is a hypertree. This follows from the observation that any vector of marginals in the local marginal polytope are the marginals of some tree structured distribution. Let $\tau \in \mathcal{T}$ be a collection of locally consistent marginal distributions and define

$$q(x) = \prod_{i \in V} \tau_i(x_i) \prod_{\alpha \in \mathcal{A}} \frac{\tau_\alpha(x_\alpha)}{\prod_{j \in \alpha} \tau_j(x_j)}.$$

To show that $q_i(x_i) = \tau_i(x_i)$ for each $i \in V, x_i \in \mathcal{X}_i$ and $q_\alpha(x_\alpha) = \tau_\alpha(x_\alpha)$ for each $\alpha \in \mathcal{A}, x_\alpha \in \mathcal{X}_\alpha$, view the hypertree as being rooted at i and sum out each variable, except for x_i, by starting at the leaves and working towards the root, eliminating all child nodes before their parent node.

As a consequence, every element in the local marginal polytope is realized by some joint probability distribution that factors over the hypertree. This yields

$$\begin{aligned}
\log Z_{\mathrm{B}} &= \max_{\tau \in \mathcal{T}} F_{\mathrm{B}}(\tau; G, \phi, \psi) \\
&= \max_{\tau \in \mathcal{T}} \left[U(\tau; G, \phi, \psi) + \tilde{H}(\tau; G) \right] \\
&= \max_{\tau \in \mathcal{T}} \Bigg[\sum_{i \in V} \sum_{x_i} \tau_i(x_i) \log \phi_i(x_i) + \sum_{\alpha \in \mathcal{A}} \sum_{x_\alpha} \tau_\alpha(x_\alpha) \log \psi_\alpha(x_\alpha) + \\
&\qquad - \sum_{i \in V} \sum_{x_i} \tau_i(x_i) \log \tau_i(x_i) - \sum_{\alpha \in \mathcal{A}} \sum_{x_\alpha} \tau_\alpha(x_\alpha) \log \frac{\tau_\alpha(x_\alpha)}{\prod_{i \in \alpha} \tau_i(x_i)} \Bigg] \\
&= \max_{q \in \mathcal{P}} \Bigg[\sum_{i \in V} \sum_{x_i} q_i(x_i) \log \phi_i(x_i) + \sum_{\alpha \in \mathcal{A}} \sum_{x_\alpha} q_\alpha(x_\alpha) \log \psi_\alpha(x_\alpha) + \\
&\qquad - \sum_{i \in V} \sum_{x_i} q_i(x_i) \log q_i(x_i) - \sum_{\alpha \in \mathcal{A}} \sum_{x_\alpha} q_\alpha(x_\alpha) \log \frac{q_\alpha(x_\alpha)}{\prod_{i \in \alpha} q_i(x_i)} \Bigg] \\
&= \log Z.
\end{aligned}$$

□

5.2.1 Convex and reweighted free energies

The nonconcavity of the Bethe free energy is due to the entropy approximation $\tilde{H}$. Other entropy approximations have been proposed that make the resulting optimization problem concave [7, 22, 13, 9, 29, 33, 14]. Substituting these alternative entropy approximations in (5.9) results in a concave optimization problem that can be maximized via standard methods (e.g., gradient ascent). Most reweighted entropy approximations introduce a vector of weights, ρ, sometimes called counting numbers, that has an entry for each $i \in V$, each $\alpha \in \mathcal{A}$, and for each pair $(\alpha \in \mathcal{A}, i \in \alpha)$. This yields an infinite family of entropy approximations.

$$\tilde{H}_\rho(\tau; G) \triangleq \sum_{i \in V, \alpha \supset i} \rho_{i\alpha} \left[H(\tau_\alpha) - H(\tau_i) \right] + \sum_{i \in V} \rho_i H(\tau_i) + \sum_{\alpha \in \mathcal{A}} \rho_\alpha H(\tau_\alpha) \tag{5.8}$$

The reweighted partition function is then obtained by replacing the tree-structured entropy in (5.6) with the reweighted entropy approximation (5.8).

$$\log Z_\rho = \max_{\tau \in \mathcal{T}} \exp \left(U(\tau; G, \phi, \psi) + \tilde{H}_\rho(\tau; G) \right) \tag{5.9}$$

The Bethe free energy is recovered from the reweighted free energy by setting $\rho_{i\alpha} = 0$ for all $i \in V, \alpha \supset i$, $\rho_\alpha = 1$ for all $\alpha \in \mathcal{A}$, and $\rho_i = 1 - |\{\alpha \in \mathcal{A} : \alpha \supset i\}|$ for all $i \in V$. Note that $|\{\alpha \in \mathcal{A} : \alpha \supset i\}|$ is the degree of node i in the hypergraph. The reweighted free energy is concave whenever $\rho \geq 0$, but this is only a sufficient condition [14, 13, 6].

While concave reweighted free energies have desirable computational properties, the optima of the Bethe free energy can provide better estimates of the partition function and marginals in practice [35]. In addition, if the reweighted free energy is not equal to the Bethe free energy, then it is no longer exact on trees. Despite this, reweighted free energies have a myriad of applications even outside the concave case [22, 21], and a number of different algorithms have been built around specific choices of the counting numbers:

- The tree-reweighted belief propagation (TRBP) algorithm picks the counting numbers so that they correspond to edge appearance probabilities with respect to some

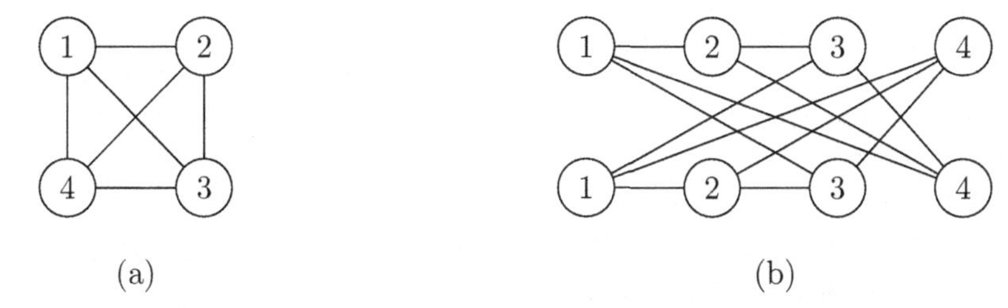

FIGURE 5.1: The complete graph on four nodes (a) and one of its 2-covers (b). The nodes in the cover are labeled for the node that they copy in the base graph.

probability distribution over spanning trees of the graph [29]. Let $G = (V, E)$ be a standard graph, let $\mathcal{S}$ be the set of all spanning trees of G, and let μ be a probability distribution over $\mathcal{S}$ such that every edge appears in at least one spanning tree $T \in \mathcal{S}$ with $\mu(T) > 0$. Let $\mu_{ij} = \sum_{T \in \mathcal{S}:(i,j) \in T} \mu(S)$ denote the edge appearance probability. For pairwise graphical models (i.e., for all $\alpha \in \mathcal{A}$, $|\alpha| \leq 2$), TRBP selects $\rho_{ie} = 0$ for all $i \in V$ and all edges e incident with i, $\rho_{ij} = \mu_{ij}$ for all $(i, j) \in E$, and $\rho_i = 1$ for all $i \in V$. The optimum of the reweighted free energy with this choice of counting numbers provably yields an upper bound on the partition function.

- Let δ be the maximum degree of any node in G. A simple choice of counting numbers similar to those of the Bethe free energy that will guarantee concavity of the entropy approximation is to choose $\rho_{i\alpha} = 0$ for all $i \in V, \alpha \supset i$, $\rho_\alpha = 1/\delta$ for all $\alpha \in \mathcal{A}$, and $\rho_i = 1 - \deg(i)/\delta$ for all $i \in V$.

- When ρ_α is an integer for each $\alpha \in \mathcal{A}$, $\rho_i = 1 - \sum_{\alpha \ni i} \rho_\alpha$, and all other counting numbers equal to zero, the reweighted entropy corresponds to the Bethe free energy of a model in which each hyperedge $\alpha \in \mathcal{A}$ in the graph has been duplicated ρ_α times [22].

5.2.2 A combinatorial characterization of the Bethe free energy

In general, understanding when the Bethe free energy approximation or one of the reweighted free energies provides a good approximation to the true partition function and/or marginals is a challenging problem. In particular, with the exception of a few special cases, even determining whether or not the Bethe partition function is an upper or a lower bound on the true partition function is non-trivial. In this regard, one final theoretical tool is worth discussing before moving on to a discussion of practical algorithms for approximate marginal inference: graph covers (sometimes called lifts of graphs). Roughly speaking, if a graph G' covers a graph G, then G' looks locally the same as G.

Definition 5.2.2. *A graph G'* ***covers*** *a graph $G = (V, E)$ if there exists a graph homomorphism $h : G' \to G$ such that for all vertices $i \in G$ and all $j \in h^{-1}(i)$, h maps the neighbors of j, denoted $N(j)$, in G' bijectively to the neighborhood $N(i)$ of i in G.*

If $h(j) = i$, then $j \in G'$ is called a copy of $i \in G$. Further, G' is said to be a k-cover of G if every vertex of G has exactly k copies in G'. In addition, the k copies of $i \in V_G$ will be denoted as $i_1, \ldots, i_k \in V_{G'}$. An example 2-cover is pictured in Figure 5.1.

Graph covers of hypergraphs can be obtained as graph covers of the corresponding factor graph: a representation of the hypergraph G as a standard graph in which the vertex set consists of both the vertices and hyperedges of G with an edge joining a hyperedge to each vertex on which it depends. In terms of graphical models, if $G = (V, \mathcal{A})$ is a hypergraph and probability distribution p factorizes over G as in (5.1), then to any graph cover $G' = (V', \mathcal{A}')$ of G under the homomorphism h define the probability distribution

$$p'(x_{V'}) = \frac{1}{Z'} \prod_{i \in V'} \phi_{h(i)}(x_i) \prod_{\alpha \in \mathcal{A}'} \psi_{h(\alpha)}(x_\alpha).$$

Notice that p' factorizes over G' and that the potentials associated to the vertices and hyperedges in G' are given by the corresponding potentials from the factorization of p over G. In the context of graphical models, one graphical model will be said to cover another whenever their factor graphs satisfy the graph cover relation and the potentials on the cover are chosen as above.

A somewhat surprising result is that the Bethe partition function, i.e., the solution to optimization problem (5.7), has an equivalent formulation as a limit of exact counting problems on covers of G. For clarity, let $Z(G)$ denote the partition function of the graph G and $Z(G')$ denote the partition function corresponding to a cover of the graph G.

Theorem 5.2.3 (Theorem 33 [28]).

$$Z(G) = \lim \sup_{k \to \infty} \sqrt[k]{\frac{\sum_{G' \in \mathcal{C}_k(G)} Z(G')}{|\mathcal{C}_k(G)|}}$$

where $\mathcal{C}_k(G)$ is the set of all k-covers of G.

Example 5.2.4 (Bethe Partition Function of a Single Cycle). The Bethe partition function has a simple expression for pairwise graphical models, i.e., graphical models in which $|\alpha| = 2$ for all $\alpha \in \mathcal{A}$, consisting of a single cycle. The approximate marginals are easy to characterize in this case [31]. The goal of this example is to show how Theorem 5.2.3 can be used to provide insight into the relationship between Z and Z_{B}.

Let C_n denote the simple cycle on n nodes. Suppose that $G = C_n$, $\mathcal{X}_i = \mathcal{X}_j$ for all $i, j \in V$, and p factorizes over G as

$$p(x) = \frac{1}{Z} \prod_{(i,j) \in E} \psi_{ij}(x_i, x_j),$$

where the potential functions are symmetric in their arguments. Each potential function ψ_{ij} can be represented by a $\mathcal{X} \times \mathcal{X}$ matrix A^{ij} whose x, y entry, $A^{ij}_{x,y}$, is given by $\psi_{ij}(x, y)$. Label the vertices of G from i_1 to i_n clockwise around the cycle. With this labeling,

$$\begin{aligned} Z &= \sum_x \prod_{(i,j) \in E} \psi_{ij}(x_i, x_j) \\ &= \sum_x A^{i_1,i_2}_{x_1,x_2} A^{i_2,i_3}_{x_2,x_3} \cdot \ldots \cdot A^{i_{n-1},i_n}_{x_{n-1},x_n} \cdot A^{i_n,i_1}_{x_n,x_1} \\ &= \mathtt{tr}\left(A^{i_1,i_2} \cdot A^{i_2,i_3} \cdot \ldots \cdot A^{i_{n-1},i_n} \cdot A^{i_n,i_1}\right). \end{aligned}$$

Let $A \triangleq A^{i_1,i_2} \cdot A^{i_2,i_3} \cdot \ldots \cdot A^{i_{n-1},i_n} \cdot A^{i_n,i_1}$, and let $\lambda_1, \ldots, \lambda_{|\mathcal{X}|}$ denote the eigenvalues of A. Every graph cover of a single cycle must be a disjoint union of single cycles. As an example, all 2-covers of C_n are isomorphic to C_{2n} or two disjoint copies of C_n. The corresponding partition functions are $\mathtt{tr}(A^2)$ and $\mathtt{tr}(A)^2$ respectively. More generally, the only connected covers of C_n are the cycles C_{kn} for some positive integer k whose partition functions are given by $\mathtt{tr}(A^k)$. In the simplest case, if all of the eigenvalues of A are nonnegative real, then $\mathtt{tr}(A^k) = \sum_{i=1}^{|\mathcal{X}|} \lambda_i^k \leq \left(\sum_{i=1}^{|\mathcal{X}|} \lambda_i\right)^k = \mathtt{tr}(A)^k$. Theorem 5.2.3 can then be used to conclude that $Z_{\mathrm{B}} \leq Z$ in this case. □

Theorem 5.2.3 can be used to translate questions about $Z_{\mathrm{B}}(G)$ into questions about $Z(G')$ for some cover G' of G. This turns out to be a useful theoretical tool. For example, the Bethe partition function can be shown to be a lower bound on Z in a variety of models including the ferromagnetic Ising model [18], the ferromagnetic Potts model with a uniform external field [19], weighted graph homomorphism problems in which the target can be represented as the sum of two nonnegative rank one matrices [19], weight enumerators of linear codes [19], and the permanent of nonnegative matrices [5], among others [25, 34]. Most of these results, argue that $Z(G)^k \geq Z(G')$ whenever G' is a k-cover of G. The inequality then follows from Theorem 5.2.3.

5.3 Algorithms for Approximate Marginal Inference

A variety of algorithms have been proposed for approximate marginal inference. The most direct methods simply perform coordinate or gradient ascent directly on the free energy approximation in order to obtain a locally optimal solution. Other approaches implement local message-passing algorithms over the graph G while others use sampling based methods to approximate the counting problem. Each of these approaches has advantages and disadvantages.

5.3.1 Loopy belief propagation

One simple algorithm for approximate inference is a local message-passing algorithm known as belief propagation. The standard belief propagation algorithm is simply variable elimination on a tree-structured model in which the variable elimination algorithm proceeds from the leaf nodes to the root of the tree, eliminating all children of a node before the node itself. This can be implemented as a message-passing algorithm in which the messages correspond to the new factors created as part of the elimination algorithm. Specifically, belief propagation iterates the following message-passing updates until convergence.

$$m^t_{i\to\alpha}(x_i) \triangleq \kappa \cdot \phi_i(x_i) \cdot \prod_{\beta\supset i:\beta\neq\alpha} m^{t-1}_{\beta\to i}(x_i) \tag{5.10}$$

$$m^t_{\alpha\to i}(x_i) \triangleq \kappa \cdot \sum_{x_{\alpha\setminus i}} \left[\psi_\alpha(x_\alpha) \cdot \prod_{k\in\alpha\setminus i} m^{t-1}_{k\to\alpha}(x_k)\right] \tag{5.11}$$

Here, the constants κ, which can be different for each message update, are used to keep the messages bounded (e.g., κ could be chosen so that $m^t_{i\to\alpha}$ is a probability distribution). This is necessary to avoid numerical issues on loopy graphs. The messages are used to construct approximate marginals, called beliefs.

$$b^t_i(x_i) \triangleq \kappa \cdot \phi_i(x_i) \cdot \prod_{\alpha\supset i} m^t_{\alpha\to i}(x_i) \tag{5.12}$$

$$b^t_\alpha(x_\alpha) \triangleq \kappa \cdot \psi_\alpha(x_\alpha) \cdot \prod_{i\in\alpha} m^t_{i\to\alpha}(x_i). \tag{5.13}$$

After the algorithm converges, the appropriately normalized beliefs, b^*, are guaranteed to satisfy the following marginalization condition for each $\alpha \in \mathcal{A}, i \in \alpha, x_i \in \mathcal{X}_V$.

$$\sum_{x_{\alpha\setminus\{i\}}} b^*_\alpha(x_\alpha) = b^*_i(x_i)$$

While the belief propagation algorithm is exact on trees (i.e., the converged beliefs are equal to the true marginals), it is often applied in the "loopy" setting as well. That is, the same message-passing algorithm that was formulated to provide the exact solution on trees is simply run on graphs with loops until a fixed point of the message-passing equations is obtained. Loopy belief propagation is not exact on graphs with cycles, where it only provides an approximation of the marginals and partition function. Additionally, in the loopy case, belief propagation may or may not converge in a finite number of iterations. There exist conditions on the potential functions that are sufficient to guarantee that the message updates yield a contractive mapping and hence converge, but in general, such conditions are difficult to verify [16].

In practice, the updates are normally damped to eliminate oscillations that are common with these types of iterative methods. The damped message updates, with damping factor $\epsilon \in (0,1)$, have the following form.

$$\tilde{m}^t_{i\to\alpha}(x_i) \triangleq (1-\epsilon)\tilde{m}^{t-1}_{i\to\alpha}(x_i) + \epsilon\left[\kappa\cdot\phi_i(x_i)\cdot\prod_{\beta\supset i:\beta\neq\alpha} m^{t-1}_{\beta\to i}(x_i)\right] \tag{5.14}$$

$$\tilde{m}^t_{\alpha\to i}(x_i) \triangleq (1-\epsilon)\tilde{m}^{t-1}_{\alpha\to i}(x_i) + \epsilon\left[\kappa\cdot\sum_{x_{\alpha\setminus i}}\left[\psi_\alpha(x_\alpha)\cdot\prod_{k\in\alpha\setminus i} m^{t-1}_{k\to\alpha}(x_k)\right]\right] \tag{5.15}$$

Both the damped and undamped versions have the same fixed points. However, the damped updates often converge in practice when the undamped version does not.

There are a number of important connections between belief propagation and the Bethe free energy, graph covers, and belief propagation. First, once a set of beliefs that satisfy the local marginalization conditions are obtained, the partition function can be estimated by plugging the beliefs into the Bethe free energy. There is a correspondence between fixed points of the belief propagation algorithm and the Lagrangian of the Bethe free energy [37]: local optima of the Lagrangian correspond to fixed points of belief propagation and vice versa. As a result, belief propagation can be used to find local optima of the Bethe free energy. Second, graph covers have been shown to be important for understanding the behavior of message-passing algorithms on networks [1]. The primary observation in these contexts is that if G' is a graph cover of G, then G and G' are locally indistinguishable in the following sense: Any local message-passing algorithm on G can be realized as a local message-passing algorithm on G'. In particular, for loopy belief propagation, this means that any fixed point of loopy belief propagation on G is also a fixed point of loopy belief propagation on G'.

5.3.2 Reweighted message-passing algorithms

Message-passing schemes similar to belief propagation can also be derived in the reweighted entropy case. For example, when $\rho_{i\alpha} = 1$ for all $i \in V, \alpha \supset I$, the fixed points of the following message-passing scheme correspond to local optima of the reweighted free energy.

$$m^t_{i\to\alpha}(x_i) \triangleq \kappa\cdot\phi_i(x_i)^{1/\rho_i}\cdot m^{t-1}_{\alpha\to i}(x_i)^{\rho_\alpha-1}\cdot\prod_{\beta\supset i:\beta\neq\alpha} m^{t-1}_{\beta\to i}(x_i)^{\rho_\alpha}$$

$$m^t_{\alpha\to i}(x_i) \triangleq \kappa\cdot\sum_{x_{\alpha\setminus i}}\left[\psi_\alpha(x_\alpha)^{1/\rho_\alpha}\cdot m^{t-1}_{i\to\alpha}(x_k)^{\rho_i-1}\cdot\prod_{k\in\alpha\setminus i} m^{t-1}_{k\to\alpha}(x_k)^{\rho_k}\right]$$

Fixed points of the above message-passing scheme correspond to local optima of the Lagrangian for the reweighted free energy. As in the case of belief propagation, the message updates are typically damped to help with convergence.

Beliefs are constructed from the converged messages, and then plugged into the reweighted free energy approximation to obtain an approximation of the partition function.

$$b_i^t(x_i) \triangleq \kappa \cdot \phi_i(x_i)^{1/\rho_i} \cdot \prod_{\alpha \supset i} m_{\alpha \to i}^{t-1}(x_i)^{\rho_\alpha}$$

$$b_\alpha^t(x_\alpha) \triangleq \kappa \cdot \psi_\alpha(x_\alpha)^{1/\rho_\alpha} \cdot \prod_{k \in \alpha} m_{k \to \alpha}^{t-1}(x_k)^{\rho_k}$$

Even in the case that the reweighted entropy approximation is concave, the above message passing scheme is not guaranteed to converge. However, if the message updates are not performed synchronously but rather asynchronously, and perhaps in a specific order, then convergence can be guaranteed in the concave case in a variety of settings [11, 13, 21, 6]. Most proofs of convergence demonstrate that these message-passing schemes perform coordinate/gradient ascent on the reweighted free energy.

5.3.3 Gradient ascent

Performing gradient ascent directly on the Bethe free energy provides one alternative to message-passing algorithms. The advantage of gradient based methods is that they can guarantee convergence to a local optimum under mild restrictions. In practice, when the message-passing algorithms work well, they tend to outperform gradient ascent. In particular, if the graph is a tree, then belief propagation will converge after a finite number of iterations that depends only on the diameter of the graph while gradient methods may still take a large number of iterations to converge.

The primary difficulty when applying gradient methods to the problem of maximizing the Bethe approximation is that standard gradient descent cannot handle constraints. In special cases, for example pairwise binary graphical models [36], the constraints can be removed analytically, but in general, projected gradient methods (or barrier methods) may be needed to keep the iterates inside of the constraint set. Other methods such as the convex-concave procedure [38] carefully integrate the constraints into the objective. The Frank-Wolf algorithm can also be used as an alternative to projected gradient methods and is particularly efficient whenever the underlying graphical model admits a fast approximate MAP routine [12, 27].

5.3.4 Naive mean field

While the Bethe partition function does not provide an upper or a lower bound on the true partition function in general, there exist simple methods that achieve lower bounds. The mean field method considers the optimization problem (5.2), but restricts the set of allowable probability distributions $q \in \mathcal{P}$ to only those distributions that factorize completely. That is,

$$q(x) = \prod_{i \in V} q_i(x_i),$$

where each $q_i : \mathcal{X}_i \to [0,1]$ is a probability distribution. Plugging distributions of this form into (5.2) yields

$$\log Z_{\mathrm{MF}} = \sup_{q_i \in V} \Bigg[\sum_{i \in V} H(q_i) + \sum_{i \in V} \sum_{x_i \in \mathcal{X}_i} q_i(x_i) \log \phi_i(x_i) + \sum_{\alpha \in \mathcal{A}} \sum_{x_\alpha \in \mathcal{X}_\alpha} \prod_{i \in \alpha} q_i(x_i) \log \psi_\alpha(x_\alpha) \Bigg] \tag{5.16}$$
$$\leq \min\{\log Z, \log Z_{\mathrm{B}}\}.$$

The inequality follows from two observations. First, $Z_{\mathrm{MF}} \leq Z$ as it optimizes the same objective over a restricted set of probability distributions. Second, setting $\tau_i(x_i) = q_i(x_i)$ and $\tau_\alpha = \prod_{i \in \alpha} q_i(x_i)$ yields a set of locally consistent marginals, i.e., $\tau \in \mathcal{T}$. This shows that the mean field objective is also a special case of the Bethe free energy. If the Bethe partition function is smaller than the true partition function, then it must attain at least as good of an estimate of the partition function as the mean field estimate. In practice, the mean field partition function can often produce very poor approximations of the true partition function as real distributions are unlikely to be close to fully factorized distributions. Still, performing gradient ascent on (5.16) is simpler than in the more general case and provably yields a lower bound.

5.3.5 Sampling methods

As an alternative to framing inference as an optimization problem, the partition function and marginals can be estimated via sampling. For example, given D independent samples, $x^{(1)}, \ldots x^{(D)}$, from a probability distribution p, the marginal distribution for the ith variable can be estimated as

$$p_i(x_i) \approx \sum_{d=1}^{D} \frac{1_{x_i^{(d)} = x_i}}{D},$$

where $1_{x_i^{(d)} = x_i}$ is equal to one if $x_i^{(d)} = x_i$ and zero otherwise. As D tends to infinity, the estimate converges to the true marginal. The primary difficulty then is to generate D such independent samples from p. For univariate probability distributions over a finite state space, this is relatively straightforward given the ability to sample uniformly from the interval $[0,1]$: carve the interval $[0,1]$ into k buckets, one for each of the possible assignments $\{1, \ldots, k\}$, such that length of the ith bucket corresponds to the probability of assignment i. Next, sample a random number $r \in [0,1]$ and return the assignment corresponding to the bucket that contains r.

Sampling from multivariate distributions requires more care if it is to be done efficiently as the simple method to sample from a univariate distribution could require carving the unit interval up into exponentially many pieces. In addition, using this strategy to sample from a probability distribution of the form (5.1) would require computing the partition function, Z. A variety of alternative sampling procedures that avoid both of the above difficulties have been developed. One of simplest such procedures is a Markov chain Monte Carlo (MCMC) method knows as the Gibbs sampling algorithm. Gibbs sampling iteratively constructs a sequence of samples in which each sample depends on the previous sample in the sequence. The method is efficient in that it only requires sampling from local probability distributions. Specifically, fix an assignment $\tilde{x} \in \prod_i \mathcal{X}_i$. To generate a new assignment, the Gibbs sampling algorithm iterates through the variables one at a time (or picks one at random). When it reaches the ith variable, it samples a new value, x_i', for that variable from the conditional probability distribution $p(x_i|\tilde{x}_{N(i)})$, where $N(i)$ is the set of neighbors of node i, and updates $x_i \leftarrow x_i'$. The algorithm then proceeds to the next variable.

The reason that this algorithm is easy to implement for graphical models is that $p(x_j|\tilde{x}_{N(j)})$ is simple to compute and, as it is a univariate distribution over a finite state space, easy to sample from.

$$\begin{aligned}
p(x_j|\tilde{x}_{N(j)}) &= \frac{p(x_j, \tilde{x}_{N(j)})}{p(\tilde{x}_{N(j)})} \\
&= \frac{\frac{1}{Z}\sum_{x':x'_{N(j)}=\tilde{x}_{N(j)},x'_j=x_j} \prod_{i\in V}\phi_i(x'_i)\prod_{\alpha\in\mathcal{A}}\psi_\alpha(x'_\alpha)}{\frac{1}{Z}\sum_{x':x'_{N(j)}=\tilde{x}_{N(j)}} \prod_{i\in V}\phi_i(x'_i)\prod_{\alpha\in\mathcal{A}}\psi_\alpha(x'_\alpha)} \\
&= \frac{g(\tilde{x}_{N(j)})\phi_j(x_j)\prod_{\alpha\supset j}\psi_\alpha(x_j, \tilde{x}_{\alpha\setminus j})}{\sum_{x'_j} g(\tilde{x}_{N(j)})\phi_j(x'_j)\prod_{\alpha\supset j}\psi_\alpha(x'_j, \tilde{x}_{\alpha\setminus j})} \\
&\propto \phi_j(x_j)\prod_{\alpha\supset j}\psi_\alpha(x_j, \tilde{x}_{\alpha\setminus j})
\end{aligned}$$

Here, $g(x_{N(j)}) = \sum_{x':x'_{N(j)}=x_{N(j)}} \prod_{i\neq j}\phi_i(x'_i)\prod_{\alpha\in\mathcal{A}:j\notin\alpha}\psi_\alpha(x'_\alpha)$ is the function obtained by performing the sum over potential functions that do not depend on the variable x_j. Note that sampling from this conditional distribution does not require computing Z as it appears in both the numerator and the denominator.

Whenever all of the potential functions are strictly positive, the Gibbs sampler is guaranteed to eventually produce unbiased samples from $p(x)$. However, the number of iterations of Gibbs sampling needed to converge to the distribution p may be quite large. Further, for probability distributions that are not strictly positive, the Gibbs sampler is not guaranteed to converge to the correct limiting distribution. Despite this, MCMC sampling is useful in certain settings where it is known to converge quickly or as a subroutine in nonparametric methods for inference in continuous graphical models [26].

5.4 Approximate Learning

As opposed to approximate marginal inference, which aims to recover the marginals of a fixed probability distribution, the learning task is to estimate an unknown probability distribution from data. That is, given a collection of data points that are sampled independently from an unknown probability distribution p that factorizes over the hypergraph G, the goal is to recover, i.e., learn, the potential functions that describe the distribution from which the data points were sampled.

5.4.1 Log-linear models

So far, the potential functions $\phi_{i\in V}$ and $\psi_{\alpha\in\mathcal{A}}$ were allowed to be arbitrary nonnegative functions over some discrete space. These functions are completely determined by their outputs over every possible input. This is often referred to as the overcomplete model: there are exactly $|\mathcal{X}_i|$ parameters to be learned for each $i \in V$ and $|\mathcal{X}_\alpha|$ parameters to be learned for each $\alpha \in \mathcal{A}$. In this case, the number of parameters needed to represent the model grows exponentially in the size of the largest hyperedge in G. Many interesting families of graphical models can be expressed with far fewer parameters.

Example 5.4.1 (The Ising Model). Let $G = (V, E)$ be a simple graph. The Ising model selects $\mathcal{X}_i = \{-1, +1\}$ and potential functions of the form

$$\phi_i(x_i) = \exp(h_i x_i)$$
$$\psi_\alpha(x_i, x_j) = \exp(J_{ij} x_i x_j).$$

The model has one parameter for each vertex and one parameter for each edge of G (the overcomplete representation would have four parameters per edge and 2 parameters per vertex). When all of the J parameters are nonnegative, the model is said to be ferromagnetic (attractive).

Ising models arose in statistical physics as a simplistic model of particles in a two dimensional grid. They are often used in computer vision applications (e.g., image segmentation where $+1$ may correspond to foreground pixels and -1 to background pixels). □

The Ising model has the mathematically convenient property that the model is log-linear in the parameters. More generally, a log-linear model with parameter vector $\theta \in \mathbb{R}^K$ that factorizes over a graph G has potential functions of the following form.

$$\phi_i(x_i|\theta) = \exp(\langle \theta, f_i(x_i) \rangle)$$
$$\psi_\alpha(x_\alpha|\theta) = \exp(\langle \theta, f_\alpha(x_\alpha) \rangle),$$

where $f_{i \in V} : \mathcal{X}_i \to \mathbb{R}^K$, $f_{\alpha \in \mathcal{A}} : \mathcal{X}_\alpha \to \mathbb{R}^K$, and $\langle \cdot, \cdot \rangle$ denotes the inner product between two vectors. Allowing extended value functions, every nonnegative function can be represented in this form. As an example, for the Ising model, θ could be a $|V| + |E|$ dimensional vector whose first $|V|$ components correspond to each h_i for $i \in V$ followed by $|E|$ components corresponding to J_{ij} for each $(i, j) \in E$. Picking $f_i(x_i)$ to be the vector of all zeros except for an x_i in the ith position would guarantee that $\langle \theta, f_i(x_i) \rangle = \theta_i x_i = h_i x_i$. The overcomplete representation can be expressed using a similar construction. To emphasize the dependence on the parameter vector θ, the probability distributions in this section will be assumed to be log-linear models and the dependence on θ will be made explicit by writing

$$p(x|\theta) = \frac{1}{Z(\theta)} \prod_{i \in V} \phi_i(x_i|\theta) \prod_{\alpha \in \mathcal{A}} \psi_\alpha(x_\alpha|\theta).$$

Log-linear models are not necessarily identifiable. That is, there may exist $\theta \neq \theta' \in \mathbb{R}^K$ such that $p(x|\theta) = p(x|\theta')$ for all $x \in \mathcal{X}_V$. When the model is not identifiable, then it is not possible to reconstruct the exact parameters from the sampled data alone. This is not problematic in practice as the learned models are typically used for prediction and the true underlying distribution is often not a member of the family of parameterized models being learned.

5.4.2 Maximum likelihood estimation (MLE)

Given a collection of data points that are assumed to be sampled independently from an unknown probability distribution p, the goal is to find a parameter vector θ that maximizes the probability of seeing these samples. Formally, given $x^{(1)}, \ldots, x^{(D)} \in \mathcal{X}_V$, the maximum likelihood estimation problem is to find the parameter vector θ that maximizes the likelihood of the data:

$$\prod_{d=1}^{D} p(x^{(d)}|\theta).$$

Maximum likelihood estimation produces a consistent estimator of the parameter vector θ: if the model is identifiable and the data is sampled from $p(\cdot|\theta^*)$ for some $\theta^* \in \mathbb{R}^K$, then in the limit of infinite data, the parameter vector θ that maximizes the likelihood will be θ^*.

In the sequel, it will be more mathematically convenient to maximize the normalized log-likelihood, denoted $\ell(\theta)$, instead of the likelihood.

$$\begin{aligned}
\log \ell(\theta) &= \frac{1}{D}\sum_{d=1}^{D} \log p\left(x^{(d)}|\theta\right) \\
&= \frac{1}{D}\sum_{d=1}^{D}\left[-\log Z(\theta) + \sum_{i\in V}\log\phi_i\left(x_i^{(d)}|\theta\right) + \sum_{\alpha\in\mathcal{A}}\log\psi_\alpha\left(x_\alpha^{(d)}|\theta\right)\right] \\
&= \frac{1}{D}\sum_{d=1}^{D}\left[-\log Z(\theta) + \sum_{i\in V}\left\langle\theta, f_i\left(x_i^{(d)}\right)\right\rangle + \sum_{\alpha\in\mathcal{A}}\left\langle\theta, f_\alpha\left(x_\alpha^{(d)}\right)\right\rangle\right] \\
&= -\log Z(\theta) + \left\langle\theta, \frac{1}{D}\sum_{d=1}^{D}\left[\sum_{i\in V} f_i\left(x_i^{(d)}\right) + \sum_{\alpha\in\mathcal{A}} f_\alpha\left(x_\alpha^{(d)}\right)\right]\right\rangle
\end{aligned} \tag{5.17}$$

A regularizer is often added to the maximum likelihood objective. For example, an ℓ_2 regularizer would add a penalty term of the form $-\frac{\lambda}{2}||\theta||_2^2$, corresponding to a zero-mean Gaussian prior over the parameters in the Bayesian interpretation, yielding the following regularized maximum likelihood objective.

$$\ell_{\text{reg}}(\theta) = -\log Z(\theta) + \left\langle\theta, \frac{1}{D}\sum_{d=1}^{D}\left[\sum_{i\in V} f_i\left(x_i^{(d)}\right) + \sum_{\alpha\in\mathcal{A}} f_\alpha\left(x_\alpha^{(d)}\right)\right]\right\rangle - \frac{\lambda}{2}||\theta||_2^2 \tag{5.18}$$

The regularizer serves two purposes. First, it helps to prevent overfitting and improve generalization. Second, it helps to combat lack of identifiability (sufficient regularization can always be used to guarantee that there is a unique parameter vector that maximizes the log-likelihood).

Lemma 5.4.2. *$\log Z(\theta)$ is a convex function of θ.*

Proof. From (5.2),

$$\begin{aligned}
\log Z(\theta) &= \sup_{q\in\mathcal{P}}\left[H(q) + \sum_{x\in\mathcal{X}_V}\sum_{i\in V} q(x)\log\phi_i(x_i|\theta) + \sum_{x\in\mathcal{X}_V}\sum_{\alpha\in\mathcal{A}} q(x)\log\psi_\alpha(x_\alpha|\theta)\right] \\
&= \sup_{q\in\mathcal{P}}\left[H(q) + \sum_{x\in\mathcal{X}_V}\sum_{i\in V} q(x)\langle\theta, f_i(x_i)\rangle + \sum_{x\in\mathcal{X}_V}\sum_{\alpha\in\mathcal{A}} q(x)\langle\theta, f_\alpha(x_\alpha)\rangle\right].
\end{aligned} \tag{5.19}$$

The terms inside of the brackets in (5.19) are linear, hence convex, in θ for each fixed $q \in \mathcal{P}$. As the supremum of convex functions is itself a convex function [2], $\log Z$ must be convex in θ. □

As a result of Lemma 5.4.2, the regularized log-likelihood (5.18) is a concave function of the parameter vector θ. Consequently, gradient ascent could be used to maximize the log-likelihood. The primary difficulty is that computing the gradient of $\ell(\theta)$ requires computing the gradient of $\log Z(\theta)$. Consider the partial derivative of $\log Z(\theta)$ with respect to a single component of the vector θ.

$$\begin{aligned}
\frac{\partial \log Z(\theta)}{\partial \theta_k} &= \frac{1}{Z(\theta)} \frac{\partial Z(\theta)}{\partial \theta_k} \\
&= \frac{1}{Z(\theta)} \frac{\partial}{\partial \theta_k} \left[\sum_x \prod_{i \in V} \phi_i(x_i|\theta) \prod_{\alpha \in \mathcal{A}} \phi_\alpha(x_\alpha|\theta) \right] \\
&= \frac{1}{Z(\theta)} \frac{\partial}{\partial \theta_k} \left[\sum_x \exp\left(\sum_{i \in V} \langle \theta, f_i(x_i) \rangle + \sum_{\alpha \in \mathcal{A}} \langle \theta, f_\alpha(x_\alpha) \rangle \right) \right] \\
&= \frac{1}{Z(\theta)} \sum_x \left[\exp\left(\sum_{i \in V} \langle \theta, f_i(x_i) \rangle + \sum_{\alpha \in \mathcal{A}} \langle \theta, f_\alpha(x_\alpha) \rangle \right) \left(\sum_{i \in V} f_i(x_i)_k + \sum_{\alpha \in \mathcal{A}} f_\alpha(x_\alpha)_k \right) \right] \\
&= \sum_x \left[\sum_{i \in V} p(x|\theta) f_i(x_i)_k + \sum_{\alpha \in \mathcal{A}} p(x|\theta) f_\alpha(x_\alpha)_k \right] \\
&= \sum_{i \in V} \sum_{x_i} p_i(x_i|\theta) f_i(x_i)_k + \sum_{\alpha \in \mathcal{A}} \sum_{x_\alpha} p_\alpha(x_\alpha|\theta) f_\alpha(x_\alpha)_k
\end{aligned}$$

The gradient of the regularized log-likelihood is then given by

$$\begin{aligned}
\nabla \ell_{reg} = \frac{1}{D} \sum_{d=1}^{D} & \left[\sum_{i \in V} f_i\left(x_i^{(d)}\right) + \sum_{\alpha \in \mathcal{A}} f_\alpha\left(x_\alpha^{(d)}\right) \right] \\
& - \left[\sum_{i \in V} \sum_{x_i} p_i(x_i|\theta) f_i(x_i) + \sum_{\alpha \in \mathcal{A}} \sum_{x_\alpha} p_\alpha(x_\alpha|\theta) f_\alpha(x_\alpha) \right] - \lambda\theta.
\end{aligned}$$

Let $\widehat{p}$ denote the empirical probability distribution.

$$\widehat{p}(x) = \frac{1}{D} \sum_{d=1}^{D} 1_{x=x^{(d)}}$$

The gradient of (5.18) can be equivalently expressed as

$$\nabla \ell_{reg} = -\lambda\theta + \sum_x \left(\widehat{p}(x) - p(x|\theta)\right) \left(\sum_{i \in V} f_i(x_i) + \sum_{\alpha \in \mathcal{A}} f_\alpha(x_\alpha) \right). \tag{5.20}$$

When $\lambda = 0$, $\nabla \ell_{reg}(\theta)$ is equal to zero whenever the two expectations agree. This condition is known as moment matching because, e.g., in the overcomplete case, this constraint requires that the model marginals must equal the empirical marginals over every vertex and hyperedge of the graph.

The most expensive step in the computation of the gradient of the log-likelihood is computing the marginals of $p(x|\theta)$. As discussed earlier, this is an intractable problem in general. However, any of the approximate inference algorithms in Section 5.3 can be used to estimate the marginal distributions (e.g., loopy belief propagation, Gibbs sampling, reweighted loopy belief propagation, etc.). This leads to what are sometimes referred to as double loop algorithms for learning: the outer loop performs gradient ascent while the inner loop estimates the marginals. Speed is of the utmost importance as the marginals must be estimated once per iteration of gradient ascent. In practice, loopy belief propagation/Gibbs sampling often take much too long to converge [16], limiting maximum likelihood estimation to relatively small graphical models. One alternative is to run the approximate inference algorithms for a fixed number of iterations, warm starting from the previous iteration, of the outer loop [3]. If the number of chosen iterations is too small, then the entire learning procedure may fail to converge while if it is too large, the learning will be time consuming. Experimentally, only a small number of iterations seem to be needed for good performance [3]. However, there does not exist a procedure, other than trial and error, to pick the number of iterations in advance.

5.4.2.1 Approximate maximum likelihood

Using the variational approach, the log-likelihood can be approximated as an optimization problem over both the parameters θ and the approximate marginals τ. Fix a vector of counting numbers, ρ, and consider the log-likelihood in which the partition function is approximated using the reweighted free energy.

$$
\begin{aligned}
\log \ell_\rho(\theta) &\triangleq -\log Z_\rho(\theta) + \left\langle \theta, \frac{1}{D}\sum_{d=1}^{D}\left[\sum_{i\in V} f_i\left(x_i^{(d)}\right) + \sum_{\alpha\in\mathcal{A}} f_\alpha\left(x_\alpha^{(d)}\right)\right]\right\rangle - \frac{\lambda}{2}||\theta||_2^2 \\
&= -\max_{\tau\in\mathcal{T}}\left[U(\tau;G,\phi,\psi) + \tilde{H}_\rho(\tau;G)\right] \\
&\qquad + \left\langle \theta, \frac{1}{D}\sum_{d=1}^{D}\left[\sum_{i\in V} f_i\left(x_i^{(d)}\right) + \sum_{\alpha\in\mathcal{A}} f_\alpha\left(x_\alpha^{(d)}\right)\right]\right\rangle - \frac{\lambda}{2}||\theta||_2^2 \\
&= -\max_{\tau\in\mathcal{T}}\left[\sum_{i\in V}\sum_{x_i}\tau_i(x_i)\langle\theta, f_i(x_i)\rangle + \sum_{\alpha\in\mathcal{A}}\sum_{x_\alpha}\tau_\alpha(x_\alpha)\langle\theta, f_\alpha(x_\alpha)\rangle + \tilde{H}_\rho(\tau;G)\right] \\
&\qquad + \left\langle \theta, \frac{1}{D}\sum_{d=1}^{D}\left[\sum_{i\in V} f_i\left(x_i^{(d)}\right) + \sum_{\alpha\in\mathcal{A}} f_\alpha\left(x_\alpha^{(d)}\right)\right]\right\rangle - \frac{\lambda}{2}||\theta||_2^2
\end{aligned}
$$

As was the case for the exact log-likelihood, the reweighted log-likelihood is also a concave function of τ for appropriate choices of the reweighting parameters, ρ. However, ℓ_ρ may not be differentiable everywhere. Despite this, supergradient ascent can be employed to maximize ℓ_ρ. The supergradient of a concave function g at a point x is the set of all linear functions that are at least as large as g everywhere on its domain and agree with g at the point x. If the supergradient of g at x contains the zero vector, then x must be a global optimum of g. The supergradient of ℓ_ρ at the vector θ, like the gradient of the log-likelihood, also enforces a moment matching constraint [8]. Specifically, a vector γ is in the supergradient of ℓ_ρ at θ if

$$
\begin{aligned}
\gamma = &\frac{1}{D}\sum_{d=1}^{D}\left[\sum_{i\in V} f_i\left(x_i^{(d)}\right) + \sum_{\alpha\in\mathcal{A}} f_\alpha\left(x_\alpha^{(d)}\right)\right] \\
&\quad - \left[\sum_{i\in V}\sum_{x_i}\tau_i^*(x_i|\theta) f_i(x_i) + \sum_{\alpha\in\mathcal{A}}\sum_{x_\alpha}\tau_\alpha^*(x_\alpha|\theta) f_\alpha(x_\alpha)\right] - \lambda\theta
\end{aligned}
$$

for some $\tau^* \in \texttt{convex hull}(\mathcal{T}_\theta)$ where

$$
\mathcal{T}_\theta \triangleq \arg\max_\tau \left[\sum_{i\in V}\sum_{x_i}\tau_i(x_i)\langle\theta, f_i(x_i)\rangle + \sum_{\alpha\in\mathcal{A}}\sum_{x_\alpha}\tau_\alpha(x_\alpha)\langle\theta, f_\alpha(x_\alpha)\rangle + \tilde{H}_\rho(\tau;G)\right].
$$

Whenever ρ is chosen so that the reweighted entropy approximation is concave, the convex hull of $\mathcal{T}_\theta$ will be equal to $\mathcal{T}_\theta$ as the set of optima of a concave function must form a convex set. If the reweighted entropy approximation is not concave, then there may exist points in the convex hull which are not elements of the arg max. In principle, this means that there could exist some empirical marginals that are difficult, if not impossible, to learn in practice when the entropy approximation is not concave (for an example of such a case, see [8]). As a result, concave entropy approximations are often preferred in practice (even if they may produce poorer estimates of the partition function).

If the reweighted entropy approximation is concave, the joint optimization problem in τ and θ is a convex-concave optimization problem. Let

$$\ell_\rho(\theta, \tau) \triangleq \left[\left\langle \theta, \frac{1}{D} \sum_{d=1}^{D} \left[\sum_{i \in V} f_i\left(x_i^{(d)}\right) + \sum_{\alpha \in \mathcal{A}} f_\alpha\left(x_\alpha^{(d)}\right) \right] \right\rangle - \frac{\lambda}{2} ||\theta||_2^2 \right.$$
$$\left. - \left[\sum_{i \in V} \sum_{x_i} \tau_i(x_i) \langle \theta, f_i(x_i) \rangle + \sum_{\alpha \in \mathcal{A}} \sum_{x_\alpha} \tau_\alpha(x_\alpha) \langle \theta, f_\alpha(x_\alpha) \rangle + \tilde{H}_\rho(\tau; G) \right] \right]. \quad (5.21)$$

The optimal solution to the minimax problem $\max_{\theta \in \mathbb{R}^K} \min_{\tau \in \mathcal{T}} \ell_\rho(\theta, \tau)$ is a saddle point.

5.4.3 Maximum entropy

In general, swapping the order of optimization in a minimax problem yields an inequality. In the case of the approximate maximum likelihood objective,

$$\max_{\theta \in \mathbb{R}^K} \min_{\tau \in \mathcal{T}} \ell_\rho(\theta, \tau) \leq \min_{\tau \in \mathcal{T}} \max_{\theta \in \mathbb{R}^K} \ell_\rho(\theta, \tau).$$

A similar inequality for ℓ can be obtained by replacing $\log Z$ by its variational optimization problem (5.2). In the particular case that the entropy approximation is concave, Sion's minimax theorem guarantees that the inequality is tight [23]. In the dual problem, $\min_{\tau \in \mathcal{T}} \max_{\theta \in \mathbb{R}^K} \ell_\rho(\theta, \tau)$, the inner maximization over θ has a closed form solution.

$$\theta(\tau) \triangleq \frac{1}{\lambda} \left[\frac{1}{D} \sum_{d=1}^{D} \left[\sum_{i \in V} f_i\left(x_i^{(d)}\right) + \sum_{\alpha \in \mathcal{A}} f_\alpha\left(x_\alpha^{(d)}\right) \right] \right.$$
$$\left. - \sum_{i \in V} \sum_{x_i} \tau_i(x_i) f_i(x_i) - \sum_{\alpha \in \mathcal{A}} \sum_{x_\alpha} \tau_\alpha(x_\alpha) f_\alpha(x_\alpha) \right] \quad (5.22)$$

Plugging into $\ell_\rho(\theta, \tau)$ yields

$$\min_{\tau \in \mathcal{T}} \max_{\theta \in \mathbb{R}^K} \ell_\rho(\theta, \tau) = \min_{\tau \in \mathcal{T}} \ell_\rho(\theta(\tau), \tau)$$
$$= \min_{\tau \in \mathcal{T}} \left[\frac{1}{2\lambda} ||\theta(\tau)||_2^2 - \tilde{H}_\rho(\tau; G) \right].$$

Equivalently,

$$\max_{\tau \in \mathcal{T}} \left[-\frac{1}{2\lambda} ||\theta(\tau)||_2^2 + \tilde{H}_\rho(\tau; G) \right].$$

In other words, the dual of the regularized log-likelihood is a regularized maximum entropy problem. Note that this does not apply in the special case in which $\lambda = 0$. However, the dual problem is still a maximum entropy problem, in this case with constraints that enforce moment matching.

Again, a variety of standard optimization techniques can be applied to maximize the reweighted entropy objective. As the objective function is constrained, projected gradient methods, which alternate between performing a gradient ascent/descent step and projecting the result back into the set of constraints are one possibility. However, these methods can be slow in this case as they require projecting onto the local marginal polytope, for which no particularly efficient algorithms are known. As an alternative, the Frank-Wolfe algorithm, sometimes known as the method of conditional gradients, yields a practical optimization strategy for the dual objective in the case that an efficient approximate MAP inference solver is available for the models being considered [27].

5.4.4 Pseudolikelihood learning

Finally, maximum likelihood optimization (even approximate) is often time consuming in practice. One alternative, maximizing the pseudolikelihood enjoys some of the same properties as the maximum likelihood estimator (e.g., it is a consistent estimator) while being much simpler to optimize in practice. The pseudolikelihood is an approximate version of the likelihood based on the chain rule. That is, let p be a joint probability distribution over n random variables.

$$\begin{aligned} p(x_1, \ldots, x_n|\theta) &= \prod_{i=1}^{n} p(x_i|x_1, \ldots, x_{i-1}, \theta) \\ &\approx \prod_{i=1}^{n} p(x_i|x_1, \ldots, x_{i-1}, x_{i+1}, \ldots, x_n, \theta) \end{aligned} \tag{5.23}$$

The approximation (5.23) is relatively easy to compute for probability distributions of the form (5.1) as, using the conditional independence properties of these models (see Chapter 1),

$$p(x_i|x_1, \ldots, x_{i-1}, x_{i+1}, \ldots, x_n, \theta) = p(x_i|x_{N(i)}, \theta).$$

As in the case of Gibbs sampling, these conditional probability distributions can be computed without first computing the partition function. We note that pseudolikelihood also underlies the technique of neighborhood selection discussed in Chapter 12.

For a probability distribution of the form (5.1), given $x^{(1)}, \ldots, x^{(D)} \in \mathcal{X}_V$, the maximum pseudolikelihood estimation problem is to find the parameter vector θ that maximizes the pseudolikelihood of the data,

$$\prod_{d=1}^{D} \prod_{i \in V} p(x_i|x_{N(i)}, \theta),$$

or, equivalently maximizing the log-pseudolikelihood:

$$\begin{aligned} \log \ell_{\text{PL}}(\theta) &\equiv \sum_{d=1}^{D} \sum_{i \in V} \log p(x_i^{(d)}|x_{N(i)}^{(d)}, \theta) \\ &= \sum_{d=1}^{D} \sum_{i \in V} \left[\log p(x_i^{(d)}, x_{N(i)}^{(d)}|\theta) - \log \sum_{x_i} p(x_i, x_{N(i)}^{(d)}|\theta) \right] \\ &= \sum_{d=1}^{D} \sum_{i \in V} \Bigg[\left\langle \theta, \left[f_i\left(x_i^{(d)}\right) + \sum_{\alpha \in \mathcal{A}: i \in \alpha} f_\alpha\left(x_\alpha^{(d)}\right) \right] \right\rangle \\ &\qquad - \log \sum_{x_i} \exp \left\langle \theta, \left[f_i\left(x_i\right) + \sum_{\alpha \in \mathcal{A}: i \in \alpha} f_\alpha\left(x_i, x_{\alpha \setminus i}^{(d)}\right) \right] \right\rangle \Bigg]. \end{aligned}$$

Like the log-likelihood, the log-pseudolikelihood is a concave function of the parameter vector θ, but the log-pseudolikelihood has the notable advantage that the log-partition function is not required to compute the pseudolikelihood. The pseudolikelihood is a consistent estimator for identifiable model: if the data was generated from a probability distribution of the form (5.1), then as the number of data points goes to infinity, the parameters that maximize the pseudolikelihood are equal to the true parameters with probability one. In a data limited setting or if the data was not generated from a model of the form (5.1), the pseudolikelihood may not yield a better estimate of the parameters than the one produced by maximizing the likelihood.

5.5 Conclusion

This chapter has presented a variety of algorithms for exact and approximate inference and learning for Markov random fields (i.e., probability distributions of the form (5.1)). While the exact methods are well-understood theoretically, limited computational power necessitates the use of their approximate counterparts in most practical settings. While some theoretical analysis is available for the approximate methods, bounding the errors introduced by these methods for problems of interest remains, for the most part, an open problem. In addition, faster/scalable methods are needed to make likelihood learning possible on large datasets and models. Last, but not least, all of the material presented in this chapter can be extended, for the most part, to models over continuous state spaces (e.g., $\mathcal{X}_i = \mathbb{R}$) including the Gaussian models treated in Chapters 9 and 10. However, even implementing belief propagation in a continuous setting requires care and often additional approximations beyond what is required for the finite state space setting (see [15, 20, 30, 4, 32, 24, 17, 10]).

Appendix: Marginal Reparameterization of a Tree-Structured Distribution

This appendix proves that each tree-structured distribution admits a factorization in terms of its marginal distributions as in (5.3). For the base case, a hypergraph with one vertex and no edges, the result is trivial. The result for hypertrees with more than one variable node follows by induction after summing out a single leaf node. Formally, suppose $|V| = n$. Without loss of generality, assume that x_1 is the variable corresponding to a leaf of the tree and that β is the only hyperedge containing it.

$$p(x_1, \dots, x_n) = p(x_2, \dots, x_n)p(x_1|x_2, \dots, x_n)$$

By induction, as $p(x_2, \dots, x_n)$ factorizes over a hypergraph with one less vertex and the hyperedge β replaced by the hyperedge $\beta \setminus 1$,

$$p(x_2, \dots, x_n) = \left[\prod_{i=2}^{n} p_i(x_i) \prod_{\alpha \in \mathcal{A} \setminus \{\beta\}} \frac{p_\alpha(x_\alpha)}{\prod_{j \in \alpha} p_j(x_j)}\right] \frac{p_{\beta \setminus 1}(x_{\beta \setminus 1})}{\prod_{j \in \beta \setminus 1} p_j(x_j)}.$$

Note that, as vertex 1 was a leaf node, $p(x_1|x_2, \ldots, x_n) = p(x_1|x_{\beta\setminus 1})$ (see I.1 and I.2). Putting all of this together gives

$$\begin{aligned} p(x_1, \ldots, x_n) &= \left[\prod_{i=2}^{n} p_i(x_i) \prod_{\alpha \in \mathcal{A}\setminus\{\beta\}} \frac{p_\alpha(x_\alpha)}{\prod_{j\in\alpha} p_j(x_j)}\right] \frac{p_{\beta\setminus 1}(x_{\beta\setminus 1})}{\prod_{j\in\beta\setminus 1} p_j(x_j)} p(x_1|x_{\beta\setminus 1}) \frac{p_1(x_1)}{p_1(x_1)} \\ &= \left[\prod_{i=2}^{n} p_i(x_i) \prod_{\alpha \in \mathcal{A}\setminus\{\beta\}} \frac{p_\alpha(x_\alpha)}{\prod_{j\in\alpha} p_j(x_j)}\right] \frac{p_\beta(x_\beta)}{\prod_{j\in\beta} p_j(x_j)} p_1(x_1) \end{aligned}$$

as desired.

Bibliography

[1] D. Angluin and A. Gardiner. Finite common coverings of pairs of regular graphs. *Journal of Combinatorial Theory, Series B*, 30(2):184–187, 1981.

[2] S. Boyd and L. Vandenberghe. *Convex Optimization.* Cambridge University Press, New York, NY, 2004.

[3] J. Domke. Learning graphical model parameters with approximate marginal inference. *PAMI*, 35(10):2454–2467, 2013.

[4] A. Frank, P. Smyth, and A. T. Ihler. Particle-based variational inference for continuous systems. In *Advances in Neural Information Processing Systems (NIPS)*, 826–834, 2009.

[5] L. Gurvits. Unleashing the power of Schrijver's permanental inequality with the help of the Bethe approximation. arXiv:1106.2844, June 2011.

[6] T. Hazan and A. Shashua. Convergent message-passing algorithms for inference over general graphs with convex free energy. In *The 24th Conference on Uncertainty in Artificial Intelligence (UAI)*, 2008.

[7] T. Hazan and A. Shashua. Norm-product belief propagation: Primal-dual message-passing for approximate inference. *Information Theory, IEEE Transactions on*, 56(12):6294–6316, Dec. 2010.

[8] U. Heinemann and A. Globerson. What cannot be learned with Bethe approximations. In *Uncertainty in Artificial Intelligence*, 2011.

[9] T. Heskes. Convexity arguments for efficient minimization of the Bethe and Kikuchi free energies. *Journal of Artificial Intelligence Research (JAIR)*, 26:153–190, 2006.

[10] A. T. Ihler and D. A. McAllester. Particle belief propagation. In *Twelfth International Conference on Artificial Intelligence and Statistics (AISTATS)*, 256–263, 2009.

[11] V. Kolmogorov. Convergent tree-reweighted message passing for energy minimization. *Transactions on Pattern Analysis and Machine Intelligence*, 28(10):1568–1583, 2006.

[12] V. Kolmogorov and C. Rother. Minimizing nonsubmodular functions with graph cuts-a review. *Pattern Analysis and Machine Intelligence, IEEE Transactions on*, 29(7):1274–1279, July 2007.

[13] T. Meltzer, A. Globerson, and Y. Weiss. Convergent message passing algorithms: a unifying view. In *Proceedings of the 25th Uncertainty in Artifical Intelligence (UAI)*, Montreal, Canada, 2009.

[14] O. Meshi, A. Jaimovich, A. Globerson, and N. Friedman. Convexifying the Bethe free energy. In *Proceedings of the 25th Conference on Uncertainty in Artificial Intelligence (UAI)*, 402–410. AUAI Press, 2009.

[15] T. P. Minka. Expectation propagation for approximate Bayesian inference. In *Proceedings of the 17th Conference on Uncertainty in Artificial Intelligence (UAI)*, 362–369, 2001.

[16] J. M. Mooij and H. J. Kappen. Sufficient conditions for convergence of loopy belief propagation. In *Proceedings of the 21st Conference on Uncertainty in Artificial Intelligence*, 396–403, 2005.

[17] N. Noorshams and M. J. Wainwright. Belief propagation for continuous state spaces: stochastic message-passing with quantitative guarantees. *Journal of Machine Learning Research (JMLR)*, 14(1):2799–2835, 2013.

[18] N. Ruozzi. The Bethe partition function of log-supermodular graphical models. In *Neural Information Processing Systems (NIPS)*, Lake Tahoe, NV, Dec. 2012.

[19] N. Ruozzi. Beyond log-supermodularity: Lower bounds and the Bethe partition function. In *Proceedings of the 29th Annual Conference on Uncertainty in Artificial Intelligence (UAI)*, 546–555, Corvallis, Oregon, 2013. AUAI Press.

[20] N. Ruozzi. Exactness of approximate MAP inference in continuous MRFs. In *Advances in Neural Information Processing Systems (NIPS)*, 2332–2340, 2015.

[21] N. Ruozzi and S. Tatikonda. Message-passing algorithms for quadratic minimization. *Journal of Machine Learning Research (JMLR)*, 14:2287–2314, 2013.

[22] N. Ruozzi and S. Tatikonda. Message-passing algorithms: Reparameterizations and splittings. *IEEE Transactions on Information Theory*, 59(9):5860–5881, Sept. 2013.

[23] M. Sion. On general minimax theorems. *Pacific Journal of Mathematics*, 8(1):171–176, 1958.

[24] E. B. Sudderth, A. T. Ihler, M. Isard, W. T. Freeman, and A. S. Willsky. Nonparametric belief propagation. In *Computer Vision and Pattern Recognition (CVPR), IEEE Computer Society Conference on Computer Science*, 2003.

[25] E. B. Sudderth, M. J. Wainwright, and A. S. Willsky. Loop series and Bethe variational bounds in attractive graphical models. In *Neural Information Processing Systems (NIPS)*, Vancouver, BC, Canada, Dec. 2007.

[26] E.B. Sudderth, A.T. Ihler, W.T. Freeman, and A.S. Willsky. Nonparametric belief propagation. In *Computer Vision and Pattern Recognition, 2003. Proceedings. 2003 IEEE Computer Society Conference on Computer Science*, volume 1, I–605. IEEE, 2003.

[27] K. Tang, N. Ruozzi, D. Belanger, and T. Jebara. Bethe learning of graphical models via map decoding. In *Proceedings of the 19th International Conference on Artificial Intelligence and Statistics (AISTATS)*, 1096–1104, 2016.

[28] P. O. Vontobel. Counting in graph covers: A combinatorial characterization of the Bethe entropy function. *Information Theory, IEEE Transactions on Information Theory*, Jan. 2013.

[29] M. J. Wainwright, T. S. Jaakkola, and A. S. Willsky. A new class of upper bounds on the log partition function. *IEEE Transactions on Information Theory*, 51(7):2313–2335, 2005.

[30] S. Wang, A. Schwing, and R. Urtasun. Efficient inference of continuous Markov random fields with polynomial potentials. In *Advances in Neural Information Processing Systems (NIPS)*, 936–944, 2014.

[31] Y. Weiss. Correctness of local probability propagation in graphical models with loops. *Neural Computation*, 12(1):1–41, January 2000.

[32] Y. Weiss and W. T. Freeman. Correctness of belief propagation in Gaussian graphical models of arbitrary topology. *Neural Computation*, 13(10):2173–2200, 2001.

[33] Y. Weiss, C. Yanover, and T. Meltzer. MAP estimation, linear programming and belief propagation with convex free energies. In *Proceedings of the 23rd Conference on Uncertainty in Artificial Intelligence (UAI)*, 2007.

[34] A. Weller and T. Jebara. Clamping variables and approximate inference. In *Advances in Neural Information Processing Systems (NIPS)*, 909–917, 2014.

[35] A. Weller, K. Tang, T. Jebara, and D. Sontag. Understanding the Bethe approximation: when and how can it go wrong? In *The 30th Conference on Uncertainty in Artificial Intelligence (UAI)*, 868–877, 2014.

[36] M. Welling and Y. Teh. Belief optimization for binary networks: A stable alternative to loopy belief propagation. In *Proceedings of the 17th Conference on Uncertainty in Artificial Intelligence (UAI)*, pages 554–561, 2001.

[37] J. S. Yedidia, W. T. Freeman, and Y. Weiss. Understanding belief propagation and its generalizations. *Exploring Artificial Intelligence in the New Millennium*, 8:236–239, 2003.

[38] A. L. Yuille and A. Rangarajan. The concave-convex procedure (cccp). *Advances in Neural Information Processing Systems (NIPS)*, 2:1033–1040, 2002.

6

MAP Estimation: Linear Programming Relaxation and Message-Passing Algorithms

Ofer Meshi

Google

Alexander G. Schwing

Department of Electrical and Computer Engineering, University of Illinois at Urbana-Champaign

CONTENTS

6.1 Introduction

Probabilistic graphical models provide a compelling framework for reasoning about multiple variables with structured dependencies. One of the central tasks in many applications of graphical models is finding the most probable assignment for the variables. This task is commonly referred to as the *maximum a-posteriori* (MAP) inference problem.

Computing the MAP assignment is tractable in some special cases. It can, for example, be solved efficiently for tree-structured graphs using a dynamic programming approach. However, for general discrete graphical models MAP inference is equivalent to a combinatorial optimization problem, which is typically NP-hard, and is even hard to approximate efficiently [56, 63, 32].

Despite this theoretical barrier, in recent years, it has been shown that approximate inference methods often obtain surprisingly accurate solutions in practice. In this chapter we focus on a family of approximations which are based on a *linear programming* (LP) relaxation. Hereby, the combinatorial program of MAP inference is first cast as an intractable (integer) linear program, and some of the constraints are then relaxed for tractability.

Although tractable in principle, LP relaxations still pose a serious computational challenge, especially for large-scale problems where the LP consists of many variables and many constraints. In such cases, standard off-the-shelf LP solvers are observed to be slow [80], and a significant research effort has been put into designing more efficient solvers that exploit the special structure of the MAP inference problem.

Many iterative solvers for MAP inference are often interpreted as performing *message-passing* on a graphical representation of the problem. Specifically, the structure of the MAP inference problem resembles a graph with sparse connectivity, and many of the devised solvers cast MAP inference as solving for variables that are indexed by edges on this graph. The goal of this chapter is to introduce the LP relaxation approaches for MAP estimation, and to present recent progress on the design and analysis of efficient solvers for this important problem.

6.2 The MAP Estimation Problem

In this section we formalize the MAP inference task for graphical models and review some strategies to solve it. To formalize the MAP problem, consider a set of n discrete variables[1] $X_1, \ldots, X_n$, and let us denote by x_i a particular assignment to variable X_i. Let $r \subseteq \{1, \ldots, n\}$ denote a subset of the variables, also known as a *region*, and we refer to the set of all regions which are used to describe the problem via the set $\mathcal{R}$. Each subset $r \in \mathcal{R}$ is associated with a local function $\varphi_r(x_r)$, which depends on the assignments $(x_i)_{i \in r}$. Using this notation, the joint probability distribution of a configuration x is defined as:

$$p(x) \propto \prod_{r \in \mathcal{R}} \varphi_r(x_r) \quad = \exp\left(\sum_{r \in \mathcal{R}} \theta_r(x_r)\right) , \tag{6.1}$$

where the local score function $\theta_r(x_r) = \log(\varphi_r(x_r))$ is often also referred to as a *potential* or *factor* [70, 32]; compare (5.1) in the preceding chapter and Section 1.7.3 in Chapter 1. Probability distributions that factorize in this way are often called *Gibbs distributions*, or *Markov Random Fields* (MRFs).

The MAP estimation problem is to find a mode of the distribution, *i.e.*, an assignment with highest probability $\operatorname{argmax}_x p(x)$, which can be written equivalently as a combinatorial optimization problem:

$$\operatorname*{argmax}_x F(x) := \sum_{r \in \mathcal{R}} \theta_r(x_r) , \tag{6.2}$$

[1] Analysis for continuous variables is beyond the scope of this chapter. The interested reader may refer to Wainwright and Jordan [70] for treatment of the Gaussian case.

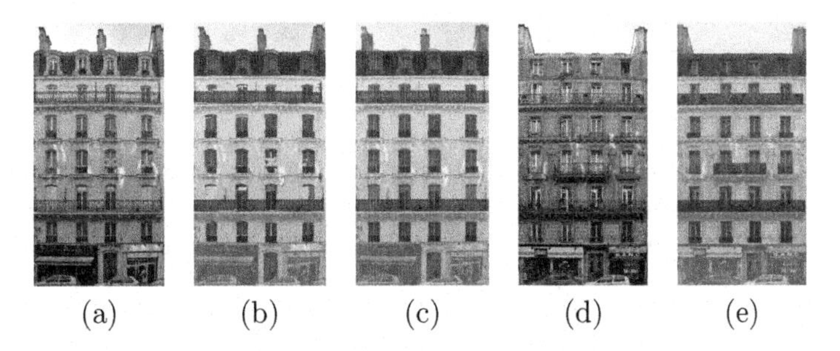

FIGURE 6.1: Semantic segmentation of the image in (a) into 8 categories using local methods (b) or more structured approaches (c). The results for the image in (d) are visualized in (e).

where $F(x)$ is often referred to as a *score function* or the *negative energy*.

A special case of this program that is commonly used in applications is a pairwise MRF. In a pairwise MRF, potentials are defined for single variables and for pairs of variables, which correspond to nodes and edges of an undirected graph G, respectively. The MAP estimation problem of this special case takes the following form:

$$\operatorname*{argmax}_x \quad \sum_{i \in V(G)} \theta_i(x_i) + \sum_{ij \in E(G)} \theta_{ij}(x_i, x_j) \ .$$

Hereby, the set $V(G)$ subsumes the vertices of the graph, while $E(G)$ refers to the edges. Correspondingly, the sets of functions $\{\theta_i\}_{i \in V(G)}$ and $\{\theta_{ij}\}_{ij \in E(G)}$ that describe the problem depend on a single variable (unary functions) and two variables (pairwise functions).

Example applications of this type include semantic image segmentation in computer vision, part-of-speech tagging in natural language processing, protein structure prediction in computational biology, to name a few. To illustrate the aforementioned concepts in the context of a concrete example, let us consider semantic image segmentation more closely. An assignment x_i refers to the configuration of a (super-)pixel X_i and encodes the predicted semantic label. The unary score functions $\theta_i(x_i)$ denote the local assignment preference for each (super-)pixel i. Note that the local score function $\theta_i(x_i)$ can differ for each pixel i, hence the subscript for both, the function θ_i and the assignment x_i. The pairwise score functions $\theta_{ij}(x_i, x_j)$ are traditionally used to encode smoothness or co-occurrence statistics between neighboring (super-)pixels i and j. We illustrate semantic segmentation of facades of Haussmannian buildings in Fig. 6.1 and refer the interested reader to Cohen et al. [9] for details.

Due to its generality, many NP-complete problems can be cast in the form of (6.2), for example, MAX-CUT, valued constraint satisfaction, independent set, and vertex cover. This equivalence suggests the general intractability of problem (6.2), and motivates the use of approximation schemes [56, 63, 32].

In the remainder of this chapter we cover several common approaches to approximate MAP inference. In particular, sampling and search-based techniques are reviewed in Section 6.3, then in Sections 6.4 and 6.5 we focus on linear programming relaxations and their optimization. Afterwards, we outline the connection to more traditional message-passing techniques in Section 6.6, and discuss the problem of extracting a MAP solution from an approximate solution in Section 6.7.

6.3 Sampling and Search-Based Methods

We begin by reviewing sampling and search-based approaches to MAP inference. These methods try to find an assignment x that has high objective value $F(x)$ by searching over the set of all possible assignments, or by transitioning between assignments in a stochastic or greedy manner.

6.3.1 Sampling

One possible approach to MAP inference is based on *sampling.* In this case, a Markov chain is simulated to generate samples from a stationary distribution corresponding to a temperature-controlled variant of $p(x)$ defined by $p_\gamma(x) \propto \exp\left(\frac{1}{\gamma}\sum_{r\in\mathcal{R}}\theta_r(x_r)\right)$ [15]. When the temperature γ is decreased the distribution becomes more peaked, and when it reaches zero, samples are drawn only from the mode of the distribution, corresponding to MAP assignments [30]. The main shortcomings of this approach are long mixing time and a slow annealing schedule that are required to guarantee the quality of the solution, making this approach less practical in general.

6.3.2 Global search methods

Another prominent approach to the MAP inference problem is based on *global search* techniques. Algorithms in this class search for an optimal solution in a systematic manner until one is obtained, along with a certificate of optimality.

Branch-and-bound is such a systematic search procedure which has been widely used to solve combinatorial optimization problems. Consider the set of all possible assignments x, denoted $\mathcal{X}$. The procedure applies two main steps: **branch:** partition the assignment set into two or more subsets,[2] say $\mathcal{X}^1$ and $\mathcal{X}^2$; **bound:** compute lower and upper bounds on the optimal value in each subset[3] (e.g., $\mathcal{L}^1 \leq \max_{x\in\mathcal{X}^1} F(x) \leq \mathcal{U}^1$). If, for example, we have that $\mathcal{L}^1 > \mathcal{U}^2$, then we can safely discard all assignments in $\mathcal{X}^2$, and apply this process recursively to $\mathcal{X}^1$. When the remaining set contains a single element, that must be the optimal solution and the search is successful.

Of course, the performance of this algorithm depends heavily on the branching and bounding strategies, and there is no single strategy that always works well. For example, if the bounds are loose, few of the active sets can be discarded. For this reason, quite a few branching and bounding approaches have been proposed. Some of those approaches, such as *AND/OR* search spaces [11] and the *Mini-Bucket* heuristic [12], exploit the structure of the underlying graphical model in order to achieve better performance.

Given the hardness of the MAP estimation task, it is no surprise that global search methods have exponential runtime in the worst case and are usually suitable for smaller scale problems. This motivates local search techniques, which trade optimality for efficiency.

6.3.3 Local search methods

Local search algorithms are initialized with some solution, and iteratively attempt to improve it in a greedy manner by optimizing over subsets of the variables until no further im-

[2]A simple partitioning strategy is to condition on all possible states of one of the variables.

[3]For example, an upper bound can be obtained using linear programming relaxations, as we discuss in Section 6.4, while any assignment x provides a lower bound.

provement is possible. The definition of the search space determines the transitions between states, and various choices give rise to different algorithms. An important consideration is efficiency of computing an optimal transition out of all possible ones.

Perhaps the simplest and best known method in this family is the *iterated conditional modes (ICM)* algorithm [4]. In ICM a variable X_i is chosen in each step, and then the current solution x is updated by: $x_i \leftarrow \text{argmax}_{x'_i} F(x'_i, x_{-i})$, where the rest of the variables X_{-i} are held fixed. Rather than greedily assigning the best label, it has also been proposed to sample labels according to the conditional distribution $p(x_i|x_{-i})$ — a procedure known as *stochastic local search* [e.g., 29, 21].

Other notable local search algorithms are based on *graph-cuts* [7]. For example, in the *α-expansion* method, multiple variables are allowed to change their current assignment to a specific label α. In *α-β-swap*, another graph-cut method, all variables with labels either α or β are allowed to switch their assignment to the other label or stay the same. In both cases, α-expansion and α-β-swap, the optimal move can be computed efficiently for some models, and approximated for others, by solving a min-cut problem on a properly constructed graph.

In contrast to the global search-based methods, local search-based techniques are often very efficient and very easy to implement. However, they are prone to get stuck in local optima. This motivates the use of techniques which take a larger context into account. Integer programming and LP relaxations, which we discuss next, take a larger context into account and are therefore computationally more demanding, but they often yield appealing results.

6.4 Integer Programming and LP Relaxations

We begin by formulating the MAP problem as an integer linear program. Given a full assignment x and a region r, consider a vector of indicators μ_r^x with an entry for each local assignment x_r, where[4] $\mu_r^x(x'_r) = \mathbb{1}\{x'_r = x_r\}$. Clearly, for all assignments x,

$$\sum_{r \in \mathcal{R}} \theta_r(x_r) = \sum_{r \in \mathcal{R}} \sum_{x'_r} \mu_r^x(x'_r)\theta_r(x'_r) \quad = \mu^x \cdot \theta \ , \tag{6.3}$$

where μ^x and θ are the concatenation of $\mu_r^x(x_r)$ and $\theta_r(x_r)$ for all regions $r \in \mathcal{R}$ and assignments x_r. Due to the correspondence between μ^x and x, instead of optimizing over assignments x, we can instead optimize the linear objective in (6.3) over all indicators μ consistent with an assignment.

To ensure consistency, it is sufficient to require that indicator variables for overlapping subsets agree with each other. A simple way to formalize this is through marginalization constraints:

$$\mu_i(x_i) = \sum_{x_{r\setminus i}} \mu_r(x_r) \quad \forall\, i, x_i, r : i \in r \ , \tag{6.4}$$

where the notation $x_{r\setminus i}$ entails that the summation is over all consistent assignments $\{x'_r \mid x'_i = x_i\}$. Notice that usage of $\mu_i(x_i)$ is generally possible, as we can define $\theta_i(x_i) = 0$ even if it is absent from the model (6.2).[5] Note that we can use other constraint sets to enforce the same agreement between factor indicators. For example, intersections between all pairs of regions: $\sum_{x_{r\setminus r'}} \mu_r(x_r) = \sum_{x_{r'\setminus r}} \mu_{r'}(x_{r'}) \quad \forall r, r'$, or all subsets of a certain size.

[4]The notation $\mu_r(x_r)$ is standard in the graphical models literature, *e.g.* Wainwright and Jordan [70].
[5]This is true for any subset of the variables, not just x_i.

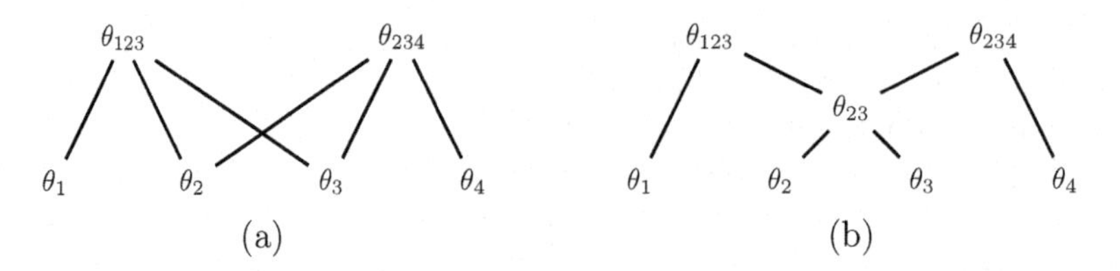

FIGURE 6.2: A factor graph in (a) and a region graph in (b).

To generally define the consistency constraints we use a *region graph*, in which each node corresponds to a region, and nodes are connected via parent-child relationships. We require that all variables in a child region are contained in the parent region [81]. For a region r, its parents in the region graph are denoted by $P(r)$, and its children by $C(r)$, and we require $r \subseteq p$ for $p \in P(r)$. The region graph defines a set of marginalization constraints,

$$\mu_r(x_r) = \sum_{x_{p\setminus r}} \mu_p(x_p) \quad \forall\, r, x_r, p \in P(r)\ . \tag{6.5}$$

A *factor graph* is a special case of a region graph that consists of nodes for model factors and for single variables. Marginalization is only enforced between factor nodes and variable nodes, as in (6.4) [37], hence a factor graph is bipartite. See Fig. 6.2 for an illustration of a factor graph and a region graph.

With the aforementioned notation at hand we can now formulate the MAP problem as the following integer linear program (ILP):

$$\max_{\mu}\ \mu\cdot\theta \quad \text{s.t.} \quad \left\{\begin{array}{ll} \mu_r(x_r) \in \{0,1\} & \forall r, x_r\ , \\ \sum_{x_r} \mu_r(x_r) = 1 & \forall r\ , \\ \mu_r(x_r) = \sum\limits_{x_{p\setminus r}} \mu_p(x_p) & \forall\, r, x_r, p \in P(r). \end{array}\right. \tag{6.6}$$

For problems of small scale, standard ILP solvers are applicable and often implement branch-and-bound like procedures.

6.4.1 LP relaxations

While solving ILPs is NP-hard in general [28], several authors [*e.g.*, 59, 36, 8, 72] proposed to obtain a *tractable approximation* by relaxing the integrality constraints ($\mu_r(x_r) \in \{0,1\}$), and replacing them with per-region simplex constraints (*i.e.*, non-negativity and normalization). This results in the following linear programming relaxation:

$$\max_{\mu\in\mathcal{M}_L}\ \mu\cdot\theta \tag{6.7}$$

$$\text{where:}\quad \mathcal{M}_L = \left\{\mu \geq 0 \;\middle|\; \begin{array}{ll} \sum_{x_r} \mu_r(x_r) = 1 & \forall r \\ \mu_r(x_r) = \sum\limits_{x_{p\setminus r}} \mu_p(x_p) & \forall\, r, x_r, p \in P(r) \end{array}\right\}\ .$$

In this relaxation, μ is no longer restricted to be a vector of indicators, and can in fact take fractional values between 0 and 1. Often, fractional solutions obtained from the LP relaxation have to be *rounded* to an integer solution for the original problem; we return to this topic in Section 6.7. The set $\mathcal{M}_L$ in (6.7) is known as the *local marginal polytope* [70]. Since the program in (6.7) was obtained by removing constraints, its optimal value is an upper bound on the optimum of (6.6) (or, equivalently, (6.2)). When basic factor-graph constraints are used (6.4), the relaxation is called the *first-order LP relaxation*, and tighter relaxations can be obtained by introducing more constraints [*e.g.*, 2, 64, 66]. As we discuss next, in some special cases the relaxation is known to be *tight*, namely the solution of the relaxed program is the same as that of the ILP.

6.4.2 Tight LP relaxations

Sometimes the computational gain from approximating the ILP comes at no cost in accuracy. Perhaps the best known case is when the region graph is acyclic, namely a *tree* (see, e.g., Fig. 6.2(b)). In this case the local marginal polytope $\mathcal{M}_L$ can be shown to have only integral vertices ($\mu_r(x_r) \in \{0,1\}$ for all r, x_r). Therefore, optimization of a linear objective over $\mathcal{M}_L$ must result in an integral point. Hence the LP relaxation solves the ILP problem given in (6.6).

Another well studied instance of tight LP relaxations is models with *supermodular* potential functions. This class of problems includes *binary pairwise models* whose factors satisfy the inequality:

$$\theta_{ij}(1,1) + \theta_{ij}(0,0) \geq \theta_{ij}(0,1) + \theta_{ij}(1,0) \ . \tag{6.8}$$

In words, these factors favor assignments that agree with each other over assignments that differ. Such models are also known as *attractive*, or *associative* [68]. In some special cases (called *balanced models*), it is possible to make the model supermodular even when not all factors satisfy (6.8) by "flipping" a subset of the variables ($x_i \leftarrow 1 - x_i$). Furthermore, a single pass over the corresponding graph is enough to determine whether the model is balanced and to find the flip-set, if it exists [18].

A third class of problems with a tight LP relaxation is *maximum-weight matching*. Given an undirected graph, the goal here is to find a subset of edges of maximum weight, such that each vertex in the graph is adjacent to at most one edge. This problem has a natural MAP formulation, and it is well known that the first-order LP relaxation is tight when the original graph is bipartite but loose otherwise [60].

It is important to note that although the LP relaxations are tight in the above cases, usage of general LP solvers is often not the most efficient way to tackle these problems. Due to their particular structure, specialized algorithms have been designed to find the MAP solutions in these cases. For tree-structured graphs the max-product algorithm is a dynamic programming approach that can compute the MAP solution in linear time [51], see Section 6.6; for supermodular potentials, the MAP solution can be computed by solving a min-cut/max-flow problem on a properly constructed graph, with cost that is cubic in the number of variables [17]; and for maximum weight matching problems, one can obtain a MAP solution *for general graphs* (not necessarily bipartite) in polynomial time using Edmond's blossom algorithm [13].

There are also cases where the first-order relaxation is not tight, but a polynomial number of additional constraints can render it tight. These include *planar graphs* with no singleton factors [1], and almost-balanced models [75]. Furthermore, even when the model does not belong to any of the tractable families above, it has been observed that for many applications, models which are learned from data in a supervised manner, often give rise to tight LP relaxations [45].

We next turn our attention to the problem of computing a solution to the relaxed linear program in (6.7). As we mentioned before, solving the LP relaxation is tractable, but practically it is still quite challenging computationally. In fact, it has been shown recently that LP relaxations for MAP inference are not easier than general LPs [52]. In the next section we present several approaches to solving the LP relaxation.

6.5 Optimization of the LP Relaxation

In this chapter we particularly emphasize algorithms that come with performance guarantees. A straightforward approach to solving the MAP LP relaxation is to use a general-purpose LP solver. Most contemporary solvers implement both the Simplex algorithm [10] and the Interior Point method [27, 79]. While the former may need an exponential number of steps in the worst case, the latter has strong polynomial time guarantees, converging to an ϵ-accurate solution with at most $O\left(\sqrt{q+m}\log(1/\epsilon)\right)$ iterations, where q is the number of variables and m the number of constraints in the LP [55]. While this is a very fast rate, it has been noticed that for many practical applications, specialized solvers that exploit the structure of the problem can often perform much better than generic LP solvers [80]. In what comes next, we present several algorithmic approaches that aim at solving the relaxed LP, or its approximations, while taking advantage of the underlying graphical representation of the problem. Specifically, these algorithms perform a sequence of local computations on the region graph that defines the LP relaxation.

6.5.1 The dual program

Many of the popular algorithms for MAP LP relaxations optimize the *dual program* of (6.7) in order to take advantage of the graph structure defined via the marginalization constraints. Using the Lagrangian formalism, it is not difficult to derive the dual problem:[6]

$$\min_{\delta} g(\delta) := \sum_{r\in\mathcal{R}} \max_{x_r} \left(\theta_r(x_r) + \sum_{p\in P(r)} \delta_{pr}(x_r) - \sum_{c\in C(r)} \delta_{rc}(x_c) \right) = \sum_{r\in\mathcal{R}} \max_{x_r} \hat{\theta}_r^{\delta}(x_r)\,. \tag{6.9}$$

Here δ are dual variables corresponding to Lagrange multipliers for the marginalization constraints in (6.7). There exists such a variable for each $r, x_r, p \in P(r)$, and we use the notation $\delta_{pr}(x_r)$. To clarify, in (6.9) x_c is the restriction of the subset assignment x_r to those variables subsumed in the child set c. Notice that the dual program given in (6.9) consists of per-factor maximization of reparameterized factors denoted $\hat{\theta}_r^{\delta}(x_r)$ [71]. Intuitively, the dual program aims at reparameterizing factors so as to ensure that local maximization is equal to the global maximization aspired in the primal program. Due to strong duality, the optimal value of the dual problem is equal to the optimal value of the primal LP relaxation, while any dual variable δ provides an upper bound $g(\delta)$ on this value.

6.5.2 Subgradient descent

(6.9) is an *unconstrained convex* optimization problem with a piecewise linear and hence *non-smooth* objective function. Perhaps the simplest way to optimize such objective functions uses the *subgradient descent* method, shown in Algorithm 6.1 [35].

Intuitively, the subgradient $\bar{\delta}$ holds information about the agreement between the factor argmax assignment $\hat{x}_r$ and its children argmax $\hat{x}_c$. If they agree then the subgradient is zero and no update occurs, however if they disagree then the update aims to encourage agreement by making the reparameterized scores closer. In fact, if all factors agree on their argmax assignments, we have a *certificate* that the current solution δ is optimal and that the LP relaxation is tight [72, 67]. In this case, composing argmax solutions from all factors yields an optimal solution to the original combinatorial problem given in (6.2).

[6] Details on the derivation leading to this dual problem can be found in previous art (e.g., Werner [77], Sontag et al. [67], Komodakis et al. [35]), and for completeness we also include it in the appendix.

Algorithm 6.1 Subgradient Descent for dual MAP LP relaxation

1: Initialize: $\delta^0 = 0$
2: **for** t = 0,...,T **do**
3: $\quad \hat{x}_r = \text{argmax}_{x_r} \hat{\theta}_r^{\delta^t}(x_r)$ for all r
4: $\quad$ Subgradient: $\bar{\delta} = 0$
5: $\quad\quad \bar{\delta}_{pr}(\hat{x}_r) += 1$ for all $p \in P(r)$
6: $\quad\quad \bar{\delta}_{rc}(\hat{x}_c) -= 1$ for all $c \in C(r)$
7: $\quad$ Update: $\delta^{t+1} = \delta^t - \eta_t \bar{\delta}$
8: **end for**

Algorithm 6.2 Star Block Coordinate Minimization for dual MAP LP relaxation

1: Initialize: $\delta^0 = 0$
2: **while** not converged **do**
3: $\quad$ Choose a region r
4: $\quad$ For all x_r and $p \in P(r)$:
5: $$\delta_{pr}^{t+1}(x_r) = \delta_{pr}^t(x_r) + \max_{x_{p \setminus r}} \hat{\theta}_p^{\delta^t}(x_p) - \frac{1}{|P(r)|}\left(\hat{\theta}_r^{\delta^t}(x_r) + \sum_{p' \in P(r)} \max_{x_{p' \setminus r}} \hat{\theta}_{p'}^{\delta^t}(x_{p'})\right)$$
6: **end while**

An important property of the dual form in (6.9) and the subgradient algorithm is that they only require computing per-factor maximization of the reparameterization. This opens up the possibility to use large structured factors for which maximization is efficient to compute. This is not possible in the primal formulation because of the need to explicitly represent all factor assignments x_r, whose number grows exponentially with the size of the factor. In terms of convergence rate, under a suitable choice of the step size η_t, such as $\eta_t = 1/(\sqrt{t}+1)$ (see Komodakis et al. [35] for more options), this algorithm is guaranteed to converge to an ϵ-accurate solution of the dual problem after at most $O(1/\epsilon^2)$ iterations [49].

6.5.3 Block coordinate minimization

Another approach that has been proposed for optimizing the dual problem in (6.9) is based on *coordinate minimization.* This is a sequential method, where at each step a subset of the variables is optimized while the others are held fixed.[7] Note that the resulting sequence of objective values is non-increasing. A key requirement for these methods is that the updates are efficient to compute, which dictates the types of possible subsets that are optimized over. Different choices lead to different algorithms in this family. Here we provide the so called *Star* update in Algorithm 6.2, which updates dual variables $\delta_{pr}(x_r)$ for all parents $p \in P(r)$ and values x_r for a given region r. Other examples of blocks include MPLP [16], Max-Sum Diffusion [77], TRW-S [33], and tree blocks [65] (see also Meltzer et al. [41]). It is worth noting that the updates in coordinate minimization algorithms require computing max-marginals (*e.g.*, $\max_{x_{p \setminus r}} \hat{\theta}_p^{\delta^t}(x_p)$ in Algorithm 6.2), which may sometimes be more expensive than per-factor maximization used in subgradient descent. For a comprehensive review and comparison to subgradient descent see Sontag et al. [67].

Unfortunately, from a theoretical point of view, coordinate minimization methods suffer from a major caveat: for non-smooth objective functions, such as (6.9), they may converge

[7]Here our interest is in coordinate minimization methods which are different than coordinate descent methods. The latter consist of taking a step in the direction of the gradient rather than full optimization of the coordinate block.

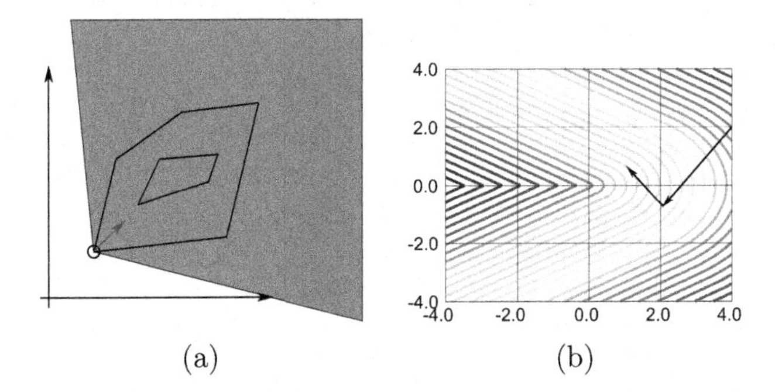

FIGURE 6.3: (a) Iso-lines of a 2-dimensional optimization problem for which coordinate descent will get stuck at the iterate highlighted by the circle. The set of subgradient directions at the iterate is highlighted in blue. Note that those are not necessarily descending directions. The red arrow highlights the steepest descending direction. (b) Iso-lines of a 2-dimensional program for which steepest descent is not globally convergent.

to a suboptimal point (except for some special cases [16, 34]). In fact, it is not too hard to find problems that lead to such behavior [*e.g.*, 33].

6.5.4 ϵ-descent

Convergence of block-coordinate minimization to a sub-optimal point for non-smooth objectives is due to the fact that block-coordinate descent techniques may get stuck in "corner" cases. To see this, consider the 2-dimensional optimization task illustrated in Fig. 6.3 (a), where iso-lines have sharp corners and steps are restricted to be axis-aligned while requiring improvement. A second drawback of block-coordinate minimization is that it does not always produce a primal optimal solution. Even if we reach the optimum of the dual program, a consistent primal solution may not be reconstructed from the dual variables (see also Section 6.7). To fix both issues, steepest subgradient descent may be considered. Although finding the subgradient direction which descends fastest is feasible for the aforementioned LP relaxation [61], it does not lead to an algorithm that is globally convergent for non-smooth functions. A counter example which was originally investigated by Wolfe [78] is given in Fig. 6.3 (b). The issue is due to the fact that the set of subgradients may change drastically from one point in the domain to the other. To fix this issue Schwing et al. [61, 62] find the steepest descending direction within an ϵ-neighborhood of the current iterate [3].

Let us consider the set of ϵ-subdifferentials, *i.e.*, the set of subgradients in an ϵ-neighborhood of our current iterate for the LP relaxation more formally. It was shown by Schwing et al. [61] that this set of ϵ-subdifferentials can be characterized for the dual program given in (6.9). In addition, it is possible to find the steepest descending direction within the set of ϵ-subdifferentials by solving the following quadratic program:

$$\min_{\mu} \sum_{r,x_r,p\in P(r)} \left(\sum_{x_{p\setminus r}} \mu_p(x_p) - \mu_r(x_r) \right)^2 \qquad (6.10)$$

$$\text{s.t.} \quad \begin{cases} \mu_r(x_r) \geq 0 & \forall r, x_r \\ \sum_{x_r} \mu_r(x_r) = 1 & \forall r \\ \sum_{x_r} \mu_r(x_r)\hat{\theta}_r^\delta(x_r) \geq \max_{x_r} \hat{\theta}_r^\delta(x_r) - \epsilon & \forall r \end{cases} .$$

Algorithm 6.3 ϵ-descent minimization for dual MAP LP relaxation

1: Initialize: $\delta^0 = 0$
2: **while** not converged **do**
3: Solve program given in (6.10) to obtain μ^*
4: Compute disagreement $d_{pr}^{t+1}(x_r) = \sum_{x_{p\setminus r}} \mu_p^*(x_p) - \mu_r^*(x_r)$
5: Update dual variables $\delta_{pr}^{t+1} = \delta_{pr}^t(x_r) + \eta d_{pr}^{t+1}$
6: **end while**

This program can be derived by considering the set of ϵ-subdifferentials for the dual program given in (6.9) and finding the direction of steepest descent. By summing the last constraint over all regions $r \in \mathcal{R}$, we observe that a cost function value of zero ensures $|\mathcal{R}|\epsilon$ optimality while the original LP relaxation is primal feasible.

An ϵ-descent based minimization algorithm iterates three steps. We first solve the program given in (6.10) to obtain the direction of steepest descent in an ϵ neighborhood of the current iterate. We then use the obtained solution to compute the marginal disagreement which is employed in a third step to update the dual variables. The updated dual variables result in a different reparameterization which changes the direction of steepest descent in the next iteration. We summarize this procedure in Algorithm 6.3.

Solving a quadratic program in every iteration may be expensive. To this end Schwing et al. [62] design an optimization scheme based on conditional gradient (a.k.a. the Frank-Wolfe algorithm), which takes advantage of the fact that the constraints of the program given in (6.10) decompose per region. The resulting algorithm is trivially parallelizable, monotonic and globally convergent. Although very effective in practice the best provable rate to achieve ϵ accuracy is $O(1/\epsilon^2)$.

6.5.5 The smoothed dual

To overcome some of the limitations mentioned above (*i.e.*, suboptimality or slow rate), several authors proposed to *smooth* the dual objective function [23, 19, 76]. This is often done by replacing the local max operation in (6.9) with a *soft-max*, which results in the following program:

$$\min_{\delta} g_\gamma(\delta) := \sum_{r \in \mathcal{R}} \gamma \log \sum_{x_r} \exp\left(\frac{\hat{\theta}_r^\delta(x_r)}{\gamma}\right), \tag{6.11}$$

where γ is a temperature parameter controlling the amount of smoothing (larger means smoother). This parameter can be chosen in advance to achieve some desired precision, or decreased gradually while approaching the optimum [see 58]. The dual form in (6.11) can be obtained by adding local entropy terms to the primal given in (6.7), *i.e.*,

$$\max_{\mu \in \mathcal{M}_L} \sum_{r, x_r} \mu_r(x_r)\theta_r(x_r) + \gamma \sum_r H(\mu_r), \tag{6.12}$$

where $H(\mu_r) = -\sum_{x_r} \mu_r(x_r) \log \mu_r(x_r)$ denotes the local entropy. The following guarantee holds for the smooth optimal value[8] g_γ^*:

$$g^* \leq g_\gamma^* \leq g^* + \gamma \sum_r \log V_r\ , \tag{6.13}$$

where g^* is the optimal value of the non-smooth dual program given in (6.9), and V_r denotes the number of variables in region r.

[8]The same guarantee holds for the primal in (6.12).

Algorithm 6.4 Star block coordinate minimization for the smooth dual

1: Initialize: $\delta^0 = 0$
2: **while** not converged **do**
3: Choose a region r
4: For all x_r and $p \in P(r)$:
5: $\delta_{pr}^{t+1}(x_r) = \delta_{pr}^t(x_r) + \gamma \log b_p(x_r) - \frac{1}{|P(r)|+1}\gamma \log \left(b_r(x_r) \cdot \prod_{p' \in P(r)} b_{p'}(x_r) \right)$
6: **end while**

The *gradient descent* algorithm can be readily applied to the smooth dual in (6.11). In this case the algorithm repeatedly applies the update $\delta^{t+1} = \delta^t - \frac{1}{L}\nabla g_\gamma(\delta^t)$, where $L = \frac{1}{\gamma}\sum_r V_r$ is the Lipschitz constant, and the gradient takes the simple form:

$$\nabla_{\delta_{pr}(x_r)} g_\gamma = \left(b_r(x_r) - \sum_{x_{p \setminus r}} b_p(x_p) \right), \text{ with } b_r(x_r) \propto \exp\left(\frac{\hat{\theta}_r^\delta(x_r)}{\gamma} \right). \tag{6.14}$$

The convergence rate of this algorithm is known to be $O(1/\gamma\epsilon)$, which can be improved using Nesterov's acceleration scheme to obtain an $O(1/\sqrt{\gamma\epsilon})$ rate [24, 57].

In contrast to the non-smooth case, for the smooth optimization problem in (6.11), coordinate minimization algorithms are globally convergent (to the smooth optimum). As an example, in Algorithm 6.4 we show the smooth counterpart of Algorithm 6.2, where with some abuse of notation, we use $b_p(x_r) = \sum_{x_{p \setminus r}} b_p(x_p)$. Notice that in Algorithm 6.4 we do not specify how the next block to update is chosen. Commonly used schedules include updating blocks in random order, updating in cyclic order, and a greedy approach that updates at each step the most "promising" block under some criterion. Meshi et al. [43] provide convergence analysis for random and greedy scheduling, showing that their convergence rate can be bounded by $O(1/\gamma\epsilon)$, which is similar to gradient descent.

Quadratic smoothing

Soft-max is not the only way to smooth the dual objective in (6.9). Recall that soft-max can be seen as the result of adding local entropy terms in the primal LP (*i.e.*, (6.12)). In fact, this is a consequence of the well known duality between strong convexity and smoothness, where the entropy terms serve to introduce strong-convexity in the primal [*e.g.*, 50]. One can thus consider other terms that induce strong-convexity of the primal. Meshi et al. [44] propose to use a simple quadratic term:

$$\max_{\mu \in \mathcal{M}_L} f_\gamma(\mu) := \mu^\top \theta - \frac{\gamma}{2}\|\mu\|^2 . \tag{6.15}$$

It turns out that the corresponding (smooth) dual function takes the form:

$$\min_\delta \tilde{g}_\gamma(\delta) := \sum_r \left(\frac{\gamma}{2} \left\| \frac{\hat{\theta}_r^\delta}{\gamma} \right\|^2 - \min_{u \in \Delta} \frac{\gamma}{2} \left\| u - \frac{\hat{\theta}_r^\delta}{\gamma} \right\|^2 \right) . \tag{6.16}$$

Thus the dual objective involves scaling the factor reparameterization $\hat{\theta}_r^\delta$ by $1/\gamma$, and then projecting the resulting vector onto the probability simplex (denoted Δ). We call the outcome of this projection u_r (or just u when clear from context). The dual stated in (6.16) is smooth with Lipschitz constant $L = q/\gamma$, where $q = |\mathcal{R}|$ is the total number of regions. Calculating the objective requires computing the projection u_r onto the simplex for all factors, which can be done efficiently [see 44].

The difference between the smooth dual optimum and the original one can be bounded as follows:

$$g^* - \frac{\gamma}{2} q \leq \tilde{g}^*_\gamma \leq g^* \ .$$

Notice that this bound is better than the corresponding soft-max bound stated in (6.13), since it does not depend on the size of regions, *i.e.*, V_r. This point is illustrated in Fig. 6.4.

To solve the dual program given in (6.16) one can use gradient-based algorithms. The gradient takes the form:

$$\nabla_{\delta_{pr}(x_r)} \tilde{g}_\gamma = \left(u_r(x_r) - \sum_{x_{p\setminus r}} u_p(x_p) \right) ,$$

which only requires computing the projection u_r, as in the objective function. Notice that this form is very similar to the soft-max gradient (6.14), with projections u taking the role of beliefs b. However, unlike b_r, the projection u_r will most likely be sparse, since only a few elements tend to have scores close to the maximum and hence non-zero value.

The convergence rate of gradient descent for the smooth dual is $O(1/\gamma\epsilon)$, which is similar to the soft-max rate [44], and again, Nesterov's accelerated gradient method achieves a better $O(1/\sqrt{\gamma\epsilon})$ rate.

Interestingly, it is not clear how to derive efficient coordinate minimization updates for the dual in (6.16), since the projection u_r depends on the dual variables δ in a non-linear manner.

6.5.6 The strongly-convex dual

Recall that the dual given in (6.9) is a piecewise linear (non-smooth) function, which makes it challenging to optimize. Introducing strong convexity to control the convergence rate, is an alternative to smoothing. Fig. 6.4 illustrates this idea and compares it to the smoothing approach presented above. Meshi et al. [44] propose to introduce strong convexity by simply adding the L_2 norm of the variables to the dual program given in (6.9), *i.e.*,

$$\min_\delta \breve{g}_\lambda(\delta) := g(\delta) + \frac{\lambda}{2}\|\delta\|^2 \ . \tag{6.17}$$

The corresponding primal objective is then:

$$\max_{\mu \in \Delta^\times} f_\lambda(\mu) := \mu^\top \theta - \frac{1}{2\lambda} \sum_{r, x_r, p \in P(r)} \left(\sum_{x_{p\setminus r}} \mu_p(x_p) - \mu_r(x_r) \right)^2 = \mu^\top \theta - \frac{\lambda}{2}\|A\mu\|^2 , \tag{6.18}$$

where $\Delta^\times$ preserves only the separable per-region simplex constraints in $\mathcal{M}_L$ (non-negativity and normalization), and for convenience we define $(A\mu)_{r,x_r,p} = \frac{1}{\lambda}\left(\sum_{x_{p\setminus r}} \mu_p(x_p) - \mu_r(x_r)\right)$. Importantly, this primal program is similar to the original primal given in (6.7), but the non-separable marginalization constraints in $\mathcal{M}_L$ are enforced softly – via a penalty term in the objective.

Note that the second part of the primal in (6.18) resembles the objective function obtained by the ϵ-steepest descent approach discussed in Section 6.5.4. Similar to Schwing et al. [62], the algorithm below is also based on conditional gradient.

The following guarantee holds for the optimum of the strongly convex dual:

$$g^* \leq \breve{g}^*_\lambda \leq g^* + \frac{\lambda}{2} h , \tag{6.19}$$

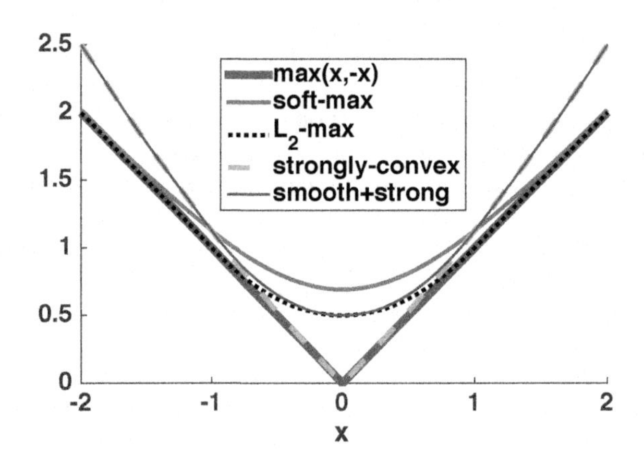

FIGURE 6.4: Illustration of smooth and/or strongly-convex approximations to the piecewise linear function $\max\{x, -x\}$. We compare: a smooth approximation based on soft-max (Section 6.5.5), a smooth approximation based on L_2 (Section 6.5.5), a strongly-convex approximation (Section 6.5.6), and a smooth and strongly-convex approximation (Section 6.5.6).

Algorithm 6.5 Block-coordinate Frank-Wolfe for soft-constrained primal

1: Initialize: $\mu_r(x_r) = \mathbb{1}\{x_r = \text{argmax}_{x'_r} \hat{\theta}_r^{\delta(\mu)}(x'_r)\}$ for all r, x_r
2: **while** not converged **do**
3: Pick r at random
4: Let $s_r(x_r) = \mathbb{1}\{x_r = \text{argmax}_{x'_r} \hat{\theta}_r^{\delta(\mu)}(x'_r)\}$ for all x_r
5: Let $\eta = \frac{\left(\hat{\theta}_r^{\delta(\mu)}\right)^\top (s_r - \mu_r)}{\frac{1}{\lambda}|P(r)|\|s_r-\mu_r\|^2 + \frac{1}{\lambda}\sum_{c\in C(r)} \|A_{rc}(s_r-\mu_r)\|^2}$, and clip to $[0, 1]$
6: Update $\mu_r \leftarrow (1-\eta)\mu_r + \eta s_r$
7: **end while**

where h is chosen such that $\|\delta^*\|^2 \leq h$. It can be shown that $h = (4Mq\|\theta\|_\infty)^2$, where M is the maximal number of configurations of any region. Notice that this bound is worse than the soft-max bound stated in (6.13) due to the dependence on the magnitude of the parameters θ and the number of configurations M.

In terms of optimization, it is easy to modify the subgradient algorithm to optimize the strongly-convex dual objective given in (6.17). It only requires adding the term $\lambda\delta$ to the subgradient. Since the objective is non-smooth and strongly convex, its convergence rate is $O(1/\lambda\epsilon)$ [48]. Note that coordinate descent algorithms are still non-convergent for (6.17), since the objective is still non-smooth. Instead, Meshi et al. [44] propose to optimize the primal given in (6.18) via a conditional gradient algorithm [14]. Specifically, Algorithm 6.5 implements the block-coordinate Frank-Wolfe algorithm proposed by Lacoste-Julien et al. [38]. In Algorithm 6.5 we define $\delta(\mu)$ as $\delta_{pr}(x_r) = \frac{1}{\lambda}\left(\sum_{x_{p\setminus r}} \mu_p(x_p) - \mu_r(x_r)\right)$, and $A_{rc}\mu_r = \sum_{x_{r\setminus c}} \mu_r(x_r)$.

Meshi et al. [44] show that the convergence rate of Algorithm 6.5 is $O(1/\lambda\epsilon)$, similar to subgradient in the strongly-convex dual. However, Algorithm 6.5 has several advantages over subgradient, including optimal step-size calculation and a sound stopping criterion [see 44, for details]. Notice also that the basic operation in the update is maximization over the

TABLE 6.1
Summary of objective functions, algorithms and rates. Row and column headers pertain to the dual objective.

	Convex	Strongly-convex
Non-smooth	**Dual:** $\min_\delta g(\delta) := \sum_r \max_{x_r} \hat{\theta}_r^\delta(x_r)$ Subgradient $O(1/\epsilon^2)$ Coordinate minimization (non-convergent) **Primal:** $\max_{\mu \in \mathcal{M}_L} \mu^\top \theta$ Proximal projections	**Dual:** $\min_\delta g(\delta) + \frac{\lambda}{2}\|\delta\|^2$ Subgradient $O(1/\lambda\epsilon)$ **Primal:** $\max_{\mu \in \Delta^\times} \mu^\top \theta - \frac{\lambda}{2}\|A\mu\|^2$ Conditional Gradient $O(1/\lambda\epsilon)$
Soft-max	**Dual:** $\min_\delta g_\gamma(\delta) := \sum_r \gamma \log \sum_{x_r} \exp\left(\frac{\hat{\theta}_r^\delta(x_r)}{\gamma}\right)$ Gradient $O(1/\gamma\epsilon)$ Accelerated $O(1/\sqrt{\gamma\epsilon})$ Coordinate minimization $O(1/\gamma\epsilon)$ **Primal:** $\max_{\mu \in \mathcal{M}_L} \mu^\top \theta + \gamma \sum_r H(\mu_r)$	**Dual:** $\min_\delta g_\gamma(\delta) + \frac{\lambda}{2}\|\delta\|^2$ Gradient $O(\frac{1}{\gamma\lambda}\log(\frac{1}{\epsilon}))$ Accelerated $O(\frac{1}{\sqrt{\gamma\lambda}}\log(\frac{1}{\epsilon}))$ **Primal:** $\max_{\mu \in \Delta^\times} \mu^\top \theta - \frac{\lambda}{2}\|A\mu\|^2 + \gamma \sum_r H(\mu_r)$
L_2-max	**Dual:** $\min_\delta \tilde{g}_\gamma(\delta) := \sum_r \max_{u \in \Delta}\left(u^\top \hat{\theta}_r^\delta - \frac{\gamma}{2}\|u\|^2\right)$ Gradient $O(1/\gamma\epsilon)$ Accelerated $O(1/\sqrt{\gamma\epsilon})$ **Primal:** $\max_{\mu \in \mathcal{M}_L} \mu^\top \theta - \frac{\gamma}{2}\|\mu\|^2$	**Dual:** $\min_\delta \tilde{g}_\gamma(\delta) + \frac{\lambda}{2}\|\delta\|^2$ Gradient $O(\frac{1}{\gamma\lambda}\log(\frac{1}{\epsilon}))$ Accelerated $O(\frac{1}{\sqrt{\gamma\lambda}}\log(\frac{1}{\epsilon}))$ **Primal:** $\max_{\mu \in \Delta^\times} \mu^\top \theta - \frac{\lambda}{2}\|A\mu\|^2 - \frac{\gamma}{2}\|\mu\|^2$ SDCA $O((1+\frac{1}{\gamma\lambda})\log(\frac{1}{\epsilon}))$

reparameterization ($\max_{x_r} \hat{\theta}_r^\delta(x_r)$), which is similar to a subgradient computation (Algorithm 6.1). As mentioned above, this operation may sometimes be cheaper than computing max-marginals, which is required by coordinate minimization.

To achieve higher accuracy, a natural idea is to nest the strongly-convex problem in an outer loop that tightens the approximation at each iteration. A simple way to achieve that is to start with a large strong-convexity parameter λ and decrease it gradually, using the previous optimum as a warm-start, until a desired level of accuracy is achieved. This approach is known as the *penalty method* and has been explored extensively in the optimization literature [see 6, for a survey]. An alternative approach is based on the *proximal point method*, where the strong-convexity term is defined with respect to the previous optimal solution, *i.e.*,

$$\delta^{(t+1)} = \operatorname*{argmin}_\delta \; g(\delta) + \frac{\lambda}{2}\|\delta - \delta^{(t)}\|^2 \; .$$

This problem is very similar to the one in (6.17) and can be solved by Algorithm 6.5 with minor adjustments.

Smooth and strong

It is possible to obtain a *smooth and strongly convex* objective function by adding an L_2 term to the smooth program given in (6.16) (similarly possible for the soft-max dual in (6.11)). Gradient-based algorithms have *linear convergence rate* in this case, *i.e.*, $O(\frac{1}{\gamma\lambda}\log\frac{1}{\epsilon})$ [49, 44]. Equivalently, one can add an L_2 term to the primal in (6.18). For more details, see Meshi et al. [44]. To conclude this section, we summarize objective functions and algorithms in Table 6.1.

Algorithm 6.6 Max-product belief propagation with a factor graph

1: Initialize: $\delta_{r\to i}(x_i) = 0$ and $\delta_{i\to r}(x_i) = 0$ for all $r, i \in C(r), x_i$,
2: **for** $t = 1, \ldots, T$ **do**
3: Pass a message from variable i to factor r:
4: $\delta_{i\to r}(x_i) = \theta_i(x_i) + \sum_{\substack{r' \in P(i) \\ r' \neq r}} \delta_{r'\to i}(x_i)$,
5: Or, pass a message from factor r to variable i:
6: $\delta_{r\to i}(x_i) = \max_{x_{r\setminus i}} \left(\theta_r(x_r) + \sum_{\substack{j \in C(r) \\ j \neq i}} \delta_{j\to r}(x_j) \right)$
7: **end for**

6.5.7 Other optimization schemes

It is important to note that the list of methods presented here is by no means exhaustive. Due to the importance of the MAP estimation problem many other methods have been proposed. Noteworthy globally convergent methods for solving the LP relaxation include augmented Lagrangian methods [39, 42, 40], bundle methods [25], and a proximal point algorithm applied to the primal LP relaxation [54]. Unfortunately, the convergence rates of some of these methods are still not well understood in the context of the MAP problem, making them hard to compare to other algorithms. For a recent survey covering additional algorithms as well as an empirical comparison of several implementations see Kappes et al. [26].

6.6 Relation to Message-Passing Algorithms

A rich line of work in graphical models is dedicated to developing efficient local message-passing schemes for inference tasks, including MAP estimation. These algorithms date back to Pearl's celebrated *max-product belief-propagation (BP)* algorithm [51], which is a dynamic programming approach to exactly compute the MAP assignment for graphs without cycles. In Algorithm 6.6 we show a version of this algorithm that can be viewed as passing messages along edges of a factor graph [37]. The same updates are sometimes used iteratively even when the graph does contain cycles (known as *loopy BP*), but in that case the procedure is not guaranteed to find the exact optimum and may not even converge (although in practice it often does[9] [47]). It is beyond the scope of this chapter to review convergence conditions of loopy BP. Instead, we refer the reader to work by Ihler et al. [22], Weiss [73], Tatikonda and Jordan [69], Heskes [20], Mooij and Kappen [46] and references therein for a more detailed discussion.

A number of variants on the basic BP algorithm have been proposed. In their seminal work, Yedidia et al. [81] present a *generalized BP* algorithm, which can operate on a region graph representation. They also reveal a correspondence between fixed points of BP and stationary points of an approximate free energy functional resulting from a variational inference framework.

Another seminal work, by Wainwright et al. [72], proposes a variant of BP called *tree-reweighted BP*, which is based on a free energy construction via a distribution over spanning

[9] Convergence of BP may require some numerical caution, for example by applying message damping and normalization.

trees of the original graph. They also discover intriguing connections between message-passing algorithms and the first-order LP relaxation of (6.7). In particular, it is shown that for tree-structured graphs, fixed points of ordinary max-product BP correspond to optimal solutions of a dual LP problem. However, for general loopy graphs this is no longer the case. On the other hand, when using tree-reweighted BP, or in fact any other *convex* BP, then a fixed-point corresponding to the LP solution is guaranteed to exist. Unfortunately, not every fixed-point necessarily corresponds to an LP solution, and (excluding special cases) unless the LP relaxation is tight and messages satisfy certain consistency conditions, it is generally hard to tell whether a given fixed-point corresponds to an LP solution or not [34, 74]. Weiss et al. [74] show that *sum-product convex BP* may be preferable to its max-product counterpart, since with sufficiently small temperatures, the beliefs will indeed approach the relaxed LP solution. This result is closely related to the smoothing approach discussed in Section 6.5.5.

The methods presented in Section 6.5 also have a local nature similar to message-passing algorithms. Despite the fact that we motivated this chapter from the optimization viewpoint, the dual variables δ_{pr} can be interpreted as messages that are passed from a parent region p to region r, locally reparameterizing the score. Intuitively, the dual problem specified in (6.9) attempts to find a reparameterization which ensures that local maximization of the reparameterized score is equivalent to global maximization of the original score. It is therefore not surprising that dynamic programming, i.e., message passing on a tree, involves steps identical to those of solving the corresponding instance of the program given in (6.9). Discovering this intricate relationship drove much of the message-passing research which dates back to early work on Ising models [5].

6.7 Rounding Schemes

Most of the methods discussed above address the LP relaxation rather than the original MAP task, i.e., (6.2). Hence, after solving the relaxation we end up with a solution to either the primal or dual programs. If we manage to obtain an integral optimal solution to the primal LP relaxation then it is guaranteed to be optimal also w.r.t. the integer program, the relaxation is tight, and a MAP assignment can be easily extracted. However, when we have a fractional primal solution at hand, or a solution to the dual LP relaxation, then we may often be interested in extracting from that an approximate MAP configuration. This is called the *rounding*, or *decoding* problem.

Rounding is a natural strategy to retrieve an integral result from a fractional solution of the primal LP relaxation [53]. The quality of the result can then be measured by its MAP value (6.2), which provides a lower bound on the optimal value. Performance guarantees of various rounding schemes applied to certain special cases of MAP problems have been investigated, *e.g.*, by Raghavan and Thompson [53], Kleinberg and Tardos [31], Chekuri et al. [8], Ravikumar et al. [54]. In the following we briefly review deterministic and randomized rounding schemes.

Among the deterministic rounding schemes, node-based rounding is the simplest and most popular procedure. In this case we choose $x_i \in \operatorname{argmax}_{x'_i} \mu_i(x'_i)$. Note that a solution obtained in this manner need not be consistent with the factor terms, *i.e.*, for $i \in r$, $\hat{x}_r \in \operatorname{argmax}_{x'_r} \mu_r(x'_r)$ and $x_i \neq \hat{x}_i$. Rounding schemes taking a more global configuration such as neighboring variables or even entire trees into account have been considered, *e.g.*, by Ravikumar et al. [54]. For a detailed review of optimality conditions see work by Weiss et al. [74], Ravikumar et al. [54].

Assuming every value in the primal solution μ to be unique, deterministic rounding schemes yield identical results when given the same input. A different approach is to use a randomized rounding procedure. In analogy to the deterministic case, one of the easiest approaches is to sample the assignment x_i with probability proportional to $\mu_i(x_i)$. Graph structured generalizations have been considered by Ravikumar et al. [54].

If we obtain a solution to the dual LP relaxation rather than the primal, then there are two options. On the one hand, we can try to map the dual solution to a feasible primal one (*e.g.*, using the algorithm proposed in Meshi et al. [43]). This can then be used to perform rounding as above, however a primal (possibly fractional) solution can be useful in itself since it gives a bound on the suboptimality of the current solution, which provides a natural stopping criterion for optimization algorithms, and can also be used for branch-and-bound procedures (Section 6.3). On the other hand, one can decode an approximate MAP assignment directly from the dual solution. Similar to primal decoding, a simple way to do this is to use the singleton reparameterizations and return $x_i \in \text{argmax}_{x'_i} \hat{\theta}_i^{\delta}(x'_i)$. Natural randomized or graph-based variants of this approach can also be applied.

6.8 Conclusion

In this chapter we reviewed the MAP estimation problem along with several solution strategies. Due to its general intractability we discussed special cases that can be solved exactly in an efficient manner, as well as approximate methods that can be applied more broadly. In general, such approximations explore an inevitable *trade-off between accuracy and efficiency*. For example, one can tighten an LP relaxation by adding constraints to the LP, which in turn increases the runtime of solvers. Similarly, we have shown that it is possible to (slightly) modify the LP objective in order to obtain optimization problems with improved convergence guarantees. Ultimately, the choice of where to be on the accuracy-efficiency trade-off depends on the specifics of an application.

Appendix: The Dual LP Relaxation

In this section we provide the detailed derivation of the Lagrangian dual of the LP in (6.7). Similar derivations can be found in Werner [77], Sontag et al. [67], Komodakis et al. [35].

We begin by formulating the Lagrangian function:

$$
\begin{aligned}
L(\mu, \alpha, \beta, \delta) = & \sum_r \sum_{x_r} \mu_r(x_r)\theta_r(x_r) \\
& + \sum_r \lambda_r \left(1 - \sum_{x_r} \mu_r(x_r)\right) && \text{[normalization]} \\
& + \sum_r \sum_{x_r} \alpha_r(x_r)\mu_r(x_r) && \text{[non-negativity]} \\
& + \sum_r \sum_{x_r} \sum_{p \in P(r)} \delta_{pr}(x_r) \left(\mu_r(x_r) - \sum_{x_{p \setminus r}} \mu_p(x_p)\right) && \text{[marginalization]}
\end{aligned}
$$

Next, we derive the optimality conditions:

$$\frac{\partial L}{\partial \mu_r(x_r)} = \theta_r(x_r) - \lambda_r + \alpha_r(x_r) + \sum_{p \in P(r)} \delta_{pr}(x_r) - \sum_{c \in C(r)} \delta_{rc}(x_c) = 0$$

This yields the dual LP:

$$\min_{\alpha \geq 0, \delta, \lambda} \sum_r \lambda_r \quad \text{s.t.}$$
$$\lambda_r = \theta_r(x_r) + \sum_{p \in P(r)} \delta_{pr}(x_r) - \sum_{c \in C(r)} \delta_{rc}(x_c) + \alpha_r(x_r) \quad \forall x_r$$

Since α does not appear in the objective we can discard it and replace the equality constraints with inequalities:

$$\min_{\delta, \lambda} \sum_r \lambda_r \quad \text{s.t.}$$
$$\lambda_r \geq \theta_r(x_r) + \sum_{p \in P(r)} \delta_{pr}(x_r) - \sum_{c \in C(r)} \delta_{rc}(x_c) \quad \forall x_r$$

Finally, since λ is minimized over, it will achieve the maximum value for all the corresponding constraints. Therefore, this problem simplifies to (6.9).

Bibliography

[1] F. Barahona. On cuts and matchings in planar graphs. *Math. Program.*, 60:53–68, 1993.

[2] F. Barahona and A.R. Mahjoub. On the cut polytope. *Math. Program.*, 36:157–173, 1986.

[3] D.P. Bertsekas. *Nonlinear Programming.* Athena Scientific, Belmont, MA, 1995.

[4] J. Besag. On the Statistical Analysis of Dirty Pictures. *J. Royal Stat. Soc. Series B (Methodological)*, 48(3):259–302, 1986.

[5] H.A. Bethe. Statistical Theory of Superlattices. *Royal Society of London, Series A, Mathematical and Physical Sciences*, 1935.

[6] D. Boukari and A.V. Fiacco. Survey of penalty, exact-penalty and multiplier methods from 1968 to 1993. *Optimization*, 32(4):301–334, 1995.

[7] Y. Boykov, O. Veksler, and R. Zabih. Fast approximate energy minimization via graph cuts. *IEEE Trans Pattern Anal Mach Intell*, 2001.

[8] C. Chekuri, S. Khanna, J. Naor, and L. Zosin. A linear programming formulation and approximation algorithms for the metric labeling problem. *SIAM J. on Discrete Math*, 18(3):608–625, 2005.

[9] A. Cohen, A.G. Schwing, and M. Pollefeys. Efficient Structured Parsing of Facades Using Dynamic Programming. In *Proc. CVPR*, 2014.

[10] G. B. Dantzig. *Origins of the simplex method.* ACM, 1990.

[11] R. Dechter and R. Mateescu. And/or search spaces for graphical models. *Artif Intell*, 171(2-3):73–106, 2007.

[12] R. Dechter and I. Rish. Mini-buckets: A general scheme for bounded inference. *J. ACM*, 50(2):107–153, 2003.

[13] J. Edmonds. Paths, trees, and flowers. *Can J Math*, 17(3):449–467, 1965.

[14] M. Frank and P. Wolfe. *An algorithm for quadratic programming*, volume 3, Wiley Periodicals, 95–110. 1956.

[15] S. Geman and D. Geman. Stochastic relaxation, Gibbs distributions, and the Bayesian restoration of images. *IEEE Trans. PAMI*, 6(6):721–741, 1984.

[16] A. Globerson and T. Jaakkola. Fixing max-product: Convergent message passing algorithms for MAP LP-relaxations. In *NIPS*. MIT Press, 2008.

[17] D.M. Greig, B.T. Porteous, and A.H. Seheult. Exact maximum a posteriori estimation for binary images. *J Royal Stat Soc. Series B (Methodological)*, 51(2):pp. 271–279, 1989.

[18] F. Harary et al. On the notion of balance of a signed graph. *Mich Math J*, 2(2):143–146, 1953.

[19] T. Hazan and A. Shashua. Norm-product belief propagation: Primal-dual message-passing for approximate inference. *IEEE Trans. Inf. Theory*, 56(12):6294–6316, 2010.

[20] T. Heskes. On the uniqueness of loopy belief propagation fixed points. *Neural Comput*, 2004.

[21] F. Hutter, H.H. Hoos, and T. Stützle. Efficient stochastic local search for MPE solving. In *Proceedings of the 19th International Joint Conference on Artificial Intelligence*, IJCAI'05, 2005.

[22] A.T. Ihler, J.W. Fisher III, and A.S. Willsky. Loopy Belief Propagation: Convergence and Effects of Message Errors. *JMLR*, 2005.

[23] J. Johnson. *Convex Relaxation Methods for Graphical Models: Lagrangian and Maximum Entropy Approaches.* PhD thesis, EECS, MIT, 2008.

[24] V. Jojic, S. Gould, and D. Koller. Fast and smooth: Accelerated dual decomposition for MAP inference. In *Proceedings of International Conference on Machine Learning (ICML)*, 2010.

[25] J.H. Kappes, B. Savchynskyy, and C. Schnörr. A bundle approach to efficient MAP-inference by Lagrangian relaxation. In *CVPR*, 2012.

[26] J.H. Kappes, B. Andres, Fred A. Hamprecht, C. Schnörr, S. Nowozin, D. Batra, S. Kim, B. X. Kausler, T. Kröger, J. Lellmann, N. Komodakis, B. Savchynskyy, and C. Rother. A comparative study of modern inference techniques for structured discrete energy minimization problems. *Int J Comput Vision*, 115(2):155–184, 2015.

[27] N. Karmarkar. A new polynomial-time algorithm for linear programming. In *Proceedings of the 16th Annual ACM Symposium on Theory of Computing*, 302–311. ACM, 1984.

[28] R.M. Karp. Reducibility among combinatorial problems. In *Complexity of Computer Computations*, 85–103. Springer, 1972.

[29] K. Kask and R. Dechter. Stochastic local search for Bayesian network. In *AISTATS*, 1999.

[30] S. Kirkpatrick, C.D. Gelatt, and M.P. Vecchi. Optimization by simulated annealing. *Science*, 220:671–680, 1983.

[31] J. Kleinberg and E. Tardos. Approximation algorithms for classification problems with pairwise relationships: Metric labeling and Markov random fields. In *IEEE Symposium on Foundations in Computer Science*, 1999.

[32] D. Koller and N. Friedman. *Probabilistic Graphical Models: Principles and Techniques.* MIT Press, 2009.

[33] V. Kolmogorov. Convergent tree-reweighted message passing for energy minimization. *IEEE T Pattern Anal*, 28(10):1568–1583, 2006.

[34] V. Kolmogorov and M.J. Wainwright. On the optimality of tree-reweighted max-product message-passing. In *Proceedings of the 21st Conference on Uncertainty in Artificial Intelligence*, 316–323, Arlington, Virginia, 2005. AUAI Press.

[35] N. Komodakis, N. Paragios, and G. Tziritas. MRF energy minimization and beyond via dual decomposition. *IEEE Transactions on Pattern Analysis and Machine Intelligence*, 33:531–552, March 2011.

[36] A. Koster, S.P.M. van Hoesel, and A.W.J. Kolen. The partial constraint satisfaction problem: Facets and lifting theorems. *Oper Res Lett*, 23:89–97, 1998.

[37] F.R. Kschischang, B.J. Frey, and H.-A. Loeliger. Factor graphs and the sum-product algorithm. *IEEE T Inform Theory*, 2001.

[38] S. Lacoste-Julien, M. Jaggi, M. Schmidt, and P. Pletscher. Block-coordinate Frank-Wolfe optimization for structural SVMs. In *ICML*, 53–61, 2013.

[39] A.L. Martins, M.A.T. Figueiredo, P.M.Q. Aguiar, N.A. Smith, and E.P. Xing. An augmented Lagrangian approach to constrained MAP inference. In *ICML*, pages 169–176, 2011.

[40] A.F.T. Martins, Mário A.T. Figueiredo, P.M.Q. Aguiar, N.A. Smith, and E.P. Xing. Ad3: Alternating directions dual decomposition for map inference in graphical models. *J Mach Learn Res*, 16:495–545, 2015.

[41] T. Meltzer, A. Globerson, and Y. Weiss. Convergent message passing algorithms: a unifying view. In *UAI*, pages 393–401. AUAI Press, 2009.

[42] O. Meshi and A. Globerson. An alternating direction method for dual MAP LP relaxation. In *ECML*, 2011.

[43] O. Meshi, T. Jaakkola, and A. Globerson. Convergence rate analysis of MAP coordinate minimization algorithms. In *NIPS*, 3023–3031, 2012.

[44] O. Meshi, M. Mahdavi, and A.G. Schwing. Smooth and strong: MAP inference with linear convergence. In *NIPS*, 2015.

[45] O. Meshi, M. Mahdavi, A. Weller, and D. Sontag. Train and test tightness of LP relaxations in structured prediction. In *ICML*, 2016.

[46] J.M. Mooij and H.J. Kappen. Sufficient conditions for convergence of Loopy Belief Propagation. In *Proc. UAI*, 2005.

[47] K.P. Murphy, Y. Weiss, and M.I. Jordan. Loopy belief propagation for approximate inference: an empirical study. In *Proc. UAI*, 1999.

[48] A.S. Nemirovski and D.B. Yudin. *Problem Complexity and Method Efficiency in Optimization.* Wiley, 1983.

[49] Y. Nesterov. *Introductory Lectures on Convex Optimization: A Basic Course*, volume 87. Kluwer Academic Publishers, 2004.

[50] Y. Nesterov. Smooth minimization of non-smooth functions. *Math. Prog.*, 103(1): 127–152, 2005.

[51] J. Pearl. *Probabilistic Reasoning in Intelligent Systems: Networks of Plausible Inference.* Morgan Kaufmann, 1988.

[52] D. Prusa and T. Werner. Universality of the local marginal polytope. In *CVPR*, 1738–1743. IEEE, 2013.

[53] P. Raghavan and C.D. Thompson. Randomized rounding: A technique for provably good algorithms and algorithmic proofs. *Combinatorica*, 1987.

[54] P. Ravikumar, A. Agarwal, and M.J. Wainwright. Message-passing for graph-structured linear programs: Proximal methods and rounding schemes. *JMLR*, 11:1043–1080, 2010.

[55] J. Renegar. A polynomial-time algorithm, based on Newton's method, for linear programming. *Math. Prog.*, 40(1-3):59–93, 1988.

[56] D. Roth. On the hardness of approximate reasoning. *Artif Intell*, 82, 1996.

[57] B. Savchynskyy, S. Schmidt, J. Kappes, and C. Schnorr. A study of Nesterov's scheme for Lagrangian decomposition and MAP labeling. *CVPR*, 2011.

[58] B. Savchynskyy, S. Schmidt, J.H. Kappes, and C. Schnörr. Efficient MRF energy minimization via adaptive diminishing smoothing. In *UAI*, 2012.

[59] M.I. Schlesinger. Syntactic analysis of two-dimensional visual signals in noisy conditions. *Kibernetika*, 4:113–130, 1976.

[60] A. Schrijver. *Combinatorial Optimization: Polyhedra and Efficiency*, volume 24. Springer, 2002.

[61] A.G. Schwing, T. Hazan, M. Pollefeys, and R. Urtasun. Globally Convergent Dual MAP LP Relaxation Solvers using Fenchel-Young Margins. In *Proc. NIPS*, 2012.

[62] A.G. Schwing, T. Hazan, M. Pollefeys, and R. Urtasun. Globally Convergent Parallel MAP LP Relaxation Solver Using the Frank-Wolfe Algorithm. In *Proc. ICML*, 2014.

[63] Y. Shimony. Finding the MAPs for belief networks is NP-hard. *Artif Intell*, 68(2): 399–410, 1994.

[64] D. Sontag and T. Jaakkola. New outer bounds on the marginal polytope. In J.C. Platt, D. Koller, Y. Singer, and S. Roweis, editors, *Adv Neur In 20*, 1393–1400. MIT Press, 2008.

[65] D. Sontag and T. Jaakkola. Tree block coordinate descent for MAP in graphical models. In *AISTATS*, volume 9, 544–551. JMLR: W&CP, 2009.

[66] D. Sontag, T. Meltzer, A. Globerson, T. Jaakkola, and Y. Weiss. Tightening LP relaxations for MAP using message passing. In *UAI*, 503–510, 2008.

[67] D. Sontag, A. Globerson, and T. Jaakkola. Introduction to dual decomposition for inference. In *Optimization for Machine Learning*, pages 219–254. MIT Press, 2011.

[68] B. Taskar, V. Chatalbashev, and D. Koller. Learning associative Markov networks. In *Proc. ICML*. ACM Press, 2004.

[69] S. Tatikonda and M. Jordan. Loopy belief propagation and Gibbs measures. In *Proc. UAI*, 2002.

[70] M. Wainwright and M.I. Jordan. *Graphical Models, Exponential Families, and Variational Inference.* Now Publishers Inc., Hanover, MA, 2008.

[71] M. Wainwright, T. Jaakkola, and A. Willsky. Tree-based reparameterization framework for analysis of sum-product and related algorithms. *IEEE Trans Inf Theory*, 49(5): 1120–1146, 2003.

[72] M. Wainwright, T. Jaakkola, and A. Willsky. MAP estimation via agreement on trees: message-passing and linear programming. *IEEE Trans Inf Theory*, 51(11):3697–3717, 2005.

[73] Y. Weiss. Correctness of local probability propagation in graphical models with loops. *Neural Computation*, 2000.

[74] Y. Weiss, C. Yanover, and T. Meltzer. MAP estimation, linear programming and belief propagation with convex free energies. In *Proceedings of the 23rd Conference on Uncertainty in Artificial Intelligence*, 416–425, Arlington, Virginia, 2007. AUAI Press.

[75] A. Weller, M. Rowland, and D. Sontag. Tightness of LP relaxations for almost balanced models. In *AISTATS*, 2016.

[76] T. Werner. Revisiting the decomposition approach to inference in exponential families and graphical models. Technical Report CTU-CMP-2009-06, Czech Technical University, 2009.

[77] T. Werner. Revisiting the linear programming relaxation approach to Gibbs energy minimization and weighted constraint satisfaction. *IEEE PAMI*, 32(8):1474–1488, 2010.

[78] P. Wolfe. A method of conjugate subgradients for minimizing nondifferentiable functions. *Mathematical Programming Study*, 1975.

[79] S.J. Wright. *Primal-dual interior-point methods.* SIAM, 1997.

[80] C. Yanover, T. Meltzer, and Y. Weiss. Linear programming relaxations and belief propagation–an empirical study. *J. of Mach Learn Res*, 7:1887–1907, 2006.

[81] J.S. Yedidia, W.T. Freeman, and Y. Weiss. Constructing free-energy approximations and generalized belief propagation algorithms. *IEEE Trans Inf Theory*, 51(7):2282–2312, 2005.

7

Sequential Monte Carlo Methods

Arnaud Doucet
Department of Statistics, University of Oxford

Anthony Lee
School of Mathematics, University of Bristol

CONTENTS

7.1 Introduction

Inference in probabilistic graphical models is generally analytically intractable for non-Gaussian, continuous-valued models and too computationally expensive for high-dimensional discrete-valued models. In these scenarios, a popular approach to carry out approximate inference consists of using Monte Carlo techniques, as discussed in Section 5.3.5 of Chapter 5. We focus here on Sequential Monte Carlo (SMC) and some related Markov chain Monte Carlo (MCMC) methods. The chapter is structured with a particular application as its main motivating example: inference in the context of discrete state-space hidden Markov models (HMMs). SMC methods are applicable much more generally, for example to general state-space HMMs and to fairly arbitrary models as outlined in Section 7.7. We have chosen to focus on discrete state-space HMMs to emphasize the important ideas: SMC methods for general state-space HMMs are essentially similar but involve notational complications that we would prefer to avoid.

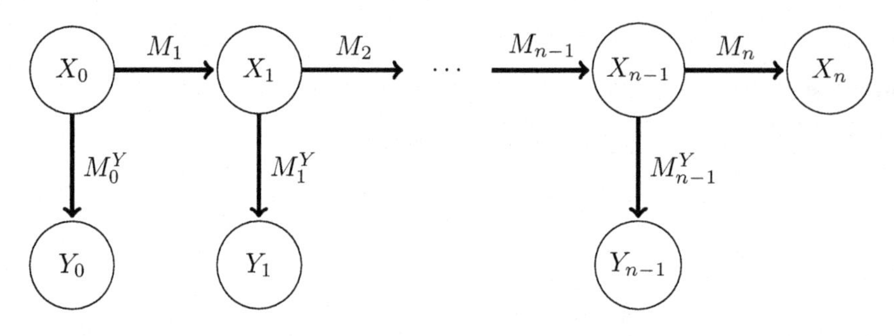

FIGURE 7.1: A graphical model of a hidden Markov model.

The rest of this chapter is organized as follows. In Section 7.2, we introduce HMMs and associated exact inference methods. Section 7.3 describes particle approximations of these inference techniques while Section 7.4 outlines the general principles behind this class of methods and presents a few convergence results. Section 7.5 discusses various extensions to the basic algorithms discussed earlier while Section 7.6 focuses on how to use SMC methods within MCMC schemes. Finally, Section 7.7 briefly describes how these ideas have been applied to a much wider class of graphical models than HMMs. This chapter complements the recent tutorial of Doucet and Johansen [28] which presents in detail more elaborate particle filtering methods but does not discuss recent techniques to parallelize these algorithms, variance estimation schemes or how to combine SMC with MCMC methods.

7.2 Hidden Markov Models

For a countable set E, we call μ a distribution on E if it is a non-negative function $\mu : \mathsf{E} \to [0,1]$ with $\sum_{x\in\mathsf{E}} \mu(x) = 1$. An unnormalized distribution is a distribution where the last requirement is weakened to $0 < \sum_{x\in\mathsf{E}} \mu(x) < \infty$. A Markov transition density is a function $M : \mathsf{E} \times \mathsf{E} \to [0,1]$ where for each $x \in \mathsf{E}$, the function $M(x,\cdot) := y \mapsto M(x,y)$ is a distribution. For an unnormalized distribution μ, we define $\nu = \mu M$ to be the unnormalized distribution satisfying $\nu(i) = \sum_{j\in E} \mu(j)M(j,i)$, so that if E is finite then computing μM is equivalent to vector-matrix multiplication. For a real-valued function f we define $M(f)(x) := \sum_{x'\in\mathsf{E}} M(x,x')f(x')$. For an E-valued random variable X, we write $X \sim \mu$ to mean $\Pr(X = x) = \mu(x)$ for all $x \in \mathsf{E}$. For two real-valued functions f and g with domain E, we denote by $f \cdot g$ the pointwise product $x \mapsto f(x)g(x)$ and f/g the pointwise division $x \mapsto f(x)/g(x)$. For an unnormalized distribution μ and real-valued function f, we define

$$\mu(f) := \sum_{x\in\mathsf{E}} \mu(x)f(x).$$

Finally, if x is a vector and $p, q \in \mathbb{Z}$ we write $x_{p:q} := (x_p, \ldots, x_q)$, with the convention that $x_{p:q} = \emptyset$ if $q < p$.

A discrete state-space HMM, shown graphically in Figure 7.1, is a possibly inhomogeneous bivariate Markov chain where $X_{0:n}$ is itself a Markov chain and $Y_{0:n-1}$ is a vector of random variables with a particular conditional independence structure. Specifically, for $p \in \{0, \ldots, n-1\}$, Y_p is conditionally independent of all other random variables given X_p. The law of $X_{0:n}$ is defined by an initial distribution μ and a sequence of Markov transition densities $M_1, \ldots, M_n$:

$$X_0 \sim \mu \text{ and for } p \in \{1, \ldots, n\}, \quad X_p \mid \{X_{p-1} = x\} \sim M_p(x, \cdot). \tag{7.1}$$

The conditional law of $Y_{0:n-1}$ given $X_{0:n}$ is defined by another sequence of Markov transition densities $M_0^Y, \ldots, M_{n-1}^Y$, with

$$Y_p \mid \{X_p = x\} \sim M_p^Y(x, \cdot), \qquad p \in \{0, \ldots, n-1\}. \tag{7.2}$$

We denote by $\mathbb{P}$ the law of $(X_{0:n}, Y_{0:n-1})$ and by $\mathbb{E}$ expectations w.r.t. $\mathbb{P}$. To keep the notation light, we shall assume that for some set X (resp. Y), the random variables $X_0, \ldots, X_n$ (resp. $Y_0, \ldots, Y_{n-1}$) are all X-valued (resp. Y-valued), where X (resp. Y) is countable. In addition, we will assume for simplicity that $M_p^Y(x, y) > 0$ for all $(x, y) \in \mathsf{X} \times \mathsf{Y}$. The Markov chain $X_{0:n}$ is typically referred to as the hidden or latent process while $Y_{0:n-1}$ is referred to as the observation process: this is because in many statistical applications where HMMs are employed, inference is conducted on the basis of observed values of $Y_{0:n-1}$, while $X_{0:n}$ are unobserved.

Three common statistical questions associated with HMMs can be phrased as follows, where p is an appropriate value in $\{0, \ldots, n\}$:

Q1. What is the conditional distribution of X_p given $Y_{0:p} = y_{0:p}$ or $Y_{0:p-1} = y_{0:p-1}$?

Q2. How likely are the observations $Y_{0:n-1} = y_{0:n-1}$?

Q3. What is the conditional distribution of X_p, for $p < n$, given $Y_{0:n-1} = y_{0:n-1}$?

These questions essentially concern quantities that we would like to calculate given observations. The first corresponds roughly to the prediction or filtering problem, where we seek to deduce the distribution of the random variable X_p conditional upon observations up to time $p-1$ or p; this is important in scenarios where one is interested in tracking the plausible values of X_p over time p as data is collected sequentially. The second is the probability of the observations, and is important for comparing the fit of two or more different HMMs to the data. The third corresponds to the smoothing problem, where we are interested in the distribution of X_p in the light of all data that has been collected up to time $n-1$, and not only up to time $p-1$ or p.

To define the quantities explicitly, we first define a sequence of positive potential functions $G_0, \ldots, G_{n-1}$ by

$$G_p(x) := M_p^Y(x, y_p), \qquad p \in \{0, \ldots, n-1\},$$

and references to the observations $y_{0:n-1}$ can therefore be made implicitly via these functions. To address Q1, we build up recursively a sequence of predictive and filtering distributions: the predictive distribution at time p, which we shall denote η_p, is the conditional distribution of X_p given $Y_{0:p-1} = y_{0:p-1}$ while the filtering distribution at time p, $\hat{\eta}_p$, is the conditional distribution of X_p given $Y_{0:p} = y_{0:p}$. As a convention, we define $\eta_0 := \mu$. From Bayes' rule and the definition of HMMs, for each $p \in \{0, \ldots, n-1\}$ and $x \in \mathsf{X}$,

$$\begin{aligned}
\hat{\eta}_p(x) &:= \mathbb{P}(X_p = x \mid Y_{0:p} = y_{0:p}) \\
&= \frac{\mathbb{P}(X_p = x \mid Y_{0:p-1} = y_{0:p-1})\mathbb{P}(Y_p = y_p \mid X_p = x, Y_{0:p-1} = y_{0:p-1})}{\mathbb{P}(Y_p = y_p \mid Y_{0:p-1} = y_{0:p-1})} \\
&= \frac{\mathbb{P}(X_p = x \mid Y_{0:p-1} = y_{0:p-1})\mathbb{P}(Y_p = y_p \mid X_p = x)}{\sum_{x' \in \mathsf{X}} \mathbb{P}(X_p = x' \mid Y_{0:p-1} = y_{0:p-1})\mathbb{P}(Y_p = y_p \mid X_p = x')} \\
&= \frac{\eta_p(x) G_p(x)}{\sum_{x' \in \mathsf{X}} \eta_p(x') G_p(x')},
\end{aligned}$$

Algorithm 7.1 Exact forward algorithm

1. Set $\eta_0 \leftarrow \mu$ and $Z_0 \leftarrow 1$.
2. For $p = 1, \ldots, n$: set

$$Z_p \leftarrow Z_{p-1}\eta_{p-1}(G_{p-1}), \quad \hat{\eta}_{p-1} \leftarrow \frac{\eta_{p-1} \cdot G_{p-1}}{\eta_{p-1}(G_{p-1})}, \quad \eta_p \leftarrow \hat{\eta}_{p-1}M_p.$$

---

and so

$$\hat{\eta}_p = \frac{\eta_p \cdot G_p}{\eta_p(G_p)}, \qquad p \in \{0, \ldots, n-1\}. \tag{7.3}$$

Similarly, for each $p \in \{1, \ldots, n\}$ and $x \in \mathsf{X}$, the definition of HMMs implies

$$\begin{aligned}
\eta_p(x) &:= \mathbb{P}(X_p = x \mid Y_{0:p-1} = y_{0:p-1}) \\
&= \sum_{x' \in \mathsf{X}} \mathbb{P}(X_{p-1} = x' \mid Y_{0:p-1} = y_{0:p-1})\mathbb{P}(X_p = x \mid X_{p-1} = x', Y_{0:p-1} = y_{0:p-1}) \\
&= \sum_{x' \in \mathsf{X}} \mathbb{P}(X_{p-1} = x' \mid Y_{0:p-1} = y_{0:p-1})\mathbb{P}(X_p = x \mid X_{p-1} = x') \\
&= \sum_{x' \in \mathsf{X}} \hat{\eta}_{p-1}(x')M_p(x', x),
\end{aligned}$$

so we can write

$$\eta_p = \hat{\eta}_{p-1}M_p, \qquad p \in \{1, \ldots n\}. \tag{7.4}$$

Henceforth, we will focus primarily on the predictive distributions $\eta_0, \ldots, \eta_n$, which can be defined recursively via $\eta_0 = \mu$ and

$$\eta_p = \frac{\eta_{p-1} \cdot G_{p-1}}{\eta_{p-1}(G_{p-1})}M_p, \qquad p \in \{1, \ldots, n\}, \tag{7.5}$$

by combining (7.3) and (7.4).

To address Q2, with the conventions $\prod_{i=0}^{-1} v_i = 1$ and $\mathbb{P}(Y_0 = y_0 \mid Y_{0:-1} = y_{0:-1}) = \mathbb{P}(Y_0 = y_0)$, we define the marginal likelihood associated with the first p observations

$$Z_p := \mathbb{P}(Y_{0:p-1} = y_{0:p-1}) = \prod_{q=0}^{p-1} \mathbb{P}(Y_q = y_q \mid Y_{0:q-1} = y_{0:q-1}),$$

for $p \in \{0, \ldots, n\}$. From the decomposition

$$\mathbb{P}(Y_p = y_p \mid Y_{0:p-1} = y_{0:p-1}) = \sum_{x \in \mathsf{X}} \mathbb{P}(Y_p = y_p \mid X_p = x)\mathbb{P}(X_p = x_p \mid Y_{0:p-1} = y_{0:p-1}),$$

we obtain $\mathbb{P}(Y_p = y_p \mid Y_{0:p-1} = y_{0:p-1}) = \eta_p(G_p)$, and so $Z_p = \prod_{q=0}^{p-1} \eta_q(G_q)$. At this point, it is instructive to consider the Forward Algorithm, Algorithm 7.1, which computes exact answers to Q1 and Q2 when X is a finite set. The key step is embodied in (7.5), and the time complexity of the algorithm is $\mathcal{O}(n\,|\mathsf{X}|^2)$, the dominant operation being vector-matrix multiplication.

Finally, to address Q3, we define the smoothing distribution at time p, denoted $\bar{\eta}_p$, which

Algorithm 7.2 Exact backward algorithm

1. Set $\bar{\eta}_n \leftarrow \eta_n$.
2. For $p = n-1, \ldots, 0$: set
$$\bar{\eta}_p \leftarrow \bar{\eta}_{p+1}\bar{M}_p.$$

Algorithm 7.3 Exact backward sampling algorithm

1. Sample $X_n \sim \eta_n$.
2. For $p = n-1, \ldots, 0$: sample $X_p \sim \bar{M}_p(X_{p+1}, \cdot)$.

is the conditional distribution of X_p given $Y_{0:n-1} = y_{0:n-1}$. Using the definition of HMMs, we have $\bar{\eta}_n = \eta_n$ and for each $p \in \{0, \ldots, n-1\}$ and $x \in \mathsf{X}$,

$$\begin{aligned}
\bar{\eta}_p(x) &= \sum_{x' \in \mathsf{X}} \mathbb{P}(X_p = x, X_{p+1} = x' \mid Y_{0:n-1} = y_{0:n-1}) \\
&= \sum_{x' \in \mathsf{X}} \mathbb{P}(X_p = x \mid Y_{0:n-1} = y_{0:n-1}, X_{p+1} = x')\bar{\eta}_{p+1}(x') \\
&= \sum_{x' \in \mathsf{X}} \mathbb{P}(X_p = x \mid Y_{0:p} = y_{0:p}, X_{p+1} = x')\bar{\eta}_{p+1}(x') \\
&= \sum_{x' \in \mathsf{X}} \frac{\hat{\eta}_p(x) M_{p+1}(x, x')}{\eta_{p+1}(x')} \bar{\eta}_{p+1}(x'),
\end{aligned}$$

so we can write $\bar{\eta}_p = \bar{\eta}_{p+1}\bar{M}_p$, where

$$\bar{M}_p(x, x') := \frac{\hat{\eta}_p(x') M_{p+1}(x', x)}{\eta_{p+1}(x)}. \tag{7.6}$$

Algorithm 7.2 can be run after Algorithm 7.1 to compute exact answers to Q3, and together they are known as the Forward-Backward Algorithm, which is an instance of a belief propagation message-passing algorithm (Chapters 4 and 5 in this book discuss this algorithm). Alternatively, one might be interested only in obtaining a sample $X_{0:n}$ from the joint smoothing distribution

$$\mathbb{P}(X_{0:n} = x_{0:n} \mid Y_{0:n-1} = y_{0:n-1}) = \eta_n(x_n) \prod_{p=0}^{n-1} \bar{M}_p(x_{p+1}, x_p) \tag{7.7}$$

using Algorithm 7.3. The expression (7.7) shows that $X_{0:n} = x_{0:n}$ conditioned on $Y_{0:n-1} = y_{0:n-1}$ is a reverse-time inhomogeneous Markov chain of initial distribution η_n and Markov transition densities $\bar{M}_{n-1}, \ldots, \bar{M}_0$ defined in (7.6).

7.3 Particle Filtering and Smoothing

When $|\mathsf{X}|$ is very large, it can be infeasible or impractical to perform Algorithm 7.1 or Algorithm 7.2. When X is a continuous state-space, it is usually only possible to perform exact

Algorithm 7.4 A particle filter

1. Set $Z_0^N \leftarrow 1$, sample $\zeta_0^i \overset{\text{i.i.d.}}{\sim} \mu$ for $i \in \{1, \ldots, N\}$ and set $\eta_0^N \leftarrow \frac{1}{N}\sum_{i=1}^N \delta_{\zeta_0^i}$.
2. For $p = 1, \ldots, n$: set

$$Z_p^N \leftarrow Z_{p-1}^N \eta_{p-1}^N(G_{p-1}), \qquad \hat{\eta}_{p-1}^N \leftarrow \frac{\eta_{p-1}^N \cdot G_{p-1}}{\eta_{p-1}^N(G_{p-1})},$$

sample

$$\zeta_p^i \overset{\text{i.i.d.}}{\sim} \hat{\eta}_{p-1}^N M_p = \frac{\sum_{j=1}^N G_{p-1}(\zeta_{p-1}^j) M_p(\zeta_{p-1}^j, \cdot)}{\sum_{j=1}^N G_{p-1}(\zeta_{p-1}^j)}, \quad i \in \{1, \ldots, N\} \quad (\star)$$

and set $\eta_p^N \leftarrow \frac{1}{N}\sum_{i=1}^N \delta_{\zeta_p^i}$.

---

calculations when the HMM is linear Gaussian as then each η_p is a Gaussian distribution whose mean vector and covariance matrix can be computed using the celebrated Kalman filter. Particle filtering methods provide a way to instead approximate each $\eta_0, \ldots, \eta_n$ through time in a way that is closely related to Algorithm 7.1.

The key idea is that instead of calculating each η_p exactly, they are approximated by a random distribution, or empirical measure, associated with N conditionally independent and identically distributed (i.i.d.) random variables. This idea was developed independently by Stewart and McCarty Jr [70], Gordon et al. [31] and Kitagawa [46]. While there are many variants, some of which will be discussed in the sequel, we now define a particular sequence of particle approximations. Algorithm 7.4 presents a particle filter, where for $x \in \mathsf{X}$, we let δ_x define the distribution satisfying

$$\delta_x(y) = \begin{cases} 1 & x = y, \\ 0 & x \neq y. \end{cases}$$

This algorithm mirrors Algorithm 7.1, but differs in two computationally crucial aspects. First, it replaces the generally intractable $\hat{\eta}_{p-1}$ with its particle approximation $\hat{\eta}_{p-1}^N$ and second, η_p^N is an empirical distribution associated with conditionally i.i.d draws from the generally intractable $\hat{\eta}_{p-1}^N M_p$. The algorithm is also known as the bootstrap particle filter.

Regarding time complexity, the first observation we can make is that the algorithm can be implemented even in cases where $|\mathsf{X}|$ is infinite, and the time complexity is linear in n and increasing in N. Most of the computation is in the simulation of $(\star)$, which involves independently sampling N times from a weighted mixture of N distributions. A straightforward approach to sampling from such a mixture is to split the process into two stages: first one samples a mixture parameter, ζ_{p-1}^k, with probability proportional to $G_{p-1}(\zeta_{p-1}^k)$ and then one samples from $M_p(\zeta_{p-1}^k, \cdot)$. This two-stage process is exactly how the algorithm is implemented in practice and additionally provides a genealogical interpretation of the particles that are produced by the algorithm. For $p \in \{1, \ldots, n\}$ and $i \in \{1, \ldots, N\}$, we denote by A_{p-1}^i the ancestor index of particle ζ_p^i. The procedure is outlined in Algorithm 7.5 and in order for the vector of ancestor indices $(A_{p-1}^1, \ldots, A_{p-1}^N)$ to be simulated in $\mathcal{O}(N)$ time one either uses the inverse transform method combined with a method for generating sorted uniform random numbers [see, e.g., 22, Section V.3] or the Alias algorithm [72, 73]. The former approach lends itself naturally to alternative resampling schemes, mentioned

Algorithm 7.5 Ancestral simulation of $(\star)$

1. For $i \in \{1, \ldots, N\}$, sample

$$A_p^i \overset{\text{i.i.d.}}{\sim} \text{Categorical}\left(G_{p-1}(\zeta_{p-1}^1), \ldots, G_{p-1}(\zeta_{p-1}^N)\right).$$

2. For $i \in \{1, \ldots, N\}$, sample independently $\zeta_p^i \sim M_p(\zeta_{p-1}^{A_{p-1}^i}, \cdot)$.

Algorithm 7.6 A particle smoother

1. Set $\bar{\eta}_n^N \leftarrow \eta_n^N$.
2. For $p = n-1, \ldots, 0$: set

$$\bar{\eta}_p^N \leftarrow \bar{\eta}_{p+1}^N \bar{M}_p^N$$

where

$$\bar{M}_p^N(x, x') = \frac{\hat{\eta}_p^N(x') M_{p+1}(x', x)}{\sum_{z \in \mathsf{X}} \hat{\eta}_p^N(z) M_{p+1}(z, x)} \mathbb{I}\left(x' \in \{\zeta_p^1, \ldots, \zeta_p^N\}\right). \tag{7.8}$$

in Section 7.5.1 below. This gives an overall time complexity of the particle filter in Algorithm 7.4 of $\mathcal{O}(Nn)$. In the sequel we denote by P the law of the random variables produced by Algorithm 7.4, where Algorithm 7.5 is used to implement $(\star)$, and by E expectations w.r.t. P.

The ancestral indices allow us to view Algorithm 7.4 as a kind of evolutionary system where at time $p > 0$ each particle has exactly one parent and at time $p < n$ each particle has some integer number of offspring. Denoting the number of offspring of particle ζ_{p-1}^i by

$$O_{p-1}^i := \sum_{j=1}^N \mathbb{I}(A_{p-1}^j = i),$$

the fact that the ancestor indices $A_{p-1}^1, \ldots, A_{p-1}^N$ are conditionally i.i.d. and categorical implies that $(O_{p-1}^1, \ldots, O_{p-1}^N)$ is a multinomial random variable. For this reason, the algorithm described so far is known as the particle filter with multinomial resampling.

Once one has obtained particle approximations of the filtering distributions, it is easy to obtain particle approximations of the smoothing distributions by replacing each distribution $\hat{\eta}_p$ in Algorithm 7.2 with its corresponding particle approximation $\hat{\eta}_p^N$ as well as η_n by η_n^N. This defines the particle smoother in Algorithm 7.6, whose time complexity is $\mathcal{O}(N^2 n)$. If one is interested only in obtaining an approximate sample from the joint smoothing distribution (7.7), then an analogue of Algorithm 7.3 giving such a sample is provided by Algorithm 7.7, whose time complexity is $\mathcal{O}(Nn)$. It is sometimes possible to reduce this

Algorithm 7.7 A particle path sampler

1. Sample $X_n \sim \eta_n^N$.
2. For $p = n-1, \ldots, 0$: sample $X_p \sim \bar{M}_p^N(X_{p+1}, \cdot)$.

Algorithm 7.8 Sampling an ancestral line

1. Sample $K_n \sim \text{Uniform}(\{1,\ldots,N\})$ and set $X_n \leftarrow \zeta_n^{K_n}$.
2. For $p = n-1,\ldots,0$: set $K_p \leftarrow A_p^{K_{p+1}}$ and $X_p \leftarrow \zeta_p^{K_p}$.

computational complexity using rejection sampling [24]. An alternative when Algorithm 7.5 is performed to implement ($\star$) in Algorithm 7.4 is to use an ancestral line as described in Algorithm 7.8, which has time complexity $\mathcal{O}(n)$ and does not require evaluation of transition densities. The conditional distribution given the particle system alone (i.e., excluding the ancestor indices) of a single path $X_{0:n}$ obtained from Algorithm 7.7 is the same as that associated with Algorithm 7.8, since for $p \in \{0,\ldots,n-1\}$,

$$\mathsf{P}\left(A_p^k = i \mid \zeta_p^1,\ldots,\zeta_p^N,\zeta_{p+1}^k\right) = \frac{G_p(\zeta_p^i)M_p(\zeta_p^i,\zeta_{p+1}^k)}{\sum_{j=1}^N G_p(\zeta_p^j)M_p(\zeta_p^j,\zeta_{p+1}^k)}$$

so that for $p \in \{0,\ldots,n-1\}$,

$$\begin{aligned}\mathsf{P}\Big(\zeta_p^{K_p}=x' \,|\, K_{p+1}=k,\zeta_p^1,\ldots,\zeta_p^N,\zeta_{p+1}^{K_{p+1}}=x\Big) &= \sum_{i=1}^N \mathsf{P}\left(A_p^k = i, \zeta_p^i = x' \mid \zeta_p^1,\ldots,\zeta_p^N,\zeta_{p+1}^k = x\right)\\ &= \bar{M}_p^N(x,x').\end{aligned}$$

We note, however, that the joint distribution of multiple paths obtained by running Algorithm 7.7 multiple times is not the same as that associated with running Algorithm 7.8 multiple times. In particular, the distribution of $X_{0:n}$ defined by Algorithm 7.8 is concentrated on at most N distinct paths, whereas the corresponding distribution defined by Algorithm 7.7 is concentrated on N^n paths. We observe that the space complexity of obtaining a draw from $\bar{\eta}_0^N$ is larger than obtaining a draw from η_n^N: the former requires $\mathcal{O}(Nn)$ space using Algorithm 7.7, or less by using Algorithm 7.8 in combination with techniques to compress the tree of ancestral lineages defined by Algorithms 7.4–7.5 [see 41, for details]. In contrast, constructing η_n^N requires only $\mathcal{O}(N)$ space as the realizations of the random variables $A_{p-1}^1,\ldots,A_{p-1}^N$ and $\zeta_{p-1}^1,\ldots,\zeta_{p-1}^N$ can be discarded at the end of the pth iteration of the for loop in Algorithm 7.4.

7.4 Sequential Monte Carlo

7.4.1 A general construction

Inspection of Algorithms 7.1–7.4 suggests a more abstract foundation than HMM. Indeed, one can ask whether it is actually necessary that the sequence of potential functions correspond to any observation process, and the answer is no. The algorithm is well-defined for any inhomogeneous Markov chain and sequence of positive potential functions. To clarify what we mean, let $G_0,\ldots,G_{n-1}$ be a sequence of positive, upper-bounded functions. Now define $\gamma_0 := \mu$ and for each $p \in \{1,\ldots,n\}$,

$$\gamma_p := (\gamma_{p-1} \cdot G_{p-1})\, M_p. \tag{7.9}$$

By construction, none of the γ_p are the zero function and we can define

$$\eta_p := \frac{\gamma_p}{\gamma_p(1)}, \qquad p \in \{0, \dots, n\}, \tag{7.10}$$

where 1 denotes the constant function $x \mapsto 1$. We see that $\eta_0, \dots, \eta_n$ defines a sequence of distributions, normalized versions of $\gamma_0, \dots, \gamma_n$ respectively. Moreover, we have $\gamma_0(1) = 1$ and can deduce from (7.9) and (7.10) that, for $p \in \{1, \dots, n\}$,

$$\gamma_p(1) = \gamma_{p-1}(1)\,(\eta_{p-1} \cdot G_{p-1})\,M_p(1) = \gamma_{p-1}(1)\eta_{p-1}(G_{p-1}),$$

from which it follows, again from (7.9) and (7.10), that $\gamma_p(1) = \prod_{q=0}^{p-1} \eta_q(G_q) = Z_p$, so $\eta_0 = \mu$ and for $p \in \{1, \dots, n\}$,

$$\eta_p = \frac{(\gamma_{p-1} \cdot G_{p-1})\,M_p}{\gamma_p(1)} = \frac{\gamma_{p-1}(1)}{\gamma_p(1)}\,(\eta_{p-1} \cdot G_{p-1})\,M_p = \frac{\eta_{p-1} \cdot G_{p-1}}{\eta_{p-1}(G_{p-1})} M_p.$$

This additional generality may not seem especially significant, but we will see in Section 7.5.4 that it can be exploited to define approximations of particle approximations of arbitrary distributions. This allows one to perform inference in arbitrary graphical models, in principle, using SMC.

7.4.2 Convergence results

Having defined a more general form of the algorithm we can now provide the basic convergence results that justify the use of Algorithm 7.4 in the regime where n is fixed and $N \to \infty$. We limit ourselves here to the presentation of simple results and refer the interested reader to the book by Del Moral [17] for a comprehensive review of convergence results. To state the results, we define

$$\gamma_n^N := Z_n^N \eta_n^N,$$

which is a particle approximation of the unnormalized distribution γ_n defined by (7.9). We also define

$$\begin{aligned} Q_p(x_{p-1}, x_p) \quad &:= \quad G_{p-1}(x_{p-1})M_p(x_{p-1}, x_p), \quad p \geq 1 \\ Q_{n,n} := \mathrm{Id}, \quad &Q_{p,n} := Q_{p+1} \cdots Q_n, \quad 0 \leq p < n, \end{aligned}$$

and

$$\sigma^2(f) := \sum_{p=0}^{n} v_p(f), \tag{7.11}$$

where

$$v_p(f) := \frac{\eta_p(Q_{p,n}(f)^2)}{\eta_p Q_{p,n}(1)^2} - \eta_n(f)^2.$$

We denote variance by var, almost sure convergence by $\xrightarrow[N\to\infty]{a.s.}$ and weak convergence by $\xrightarrow[N\to\infty]{L}$ as $N \to \infty$, all w.r.t. the law P of the particle filter in Algorithm 7.4. We have the following results, which are implied by Del Moral [17, Theorem 7.4.2 and Section 9.4] and Cérou et al. [11, Theorem 1.4]. The first two statements indicate that $\gamma_n^N(f)$ is an unbiased approximation of $\gamma_n(f)$ for any N, that $\gamma_n^N(f)$ and $\eta_n^N(f)$ converge almost surely to $\gamma_n(f)$ and $\eta_n(f)$, respectively as $N \to \infty$. The third states that a central limit theorem holds for $\gamma_n^N(f)$ and $\eta_n^N(f)$, and that the limiting variances in these central limit theorems are equivalent to asymptotic variances defined by the functional (7.11).

Theorem 7.4.1. *Let $f : \mathsf{X} \to \mathbb{R}$ be bounded. Then, with $\bar{f} = f - \eta_n(f)$,*

1. $\mathsf{E}\left[\gamma_n^N(f)\right] = \gamma_n(f)$.
2. $\gamma_n^N(f) \xrightarrow[N\to\infty]{a.s.} \gamma_n(f)$ and $\eta_n^N(f) \xrightarrow[N\to\infty]{a.s.} \eta_n(f)$.
3. $\sqrt{N}\left[\frac{\gamma_n^N(f)-\gamma_n(f)}{\gamma_n(1)}\right] \xrightarrow[N\to\infty]{L} \mathcal{N}(0,\sigma^2(f))$ and $\sqrt{N}\left[\eta_n^N(f) - \eta_n(f)\right] \xrightarrow[N\to\infty]{L} \mathcal{N}(0,\sigma^2(\bar{f}))$.
4. $N\mathrm{var}\left(\frac{\gamma_n^N(f)}{\gamma_n(1)}\right) \xrightarrow[N\to\infty]{} \sigma^2(f)$ and $N\mathsf{E}\left[\left\{\eta_n^N(f) - \eta_n(f)\right\}^2\right] \xrightarrow[N\to\infty]{} \sigma^2(\bar{f})$.

We can interpret the asymptotic variance terms $v_p(f)$ in the HMM setting, observing that

$$\gamma_n(x) = \mathbb{P}(X_n = x, Y_{0:n-1} = y_{0:n-1}),$$

and if we define $g_n(x) = 1$,

$$g_p(x) := \frac{\mathbb{P}(Y_{p:n-1} = y_{p:n-1} \mid X_p = x)}{\mathbb{P}(Y_{p:n-1} = y_{p:n-1} \mid Y_{0:p-1} = y_{0:p-1})}, \qquad p \in \{0,\ldots,n-1\},$$

$f_n(x) = f(x)$ and

$$f_p(x) := \mathbb{E}\left[f(X_n) \mid X_p = x, Y_{p+1:n-1} = y_{p+1:n-1}\right], \qquad p \in \{0,\ldots,n-1\},$$

then

$$v_p(f) = \frac{\eta_p(Q_{p,n}(f)^2)}{\eta_p Q_{p,n}(1)^2} - \eta_n(f)^2 = \mathbb{E}\left[g_p(X_p)^2 f_p(X_p)^2 \mid Y_{0:p-1} = y_{0:p-1}\right] - \eta_n(f)^2. \quad (7.12)$$

This expression is fairly instructive when considering the asymptotic variance functional (7.11). In particular, it tells us that SMC methods in HMMs have particularly nice properties when the underlying HMM exhibits specific types of weak dependency. For example, under strong but standard assumptions about weak dependence in HMMs there exists $\rho \in (0,1)$ such that for bounded f satisfying $\eta_n(f) = 0$, $f_p(x)^2 \le C_f \rho^{n-p}$ for all $p \in \{0,\ldots,n\}$. Second, there exists some $C_g > 0$ such that for all $p \in \{0,\ldots,n-1\}$, $g_p(x)^2 \le C_g$. In such cases, we can then consider $\sigma^2(f)$ through (7.12). First, recall that $Z_n = \gamma_n(1)$, so the asymptotic relative variance of the approximation Z_n^N of Z_n is

$$\sigma^2(1) = \sum_{p=0}^{n} \left\{\mathbb{E}\left[g_p(X_p)^2 \mid Y_{0:p-1} = y_{0:p-1}\right] - 1\right\} \le n(C_g - 1),$$

which grows only linearly with n. Now, consider the estimate $\eta_n^N(f)$ of $\eta_n(f)$ for bounded f, and without loss of generality assume that $\eta_n(f) = 0$. We see that

$$\begin{aligned}\sigma^2(f) &= \sum_{p=0}^{n} \mathbb{E}\left[g_p(X_p)^2 f_p(X_p)^2 \mid Y_{0:p-1} = y_{0:p-1}\right] \\ &\le C_g \sum_{p=0}^{n} C_f \rho^{n-p} \le C_g C_f/(1-\rho),\end{aligned}$$

so the asymptotic variance is bounded independent of n.

Much stronger results of this type are formalized rigorously in Cérou et al. [11] and Del Moral [17], and recent extensions include Whiteley [75], Bérard et al. [5], Douc et al. [25].

In particular it is known that, under certain assumptions, the non-asymptotic mean-squared error of $\eta_n^N(f)$ is uniformly bounded in n and of order $\mathcal{O}(N^{-1})$ for bounded functions and the non-asymptotic relative variance of Z_n^N can be controlled by taking N proportional to n. These phenomena demonstrate that, in favorable scenarios, SMC methods outperform simpler approaches like importance sampling by orders of magnitude. An importance sampling approach can be viewed as a modification of Algorithm 7.4 in which $\zeta_p^i \overset{\text{ind}}{\sim} M_p(\zeta_{p-1}^i, \cdot)$ for all $i \in \{1, \ldots, N\}$, instead of being distributed according $\hat{\eta}_{p-1}^N M_p$: the particle approximation of η_n becomes then a *weighted* empirical measure; see Section 7.5.3 below for more details. This lack of interaction can severely compromise the quality of the particle approximations; for example, the relative variance of the corresponding importance sampling approximation of Z_n typically grows exponentially in n, even in the favorable scenarios mentioned above.

7.4.3 Variance estimation

SMC methods are computationally intensive, so in many practical applications one simulates a single realization of $\gamma_n^N(f)$ or $\eta_n^N(f)$ and uses these to approximate $\gamma_n(f)$ or $\eta_n(f)$. Assessing empirically the adequacy of these approximations is non-trivial: the dependent nature of the random variables produced by the algorithm precludes a simple sample variance calculation. In addition, the asymptotic variance functional σ^2 defined in (7.11) is not tractable in general, which complicates the construction of asymptotic confidence intervals. We mention here recent developments that address these variance estimation issues.

Key random variables appearing in the variance estimator expressions below are the indices associated with ancestral lineages defined by Algorithms 7.4–7.5. That is, associated with each random variable ζ_n^i is a time 0 ancestor $\zeta_0^{E_n^i}$ where $E_0^i = i$ and for $p \in \{1, \ldots, n\}$, $E_p^i = E_{p-1}^{A_{p-1}^i}$. The first consistent estimator of $\sigma^2(f - \eta_n(f))$ that can be computed as a by-product of running Algorithm 7.4 was proposed by Chan and Lai [12],

$$\frac{1}{N}\left[\sum_{i=1}^{N}\left\{\sum_{j:E_n^j=i} f(\zeta_n^j) - \eta_n^N(f)\right\}^2\right]. \tag{7.13}$$

An alternative approach, based on identifying an analogue of the sample variance in the SMC setting, was proposed by Lee and Whiteley [48] who defined

$$V_n^N(f) := \eta_n^N(f)^2 - \left(\frac{N}{N-1}\right)^n \left[\frac{N}{N-1}\eta_n^N(f)^2 - \frac{1}{N(N-1)}\sum_{i=1}^{N}\left\{\sum_{j:E_n^j=i} f(\zeta_n^j)\right\}^2\right],$$

which satisfies

$$\mathsf{E}\left[\gamma_n^N(1)^2 V_n^N(f)\right] = \text{var}\left[\gamma_n^N(f)\right], \qquad N V_n^N(f) \xrightarrow[N\to\infty]{P} \sigma^2(f),$$

and

$$N V_n^N(f - \eta_n^N(f)) \xrightarrow[N\to\infty]{P} \sigma^2(f - \eta_n(f)),$$

where $\xrightarrow[N\to\infty]{P}$ denotes convergence in probability as $N \to \infty$. The expression $NV_n^N(f - \eta_n^N(f))$ can be simplified, since $\eta_n^N(f - \eta_n^N(f)) = 0$, giving

$$N V_n^N(f - \eta_n^N(f)) = \left(\frac{N}{N-1}\right)^n \frac{1}{(N-1)}\left[\sum_{i=1}^{N}\left\{\sum_{j:E_n^j=i} f(\zeta_n^j) - \eta_n^N(f)\right\}^2\right], \tag{7.14}$$

which is a deterministic and asymptotically negligible modification of the estimator of Chan and Lai [12]. It is also possible to estimate the individual terms $v_p(f)$ appearing in (7.11) on the basis of a single SMC algorithm [48], and to obtain unbiased and consistent estimates of var $\left[\gamma_n^N(f)\right]$ by running a different SMC algorithm [47].

We observe that both (7.13) and (7.14) will be 0 if $E_n^1 = \cdots = E_n^N$, an event that will occur eventually with probability 1 as n increases with N fixed, irrespective of the mean-squared error of $\eta_n^N(f)$ as an approximation of $\eta_n(f)$. To counteract this phenomenon, Olsson and Douc [63] propose a fixed-lag modification of (7.13), and show that under assumptions of weak dependence in HMMs similar to those described in Section 7.4.2 their fixed-lag variance estimator is able to accurately approximate the asymptotic mean-squared error of $\eta_n^N(f)$ for values of N such that (7.13) and (7.14) would both be inaccurate.

7.5 Methodological Innovations

We present in this section a number of modifications of the SMC methodology described so far, most of which are used in practice to reduce the asymptotic variance associated with the SMC approximations, to make the algorithm more suitable for parallel or distributed implementation, or to extend the scope of the algorithm to almost arbitrary settings.

7.5.1 Resampling schemes

One way in which the algorithm can be modified is to allow $(A_{p-1}^1, \ldots, A_{p-1}^N)$ to be dependent so that $\left(\zeta_p^1, \ldots, \zeta_p^N\right)$ are no longer conditionally i.i.d. This leads to the offspring vector $(O_{p-1}^1, \ldots, O_{p-1}^N)$ no longer being multinomially distributed. In practice, a few resampling schemes are popular and can all be viewed as a variation on how the inverse transform method is used to generate $(A_{p-1}^1, \ldots, A_{p-1}^N)$. More precisely, in Algorithm 7.5 we can set $A_{p-1}^i = F_{p-1}^{-1}(U_{p-1}^i)$ for $i \in \{1, \ldots, N\}$, where $U_{p-1}^1, \ldots, U_{p-1}^N$ are independent and uniformly distributed on $(0,1]$ and F_{p-1}^{-1} is the generalized inverse distribution function of the Categorical distribution defined therein. Alternative resampling schemes correspond to making $U_{p-1}^1, \ldots, U_{p-1}^N$ statistically dependent. For example, stratified resampling partitions the interval $(0,1]$ into sub-intervals $((k-1)/N, k/N]$ for $k = 1, \ldots, N$ and samples U_{p-1}^k uniformly on the k'th sub-interval [46] while systematic resampling draws a single random variable U_{p-1}^1 uniformly on $(0, 1/N]$ then sets $U_{p-1}^k = U_{p-1}^1 + (k-1)/N$ for $k = 2, \ldots, N$ [10]. These schemes are easy to implement and satisfy like multinomial resampling

$$\mathsf{E}(O_{p-1}^i = k) = \frac{N G_{p-1}(\zeta_{p-1}^k)}{\sum_{j=1}^N G_{p-1}(\zeta_{p-1}^j)}, \qquad i, k \in \{1, \ldots, N\}. \tag{7.15}$$

Under property (7.15), $\mathsf{E}\left[\gamma_n^N(f)\right] = \gamma_n(f)$ holds, which is key to ensure the validity of the methods presented in Section 7.6. More detailed theoretical properties of these resampling schemes can be found in [23]. More recently, a resampling scheme that takes not only the potentials but also the values of the particles into account has been devised using optimal transport techniques [67], although it has time complexity in $\mathcal{O}(N^3)$. Related ideas are at play in the resampling mechanisms proposed by Gerber and Chopin [29] and Jacob et al. [42].

7.5.2 Auxiliary particle filters

The general construction of SMC algorithms in Section 7.4.1 involves defining the unnormalized distribution γ_n via an initial distribution μ and sequences of Markov transition densities and positive potential functions $(M_p, G_{p-1})_{p=1}^n$. It should come as no surprise that there are multiple such sequences that define the same γ_n, and this permits one to ask the question: which sequences are optimal in terms of the behavior of the particle approximations? To keep the presentation simple, and make clear precisely what we mean by optimal, we consider the particle approximation Z_n^N of Z_n and seek to define an SMC algorithm for which this approximation is exact.

We consider a modified SMC algorithm defined by a sequence of positive functions $\boldsymbol{\psi} = (\psi_0, \ldots, \psi_n)$ that transform the ingredients described above. That is, we define an initial distribution $\mu^{\boldsymbol{\psi}}$, and sequences of Markov transition densities and positive potential functions $(M_p^{\boldsymbol{\psi}}, G_{p-1}^{\boldsymbol{\psi}})_{p=1}^n$. We define

$$\mu^{\boldsymbol{\psi}} := \frac{\mu \cdot \psi_0}{\mu(\psi_0)}, \qquad G_0^{\boldsymbol{\psi}}(x) := \frac{G_0(x) M_1(\psi_1)(x)}{\psi_0(x)} \mu(\psi_0),$$

and for $p \in \{1, \ldots, n-1\}$,

$$M_p^{\boldsymbol{\psi}}(x, x') := \frac{M_p(x, x')\psi_p(x')}{M_p(\psi_p)(x)}, \qquad G_p^{\boldsymbol{\psi}}(x) = \frac{G_p(x) M_{p+1}(\psi_{p+1})(x)}{\psi_p(x)},$$

while $M_n^{\boldsymbol{\psi}}(x, x') = M_n(x, x')$ for all $x, x' \in \mathsf{X}$. Straightforward calculations show then that

$$\gamma_0^{\boldsymbol{\psi}} = \frac{\gamma_0 \cdot \psi_0}{\mu(\psi_0)}, \qquad \gamma_p^{\boldsymbol{\psi}} = \gamma_p \cdot \psi_p, \qquad p \in \{1, \ldots, n\},$$

and in particular $\gamma_n^{\boldsymbol{\psi}} = \gamma_n$ for any sequence of positive functions $\boldsymbol{\psi}$ such that $\psi_n \equiv 1$. Moreover, if the functions chosen satisfy

$$\psi_p(x) \propto Q_{p,n}(1)(x), \qquad p \in \{0, \ldots, n\}, \qquad x \in \mathsf{X},$$

then one obtains that the SMC approximation of $Z_n^{\boldsymbol{\psi}}$ is almost surely equal to Z_n. In the context of HMMs, this corresponds to choosing

$$\psi_p(x) \propto \mathbb{P}\left(Y_{p:n-1} = y_{p:n-1} \mid X_p = x\right), \qquad p \in \{0, \ldots, n-1\},$$

and hence is related to exact look-ahead procedures which are reviewed in Lin et al. [53].

Although the optimal sequence of functions is typically intractable and cannot be used in practical applications, any sequence of positive functions approximating it can be used to define a valid SMC algorithm. A notable example is the fully-adapted auxiliary particle filter proposed in Pitt and Shephard [64], which corresponds to taking

$$\psi_p(x) = \mathbb{P}(Y_p = y_p \mid X_p = x), \qquad p \in \{0, \ldots, n-1\},$$

and so any $\boldsymbol{\psi}$-modulated SMC algorithm can be viewed as a generalization of the auxiliary particle filter. Guarniero et al. [32] present an approach to approximating the optimal sequence of functions iteratively using SMC, adding to a literature on approximating the exact look-ahead procedure [26, 53, 45]. These auxiliary particle filter ideas involve twisting in some sense the Markov transition densities and potential functions defining the SMC algorithm. A related idea is to twist instead the dynamics of the particle system defined by the SMC algorithm [76].

Algorithm 7.9 αSMC

1. Sample $\zeta_0^i \overset{\text{iid}}{\sim} \mu$ for $i \in \{1, \ldots, N\}$.
2. For $p = 1, \ldots, n$: choose α_{p-1}, set

$$W_p^i \leftarrow \sum_{j=1}^{N} \alpha_{p-1}^{ij} G_{p-1}(\zeta_{p-1}^j) W_{p-1}^j, \qquad i \in \{1, \ldots, N\}$$

   and sample

$$\zeta_p^i \overset{\text{ind}}{\sim} \frac{\sum_{j=1}^N \alpha_{p-1}^{ij} G_{p-1}(\zeta_{p-1}^j) W_{p-1}^j M_p(\zeta_{p-1}^j, \cdot)}{W_p^i}, \quad i \in \{1, \ldots, N\}.$$

3. Set $\eta_n^N \leftarrow \sum_{i=1}^N W_n^i \delta_{\zeta_n^i} / \sum_{j=1}^N W_n^j$ and $Z_n^N \leftarrow N^{-1} \sum_{i=1}^N W_n^i$.

7.5.3 Reducing interaction for distributed implementation

An important computational feature of SMC algorithms is that they are well-suited to parallel implementation, i.e. implementation in an architecture with multiple processing cores that have access to shared memory. Indeed, two of the main operations involve evaluating N positive functions and sampling N times from distributions defined by the Markov transition densities, which are embarrassingly parallel. The only operation that presents difficulty in parallel implementation is the simulation of ancestor indices in Algorithm 7.5 [51, 60].

In a distributed computing environment, i.e. where the processing cores are distributed over a network and memory is no longer shared, this interaction step poses more serious computational problems. In particular, one wishes to avoid sending all of the particles over the network as this can lead to computation time being dominated by data transfer. This has lead to the development of SMC algorithms with less interaction than Algorithm 7.4 but that still enjoy the theoretical properties that motivate the use of SMC over importance sampling, as discussed at the end of Section 7.4.2.

Early work in this direction is Bolic et al. [8], which is generalized and theoretically studied in Whiteley et al. [77]; the resulting αSMC approach is described in Algorithm 7.9. The main principle is to modulate the interaction between particles via an implicitly defined sparse Markov transition matrix α_{p-1} on $\{1, \ldots, N\}$ at each time $p \in \{1, \ldots, n\}$, whose (i, j)'th element is denoted α_{p-1}^{ij}. The sparsity of α_{p-1} ensures that ζ_p^i can be simulated by considering only the subset of particles $\{\zeta_{p-1}^j : \alpha_{p-1}^{ij} > 0\}$. This reduced interaction is then counterbalanced by the fact that the particle approximations are weighted. One simple choice is $\alpha_{p-1} = I_N$, the $N \times N$ identity matrix, and if this choice is made for all $p \in \{1, \ldots, n\}$ then the resulting αSMC algorithm corresponds exactly to (sequential) importance sampling. If $\alpha_{p-1} = \mathbf{1}_{1/N}$, the matrix where $\alpha_{p-1}^{ij} = N^{-1}$ for all $i, j \in \{1, \ldots, N\}$, for all $p \in \{1, \ldots, n\}$ then the αSMC algorithm corresponds exactly to Algorithm 7.4. Hence, the choices of α_{p-1} allow the algorithm to use degrees of interaction between these two extremes. Since selection from smaller pools of particles leads to weighted particle approximations, choosing an appropriate sequence of sparse matrices involves balancing in an appropriate sense the tradeoff between degree of interaction and variability of the weights in the approximation. The theoretical guidance provided by Whiteley et al. [77] in this regard is used to inform the development of practical implementations in Lee and Whiteley

[50]. Related work in this area includes [36, 35, 59]. We note that even in a non-distributed context, it can be beneficial to consider choosing $\alpha_{p-1} = I_N$ for some p and $\alpha_{p-1} = \mathbf{1}_{1/N}$ for others. In [57] it is proposed to use $\alpha_{p-1} = \mathbf{1}_{1/N}$ if and only if the random variables $\{W^i_{p-1} G_{p-1}(\zeta^j_{p-1})\}_{i=1}^N$ vary too much according to a specific measure of variability.

Another direction of research, due to Vergé et al. [71], involves viewing an SMC algorithm with N particles as evolving according to an initial distribution $\boldsymbol{\mu}$ and sequence of Markov transition densities $(\mathbf{M}_p)_{p=1}^n$ on X^N, one for each step of the algorithm, and defining an appropriate sequence of potential functions $(\mathbf{G}_p)_{p=0}^{n-1}$ by $\mathbf{G}_p(x_1, \ldots, x_N) = N^{-1} \sum_{i=1}^N G_p(x_i)$. These ingredients are then used to define the "double bootstrap", an SMC algorithm with M particles using $\boldsymbol{\mu}$ and $(\mathbf{M}_p, \mathbf{G}_{p-1})_{p=1}^n$ in which each of the M X^N-valued particles is a collection of N X-valued particles that evolve according to Algorithm 7.4. One benefit in terms of distributed implementation is that blocks of N particles are selected during the resampling step, leading to larger packet sizes being transmitted.

7.5.4 SMC samplers

SMC samplers were proposed by Del Moral et al. [18] to define SMC approximations of an arbitrary target distribution π. This methodology built upon related ideas by Crooks [16] and Neal [62], and the simple but generic variant we describe here was proposed by Gilks and Berzuini [30] and Chopin [13] in slightly different settings. Let μ be some distribution one can obtain i.i.d. samples from and define a sequence of strictly positive unnormalized distributions $\nu_0, \ldots, \nu_n$ where $\nu_0 \propto \mu$ and $\nu_n \propto \pi$. Now let $M_1, \ldots, M_n$ be a sequence of Markov transition densities and $G_0, \ldots, G_{n-1}$ be a sequence of positive functions satisfying

$$\nu_p M_p = \nu_p, \qquad G_{p-1} = \nu_p / \nu_{p-1}, \qquad p \in \{1, \ldots, n\}.$$

The first of these means that each M_p is ν_p-invariant, and the second provides that if we define $\gamma_0 = \nu_0$ then $\gamma_p = \nu_p$ for all $p \in \{0, \ldots, n\}$ by (7.9) and an inductive argument. The inductive step amounts to observing that if $\gamma_{p-1} = \nu_{p-1}$ then

$$\gamma_p = (\gamma_{p-1} \cdot G_{p-1}) \, M_p = (\nu_{p-1} \cdot \nu_p / \nu_{p-1}) \, M_p = \nu_p M_p = \nu_p.$$

One also obtains $\gamma_p(1) = \nu_p(1)/\nu_0(1)$ for all $p \in \{0, \ldots, n\}$. Hence, Theorem 7.4.1 provides that $\gamma_n^N(f)$ and $\eta_n^N(f)$ are respectively approximations of $\nu_n(f)$ and $\pi(f)$ that are accurate for large enough N. The choice of $\nu_1, \ldots, \nu_{n-1}$ is in some sense arbitrary, although typical strategies involve tempering. For example, one can define first $\nu_0 \propto \mu$ and $\nu_n \propto \pi$ and then the intermediate unnormalized distributions via $\nu_p(x) = \nu_0(x)^{1-\beta_p} \nu_n(x)^{\beta_p}$, where $0 < \beta_1 < \cdots < \beta_{n-1} < 1$ is a sequence of inverse "temperatures". Improvements can be obtained in principle, and in practice in some settings [18].

The practical implication of this is that by defining a sequence of distributions $\nu_0, \ldots, \nu_n$ that move from μ to π, as well as Markov transition densities $M_1, \ldots, M_n$ using, e.g., the Metropolis–Hastings algorithm [58, 34], an SMC approximation of an arbitrary distribution π is obtained simply by running Algorithm 7.4. Therefore, SMC methods can be used to construct a family of approximate methods for calculating marginals and likelihoods, and hence complement the methods discussed in Chapter 5 in this book. The accuracy of the resulting approximation for moderate values of N will typically improve if $G_0, \ldots, G_{n-1}$ are not very variable, which can be deduced from the convergence results in Section 7.4.2, and corresponds to ν_{p-1} being close to ν_p in a particular sense for all $p \in \{1, \ldots, n\}$. For example, the sequence $\nu_0, \ldots, \nu_n$ could correspond to the gradual introduction and strengthening of dependencies between random variables in a given graphical model.

7.5.5 Alternative perspectives

The perspective adopted throughout this chapter, particularly the emphasis on sequentially approximating a sequence of distributions of X-valued random variables, is by no means the only useful one. In some settings it is beneficial to define η_p to be a distribution on X_p, $M_p : \mathsf{X}_{p-1} \times \mathsf{X}_p \to [0,1]$ to be a Markov kernel whose source and target differ and G_p a non-negative function with domain X_p. This allows one to define SMC algorithms where, e.g., X_p=X^p and $\eta_p(x_{0:p}) = \mathbb{P}(X_{0:p} = x_{0:p} \mid Y_{0:p-1} = y_{0:p-1})$ for $p \in \{1, \ldots, n\}$, by setting

$$M_p(x_{0:p-1}, x'_{0:p}) = \mathbb{P}(X_p = x'_p \mid X_{p-1} = x_{p-1})\mathbb{I}\left(x'_{0:p-1} = x_{0:p-1}\right),$$

and $G_p(x_{0:p}) = \mathbb{P}(Y_p = y_p \mid X_p = x_p)$. This also serves as a justification of Algorithm 7.8. In this setting there are even more ways to modify the transitions and potential functions $(M_p, G_{p-1})_{p=0}^n$ without changing γ_n or η_n, as in Section 7.5.2; some of these are discussed in [28]. It is also easier in this setting to consider SMC methods for non-Markovian systems.

There are a variety of innovations that aim to reduce the error of approximations delivered by SMC algorithms. One approach to this is a Monte Carlo variance reduction technique known as Rao–Blackwellization, in which one computes some expectations exactly rather than approximating them using random variables. This is discussed in [28]. A recent departure from standard SMC methodology is the use of quasi-random numbers in lieu of random or pseudo-random numbers [29]. This approach requires a substantially different type of analysis, but associated approximations can have better error bounds than standard SMC methods. Finally, the behavior of SMC methods in HMMs often deteriorates rapidly with the dimension of the latent states, i.e. with d when $\mathsf{X} = \mathbb{Z}^d$. Although in some settings approximate look-ahead techniques can be helpful in mitigating the effects of this, in general this remains a significant obstacle to the use of SMC in data assimilation problems. Recently, Rebeschini and Van Handel [65] showed that for spatial models exhibiting a decay of correlations which is a spatio-temporal generalization of the weak dependency properties described in Section 7.4.2, it is possible in principle to design modified SMC methods that exploit a bias-variance tradeoff to obtain a mean-squared error that can be controlled at a computational cost that grows slower than exponentially in d. An alternative approach is to introduce auxiliary intermediate distributions to account for the dimension of the latent states [6]. Finally, it is worth noting that there is no explicit dependence on dimension in the asymptotic variance (7.11), and indeed there are situations where the approximations have dimension-independent error bounds [see, e.g., 7].

7.6 Particle MCMC

In many applications, the initial distribution μ, the Markov transition densities $M_1, \ldots, M_n$ and $M_0^Y, \ldots, M_{n-1}^Y$ of the HMM defined through (7.1)–(7.2) are unknown so we define instead a family of HMMs with a statistical parameter $\vartheta \in \Theta$, that is

$$X_0 \sim \mu_\vartheta \text{ and for } p \in \{1, \ldots, n\}, \quad X_p \mid \{X_{p-1} = x\} \sim M_{p,\vartheta}(x, \cdot),$$

while

$$Y_p \mid \{X_p = x\} \sim M_{p,\vartheta}^Y(x, \cdot), \qquad p \in \{0, \ldots, n-1\}.$$

We write $G_{p,\vartheta}(x) := M_{p,\vartheta}^Y(x, y_p)$ and denote by $\mathbb{P}_\vartheta$ the law of the random variables $(X_{0:n}, Y_{0:n-1})$. Given observed values of $Y_{0:n-1}$, we are then interested in inferring the parameter ϑ and the states $X_{0:n}$. For ease of presentation, we assume that Θ is a countable

---

Algorithm 7.10 Sampling from the marginal MH kernel $K_{\mathrm{MH}}((\theta, x_{0:n}), \cdot)$

---

1. Sample $\theta' \sim Q(\theta, \cdot)$.
2. For parameter $\vartheta = \theta'$, execute Algorithm 7.1 to obtain $Z_{n,\theta'}$ then Algorithm 7.3 to obtain a sample $X'_{0:n} \sim \mathbb{P}_{\theta'}(X_{0:n} = \cdot \mid Y_{0:n-1} = y_{0:n-1})$.
3. With probability
$$1 \wedge \frac{\nu(\theta')\, Q(\theta', \theta)\, Z_{n,\theta'}}{\nu(\theta)\, Q(\theta, \theta')\, Z_{n,\theta}},$$
output $(\theta', X'_{0:n})$. Otherwise, output $(\theta, x_{0:n})$.

---

set. We follow here a Bayesian approach where ϑ is a random variable to which we assign a prior distribution ν. The resulting posterior distribution π of the states $X_{0:n}$ and parameter ϑ is therefore
$$\pi(\theta, x_{0:n}) \propto \nu(\theta)\, \mathbb{P}_\theta(X_{0:n} = x_{0:n}, Y_{0:n} = y_{0:n-1}), \tag{7.16}$$
which implies that the θ-marginal satisfies $\pi_\vartheta(\theta) \propto \nu(\theta)\, Z_{n,\theta}$, where $Z_{n,\theta} := \mathbb{P}_\theta(Y_{0:n-1} = y_{0:n-1})$. We consider asymptotically sampling from π and approximating posterior expectations using MCMC schemes in which SMC algorithms play a key role. In general graphical models, ϑ could be a parameter governing the statistical dependencies between random variables in the model, but we do not pursue this level of generality further here.

An MCMC algorithm on a discrete state-space with target distribution $\varpi : \mathsf{E} \to [0, 1]$ simulates a Markov chain $\xi_0, \xi_1, \ldots$ with invariant distribution ϖ. If the chain is ergodic, i.e. aperiodic and irreducible, then the distribution of ξ_m converges to ϖ as $m \to \infty$ irrespective of the initial distribution of ξ_0. In such cases, ϖ is the unique stationary distribution for the chain and the algorithm provides in an asymptotic sense samples from ϖ. Moreover, a strong law of large numbers holds: $m^{-1} \sum_{i=1}^m f(\xi_i)$ converges almost surely to $\varpi(f)$ as $m \to \infty$ for any $f : \mathsf{E} \to \mathbb{R}$ such that $\varpi(|f|) < \infty$. Hence, we can think of $m^{-1} \sum_{i=1}^m f(\xi_i)$ as an MCMC approximation of $\varpi(f)$ for finite m.

Designing efficient MCMC algorithms with target π defined in (7.16) is challenging especially when n is large; see Andrieu et al. [2, Section 2.3] for a discussion. If it were possible to evaluate pointwise π_ϑ up to a normalizing constant and sample from $\mathbb{P}_\theta(X_{0:n} = \cdot \mid Y_{0:n} = y_{0:n-1})$, we could use an idealized Metropolis–Hastings (MH) algorithm which simulates under weak assumptions an ergodic, π-invariant Markov chain. The transition density K_{MH} of this Markov chain depends on an auxiliary proposal Markov transition density Q; pseudocode for sampling from K_{MH} is presented in Algorithm 7.10. We refer to this algorithm as the marginal MH algorithm because if Step 2 of Algorithm 7.10 is omitted, it is a standard MH algorithm targeting the marginal posterior distribution π_ϑ. Except in scenarios where $\theta \mapsto Z_{n,\theta}$ can be evaluated pointwise, this algorithm cannot be implemented. The particle marginal MH algorithm is an algorithmically simple approximation of the marginal MH algorithm which replaces the intractable terms of the form $Z_{n,\theta}$ by their corresponding particle estimates $Z_{n,\theta}^N$ [2]. It defines a Markov transition density K_{PMMH}^N which can be sampled from as described in Algorithm 7.11. In practice, using Algorithm 7.8 in Step 2 of Algorithm 7.11 is preferable to using Algorithm 7.7 as it is probabilistically equivalent, computationally cheaper and does not require the evaluation of the transition densities $(M_p)_{p=1}^n$, which may not be tractable in some applications.

The particle marginal MH algorithm defines a π-invariant Markov chain for any number of particles N that is ergodic under weak additional assumptions [2, Theorem 4]. This is established by showing that this algorithm is a standard MH algorithm targeting a dis-

Algorithm 7.11 Sampling from the particle marginal Metropolis–Hastings kernel $K^N_{\text{PMMH}}((\theta, Z^N_{n,\theta}, x_{0:n}), \cdot)$

1. Sample $\theta' \sim Q(\theta, \cdot)$.
2. For parameter $\vartheta = \theta'$, execute Algorithm 7.4 to obtain $Z^N_{n,\theta'}$ then Algorithm 7.7 or Algorithm 7.8 to obtain a sample $X'_{0:n}$ approximately distributed according to $\mathbb{P}_{\theta'}(X_{0:n} = \cdot \mid Y_{0:n-1} = y_{0:n-1})$.
3. With probability
$$1 \wedge \frac{\nu(\theta')\, Q(\theta', \theta)\, Z^N_{n,\theta'}}{\nu(\theta)\, Q(\theta, \theta')\, Z^N_{n,\theta}},$$
output $(\theta', Z^N_{n,\theta'}, X'_{0:n})$. Otherwise, output $(\theta, Z^N_{n,\theta}, x_{0:n})$.

Algorithm 7.12 Sampling from the Gibbs kernel $K_{\text{G}}((\theta, x_{0:n}), \cdot)$

1. Sample $X'_{0:n} = x'_{0:n} \sim \mathbb{P}_\theta(X_{0:n} = \cdot \mid Y_{0:n-1} = y_{0:n-1})$.
2. Sample $\theta' \sim \pi(\cdot \mid x'_{0:n})$ and output $(\theta', x'_{0:n})$

tribution on an extended space admitting π as a marginal. A sequential version of this algorithm has been proposed in Chopin et al. [15]. Particle marginal MH is an instance of a pseudo-marginal algorithm, and these have been recently studied in some detail. Properties of the collection of random variables $\{Z^N_{n,\theta}/Z_{n,\theta} : \theta \in \Theta\}$ are critical to such analyses [1, 4, 49, 27, 68, 20], and characterize in some cases the behavior of the Markov chain and/or associated Monte Carlo approximations. Recently, an algorithmic variant of the approach presents the possibility of attempting to control in a specific sense the quantity $Z^N_{n,\theta'}/Z^N_{n,\theta}$ appearing in the acceptance ratio by introducing dependency through auxiliary random variables [21].

The particle marginal MH can be thought of as a particle approximation to the marginal MH kernel. Similarly, it is possible to come up with a particle approximation of the Gibbs sampler (from Section 5.3.5 of this book) targeting the joint posterior distribution of the parameter and states given $y_{0:n-1}$ described in Algorithm 7.12. A naive particle approximation of this Gibbs algorithm obtained by replacing Step 1 of Algorithm 7.12 by a draw approximately distributed according to $\mathbb{P}_\theta(X_{0:n} = \cdot \mid Y_{0:n} = y_{0:n-1})$ obtained by running Algorithm 7.4 then Algorithm 7.7 or 7.8 for parameter θ does not leave π invariant. A valid approximation of this algorithm requires the use of a variant of particle filter known as the conditional particle filter. For parameter θ, the conditional particle filter is a Markov transition kernel on X^{n+1} which is ergodic and admits $\mathbb{P}_\theta(X_{0:n} = \cdot \mid Y_{0:n} = y_{0:n-1})$ as invariant distribution whenever $N \geq 2$ [2, Theorem 5]. The algorithm is presented in Algorithm 7.13. In contrast to Algorithm 7.4, the law of paths produced by Algorithm 7.13 in conjunction with Algorithm 7.7 or Algorithm 7.8 do depend on which of the two path samplers are used. Despite increased computational time, Algorithm 7.13 is in many applications empirically far superior when one executes Algorithm 7.7 in Step 4, as suggested by Whiteley [74], instead of Algorithm 7.8 as initially proposed in Andrieu et al. [2]; see also [54] for an alternative to Step 4. The particle Gibbs sampler proceeds as detailed in Algorithm 7.14, and under very weak assumptions it generates an ergodic Markov chain of invariant distribution

Algorithm 7.13 Sampling from the conditional particle filter $K^N_{\theta,\mathrm{CSMC}}\left(x_{0:n},\cdot\right)$

1. Set $\zeta^1_p \leftarrow x_p$ for $p \in \{0,\ldots,n\}$.
2. Sample $\zeta^i_0 \overset{\text{iid}}{\sim} \mu$ for $i \in \{2,\ldots,N\}$ and set $\eta^N_{0,\theta} \leftarrow \frac{1}{N}\sum_{i=1}^N \delta_{\zeta^i_0}$.
3. For $p = 1,\ldots,n$: set

$$\hat{\eta}^N_{p-1,\theta} \leftarrow \frac{\eta^N_{p-1,\theta}\cdot G_{p-1,\theta}}{\eta^N_{p-1,\theta}\left(G_{p-1,\theta}\right)},$$

sample

$$\zeta^i_p \overset{\text{iid}}{\sim} \hat{\eta}^N_{p-1,\theta}M_{p,\theta} = \frac{\sum_{j=1}^N G_{p-1,\theta}(\zeta^j_{p-1})M_{p,\theta}(\zeta^j_{p-1},\cdot)}{\sum_{j=1}^N G_{p-1,\theta}(\zeta^j_{p-1})}, \qquad i \in \{2,\ldots,N\},$$

and set $\eta^N_{p,\theta} \leftarrow \frac{1}{N}\sum_{i=1}^N \delta_{\zeta^i_p}$.

4. For parameter $\vartheta = \theta$, execute Algorithm 7.7 or Algorithm 7.8 to return a sample $X_{0:n}$.

Algorithm 7.14 Sampling from the particle Gibbs kernel $K^N_{\mathrm{PG}}\left((\theta, x_{0:n}),\cdot\right)$

1. Sample $X'_{0:n} = x'_{0:n} \sim K^N_{\theta,\mathrm{CSMC}}\left(x_{0:n},\cdot\right)$ using Algorithm 7.13.
2. Sample $\theta' \sim \pi(\cdot \mid x'_{0:n})$ and output $(\theta', x'_{0:n})$.

π whenever $N \geq 2$ [2, Theorem 5]. Sharper theoretical results for Algorithms 7.13 and 7.14 are also available [14, 3, 55, 69, 19].

7.7 Discussion

In this chapter, we have mostly limited ourselves to the applications of SMC methods to HMMs but their scope of application, as exemplified by Section 7.5.4, is much broader and they have already found many applications to more general graphical models. For example, SMC methods have been used to sample approximately from the posterior distribution associated with pairwise undirected graphs and estimate the associated marginal likelihood using either the SMC sampler approach [33], SMC approximations of loopy belief propagation when the graphs contain cycles [9, 38, 40, 52] or SMC methods introducing a sequence of auxiliary graphical models of increasing dimension [61]. More recently, more elaborate SMC algorithms based on a divide-and-conquer approach exploiting an auxiliary tree-structure decomposition of the graphical model of interest have been explored in Lindsten et al. [56]. Although these methods can perform empirically well, their properties are not well understood, in contrast to SMC methods for HMMs and would be worth investigating.

Many recent developments in SMC methodology have not been covered in this chapter. An alternative to Bayesian inference for parameter estimation in HMMs outlined in Section 7.6, is to use maximum likelihood estimators based on SMC; many of these are reviewed by Kantas et al. [44]. Novel techniques continue to be developed, e.g. Ionides et al. [39] propose an original SMC method with artificial dynamics on ϑ that approximate the maximum likelihood estimate. Ideas from optimal transport have also been explored, e.g. by Reich [66], Yang et al. [78] and Heng et al. [37], which provide characteristically different types of particle approximations. Sophisticated multilevel Monte Carlo ideas have also been proposed in the SMC context for partially observed diffusions [43].

In closing, we reiterate that the presentation here has focused on discrete state-spaces to avoid heavy notation that obscures the main ideas. SMC methods for general state-spaces are practically identical to those described here, with probability measures replacing distributions and Markov kernels replacing Markov transition densities. It is also unnecessary to insist upon potential functions that are strictly positive and bounded; non-negative and unbounded potential functions do not present any fundamental issues so long as the sequence of distributions $\eta_1, \ldots, \eta_n$ remain well-defined [see, e.g., 17].

Bibliography

[1] C. Andrieu and G. O. Roberts. The pseudo-marginal approach for efficient Monte Carlo computations. *Ann. Statist.*, 37(2):697–725, 2009.

[2] C. Andrieu, A. Doucet, and R. Holenstein. Particle Markov chain Monte Carlo methods. *J. R. Stat. Soc. Ser. B Stat. Methodol.*, 72(3):269–342, 2010.

[3] C. Andrieu, A. Lee, and M. Vihola. Uniform ergodicity of the iterated conditional SMC and geometric ergodicity of particle Gibbs samplers. *Bernoulli*, 24(2):843–872, 2018.

[4] C. Andrieu and M. Vihola. Convergence properties of pseudo-marginal Markov chain Monte Carlo algorithms. *Ann. Appl. Probab.*, 25(2):1030–1077, 2015.

[5] J. Bérard, P. Del Moral, and A. Doucet. A lognormal central limit theorem for particle approximations of normalizing constants. *Electron. J. Probab.*, 19(94):1–28, 2014.

[6] A. Beskos, D. Crisan, A. Jasra, K. K., and Y. Zhou. A stable particle filter in high-dimensions. *Adv. Appl. Probab.*, 49(1):24–48, 2017.

[7] A. Beskos, A. Jasra, E. A. Muzaffer, and A. M. Stuart. Sequential Monte Carlo methods for Bayesian elliptic inverse problems. *Stat. Comput.*, 25(4):727–737, 2015.

[8] M. Bolic, P. M. Djuric, and S. Hong. Resampling algorithms and architectures for distributed particle filters. *IEEE Trans. Signal Process.*, 53(7):2442–2450, 2005.

[9] M. Briers, A. Doucet, and S. S. Singh. Sequential auxiliary particle belief propagation. In *Proceedings of the 7th International Conference on Information Fusion*, volume 1, 705–711, 2005.

[10] J. Carpenter, P. Clifford, and P. Fearnhead. Improved particle filter for nonlinear problems. *IEE Proceedings-Radar, Sonar and Navigation*, 146(1):2–7, 1999.

[11] F. Cérou, P. Del Moral, and A. Guyader. A nonasymptotic theorem for unnormalized Feynman–Kac particle models. *Ann. Inst. H. Poincaré B*, 47(3):629–649, 2011.

[12] H. P. Chan and T. L. Lai. A general theory of particle filters in hidden Markov models and some applications. *Ann. Statist.*, 41(6):2877–2904, 2013.

[13] N. Chopin. A sequential particle filter method for static models. *Biometrika*, 89(3): 539–552, 2002.

[14] N. Chopin and S. S. Singh. On particle Gibbs sampling. *Bernoulli*, 21(3):1855–1883, 2015.

[15] N. Chopin, P. E. Jacob, and O. Papaspiliopoulos. SMC^2: an efficient algorithm for sequential analysis of state space models. *J. R. Stat. Soc. Ser. B Stat. Methodol.*, 75 (3):397–426, 2013.

[16] G. E. Crooks. Nonequilibrium measurements of free energy differences for microscopically reversible Markovian systems. *J. Stat. Phys.*, 90(5-6):1481–1487, 1998.

[17] P. Del Moral. *Feynman-Kac Formulae: Genealogical and Interacting Particle Systems with Applications.* Springer Verlag, 2004.

[18] P. Del Moral, A. Doucet, and A. Jasra. Sequential Monte Carlo samplers. *J. R. Stat. Soc. Ser. B Stat. Methodol.*, 68(3):411–436, 2006.

[19] P. Del Moral, R. Kohn, and F. Patras. On Feynman-Kac and particle Markov chain Monte Carlo models. *Ann. I. H. Poincaré B*, 52(4):1687–1733, 2016.

[20] G. Deligiannidis and A. Lee. Which ergodic averages have finite asymptotic variance? *Ann. Appl. Prob.*, to appear.

[21] G. Deligiannidis, A. Doucet, and M. K. Pitt. The correlated pseudo-marginal method. *arXiv:1511.04992*, 2015.

[22] L. Devroye. *Non-Uniform Random Variate Generation.* Springer Verlag, 1986.

[23] R. Douc, O. Cappé, and E. Moulines. Comparison of resampling schemes for particle filtering. In *Proceedings of the 4th International Symposium on Image and Signal Processing and Analysis*, 64–69, 2005.

[24] R. Douc, A. Garivier, E. Moulines, and J. Olsson. Sequential Monte Carlo smoothing for general state space hidden Markov models. *Ann. Appl. Probab.*, 21(6):2109–2145, 2011.

[25] R. Douc, E. Moulines, and J. Olsson. Long-term stability of sequential Monte Carlo methods under verifiable conditions. *Ann. Appl. Probab.*, 24(5):1767–1802, 2014.

[26] A. Doucet, M. Briers, and S. Sénécal. Efficient block sampling strategies for sequential Monte Carlo methods. *J. Comput. Graph. Statist.*, 15(3):693–711, 2006.

[27] A. Doucet, M. K. Pitt, G. Deligiannidis, and R. Kohn. Efficient implementation of Markov chain Monte Carlo when using an unbiased likelihood estimator. *Biometrika*, 102(2):295–313, 2015.

[28] A. Doucet and A. M. Johansen. A tutorial on particle filtering and smoothing: Fifteen years later. In D. Crisan and B. Rozovsky, editors, *The Oxford Handbook of Nonlinear Filtering*, 656–704. Oxford University Press, 2011.

[29] M. Gerber and N. Chopin. Sequential quasi Monte Carlo. *J. R. Stat. Soc. Ser. B Stat. Methodol.*, 77(3):509–579, 2015.

[30] W. R. Gilks and C. Berzuini. Following a moving target—Monte Carlo inference for dynamic Bayesian models. *J. R. Stat. Soc. Ser. B Stat. Methodol.*, 63(1):127–146, 2001.

[31] N. J. Gordon, D. J. Salmond, and A. F. M. Smith. Novel approach to nonlinear/non-Gaussian Bayesian state estimation. *Radar and Signal Processing, IEE Proceedings F*, 140(2):107–113, 1993.

[32] P. Guarniero, A. M. Johansen, and A. Lee. 2017. The iterated auxiliary particle filter. *J. Amer. Statist. Assoc.*, 112(520):1636–1647.

[33] F. Hamze and N. de Freitas. Hot coupling: A particle approach to inference and normalization on pairwise undirected graphs of arbitrary topology. *Adv Neur In*, 18: 1–8, 2005.

[34] W. K. Hastings. Monte Carlo sampling methods using Markov chains and their applications. *Biometrika*, 57(1):97–109, 1970.

[35] K. Heine and N. Whiteley. Fluctuations, stability and instability of a distributed particle filter with local exchange. *Stoch. Proc. Appl.*, 127(8):2508–2541.

[36] K. Heine, N. Whiteley, A. Taylan Cemgil, and H. Guldas. Butterfly resampling: asymptotics for particle filters with constrained interactions. *arXiv:1411.5876*, 2014.

[37] J. Heng, A. Doucet, and Y. Pokern. Gibbs flow for approximate transport with applications to Bayesian computation. *arXiv:1509.08787*, 2015.

[38] A. T. Ihler and D. A. McAllester. Particle belief propagation. In *Proceedings of AISTATS*, 256–263, 2009.

[39] E. L. Ionides, D. Nguyen, Y. Atchadé, S. Stoev, and A. A. King. Inference for dynamic and latent variable models via iterated, perturbed Bayes maps. *Proc. Natl. Aca. Sci.*, 112(3):719–724, 2015.

[40] M. Isard. Pampas: Real-valued graphical models for computer vision. In *Proceedings of the Conference on CVPR*, volume 1, 613–620, 2003.

[41] P. E. Jacob, L. M. Murray, and S. Rubenthaler. Path storage in the particle filter. *Stat. Comput.*, 25(2):487–496, 2015.

[42] P. E. Jacob, F. Lindsten, and T. B. Schön. Coupling of particle filters. *arXiv:1606.01156*, 2016.

[43] A. Jasra, K. Kamatani, K. J. H. Law, and Y. Zhou. Multilevel particle filter. *SIAM Journ. on Num. Anal*, 55(6):3068–3096, 2017.

[44] N. Kantas, A. Doucet, S. S. Singh, J. Maciejowski, and N. Chopin. On particle methods for parameter estimation in state-space models. *Statist. Sci.*, 30(3):328–351, 2015.

[45] H. J. Kappen and H. C. Ruiz. Adaptive importance sampling for control and inference. *J. Stat. Phys.*, 162(5):1244–1266, 2016.

[46] G. Kitagawa. A Monte Carlo filtering and smoothing method for non-Gaussian nonlinear state space models. In *Proceedings of the 2nd US-Japan Joint Seminar on Statistical Time Series Analysis*, 110–131, 1993.

[47] S. Kostov and N. Whiteley. An algorithm for approximating the second moment of the normalizing constant estimate from a particle filter. *Methodol. Comput. Appl. Probab.*, 19(3), 799–818, 2017.

[48] A. Lee and N. Whiteley. Variance estimation in the particle filter. *Biometrika*, to appear. 2015.

[49] A. Lee and K. Łatuszyński. Variance bounding and geometric ergodicity of Markov chain Monte Carlo kernels for approximate Bayesian computation. *Biometrika*, 101(3): 655–671, 2014.

[50] A. Lee and N. Whiteley. Forest resampling for distributed sequential Monte Carlo. *Stat. Anal. Data Min.*, 9(4):230–248, 2016.

[51] A. Lee, C. Yau, M. B. Giles, A. Doucet, and C. C. Holmes. On the utility of graphics cards to perform massively parallel simulation of advanced Monte Carlo methods. *J. Comput. Graph. Statist.*, 19(4):769–789, 2010.

[52] T. Lienart, Y. W. Teh, and A. Doucet. Expectation particle belief propagation. In *Adv Neur In*, 3609–3617, 2015.

[53] M. Lin, R. Chen, and J. S. Liu. Lookahead strategies for sequential Monte Carlo. *Statist. Sci.*, 28(1):69–94, 2013.

[54] F. Lindsten, M. I. Jordan, and T. B. Schön. Particle Gibbs with ancestor sampling. *J. Mach. Learn. Res.*, 15(1):2145–2184, 2014.

[55] F. Lindsten, R. Douc, and E. Moulines. Uniform ergodicity of the particle Gibbs sampler. *Scand. J. Statist.*, 42(3):775–797, 2015.

[56] F. Lindsten, A. M. Johansen, C. A. Naesseth, B. Kirkpatrick, T. B. Schön, J. Aston, and A. Bouchard-Côté. Divide-and-conquer with sequential Monte Carlo. *J. Comput. Graph. Statist.*, 26(2):445–458, 2017.

[57] J. S. Liu and R. Chen. Blind deconvolution via sequential imputations. *J. Amer. Statist. Assoc.*, 90:567–576, 1995.

[58] N. Metropolis, A. W. Rosenbluth, M. N. Rosenbluth, A. H. Teller, and E. Teller. Equation of state calculations by fast computing machines. *J. Chem. Phys.*, 21(6): 1087–1092, 1953.

[59] J. Míguez and M. A. Vázquez. A proof of uniform convergence over time for a distributed particle filter. *Signal Process*, 122:152–163, 2016.

[60] L. M. Murray, A. Lee, and P. E. Jacob. Parallel resampling in the particle filter. *J. Comput. Graph. Statist.*, 25(3):789–805, 2016.

[61] C. A. Naesseth, F. Lindsten, and T. B. Schön. Sequential Monte Carlo for graphical models. In *Adv Neur Inf*, 1862–1870, 2014.

[62] R. M. Neal. Annealed importance sampling. *Stat. Comput.*, 11(2):125–139, 2001.

[63] J. Olsson and R. Douc. Numerically stable online estimation of variance in particle filters. *Bernoulli*, to appear.

[64] M. K. Pitt and N. Shephard. Filtering via simulation: Auxiliary particle filters. *J. Amer. Statist. Assoc.*, 94(446):590–599, 1999.

[65] P. Rebeschini and R. Van Handel. Can local particle filters beat the curse of dimensionality? *Ann. Appl. Probab.*, 25(5):2809–2866, 2015.

[66] S. Reich. A dynamical systems framework for intermittent data assimilation. *BIT Numerical Mathematics*, 51(1):235–249, 2011.

[67] S. Reich. A nonparametric ensemble transform method for Bayesian inference. *SIAM J Sci Comput*, 35(4):A2013–A2024, 2013.

[68] C. Sherlock, A. H. Thiery, G. O. Roberts, and J. S. Rosenthal. On the efficiency of pseudo-marginal random walk Metropolis algorithms. *Ann. Statist.*, 43(1):238–275, 02 2015.

[69] S. S. Singh, F. Lindsten, and E. Moulines. Blocking strategies and stability of particle Gibbs samplers. *Biometrika*, 104(4):953–969, 2015.

[70] L. Stewart and P. McCarty Jr. Use of Bayesian belief networks to fuse continuous and discrete information for target recognition, tracking, and situation assessment. In *Aerospace Sensing*, 177–185. International Society for Optics and Photonics, 1992.

[71] C. Vergé, C. Dubarry, P. Del Moral, and E. Moulines. On parallel implementation of sequential Monte Carlo methods: the island particle model. *Stat. Comput.*, 25(2): 243–260, 2015.

[72] A. J. Walker. New fast method for generating discrete random numbers with arbitrary frequency distributions. *Electronics Letters*, 10(8):127–128, 1974.

[73] A. J. Walker. An efficient method for generating discrete random variables with general distributions. *ACM Trans. Math. Software*, 3(3):253–256, 1977.

[74] N. Whiteley. Discussion of Particle Markov chain Monte Carlo methods. *J. R. Stat. Soc. Ser. B Stat. Methodol.*, 72(3):306–307, 2010.

[75] N. Whiteley. Stability properties of some particle filters. *Ann. Appl. Probab.*, 23(6): 2500–2537, 2013.

[76] N. Whiteley and A. Lee. Twisted particle filters. *Ann. Statist.*, 42(1):115–141, 2014.

[77] N. Whiteley, A. Lee, and K. Heine. On the role of interaction in sequential Monte Carlo algorithms. *Bernoulli*, 22(1):494–529, 2016.

[78] T. Yang, P. G. Mehta, and S. P. Meyn. Feedback particle filter. *IEEE Trans. Automat. Control*, 58(10):2465–2480, 2013.

Part III

Statistical inference

8

Discrete Graphical Models and Their Parameterization

Luca La Rocca
Department of Physics, Informatics and Mathematics, University of Modena and Reggio Emilia

Alberto Roverato
Department of Statistical Sciences, University of Bologna

CONTENTS

8.1 Introduction

This chapter is devoted to graphical models in which the observed variables are *categorical*, that is, whose state space consists of a finite number of values. Although the interpretation of a graphical model and the inferential questions of interest usually do not depend on the kind of variables involved in the analysis, the effective specification of the problems and the techniques for statistical analysis differ from case to case. An essential ingredient for the implementation of statistical techniques is the specification of a suitable parameterization of the model, and indeed the present chapter is mostly focused on this issue.

The restrictions characterizing a graphical model reflect the independence relationships stated by a graph through a Markov property. The joint distribution of a vector of categorical variables can be given in the form of a probability mass function on the cells of a cross-classified table. Cell probabilities are easy to interpret, but they have the drawback that independence relationships involve nonlinear constraints on these parameters. Hence, research has focused on the development of alternative parameterizations, which may allow one to deal more effectively with the restrictions defining the model. A number of features should be considered when comparing different parameterizations, but it is typically easier to deal with parameterizations where submodels of interest correspond to linear subspaces of the parameter space of the saturated model. Indeed, it is desirable that any constraint of interest corresponds to the vanishing of certain parameters. Another desirable feature of a parameterization is *variation independence*, which holds when the permissible values of any parameter are not affected by the value of the others. Furthermore, with specific reference to categorical variables, the existence of a closed form expression to compute the cell probabilities from the parameters can be considered an advantage.

A general difficulty when dealing with categorical variables is that the dimension of the parameter space increases exponentially with the number of variables. In this respect, one may wish to follow the *principle of parsimony*, requiring that models should have as few parameters as possible. The implementation of this principle in the context of graphical models requires the specification of additional constraints with respect to those encoded by the relevant independence graph; this produces a non-graphical model and raises an issue concerning the interpretation of the additional constraints. When interest lies in independence relationships, it is somehow natural to look at *context specific independencies*, i.e. conditional independencies that are not satisfied for every value of the conditioning variables but only hold in specific contexts; see [7, 8, 27, 29, 44] and references therein.

The following is an annotated outline of the chapter. After introducing notation and terminology (Section 8.2), we start with an overview of discrete graphical models focused on chain graph models (recall Chapter 2) and, more specifically, on the role played by the class of regression graph models within the existing families of graphical models (Section 8.3). We focus on regression graph models because this family of models allows us to approach discrete graphical models with a sufficient degree of generality. Indeed, regression graph models include as special cases the most relevant families of graphical models, that is, undirected, bidirected and directed acyclic graph models. The chapter continues with its core contents. We first give some general properties concerning conditional independence and Möbius inversion (Section 8.4). Next, we exploit these basic results to provide a unified approach to the parameterization of undirected, bidirected and regression graph models (Sections 8.5, 8.6 and 8.7). The three parameterizations are derived by applying the same Möbius inversion formula to three different vectors of log-probabilities. In the undirected case this procedure leads to the usual corner-constrained parameterization of the class of *log-linear models* for contingency tables; see e.g. Agresti [1]. This is the standard parameterization of undirected graph models, and we show that some well-known properties of this parameterization, such as the connection between vanishing terms and independence relationships, as well as the capability of defining context specific independencies, follow directly from the constructing procedure. In this way, we obtain that these properties automatically hold true also for other parameterizations based on the same constructing procedure. We exploit this feature to present the theory of the three classes of models in a common framework. For notational simplicity, we present the material for the special case of binary data. The extension of our approach to the case of categorical variables with an arbitrary number of levels is briefly discussed at the end of the chapter, together with likelihood inference for the models examined (Section 8.8).

8.2 Notation and Terminology

Let $X_V = (X_v)_{v \in V}$ be a vector of categorical random variables indexed by a finite set V. We denote by $\mathcal{I}_v = \{0, 1, \ldots, \ell_v - 1\}$ the state space of X_v, for $v \in V$, and by X_A the subvector of X_V indexed by the non-empty subset $A \subseteq V$. Hence, the marginal vector X_A

takes values in the *cross-classified table* $\mathcal{I}_A = \times_{v \in A} \mathcal{I}_v$, and each element $i_A \in \mathcal{I}_A$ is called a *cell* of the table. In order to lighten notation, we drop brackets and commas from the subscripts so that, for example, if $A = \{u, v, w\}$ we write $X_A = X_{uvw}$.

We restrict our attention to the binary case so that $\ell_v = 2$ and, for any non-empty margin $A \subseteq V$, we can write $\mathcal{I}_A = \{\epsilon_{A'}\}_{A' \subseteq A}$, where $(\epsilon_{A'})_v = 1$ if $v \in A'$ and $(\epsilon_{A'})_v = 0$ if $v \notin A'$. The distribution of X_A is then described by the probability mass function

$$\mathrm{P}(X_A = \epsilon_{A'}) = \mathrm{P}(X_{A'} = 1, X_{A \setminus A'} = 0), \quad \text{for every } A' \subseteq A,$$

where 1 and 0 denote suitably indexed vectors of ones and zeros. As a limiting case, we also consider the empty margin $\emptyset$, reading $\mathrm{P}(X_\emptyset = \epsilon_\emptyset) = 1$ and $\mathrm{P}(X_A = \epsilon_{A'}, X_\emptyset = \epsilon_\emptyset) = \mathrm{P}(X_{A \cup \emptyset} = \epsilon_{A' \cup \emptyset}) = \mathrm{P}(X_A = \epsilon_{A'})$ for every $A' \subseteq A$.

We assume a positive distribution for X_V, that is, we let $\mathrm{P}(X_V = \epsilon_D) > 0$ for every $D \subseteq V$. Under this assumption, which paves the way for using logarithms, all conditional probabilities $\mathrm{P}(X_A = \epsilon_{A'} \mid X_B = \epsilon_{B'})$ are uniquely defined, if A and B are non-empty disjoint subsets of V, while $A' \subseteq A$ and $B' \subseteq B$. Then, on grounds of convention, we interpret $\mathrm{P}(X_A = \epsilon_{A'} \mid X_\emptyset = \epsilon_\emptyset)$ as $\mathrm{P}(X_A = \epsilon_{A'})$ for every $A \neq \emptyset$ and $A' \subseteq A$. If one is interested in dropping the positive distribution assumption, implicit model descriptions can be useful; see Geiger, Meek and Sturmfels [24].

The variables forming X_V are in one-to-one correspondence with the vertices of a graph $\mathcal{G}$, which therefore has *vertex set* V. We consider graphs $\mathcal{G} = (V, \mathcal{U}, \mathcal{B}, \mathcal{D})$ with three sets of edges: a set $\mathcal{U}$ of *undirected edges* (—), a set $\mathcal{B}$ of *bidirected edges* ($\leftrightarrow$), and a set $\mathcal{D}$ of *directed edges* ($\rightarrow$). Furthermore, the graphs we consider are *simple*, i.e. they have no edge joining a vertex with itself and no multiple edges joining the same pair or vertices. A non-empty subset $A \subseteq V$ is *complete* when every pair of distinct vertices in A are joined by an edge; otherwise it is *incomplete*. The *subgraph* of $\mathcal{G}$ induced by the non-empty subset $A \subseteq V$, denoted by $\mathcal{G}_A$, is the graph having vertex set A and exactly those edges of $\mathcal{G}$ that join two vertices in A. A *path* (of length $k \geq 1$) from v to w is a sequence $v = v_0, \ldots, v_k = w$ of distinct vertices such that v_{i-1} and v_i are joined by an edge for all $i = 1, \ldots, k$. A graph $\mathcal{G}$ is *connected* when every pair of distinct vertices in V is joined by a path, and a non-empty subset $C \subseteq V$ is a *connected set* in $\mathcal{G}$ when $\mathcal{G}_C$ is connected. Given a graph $\mathcal{G}$, every non-empty subset $A \subseteq V$ can be partitioned uniquely into inclusion maximal *connected sets*, $A = C_1 \uplus \cdots \uplus C_r$, where the symbol $\uplus$ denotes a union of pairwise disjoint sets; the sets $C_1, \ldots, C_r$ are called the *connected components* of A in $\mathcal{G}$. We have $r \geq 2$ when $\mathcal{G}_A$ is *disconnected*. Two non-empty, disjoint, subsets $A, B \subseteq V$ are *separated* in $\mathcal{G}$ by a third subset $C \subseteq V$ when every path in $\mathcal{G}$ from a vertex in A to a vertex in B contains a vertex in C. A *cycle* is defined like a path, but with $v = w$; a *directed cycle* is a cycle such that $\mathcal{G}$ contains $v_{i-1} \rightarrow v_i$ for all $i = 1, \ldots, k$.

Undirected graphs (UGs) and *bidirected graphs* (BGs) are graphs with only undirected and bidirected edges, i.e. lines and arrows with two heads, respectively. *Directed acyclic graphs* (DAGs) are graphs with only directed edges, i.e. arrows, and no directed cycle. We define a *chain graph* (CG) as a graph whose vertex set can be partitioned into an ordered sequence of pairwise disjoint blocks such that the edges joining vertices belonging to different blocks are all directed and point from a block in a lower position to a block in a higher position in the sequence. Furthermore, directed edges are not allowed within blocks and, more specifically, we consider CGs where every block induces a subgraph that is either undirected or bidirected. The connected components of the blocks are called the *chain components* of the CG so that a CG where all chain components are singletons is a DAG. We remark that the same CG can possibly come with different block orderings of its vertices, but we always assume that one of them has been selected on the basis of subject matter knowledge. A *regression graph* (RG) is a CG that admits an ordered sequence of blocks $(U, T_1, \ldots, T_k)$ where the first block U induces an UG whereas each of the remaining

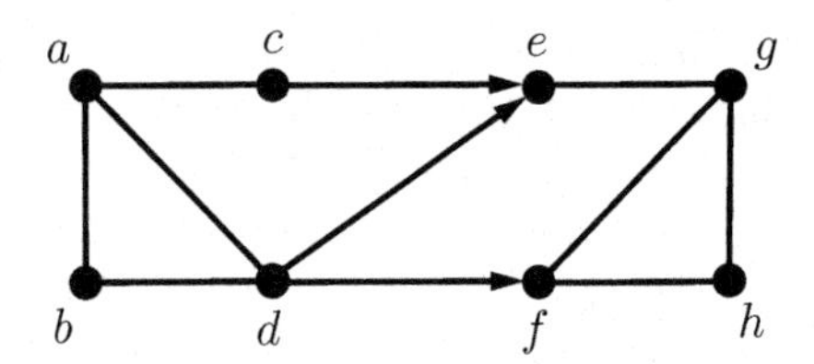

FIGURE 8.1: Example of chain graph with two blocks, each inducing an undirected graph.

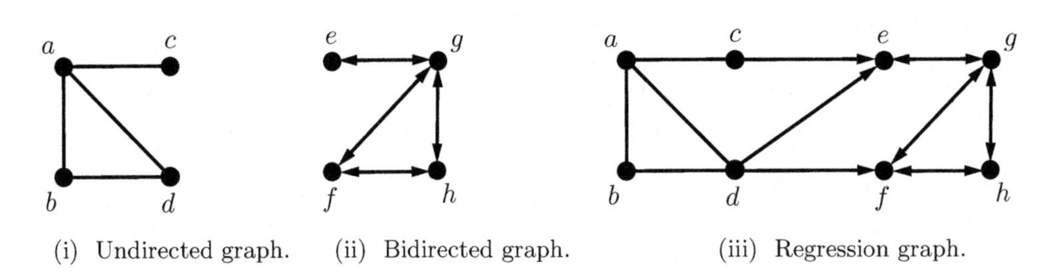

FIGURE 8.2: Example of a regression graph (iii) together with the undirected graph (i) and the bidirected graph (ii) induced by its blocks.

blocks T_i, for $i = 1, \dots, k$, induces a BG. Examples of different types of graphs are depicted in Figures 8.1 and 8.2.

For a non-empty subset $A \subseteq V$ we denote by $\mathrm{pa}_\mathcal{G}(A)$ the set of *parents* of A in $\mathcal{G}$, that is, the set of vertices $v \in V \setminus A$ such that $\mathcal{G}$ contains a directed edge $v \to w$ with $w \in A$. Furthermore, if $\mathcal{G}$ is a RG with block ordering $(U, T_1, \dots, T_k)$, we define the sets of *predecessors* of the bidirected blocks of $\mathcal{G}$ as $\mathrm{pre}_\mathcal{G}(T_1) = U$ and $\mathrm{pre}_\mathcal{G}(T_i) = U \uplus T_1 \uplus \cdots \uplus T_{i-1}$ for $i = 2, \dots, k$. Finally, we write $\mathrm{pa}(A)$ for $\mathrm{pa}_\mathcal{G}(A)$, and $\mathrm{pre}(T_i)$ for $\mathrm{pre}_\mathcal{G}(T_i)$, $i = 1, \dots, k$, if the graph $\mathcal{G}$ is clearly identifiable from the context.

8.3 Overview of Discrete Graphical Models

We present here a historical overview of the most relevant families of graphical models for categorical variables. In a graphical model the graph is used to represent independence relationships among variables. More precisely, the vertices of the graph are associated with the variables and conditional independencies are implied by the missing edges. Different types of missing edges typically imply different types of independencies. The rules that translate properties of the graph into conditional independence statements are called Markov properties. Special attention is paid in this section to the use of log-linear parameterizations in graphical modeling, and to models specified by means of RGs with an emphasis on their connection with other relevant families of graphical models.

The seminal paper [12] by Darroch, Lauritzen and Speed showed that discrete graphical models based on UGs form a subclass of hierarchical *log-linear models* for contingency tables. Since then, the log-linear parameterization has become the standard tool to deal with this class of models. More generally, parameterizations based on log-linear expansions of cell probabilities play a fundamental role in most of the existing classes of discrete graphical models. A second long-established class of graphical models is that of DAG models, which is suited to model recursive sequences of univariate regressions; see e.g. [11, 36, 61]. The parameterization of these models is typically obtained from the parameterizations of the single regression components. In the discrete case, these can be represented in the form of log-linear models as is illustrated, for example, in Wermuth and Lauritzen [58] and Agresti [1, Sect. 9.5]. Both UG models and DAG models are special cases of CG models, which we present below in some detail.

In CG models the variables can be arranged in a sequence of blocks ordered on the basis of subject matter considerations. The variables within each block are considered to be on an equal standing and, consequently, all edges within a block are of a same type and symmetric (either lines or arrows with two heads). As clearly pointed out by Drton [14], a CG can be regarded as a *DAG of chain components*, and CG models generalize DAG models accordingly. More specifically, every block contains variables which are *multiple responses* of the previous blocks and *explanatory* of the following blocks. The first block has no incoming arrows, and therefore it contains purely explanatory variables, also called *context variables.* Likewise, the last block contains the *primary responses.*

The classical Markov property for CGs was introduced by Frydenberg [23] and by Wermuth and Lauritzen [59] and, following the standard literature in this field, we will refer to it as the *LWF Markov property.* An alternative Markov property for CGs, hereafter referred to as the *AMP Markov property,* was introduced by Andersson, Madigan and Perlman [3]. These two Markov properties differ with respect to the independence relationships implied by missing arrows. However, under both interpretations, conditionally on the previous blocks, every block defines an UG model and, accordingly, every edge joining two vertices belonging to the same block is depicted as a line. In this way, the notion of variables in the same block being on an equal standing acquires a more specific meaning; see Lauritzen and Richardson [37]. For example, the graph in Figure 8.1 is a CG with two blocks, a block of context variables $\{a, b, c, d\}$ and a block of primary responses $\{e, f, g, h\}$. Under both the LWF and the AMP Markov properties the marginal distribution of X_{abcd} belongs to the model defined by the UG induced by $\{a, b, c, d\}$, also given in Figure 8.2(i), whereas the conditional distribution $X_{efgh} \mid X_{abcd}$ belongs to the model defined by the UG induced by $\{e, f, g, h\}$, for all values of X_{abcd}. The log-linear parameterization of UG models can be naturally extended to parameterize LWF models, as shown by Lauritzen and Wermuth [38, 59]. On the other hand, Drton [14] showed that the class of discrete AMP models comprises also non-smooth models, which renders discrete AMP models less interesting in practice.

More recently, several classes of graphical models have been introduced to represent the independence relationships implied by marginalization and conditioning of DAG models; see [34, 46, 47, 56] and references therein. The number of available classes of graphical models has thus greatly increased, and Sadeghi and Lauritzen [52] provided a unification of the theory, thereby clarifying relevant connections existing between most of such classes. A common feature of these more recent models is that they are specified by *mixed graphs*, which in addition to lines and arrows may include a third type of edge to express a second type of symmetric relationship between variables. These additional edges are usually represented by either a dashed line or an arrow with two heads, and we here adopt the latter convention.

An example of mixed graph is given in Figure 8.2(iii). Models defined by mixed graphs with only undirected edges are UG models, whereas models defined by mixed graphs with only directed edges are DAG models. Furthermore, models defined by mixed graphs with only bidirected edges form the family of BG models. This is a class of models of marginal independence introduced by Kauermann [31]. Independencies can be read off a bidirected graph through the *dual global Markov property,* which means for instance that X_{fh} and X_e are independent in the model defined by the graph in Figure 8.2(ii). This should be compared with X_{bd} and X_c being conditionally independent given X_a in the UG model for the graph of Figure 8.2(i).

Interestingly, neither the models defined by the LWF Markov property nor those defined by the AMP Markov property form a subclass of the models specified by mixed graphs. The family of CGs resulting as a special case of this framework have bidirected edges within chain components; they are named *multivariate regression chain graphs* (MRCGs) and the corresponding models were first introduced by Cox and Wermuth [10, 11]. Accordingly, the relative Markov property was named the *MRCG Markov property* by Drton [14]. Here it deserves noting that MRCG models are a subclass of *acyclic directed mixed graph* (ADMG) models; see e.g. Richardson [46].

Wermuth and Sadeghi [60] considered an extension of MRCG models, which they called *regression graph* (RG) models, obtained by assuming an UG model, rather than a BG model, for the context variables; see also [10, 11, 57]. An example of RG is given in Figure 8.2(iii). Like the LWF model for the graph in Figure 8.1, the RG model for the graph in Figure 8.2(iii) has a block of context variables given by X_{abcd}, and a block of primary responses given by X_{efgh}; furthermore the marginal distribution of X_{abcd} belongs to the UG model defined by the graph in Figure 8.2(i). On the other hand, under the RG Markov property, the conditional distribution of $X_{edfg} \mid X_{abcd}$ belongs to the BG model defined by the graph in Figure 8.2(ii), for all values of X_{abcd}.

It is useful to compare LWF and RG models with respect to independence relationships involving variables in different blocks. To fix ideas, consider the case where we are interested in the independence relationships involving the response X_e. The RG Markov property focuses on the marginal distributions of responses. Hence, we can consider the regression with response X_e and read from the graph in Figure 8.2(ii) that X_e is independent of X_{ab} given X_{cd}. On the other hand, the LWF Markov property focuses on the joint distribution of responses. Thus, in this case, we must consider the regression with response X_{efgh}, and this leads us to read from the graph in Figure 8.1 that X_e is independent of X_{ab} given X_{cdg}. The substantial difference is that the former statement does not involve the response X_g, whereas the latter statement does. It is worth remarking that a similar interpretation of missing arrows holds under the AMP Markov property. However, unlike discrete AMP models, discrete RG models have the advantage of being smooth; see Drton [14].

For the sake of completeness, we remark that by combining the interpretation of missing arrows under the LWF Markov property with BG models within blocks a fourth family of chain graph models can be obtained, which also contains non-smooth models; see again [14]. We are not aware of notable applications for this class of models.

We conclude that RG models are an appealing family of graphical models. They form a flexible class that includes, as special cases, UG, BG, DAG and MRCG models, so that dealing with discrete RG models allows us to provide an overview of some of the most relevant existing families of discrete graphical models. Furthermore, whenever it is reasonable to assume the existence of a partial ordering of the variables, RG models play a central role, among the existing families of CG models, in both observational and intervention studies; we refer to Wermuth and Cox [57] for relevant discussion on this issue. A crucial aspect is that these models allow one to focus on the marginal distribution of the response variables; see also Whittaker [61, Sect. 10.5]. Conversely, the association among context variables is modeled through an UG, thereby providing insight into the conditional independence relationships among these variables.

There is no straightforward way to extend the log-linear parameterization of discrete UG models to RG models. Some researchers exploited the theory of *marginal log-linear models* of Bergsma and Rudas [6] to parameterize ADMG and MRCG models [5, 20, 42, 51]. On the other hand, it was shown in [14, 19, 21] that ADMG and therefore MRCG models can be parameterized by means of a collection of marginal and conditional probabilities called the *(generalized) Möbius parameters.* Subsequently, the *log-mean linear parameterization* was introduced in [49, 50] as a log-linear expansion of the Möbius parameters, and used to parameterize BG models.

In this chapter we extend the log-mean linear parameterization of BG models to RG models in such a way that some fundamental properties of this parameterization follow immediately from the same properties of the log-linear parameterization of UG models. We deem that this may represent a convenient way to approach RG models, because it provides insight into their connection with classical log-linear models and, furthermore, it allows us to overview the theory concerning the parameterization of three prominent families of models in an efficient manner. Indeed, technical proofs are only required for the basic results of Section 8.4; every other result follows in a straightforward way from those.

8.4 Basic Lemmas

We give here some results which provide the basis for the development of the parameterizations of the three families of models considered in the forthcoming sections. Initially, we give three alternative, but related, rules to asses the (conditional) independence of two random vectors. Next, we introduce the Möbius inversion formula, which we will apply to obtain a log-linear expanded parameterization for each of the three families of models.

8.4.1 Establishing independence relationships

We are interested in conditional independencies between subvectors of the binary random vector X_V. Recall that, given three pairwise disjoint margins $A, B, C \subseteq V$ with $A, B \neq \emptyset$, X_A and X_B are conditionally independent given X_C, in symbols $X_A \perp\!\!\!\perp X_B \mid X_C$, when

$$P(X_A = \epsilon_{A'} \mid X_B = \epsilon_{B'}, X_C = \epsilon_{C'}) = P(X_A = \epsilon_{A'} \mid X_C = \epsilon_{C'}) \tag{8.1}$$

for every $A' \subseteq A$, $B' \subseteq B$ and $C' \subseteq C$. It is immediate to check that (8.1) holds if and only if it holds with A and B exchanging roles. As a special case, when $C = \emptyset$, X_A and X_B are marginally independent (or simply independent); in symbols $X_A \perp\!\!\!\perp X_B$. We write $X_A \perp\!\!\!\perp X_B \mid X_C = \epsilon_{C'}$ when (8.1) holds for a specific $C' \subseteq C$, that is, when X_A and X_B are independent conditionally on the event $\{X_C = \epsilon_{C'}\}$.

The following result translates the statement $X_A \perp\!\!\!\perp X_B \mid X_C = \epsilon_{C'}$ into three alternative rules, each suited to a different scenario. Informally, the differences between the three rules can be appreciated as follows. The first rule involves probabilities from the full distributions of both X_A and X_B, and it will be applied to the parameterization of UG models. The second rule involves probabilities from the marginal distributions of X_A and the full distribution of X_B, and it will be applied to RG models. The third rule involves probabilities from the marginal distributions of both X_A and X_B, and it will be used for BG models.

Lemma 8.4.1. *Let X_V be a binary random vector indexed by a finite set V, and let A, B, C be a triple of pairwise disjoint subsets of V with $A, B \neq \emptyset$. For all $C' \subseteq C$, the following statements are equivalent:*

(i) $X_A \perp\!\!\!\perp X_B \mid X_C = \epsilon_{C'}$;

(ii) $P(X_A = \epsilon_{A'} \mid X_B = \epsilon_{B'}, X_C = \epsilon_{C'}) = P(X_A = \epsilon_{A'} \mid X_B = 0, X_C = \epsilon_{C'})$ *for every $A' \subseteq A$ and $B' \subseteq B$ with $A', B' \neq \emptyset$;*

(iii) $\mathrm{P}(X_{A'} = 1 \mid X_B = \epsilon_{B'}, X_C = \epsilon_{C'}) = \mathrm{P}(X_{A'} = 1 \mid X_B = 0, X_C = \epsilon_{C'})$ *for every* $A' \subseteq A$ *and* $B' \subseteq B$ *with* $A', B' \neq \emptyset$*;*

(iv) $\mathrm{P}(X_{A'} = 1 \mid X_{B'} = 1, X_C = \epsilon_{C'}) = \mathrm{P}(X_{A'} = 1 \mid X_C = \epsilon_{C'})$ *for every* $A' \subseteq A$ *and* $B' \subseteq B$ *with* $A', B' \neq \emptyset$*.*

Then, the statement $X_A \perp\!\!\!\perp X_B \mid X_C$ *is equivalent to any of (ii), (iii) or (iv) holding for every* $C' \subseteq C$*.*

Proof. We first show that (i) and (ii) are equivalent. If (i) holds, then both sides of (ii) are equal to $\mathrm{P}(X_A = \epsilon_{A'} \mid X_C = \epsilon_{C'})$ by (8.1). Conversely, if (ii) holds, it also holds for $B' = \emptyset$, trivially, because $\epsilon_\emptyset = 0$, and for $A' = \emptyset$ because both sides sum to one over $A' \subseteq A$; then (i) follows from the law of total probability, which gives $\mathrm{P}(X_A = \epsilon_{A'} \mid X_C = \epsilon_{C'})$ as in (8.1).

We next show that both (ii) and (iii) are equivalent to

$$\tilde{\mathrm{P}}(X_{A''} = \epsilon_{A'} \mid X_B = \epsilon_{B'}) = \tilde{\mathrm{P}}(X_{A''} = \epsilon_{A'} \mid X_B = 0) \tag{8.2}$$

for every $A'' \subseteq A$, $A' \subseteq A''$ and $B' \subseteq B$, with $A'' \neq \emptyset$, $A' \neq \emptyset$ and $B' \neq \emptyset$, where $\tilde{\mathrm{P}}$ denotes P conditioned on $\{X_C = \epsilon_{C'}\}$. Clearly, both (ii) and (iii) are implied by (8.2): just let $A'' = A$ for (ii) or $A'' = A'$ for (iii). Conversely, if (ii) holds, we can use finite induction on the cardinality of $A \setminus A''$. The base case $A'' = A$ is (ii). Then, in the inductive step, we pick $a \in A \setminus A''$, let $A''_\star = A'' \cup \{a\}$, and write $\tilde{\mathrm{P}}(X_{A''} = \epsilon_{A'} \mid X_B = \epsilon_{B'})$ as the sum of $\tilde{\mathrm{P}}(X_{A''_\star} = \epsilon_{A'} \mid X_B = \epsilon_{B'})$ and $\tilde{\mathrm{P}}(X_{A''_\star} = \epsilon_{A' \cup \{a\}} \mid X_B = \epsilon_{B'})$; we conclude by the inductive hypothesis, with $|A \setminus A''_\star| = |A \setminus A''| - 1$, and recombining the two terms into $\tilde{\mathrm{P}}(X_{A''} = \epsilon_{A'} \mid X_B = 0)$. Similarly, if (iii) holds, we can use finite induction on the cardinality of $A'' \setminus A'$. The base case $A'' = A'$ is (iii). In the inductive step, we pick $a \in A'' \setminus A'$, let $A''_\star = A'' \setminus \{a\}$, and write $\tilde{\mathrm{P}}(X_{A''} = \epsilon_{A'} \mid X_B = \epsilon_{B'})$ as the difference between $\tilde{\mathrm{P}}(X_{A''_\star} = \epsilon_{A'} \mid X_B = \epsilon_{B'})$ and $\tilde{\mathrm{P}}(X_{A''} = \epsilon_{A' \cup \{a\}} \mid X_B = \epsilon_{B'})$; we then use the inductive hypothesis, with $|A''_\star \setminus A'| = |A'' \setminus (A' \cup \{a\})| = |A'' \setminus A'| - 1$, and recombine the two terms into $\tilde{\mathrm{P}}(X_{A''} = \epsilon_{A'} \mid X_B = 0)$ to conclude.

We finally show that both (iii) and (iv) are equivalent to

$$\tilde{\mathrm{P}}(X_{A'} = 1, X_{B''} = \epsilon_{B'}) = \tilde{\mathrm{P}}(X_{A'} = 1)\tilde{\mathrm{P}}(X_{B''} = \epsilon_{B'}) \tag{8.3}$$

for every $B'' \subseteq B$, $B' \subseteq B''$ and $A' \subseteq A$, with $B'' \neq \emptyset$, $B' \neq \emptyset$ and $A' \neq \emptyset$. To this aim, we remark that $\mathrm{P}(X_{A'} = 1 \mid X_B = 0, X_C = \epsilon_{C'})$ in the right hand side of (iii) can be replaced by $\mathrm{P}(X_{A'} = 1 \mid X_C = \epsilon_{C'})$ because of the law of total probability. Then, to obtain (iii) or (iv) from (8.3), we divide by $\tilde{\mathrm{P}}(X_{B''} = \epsilon_{B'})$ and let $B'' = B$ for (iii) or $B'' = B'$ for (iv). Conversely, reasoning as above, we get (8.3) from (iii) by finite induction on $|B \setminus B''|$, and (8.3) from (iv) by finite induction on $|B'' \setminus B'|$. □

8.4.2 Two properties of Möbius inversion

Now we introduce the Möbius inversion formula and prove two properties of this operation. The first property will allow us to interpret some zero patterns of a log-linear expanded parameterization in terms of conditional independencies. The second property provides a general tool to deal with context specific independencies.

The relevance of the theory of Möbius inversion in combinatorial mathematics was shown by Rota [48], who elaborated on previous work [26, 55] in the area of group theory. Although Möbius inversion plays a central role in this chapter, we only need a basic version of this result, which is given below in terms of two generic real vectors $\theta = (\theta_D)_{D \subseteq V}$ and $\omega = (\omega_D)_{D \subseteq V}$, indexed by the subsets of a finite set V, and two specific matrices M_V and Z_V, whose rows and columns are also indexed by the subsets of V.

Proposition 8.4.2. *(Möbius inversion formula) Let $\theta = (\theta_D)_{D\subseteq V}$ and $\omega = (\omega_D)_{D\subseteq V}$ be two real vectors indexed by the subsets of a finite set V. Then, it holds that*

$$\omega_D = \sum_{D'\subseteq D} \theta_{D'} \quad \text{if and only if} \quad \theta_D = \sum_{D'\subseteq D} (-1)^{|D\setminus D'|}\, \omega_{D'}, \tag{8.4}$$

where the identities are intended for all $D \subseteq V$. This statement can be written in matrix notation as

$$\omega = Z_V^\top \theta \quad \text{if and only if} \quad \theta = M_V^\top \omega,$$

where Z_V is the matrix with entries $\mathbb{I}_{\{D\subseteq E\}}$, and M_V the matrix with entries $(-1)^{|E\setminus D|}\mathbb{I}_{\{D\subseteq E\}}$, for $D, E \subseteq V$, with $\mathbb{I}_{\{\cdot\}}$ denoting the indicator function.

Proof. It is immediate to check that $Z_V = M_V^{-1}$. □

The matrix M_V is called the Möbius matrix associated with V, while the matrix Z_V is called the Zeta matrix. Hereafter, we will omit the subscript when the set V can be identified from the context, writing for instance M for M_V. A fast algorithm for Möbius inversion and a memory efficient data structure for storing the Möbius parameters are given in [33] and [43], respectively. It is also worth noting that the Möbius map can be given a Kronecker product structure; see e.g. [30]. Both properties given below exploit the fact that Möbius inversion is a recursive operation.

The following lemma establishes a connection between zero entries in the vector θ and vanishing contrasts of the entries of ω. We will see that, when ω is a vector of log-probabilities, this result will allow us to interpret some zero patterns of θ in terms of independence relationships.

Lemma 8.4.3. *Let ω be a real vector indexed by the subsets of a finite set V and set $\theta = M^\top \omega$. For any triple A, B, C of pairwise disjoint subsets of V with $A, B \neq \emptyset$, the following statements are equivalent:*

(i) $\theta_D = 0$ for every $D \subseteq A \cup B \cup C$ such that both $D \cap A \neq \emptyset$ and $D \cap B \neq \emptyset$;

(ii) $\omega_{A'\cup B'\cup C'} - \omega_{A'\cup C'} - \omega_{B'\cup C'} + \omega_{C'} = 0$ for every $A' \subseteq A$, $B' \subseteq B$ and $C' \subseteq C$, such that $A', B' \neq \emptyset$.

Proof. We first show *(i)* ⇒ *(ii)*. Since $\omega_D = \sum_{D'\subseteq D} \theta_{D'}$, for every $D \subseteq V$, then, for every non-empty $A' \subseteq A$, non-empty $B' \subseteq B$, and $C' \subseteq C$,

$$\omega_{A'\cup B'\cup C'} = \sum_{D'\subseteq A'\cup B'\cup C'} \theta_{D'}, \tag{8.5}$$

and it follows from (i) that $\theta_{D'} = 0$ for every D' such that $D' \not\subseteq A' \cup C'$ and $D' \not\subseteq B' \cup C'$. Hence, we can write (8.5) as

$$\omega_{A'\cup B'\cup C'} = \sum_{D'\subseteq A'\cup C'} \theta_{D'} + \sum_{D'\subseteq B'\cup C'} \theta_{D'} - \sum_{D'\subseteq C'} \theta_{D'}, \tag{8.6}$$

that is, $\omega_{A'\cup B'\cup C'} = \omega_{A'\cup C'} + \omega_{B'\cup C'} - \omega_{C'}$, which implies (ii). Note that the term $-\sum_{D'\subseteq C'} \theta_{D'}$ is necessary because, otherwise, every term $\theta_{D'}$ with $D' \subseteq C'$ would be added twice, once for each of the other sums in (8.6).

We now show *(ii)* ⇒ *(i)* by finite induction on the cardinality of the set $A \cup B \cup C$. We first show that the result is true for $|A \cup B \cup C| = 2$. In this case $A = \{a\}$, $B = \{b\}$, and $C = \emptyset$, because A and B cannot be empty. Hence, we have to show that $\omega_{ab} - \omega_a - \omega_b + \omega_\emptyset = 0$ implies $\theta_{ab} = 0$, and this follows immediately from the fact that $\theta_{ab} = \omega_{ab} - \omega_a - \omega_b + \omega_\emptyset$

by the definition of θ in (8.4). We now set $k > 2$ and show that if result holds true for $|A \cup B \cup C| < k$ then it also holds true for $|A \cup B \cup C| = k$. More specifically, the induction assumption implies $\theta_{D'} = 0$ for every $D' \subset A \cup B \cup C$ such that $D' \cap A \neq \emptyset$ and $D' \cap B \neq \emptyset$. This in turn implies both that it is sufficient to show that (ii) implies $\theta_{A \cup B \cup C} = 0$ and that

$$\sum_{D' \subseteq A \cup B \cup C} \theta_{D'} = \theta_{A \cup B \cup C} + \sum_{D' \subseteq A \cup C} \theta_{D'} + \sum_{D' \subseteq B \cup C} \theta_{D'} - \sum_{D' \subseteq C} \theta_{D'}. \tag{8.7}$$

Every sum in (8.7) corresponds to an entry of ω, so that (8.7) can be rewritten as

$$\omega_{A \cup B \cup C} = \theta_{A \cup B \cup C} + \omega_{A \cup C} + \omega_{B \cup C} - \omega_C.$$

This rewriting gives $\theta_{A \cup B \cup C} = 0$, as required to complete the proof, because (ii) implies $\omega_{A \cup B \cup C} - \omega_{A \cup C} - \omega_{B \cup C} + \omega_C = 0$. □

For a pair of subsets $S, W \subseteq V$ forming a partition of V, i.e. $V = S \uplus W$, we denote by $\mathrm{s}_W(\omega) = (\omega_{W \cup S'})_{S' \subseteq S}$ the subvector of ω whose entries are indexed by the supersets of W. In this context, an object of interest is the vector $\theta_{[W]} = M_S^{\top} \mathrm{s}_W(\omega)$, whose entries can be written as

$$\theta_{[W]S'} = \sum_{S'' \subseteq S'} (-1)^{|S' \setminus S''|} \omega_{W \cup S''}, \qquad \text{for every } S' \subseteq S. \tag{8.8}$$

If $\theta = M_V^{\top} \omega$, it follows directly from the definition of $\theta_{[W]}$ that $\theta_{[\emptyset]} = \theta$, but in general $\theta_{[W]}$ is not a subvector of θ, and the next lemma states the connection between $\theta_{[W]}$ and θ.

Lemma 8.4.4. *Let ω be a real vector indexed by the subsets of a finite set V and set $\theta = M_V^{\top} \omega$. Moreover, for $W \subseteq V$, let $\theta_{[W]}$ be the vector defined by (8.8) with $S = V \setminus W$. Then, it holds that*

$$\theta_{[W]S'} = \sum_{W' \subseteq W} \theta_{S' \cup W'} \quad \text{for every } S' \subseteq S. \tag{8.9}$$

Proof. Let $\theta^* = (\theta^*_{S'})_{S' \subseteq S}$ be the vector with entries $\theta^*_{S'} = \sum_{W' \subseteq W} \theta_{S' \cup W'}$. To establish (8.9) we have to show that $\theta_{[W]} = \theta^*$. Since $\omega = Z_V^{\top} \theta$, then for every $S' \subseteq S$

$$\omega_{W \cup S'} = \sum_{D \subseteq S' \cup W} \theta_D = \sum_{S'' \subseteq S'} \sum_{W' \subseteq W} \theta_{S'' \cup W'} = \sum_{S'' \subseteq S'} \theta^*_{S''}, \tag{8.10}$$

and (8.10) can be written in vector form as $\mathrm{s}_W(\omega) = Z_S^{\top} \theta^*$. On the other hand, by definition $\theta_{[W]} = M_S^{\top} \mathrm{s}_W(\omega)$, and therefore $\mathrm{s}_W(\omega) = Z_S^{\top} \theta_{[W]}$. Hence, it holds that $Z_S^{\top} \theta_{[W]} = Z_S^{\top} \theta^*$ and $\theta_{[W]} = \theta^*$ follows by noting that Z_S has full rank. □

8.5 Undirected Graph Models

We consider in this section the case where the random vector X_V is indexed by the vertex set of a graph $\mathcal{G} = (V, \mathcal{U})$ having only undirected edges (an UG). We assume that missing edges of $\mathcal{G}$ encode independence relationships between subvectors of X_V according to the *global Markov property*: $X_A \perp\!\!\!\perp X_B \mid X_C$ for every triple A, B, C of pairwise disjoint subsets of V with $A, B \neq \emptyset$ and C separating A from B in $\mathcal{G}$. In particular, letting $A = \{u\}$, $B = \{v\}$, and $C = V \setminus \{u, v\}$, it holds that $X_u \perp\!\!\!\perp X_v \mid X_{V \setminus \{u,v\}}$ for every pair u, v of distinct vertices in V such that $u - v \notin \mathcal{U}$. The latter statement is known as the *pairwise Markov property*, and it is known to be equivalent to the global Markov property under the assumption of positive cell probabilities, although in general it is a weaker requirement; see e.g. Cowell, Dawid, Lauritzen and Spiegelhalter [9, Sect. 5.2].

Since we assume a positive distribution for X_V, we simply say that X_V is Markov with respect to $\mathcal{G}$ when it satisfies either (and hence both) of the above described Markov properties. For instance, if X_V is Markov with respect to the graph in Figure 8.2(i), then $X_c \perp\!\!\!\perp X_d \mid X_{ab}$ (pairwise statement) as well as $X_c \perp\!\!\!\perp X_d \mid X_a$ and $X_{bd} \perp\!\!\!\perp X_c \mid X_a$ (global statements). Hence, the graphical model specified by $\mathcal{G}$, denoted by $\mathbb{M}_+(\mathcal{G})$, is the family of all positive probability distributions for X_V such that X_V is Markov with respect to $\mathcal{G}$. In the following, we deal with effectively parameterizing $\mathbb{M}_+(\mathcal{G})$ by means of the classical log-linear parameterization, as originally suggested by Darroch, Lauritzen and Speed [12]; see Cox and Wermuth [11, Ch. 2] for early references in the area of UG modeling, and Edwards [16, Ch. 2] for a list of relevant classical monographs.

The saturated model for a binary random vector X_V is typically parameterized by $\pi = (\pi_D)_{D\subseteq V}$, where $\pi_D = \mathrm{P}(X_V = \epsilon_D)$, under the affine constraint $\sum_{D\subseteq V} \pi_D = 1$. We note that π is a vector indexed by the subsets of V, and we call it the *probability parameter* of X_V. Pairwise independence statements on X_V result in multiplicative constraints on π. A log-linear expansion of π can be used to express such independencies as linear constraints (actually zero constraints). We here adopt the corner-constrained expansion with zero baseline: $\lambda = M^\top \log \pi$, where M is the Möbius matrix associated with V, and the logarithm is taken entrywise. It is immediate to check that $\pi = \exp(Z^\top \lambda)$, where $Z = M^{-1}$ is the corresponding Zeta matrix, and the exponential is also taken entrywise. The map $\pi \to \lambda$ is therefore a valid reparameterization (a smooth bijection) and it is worth remarking that the inverse map $\lambda \to \pi$ is available in closed form. We call λ the *log-linear parameter* of X_V.

The effective parameter space for λ can be described as follows: λ_D, $D \neq \emptyset$, are variation independent, varying in $\mathbb{R}^{2^{|V|}-1}$, while $\lambda_\emptyset$ is uniquely determined by λ_D, $D \neq \emptyset$, through the sum-to-one constraint on π. Indeed

$$\log \pi_D = \lambda_\emptyset + \sum_{D'\subseteq D: D'\neq\emptyset} \lambda_{D'}, \quad \text{for every } D \subseteq V, \tag{8.11}$$

and any choice of $\lambda_{D'}$, $D' \neq \emptyset$, determines π_D, $D \subseteq V$, up to the common normalizing factor $\exp(\lambda_\emptyset) = \mathrm{P}(X_V = 0)$. Equation (8.11) also shows that the saturated model for X_V is a regular exponential family of order $2^{|V|} - 1$, with canonical parameter $\lambda_{D'}$, $D' \neq \emptyset$, and canonical statistic $t_{D'}(\cdot)$, $D' \neq \emptyset$, where $t_{D'}(\epsilon_D) = 1$ if $D' \subseteq D$ and $t_{D'}(\epsilon_D) = 0$ if $D' \nsubseteq D$, for any choice of $D \subseteq V$. We can indeed rewrite (8.11) as $\mathrm{P}(X_V = \epsilon_D) \propto \exp\{\sum_{D'\neq\emptyset} \lambda_{D'} t_{D'}(\epsilon_D)\}$, for every $D \subseteq V$, which shows the exponential nature of the saturated model.

The factorization criterion, see e.g. [61, Sect. 2.2], applied to (8.11), shows that the independence statement $X_A \perp\!\!\!\perp X_B \mid X_{V\setminus(A\cup B)}$ corresponds to the constraint $\lambda_D = 0$ for every $D \subseteq V$ such that both $D \cap A \neq \emptyset$ and $D \cap B \neq \emptyset$. For instance, in the graph of Figure 8.2(i), the statement $X_c \perp\!\!\!\perp X_d \mid X_{ab}$ corresponds to the condition $\lambda_{cd} = \lambda_{acd} = \lambda_{bcd} = \lambda_{abcd} = 0$. Similarly, in the same graph, the statement $X_{bd} \perp\!\!\!\perp X_c \mid X_a$ corresponds to having $\lambda_{bcd} = \lambda_{abcd} = 0$. The following result derives this well-known correspondence from the basic tools introduced in Section 8.4, which are designed to deal in a unified fashion with all kinds of independence statements and parameterizations considered in this chapter.

Theorem 8.5.1. *Let $\pi > 0$ be the probability parameter of a binary random vector X_V and set $\lambda = M^\top \log \pi$. Then, for every pair A, B of non-empty disjoint subsets of V, the following statements are equivalent:*

(i) $X_A \perp\!\!\!\perp X_B \mid X_{V\setminus(A\cup B)}$;

(ii) $\log \pi_{A' \cup B' \cup C'} - \log \pi_{B' \cup C'} = \log \pi_{A' \cup C'} - \log \pi_{C'}$ *for every* $A' \subseteq A$, $B' \subseteq B$ *and* $C' \subseteq V \setminus (A \cup B)$, *such that* $A', B' \neq \emptyset$;

(iii) $\lambda_D = 0$ *for every* $D \subseteq V$ *such that both* $D \cap A \neq \emptyset$ *and* $D \cap B \neq \emptyset$.

Proof. We can rewrite (ii) as

$$\frac{\mathrm{P}(X_A = \epsilon_{A'} \mid X_B = \epsilon_{B'}, X_C = \epsilon_{C'})}{\mathrm{P}(X_A = 0 \mid X_B = \epsilon_{B'}, X_C = \epsilon_{C'})} = \frac{\mathrm{P}(X_A = \epsilon_{A'} \mid X_B = 0, X_C = \epsilon_{C'})}{\mathrm{P}(X_A = 0 \mid X_B = 0, X_C = \epsilon_{C'})}$$

for every $A' \subseteq A$, $B' \subseteq B$ and $C' \subseteq V \setminus (A \cup B)$, such that $A, B \neq \emptyset$; this rewriting is equivalent to (ii) in Lemma 8.4.1 with $C = V \setminus (A \cup B)$, because of the sum-to-one constraint over $A' \subseteq A$. Hence, the equivalence of (i) and (ii) follows from Lemma 8.4.1. On the other hand, the equivalence of (ii) and (iii) is an immediate consequence of Lemma 8.4.3 with $\omega = \log \pi$ and $C = V \setminus (A \cup B)$. □

In light of Theorem 8.5.1, if X_V is Markov with respect to $\mathcal{G} = (V, \mathcal{U})$, we have $\lambda_D = 0$ for every incomplete $D \subseteq V$. Indeed, if $u, v \in D$ with $u \neq v$ and $u{-}v \notin \mathcal{U}$, we have $X_u \perp\!\!\!\perp X_v \mid X_{V \setminus \{u,v\}}$, and therefore $\lambda_D = 0$. Conversely, if $\lambda_D = 0$ for every incomplete $D \subseteq V$, then X_V is Markov with respect to $\mathcal{G}$. Indeed, if u, v is a pair of distinct vertices in V with $u{-}v \notin \mathcal{U}$, then $\lambda_D = 0$ for every $D \subseteq V$ such that $u, v \in D$, and $X_u \perp\!\!\!\perp X_v \mid X_{V \setminus \{u,v\}}$. A parameterization for $\mathbb{M}_+(\mathcal{G})$ is therefore given by λ_D, D non-empty and complete, varying unconstrained in a certain number of dimensions. For instance, the model defined by the graph in Figure 8.2(i) is parameterized by $\lambda_a, \lambda_b, \lambda_c, \lambda_d, \lambda_{ab}, \lambda_{ac}, \lambda_{ad}, \lambda_{bd}, \lambda_{abd}$; it is therefore a nine dimensional model, whereas the saturated model is fifteen dimensional.

We remark that the pairwise Markov property suffices to obtain $\lambda_D = 0$ for every incomplete $D \subseteq V$. On the other hand, the latter condition implies, through Theorem 8.5.1, all global Markov statements with $C = V \setminus A \cup B$. These in turn imply all other global Markov statements, because the vertices in $V \setminus (A \cup B)$ can be assigned to either A or B, without breaking separation, and then marginalized out. Hence, the equivalence between the pairwise Markov property and the global Markov property, if the cell probabilities are positive, can be derived from Theorem 8.5.1.

Adopting the principle of parsimony, as discussed in the Introduction, there is an interest for refining a given graphical model by means of some context specific independencies. We here specifically consider independence relationships that hold on a subvector X_S of X_V given a specific event $\{X_W = \epsilon_{W'}\}$, where $W = V \setminus S$ and $W' \subseteq W$, that is, only in the context specified by $X_W = \epsilon_{W'}$. These relationships offer an opportunity for specifying parsimonious models with interpretable constraints: a graphical model refined by a context specific independence will no longer be graphical, strictly speaking, but all its constraints will nonetheless be interpretable as independence relationships. The following result, which is a direct consequence of Lemma 8.4.4, shows that context specific independencies of the above type correspond to sum-to-zero constraints on the log-linear parameter λ.

Proposition 8.5.2. *Let* $\pi > 0$ *be the probability parameter of a binary random vector* X_V *and set* $\lambda = M^\top \log \pi$. *Furthermore, let* $V = S \uplus W$ *be a partition of* V, *with* $S \neq \emptyset$, *and let* W' *be any subset of* W. *Then, for any non-empty subset* $S' \subseteq S$, *it holds that*

$$\lambda_{S'}^{S \mid X_W = \epsilon_{W'}} = \sum_{W'' \subseteq W'} \lambda_{S' \cup W''}, \tag{8.12}$$

where $\lambda^{S \mid X_W = \epsilon_{W'}}$ *denotes the log-linear parameter of* $X_S \mid X_W = \epsilon_{W'}$.

Proof. We first apply the definition of log-linear parameter to compute

$$\lambda_{S'}^{S|X_W=\epsilon_{W'}} = \sum_{S''\subseteq S'} (-1)^{|S'\setminus S''|} \log \mathrm{P}(X_{S'} = \epsilon_{S''} \mid X_W = \epsilon_{W'}) \quad (8.13)$$

$$= \sum_{S''\subseteq S'} (-1)^{|S'\setminus S''|} \log \mathrm{P}(X_{S'} = \epsilon_{S''}, X_W = \epsilon_{W'}) \quad (8.14)$$

$$= \sum_{S''\subseteq S'} (-1)^{|S'\setminus S''|} \log \pi_{S''\cup W'} = \lambda_{[W']S'} \quad (8.15)$$

if $S' \neq \emptyset$. The condition $S' \neq \emptyset$ ensures the equality of (8.13) and (8.14), because in this case the sum has the same number of positive and negative terms, and the common term $\log \mathrm{P}(X_W = \epsilon_{W'})$ cancels. The condition $V = S \uplus W$ guarantees the equality of (8.14) and (8.15). The equality in (8.15) is the equality in (8.8) with $\omega = \log \pi$ (and therefore $\theta = \lambda$). Then, to conclude, we obtain (8.12) from (8.15) by applying Lemma 8.4.4. □

In light of Proposition 8.5.2, if we want to refine the UG model defined by the graph in Figure 8.2(i) by adding the context specific independence $X_a \perp\!\!\!\perp X_b \mid X_c = 0, X_d = 1$, we will introduce the constraint $\lambda_{ab} + \lambda_{abd} = 0$ and thus obtain an eight dimensional model, which we will call a *quasi-graphical* model.

Proposition 8.5.2 implies $\lambda_{S'} = \lambda_{S'}^{S'|X_{V\setminus S'}=0}$ (letting $S = S'$ and $W' = \emptyset$). This helps us shed light on the interpretation of individual entries of λ. We have already seen that $\lambda_\emptyset$ is the logarithm of $\mathrm{P}(X_V = 0)$. Next, we can write $\lambda_v = \log\{\mathrm{P}_0(X_v = 1)/\mathrm{P}_0(X_v = 0)\}$, for any $v \in V$, where P_0 is conditional on $X_{V\setminus\{v\}} = 0$. We thus see that λ_v, the *main effect* of v, represents the log odds for $X_v = 1$, against $X_v = 0$, conditional on all other variables being zero. Then, for any $u, v \in V$ with $u \neq v$, we can write

$$\lambda_{uv} = \log \frac{\mathrm{P}_0(X_u = 1, X_v = 1)\mathrm{P}_0(X_u = 0, X_v = 0)}{\mathrm{P}_0(X_u = 1, X_v = 0)\mathrm{P}_0(X_u = 0, X_v = 1)},$$

where P_0 is now conditional on $X_{V\setminus\{u,v\}} = 0$. This shows that λ_{uv}, the *interaction* between u and v, is the log cross-product ratio (log odds ratio) between X_u and X_v conditional on all other variables being zero. Finally, the interpretation of *higher order interactions* in λ, that is, its entries λ_D with $|D| \geq 3$, is similar, but of course less immediate. For instance, the interaction λ_{uvw} is the differential log cross-product ratio between u and v when comparing the context $X_w = 1$ to the context $X_w = 0$, conditional on $X_{V\setminus\{u,v,w\}} = 0$. The same can be said with u or v exchanging role with w.

In alternative to context specific independencies, very parsimonious UG models can be specified by setting $\lambda_D = 0$ for every $D \subseteq V$ with $|D| \geq 3$. In this way λ_{uv} becomes the common value of the cross-product ratio between X_u and X_v given $X_{V\setminus\{u,v\}} = \epsilon_{W'}$, no matter which $W' \subseteq V \setminus \{u, v\}$ is chosen; see Whittaker [61, Ch. 9]. Hence, these models admit an interesting interpretation. However, they are also very restrictive, and mostly interesting for sparse tables.

8.6 Bidirected Graph Models

We present in this section the case where the random vector X_V is indexed by the vertices of a graph $\mathcal{G} = (V, \mathcal{B})$ having only bidirected edges (a BG). In this case, we assume that missing edges of $\mathcal{G}$ encode independence relationships between subvectors of X_V according to the *dual global Markov property*: $X_A \perp\!\!\!\perp X_B \mid X_C$ for every triple A, B, C of pairwise disjoint subsets of V with $A, B \neq \emptyset$ and $V \setminus (A \cup B \cup C)$ separating A from B in $\mathcal{G}$. Note the different roles played by C in the definition here and in that for the UGs. When $C = \emptyset$ the dual global Markov property implies $X_A \perp\!\!\!\perp X_B$ if no edge of $\mathcal{G}$ joins a vertex in A with a vertex in B. In particular, letting $A = \{u\}$ and $B = \{v\}$, we get $X_u \perp\!\!\!\perp X_v$ if $u \leftrightarrow v \notin \mathcal{B}$. This *dual pairwise Markov property* only concerns bivariate marginal distributions and does not imply all global Markov statements, in general, not even in the case of positive cell probabilities; see e.g. [15, Example 3].

We define the *global graphical model* specified by $\mathcal{G}$ as the family $\mathbb{M}_+(\mathcal{G})$ of all positive distributions for X_V such that X_V is dual global Markov with respect to $\mathcal{G}$, and the *pairwise graphical model* specified by $\mathcal{G}$ as the broader family $\widetilde{\mathbb{M}}_+(\mathcal{G})$ of all positive distributions for X_V such that X_V is dual pairwise Markov with respect to $\mathcal{G}$. For instance, if the distribution of X_V belongs to the pairwise model specified by the graph in Figure 8.2(ii), it holds that both $X_e \perp\!\!\!\perp X_f$ and $X_e \perp\!\!\!\perp X_h$ (dual pairwise statements), but these statements do not imply that that X_e is independent of X_{fh}. If the distribution of X_V also belongs to the global model specified by the same graph, it also holds that $X_e \perp\!\!\!\perp X_{fh}$ and $X_e \perp\!\!\!\perp X_f \mid X_h$ (dual global statements). We discuss below the parameterization of $\mathbb{M}_+(\mathcal{G})$, and $\widetilde{\mathbb{M}}_+(\mathcal{G})$, which will give rise to a more variegated picture than in the UG case. We start by presenting a third Markov property, which we do from a historical perspective.

The dual global Markov property was introduced by Kauermann [31] to specify Gaussian models by means of covariance graphs; see also [4]. Kauermann's work is part of a range of early efforts in the area of marginal modeling, which were rooted in an interest for the effect of unobserved variables. These efforts, e.g. the paper [45] by Pearl and Wermuth, are documented in [15], where Drton and Richardson unlocked discrete BG models using the *connected set Markov property* of Richardson [46]: $X_{C_1} \perp\!\!\!\perp \ldots \perp\!\!\!\perp X_{C_r}$ for every disconnected subset D of V with connected components $C_1, \ldots, C_r$, $r \geq 2$. The connected set Markov property is clearly implied by the dual global Markov property; the case $r = 2$ was considered before, while the case $r \geq 3$ follows from the case $r - 1$ by letting $A = C_1 \uplus \cdots \uplus C_{r-1}$ and $B = C_r$. Richardson [46] showed that the two properties are in fact equivalent, and this led to the parameterization of global BG models for categorical data.

We have seen in Section 8.5 that the saturated model for a binary random vector X_V is a regular exponential family with canonical statistic $t_{D'}(\cdot)$, $D' \subseteq V$ with $D' \neq \emptyset$, where $t_{D'}(\epsilon_D) = 1$ if $D' \subseteq D$ and $t_{D'}(\epsilon_D) = 0$ if $D' \nsubseteq D$, for any choice of $D \subseteq V$. The *mean parameter* of this exponential family is $\mu_{D'} = \mathrm{E}[t_{D'}(X_V)] = \mathrm{P}(X_{D'} = 1)$, $D' \subseteq V$ with $D' \neq \emptyset$. Letting $\mu_\emptyset = 1$, consistently with our empty margin convention, we find $\mu = Z\pi$. Hence $\pi = M\mu$, and the map $\pi \to \mu$ is a valid reparameterization with closed form inverse. However, the entries of μ are subject to non-trivial inequality constraints, and they are not variation independent; see [15] for an illustration of this issue.

Marginal independence statements on X_V, such as $X_u \perp\!\!\!\perp X_v$, are easier to express in terms of μ than in terms of π; for instance $X_u \perp\!\!\!\perp X_v$ corresponds to $\mu_{uv} = \mu_u \mu_v$, but it implies no factorization of π (in general). This is the route followed by Drton and Richardson in [15], working with connected set independence statements on X_V. Here, following [50], we use a log-linear expansion of μ to turn marginal independencies into zero constraints. Specifically, we define the *log-mean linear parameter* of X_V as $\gamma = M^\top \log \mu$. In this way $\gamma_\emptyset = \log \mu_\emptyset = 0$, $\gamma_v = \log \mu_v$, and $\gamma_{uv} = \log\{\mu_{uv}/(\mu_u \mu_v)\}$, for all $u, v \in V$. Hence, for instance, $X_u \perp\!\!\!\perp X_v$ corresponds to $\gamma_{uv} = 0$. The map $\mu \to \gamma$ can be inverted in closed form as $\mu = \exp Z^\top \gamma$, but the entries of γ, like those of μ, are not variation independent. The following result connects the log-mean linear parameter to marginal independencies, using the basic tools introduced in Section 8.4.

Theorem 8.6.1. *Let $\pi > 0$ be the probability parameter of a binary random vector X_V. Set*

$\mu = Z\pi$ *and* $\gamma = M^\top \log \mu$. *Then, for every pair* A, B *of non-empty disjoint subsets of* V, *the following statements are equivalent:*

(i) $X_A \perp\!\!\!\perp X_B$;

(ii) $\log \mu_{A' \cup B'} - \log \mu_{B'} = \log \mu_{A'} - \log \mu_\emptyset$ *for every* $A' \subseteq A$ *and* $B' \subseteq B$ *with* $A', B' \neq \emptyset$;

(iii) $\gamma_D = 0$ *for every* $D \subseteq A \cup B$ *such that both* $D \cap A \neq \emptyset$ *and* $D \cap B \neq \emptyset$.

Proof. Since $\mu_\emptyset = 1$, we can rewrite (ii) as

$$\mathrm{P}(X_{A'} = 1 \mid X_{B'} = 1) = \mathrm{P}(X_{A'} = 1)$$

for every $A' \subseteq A$ and $B' \subseteq B$ with $A', B' \neq \emptyset$; this is (iv) in Lemma 8.4.1 with $C = \emptyset$, which shows the equivalence of (i) and (ii). On the other hand, the equivalence of (ii) and (iii) follows directly from Lemma 8.4.3 applied with $\omega = \log \mu$ and $C = \emptyset$. □

Theorem 8.6.1, first given as Theorem 1 in [50], shows that every connected set Markov statement corresponds to a set of zero constraints on some log-mean linear parameters. Indeed, if $D = C_1 \uplus \cdots \uplus C_r$ is a disconnected subset of V partitioned into its connected components, we have $X_{C_1} \perp\!\!\!\perp \ldots \perp\!\!\!\perp X_{C_r}$ if and only if $\gamma_{D'} = 0$ for every $D' \subseteq D$ such that D' is not included in any C_j. This is Corollary 1 in [50] and can be shown by finite induction on r. For instance, in the graph of Figure 8.2(ii), the independence statement $X_e \perp\!\!\!\perp X_f$ corresponds to the constraint $\gamma_{ef} = 0$, while the statement $X_e \perp\!\!\!\perp X_{fh}$ corresponds to the condition $\gamma_{ef} = \gamma_{eh} = \gamma_{efh} = 0$.

As a consequence of the above described correspondence, if X_V is dual global Markov with respect to $\mathcal{G} = (V, \mathcal{B})$, we have $\gamma_D = 0$ for every disconnected $D \subseteq V$. This can be seen using $D' = D$ above. Conversely, if $\gamma_{D'} = 0$ for every disconnected $D' \subseteq V$, we have $X_{C_1} \perp\!\!\!\perp \ldots \perp\!\!\!\perp X_{C_r}$ for every disconnected $D = C_1 \uplus \cdots \uplus C_r \subseteq V$. Therefore, a parameterization for $\mathbb{M}_+(\mathcal{G})$ is given by γ_C, C non-empty and connected. For instance, the global model specified by the graph in Figure 8.2(ii) is parameterized by $\gamma_e, \gamma_f, \gamma_g, \gamma_h, \gamma_{eg}, \gamma_{fg}, \gamma_{fh}, \gamma_{gh}, \gamma_{efg}, \gamma_{egh}, \gamma_{fgh}, \gamma_{efgh}$. These parameters are not variation independent, but they vary in a section of the image of the map $\pi \to M^\top \log Z\pi$, which is concretely determined by the condition $\gamma_{ef} = \gamma_{eh} = \gamma_{efh} = 0$. Every choice of γ compatible with this section uniquely determines a sum-to-one $\pi = M \exp(Z^\top \gamma)$ in closed form.

We remark that the global model defined by the BG in Figure 8.2(ii) is twelve dimensional, whereas the analogous model defined by the UG in Figure 8.2(i) is nine dimensional; global BG models are less parsimonious than UG models, because every complete set is connected. Of course, pairwise BG models are even less parsimonious, because $\mathbb{M}_+(\mathcal{G})$ corresponds to a reduced set of constraints: $\lambda_D = 0$ for every disconnected $D \subseteq V$ such that $|D| = 2$, that is, $\lambda_{uv} = 0$ for every $u, v \in V$ such that $u \leftrightarrow v \notin \mathcal{B}$. Nevertheless, on top of a given global BG model, pairwise independencies can help to increase parsimony.

The log-mean linear parameters are related to the *dependence ratios* of Ekholm, Smith and McDonald [17]. Specifically, second order dependence ratios coincide with second order log-mean linear interactions, while higher order dependence ratios are defined differently, but they can replace higher order log-mean linear interactions in Theorem 8.6.1. For instance, in general $\gamma_{uvw} \neq \log\{\mu_{uvw}/(\mu_u \mu_v \mu_w)\}$, but equality holds if $\gamma_{uv} = \gamma_{uw} = \gamma_{vw} = 0$.

Two alternative parameterizations for discrete BG models were proposed by Lupparelli, Marchetti and Bergsma [40] within the general class of *marginal log-linear parameterizations* introduced by Bergsma and Rudas [6]. Specifically, these authors showed that any discrete BG model can be parameterized by the multivariate logistic parameterization of Glonek and McCullagh [25], whose importance for marginal modeling was first pointed out by Kauermann [32], or by a graph specific parameterization generated by the disconnected sets. Unlike the log-mean linear parameterization, marginal log-linear parameterizations do not guarantee a closed form inverse. However, as pointed out in [40], marginal log-linear parameterizations may lead to variation independence and, furthermore, they suggest a class of parsimonious models defined by constraining certain higher order log-linear parameters to zero, with an interpretation along the lines described at the end of Section 8.5.

Log-mean linear parameters allow for marginal context specific independencies on any subvector X_S of X_V when the context is defined by $X_W = 1$, where $W \subseteq V \setminus S$. These context specific independences can indeed be expressed as sum-to-zero constraints on γ, as shown by the following result, which is another direct consequence of Lemma 8.4.4. We remark that the resulting class of quasi-graphical models depends on the actual coding (labeling) of the variable levels, i.e. it is not invariant to permutation of levels.

Proposition 8.6.2. *Let $\pi > 0$ be the probability parameter of a binary random vector X_V and set $\gamma = M^\top \log Z\pi$. Furthermore, let $V = S \uplus W$ be a partition of V such that $S \neq \emptyset$. Then, for any non-empty subset $S' \subseteq S$, it holds that*

$$\gamma_{S'}^{S|X_W=1} = \sum_{W' \subseteq W} \gamma_{S' \cup W'}, \tag{8.16}$$

where $\gamma^{S|X_W=1}$ denotes the log-mean linear parameter of $X_S \mid X_W = 1$.

Proof. Mimicking the proof of Proposition 8.5.2, we first compute the generic log-mean linear parameter of $X_S \mid X_W = 1$ as

$$\gamma_{S'}^{S|X_W=1} = \sum_{S'' \subseteq S'} (-1)^{|S' \setminus S''|} \log \mathrm{P}(X_{S''} = 1 \mid X_W = 1) \tag{8.17}$$

$$= \sum_{S'' \subseteq S'} (-1)^{|S' \setminus S''|} \log \mathrm{P}(X_{S''} = 1, X_W = 1) = \gamma_{[W]S'}^{S \cup W} \tag{8.18}$$

if $S' \neq \emptyset$; this condition ensures the equality of (8.17) and (8.18) while the equality in (8.18) follows from the assumption $S \cap W = \emptyset$ and the equation (8.8) with $\omega = \log \mu^{S \cup W}$, hence $\theta = \gamma^{S \cup W}$, where $\mu^{S \cup W}$ is the mean parameter of $X_{S \cup W}$. Then, we obtain (8.16) from (8.18) by applying Lemma 8.4.4 and noting that $\gamma_D^{S \cup W} = \gamma_D^V = \gamma_D$ for all $D \subseteq S \cup W$. □

In light of Proposition 8.6.2, starting from the global model defined by the graph in Figure 8.2(ii), we may like to impose the context specific independence $X_e \perp\!\!\!\perp X_g \mid X_h = 1$. This will result in the additional constraint $\gamma_{eg} + \gamma_{egh} = 0$, obtained by setting either $S = \{e, g\}$ and $W = \{h\}$ or $S = \{e, f, g\}$ and $W = \{h\}$. Notice the *upward compatibility* property of log-mean linear parameters: $\gamma_D^{S|X_W=1} = \gamma_D^{D|X_W=1}$ for all $S \subseteq V \setminus W$ such that $D \subseteq S$. This is a typical feature of marginal parameterizations, which is not shared by the log-linear parameters, because the latter are conditional on all other variables being zero. It can be considered an advantage, because it makes the interaction indexed by a given margin independent of the variables not included in that margin. For instance, letting $W = \emptyset$ and $D = \{u, v\}$, the interaction γ_{uv} only depends on X_{uv}, and not on $X_{V \setminus \{u,v\}}$. The practical meaning is that, if we measure the association between X_u and X_v by γ_{uv}, we should not worry about which other variables we include in our study (in X_V).

8.7 Regression Graph Models

We now assume that the entries of X_V are associated with the vertices of a regression graph $\mathcal{G} = (V, \mathcal{U}, \mathcal{B}, \mathcal{D})$ compatible with a given block ordering of variables $(U, T_1, \ldots, T_k)$.

In applied work, the partition of variables into context variables, intermediate responses and primary responses, as well as the specification of a block ordering of variables, usually follows from subject matter working hypotheses, or from the temporal ordering of the variables. A certain degree of arbitrariness is inevitable in the determination of this partition, and it is therefore important to remark that, although we deem that assuming a block ordering of variables facilitates the presentation of the theory, the material in this section only depends on the structure of the RG, i.e. on its sets of vertices and edges.

The distribution of X_V obeys the *RG Markov property* with respect to $\mathcal{G}$ when

(RG1) $X_A \perp\!\!\!\perp X_{\mathrm{pre}(T_i)\setminus \mathrm{pa}(A)} \mid X_{\mathrm{pa}(A)}$ for all $A \subseteq T_i$ with $A \neq \emptyset$ and $\mathrm{pre}(T_i) \setminus \mathrm{pa}(A) \neq \emptyset$ and every $i = 1, \ldots, k$;

(RG2) the distribution of $X_{T_i} \mid X_{\mathrm{pre}(T_i)}$ obeys the dual global Markov property with respect to $\mathcal{G}_{T_i}$, for every $i = 1, \ldots, k$;

(RG3) the distribution of X_U obeys the global Markov property with respect to $\mathcal{G}_U$.

The RG Markov property is closely related to other existing Markov properties. A MRCG is a RG without the undirected block, i.e. with $U = \emptyset$, and conditions (RG1) and (RG2) above define the MRCG Markov property given e.g. in Drton [14]. More precisely, Marchetti and Lupparelli [42] showed that the Markov property defined by (RG1) and (RG2) is equivalent to the MRCG Markov property. This also implies that models defined by these two conditions do not depend on the chosen block ordering of variables, but only on the partial ordering implied by the RG. It is also worth noting that RG models are special cases of more general classes of models specified by mixed graphs, which use the tool of *m-separation* to encode conditional independence relationships; see e.g. Sadeghi and Lauritzen [52] and Richardson [46]. Finally, we remark that the RG Markov property considered here is a global property, and we refer to Sadeghi and Wermuth [53] for the relevant pairwise properties.

The graphical model defined by a RG can be dealt with by considering an UG model for the undirected block and then a MRCG model for the bidirected blocks conditional on the undirected block. However, we here follow a joint approach, and we start by considering the basic case where there are only two blocks, i.e. $V = U \uplus T$ with X_U the context variables and X_T the primary responses. The graph depicted in Figure 8.2(iii) is an instance of this framework, with $U = \{a, b, c, d\}$ and $T = \{e, f, g, h\}$. In this case, we define the *hybrid probability-mean parameter*, or *hybrid parameter* for short, as the vector $\pi^{(U,T)}$ with entries

$$\pi_D^{(U,T)} = \mathrm{P}(X_D = 1, X_{U\setminus D} = 0), \qquad D \subseteq V. \tag{8.19}$$

The name 'hybrid' comes from the fact that $\pi^{(U,T)}$ combines the features of the parameterizations π and μ of X_V used in Sections 8.5 and 8.6, respectively. Furthermore, the latter are special cases of $\pi^{(U,T)}$, because $\pi = \pi^{(V,\emptyset)}$ and $\mu = \pi^{(\emptyset,V)}$.

The generic entry μ_D of the mean parameter is a probability marginalized over $X_{V\setminus D}$. The entries of μ are thus computed by summing the relevant entries of π, and this is done by the matrix Z_V through the equation $\mu = Z_V \pi$. For a given subset $A \subseteq V$, let I_A be the identity matrix indexed by the subsets of A, that is, the matrix with entries $\mathbb{I}_{\{D=E\}}$, for $D, E \subseteq A$. The generic entry $\pi_D^{(U,T)}$ of the hybrid parameter is a probability marginalized over $X_{T\setminus D}$, but not over $X_{U\setminus D}$, and it can be shown that $\pi^{(U,T)}$ can be computed from π as $\pi^{(U,T)} = (I_U \otimes Z_T)\pi$, where $\otimes$ denotes the Kronecker product of matrices. Consequently, π can be computed in closed form from $\pi^{(U,T)}$ as $\pi = (I_U \otimes M_T)\pi^{(U,T)}$, and this shows that $\pi^{(U,T)}$ is a valid reparameterization of X_V.

We now apply the usual log-linear expansion to $\pi^{(U,T)}$ and obtain the *log-hybrid linear parameter*

$$\varphi^{(U,T)} = M^\top \log \pi^{(U,T)}, \tag{8.20}$$

which extends both λ and γ because $\varphi^{(V,\emptyset)} = \lambda$ and $\varphi^{(\emptyset,V)} = \gamma$. We will write φ for $\varphi^{(U,T)}$ when (U,T) is clear from the context. We next show that φ is suited to parameterize basic RG models with a single bidirected block. The following theorem connects conditional independencies between variables in different blocks with vanishing entries of φ; it runs parallel to Theorems 8.5.1 and 8.6.1.

Theorem 8.7.1. *Let $\pi > 0$ be the probability parameter of a binary random vector X_V. Set $\pi^{(U,T)} = (I_U \otimes Z_T)\pi$ and $\varphi = M^\top \log \pi^{(U,T)}$, where $V = U \uplus T$ is a partition of V. Then, for every $A \subseteq T$ and $B \subseteq U$ with $A, B \neq \emptyset$, the following statements are equivalent:*

(i) $X_A \perp\!\!\!\perp X_B | X_{U\setminus B}$;

(ii) $\log \pi^{(U,T)}_{A'\cup B'\cup C'} - \log \pi^{(U,T)}_{B'\cup C'} = \log \pi^{(U,T)}_{A'\cup C'} - \log \pi^{(U,T)}_{C'}$ *for every $A' \subseteq A$, $B' \subseteq B$ and $C' \subseteq U \setminus B$, such that $A', B' \neq \emptyset$;*

(iii) $\varphi_D = 0$ *for every $D \subseteq A \cup U$ such that both $D \cap A \neq \emptyset$ and $D \cap B \neq \emptyset$.*

Proof. If we rewrite (ii) as

$$\mathrm{P}(X_{A'} = 1 \mid X_B = \epsilon_{B'}, X_C = \epsilon_{C'}) = \mathrm{P}(X_{A'} = 1 \mid X_B = 0, X_C = \epsilon_{C'})$$

for every $A' \subseteq A$, $B' \subseteq B$ and $C' \subseteq U \setminus B$, such that $A', B' \neq \emptyset$, we obtain point (iii) of Lemma 8.4.1, and this shows the equivalence of (i) and (ii). On the other hand, the equivalence of (ii) and (iii) follows directly from Lemma 8.4.3 with $\omega = \log \pi^{(U,T)}$ and $C = U \setminus B$. □

Assume, for example, that the distribution of X_V obeys the RG Markov property with respect to the graph in Figure 8.2(iii). Then, it holds that $X_e \perp\!\!\!\perp X_{ab} \mid X_{cd}$. Theorem 8.7.1 tells us that this statement is satisfied if and only if $\varphi_D = 0$ for all $D \subseteq \{a,b,c,d,e\}$ such that $e \in D$ and $D \cap \{a,b\} \neq \emptyset$. The graph also implies $X_{gh} \perp\!\!\!\perp X_{abcd}$ and this independence holds if and only if $\varphi_D = 0$ for all $D \subseteq \{a,b,c,d,g,h\}$ such that $D \cap \{a,b,c,d\} \neq \emptyset$ and $D \cap \{g,h\} \neq \emptyset$. It follows that the zero structure of φ characterizes the family of distributions for X_V satisfying (RG1) for any basic RG with two blocks, as detailed by the next result.

Corollary 8.7.2. *In the setting of Theorem 8.7.1, the distribution of X_V satisfies condition (RG1) with respect to a RG with block ordering (U,T) if and only if $\varphi_D = 0$ for every $D \subseteq V$ such that $D \cap T \neq \emptyset$ and $D \cap U \nsubseteq \mathrm{pa}(D \cap T)$.*

Proof. If the distribution of X_V satisfies condition (RG1), we have $X_{D\cap T} \perp\!\!\!\perp X_{U\setminus \mathrm{pa}(D\cap T)} \mid X_{\mathrm{pa}(D\cap T)}$ for all $D \subseteq V$ such that $D \cap T \neq \emptyset$ and $U \setminus \mathrm{pa}(D \cap T) \neq \emptyset$. In this case, if $D \cap U \nsubseteq \mathrm{pa}(D \cap T)$ then $\varphi_D = 0$ by point (iii) of Theorem 8.7.1. Conversely, assume that $\varphi_D = 0$ for every $D \subseteq V$ such that $D \cap T \neq \emptyset$ and $D \cap U \nsubseteq \mathrm{pa}(D \cap T)$. We have to show that $X_A \perp\!\!\!\perp X_B \mid X_{U\setminus B}$ for all non-empty $A \subseteq T$ such that $B = U \setminus \mathrm{pa}(A)$ is also non-empty. By Theorem 8.7.1 it suffices to show that $\varphi_D = 0$ for every $D \subseteq A \cup U$ such that $D \cap A \neq \emptyset$ and $D \cap B \neq \emptyset$. Now $D \subseteq A \cup U$ implies $D \cap T = D \cap A \neq \emptyset$, while $D \cap B \neq \emptyset$ implies $D \cap U \nsubseteq \mathrm{pa}(A)$; since $\mathrm{pa}(D \cap A) \subseteq \mathrm{pa}(A)$ it follows that $D \cap U \nsubseteq \mathrm{pa}(D \cap A) = \mathrm{pa}(D \cap T)$. Therefore, we have $\varphi_D = 0$. □

Every subset $D \subseteq V$ with $D \cap T \neq \emptyset$ identifies two subsets of U: $\mathrm{pa}(D \cap T)$ and $D \cap U$. If $D \cap U \subseteq \mathrm{pa}(D \cap T)$, that is, all the elements of D belonging to U are parents of $D \cap T$, then the condition (RG1) imposes no constraints on φ_D; otherwise φ_D vanishes. Consider the RG model defined by the graph in Figure 8.2(iii). The set $D = \{b,d,f,h\}$ can be split into $D \cap T = \{f,h\}$ and $D \cap U = \{b,d\}$, and since $\{b,d\} \nsubseteq \mathrm{pa}(\{f,h\}) = \{d\}$ it follows that $\varphi_{bdfh} = 0$. As a second example, in the same graph, let $D = \{c,d,e,g,h\}$. In this case we find $D \cap T = \{e, g, h\}$ and $D \cap U = \{c, d\}$, so that $\{c, d\} \subseteq \text{pa}(\{e, g, h\}) = \{c, d\}$, and therefore φ_{cdegh} is not restricted by condition (RG1). Finally, the set $D = \{d, e, h\}$ is such that $D \cap T = \{e, h\}$ and $D \cap U = \{d\}$, so that $\{d\} \subseteq \text{pa}(\{e, h\}) = \{c, d\}$. Hence, φ_{deh} is not restricted by condition (RG1); however we will see below that this parameter vanishes in the RG model from Figure 8.2(iii) because of condition (RG2).

Section 8.6 describes how the the log-mean linear parameters can be used to parameterize BG models. To deal with the independence relationships encoded through condition (RG2) by the BG induced by T, we show that every log-mean linear parameter of $X_T \mid X_U$ can be obtained as a linear combination of log-hybrid linear parameters. In this way, the theory of Section 8.6 can be readily applied to RG models. Like Propositions 8.5.2 and 8.6.2, this result is a direct application of Lemma 8.4.4.

Proposition 8.7.3. *In the setting of Theorem 8.7.1, for every $T' \subseteq T$ and $U' \subseteq U$ with $T' \neq \emptyset$, it holds that*

$$\gamma_{T'}^{T|X_U=\epsilon_{U'}} = \sum_{U'' \subseteq U'} \varphi_{T' \cup U''}, \tag{8.21}$$

where $\gamma^{T|X_U=\epsilon_{U'}}$ denotes the log-mean linear parameter of $X_T \mid X_U = \epsilon_{U'}$.

Proof. The same arguments used in the proof of Proposition 8.6.2 can be used to show that

$$\gamma_{T'}^{T|X_U=\epsilon_{U'}} = \sum_{T'' \subseteq T'} (-1)^{|T' \setminus T''|} \log \text{P}(X_{T''} = 1 \mid X_U = \epsilon_{U'}) = \varphi_{[U']T'},$$

and the equality (8.21) follows from Lemma 8.4.4. □

Proposition 8.7.3 shows that, for any given subset $U' \subseteq U$, marginal independencies in the distribution of $X_T \mid X_U = \epsilon_{U'}$ correspond to linear constraints on the entries of φ. On the other hand, the marginal independencies implied by the dual Markov property in (RG2) hold true for every $U' \subseteq U$; they thus correspond to the vanishing of certain entries of φ.

Corollary 8.7.4. *In the setting of Theorem 8.7.1, let $T' \subseteq T$ with $T' \neq \emptyset$. The following conditions are equivalent:*

(i) $\gamma_{T'}^{T|X_U=\epsilon_{U'}} = 0$ *for every* $U' \subseteq U$*;*

(ii) $\varphi_{T' \cup U'} = 0$ *for every* $U' \subseteq U$.

Proof. Point (ii) implies (i) by (8.21) whereas (i) implies (ii) because, by Möbius inversion, $\varphi_{T' \cup U'} = \sum_{U'' \subseteq U'} (-1)^{|U' \setminus U''|} \gamma_{T'}^{T|X_U=\epsilon_{U''}}$ for every $U' \subseteq U$. □

In the RG model from Figure 8.2(iii) condition (RG2) implies $X_e \perp\!\!\!\perp X_h \mid X_{abcd}$, and it follows from Corollary 8.7.4 and Theorem 8.6.1 in Section 8.6 that this independence relationship is satisfied if and only if $\varphi_{\{e,h\} \cup U'} = 0$ for every $U' \subseteq \{a, b, c, d\}$; in particular we have $\varphi_{deh} = 0$, which is the term mentioned above. On the other hand, the vertices e and g are joined by an edge in the graph of Figure 8.2(iii), and therefore condition (RG2) does not imply the conditional independence of X_e and X_g given X_{abcd}. However, one can exploit Proposition 8.7.3 to identify context specific independencies. For example, the relationship $X_e \perp\!\!\!\perp X_g \mid X_{abcd} = 0$ is given by the constraint $\varphi_{eg} = 0$, whereas $\varphi_{eg} + \varphi_{deg} = 0$ is equivalent to $X_e \perp\!\!\!\perp X_g \mid X_{abc} = 0, X_d = 1$.

Finally, although we do not pursue this issue here, it is worth mentioning that Proposition 8.7.3 is useful for the interpretation of the log-hybrid parameters $\varphi_{T' \cup U'}$, $U' \subseteq U$ and $T' \subseteq T$ with $T' \neq \emptyset$, because it connects these parameters with the *log-mean linear regression parameters* introduced by Lupparelli and Roverato [41], which can be interpreted in terms of relative risks.

We now turn to condition (RG3) and show that the entries of φ indexed by subsets of U are the log-linear parameters computed on the marginal distribution of X_U so that they can be used to deal with the UG model defined by $\mathcal{G}_U$ (and possibly with related context specific independencies) as illustrated in Section 8.5.

Proposition 8.7.5. *In the setting of Theorem 8.7.1, it holds that*

$$\varphi_{U'} = \lambda^U_{U'} \quad \text{for every } U' \subseteq U,$$

where λ^U is the log-linear parameter of X_U.

Proof. This follows from the fact that $\pi^{(U,T)}_{U'} = \pi^U_{U'}$, where π^U is the probability parameter of X_U, for every $U' \subseteq U$. □

It follows from Proposition 8.7.5 that the log-hybrid linear parameter φ can be partitioned into a subvector $(\varphi_D)_{D \subseteq U} = \lambda^U$, with variation independent entries, which parameterizes the distribution of X_U, and a subvector $(\varphi_D)_{D \not\subseteq U}$, whose entries are typically not variation independent, which parameterizes the distribution of $X_T \mid X_U$. The RG we consider here is a CG, and the parameterization of this kind of model is typically approached by explicitly considering the conditional distribution of $X_T \mid X_U$ and then the marginal distribution of X_U. Our approach is slightly different. The hybrid parameter $\pi^{(U,T)}$ parameterizes the joint distribution of $X_{U \cup T}$, and it is constructed in such a way that the independence relationships among variables in X_U are modeled by an UG, whereas those among variables in X_T by a BG. The direction of the arrows takes no role in the computation of $\pi^{(U,T)}$, and therefore of φ. The fact that a subvector of φ parameterizes the distribution of $X_T \mid X_U$ follows from the Möbius inversion formula applied to $\log \pi^{(U,T)}$, and for this reason the properties of φ follow from the general properties of Möbius inversion given in Section 8.4.2. Indeed, although we do not pursue this aspect here, by Lemma 8.4.4 other kinds of context specific independencies may be specified by means of linear constraints on φ. These include context specific independencies between variables in different blocks or between subsets of context variables conditionally on the responses.

In summary, in a basic RG model with two blocks, the only φ-terms which are not constrained to vanish are those indexed by subsets $D \subseteq V$ such that either $D \subseteq U$ and D is complete or $D \cap T \neq \emptyset$ and both $D \cap T$ is connected and $D \cap U \subseteq \mathrm{pa}(D \cap T)$.

We now sketch the general case of a RG model with an arbitrary number of blocks; let $(U, T_1, \ldots, T_k)$ be the block ordering of its variables. The distribution of X_V is characterized by the collection of conditional distributions $X_{T_i} \mid X_{\mathrm{pre}(T_i)}$, $i = 1, \ldots, k$, together with the marginal distribution of X_U. Accordingly, the φ-terms for the distribution of X_V can be obtained by noting that, for every $i = 1, \ldots, k$, the vector $\varphi^{(\mathrm{pre}(T_i),T_i)}$ parameterizes the distribution of $X_{\mathrm{pre}(T_i) \cup T_i}$ in such a way that the subvector $(\varphi^{(\mathrm{pre}(T_i),T_i)}_D)_{D \subseteq \mathrm{pre}(T_i)} = \lambda^{\mathrm{pre}(T_i)}$ parameterizes the marginal distribution of $X_{\mathrm{pre}(T_i)}$, while the complementary subvector $(\varphi^{(\mathrm{pre}(T_i),T_i)}_D)_{D \not\subseteq \mathrm{pre}(T_i)}$ parameterizes the conditional distribution of $X_{T_i} \mid X_{\mathrm{pre}(T_i)}$.

8.8 Non-binary Variables and Likelihood Inference

The approach we have followed to deal with RG models, and therefore with the special cases of UG and BG models, for binary data can be readily extended to the general case of categorical variables with an arbitrary number of levels. This can be done by introducing a version of the cross-classified table deprived of one level for every variable. Formally, the *restricted state space* of X_v is defined as $\mathcal{J}_v = \mathcal{I}_v \setminus \{0\} = \{1, \ldots, \ell_v - 1\}$, so that the restricted state space of X_V is given by $\mathcal{J}_V = \times_{v \in V} \mathcal{J}_v$. It is worth remarking that the choice of the level to be removed is arbitrary; see also Drton [14]. For every non-empty subset $A \subseteq V$, and every cell $j \in \mathcal{J}_V$, we denote by $j_A \in \mathcal{J}_A$ the marginalization of j to the restricted state space of X_A. Then, we can generalize the vector $\pi^{(U,T)}$ in (8.19) to the collection of vectors $\pi^{(U,T,j)}$, $j \in \mathcal{J}_V$, where

$$\pi_D^{(U,T,j)} = \mathrm{P}(X_D = j_D, X_{U \setminus D} = 0), \qquad \text{for every } D \subseteq V.$$

It can be shown that the above collection of vectors parameterizes the distribution of X_V, and therefore we can generalize the log-hybrid parameter of X_V by computing the collection of vectors $\varphi^{(U,T,j)} = M^\top \log \pi^{(U,T,j)}$, $j \in \mathcal{J}_V$. Clearly, the case $T = \emptyset$ gives a collection of log-linear vectors, whereas the case $U = \emptyset$ gives a collection of log-mean linear vectors, whose properties have been investigated in [49]. It follows that the results on Möbius inversion given in Section 8.4.2 directly apply to $\varphi^{(U,T,j)}$, for each $j \in \mathcal{J}_V$; see also Letac and Massam [39]. Hence, for the formal generalization of most of the results presented in this chapter, it is sufficient to give a more general version of Lemma 8.4.1 for non-binary variables. This is not given here, due to space constraints, but we note that its proof follows the same lines as that of Lemma 8.4.1.

We conclude the chapter with a discussion of the inferential issues raised by a random sample from a distribution in a discrete quasi-graphical model of UG, BG or RG type. We take a likelihood based approach, and we consider as data a table of multinomial cross-classified counts, which is a sufficient statistic in this setting. The reader is referred to Lauritzen [36, Sect. 4.2.1] for an illustration of the link to Poisson sampling. We focus for simplicity on the binary case, but we note that this restriction is not at all essential.

We first deal with point estimation, that is, likelihood maximization. Consider a model specified by a set of independence constraints including a mixture of conditional, marginal, and context-specific relationships. Working with the probability parameter, any such *independence model* is as a subset of the standard $2^{|V|} - 1$ dimensional simplex. The standard simplex is bounded, and thus the model is relatively compact. Since the likelihood is a continuous function of cell probabilities, it follows that its constrained supremum is attained within the closure of the model. This means that independence models always admit an *extended* maximum likelihood estimator (MLE). Note however that, in general, there is no guarantee of uniqueness. If all counts are positive, any extended MLE will have positive cell probabilities, and it will thus be a proper MLE, because in the discrete case independence relationships are preserved in the limit; see Lauritzen [36, Prop. 3.12]. This will happen with probability tending to one as the sample size tends to infinity, when sampling from a positive distribution. We can therefore count on the existence of a MLE for the models presented in this chapter. The reader is referred to the work of Fienberg and Rinaldo [22] for an in-depth study of MLE existence in log-linear models; see also the recent work of Wang, Rauh and Massam [54].

The actual computation of a MLE for discrete independence models is a problem in constrained optimization. We here content ourselves with mentioning a couple of landmark references in the statistical literature. Aitchison and Silvey [2] gave a general iterative algorithm for constrained likelihood maximization, which was later specialized to categorical data by Lang [35]. A recent discussion of this approach is offered by Evans and Forcina [18]. Lang's algorithm encompasses all models presented in this chapter, and thus provides a unified approach to their fitting. However, its output is a *local* maximum, and therefore one should try different starting points to explore a multimodal likelihood. For RG models the performance of this approach should be compared with that of a couple of dedicated algorithms, which we briefly mention at the end of this section.

We now turn to confidence regions and model comparison. A key fact here is that all models presented in this chapter are exponential families. Discrete UG models are regular exponential families, which is apparent from their parameterization in terms of log-linear parameters. Discrete BG models are curved exponential families, as shown by Drton and Richardson [15, Corollary 3]. The same result can be obtained by studying the bijection between γ and λ. Any discrete RG model is the product of an UG model and a MRCG model. The latter is a product of conditional BG models and thus a curved exponential family; see Drton [14, Corollary 10]. Finally, context-specific independencies expressed by sum-to-zero constraints give rise to curved exponential families. For large samples from both regular and curved exponential families, there is a unique MLE, which is normally distributed, and model deviance with respect to the saturated model is chi-square distributed. These results can be used to deal with uncertainty on the model and its parameters. For a review of the general theory of exponential families the reader is referred to Lauritzen [36, Appendix D] and references therein.

Finally, we provide more information on the MLE for RG models. For regular exponential families the MLE is such that the mean parameter equals the mean canonical statistic. In the UG case this means that marginal cell probabilities should equal marginal relative frequencies for all non-empty complete margins. This characterization is exploited by the iterative proportional fitting/scaling (IPF/IPS) algorithm, which is discussed for instance by Højsgaard, Edwards and Lauritzen [28, Sect. 2.3.4]. A dual of this algorithm, named the iterative conditional fitting (ICF) algorithm, was developed in [15] by Drton and Richardson to fit discrete BG models, and then extended by Drton [13] to fit the bidirected components of regression graph models. Remarkably, in the special case of *decomposable* UG models the MLE is available in closed form, and the IPS algorithm converges in at most two iterations. Recall that an UG is decomposable when all its cycles induce a subgraph with at least one extra edge; see e.g. Lauritzen [36, Sect. 2.1.2].

Acknowledgments

The authors are grateful to Monia Lupparelli and Kayvan Sadeghi, for useful discussion, as well as to two anonymous referees, for their insightful comments. Work of Alberto Roverato was supported by the Air Force Office of Scientific Research under award number FA9550-17-1-0039.

Bibliography

[1] Alan Agresti. *Categorical Data Analysis.* John Wiley and Sons, New York, 3rd edition, 2013.

[2] John Aitchison and S. D. Silvey. Maximum-likelihood estimation of parameters subject to restraints. *Annals of Mathematical Statistics*, 29(3):813–828, 1958.

[3] Steen A. Andersson, David Madigan, and Michael D. Perlman. Alternative Markov properties for chain graphs. *Scandinavian Journal of Statistics*, 28(1):33–85, 2001.

[4] Moulinath Banerjee and Thomas S. Richardson. On a dualization of graphical Gaussian models: A correction note. *Scandinavian Journal of Statistics*, 30(4):817–820, 2003.

[5] Francesco Bartolucci, Roberto Colombi, and Antonio Forcina. An extended class of marginal link functions for modelling contingency tables by equality and inequality constraints. *Statistica Sinica*, 17(2):691–711, 2007.

[6] Wicher P. Bergsma and Tamas Rudas. Marginal models for categorical data. *Annals of Statistics*, 30(1):140–159, 2002.

[7] Craig Boutilier, Nir Friedman, Moises Goldszmidt, and Daphne Koller. Context-specific independence in Bayesian networks. In *Proceedings of the Twelfth International Conference on Uncertainty in Artificial Intelligence*, 115–123. Morgan Kaufmann Publishers, 1996.

[8] Jukka Corander. Labelled graphical models. *Scandinavian Journal of Statistics*, 30(3):493–508, 2003.

[9] Robert G. Cowell, A. Philip Dawid, Steffen L. Lauritzen, and David J. Spiegelhalter. *Probabilistic Networks and Expert Systems.* Springer-Verlag, New York, 1999.

[10] David R. Cox and Nanny Wermuth. Linear dependencies represented by chain graphs. *Statistical Science*, 8(3):204–218, 1993.

[11] David R. Cox and Nanny Wermuth. *Multivariate Dependencies: Models, Analysis, and Interpretation.* Chapman & Hall, Boca Raton, 1996.

[12] John N. Darroch, Steffen L. Lauritzen, and Terry P. Speed. Markov fields and log-linear interaction models for contingency tables. *Annals of Statistics*, 8(3):522–539, 1980.

[13] Mathias Drton. Iterative conditional fitting for discrete chain graph models. In Paula Brito, editor, *COMPSTAT 2008 – Proceedings in Computational Statistics*, pages 93–104. Springer, 2008.

[14] Mathias Drton. Discrete chain graph models. *Bernoulli*, 15(3):736–753, 2009.

[15] Mathias Drton and Thomas S. Richardson. Binary models for marginal independence. *Journal of the Royal Statistical Society: Series B (Statistical Methodology)*, 70(2):287–309, 2008.

[16] David Edwards. *Introduction to Graphical Modelling.* Springer-Verlag, New York, 2nd edition, 2000.

[17] Anders Ekholm, Peter W. F. Smith, and John W. McDonald. Marginal regression analysis of a multivariate binary response. *Biometrika*, 82(4):847–854, 1995.

[18] Robin J. Evans and Antonio Forcina. Two algorithms for fitting constrained marginal models. *Computational Statistics & Data Analysis*, 66:1–7, 2013.

[19] Robin J. Evans and Thomas S. Richardson. Maximum likelihood fitting of acyclic directed mixed graphs to binary data. In P. Grunwald and P. Spirtes, editors, *Proceedings of the 26th Conference on Uncertainty in Artificial Intelligence (UAI 2010)*, 177–184. AUAI Press, 2010.

[20] Robin J. Evans and Thomas S. Richardson. Marginal log-linear parameters for graphical Markov models. *Journal of the Royal Statistical Society: Series B (Statistical Methodology)*, 75(4):743–768, 2013.

[21] Robin J. Evans and Thomas S. Richardson. Markovian acyclic directed mixed graphs for discrete data. *Annals of Statistics*, 42(4):1452–1482, 2014.

[22] Stephen E. Fienberg and Alessandro Rinaldo. Maximum likelihood estimation in log-linear models. *Annals of Statistics*, 40(2):996–1023, 2012.

[23] Morten Frydenberg. The chain graph Markov property. *Scandinavian Journal of Statistics*, 17(4):333–353, 1990.

[24] Dan Geiger, Christopher Meek, and Bernd Sturmfels. On the toric algebra of graphical models. *Annals of Statistics*, 34(3):1463–1492, 2006.

[25] Gary F. V. Glonek and Peter McCullagh. Multivariate logistic models. *Journal of the Royal Statistical Society: Series B (Methodological)*, 57(3):533–546, 1995.

[26] Philip Hall. A contribution to the theory of groups of prime-power order. *Proceedings of the London Mathematical Society*, 2(1):29–95, 1934.

[27] Søren Højsgaard. Statistical inference in context specific interaction models for contingency tables. *Scandinavian Journal of Statistics*, 31(1):143–158, 2004.

[28] Søren Højsgaard, David Edwards, and Steffen L. Lauritzen. *Graphical Models with R*. Springer Science+Business Media, New York, 2012.

[29] James E. Johndrow, Anirban Bhattacharya, and David B. Dunson. Tensor decompositions and sparse log-linear models. *Annals of Statistics*, 45(1):1–38, 2017.

[30] Jukka Jokinen. Fast estimation algorithm for likelihood-based analysis of repeated categorical responses. *Computational Statistics & Data Analysis*, 51(3):1509–1522, 2006.

[31] Göran Kauermann. On a dualization of graphical Gaussian models. *Scandinavian Journal of Statistics*, 23(1):105–116, 1996.

[32] Göran Kauermann. A note on multivariate logistic models for contingency tables. *Australian Journal of Statistics*, 39(3):261–276, 1997.

[33] Robert Kennes and Philippe Smets. Computational aspects of the Möbius transform. In P. Bonissone, M. Henrion, L. Kanal, and J. Lemmer, editors, *Proceedings of the 6th Conference on Uncertainty in Artificial Intelligence (UAI1990)*, 401–416. North-Holland, 1991.

[34] Jan T.A. Koster. Marginalizing and conditioning in graphical models. *Bernoulli*, 8(6):817–840, 2002.

[35] Joseph B. Lang. Maximum likelihood methods for a generalized class of log-linear models. *Annals of Statistics*, 24(2):726–752, 1996.

[36] Steffen L. Lauritzen. *Graphical Models.* Oxford University Press, Oxford, 1996.

[37] Steffen L. Lauritzen and Thomas S. Richardson. Chain graph models and their causal interpretations. *Journal of the Royal Statistical Society: Series B (Statistical Methodology)*, 64(3):321–348, 2002.

[38] Steffen L. Lauritzen and Nanny Wermuth. Graphical models for associations between variables, some of which are qualitative and some quantitative. *Annals of Statistics*, 17(1):31–57, 1989.

[39] Gérard Letac and Hélène Massam. Bayes factors and the geometry of discrete hierarchical loglinear models. *Annals of Statistics*, 40(2):861–890, 2012.

[40] Monia Lupparelli, Giovanni M. Marchetti, and Wicher P. Bergsma. Parameterizations and fitting of bi-directed graph models to categorical data. *Scandinavian Journal of Statistics*, 36(3):559–576, 2009.

[41] Monia Lupparelli and Alberto Roverato. Log-mean linear regression models for binary responses with an application to multimorbidity. *Journal of the Royal Statistical Society: Series C (Applied Statistics)*, 66(2):227–252, 2017.

[42] Giovanni M. Marchetti and Monia Lupparelli. Chain graph models of multivariate regression type for categorical data. *Bernoulli*, 17(3):827–844, 2011.

[43] Andrew Moore and Mary Soon Lee. Cached sufficient statistics for efficient machine learning with large datasets. *Journal of Artificial Intelligence Research*, 8:67–91, 1998.

[44] Henrik Nyman, Johan Pensar, Timo Koski, and Jukka Corander. Stratified graphical models - context-specific independence in graphical models. *Bayesian Analysis*, 9(4):883–908, 2014.

[45] Judea Pearl and Nanny Wermuth. When can association graphs admit a causal interpretation? In P. Cheeseman and R. W. Oldford, editors, *Selecting Models from Data: Artificial Intelligence and Statistics IV*, 205–214. Springer, New York, 1994.

[46] Thomas S. Richardson. Markov properties for acyclic directed mixed graphs. *Scandinavian Journal of Statistics*, 30(1):145–157, 2003.

[47] Thomas S. Richardson and Peter Spirtes. Ancestral graph Markov models. *Annals of Statistics*, 30(4):962–1030, 2002.

[48] Gian-Carlo Rota. On the foundations of combinatorial theory I. Theory of Möbius functions. *Probability Theory and Related Fields*, 2(4):340–368, 1964.

[49] Alberto Roverato. Log-mean linear parameterization for discrete graphical models of marginal independence and the analysis of dichotomizations. *Scandinavian Journal of Statistics*, 42(2):627–648, 2015.

[50] Alberto Roverato, Monia Lupparelli, and Luca La Rocca. Log-mean linear models for binary data. *Biometrika*, 100(2):485–494, 2013.

[51] Tamás Rudas, Wicher P. Bergsma, and Renáta Németh. Marginal log-linear parameterization of conditional independence models. *Biometrika*, 97:1006–1012, 2010.

[52] Kayvan Sadeghi and Steffen L. Lauritzen. Markov properties for mixed graphs. *Bernoulli*, 20(2):676–696, 2014.

[53] Kayvan Sadeghi and Nanny Wermuth. Pairwise Markov properties for regression graphs. *Stat*, 2016.

[54] Nanwei Wang, Johannes Rauh, and Hélène Massam. Approximating faces of marginal polytopes in discrete hierarchical models. *arXiv:1603.04843*, 2016.

[55] Louis Weisner. Abstract theory of inversion of finite series. *Transactions of the American Mathematical Society*, 38(3):474–484, 1935.

[56] Nanny Wermuth. Probability distributions with summary graph structure. *Bernoulli*, 17(3):845–879, 2011.

[57] Nanny Wermuth and David R. Cox. Graphical Markov models: overview. In J. D. Wright, editor, *International Encyclopedia of the Social and Behavioral Sciences*, volume 10, pages 341–350. Elsevier, Oxford, 2nd edition, 2015.

[58] Nanny Wermuth and Steffen L. Lauritzen. Graphical and recursive models for contigency tables. *Biometrika*, 70(3):537–552, 1983.

[59] Nanny Wermuth and Steffen L. Lauritzen. On substantive research hypotheses, conditional independence graphs and graphical chain models. *Journal of the Royal Statistical Society: Series B (Statistical Methodology)*, 52(1):21–50, 1990.

[60] Nanny Wermuth and Kayvan Sadeghi. Sequences of regressions and their independences. *TEST*, 21(2):215–252, 2012.

[61] Joe Whittaker. *Graphical Models in Applied Multivariate Statistics*. John Wiley and Sons, Chichester, 1990.

9

Gaussian Graphical Models

Caroline Uhler
Department of Electrical Engineering and Computer Science, Massachusetts Institute of Technology

CONTENTS

After the discussion of discrete graphical models in the preceding Chapter 8, we now turn to the continuous setting and introduce Gaussian graphical models. As we will see in this chapter, assuming Gaussianity leads to a rich geometric structure that can be exploited for parameter estimation. However, Gaussianity is not only assumed for mathematical simplicity. Gaussian distributions are commonly used for modeling continuous phenomena. As a consequence of the central limit theorem, physical quantities that are expected to be the sum of many independent contributions often follow approximately a Gaussian distribution. For example, people's height is approximately normally distributed; height is believed to be the sum of many independent contributions from various genetic and environmental factors.

Another reason for assuming normality is that the Gaussian distribution has maximum entropy among all real-valued distributions with a specified mean and covariance. Hence, assuming Gaussianity imposes the least number of structural constraints beyond the first and second moments. So another reason for assuming Gaussianity is that it is the least-informative distribution. In addition, many physical systems tend to move towards maximal entropy configurations over time.

Throughout this chapter, we denote by $\mathbb{S}^p$ the vector space of real symmetric $p \times p$ matrices. This vector space is equipped with the *trace inner product* $\langle A, B\rangle := \operatorname{tr}(AB)$. In addition, we denote by $\mathbb{S}^p_{\succeq 0}$ the convex cone of positive semidefinite matrices. Its interior is the open cone $\mathbb{S}^p_{\succ 0}$ of positive definite matrices.

A random vector $X \in \mathbb{R}^p$ is distributed according to the *multivariate Gaussian distribution* $\mathcal{N}(\mu, \Sigma)$ with parameters $\mu \in \mathbb{R}^p$ (the *mean*) and $\Sigma \in \mathbb{S}^p_{\succ 0}$ (the *covariance matrix*), if it has density function

$$f_{\mu,\Sigma}(x) = (2\pi)^{-p/2}(\det \Sigma)^{-1/2} \exp\left\{-\frac{1}{2}(x-\mu)^T\Sigma^{-1}(x-\mu)\right\}, \quad x \in \mathbb{R}^p.$$

In the following, we denote the inverse covariance matrix, also known as the *precision matrix* or the *concentration matrix*, by K. In terms of $K = \Sigma^{-1}$ and using the trace inner product on $\mathbb{S}^p$, the density $f_{\mu,\Sigma}$ can equivalently be formulated as:

$$f_{\mu,K}(x) = \exp\left\{\mu^T K x - \langle K, \frac{1}{2}xx^T\rangle - \frac{p}{2}\log(2\pi) + \frac{1}{2}\log\det(K) - \frac{1}{2}\mu^T K\mu\right\}.$$

Hence, the Gaussian distribution is an *exponential family* with *canonical parameters* $(-\mu^T K, K)$, *sufficient statistics* $(x, \frac{1}{2}xx^T)$ and *log-partition function* (also known as the *cumulant generating function*) $\frac{p}{2}\log(2\pi) - \frac{1}{2}\log\det(K) + \frac{1}{2}\mu^T K\mu$; see [5, 11] for an introduction to exponential families.

Let $G = (V, E)$ be an undirected graph with vertices $V = [p]$ and edges E, where $[p] = \{1, \ldots, p\}$. A random vector $X \in \mathbb{R}^p$ is said to *satisfy the (undirected) Gaussian graphical model with graph* G, if X has a multivariate Gaussian distribution $\mathcal{N}(\mu, \Sigma)$ with

$$\left(\Sigma^{-1}\right)_{i,j} = 0 \quad \text{for all } (i,j) \notin E.$$

Hence, the graph G describes the sparsity pattern of the concentration matrix. This explains why G is also known as the *concentration graph*. As we will see in Section 9.1, missing edges in G also correspond to conditional independence relations in the corresponding Gaussian graphical model. Hence, sparser graphs correspond to simpler models with fewer canonical parameters and more conditional independence relations.

Gaussian graphical models are the continuous counter-piece to the Ising models that were discussed in Section 5.4.1 (Ising models are special versions of the log-linear models from Chapter 8). Like Ising models, Gaussian graphical models are quadratic exponential families. These families only model pairwise interactions between nodes, i.e., interactions are only on the edges of the underlying graph G. But nevertheless, Ising models and Gaussian graphical models are extremely flexible models; in fact, they can capture any pairwise correlation structure that can be constructed for binary or for continuous data.

This chapter discusses maximum likelihood (ML) estimation for Gaussian graphical models. There are two problems of interest in this regard: (1) to estimate the edge weights, i.e. the canonical parameters, given the graph structure, and (2) to learn the underlying graph structure. This chapter is mainly focused with the first problem (Sections 9.2-9.6), while the second problem is only discussed in Section 9.7. The second problem is particularly important in the high-dimensional setting when the number of samples n is smaller than the number of variables p. This problem is the subject of Chapters 12 and 14.

The remainder of this chapter is structured as follows: In Section 9.1, we examine conditional independence relations for Gaussian distributions. Then, in Section 9.2, we introduce the Gaussian likelihood. We show that ML estimation for Gaussian graphical models is a convex optimization problem and we describe its dual optimization problem. In Section 9.3, we analyze this dual optimization problem and explain the close links to positive definite matrix completion problems studied in linear algebra. In Section 9.4, we develop a geometric picture of ML estimation for Gaussian graphical models that complements the point of view of convex optimization. The combination of convex optimization, positive definite matrix completion, and convex geometry allows us to obtain results about the existence of the maximum likelihood estimator (MLE) and algorithms for computing the MLE. These are presented in Section 9.5 and in Section 9.6, respectively. Gaussian graphical models are defined by zero constraints on the concentration matrix K. In Section 9.7, we describe methods for learning the underlying graph, or equivalently, the zero pattern of K. Finally, in Section 9.8, we end with a discussion of other Gaussian models with linear constraints on the concentration matrix or the covariance matrix.

9.1 The Gaussian Distribution and Conditional Independence

We start this section by reviewing some of the extraordinary properties of Gaussian distributions. The following result shows that the Gaussian distribution is closed under marginalization and conditioning. We here only provide proofs that will be useful in later sections of the chapter. A complete proof of the following well-known result can be found for example in [4, 17].

Proposition 9.1.1. *Let $X \in \mathbb{R}^p$ be distributed as $\mathcal{N}(\mu, \Sigma)$ and partition the random vector X into two components $X_A \in \mathbb{R}^a$ and $X_B \in \mathbb{R}^b$ such that $a + b = p$. Let μ and Σ be partitioned accordingly, i.e.,*

$$\mu = \begin{pmatrix} \mu_A \\ \mu_B \end{pmatrix} \quad \text{and} \quad \Sigma = \begin{pmatrix} \Sigma_{A,A} & \Sigma_{A,B} \\ \Sigma_{B,A} & \Sigma_{B,B} \end{pmatrix},$$

where, for example, $\Sigma_{B,B} \in \mathbb{S}^b_{\succ 0}$. Then,

(a) the marginal distribution of X_A is $\mathcal{N}(\mu_A, \Sigma_{A,A})$;

(b) the conditional distribution of $X_A \mid X_B = x_B$ is $\mathcal{N}(\mu_{A|B}, \Sigma_{A|B})$, where

$$\mu_{A|B} = \mu_A + \Sigma_{A,B}\Sigma_{B,B}^{-1}(x_B - \mu_B) \quad \text{and} \quad \Sigma_{A|B} = \Sigma_{A,A} - \Sigma_{A,B}\Sigma_{B,B}^{-1}\Sigma_{B,A}.$$

Proof. We only prove (b) to demonstrate the importance of Schur complements when working with Gaussian distributions. Fixing x_B, we find by direct calculation that the conditional density $f(x_A \mid x_B)$ is proportional to:

$$\begin{aligned} f(x_A \mid x_B) &\propto \exp\Big\{ -\frac{1}{2}(x_A - \mu_A)^T K_{A,A}(x_A - \mu_A) - (x_A - \mu_A)^T K_{A,B}(x_B - \mu_B)\Big\} \\ &\propto \exp\Big\{ -\frac{1}{2}\big(x_A - \mu_A - K_{A,A}^{-1}K_{A,B}(x_B - \mu_B)\big)^T K_{A,A} \\ &\qquad\qquad \times \big(x_A - \mu_A - K_{A,A}^{-1}K_{A,B}(x_B - \mu_B)\big)\Big\}, \end{aligned} \tag{9.1}$$

where we used the same partitioning for K as for Σ. Using Schur complements, we obtain

$$K_{A,A}^{-1} = \Sigma_{A,A} - \Sigma_{A,B}\Sigma_{B,B}^{-1}\Sigma_{B,B},$$

and hence $K_{A,A} = \Sigma_{A|B}^{-1}$. Similarly, we obtain $K_{A,A}^{-1}K_{A,B} = -\Sigma_{A,B}\Sigma_{B,B}^{-1}$. Combining these two identities with the conditional density in (9.1) completes the proof. □

These basic properties of the multivariate Gaussian distribution have interesting implications with respect to the interpretation of zeros in the covariance and the concentration matrix. Namely, as described in the following corollary, zeros correspond to (*conditional*) *independence relations*. For disjoint subsets $A, B, C \subset [p]$ we denote the statement that X_A is conditionally independent of X_B given X_C by $X_A \perp\!\!\!\perp X_B \mid X_C$. If $C = \emptyset$, then we write $X_A \perp\!\!\!\perp X_B$.

Corollary 9.1.2. *Let $X \in \mathbb{R}^p$ be distributed as $\mathcal{N}(\mu, \Sigma)$ and let $i, j \in [p]$ with $i \neq j$. Then*

(a) $X_i \perp\!\!\!\perp X_j$ if and only if $\Sigma_{i,j} = 0$;

(b) $X_i \perp\!\!\!\perp X_j \mid X_{[p]\setminus\{i,j\}}$ if and only if $K_{i,j} = 0$ if and only if $\det(\Sigma_{[p]\setminus\{i\},[p]\setminus\{j\}}) = 0$.

Proof. Statement (a) follows directly from the expression for the conditional mean in Proposition 9.1.1 (b). From the expression for the conditional covariance in Proposition 9.1.1 (b) it follows that $\Sigma_{\{i,j\}|([p]\setminus\{i,j\})} = (K_{\{i,j\},\{i,j\}})^{-1}$. To prove (b), note that it follows from (a) that $X_i \perp\!\!\!\perp X_j \mid X_{[p]\setminus\{i,j\}}$ if and only if the 2×2 conditional covariance matrix $\Sigma_{\{i,j\}|([p]\setminus\{i,j\})}$ is diagonal. This is the case if and only if $K_{\{i,j\},\{i,j\}}$ is diagonal, or equivalently, $K_{i,j} = 0$. This proves the first equivalence in (b). The second equivalence is a consequence of the cofactor formula for matrix inversion, since

$$K_{i,j} = (\Sigma^{-1})_{i,j} = (-1)^{i+j} \frac{\det(\Sigma_{[p]\setminus\{i\},[p]\setminus\{j\}})}{\det(\Sigma)},$$

which completes the proof. □

Corollary 9.1.2 shows that for undirected Gaussian graphical models a missing edge (i,j) in the underlying graph G (i.e. the concentration graph) corresponds to the conditional independence relation $X_i \perp\!\!\!\perp X_j \mid X_{[p]\setminus\{i,j\}}$. Corollary 9.1.2 can be generalized to an equivalence between any conditional independence relation and the vanishing of a particular almost principal minor of Σ or K. This is shown in the following proposition; see also Proposition 3.3.10.

Proposition 9.1.3. *Let $X \in \mathbb{R}^p$ be distributed as $\mathcal{N}(\mu, \Sigma)$. Let $i, j \in [p]$ with $i \neq j$ and let $S \subseteq [p] \setminus \{i,j\}$. Then the following statements are equivalent:*

(a) $X_i \perp\!\!\!\perp X_j \mid X_S$;

(b) $\det(\Sigma_{iS,jS}) = 0$, *where* $iS = \{i\} \cup S$;

(c) $\det(K_{iR,jR}) = 0$, *where* $R = [p] \setminus (S \cup \{i,j\})$.

Proof. By Proposition 9.1.1 (a), the marginal distribution of $X_{S\cup\{i,j\}}$ is Gaussian with covariance matrix $\Sigma_{ijS,ijS}$. Then Corollary 9.1.2 (b) implies the equivalence between (a) and (b). Next we show the equivalence between (a) and (c): It follows from Proposition 9.1.1 (b) that the inverse of $K_{ijR,ijR}$ is equal to the conditional covariance $\Sigma_{ijR|S}$. Hence by Corollary 9.1.2 (a), the conditional independence statement in (a) is equivalent to $((K_{ijR,ijR})^{-1})_{ij} = 0$, which by the cofactor formula for matrix inversion is equivalent to (c). This completes the proof. □

9.2 The Gaussian Likelihood and Convex Optimization

Given n i.i.d. observations $X^{(1)}, \ldots, X^{(n)}$ from $\mathcal{N}(\mu, \Sigma)$, we define the *sample covariance matrix* as

$$S = \frac{1}{n}\sum_{i=1}^{n}(X^{(i)} - \bar{X})(X^{(i)} - \bar{X})^T,$$

where $\bar{X} = \frac{1}{n}\sum_{i=1}^{n} X^{(i)}$ is the *sample mean*. We will see that $\bar{X}$ and S are sufficient statistics for the Gaussian model and hence we can write the log-likelihood function in terms of these quantities. Ignoring the normalizing constant, the Gaussian log-likelihood expressed as a function of (μ, Σ) is

$$\begin{aligned}
\ell(\mu, \Sigma) &\propto -\frac{n}{2} \log\det(\Sigma) - \frac{1}{2}\sum_{i=1}^{n}(X^{(i)} - \mu)^T\Sigma^{-1}(X^{(i)} - \mu) \\
&= -\frac{n}{2} \log\det(\Sigma) - \frac{1}{2}\operatorname{tr}\left(\Sigma^{-1}\Big(\sum_{i=1}^{n}(X^{(i)} - \mu)(X^{(i)} - \mu)^T\Big)\right) \\
&= -\frac{n}{2} \log\det(\Sigma) - \frac{n}{2}\operatorname{tr}(S\Sigma^{-1}) - \frac{n}{2}(\bar{X} - \mu)^T\Sigma^{-1}(\bar{X} - \mu),
\end{aligned}$$

where for the last equality we expanded $X^{(i)} - \mu = (X^{(i)} - \bar{X}) + (\bar{X} - \mu)$ and used the fact that $\sum_{i=1}^{n}(X^{(i)} - \bar{X}) = 0$. Hence, it can easily be seen that in the *saturated* (unconstrained) *model* where $(\mu, \Sigma) \in \mathbb{R}^p \times \mathbb{S}^p_{\succ 0}$, the MLE is given by

$$\hat{\mu} = \bar{X} \quad \text{and} \quad \hat{\Sigma} = S,$$

assuming that $S \in \mathbb{S}^p_{\succ 0}$.

ML estimation under general constraints on the parameters (μ, Σ) can be complicated. Since Gaussian graphical models only pose constraints on the covariance matrix, we will restrict ourselves to models where the mean μ is unconstrained, i.e. $(\mu, \Sigma) \in \mathbb{R}^p \times \Theta$, where $\Theta \subseteq \mathbb{S}^p_{\succ 0}$. In this case, $\hat{\mu} = \bar{X}$ and the ML estimation problem for Σ boils down to the optimization problem

$$\begin{aligned}
\underset{\Sigma}{\text{maximize}} \quad & -\log\det(\Sigma) - \operatorname{tr}(S\Sigma^{-1}) \\
\text{subject to} \quad & \Sigma \in \Theta.
\end{aligned} \tag{9.2}$$

While this objective function as a function of the covariance matrix Σ is in general not concave over the whole cone $\mathbb{S}^p_{\succ 0}$, it is concave over a large region of the cone, namely for all $\Sigma \in \mathbb{S}^p_{\succ 0}$ such that $\Sigma - 2S \in \mathbb{S}^p_{\succ 0}$ (see [10, Exercise 7.4]).

Gaussian graphical models are given by linear constraints on K. So it is convenient to write the optimization problem (9.2) in terms of the concentration matrix K:

$$\begin{aligned}
\underset{K}{\text{maximize}} \quad & \log\det(K) - \operatorname{tr}(SK) \\
\text{subject to} \quad & K \in \mathcal{K},
\end{aligned} \tag{9.3}$$

where $\mathcal{K} = \Theta^{-1}$. In particular, for a Gaussian graphical model with graph $G = (V, E)$ the constraints are given by $K \in \mathcal{K}_G$, where

$$\mathcal{K}_G := \{K \in \mathbb{S}^p_{\succ 0} \mid K_{i,j} = 0 \text{ for all } i \neq j \text{ with } (i,j) \notin E\}.$$

In the following, we show that the objective function in (9.3), i.e. as a function of K, is concave over its full domain $\mathbb{S}^p_{\succ 0}$. Since $\mathcal{K}_G$ is a convex cone, this implies that ML estimation for Gaussian graphical models is a convex optimization problem.

Proposition 9.2.1. *The function $f(Y) = \log\det(Y) - \operatorname{tr}(SY)$ is concave on its domain $\mathbb{S}^p_{\succ 0}$.*

Proof. Since $\operatorname{tr}(SY)$ is linear in Y it suffices to prove that the function $\log\det(Y)$ is concave over $\mathbb{S}^p_{\succ 0}$. We prove this by showing that the function is concave on any line in $\mathbb{S}^p_{\succ 0}$. Let $Y \in \mathbb{S}^p_{\succ 0}$ and consider the line $Y + tV$, $V \in \mathbb{S}^p$, that passes through Y. It suffices to prove that $g(t) = \log\det(Y + tV)$ is concave for all $t \in \mathbb{R}$ such that $Y + tV \in \mathbb{S}^p_{\succ 0}$. This can be seen from the following calculation:

$$\begin{aligned} g(t) &= \log\det(Y + tV) \\ &= \log\det(Y^{1/2}(I + tY^{-1/2}VY^{-1/2})Y^{1/2}) \\ &= \log\det(Y) + \sum_{i=1}^{p} \log(1 + t\lambda_i), \end{aligned}$$

where I denotes the identity matrix and λ_i are the eigenvalues of $Y^{-1/2}VY^{-1/2}$. This completes the proof, since $\log\det(Y)$ is a constant and $\log(1 + t\lambda_i)$ is concave in t. □

As a consequence of Proposition 9.2.1, we can study the dual of (9.3) with $\mathcal{K} = \mathcal{K}_G$. See e.g. [10] for an introduction to convex optimization and duality theory. The Lagrangian of this convex optimization problem is given by:

$$\begin{aligned} \mathcal{L}(K, \nu) &= \log\det(K) - \mathrm{tr}(SK) - 2 \sum_{(i,j)\notin E, i\neq j} \nu_{i,j} K_{i,j} \\ &= \log\det(K) - \sum_{i=1}^{p} S_{i,i}K_{i,i} - 2 \sum_{(i,j)\in E} S_{i,j}K_{i,j} - 2 \sum_{(i,j)\notin E,\, i\neq j} \nu_{i,j}K_{i,j}, \end{aligned}$$

where $\nu = (\nu_{i,j})_{(i,j)\notin E}$ are the Lagrangian multipliers. To simplify the calculations, we omit the constraint $K \in \mathbb{S}^p_{\succ 0}$. This can be done, since it is assumed that K is in the domain of $\mathcal{L}$. Maximizing $\mathcal{L}(K, \nu)$ with respect to K gives

$$(\hat{K}^{-1})_{i,j} = \begin{cases} S_{i,j} & \text{if } i = j \text{ or } (i, j) \in E \\ \nu_{i,j} & \text{otherwise.} \end{cases}$$

The Lagrange dual function is obtained by plugging in $\hat{K}$ for K in $\mathcal{L}(K, \nu)$, which results in

$$g(\nu) = \log\det(\hat{K}) - \mathrm{tr}(\hat{K}^{-1}\hat{K}) = \log\det(\hat{K}) - p.$$

Hence, the dual optimization problem to ML estimation in Gaussian graphical models is given by

$$\begin{aligned} &\underset{\Sigma\in\mathbb{S}^p_{\succ 0}}{\text{minimize}} \quad -\log\det\Sigma - p \\ &\text{subject to} \quad \Sigma_{i,j} = S_{i,j} \ \text{ for all } \ i = j \text{ or } (i, j) \in E. \end{aligned} \tag{9.4}$$

Note that this optimization problem corresponds to *entropy maximization* for fixed sufficient statistics. In fact, this dual relationship between likelihood maximization and entropy maximization holds more generally for exponential families; see [46].

Sections 9.4 and 9.5 are centered around the existence of the MLE. We say that the MLE does not exist if the likelihood does not attain the global maximum. Note that the identity matrix is a strictly feasible point for (9.3) with $\mathcal{K} = \mathcal{K}_G$. Hence, the MLE does not exist if and only if the likelihood is unbounded. *Slater's constraint qualification* states that the existence of a strictly primal feasible point is sufficient for strong duality to hold for a convex optimization problem (see e.g. [10] for an introduction to convex optimization). Since the identity matrix is a strictly feasible point for (9.3), strong duality holds for the optimization problems (9.3) with $\mathcal{K} = \mathcal{K}_G$ and (9.4), and thus we can equivalently study the dual problem (9.4) to obtain insight into ML estimation for Gaussian graphical models. In particular, the MLE does not exist if and only if there exists no feasible point for the dual optimization problem (9.4). In the next section, we give an algebraic description of this property. A generalization of this characterization for the existence of the MLE holds also more generally for regular exponential families; see [5, 11].

9.3 The MLE as a Positive Definite Completion Problem

To simplify notation, we use $E^* = E \cup \{(i,i) \mid i \in V\}$. We introduce the projection on the augmented edge set E^*, namely

$$\pi_G : \mathbb{S}^p_{\succeq 0} \to \mathbb{R}^{|E^*|}, \quad \pi_G(S) = \{S_{i,j} \mid (i,j) \in E^*\}.$$

Note that $\pi_G(S)$ can be seen as a partial matrix, where the entries corresponding to missing edges in the graph G have been removed (or replaced by question marks as shown in (9.5) for the case where G is the 4-cycle). In the following, we use S_G to denote the partial matrix corresponding to $\pi_G(S)$. Using this notation, the constraints in the optimization problem (9.4) become $\Sigma_G = S_G$. Hence, existence of the MLE in a Gaussian graphical model is a *positive definite matrix completion problem*: The MLE exists if and only if the partial matrix S_G can be completed to a positive definite matrix. In that case, the MLE $\hat{\Sigma}$ is the unique positive definite completion that maximizes the determinant. And as a consequence of strong duality, we obtain that $(\hat{\Sigma}^{-1})_{i,j} = 0$ for all $(i,j) \notin E^*$.

Positive definite completion problems have been widely studied in the linear algebra literature [6, 7, 22, 30]. Clearly, if a partial matrix has a positive definite completion, then every specified (i.e., with given entries) principal submatrix is positive definite. Hence, having a positive definite completion imposes some obvious necessary conditions. However, these conditions are in general not sufficient as seen in the following example, where the graph G is the 4-cycle:

$$S_G = \begin{pmatrix} 1 & 0.9 & ? & -0.9 \\ 0.9 & 1 & 0.9 & ? \\ ? & 0.9 & 1 & 0.9 \\ -0.9 & ? & 0.9 & 1 \end{pmatrix}. \tag{9.5}$$

It can easily be checked that this partial matrix does not have a positive definite completion, although all the specified 2×2-minors are positive. Hence, the MLE does not exist for the sufficient statistics given by S_G.

This example leads to the question if there are graphs for which the obvious necessary conditions are also sufficient for the existence of a positive definite matrix completion. The following remarkable theorem proven in [22] answers this question.

Theorem 9.3.1. *For a graph G the following statements are equivalent:*

(a) A G-partial matrix $M_G \in \mathbb{R}^{|E^*|}$ has a positive definite completion if and only if all completely specified submatrices in M_G are positive definite.

(b) G is chordal *(also known as* triangulated*), i.e. every cycle of length 4 or larger has a chord.*

The proof in [22] is constructive. It makes use of the fact that any chordal graph can be turned into a complete graph by adding one edge at a time in such a way, that the resulting graph remains chordal at each step. Following this ordering of edge additions, the partial matrix is completed entry by entry in such a way as to maximize the determinant of the largest complete submatrix that contains the missing entry. Hence the proof in [22] can be turned into an algorithm for finding a positive definite completion for partial matrices on chordal graphs.

We will see in Section 9.5 how to make use of positive definite completion results to determine the minimal number of observations required for existence of the MLE in a Gaussian graphical model.

9.4 ML Estimation and Convex Geometry

After having introduced the connections to positive definite matrix completion problems, we now discuss how convex geometry enters the picture for ML estimation in Gaussian graphical models. We already introduced the set

$$\mathcal{K}_G := \{K \in \mathbb{S}^p_{\succ 0} \mid K_{i,j} = 0 \text{ for all } (i,j) \notin E^*\}.$$

Note that $\mathcal{K}_G$ is a convex cone obtained by intersecting the convex cone $\mathbb{S}^p_{\succ 0}$ with a linear subspace. We call $\mathcal{K}_G$ the *cone of concentration matrices.*

A second convex cone that plays an important role for ML estimation in Gaussian graphical models is the *cone of sufficient statistics* denoted by $\mathcal{S}_G$. It is defined as the projection of the positive semidefinite cone onto the entries E^*, i.e.,

$$\mathcal{S}_G := \pi_G(\mathbb{S}^p_{\succeq 0}).$$

In the following proposition, we show how these two cones are related to each other.

Proposition 9.4.1. *Let G be an undirected graph. Then the cone of sufficient statistics $\mathcal{S}_G$ is the dual cone to the cone of concentration matrices $\mathcal{K}_G$, i.e.*

$$\mathcal{S}_G = \left\{ S_G \in \mathbb{R}^{|E^*|} \mid \langle S_G, K\rangle \geq 0 \textit{ for all } K \in \mathcal{K}_G \right\}. \tag{9.6}$$

Proof. Let $\mathcal{K}_G^\vee$ denote the dual of $\mathcal{K}_G$, i.e. the right-hand side of (9.6). Let

$$\mathcal{L}_G := \{K \in \mathbb{S}^p \mid K_{i,j} = 0 \text{ for all } (i,j) \notin E^*\}$$

denote the linear subspace defined by the graph G. We denote by $\mathcal{L}_G^\perp$ the orthogonal complement of $\mathcal{L}_G$ in $\mathbb{S}^p$. Using the fact that the dual of the full-dimensional cone $\mathbb{S}^p_{\succ 0}$ is $\mathbb{S}^p_{\succeq 0}$, i.e. $(\mathbb{S}^p_{\succ 0})^\vee = \mathbb{S}^p_{\succeq 0}$, general duality theory for convex cones (see e.g. [8]) implies:

$$\mathcal{K}_G^\vee = (\mathbb{S}^p_{\succ 0} \cap \mathcal{L}_G)^\vee = (\mathbb{S}^p_{\succeq 0} + \mathcal{L}_G^\perp)/\mathcal{L}_G^\perp = \mathcal{S}_G,$$

which completes the proof. □

It is clear from this proof that the geometric picture we have started to draw holds more generally for any Gaussian model that is given by linear constraints on the concentration matrix. We will therefore use $\mathcal{L}$ to denote any linear subspace of $\mathbb{S}^p$ and we assume that $\mathcal{L}$ intersects the interior of $\mathbb{S}^p_{\succeq 0}$. Hence, $\mathcal{L}_G$ is a special case defined by zero constraints given by missing edges in the graph G. Then,

$$\mathcal{K}_\mathcal{L} = \mathcal{L} \cap \mathbb{S}^p_{\succ 0}, \quad \mathcal{S}_\mathcal{L} = \pi_\mathcal{L}(\mathbb{S}^p_{\succeq 0}) = \mathcal{K}_\mathcal{L}^\vee,$$

where $\pi_\mathcal{L} : \mathbb{S}^p \to \mathbb{S}^p/\mathcal{L}^\perp$. Note that given a basis $K_1, \ldots, K_d$ for $\mathcal{L}$, this map can be identified with

$$\pi_\mathcal{L} : \mathbb{S}^p \to \mathbb{R}^d, \quad S \mapsto \big(\langle S, K_1\rangle, \ldots, \langle S, K_d\rangle\big).$$

A *spectrahedron* is a convex set that is defined by linear matrix inequalities. Given a sample covariance matrix S, we define the spectrahedron

$$\text{fiber}_\mathcal{L}(S) = \{\Sigma \in \mathbb{S}^p_{\succ 0} \mid \langle \Sigma, K\rangle = \langle S, K\rangle \text{ for all } K \in \mathcal{L}\}.$$

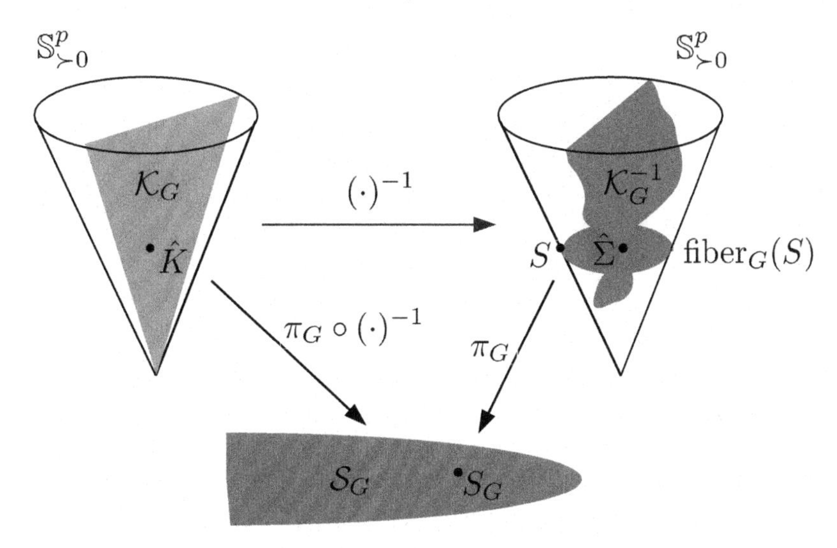

FIGURE 9.1: Geometry of maximum likelihood estimation in Gaussian graphical models. The cone $\mathcal{K}_G$ consists of all concentration matrices in the model and $\mathcal{K}_G^{-1}$ is the corresponding set of covariance matrices. The cone of sufficient statistics $\mathcal{S}_G$ is defined as the projection of $\mathbb{S}^p_{\succeq 0}$ onto the (augmented) edge set E^* of G. It is dual and homeomorphic to $\mathcal{K}_G$. Given a sample covariance matrix S, $\text{fiber}_G(S)$ consists of all positive definite completions of the G-partial matrix S_G, and it intersects $\mathcal{K}_G^{-1}$ in at most one point, namely the MLE $\hat{\Sigma}$.

For a Gaussian graphical model with underlying graph G this spectrahedron consists of all positive definite completions of S_G, i.e.

$$\text{fiber}_G(S) \quad = \quad \left\{\Sigma \in \mathbb{S}^p_{\succ 0} \mid \Sigma_G \; = \; S_G\right\}.$$

The following theorem combines the point of view of convex optimization developed in Section 9.2, the connection to positive definite matrix completion discussed in Section 9.3, and the link to convex geometry described in this section into a result about the existence of the MLE in Gaussian models with linear constraints on the concentration matrix, which includes Gaussian graphical models as a special case. This result is essentially also given in [22, Theorem 2].

Theorem 9.4.2. *Consider a Gaussian model with linear constraints on the concentration matrix defined by $\mathcal{L}$ with $\mathcal{L} \cap \mathbb{S}^p_{\succ 0} \neq \emptyset$. Then the MLEs $\hat{\Sigma}$ and $\hat{K}$ exist for a given sample covariance matrix S if and only if* $\text{fiber}_{\mathcal{L}}(S)$ *is non-empty, in which case* $\text{fiber}_{\mathcal{L}}(S)$ *intersects $\mathcal{K}_{\mathcal{L}}^{-1}$ in exactly one point, namely the MLE $\hat{\Sigma}$. Equivalently, $\hat{\Sigma}$ is the unique maximizer of the determinant over the spectrahedron* $\text{fiber}_{\mathcal{L}}(S)$.

Proof. This proof is a simple exercise in convex optimization; see [10] for an introduction. The ML estimation problem for Gaussian models with linear constraints on the concentration matrix is given by

$$\begin{aligned} &\underset{K}{\text{maximize}} \quad && \log\det K - \text{tr}(SK) \\ &\text{subject to} \quad && K \in \mathcal{K}_{\mathcal{L}}. \end{aligned}$$

Its dual is

$$\begin{aligned} &\underset{\Sigma}{\text{minimize}} \quad && -\log\det\Sigma - p \\ &\text{subject to} \quad && \Sigma \in \text{fiber}_{\mathcal{L}}(S). \end{aligned}$$

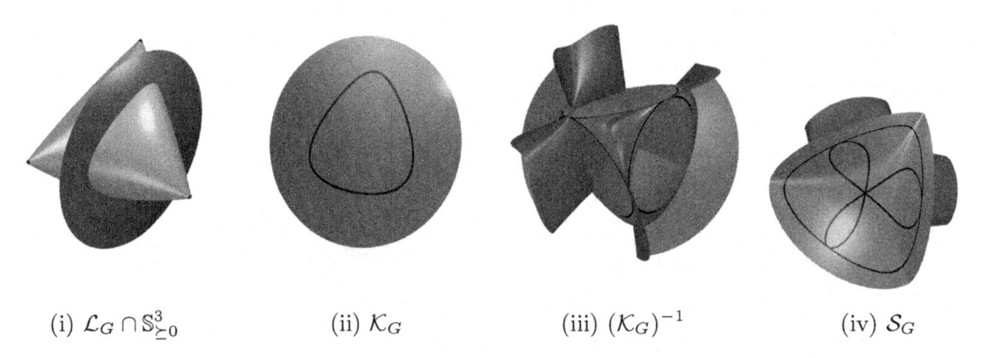

(i) $\mathcal{L}_G \cap \mathbb{S}^3_{\succeq 0}$ (ii) $\mathcal{K}_G$ (iii) $(\mathcal{K}_G)^{-1}$ (iv) $\mathcal{S}_G$

FIGURE 9.2: Geometry of Gaussian graphical models for $p = 3$. The tetrahedral-shaped pillow in (i) corresponds to the set of all 3×3 concentration matrices with ones on the diagonal. The linear subspace in (i) is defined by the missing edges in G. The resulting cone of concentration matrices is shown in (ii). The corresponding set of covariance matrices is shown in (iii), and the cone of sufficient statistics $\mathcal{S}_G$, dual to $\mathcal{K}_G$, is shown in (iv).

Since by assumption the primal problem is strictly feasible, strong duality holds by Slater's constraint qualification with the solutions satisfying $\hat{\Sigma} = \hat{K}^{-1}$. The MLE exists, i.e. the global optimum of the two optimization problems is attained, if and only if the dual is feasible, i.e. $\text{fiber}_{\mathcal{L}}(S)$ is non-empty. Let $\Sigma \in \text{fiber}_{\mathcal{L}}(S) \cap \mathcal{K}_{\mathcal{L}}^{-1}$. Then (Σ^{-1}, Σ) satisfies the KKT conditions, namely stationarity, primal and dual feasibility, and complimentary slackness. Hence, this pair is primal and dual optimal. Thus, if $\text{fiber}_{\mathcal{L}}(S)$ is non-empty, then $\text{fiber}_{\mathcal{L}}(S)$ intersects $\mathcal{K}_{\mathcal{L}}^{-1}$ in exactly one point, namely the MLE $\hat{\Sigma}$, which is the dual optimal solution. This completes the proof. □

The geometry of ML estimation in Gaussian models with linear constraints on the concentration matrix is summarized in Figure 9.1 for the special case of Gaussian graphical models. The geometric picture for general linear concentration models is completely analogous. The convex geometry of Gaussian graphical models on 3 nodes is shown in Figure 9.2. Since a general covariance matrix on 3 nodes lives in a 6-dimensional space, we show the picture for correlation matrices instead, which live in the 3-dimensional space.

Theorem 9.4.2 was first proven for Gaussian graphical models by Dempster [15] and later more generally for regular exponential families in [5, 11]. One can show that the map

$$\pi_G \circ (\cdot)^{-1} : \mathcal{K}_G \to \mathcal{S}_G$$

in Figure 9.1 corresponds to the gradient of the log-partition function. To embed this result into the theory of regular exponential families, we denote canonical parameters by θ, minimal sufficient statistics by $t(X)$, and the log-partition function of a regular exponential family by $A(\theta)$. Then the theory of regular exponential families (see e.g. [5, 11]) implies that the gradient of the log-partition function $\nabla A(\cdot)$ defines a homeomorphism between the space of canonical parameters and the relative interior of the convex hull of sufficient statistics, and it is defined by $\nabla A(\theta) = \mathbb{E}_\theta(t(X))$. For Gaussian models we have $A(\theta) = \log\det(\theta)$; the algebraic structure in maximum likelihood estimation for Gaussian graphical models is a consequence of the fact that $\nabla A(\cdot)$ is a rational function.

The geometric results and duality theory that hold for Gaussian graphical models can be extended to all regular exponential families [5, 11]. The algebraic picture can be extended to exponential families where $\nabla A(\cdot)$ is a rational function. This was shown in [36], where it was proven that such exponential families are defined by *hyperbolic polynomials.*

The problem of existence of the MLE can be studied at the level of sufficient statistics, i.e. in the cone $\mathcal{S}_G$, or at the level of observations. As explained in Section 9.3, the MLE exists if and only if the sufficient statistics S_G lie in the interior of the cone $\mathcal{S}_G$. Hence, analyzing existence of the MLE at the level of sufficient statistics requires analyzing the boundary of the cone $\mathcal{S}_G$. The boundary of $\mathcal{K}_G$ is defined by the hypersurface $\det(K) = 0$ with $K_{i,j} = 0$ for all $(i,j) \notin E^*$. It has been shown in [42] that the boundary of the cone $\mathcal{S}_G$ can be obtained by studying the dual of the variety defined by $\det(K) = 0$. This algebraic analysis results in conditions that characterize existence of the MLE at the level of sufficient statistics.

But perhaps more interesting from a statistical point of view, is a characterization of existence of the MLE at the level of observations. Note that if $\text{rank}(S) < p$ then it can happen that $\text{fiber}_{\mathcal{L}}(S)$ is empty, in which case the MLE does not exist for $(\mathcal{L}, S)$. In the next section, we discuss conditions on the number of observations n, or equivalently on the rank of S, that ensure existence of the MLE with probability 1 for particular classes of graphs.

9.5 Existence of the MLE for Various Classes of Graphs

Since the Gaussian density is strictly positive, $\text{rank}(S) = \min(n,p)$ with probability 1. The *maximum likelihood threshold* of a graph G, denoted $\text{mlt}(G)$, is defined as the minimum number of observations n such that the MLE in the Gaussian graphical model with graph G exists with probability 1. This is equivalent to the smallest integer n such that for all generic positive semidefinite matrices S of rank n there exists a positive definite matrix Σ with $S_G = \Sigma_G$. Although in this section we only consider Gaussian graphical models, note that this definition can easily be extended to general linear Gaussian concentration models.

The maximum likelihood threshold of a graph was introduced by Gross and Sullivant in [23]. Ben-David [9] introduced a related but different notion, the *Gaussian rank* of a graph, namely the smallest n such that the MLE exists for every positive semidefinite matrix S of rank n for which every $n \times n$ principal submatrix is non-singular. Note that with probability 1 every $n \times n$ principal submatrix of a sample covariance matrix based on n i.i.d. samples from a Gaussian distribution is non-singular. Hence, the Gaussian rank of G is an upper bound on $\text{mlt}(G)$. Since a sample covariance matrix of size $p \times p$ based on $n \leq p$ observations from a Gaussian population is of rank n with probability 1, we here concentrate on the maximum likelihood threshold of a graph.

A *clique* in a graph G is a completely connected subgraph of G. We denote by $q(G)$ the maximal clique-size of G. It is clear that the MLE cannot exist if $n < q(G)$, since otherwise the partial matrix S_G would contain a completely specified submatrix that is not positive definite (the submatrix corresponding to the maximal clique). This results in a lower bound for the maximum likelihood threshold of a graph, namely

$$\text{mlt}(G) \geq q(G).$$

For chordal graphs, Theorem 9.3.1 shows that the MLE exists with probability 1 if and only if $n \geq q(G)$. Hence for chordal graphs it holds that $\text{mlt}(G) = q(G)$. However, this is not the case in general as shown by the following example.

Example 9.5.1. Let G be the 4-cycle with edges $(1,2)$, $(2,3)$, $(3,4)$, and $(1,4)$. Then $q(G) = 2$. We define $X \in \mathbb{R}^{4\times 2}$ consisting of 2 samples in $\mathbb{R}^4$ and the corresponding sample covariance matrix $S = XX^T$ by

$$X = \begin{pmatrix} 1 & 0 \\ \frac{1}{\sqrt{2}} & \frac{1}{\sqrt{2}} \\ 0 & 1 \\ -\frac{1}{\sqrt{2}} & \frac{1}{\sqrt{2}} \end{pmatrix} \quad \text{and hence} \quad S = \begin{pmatrix} 1 & \frac{1}{\sqrt{2}} & 0 & -\frac{1}{\sqrt{2}} \\ \frac{1}{\sqrt{2}} & 1 & \frac{1}{\sqrt{2}} & 0 \\ 0 & \frac{1}{\sqrt{2}} & 1 & \frac{1}{\sqrt{2}} \\ -\frac{1}{\sqrt{2}} & 0 & \frac{1}{\sqrt{2}} & 1 \end{pmatrix}.$$

One can check that S_G cannot be completed to a positive definite matrix. In addition, there exists an open ball around X for which the MLE does not exist. This shows that in general for non-chordal graphs $\text{mlt}(G) > q(G)$.

From Theorem 9.3.1 we can determine an upper bound on $\text{mlt}(G)$ for general graphs. For a graph $G = (V, E)$ we denote by $G^+ = (V, E^+)$ a *chordal cover* of G, i.e. a chordal graph satisfying $E \subseteq E^+$. We denote the maximal clique size of G^+ by q^+. A *minimal chordal cover*, denoted by $G^\# = (V, E^\#)$, is a chordal cover of G, whose maximal clique size $q^\#$ achieves $q^\# = min(q^+)$ over all chordal covers of G. The quantity $q^\#(G) - 1$ is also known as the *treewidth* of G. It follows directly from Theorem 9.3.1 that

$$\text{mlt}(G) \leq q^\#(G),$$

since if $S_{G^\#}$ can be completed to a positive definite matrix, so can S_G.

If G is a cycle, then $q(G) = 2$ and $q^\#(G) = 3$. Hence the MLE does not exist for $n = 1$ and it exists with probability 1 for $n = 3$. From Example 9.5.1 we can conclude that for cycles $\text{mlt}(G) = 3$. Buhl [12] shows that for $n = 2$ the MLE exists with probability in $(0, 1)$. More precisely, for $n = 2$ we can view the two samples as vectors $x_1, \ldots, x_p \in \mathbb{R}^2$. We denote by $\ell_1, \ldots, \ell_p$ the lines defined by $x_1, \ldots, x_p$. Then Buhl [12] shows using an intricate trigonometric argument that the MLE for the p-cycle for $n = 2$ exists if and only if the lines $\ell_1, \ldots, \ell_p$ do not occur in one of the two sequences conforming with the ordering in the cycle G as shown in Figure 9.3. In the following, we give an algebraic proof of this result by using an intriguing characterization of positive definiteness for 3×3 symmetric matrices given in [7].

Proposition 9.5.2 (Barrett et al. [7]). *The matrix*

$$\begin{pmatrix} 1 & \cos(\alpha) & \cos(\beta) \\ \cos(\alpha) & 1 & \cos(\gamma) \\ \cos(\beta) & \cos(\gamma) & 1 \end{pmatrix}$$

with $0 < \alpha, \beta, \gamma < \pi$ *is positive definite if and only if*

$$\alpha < \beta + \gamma, \quad \beta < \alpha + \gamma, \quad \gamma < \alpha + \beta, \quad \alpha + \beta + \gamma < 2\pi.$$

Let G denote the p-cycle. Then, as shown in [7], this result can be used to give a characterization for completability of a G-partial matrix to a positive definite matrix through induction on the cycle length p.

Corollary 9.5.3 (Barrett et al. [7]). *Let G be the p-cycle. Then the G-partial matrix*

$$\begin{pmatrix} 1 & \cos(\theta_1) & & & \cos(\theta_p) \\ \cos(\theta_1) & 1 & \cos(\theta_2) & ? & \\ & \cos(\theta_2) & 1 & & \\ & ? & & \ddots & \cos(\theta_{p-1}) \\ \cos(\theta_p) & & & \cos(\theta_{p-1}) & 1 \end{pmatrix}$$

with $0 < \theta_1, \theta_2, \dots \theta_p < \pi$ *has a positive definite completion if and only if for each* $S \subseteq [p]$ *with* $|S|$ *odd,*

$$\sum_{i \in S} \theta_i < (|S| - 1)\pi + \sum_{j \notin S} \theta_j.$$

Buhl's result [12] can easily be deduced from this algebraic result about the existence of positive definite completions: For $n = 2$ we view the observations as vectors $x_1, \dots, x_p \in \mathbb{R}^2$. Note that we can rescale and rotate the data vectors $x_1, \dots, x_p$ (i.e. perform an orthogonal transformation) without changing the problem of existence of the MLE. So without loss of generality we can assume that the vectors $x_1, \dots, x_p \in \mathbb{R}^2$ have length one, lie in the upper unit half circle, and $x_1 = (1, 0)$. Now we denote by θ_i the angle between x_i and x_{i+1}, where $x_{p+1} := x_1$. One can show that the angular conditions in Corollary 9.5.3 are equivalent to requiring that the vectors $x_1, \dots, x_p \in \mathbb{R}^2$ do not occur in one of the two sequences conforming with the ordering in the cycle G as shown in Figure 9.3.

Hence, for a G-partial matrix to be completable to a positive definite matrix, it is necessary that every submatrix corresponding to a clique in the graph is positive definite and every partial submatrix corresponding to a cycle in G satisfies the conditions in Corollary 9.5.3. Barrett et al. [6] characterized the graphs for which these conditions are sufficient for existence of a positive definite completion. They showed that this is the case for graphs that have a chordal cover with no new 4-cliques. Such graphs can be obtained as a *clique sum* of chordal graphs and series-parallel graphs (i.e. graphs G with $q^{\#}(G) \leq 3$) [28]. To be more precise, for such graphs $G = (V, E)$ the vertex set can be decomposed into three disjoint subsets $V = V_1 \cup V_2 \cup V_3$ such that there are no edges between V_1 and V_3, the subgraph induced by V_2 is a clique, and the subgraphs induced by $V_1 \cup V_2$ and $V_2 \cup V_3$ are either chordal or series-parallel graphs or can themselves be decomposed as a clique sum of chordal or series-parallel graphs. For such graphs it follows that

$$\mathrm{mlt}(G) = \max(3, q(G)) = q^{\#}(G).$$

This raises the question whether there exist graphs for which $\mathrm{mlt}(G) < q^{\#}(G)$, i.e., graphs

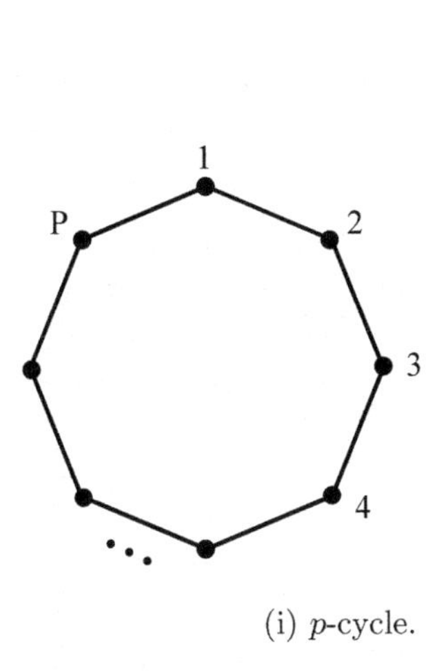

(i) p-cycle.

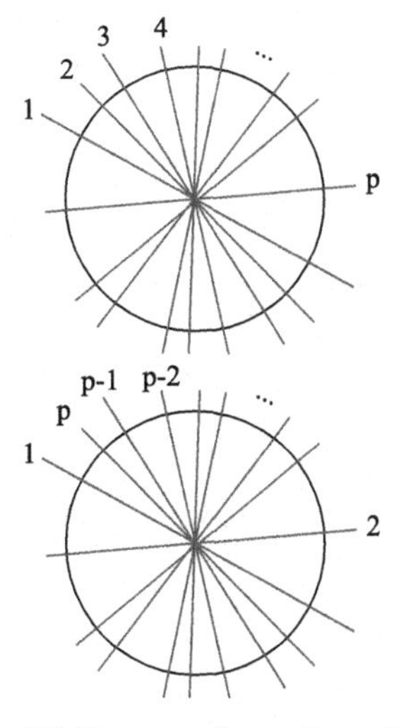

(ii) Line configurations for which the MLE does not exist.

FIGURE 9.3: Buhl's geometric criterion [12] for existence of the MLE for $n = 2$ in a Gaussian graphical model on the p-cycle.

for which the MLE exists with probability 1 even if the number of observations is strictly smaller than the maximal clique size in a minimal chordal cover of G. This question has been answered to the positive for 3×3 grids using an algebraic argument in [45] and more generally for grids of size $m \times m$ using a combinatorial argument in [23]. In particular, let G be a grid of size $m \times m$. Then $q^{\#}(G) = m + 1$, but it was shown in [23] that the MLE exists with probability 1 for $n = 3$, independent of the grid size m. Grids are a special class of planar graphs. Gross and Sullivant [23] more generally proved that for any planar graph it holds that $\text{mlt}(G) \leq 4$.

9.6 Algorithms for Computing the MLE

After having discussed when the MLE exists, we now turn to the question of how to compute the MLE for Gaussian graphical models. As described in Section 9.2, determining the MLE in a Gaussian model with linear constraints on the inverse covariance matrix is a convex optimization problem. Hence, it can be solved in polynomial time for instance using interior point methods [10]. These are implemented for example in `cvx`, a user-friendly `matlab` software for disciplined convex programming [21].

Although interior point methods run in polynomial time, for very large Gaussian graphical models it is usually more practical to apply coordinate descent algorithms. The idea of using coordinate descent algorithms for computing the MLE in Gaussian graphical models was already present in the original paper by Dempster [15]. Coordinate descent on the entries of Σ was first implemented by Wermuth and Scheidt [47] and is shown in Algorithm 9.1. In this algorithm, we start with $\Sigma^0 = S$ and iteratively update the entries $(i, j) \notin E^*$ by maximizing the log-likelihood in direction $\Sigma_{i,j}$ and keeping all other entries fixed.

Note that step (2) in Algorithm 9.1 can be given in closed-form: Let $A = \{u, v\}$ and $B = V \setminus A$. We now show that the objective function in step (2) of Algorithm 9.1 can be written in terms of the 2×2 Schur complement $\Sigma' = \Sigma_{A,A} - \Sigma_{A,B}\Sigma_{B,B}^{-1}\Sigma_{B,A}$. To do this, note that $\det(\Sigma) = \det(\Sigma')\det(\Sigma_{B,B})$. Since $\Sigma_{B,B}$ is held constant in the optimization problem, then up to an additive constant it holds that

$$\log\det(\Sigma) = \log\det(\Sigma').$$

Thus, the optimization problem in step (2) of Algorithm 9.1 is equivalent to

$$\begin{aligned} &\underset{\Sigma' \succeq 0}{\text{maximize}} && \log\det(\Sigma') \\ &\text{subject to} && \Sigma'_{i,i} = \Sigma^0_{i,i} - \Sigma^0_{i,B}(\Sigma^0_{B,B})^{-1}\Sigma^0_{B,i}, \quad i \in A, \end{aligned}$$

and the global maximum is attained by $\Sigma'_{u,v} = 0$. Hence, the solution to the univariate optimization problem in step (2) of Algorithm 9.1 is

$$\Sigma_{u,v} = \Sigma_{u,B}\Sigma_{B,B}^{-1}\Sigma_{B,v},$$

forcing the corresponding entry of Σ^{-1} to be equal to zero.

Dual to this algorithm, one can define an equivalent algorithm that cycles through entries of the concentration matrix corresponding to $(i, j) \in E$, starting in the identity matrix. This procedure is shown in Algorithm 9.2. Similarly as for Algorithm 9.1, the solution to the optimization problem in step (2) can be given in closed-form. Defining as before, $A = \{u, v\}$ and $B = V \setminus A$, then analogously as in the derivation above, one can show that the solution

Algorithm 9.1 Coordinate descent on Σ

Input: Graph $G = (V, E)$, sample covariance matrix S, and precision ϵ.
Output: MLE $\hat{\Sigma}$.

1: Let $\Sigma^0 = S$.
2: Cycle through $(u, v) \notin E^*$ and solve the following optimization problem:

$$\begin{aligned} \underset{\Sigma \succeq 0}{\text{maximize}} \quad & \log\det(\Sigma) \\ \text{subject to} \quad & \Sigma_{i,j} = \Sigma^0_{i,j} \quad \text{for all } (i,j) \neq (u,v) \end{aligned}$$

and update $\Sigma^1 := \Sigma$.
3: **if** $\|\Sigma^0 - \Sigma^1\|_1 < \epsilon$ **then**
4: let $\hat{\Sigma} := \Sigma^1$
5: **else**
6: let $\Sigma^0 := \Sigma^1$ and return to line 2.
7: **end if**

to the optimization problem in step (2) of Algorithm 9.2 is

$$K_{A,A} = (S_{A,A})^{-1} + K_{A,B}K^{-1}_{B,B}K_{B,A},$$

forcing $\Sigma_{A,A}$ to be equal to $S_{A,A}$. This algorithm, which tries to match the sufficient statistics, is analogous to *iterative proportional scaling* for computing the MLE in contingency tables [24]. Convergence proofs for both algorithms were given by Speed and Kiiveri [41].

In general the MLE must be computed iteratively. However, in some cases estimation can be made in closed form. A trivial case when the MLE of a Gaussian graphical model can be given explicitly is for complete graphs: In this case, assuming that the MLE exists, i.e. S is non-singular, then $\hat{K} = S^{-1}$. In [32, Section 5.3.2], Lauritzen showed that also for chordal graphs the MLE has a closed-form solution. This result is based on the fact that any chordal graph $G = (V, E)$ is a clique sum of cliques, i.e., the vertex set can be decomposed into three disjoint subsets $V = A \cup B \cup C$ such that there are no edges between A and C, the subgraph induced by B is a clique, and the subgraphs induced by $A \cup B$ and $B \cup C$

Algorithm 9.2 Coordinate descent on K

Input: Graph $G = (V, E)$, sample covariance matrix S, and precision ϵ.
Output: MLE $\hat{K}$.

1: Let $K^0 = \text{Id}$.
2: Cycle through $(u, v) \in E$ and solve the following optimization problem:

$$\begin{aligned} \underset{K \succeq 0}{\text{maximize}} \quad & \log\det(K) - \text{trace}(KS) \\ \text{subject to} \quad & K_{i,j} = K^0_{i,j} \quad \text{for all } (i,j) \in (V \times V) \setminus \{(u,u),(v,v),(u,v)\} \end{aligned}$$

and update $K^1 := K$.
3: **if** $\|K^0 - K^1\|_1 < \epsilon$ **then**
4: let $\hat{K} := K^1$
5: **else**
6: let $K^0 := K^1$ and return to line 2.
7: **end if**

are either cliques or can themselves be decomposed as a clique sum of cliques. In such a decomposition, B is known as a *separator*. In [32, Proposition 5.9], Lauritzen shows that, assuming existence of the MLE, then the MLE for a chordal Gaussian graphical model is given by

$$\hat{K} = \sum_{C\in\mathcal{C}} \left[(S_{C,C})^{-1}\right]^{\text{fill}} - \sum_{B\in\mathcal{B}} \left[(S_{B,B})^{-1}\right]^{\text{fill}}, \tag{9.7}$$

where $\mathcal{C}$ denotes the maximal cliques in G, $\mathcal{B}$ denotes the separators in the clique decomposition of G (with multiplicity, i.e., a clique could appear more than once), and $[A_{HH}]^{\text{fill}}$ denotes a $p \times p$ matrix, where the submatrix corresponding to $H \subset V$ is given by A and all the other entries are filled with zeros.

To gain more insight into the formula (9.7), consider the simple case where the subgraphs corresponding to $A \cup B$ and $B \cup C$ are cliques. Then (9.7) says that the MLE is given by

$$\hat{K} = \left[S_1^{-1}\right]^{\text{fill}} + \left[S_2^{-1}\right]^{\text{fill}} - \left[S_B^{-1}\right]^{\text{fill}}, \tag{9.8}$$

where we simplified notation by setting $S_1 = S_{AB,AB}$, $S_2 = S_{BC,BC}$, and $S_B = S_{B,B}$, also to clarify that we first take the submatrix and then invert it. To prove (9.8), it suffices to show that $\left(\hat{K}^{-1}\right)_G = S_G$, since $\hat{K}_{i,j} = 0$ for all $(i,j) \notin E^*$. We first expand $\hat{K}$ and then use Schur complements to compute its inverse:

$$\hat{K} = \begin{pmatrix} \left(S_1^{-1}\right)_{A,A} & \left(S_1^{-1}\right)_{A,B} & 0 \\ \left(S_1^{-1}\right)_{B,A} & \left(S_1^{-1}\right)_{B,B} + \left(S_2^{-1}\right)_{B,B} - S_B^{-1} & \left(S_2^{-1}\right)_{B,C} \\ 0 & \left(S_2^{-1}\right)_{C,B} & \left(S_2^{-1}\right)_{C,C} \end{pmatrix}. \tag{9.9}$$

Denoting $\hat{K}^{-1}$ by $\hat{\Sigma}$ and using Schur complements, we obtain

$$\hat{\Sigma}_{AB,AB} = \begin{pmatrix} \left(S_1^{-1}\right)_{A,A} & \left(S_1^{-1}\right)_{A,B} \\ \left(S_1^{-1}\right)_{B,A} & \left(S_1^{-1}\right)_{B,B} + \left(S_2^{-1}\right)_{B,B} - S_B^{-1} - \left(S_2^{-1}\right)_{B,C}\left((S_2^{-1})_{C,C}\right)^{-1}\left(S_2^{-1}\right)_{C,B} \end{pmatrix}^{-1}.$$

Note that by using Schur complements once again,

$$\left(S_2^{-1}\right)_{B,B} - \left(S_2^{-1}\right)_{B,C}\left((S_2^{-1})_{C,C}\right)^{-1}\left(S_2^{-1}\right)_{C,B} = S_B^{-1},$$

and hence $\hat{\Sigma}_{AB,AB} = S_1$. Analogously, it follows that $\hat{\Sigma}_{BC,BC} = S_2$, implying that $\hat{\Sigma}_G = S_G$. The more general formula for the MLE of chordal Gaussian graphical models in (9.7) is obtained by induction and repeated use of (9.9).

A stronger property than existence of a closed-form solution for the MLE is to ask which Gaussian graphical models have rational formulas for the MLE in terms of the entries of the sample covariance matrix. An important observation is that the number of critical points to the likelihood equations is constant for generic data, i.e., it is constant with probability 1 (it can be smaller on a measure zero subspace). The number of solutions to the likelihood equations for generic data, or equivalently, the maximum number of solutions to the likelihood equations, is called the *maximum likelihood degree* (*ML degree*). Hence, a model has a rational formula for the MLE if and only if it has ML degree 1. It was shown in [42] that the ML degree of a Gaussian graphical model is 1 if and only if the underlying graph is chordal. The ML degree of the 4-cycle can easily be computed and is known to be 5; see [16, Example 2.1.13] for some code on how to do the computation using the open-source computer algebra system `Singular` [14]. It is conjectured in [16, Section 7.4] that the ML degree of the cycle grows exponentially in the cycle length, namely as $(p-3)2^{p-2}+1$, where $p \geq 3$ is the cycle length.

Since the likelihood function is strictly concave for Gaussian graphical models, this implies that even when the ML degree is larger than 1, there is still a unique local maximum of the likelihood function. As a consequence, while there are multiple complex solutions to the ML equations for non-chordal graphs, there is always a unique solution that is real and results in a positive definite matrix.

9.7 Learning the Underlying Graph

Until now we have assumed that the underlying graph is given to us. In this section, we present methods for learning the underlying graph. We here only provide a short overview of some of the most prominent methods for model selection in Gaussian graphical models; for more details and for practical examples, see [25].

A popular method for performing model selection is to take a stepwise approach. We start in the empty graph (or in the complete graph) and run a forward search (or a backward search). We cycle through the possible edges and add an edge (or remove an edge) if it decreases some criterion. Alternatively, one can also search for the edge which minimizes some criterion and add (or remove) this edge, but this is considerably slower. Two popular objective functions are the *Akaike information criterion* (AIC) and the *Bayesian information criterion* (BIC) [1, 39]. These criteria are based on penalizing the likelihood according to the model complexity, i.e.

$$-2\ell + \lambda|E|, \tag{9.10}$$

where ℓ is the log-likelihood function, λ is a parameter that penalizes model complexity, and $|E|$ denotes the number of edges, or equivalently, the number of parameters in the model. The AIC is defined by choosing $\lambda = 2$, whereas the BIC is defined by setting $\lambda = \log(n)$ in (9.10).

Alternatively, one can also use significance tests for testing whether a particular partial correlation is zero and removing the corresponding edge accordingly. A hypothesis test for zero partial correlation can be built based on Fisher's z-transform [19]: For testing whether $K_{i,j} = 0$, let $A = \{i, j\}$ and $B = V \setminus A$. In Proposition 9.1.1 we saw that $K_{A,A}^{-1} = \Sigma_{A|B}$. Hence testing whether $K_{i,j} = 0$ is equivalent to testing whether the correlation $\rho_{i,j|B}$ is zero. The sample estimate of $\rho_{i,j|B}$ is given by

$$\hat{\rho}_{i,j|B} = S_{i,j} - S_{i,B} S_{B,B}^{-1} S_{B,j}.$$

Fisher's z-transform is defined by

$$\hat{z}_{i,j|B} = \frac{1}{2} \log\left(\frac{1 + \hat{\rho}_{i,j|B}}{1 - \hat{\rho}_{i,j|B}}\right).$$

Fisher [19] showed that using the test statistic $T_n = \sqrt{n - p + 2 - 3}|\hat{z}_{i,j|B}|$ with a rejection region $R_n = (-\Phi^{-1}(1 - \alpha/2), \Phi^{-1}(1 - \alpha/2))$, where Φ denotes the cumulative distribution function of $\mathcal{N}(0, 1)$, leads to a test of size α.

A problem with stepwise selection strategies is that they are impractical for large problems or only a small part of the relevant search space can be covered during the search. A simple alternative, but a seemingly naive method for model selection in Gaussian graphical models, is to set a specific threshold for the partial correlations and remove all edges corresponding to the partial correlations that are less than the given threshold. This often works well, but a disadvantage is that the resulting estimate of the inverse covariance matrix might not be positive definite.

An alternative is to use the *glasso* algorithm [20]. It is based on maximizing the ℓ_1-penalized log-likelihood function, i.e.

$$\ell_{\text{pen}}(K) = \log\det(K) - \operatorname{tr}(KS) - \lambda|K|_1,$$

where λ is a non-negative parameter that penalizes model complexity and $|K|_1$ is the sum of the absolute values of the off-diagonal elements of the concentration matrix. The use of $|K|_1$ is a convex proxy for the number of non-zero elements of K and allows efficient optimization of the penalized log-likelihood function by convex programming methods such as interior point algorithms or coordinate descent approaches similar to the ones discussed in Section 9.6; see e.g. [34]. A big advantage of using ℓ_1-penalized maximum likelihood estimation for model selection in Gaussian graphical models is that it can also be applied in the high-dimensional setting and comes with structural recovery guarantees [38]. Various alternative methods for learning high-dimensional Gaussian graphical models have been proposed that have similar guarantees, including node-wise regression with the lasso [35], a constrained ℓ_1-minimization approach for inverse matrix estimation (CLIME) [13], and a testing approach with false discovery rate control [33]. Graphical models in the high-dimensional setting are discussed in detail in Chapter 12.

9.8 Other Gaussian Models with Linear Constraints

Gaussian graphical models are Gaussian models with particular equality constraints on the concentration matrix, namely where some of the entries are set to zero. We end this chapter by giving an overview on other Gaussian models with linear constraints.

Gaussian graphical models can be generalized by introducing a vertex and edge coloring: Let $G = (V, E)$ be an undirected graph, where the vertices are colored with s different colors and the edges with t different colors. This leads to a partition of the vertex and edge set into color classes, namely,

$$V = V_1 \cup V2 \cup V_s, \quad s \leq p, \quad \text{and} \quad E = E_1 \cup E_2 \cup \cdots \cup E_t, \quad t \leq |E|.$$

An *RCON* model on G is a Gaussian graphical model on G with some additional equality constraints, namely that $K_{i,i} = K_{j,j}$ if i and j are in the same vertex color class and $K_{i,j} = K_{u,v}$ if (i, j) and (u, v) are in the same edge color class. Hence a Gaussian graphical model on a graph G is an RCON model on G, where each vertex and edge has a separate color.

Determining the MLE for RCON models leads to a convex optimization problem and the corresponding dual optimization problem can be readily computed:

$$\begin{aligned}
&\underset{\Sigma\succeq 0}{\text{minimize}} && -\log\det\Sigma - p \\
&\text{subject to} && \sum_{\alpha\in V_i} \Sigma_{\alpha,\alpha} = \sum_{\alpha\in V_i} S_{\alpha,\alpha}, && \text{for all } 1 \leq i \leq s, \\
& && \sum_{(\alpha,\beta)\in E_j} \Sigma_{\alpha,\beta} = \sum_{(\alpha,\beta)\in E_j} S_{\alpha,\beta}, && \text{for all } 1 \leq j \leq t.
\end{aligned}$$

This shows that the constraints for existence of the MLE in an RCON model on a graph G are relaxed as compared to a Gaussian graphical model on G; namely, in an RCON model the constraints are only on the sum of the entries in a color class, whereas in a Gaussian graphical model the constraints are on each entry.

RCON models were introduced by Højsgaard and Lauritzen in [26]. These models are useful for applications, where symmetries in the underlying model can be assumed. Adding symmetries reduces the number of parameters and in some cases also the number of observations needed for existence of the MLE. For example, defining G to be the 4-cycle and having only one vertex color class and one edge color class (i.e., we color each vertex in the same color and each edge in the same color), then one can show that the MLE already exists for 1 observation with probability 1. This is in contrast to the result that $\mathrm{mlt}(G) = 3$ for cycles as shown in Section 9.5. For further examples see [26, 45].

More general Gaussian models with linear equality constraints on the concentration matrix or the covariance matrix were introduced by Anderson [2]. He was motivated by the linear structure of covariance and concentration matrices resulting from various time series models. As pointed out in Section 9.2, the Gaussian likelihood as a function of Σ is not concave over the whole cone of positive definite matrices. Hence maximum likelihood estimation for Gaussian models with linear constraints on the covariance matrix in general does not lead to a convex optimization problem and has many local maxima. Anderson proposed iterative procedures for calculating the MLE for such models, such as the Newton-Raphson method [2] and a scoring method [3].

As mentioned in Section 9.2, while not being concave over the whole cone of positive definite matrices, the Gaussian likelihood as a function of Σ is concave over a large region of $\mathbb{S}^p_{\succ 0}$, namely for all Σ that satisfy $\Sigma - 2S \in \mathbb{S}^p_{\succ 0}$. This is useful, since it was shown in [48] that the MLE for Gaussian models with linear equality constraints on the covariance matrix lies in this region with high probability as long as the sample size is sufficiently large ($n \simeq 14p$). Hence in this regime, maximum likelihood estimation for linear Gaussian covariance models behaves as if it were a convex optimization problem.

Similarly as we posed the question for Gaussian graphical models in Section 9.6, one can ask when the MLE of a linear Gaussian covariance model has a closed form representation. Szatrowski showed in [43, 44] that the MLE for linear Gaussian covariance models has an explicit representation if and only if Σ and Σ^{-1} satisfy the same linear constraints. This is equivalent to requiring that the linear subspace $\mathcal{L}$, which defines the model, forms a Jordan algebra, i.e., if $\Sigma \in \mathcal{L}$ then also $\Sigma^2 \in \mathcal{L}$ [27]. Furthermore, Szatrowski proved that for this model class Anderson's scoring method [3] yields the MLE in one iteration when initiated at any positive definite matrix in the model.

Linear inequality constraints on the concentration matrix also lead to a convex optimization problem for ML estimation. An example of such models are Gaussian distributions that are *multivariate totally positive of order two* (MTP_2). This is a form of positive dependence, which for Gaussian distributions implies that $K_{i,j} \leq 0$ for all $i \neq j$. Gaussian MTP_2 distributions were studied by Karlin and Rinott [29] and more recently in [31, 40] from a machine learning and more applied perspective. It was shown in [18] that MTP_2 distributions have remarkable properties with respect to conditional independence constraints. In addition, for such models the spanning forest of the sample correlation matrix is always a subgraph of the maximum likelihood graph, which can be used to speed up graph learning algorithms [31]. Furthermore, the MLE for MTP_2 Gaussian models exists already for 2 observations with probability 1 [40]. These properties make MTP_2 Gaussian models interesting for the estimation of high-dimensional graphical models.

We end by referring to Pourahmadi [37] for a comprehensive review of covariance estimation in general and a discussion of numerous other specific covariance matrix constraints.

Bibliography

[1] H. Akaike. A new look at the statistical identification problem. *IEEE Transactions on Automatic Control*, 19:716–723, 1974.

[2] T. W. Anderson. Estimation of covariance matrices which are linear combinations or whose inverses are linear combinations of given matrices. In *Essays in Probability and Statistics*, pages 1–24. University of North Carolina Press, 1970.

[3] T. W. Anderson. Asymptotically efficient estimation of covariance matrices with linear structure. *Annals of Statistics*, 1:135–141, 1973.

[4] T. W. Anderson. *An Introduction to Multivariate Statistical Analysis*. John Wiley & Sons, third edition, 2003.

[5] O. Barndorff-Nielsen. *Information and Exponential Families in Statistical Theory*. Wiley Series in Probability and Mathematical Statistics. John Wiley & Sons Ltd., 1978.

[6] W. Barrett, C. Johnson, and R. Loewy. The real positive definite completion problem: cycle completability. *Memoirs of the American Mathematical Society*, 584:69, 1996.

[7] W. Barrett, C. Johnson, and P. Tarazaga. The real positive definite completion problem for a simple cycle. *Linear Algebra and Its Applications*, 192:3–31, 1993.

[8] A. I. Barvinok. *A Course in Convexity*, volume 54 of *Graduate Studies in Mathematics*. American Mathematical Society, Providence, 2002.

[9] E. Ben-David. Sharp lower and upper bounds for the Gaussian rank of a graph. *Journal of Multivariate Analysis*, 139:207–218, 2014.

[10] S. Boyd and L. Vandenberghe. *Convex Optimization*. Cambridge University Press, 2004.

[11] L. D. Brown. *Fundamentals of Statistical Exponential Families with Applications in Statistical Decision Theory*, volume 9 of *Institute of Mathematical Statistics Lecture Notes—Monograph Series*. Institute of Mathematical Statistics, Hayward, CA, 1986.

[12] S. L. Buhl. On the existence of maximum likelihood estimators for graphical Gaussian models. *Scandinavian Journal of Statistics*, 20:263–270, 1993.

[13] T. Cai, W. Liu, and X. Luo. A constrained ℓ_1 minimization approach to sparse precision matrix estimation. *Journal of the American Statistical Association*, 106:594–607, 2011.

[14] W. Decker, G.-M. Greuel, G. Pfister, and H. Schönemann. SINGULAR 4-0-2 — A computer algebra system for polynomial computations. `http://www.singular.uni-kl.de`, 2015.

[15] A. P. Dempster. Covariance selection. *Biometrics*, 28:157–175, 1972.

[16] M. Drton, B. Sturmfels, and S. Sullivant. *Lectures on Algebraic Statistics*, volume 39 of *Oberwolfach Seminars*. Springer, 2009.

[17] M. L. Eaton. *Multivariate Statistics. A Vector Space Approach*. John Wiley & Sons, New York, 1983.

[18] S. Fallat, S. L. Lauritzen, K. Sadeghi, C. Uhler, N. Wermuth, and P. Zwiernik. Total positivity in Markov structures. *Annals of Statistics*, 45:1152–1184, 2017.

[19] R. A. Fisher. Frequency distribution of the values of the correlation coefficient samples of an indefinitely large population. *Biometrika*, 10:507–521, 1915.

[20] J. Friedman, T. Hastie, and R. Tibshirani. Sparse inverse covariance estimation with the graphical lasso. *Biostatistics*, 9:432–441, 2008.

[21] M. Grant and S. Boyd. CVX: Matlab software for disciplined convex programming, version 2.1. `http://cvxr.com/cvx`, March 2014.

[22] R. Grone, C. R. Johnson, E. M. de Sá, and H. Wolkowicz. Positive definite completions of partial hermitian matrices. *Linear Algebra and Its Applications*, 58:109–124, 1984.

[23] E. Gross and S. Sullivant. The maximum likelihood threshold of a graph. To appear in *Bernoulli*, 2014.

[24] S. J. Haberman. *The Analysis of Frequency Data.* Statistical Research Monographs. University of Chicago Press, Chicago, 1974.

[25] S. Hojsgaard, D. Edwards, and S. L. Lauritzen. *Graphical Models with R.* Use R! Springer, New York, 2012.

[26] S. Hojsgaard and S. L. Lauritzen. Graphical Gaussian models with edge and vertex symmetries. *Journal of the Royal Statistical Society. Series B (Statistical Methodology)*, 70:1005–1027, 2008.

[27] S. T. Jensen. Covariance hypotheses which are linear in both the covariance and the inverse covariance. *Annals of Statistics*, 16(1):302–322, 1988.

[28] C. R. Johnson and T. A. McKee. Structural conditions for cycle completable graphs. *Discrete Mathematics*, 159:155–160, 1996.

[29] S. Karlin and Y. Rinott. M-matrices as covariance matrices of multinormal distributions. *Linear Algebra and Its Applications*, 52:419 – 438, 1983.

[30] M. Laurent. The real positive semidefinite completion problem for series-parallel graphs. *Linear Algebra and Its Applications*, 252:347–366, 1997.

[31] S. Lauritzen, C. Uhler, and P. Zwiernik. Maximum likelihood estimation in Gaussian models under total positivity. Preprint available at *https://arxiv.org/abs/1702.04031*, 2017.

[32] S. L. Lauritzen. *Graphical Models.* Oxford University Press, 1996.

[33] W. Liu. Gaussian graphical model estimation with false discovery rate control. *Annals of Statistics*, 41:2948–2978, 2013.

[34] R. Mazumder and T. Hastie. The graphical lasso: New insights and alternatives. *Electronic Journal of Statistics*, 6:2125–2149, 2012.

[35] N. Meinshausen and P. Bühlmann. High-dimensional graphs and variable selection with the Lasso. *Annals of Statistics*, 34:1436–1462, 2006.

[36] M. Michalek, B. Sturmfels, C. Uhler, and P. Zwiernik. Exponential varieties. *Proceedings of the London Mathematical Society*, 112:27–56, 2016.

[37] M. Pourahmadi. Covariance estimation: The GLM and regularization perspectives. *Statistical Science*, 3:369–387, 2011.

[38] P. Ravikumar, M. J. Wainwright, G. Raskutti, and B. Yu. High-dimensional covariance estimation by minimizing ℓ_1-penalized log-determinant divergence. *Electronic Journal of Statistics*, 5:935–980, 2011.

[39] G. Schwarz. Estimating the dimension of a model. *Annals of Mathematical Statistics*, 6:461–464, 1978.

[40] M. Slawski and M. Hein. Estimation of positive definite M-matrices and structure learning for attractive Gaussian Markov random field. *Linear Algebra and Its Applications*, 473:145–179, 2015.

[41] T. P. Speed and H. T. Kiiveri. Gaussian Markov distributions over finite graph. *Annals of Statistics*, 14:138–150, 1986.

[42] B. Sturmfels and C. Uhler. Multivariate Gaussians, semidefinite matrix completion, and convex algebraic geometry. *Annals of the Institute of Statistical Mathematics*, 62:603–638, 2010.

[43] T. H. Szatrowski. Necessary and sufficient conditions for explicit solutions in the multivariate normal estimation problem for patterned means and covariances. *Annals of Statistics*, 8:802–810, 1980.

[44] T. H. Szatrowski. Patterned covariances. In *Encyclopedia of Statistical Sciences*, 638–641. Wiley, New York, 1985.

[45] C. Uhler. Geometry of maximum likelihood estimation in Gaussian graphical models. *Annals of Statistics*, 40:238–261, 2012.

[46] M. J. Wainwright and M. I. Jordan. *Graphical Models, Exponential Families, and Variational Inference*, volume 1 of *Foundations and Trends in Machine Learning.* 2008.

[47] N. Wermuth and E. Scheidt. Fitting a covariance selection model to a matrix. *Journal of the Royal Statistical Society. Series C (Applied Statistics)*, 26:88–92, 1977.

[48] P. Zwiernik, C. Uhler, and D. Richards. Maximum likelihood estimation for linear Gaussian covariance models. *Journal of the Royal Statistical Society. Series B (Statistical Methodology)*, 2016.

10

Bayesian Inference in Graphical Gaussian Models

Hélène Massam

Department of Mathematics and Statistics, York University

CONTENTS

10.1 Introduction

Graphical Gaussian models are one of the main tools for the analysis of high-dimensional data with applications in a variety of disciplines. A graphical Gaussian model for the random vector $Z \in \mathbb{R}^r$ is a Gaussian model where the dependencies between the components of Z are represented by means of a graph. The parameter of a centered Gaussian model is its covariance Σ or equivalently its concentration or precision matrix $K = \Sigma^{-1} \in P_r$ where P_r is the cone of $r \times r$ positive definite matrices. Because the dependencies between the components of Z are represented by means of a graph G, the expression of Σ or K depends upon G. In a Bayesian framework, we express uncertainty on the parameters by putting prior distributions $f_2(K|G)$ and $f_3(G)$ on K given G and G respectively. Given a sample $\mathcal{D} = (Z^1, \ldots, Z^n)$ from the Gaussian distribution with precision matrix K, the joint distribution for $(\mathcal{D}, K, G)$ is $f(z^1, \ldots, z^n, K, G) = f_1(z^1, \ldots, z^n|K, G)f_2(K|G)f_3(G)$ from which we derive the posterior distribution of the model, that is G, as

$$p(G \mid \mathcal{D}) \propto f_3(G) \int f_1(z^1, \dots, z^n | K, G) f_2(K|G) dK.$$

After scanning the space of graphs, we can then select a model that has a high posterior probability or average predictions from several good models. Bayesian inference in graphical Gaussian models is concerned with the selection of the graph representing the model and the estimation of its covariance or precision parameters.

In this paper, we will give a largely chronological account of the different methods that have been developed for model selection and parameter estimation. I apologize for any work I may not have mentioned here, due to lack of space. As we shall see, the two main problems are to sample from the prior or posterior distribution of K given G and to find an efficient algorithm to traverse the space of graphs. We have tried to follow the evolution of ideas and emphasize what was innovative in each one of the papers without getting into the details of implementation of various algorithms.

The remainder of this paper is organized as follows. In Section 10.2, we recall some notions fundamental to graphical Gaussian models: graphs, Markov properties and the Wishart distribution. In Section 10.3, we discuss inference for graphical Gaussian models Markov with respect to a decomposable graph. The building blocks of inference in that case are the hyper inverse Wishart prior for Σ and the Bayes factor which allows for pairwise comparison of models and can be computed explicitly. As we shall see, the Bayes factor for the comparison of two models G_1 and G_2 is equal to the G_1/G_2 ratio of prior and posterior normalizing constants for the hyper inverse Wishart. The prior induced from the hyper inverse Wishart on the precision matrix K is the so-called G-Wishart distribution. We will see in Section 10.4 that the G-Wishart generalizes to the case where G is an arbitrary undirected graph but then, its normalizing constant cannot be computed explicitly in reasonable time. MCMC algorithms with state variables (K, G), which allow us to explore the space of graphs and simultaneously update the value of K in an efficient manner have been developed. Though these algorithms do not use the Bayes factor, it is still necessary to compute the prior and posterior normalizing constants in order to calculate the acceptance probability of a new move. More recent methods have stirred away from the G-Wishart, away from the computation of normalizing constants or away from both. In Section 10.5, we consider matrix or array variate (as opposed to vector variate) random variable. The methods in that case tend to be adaptations of methods used for the vector variate case. We note that in the methodology given in Sections 10.4 and 10.5, the model or graph structure is selected not through its posterior probability, as it was done traditionally. Rather, the selected graph(s) are the median probability graph(s) as defined by [4]. Finally in Section 10.6, we consider work using fractional Bayes factors for inference. Fractional Bayes factors have been used for DAG Markov or decomposable models but some recent work has, interestingly, combined pseudo-likelihood and fractional Bayes factor to search through the space of models Markov with respect to arbitrary undirected graphs. We end our survey of methods for Bayesian inference with a short review of two papers that are slightly off the mainstream of the methodology outlined above.

10.2 Preliminaries

10.2.1 Graphs and Markov properties

An undirected graph is a pair (V, E) where $V = \{1, \ldots, r\}$ and $E \subset \{F \subset V : |F| = 2\}$ is a set of unordered pairs $\{i, j\}$ such that $i \neq j$. Any $\{i, j\} \in E$ will be called an edge. By abuse of notation, we will denote an edge $\{i, j\}$ by (i, j).

Graphical Gaussian models are statistical models Markov with respect to graphs. We will use here the Markov properties with respect to undirected graphs, decomposable or not, and with respect to DAGs. The reader is referred to Section 1.7 of Chapter 1 for a review of Markov properties with respect to these graphs. We will only recall here the notions for graphs that are essential to what we do below. Details can be found in [27].

The class of decomposable graphs plays a special role in Bayesian inference for Gaussian graphical models. A graph is said to be decomposable if it does not contain induced cycles of length greater than or equal to four without a cord. Though, for a given set of vertices, this class forms only a relatively small percentage of the class of all undirected graphs (increasingly small as the number of vertices increases, see [24]), they are important for two reasons. First, as we shall see in the next sections, computations in the case of Gaussian models Markov with respect to a decomposable graph can be done explicitly. Second, priors for the covariance parameter in decomposable Gaussian graphical models will be the basis for priors in Gaussian models Markov with respect to arbitrary undirected graphs and also directed acyclic graphs (DAG). Let $(C_1, \ldots, C_k)$ be a given order of the cliques of an arbitrary graph G. For such a given order, we use the notation $H_1 = R_1 = C_1$, $S_1 = \emptyset$ while for $j = 2, \ldots, k$ we write

$$H_j = C_1 \cup \ldots \cup C_j, \quad S_j = H_{j-1} \cap C_j, \quad R_j = C_j \setminus H_{j-1}. \tag{10.1}$$

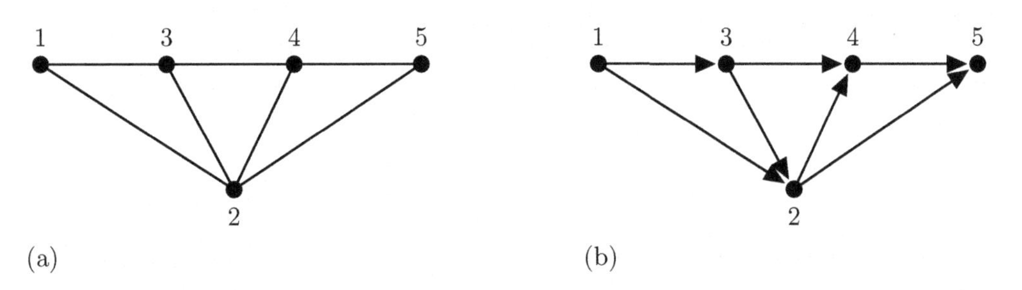

FIGURE 10.1: A decomposable graph in (a) and a Markov equivalent DAG in (b).

The order $(C_1, \ldots, C_k)$ of cliques is said to be perfect if, for every $i > 1$, there exists a $j < i$ such that $S_i \subset C_j$. Every decomposable graph admits a perfect order of its cliques. If the order $C_1, \ldots, C_k$ is perfect, $\{S_j,\ j = 2, \ldots, k\}$ is the set of minimal separators of G. Some of these separators can be identical. We let $\nu(S)$ denote the multiplicity of S that is the number of $j = 2, \ldots, k$ such that $S_j = S$. We will use $\mathcal{C}$ to denote the set of cliques of a graph G and $\mathcal{S}$ for the set of separators. It has been proven in [27] that the multiplicity $\nu(S)$ of a given minimal separator S is independent of the perfect order of the cliques considered. It can also be proven that for a given perfect order, S_j separates $H_{j-1} \setminus S_j$ from R_j in the induced graph G_{H_j} and hence $(H_{j-1} \setminus S_j, S_j, R_j)$ decomposes the induced subgraph G_{H_j}. To a perfect order of the cliques, we can attach a perfect order of the vertices: this can be done, for example by numbering the vertices in $C_1 \setminus S_2$ first, then the vertices in S_2 and next, successively, the vertices in $R_j, j = 2, \ldots, k$.

If the edges of a decomposable graph G are given a direction going from the smallest vertex to the larger one according to this perfect ordering of the vertices, we obtain a particular kind of directed acyclic graph $\mathcal{G}$ (henceforth abbreviated DAG), called a moral DAG. A DAG $\mathcal{G}$ is said to be moral if two vertices a and b such that there exists a vertex c with $a \to c$ and $b \to c$, are necessarily linked by a directed edge. The DAG $\mathcal{G}$ thus obtained is Markov equivalent to the decomposable graph G. For standard notation relative to DAGs, the reader is referred to Section 1.6 of Chapter 1. We will use, in particular, the notions of parents, descendants and non-descendants of a vertex $v \in V$ which will be denoted respectively as $pa(v), de(v)$ and $nd(v)$.

For the decomposable graph G and its Markov equivalent DAG $\mathcal{G}$ given in Figure 10.1, the cliques $C_1 = \{1,2,3\}, C_2 = \{2,3,4\}, C_3 = \{2,4,5\}$ form a perfect ordering of the cliques of G. The corresponding separators are $S_2 = \{2,3\}$ and $S_3 = \{2,4\}$ while the residuals sets are $R_2 = \{4\}$ and $R_3 = \{5\}$. The Markov properties represented by G and $\mathcal{G}$ are as follows. If $Z = (Z_i, i = 1, \ldots, 5)$ is Markov with respect to G, we have

$$Z_4 \perp\!\!\!\perp Z_1 \mid Z_{23}, \quad Z_5 \perp\!\!\!\perp Z_{13} \mid Z_{24} \,.$$

If Z is Markov with respect to $\mathcal{G}$, we have, of course, the same properties.

When an undirected graph is not decomposable, it can no longer be decomposed according to a perfect ordering of its cliques. However it can be decomposed according to a perfect ordering $\mathcal{P}_1, \ldots, \mathcal{P}_k$ of its prime components with corresponding separators $\mathcal{S}_2, \ldots, \mathcal{S}_k$ (see [27], p.14). An induced subgraph $G_{\mathcal{P}}$ of G is a prime component of G if it is a maximal (with respect to inclusion) non decomposable graph. This will imply in particular that the normalizing constant for an undirected graph G can be decomposed into a Markov ratio of the normalizing constants for $G_{\mathcal{P}_j}, j = 1, \ldots, k$ and $G_{\mathcal{S}_j}, j = 2, \ldots, k$.

10.2.2 The Wishart distribution

Let us consider a random variable Z which follows the $N_r(0, \Sigma)$ distribution where the covariance matrix Σ belongs to the cone P_r of $r \times r$ positive definite matrices and where, without loss of generality, we assume that the mean of the distribution is 0. Consider a sample $\mathcal{D}$ from this distribution. It is then well-known that, for $n \geq r$,

$$\sum_{i=1}^{n} Z^i (Z^i)^t$$

follows a Wishart distribution of dimension r with density, with respect to the Lebesgue measure on P_r, of the form

$$W_r(x \mid \frac{n}{2}, \Sigma) = \frac{|\Sigma|^{-\frac{n}{2}}}{\Gamma_r(\frac{n}{2})} |x|^{\frac{n}{2} - \frac{r+1}{2}} \exp(-\frac{1}{2}\langle x, \Sigma^{-1}\rangle) \tag{10.2}$$

where, for two symmetric matrices x and y of order r, $\langle x, y\rangle = \mathrm{tr}(xy)$ denotes their inner product, $|x|$ is the determinant of x and for $p > \frac{r-1}{2}$, $\Gamma_r(p)$ is the r-multivariate gamma function $\Gamma_r(p) = \pi^{\frac{1}{4}r(r-1)} \prod_{j=1}^{r} \Gamma(p - \frac{1}{2}(j-1))$. The matrix Σ is the scale parameter of the distribution and $\frac{n}{2}$ its shape parameter. In fact, it can be shown (see [35]) that the Wishart distribution can be defined for any shape parameter p in the Gindikin set

$$\Lambda_r = \left\{\frac{1}{2}, \ldots, \frac{r-1}{2}\right\} \cup \left(\frac{r-1}{2}, +\infty\right)$$

and any scale parameter $\sigma \in P_r$. It is then denoted as $W_r(x \mid p, \sigma)$. If X follows the $W_r(p, \sigma)$ distribution where $p > \frac{r-1}{2}$, i.e. if p belongs to the continuous part of the Gindikin set, and

$\sigma \in P_r$, then its density exists and is equal to (10.2) where Σ is replaced by σ and $\frac{n}{2}$ by p. Our assumption that the sample size satisfies $n \geq r$ thus guarantees that the density of $\sum_{i=1}^{n} Z^i(Z^i)^t$ exists. If $X \sim W_r(p, \sigma)$, its inverse $Y = X^{-1}$ follows the inverse Wishart $IW_r(p, \theta)$, $\theta = \sigma^{-1}$ distribution. Since the Jacobian of the change of variable $X \to Y$ is $|X|^{r+1}$, the density of the $IW_r(p, \theta)$, with respect to the Lebesgue measure on P_r, is

$$IW_r(y \mid p, \theta) = \frac{|\theta|^p}{\Gamma_r(p)} |y|^{-p-\frac{r+1}{2}} \exp(-\frac{1}{2}\langle y^{-1}, \theta\rangle) . \tag{10.3}$$

In Bayesian statistics, given a sample $\mathcal{D}$ from the multidimensional normal $N_r(0, \Sigma)$ distribution, the $IW_r(p, \theta)$ is the natural Diaconis-Ylvisaker conjugate prior distribution (see [17]) for the scale parameter Σ. If we let

$$u = \sum_{i=1}^{n} z^i (z^i)^t,$$

the posterior distribution for Σ is then proportional to

$$|\theta|^p |\Sigma|^{-p-\frac{n}{2}-\frac{r+1}{2}} \exp(-\frac{1}{2}\langle \Sigma^{-1}, \theta + u\rangle)$$

and it is clearly the $IW_r(p + \frac{n}{2}, \theta + u)$ distribution. Dawid [13] defined a slightly different notation for the shape parameter δ. He defined δ such that

$$p = \frac{\delta + r - 1}{2}. \tag{10.4}$$

The reason for this change in parametrization will be obvious from Proposition 10.2.1 below. To emphasize the difference, we use two slightly different notations for the Wishart distribution, namely $\mathcal{W}_r(\delta, \sigma)$ when the shape parameter is δ and $W_r(p, \sigma)$ when it is p. With the parametrization δ, the density of the Wishart and inverse Wishart distributions are respectively written as

$$\mathcal{W}_r(x \mid \delta, \sigma) = \frac{|\sigma|^{-\frac{\delta+r-1}{2}}}{\Gamma_r(\frac{\delta+r-1}{2})} |x|^{\frac{\delta-2}{2}} \exp(-\frac{1}{2}\langle x, \sigma^{-1}\rangle)$$

$$\mathcal{IW}_r(y \mid \delta, \theta) = \frac{|\theta|^{\frac{\delta+r-1}{2}}}{\Gamma_r(\frac{\delta+r-1}{2})} |y|^{-\frac{\delta+2r}{2}} \exp(-\frac{1}{2}\langle y^{-1}, \theta\rangle) \tag{10.5}$$

and the posterior distribution for Σ is the $\mathcal{IW}(\delta + n, \theta + U)$ distribution. The key to many priors, conjugate or not, for the covariance or precision matrix in Gaussian graphical models, is the decomposition of the Wishart and its inverse given in the following proposition. Consider a partition of X as $X = \begin{pmatrix} X_{11} & X_{12} \\ X_{21} & X_{22} \end{pmatrix}$ where X_{11} is a $k \times k$ matrix and a similar partition for Y. We use the notation $X_{22\cdot1} = X_{22} - X_{21}X_{11}^{-1}X_{12}$ and $\otimes$ for the Kronecker product.

Proposition 10.2.1. *Let $X \sim W_r(p, \sigma)$ so that $Y = X^{-1} \sim \mathcal{IW}_r(\delta, \theta)$ with $\theta = \sigma^{-1}$. Then*

$$X_{11} \sim W_k(p, \sigma_{11}), \qquad Y_{11} \sim \mathcal{IW}_k(\delta, \theta_{11}) \tag{10.6}$$

$$X_{22\cdot1} \sim W_{r-k}(p - \frac{k}{2}, \sigma_{22\cdot1}), \qquad Y_{22\cdot1} \sim \mathcal{IW}_{r-k}(\delta + k, \theta_{22\cdot1}) \tag{10.7}$$

$$X_{21}X_{11}^{-1} | X_{11} \sim N_{r-k,k}(\sigma_{21}\sigma_{11}^{-1}, x_{11}^{-1} \otimes \sigma_{22\cdot1}), \quad Y_{21}Y_{11}^{-1} | Y_{22\cdot1} \sim N_{r-k,k}(\theta_{21}\theta_{11}^{-1}, Y_{22\cdot1} \otimes \theta_{11}^{-1}) \tag{10.8}$$

$$(X_{21}X_{11}^{-1}, X_{11}) \perp\!\!\!\perp X_{22\cdot1}, \qquad (Y_{21}Y_{11}^{-1}, Y_{22\cdot1}) \perp\!\!\!\perp Y_{11} \tag{10.9}$$

We see that the (δ, θ) notation was chosen so that both marginals for X_{11} and Y_{11} have a simple form in terms of the parameters (p, σ) and (δ, θ) respectively. We will see below that the (δ, θ) parametrization is also very useful when we define the G-Wishart distribution on the cone of positive definite matrices with fixed zeros according to an undirected graph G.

Before closing this section, we should mention the important work of Geiger and Heckerman [21] who proved that a random matrix X follows a Wishart distribution if and only if, for every block partitioning X_1, X_{12}, X_{21}, X_2 of X, we have that $X_{11\cdot 2}$ is independent of (X_{12}, X_{22}).

10.3 Decomposable Graphs and the Hyper Inverse Wishart

10.3.1 The graphical Gaussian (or concentration graph) model

A fundamental fact, true for an arbitrary graph G, not necessarily decomposable, (see Proposition 5.2 in [27]) is that if $Z \sim N_r(0, \Sigma)$ and $K = \Sigma^{-1}$, then

$$X_i \perp\!\!\!\perp X_j \mid X_{V\setminus\{i,j\}} \Leftrightarrow K_{ij} = 0. \tag{10.10}$$

We are then led to consider the cone

$$P_G = \{Y \in P_r \mid Y_{ij} = 0, (i,j) \notin E\}.$$

The graphical Gaussian model Markov with respect to G is the family of distributions

$$\mathcal{N}_G = \{N_r(0, \Sigma) \mid K = \Sigma^{-1} \in P_G\}. \tag{10.11}$$

This model was introduced by Dempster [16] as the covariance selection model. In reference to the fact that missing edges in the graph G translate as 0's in the concentration matrix $K = \Sigma^{-1}$, it is nowadays sometimes called the concentration graph model. In a subsequent section, we will consider covariance graph models where missing edges in G translate as 0's in the covariance matrix Σ.

Let us assume now that G is decomposable and $Z \sim N_r(0, \Sigma)$. It follows from (10.10) that $K_{ij} = 0$ whenever the edge (i,j) is not in a clique C for some $C \in \mathcal{C}$. As a consequence the free elements of Σ are the entries of $\sigma_C, C \in \mathcal{C}$. We are thus led to consider the cone of partially positive definite (denoted > 0) incomplete matrices according to G. More precisely, let I_G be the linear space of symmetric incomplete matrices x with missing entries $x_{ij}, (i,j) \notin E$. Then

$$Q_G = \{x \in I_G \mid x_{C_i} > 0, i = 1, \ldots, k\}. \tag{10.12}$$

There is a 1-1 correspondence between elements of Q_G and elements of P_G. Grone et al. [23] proved that when G is decomposable, for any x in Q_G there exists a unique completion $\hat{x} \in P_r$ such that for all (i,j) in E we have $x_{ij} = \hat{x}_{ij}$ and such that $\hat{x}^{-1} \in P_G$. Here we will simply denote the completion of $\Sigma_G = (\Sigma_C, C \in \mathcal{C})$ by Σ. Thus the correspondence

$$\Sigma_G = (\Sigma_C, C \in \mathcal{C}) \in Q_G \longleftrightarrow K = \Sigma^{-1} \in P_G \tag{10.13}$$

is 1-1 and we can take as the parameter of the Gaussian graphical model either Σ_G or K. Lemma 5.5 in [27] states that

$$K = \Sigma^{-1} = \sum_{C \in \mathcal{C}} \Sigma_C^{-1} - \sum_{S \in \mathcal{S}} \Sigma_S^{-1} \text{ and } |\Sigma| = \frac{\prod_{C \in \mathcal{C}} |\Sigma_C|}{\prod_{S \in \mathcal{S}} |\Sigma_S|} \tag{10.14}$$

where, in this formula, Σ_C^{-1} is taken to be the $r \times r$ zero matrix with its $C \times C$ block replaced by the inverse of Σ_C and similarly for Σ_S^{-1}. Using (10.14) and the notation $p_C(u_C \mid \Sigma_C) = |\Sigma_C|^{-n/2} \exp(-\frac{1}{2}\langle \Sigma_C^{-1}, s_C \rangle)$, the joint distribution of a sample $(z_1, \ldots, z_n)$ from the $N(0, \Sigma) \in \mathcal{N}_G$ can be written as

$$f(z^1, \ldots, z^n | \Sigma, G) \quad \propto \quad \frac{\prod_{C \in \mathcal{C}} p_C(u_C \mid \Sigma_C)}{\prod_{S \in \mathcal{S}} p_C(u_S \mid \Sigma_S)}, \tag{10.15}$$

the Markov ratio of marginal Gaussian $N(0, \Sigma_C), C \in \mathcal{C}$ and $N(0, \Sigma_S), S \in \mathcal{S}$ distributions. The prior distribution on the parameter of the $\mathcal{N}_G$ model has to be a prior distribution on $\Sigma_G \in Q_G$ or equivalently on $K \in P_G$.

10.3.2 The hyper inverse Wishart prior

For $\delta > 0$ and $\theta \in Q_G$ given, Dawid and Lauritzen [14] consider the prior distribution of Σ_G with density equal to the Markov ratio

$$HIW(\Sigma_G \mid \delta, \theta) \quad = \quad \frac{\prod_{C \in \mathcal{C}} \mathcal{IW}_c(\delta, \theta_C; \Sigma_C)}{\prod_{S \in \mathcal{S}} \mathcal{IW}_s(\delta, \theta_S; \Sigma_S)} \tag{10.16}$$

where $\mathcal{IW}_c$ and $\mathcal{IW}_s$ are as in (10.5) and where c and s denote the cardinality of a clique $C \in \mathcal{C}$ and of a separator $S \in \mathcal{S}$ respectively. In (10.20) and other equations using (10.20) below, we will use the factorization

$$|\Sigma_C|^{-\frac{\delta+2r}{2}} = |\Sigma_C|^{-\frac{\delta+r-1}{2}} \times |\Sigma_C|^{-\frac{r+1}{2}}, \; r = c, s \tag{10.17}$$

Distribution (10.16) is called the hyper inverse Wishart. It has several remarkable properties.

(P1) It is conjugate to the Gaussian distribution Markov with respect to G.

(P2) The marginal distribution of Σ_C is $\mathcal{IW}_c(\delta, \theta_C)$, $C \in \mathcal{C}$. Similarly $\Sigma_S \sim \mathcal{IW}_s(\delta, \theta_S)$. Moreover from (10.6), we see that for any two cliques C_i, C_j with $S = C_i \cap C_j$, the marginal distribution on S induced from either of $\mathcal{IW}(\delta, \theta_{C_i}), i = 1, 2$ is the same $\mathcal{IW}(\delta, \theta_S)$.

(P3) It is strong hyper Markov. Recall that if a random variable Z has a distribution Markov with respect to G and with parameter ρ, the distribution of ρ is said to be strong hyper Markov over G if, for $(A \setminus B, A \cap B, B \setminus A)$ a decomposition of G, with $A \cap B$ complete, the parameter ρ_A of the marginal distribution of Z_A and the parameter $\rho_{B|A}$ of the conditional distribution of Z_B given Z_A are stochastically independent.

(P4) Its inverse $K = \Sigma^{-1}$ has density, with respect to the Lebesgue measure on P_G (see [5], [41] or [31]), equal to

$$f(K) = I_G(\delta, \theta)^{-1} |K|^{\frac{\delta-2}{2}} \exp(-\frac{1}{2}\langle K, \theta \rangle) \tag{10.18}$$

where the normalizing constant is

$$I_G(\delta, \theta) = \frac{\prod_{S \in \mathcal{S}} |\theta_S|^{\frac{\delta+s-1}{2}} \, \Gamma_c(\frac{\delta+c-1}{2})}{\prod_{C \in \mathcal{C}} |\theta_C|^{\frac{\delta+c-1}{2}} \, \Gamma_s(\frac{\delta+s-1}{2})}. \tag{10.19}$$

It is the Diaconis-Ylvisaker conjugate prior for the canonical parameter $K = \Sigma^{-1}$ of the Gaussian distribution Markov with respect to the decomposable graph G. As we shall see in Section 10.4, this distribution on K can be generalized to graphical models with G not necessarily decomposable and is called the G-Wishart. It is a natural exponential family and for G decomposable, it is the natural exponential family generated by the image $\nu(K) = (\mu \circ \psi^{-1})(K) = |K|^{(\delta-2)/2}$ by the mapping $\phi:\ \Sigma_G = (\Sigma_C, C \in \mathcal{C}) \to K = \Sigma^{-1}$ of the measure (see (10.16) and (10.17))

$$\mu(\Sigma_G) = \frac{\prod_{C\in\mathcal{C}} |\Sigma_C|^{-\frac{\delta+c-1}{2}}}{\prod_{S\in\mathcal{S}} |\Sigma_S|^{-\frac{\delta+s-1}{2}}} \mu_G(\Sigma_G) \text{ with } \mu_G(\Sigma_G) = \frac{\prod_{C\in\mathcal{C}} |\Sigma_C|^{-\frac{c+1}{2}}}{\prod_{S\in\mathcal{S}} |\Sigma_S|^{-\frac{s+1}{2}}}\,. \tag{10.20}$$

Indeed, the expression of $\nu(K)$ is obtained by multiplying $\mu(\Sigma_G)$ by the Jacobian (see Theorem 2.1, 4., in [31])

$$\left|\frac{d\Sigma_G}{dK}\right| = \frac{\prod_{C\in\mathcal{C}} |\Sigma_C(K)|^{c+1}}{\prod_{S\in\mathcal{S}} |\Sigma_S(K)|^{s+1}} \tag{10.21}$$

and expressing this product in terms of K. Indeed, we have

$$\frac{\prod_{C\in\mathcal{C}} |\Sigma_C|^{-\frac{\delta+c-1}{2}-\frac{c+1}{2}+(c+1)}}{\prod_{S\in\mathcal{S}} |\Sigma_S|^{-\frac{\delta+s-1}{2}-\frac{s+1}{2}+(s+1)}} = \det(K)^{\frac{\delta-2}{2}}\,.$$

When the canonical parameter of the natural exponential family is $\theta \in Q_G$, the densities in the family take the form (10.18).

The strong hyper Markov property (P3) means that if we can find a decomposition $(A \backslash B, A\cap B, B\setminus A)$ of G, then (see [14], Proposition 5.1), if $Z \sim N(0,\Sigma) \in \mathcal{N}_G$ and $\Sigma_G \sim \mathcal{IW}(\delta,\theta)$

$$(Z_A, (\Sigma_G)_A) \perp\!\!\!\perp (Z_B, (\Sigma_G)_{B|A}) \mid Z_{B\cap A}.$$

Inference can then be made layer by layer, first for $(\Sigma_G)_A$ and then for $(\Sigma_G)_{B|A}$. The decomposition (10.1) for a perfect order of the cliques $C_1,\ldots,C_k$ of G together with the global Markov property yields that $Z_{R_j} \perp\!\!\!\perp Z_{H_{j-1}} \mid Z_{S_j}$. The parameter of $Z_{R_j} \mid Z_{S_j}$ is $(\Sigma_{R_j,S_j}\Sigma_{S_j}^{-1}, \Sigma_{R_j\cdot S_j})$. By the strong hyper Markov property, it is independent of $(\Sigma_G)_{H_{j-1}}$. Continuing with the decomposition of H_{j-1} and applying Proposition 10.2.1 repeatedly we obtain the following.

Proposition 10.3.1. *Let G be decomposable, let $\Sigma_G = (\Sigma_C, C \in \mathcal{C})$ follow the hyper inverse Wishart distribution* (10.16) *with scale parameter θ and shape parameter δ. Then*
(i) the parameters

$$\{\Sigma_{C_1}, (\Sigma_{R_j,S_j}\Sigma_{S_j}^{-1}, \Sigma_{R_j\cdot S_j}),\ j = 2,\ldots k\} \tag{10.22}$$

are independent.
(ii)

$$\begin{aligned}
\Sigma_{C_1} &\sim \mathcal{IW}_{c_1}(\delta, \theta_{C_1})\\
\Sigma_{R_j\cdot S_j} &\sim \mathcal{IW}_{c_j - s_j}(\delta + s_i, \theta_{R_j\cdot S_j})\\
\Sigma_{R_j,S_j}\Sigma_{S_j}^{-1} \mid \Sigma_{S_j} &\sim N_{c_j - s_j, s_j}(\theta_{R_j,S_j}\theta_{S_j}^{-1}, \Sigma_{R_j\cdot S_j} \otimes \theta_{S_j}^{-1})\,.
\end{aligned}$$

In fact the parameters (10.22) are the building blocks of the Cholesky decomposition of Σ and K such that $K \in P_G$. This can readily be seen by writing the model as a series of regression equations of the form

$$\begin{aligned}
&Z_{C_1} = \varepsilon_{C_1}\\
&Z_{R_i,S_i} - \Sigma_{R_i,S_i}\Sigma_{S_i}^{-1} Z_{S_i} = \varepsilon_{R_i},\ i = 2,\ldots,k
\end{aligned}$$

where the $\varepsilon_{C_1}, \varepsilon_{R_i}$, $i = 2, \ldots, k$ are independent, normally distributed with variance $\Sigma_{R_i \cdot S_i}$. These equations can be written under matrix form as

$$LZ = \varepsilon \tag{10.23}$$

where L has 1 on the diagonal and R_i-row block $-\Sigma_{R_i,S_i}\Sigma_{S_i}^{-1}$ with 0 everywhere else in the row block. From (10.23), it follows that $L\Sigma L^t = D$ where the R_i-block diagonal element of D is $\Sigma_{R_j \cdot S_j}$. Thus

$$\Sigma = L^{-1}DL^{-t} \text{ and } K = L^t D^{-1} L. \tag{10.24}$$

It is easy to verify that the entries $(R_j, H_{j-1} \setminus S_j)$ of K are filled with 0's. Proposition 10.3.1 above is a generalization of the Bartlett decomposition of the inverse Wishart distribution as can be deduced from [35], Theorem 3.2.14.

Sampling from the hyper inverse Wishart

Proposition 10.3.1 also gives us the tools to sample Σ from the hyper inverse Wishart $\mathcal{IW}_r(\delta, \theta)$. This was done in [7]. The sampling scheme is defined as follows:
(1) sample $\Sigma_{C_1} \sim \mathcal{IW}_{c_1}(\delta, \theta_{C_1})$ which gives values to the submatrix Σ_{S_2}.
(2) for $i = 2, \ldots, k$, sample

$$\begin{aligned} \Sigma_{R_i \cdot S_i} &\sim \mathcal{IW}_{c_j - s_j}(\delta + s_i, \theta_{R_i \cdot S_i}) \\ U_i &\sim N_{c_i - s_i, s_i}(\theta_{R_j,S_j}\theta_{S_j}^{-1}, \Sigma_{R_j \cdot S_j} \otimes \theta_{S_j}^{-1}) \end{aligned}$$

and then directly compute the implied values of $\Sigma_{R_j,S_j} = U_i \Sigma_{S_i}$ and $\Sigma_{R_i} = \Sigma_{R_j \cdot S_j} + \Sigma_{R_i,S_i}\Sigma_{S_i}^{-1}\Sigma_{S_i,R_i}$.

This gives the entries of $\Sigma_C, C \in \mathcal{C}$. It remains to fill in the missing values in this incomplete matrix. With the notation $A_{i-1} = H_{i-1} \setminus S_i$, the standard completion operation yields

$$\Sigma_{R_i,A_{i-1}} = \Sigma_{R_i,S_i}\Sigma_{S_i}^{-1}\Sigma_{S_i,A_{i-1}}.$$

Bayes factors and model selection

In a Bayesian framework, the classical approach for choosing between two models $\mathcal{N}_G$ and $\mathcal{N}_{G'}$ is to compute their posterior probability and choose the model with the highest posterior probability. Assuming a uniform prior distribution on the space $\mathcal{G}$ of decomposable graphs on r variables, from (10.15) and (10.18) we obtain the joint distribution

$$f(z^1, \ldots, z^n, K, G) = \frac{1}{(2\pi)^{\frac{np}{2}}|\mathcal{G}|} \frac{1}{I_G(\delta, \theta)} |K|^{\frac{\delta+n-2}{2}} \exp(-\frac{1}{2}\langle K, \theta + u \rangle) .$$

Integrating out K, we obtain

$$\begin{aligned} f(z^1, \ldots, z^n, G) &= \frac{1}{(2\pi)^{\frac{np}{2}}|\mathcal{G}|} \frac{1}{I_G(\delta, \theta)} \int_{P_G} |K|^{\frac{\delta+n-2}{2}} \exp(-\frac{1}{2}\langle K, \theta + u \rangle) dK \\ &= \frac{1}{(2\pi)^{\frac{np}{2}}|\mathcal{G}|} \frac{I_G(\delta + n, \theta + u)}{I_G(\delta, \theta)} \end{aligned}$$

which is proportional to the posterior distribution of G given $(z^1, \ldots, z^n)$. The Bayes factor $B_{G,G'}$, ratio of the posterior distributions, is then

$$B_{G,G'} = \frac{I_G(\delta + n, \theta + u) I_{G'}(\delta, \theta)}{I_G(\delta, \theta) I_{G'}(\delta + n, \theta + u)} \tag{10.25}$$

and can be computed explicitly using (10.19).

It now remains to travel through the space $\mathcal{G}$ of decomposable graphs on r vertices. The basic algorithm to do so is a Metropolis-Hastings algorithm. It was used, for example in Madigan and York [32].

Algorithm 10.3.2. *Let $nbd_r(G)$ be the set of graphs obtained from G by adding or removing an edge from G. Starting from a decomposable graph G_0, repeat the following two steps for $t = 0, 1, 2, \ldots$:*

1. *sample G' from the uniform distribution on $nbd_r(G_t)$,*

2. *If G' is decomposable, accept the move $G_{t+1} = G'$ with probability*

$$min\{1, B_{G_{t+1},G_t}\}$$

where B_{G_{t+1},G_t} is as defined in (10.25). If the move is rejected or if G' is not decomposable, set $G_{t+1} = G$.

In a seminal paper for decomposable graphical Gaussian model selection, using the hyper inverse Wishart prior, Giudici and Green [22], departing from the algorithm above, proposed a reversible jump MCMC algorithm where the state variable includes the graph and the covariance or precision matrix. The methodology presented in that paper is the starting point of more modern methods of model selection using reversible jump MCMC but improving on various aspects of the algorithm and extending it to the class of arbitrary undirected graphs.

10.3.3 Priors with several shape parameters

The shape parameter δ of the hyper inverse Wishart is added up to the sample size n as we just saw in the previous section and can be used to increase the influence of the prior on inference. Often more flexibility is needed in order to reflect prior knowledge and give more importance to some variables over others. With this aim in mind and to generalize the hyper inverse Wishart, Letac and Massam [31] developed a prior with several shape parameters. In the measure (10.20), they replace the exponents

$$\frac{\delta + c - 1}{2}, C \in \mathcal{C} \text{ and } \frac{\delta + s - 1}{2}, S \in \mathcal{S}$$

with different hyper parameters $\alpha = (\alpha_C, C \in \mathcal{C})$ and $\beta = (\beta_S, s \in \mathcal{S})$ respectively. The resulting distribution has density

$$IW_{P_G}(\alpha, \beta, \theta; d\Sigma_G) = \frac{e^{-\langle \theta, \Sigma^{-1} \rangle} H_G(\alpha, \beta; \Sigma_G)}{\Gamma_{II}(\alpha, \beta) H_G(\alpha, \beta; \theta)} \mu_G(d\Sigma_G) \tag{10.26}$$

where $H_G(\alpha, \beta; \Sigma_G) = \frac{\prod_{C \in \mathcal{C}} (\det \Sigma_C)^{\alpha(C)}}{\prod_{S \in \mathcal{S}} (\det \Sigma_S)^{\nu(S)\beta(S)}}$ and $\Gamma_{II}(\alpha, \beta)$ is a known constant. There are in fact only $k+1$ free hyper parameters $(\alpha_C, C \in \mathcal{C})$ and β_{S_2}, chosen in a set B_P dependent upon the perfect order P of the cliques of G considered so that the distribution be proper and the normalizing constant be explicitly computable. The reader is referred to [31], Section 3.4 for details on the set of hyper parameters (α, β) and the expression of $\Gamma_{II}(\alpha, \beta)$ as a product of Gamma functions.

This generalized hyper inverse Wishart is called the inverse W_{P_G} distribution because its inverse, the W_{P_G} distribution, as defined in [31], is the generalization of the G-Wishart as defined in (10.18). While giving flexibility to the hyper parameters, the IW_{P_G} distribution keeps the essential properties (P1), (P3), (P4) and the distributional properties of Proposition 10.3.1 of the hyper inverse Wishart. We summarize these properties in the following. Before stating the theorem, let us recall the definition of the strong directed hyper Markov property. Given a DAG D with vertex set V, if a random variable Z follows a distribution with parameter ρ Markov with respect to D, we say that the law of ρ is strong directed hyper Markov over D if

$$\rho_{v|pa(v)} \perp\!\!\!\perp \rho_{nd(v)}$$

where, as usual, $\rho_{v|pa(v)}$ denotes the parameters of the law of Z_v given $Z_{pa(v)}$ and $\rho_{nd(v)}$ denotes the parameter of the marginal distribution of $Z_{nd(v)}$.

Theorem 10.3.3. *Let $(z^1, \ldots, z^n)$ be a sample from the $N(0, \Sigma) \in \mathcal{N}_G$ distribution and let $\Sigma_G \sim IW_{P_G}(\alpha, \beta, \theta)$. Let P be a perfect order of the cliques of G. Then the following hold:*
(i) The $IW_{P_G}(\alpha, \beta, \theta)$ is conjugate to the Gaussian distribution and the posterior distribution of Σ_G given u is $IW_{P_G}(\alpha - \frac{n}{2}, \beta - \frac{n}{2}, \theta + u)$ for appropriate values of (α, β) in B_P.
(ii) For any perfect order P of the cliques of G, the $IW_{P_G}(\alpha, \beta, \theta)$ is strong directed hyper Markov.
(iii) More precisely, if $\Sigma_G \sim IW_{P_G}(\alpha, \beta, \theta)$ with $(\alpha, \beta) \in B_P$ and $\theta \in Q_G$, then

$$\begin{aligned}
\Sigma_{R_i \cdot S_i} &\sim IW_{c_i - s_i}(-\alpha_i, \; \theta_{R_i \cdot S_i}), \; i = 1, \ldots, k \\
\Sigma_{R_1, S_2} | \Sigma_{R_1 \cdot S_2} &\sim N_{(c_1 - s_2) \times s_2}(\theta_{R_1, S_2}, \; \theta_{S_2}^{-1} \otimes \Sigma_{R_1 \cdot S_2}) \\
\Sigma_{S_2} &\sim IW_{s_2}(-(\alpha_1 + \frac{c_1 - s_2}{2} + \gamma_2), \theta_{S_2}) \\
\Sigma_{R_j, S_j} \Sigma_{S_j}^{-1} | \Sigma_{R_j \cdot S_j} &\sim N_{(c_j - s_j) \times s_j}(\theta_{R_j, S_j} \theta_{S_j}^{-1}, \; \theta_{S_j}^{-1} \otimes \Sigma_{R_j \cdot S_j}), \; j = 2, \ldots, k.
\end{aligned}$$

(iv) Also, we have that

$$\{(\Sigma_{R_1, S_2}, \Sigma_{R_1 \cdot S_2}), \; \Sigma_{S_2}, (\Sigma_{R_j, S_j} \Sigma_{S_j}^{-1}, \Sigma_{R_j \cdot S_j}), j = 2, \ldots, k\} \tag{10.27}$$

are mutually independent.
(v) The distribution of $K = \Sigma^{-1}$ is denoted the W_{P_G} Wishart and is a generalization of the G-Wishart. It is the natural exponential family generated by the image of the measure $H_G(\alpha, \beta; \Sigma_G) \mu_G(d\Sigma_G)$ under the mapping $\Sigma_G = (\Sigma_C, C \in \mathcal{C}) \to K = \Sigma^{-1}$.

For the graph G of Figure 10.1, property (iv) translates into the fact that the random variables

$$(\Sigma_{1,(23)}, \Sigma_{1 \cdot (23)}), \Sigma_{(23)}, (\Sigma_{4 \cdot (23)} \Sigma_{(23)}^{-1}, \Sigma_{4 \cdot (23)}), \Sigma_{5 \cdot (24)} \Sigma_{(24)}^{-1}, \Sigma_{5 \cdot (24)})$$

are mutually independent with distributions as given in (i), (ii), and (iii) above. The notation $\Sigma_{(23)}$ above means the submatrix of Σ obtained by taking the second and third rows and columns only.

From properties (i)-(iv) above, it follows immediately that, as for the hyper inverse Wishart, one can sample from the IW_{P_G} distribution. Moreover, the estimation of the posterior mean of K and Σ can be done explicitly. Since the W_{P_G} is a natural exponential family, the posterior mean of K is obtained by differentiating the cumulant generating function of its distribution. Due to the distributional properties and the independences in (iii) and (iv) of the theorem above, it is also straightforward to derive the expected value of Σ (see Theorem 3.1 of [40]) through the expected value of each one of the Cholesky elements listed in (iv). Each $E(\Sigma_{R_i \cdot S_i} | z^1, \ldots, z^n)$ is a linear combination of $\theta_{R_i \cdot S_i}$ and $u_{R_i \cdot S_i}$. So, the posterior mean is a shrinkage estimate of Σ and through the use of (α, β) we can shrink different layers of the Cholesky decomposition with different intensity.

Rajaratnam et al. [40] use this flexible prior to do model selection using Bayes factors in the restricted class of decomposable models with banded precision matrix. Since the IW_{P_G} is dependent upon the perfect order P of the cliques of G, the class of models considered must necessarily have a given order of the vertices and it is not a very efficient tool for model selection in the space of models Markov with respect to an undirected graph. However, as we will see, a similar idea will be used for model selection in the class of DAG Markov models.

10.3.4 Covariance graph models

Cox and Wermuth [12] introduced covariance graph models, defined as follows. Let G be an undirected graph. For $\Sigma \in P_r$, the covariance graph model Markov with respect to G is the family of distributions

$$\mathcal{K}_G = \{N_r(0,\Sigma) \mid \Sigma \in P_G\} .$$

Clearly this is a model that prescribes marginal independences rather than conditional independences as in $\mathcal{N}_G$ and the linear restriction that $\Sigma \in P_G$ is not a linear restriction on the canonical parameter of the Gaussian family which is thus a curved exponential model. There is therefore no Diaconis-Ylvisaker conjugate prior. But Khare and Rajaratnam [26] proposed a prior on Σ that is conjugate, has several shape parameters for flexibility and is such that it can be sampled from. If we let $\Sigma = LDL^t$ be the Cholesky decomposition of Σ where L is a lower triangular matrix with 1 on the diagonal and D is a diagonal matrix with $D_{ii} > 0, i = 1, \dots, r$, we can write the sampling distribution as

$$f(z^1, \dots, z^n|\Sigma) = \frac{1}{(2\pi)^{n/2}} \exp(-\frac{1}{2}\{\langle (LDL^t)^{-1}, u\rangle + \sum_{i=1}^{r} nD_{ii}\}).$$

Moreover Paulsen et al. [39] have proved that $G = (V, E)$ is decomposable if and only there is a numbering of the vertices such that, in the Cholesky decomposition of any matrix $A = LDL^t \in P_G$, L belongs to the space

$$\mathcal{L}_G = \{L \in M_r \mid L_{ij} = 0 \text{ whenever } i < j \text{ or } (i,j) \notin E\}.$$

Thus the parameter space of $\mathcal{K}_G$ is

$$\Theta_G = \{(L, D) \mid L \in \mathcal{L}_G, D_{ii} > 0, i = 1, \dots, r\} .$$

A prior for (L, D) conjugate to the $N(0, \Sigma) \in \mathcal{K}_G$ will necessarily be of the form

$$\pi_{U,\alpha}(L, D)) \propto \exp[-\frac{1}{2}\{\langle (LDL^t)^{-1}, U\rangle + \sum_{i=1}^{r} \alpha_{ii}\}] \tag{10.28}$$

where $U \in P_r$ and $\alpha = (\alpha_{ii}, i = 1, \dots, r)$ has all its components positive. A sufficient condition for the prior in (10.28) to be proper is that α_i be greater than 2 plus the number of neighbors in G preceding i in the given perfect ordering of the vertices. This prior on (L, D) has the flavor of an inverse Wishart in the sense that, if, for $v \in V$, we denote by $L_{\cdot v}$ the v-th column of L, we have that
(i) the distribution of $L_{\cdot v}$ given all the other parameters $(L \setminus L_{\cdot v}, D)$ is normal,
(ii) the diagonal elements $D_{ii}, i = 1, \dots, r$ are independent and follow an inverse Gamma distribution.
It is thus possible to sample from the prior and posterior distribution (10.28) using a block Gibbs sampler and to give an estimate of the posterior mean of $\Sigma \in P_G$. The reader is

referred to [26] for sampling details. There is, however, no attempt in [26] to do model selection in the class of covariance graph models. For the general class of directed mixed graphs, which includes as a special case the covariance graph models, Silva and Ghahramani [45] give a similar prior to that in [26], but without the multiple shape parameters, and do model selection in the larger class of acyclic directed mixed graphs.

10.4 Arbitrary Undirected Graphs and the G-Wishart

For G an arbitrary undirected graph, the definition of the Gaussian model Markov with respect to G remains as in (10.11).

10.4.1 Computing the normalizing constant of the G-Wishart

When the graph G is not necessarily decomposable, there is no equivalent to the hyper inverse Wishart but as we saw in Section 10.3.1, we still have $K \in P_G$. The joint density of $(z^1, \dots, z^n)$ is proportional to

$$|K|^{\frac{n}{2}} \exp(-\frac{1}{2}\langle K, u\rangle)$$

and the Diaconis-Ylvisaker prior with density of the same form as (10.18) but where the normalizing constant $I_G(\delta, \theta)$, unlike that in (10.19), has no closed analytic expression.

To do model search using the usual Bayesian paradigm of computing the marginal likelihood for each model G and using Bayes factor could only be used if there was a way to compute $I_G(\delta, \theta)$. Traditionally, the only way to compute the normalizing constant $I_G(\delta, \theta)$ was through the Laplace approximation. In 2005, Atay-Kayis and Massam [3] gave a simple Monte-Carlo method to compute it. They considered the Cholesky decompositions $K = \phi^t\phi$ and $\theta^{-1} = T^tT$ where ϕ and T are upper triangular matrices and made the change of variable $K \to \psi = \phi T^{-1}$. Because the entries $K_{ij}, (i,j) \notin E$ are null, the corresponding $\psi_{ij}, (i,j) \notin E$ are functions of $\psi_{ij}, (i,j) \in E, \psi_{ii}, i = 1, \dots, r$ and $\psi^E = (\psi_{ij}, (i,j) \in E, \psi_{ii}, i = 1, \dots, r)$ is the Cholesky random variable corresponding to K. With the change of variable $K \in P_G \to \psi^E$, $I_G(\delta, \theta)$ can be expressed as the product of a known constant and the expected value of

$$f_T(\psi^E) = \exp(-\frac{1}{2} \sum_{(i,j)\notin E} \psi_{ij}^2) \tag{10.29}$$

where

$$\psi_{ii}^2 \sim \chi^2_{\delta+\nu_i}, i = 1, \dots, r, \quad \phi_{ij} \sim N(0,1), (i,j) \in E \tag{10.30}$$

are mutually independent and where ν_i is the number of neighbors of i with a label $j > i$ in the perfect ordering of the vertices in V. The $\psi_{ij}, (i,j) \notin E$ are computed through the completion operation which consists in giving the explicit expression of $\psi_{ij}, (i,j) \notin E$ in terms of $\psi_{ij}, (i,j) \in E$ and entries of T (see formula (31) in [3]).

Calculations are considerably simplified by using the decomposition of any graph into its prime components. Let $P_1, \dots, P_k$ be a perfect sequence of prime components of G and let $S_i, i = 2, \dots, k$ be the corresponding separators. We note that, by definition of a prime component, such a sequence always exists. Let G_{P_i} and G_{S_i} be the corresponding induced graphs. Then

$$I_G(\delta, \theta) = \frac{\prod_{j=1}^k I_{G_{P_j}}(\delta, \theta_{P_j})}{\prod_{j=2}^k I_{G_{S_j}}(\delta, \theta_{S_j})}.$$

Roverato [42] and Dellaportas et al. [15] previously gave methods to compute $I_G(\delta, \theta)$ that used importance sampling. These were more delicate to implement and have not been much used.

Following [3], a landmark paper in graphical Gaussian model selection was Jones et al. [25]. They used the method in [3] to compute $I_G(\delta, \theta)$ and $I_G(\delta + n, \theta + S)$, and traverse the space of graphs with the add-delete Metropolis-Hastings sampler as well as a heuristic method, the "Stochastic Shotgun Search" algorithm. The priors on the space of graphs were the uniform prior but also the Erdős-Renyi prior where a graph with $|E|$ edges has prior $\beta^{|E|}(1-\beta)^{\frac{r(r+1)}{2}-|E|}$, a prior that will often be used in subsequent works. Indeed, the uniform prior distribution favors graphs of middle size which is problematic if one wants to promote sparsity. Previous to [25], Wong et al. [52] had also searched the space of arbitrary undirected graphs with a prior that gives equal probability to graphs within the class of graphs with the same number of edges and penalizes dense graphs. However, this prior does not seem to be much used. Much more used is the Erdős-Renyi prior with a hyper-prior on β, called in [43] the multiplicity correction prior. This prior allows control of the probability that an edge will appear in the graph. In [9] it is shown that, when doing model selection, it keeps the number of false positives constant as the number of vertices increase. It has more recently been adopted in [1] and [30] for DAG Markov model selection. It remains that the uniform distribution is still often favored for its convenience, for example in [34] who, though, also propose an alternative truncated Poisson prior $p(G) = \gamma^{|E|}/|E|!$ where E is the number of edges in the graph.

When the set of graphs considered is that of decomposable graphs, the uniform and Erdős-Renyi prior laws used by [25], are particular examples of a more general class of laws, defined by Byrne and Dawid [6]. There, the authors extend the hyper Markov concept to so-called structurally Markov distributions over the set of graphs. They show that, with compatible laws for the sampling variable and for the parameter of the distribution of the sampling variable, these laws are conjugate and lead to easy posterior inference. They also show that a law over the set of decomposable graphs is structurally Markov if and only if it is a member of the clique exponential family. This means that a density in that family is of the form $\pi_\omega(G) = \frac{1}{Z(\omega)} \exp(\langle \omega, t(G) \rangle)$, where for each $A \subset V$,

$$t(G) = \begin{cases} 1 & \text{if } A \in \mathcal{C} \\ -\nu(A) & \text{if } A \in \mathcal{S} \\ 0 & \text{otherwise,} \end{cases}$$

ω is a parameter in $\mathbb{R}^{2^V}$ and $Z(\omega)$ is the normalizing constant. For the uniform distribution over the set of decomposable graphs $\omega_A, A \subset V$, is a constant and for the Erdős-Renyi prior, $\omega_A = \binom{|A|}{2} \log \beta/(1-\beta)$. Clearly ω can be chosen to penalize or privilege different types of graphs. The value of this general family of distributions over classes of graphs is that they complete the fully Markov Bayesian structure of model selection. One notes that the priors proposed in [52] are not structurally Markov.

Coming back to the work of [25], though it represented a major progress in model selection, for high-dimensional data, due to the sheer size of the search space, their methodology as described above was slow and could not deal with problems of more than 150 variables.

10.4.2 Sampling from the G-Wishart

The publication of [3] was followed by a string of papers giving algorithms to sample from the G-Wishart. First, Mitsakakis et al. [33] gave an independence chain Metropolis Hastings algorithm with proposal density equal to

$$h(\psi^E) = \prod_{i=1}^{p} \sqrt{\chi^2_{\delta+\nu_i}} \times N_E(0_{|E|}, I_{|E|})$$

and acceptance probability

$$\min\left\{\frac{f_T(\psi^E_{proposed})}{f_T(\psi^E_{current})}, 1\right\},$$

where at each step, they generate all of ψ^E. The full triangular matrix ψ is obtained through the completion operation as given in equations (31) and (32) of [3]. Wang and Carvalho [49] gave a direct sampler using (10.30), a rejection sampling algorithm and the reconstruction of $K_{G_{P_j}}$ using the method in [3]. Both of these methods are inefficient in high-dimension because of the completion step of either K or ϕ. Lenkoski and Dobra [29] applied the Bayesian iterative proportional scaling algorithm due to Asci and Piccioni [2] to sample from the G-Wishart. However, to do so, the set of cliques of G needs to be enumerated and that is an NP-hard problem. Moreover, at each step it involves the inversion of a potentially large matrix. Carvalho and Scott [9] offer a model selection method for decomposable models which is original in the sense that it uses objective priors and fractional Bayes. We will come back to this paper in Section 10.6.

Dobra et al. [18] designed yet another sampler for the G-Wishart, a Metropolis-Hasting Markov chain similar to that in [33] but where only one entry of ψ^E is updated at each step. If the entry to update is a diagonal ψ_{ii}, it is sampled from a truncated normal with mean the current value of ψ_{ii} thus ensuring that it is positive, and if it is an off-diagonal element ψ_{ij}, then it is sampled from a normal with mean the current value of ψ_{ij} and a variance parameter that can be adjusted. By working with ψ rather than K, the positive definiteness of $K = \psi^t\psi$ is insured unlike what happened in [22] where entries of K were updated and one had to check after every update that K was positive definite.

10.4.3 Moving away from Bayes factors

Like [22], Dobra et al. [18] develop a reversible jump MCMC sampler to sample from the joint distribution of (G, K) thus avoiding an algorithm to traverse the space of graphs and Bayes factors. However, the graph update step still requires the computation of the prior normalizing constants for the current and proposed graph. To compute the prior normalizing constant $I_G(\delta, \theta)$ they find that the method given in [3] is very efficient even for large graphs. Matrix model graphical Gaussian models are also considered in [18] but we will come back to that in Section 10.5.

Finding that computing the normalizing constant of the G-Wishart creates a real bottleneck on the road to high-dimensional graphical Gaussian model selection, Wang and Li [50] propose a double reversible jump sampler on (G, K) without the need to evaluate the G-Wishart normalizing constants for each graph. Indeed, to compute the acceptance probability of moving from $G = (V, E)$ to $G' = (V, E')$ where $E' = E \setminus (i, j)$, their method uses the conditional posterior odds against the edge (i, j) given by

$$\frac{P(G'|K \setminus \{K_{ij}, K_{jj}\}, Z)}{P(G|K \setminus \{K_{ij}, K_{jj}\}, Z)} = \frac{P(Z, K \setminus \{K_{ij}, K_{jj}\}|G')P(G')}{P(Z, K \setminus \{K_{ij}, K_{jj}\}|G)P(G)} \tag{10.31}$$

where the normalizing constants in both $P(Z, K\setminus\{K_{ij}, K_{jj}\}|G')$ and $P(Z, K\setminus\{K_{ij}, K_{jj}\}|G)$ have explicit analytic expressions and there is thus no need for approximations. Lenkoski [28] exploits this same idea but with a new direct sampler for the G-Wishart and illustrates his method on small dimensional problems.

Mohammadi and Witt [34] exploit many features from previous papers but came up with a radically new idea to move in the space of graphs. Their Birth and Death MCMC algorithm (BDMCMC) samples from the joint posterior distribution of (G, K) given $z^1, \dots, z^n$. The prior on the space of graph is uniform or a truncated Poisson with density $p(G) \propto \frac{\gamma^{|E|}}{|E|!}$. To explore the space of graphs, starting from a graph G, they assume independent birth and death processes to add or remove an edge $e = (i, j)$ with rates

$$\beta_e(K) = \frac{P(G^{+e}, K^{+e} \setminus (K_{ij}, K_{jj})|z^1, \dots, z^n)}{P(G, K \setminus K_{jj}|z^1, \dots, z^n)} \text{ and } \delta_e(K) = \frac{P(G^{-e}, K^{-e} \setminus (K_{jj})|z^1, \dots, z^n)}{P(G, K \setminus (K_{ij}, K_{jj})|z^1, \dots, z^n)}$$

for each $e \notin E$ and each $e \in E$, respectively. Here G^{+e}, G^{-e} denote the graphs obtained from G by adding or removing the edge e. These ratios can be computed explicitly as in (10.31) using similar results to that used in [50]. To sample from the posterior distribution of K they use the direct sampler of Lenkoski [28]. They illustrate how well their sampler explores the space of graphs on several moderate size problems.

10.4.4 Moving away from Bayes factors and the G-Wishart

Meanwhile, in a linear regression context, the first formal link between the frequentist Lasso method and Bayesian inference is made. Park and Casella [38] consider the regression model $Y = \mu \mathbf{1}_n + X\beta + \epsilon$ where Y is an $n \times 1$ vector of responses, μ is the overall mean, X is an $n \times p$ matrix of standardized regressors, β is a $p \times 1$ vector of parameters and $\varepsilon \sim N_n(0, \sigma^2 I_n)$. For $\tilde{Y} = Y - \bar{Y}\mathbf{1}_n$, the Lasso estimate of β is obtained by minimizing, with respect to β,

$$(\tilde{Y} - X\beta)^t(\tilde{Y} - X\beta) + \lambda \sum_{j=1}^{p} |\beta_j|$$

for some $\lambda \geq 0$. As noticed by [46], this quantity looks like the log of the product of the Gaussian $N_n(0, \sigma^2 I_n)$ for $\tilde{Y}$ and the double exponential (or Laplace) prior on β with density

$$\pi(\beta|\sigma^2) = \prod_{j=1}^{p} \frac{\lambda}{2\sqrt{\Sigma^2}} e^{-\lambda|\beta_j|/\sqrt{\sigma^2}}.$$

The double exponential has a wonderful property. It can be viewed as the scale mixture of normals through the following formula

$$\frac{a}{2} e^{-a|z|} = \int_0^{+\infty} \frac{1}{\sqrt{2\pi s}} e^{-z^2/2s} \frac{a^2}{2} e^{-a^2 s/2} ds, \quad a > 0, \tag{10.32}$$

where the mixing density is the exponential distribution.

Following [38] and in a first attempt to move away from traversing the space of graphs, Wang [47] naturally introduces the Bayesian graphical Lasso. For $Z \sim N_r(0, K^{-1})$, giving a Bayesian interpretation to the expression

$$\log(\det K) - \frac{1}{n}\langle u, K \rangle - \lambda \sum_{1 \leq i,j \leq r} |K_{ij}|$$

he puts the following prior on K

$$p(K|\lambda) = C^{-1} \prod_{i<j} DE(K_{ij}|\lambda) \prod_{i=1}^{r} EXP(K_{ii}|\frac{\lambda}{2})\, \mathbf{1}_{P_r}(K)$$

where $DE(x|\lambda)$ is the double exponential distribution with density $p(x) = \frac{\lambda}{2}\exp(-\lambda|x|)$ and $EXP(x|\lambda)$ is the exponential distribution with density $p(x) = \lambda\exp(-\lambda x)\mathbf{1}_{R^+}(x)$. As in [38], the double exponential is expressed as a scale mixture of normals according to (10.32). The posterior distribution is then

$$p(K,\tau \mid z^1,\dots z^n,\lambda) \propto |K|^{\frac{n}{2}}\exp(-\frac{1}{2}\langle u,K\rangle) \prod_{i<j}\tau_{ij}^{-\frac{1}{2}}\exp(-\frac{K_{ij}^2}{2\tau_{ij}})\exp(-\frac{\lambda^2}{2}\tau_{ij})\prod_{i=1}^{r}\exp(-\frac{\lambda}{2}K_{ii}) \tag{10.33}$$

The aim of the paper is now to design a block Gibbs sampler for K and set to 0 entries that are too small. From (10.33), there is no obvious way to consider conditional distributions and design a Gibbs sampler. However, letting $\Upsilon = (\tau_{ij})$ be a $r\times r$ symmetric matrix with zeros on the diagonal, splitting the matrices K, u and Υ in blocks according to the last column as follows

$$K = \begin{pmatrix} \mathbf{K}_{11} & K_{12} \\ K_{21} & K_{22}\end{pmatrix}, \quad u = \begin{pmatrix} \mathbf{u}_{11} & u_{12} \\ u_{21} & u_{22}\end{pmatrix}, \quad \Upsilon = \begin{pmatrix} \mathbf{\Upsilon}_{11} & \tau_{12} \\ \tau_{21} & 0\end{pmatrix}$$

and, for $\mathbf{K}_{11}$ given, making the change of variable

$$(K_{12}, K_{22}) \to (K_{22\cdot 1}, K_{12}), \tag{10.34}$$

it follows from (10.33) that the conditional joint distribution of $(K_{2\cdot 1}, K_{12})$ is

$$\begin{aligned} p(K_{2\cdot1},K_{12}|\mathbf{K}_{11},\Upsilon,z^1,\dots,z^n,\lambda) &\propto K_{2\cdot1}^{\frac{n}{2}}\exp(-\frac{u_{22}+\lambda}{2}K_{2\cdot1}) \\ &\times\exp\Big\{-\frac{1}{2}\Big(K_{12}(D_\tau^{-1}+(u_{22}+\lambda)\mathbf{K}_{11})K_{21}+2u_{21}K_{12}\Big)\Big\}. \end{aligned} \tag{10.35}$$

The reader will notice that though we did not start with a Wishart prior on K, using the density of $(z^1,\dots,z^n)$ and borrowing from the Bayesian Lasso priors on the entries of K, we end up in (10.35) with the conditional distribution of $(K_{2\cdot1}, K_{12})$ as if we had put a Wishart prior on K (see Proposition 10.2.1) i.e. $K_{2\cdot1}$ is Gamma and K_{12} is normal and they are independent. From (10.33), we see that, conditional on $K, z^1,\dots,z^n$, the components $\tau_{ij}, 1\le i<j\le r$, of τ follow independent inverse Gaussian distributions. So, for a given Lasso coefficient λ, at each update, the Gibbs sampler samples from $(K_{2\cdot1}, K_{12})$ thus yielding the last column of K. It remains of course to choose the shrinkage parameter λ. The threshold for setting some K_{ij} to 0 in order to get a graphical model is set by a heuristic rule recommended in [8]. The method scales up to moderate dimension but in some cases takes up to 2 days for a 100 variable problem.

In a recent paper, Wang [48] replaces the continuous mixture of normal priors on K_{ij} of [47] by a continuous spike and slab prior, the mixture of just two normal priors, working on precision as well as covariance graph models. Moreover, unlike what had been done in [47], a graph structure is also included in the model under the form of a latent variable $U = (U_{ij})_{i<j} \in \{0,1\}^{\frac{r(r-1)}{2}}$. For a matrix A, variance or precision matrix, the prior distribution is thus

$$p(a) = C(\theta)^{-1}\prod_{i<j}\{(1-\pi)N(a_{ij}|0,v_0^2)+\pi N(a_{ij}|0,v_1^2)\}\prod_{i=1}^{r}EXP(a_{ii}|\frac{\lambda}{2})\mathbf{1}_{P_r}(a) \tag{10.36}$$

where $\theta = (v_0, v_1, \pi, \lambda)$. Including the latent variable U and the prior $p(U|\theta)$ on U below, the model becomes

$$p(a|U,\theta) = C(U,v_0,v_1,\lambda)^{-1}\prod_{i<j}\{N(a_{ij}|0,v^2_{u_{ij}})\} \times \prod_{i=1}^{r} EXP(a_{ii}|\frac{\lambda}{2})\mathbf{1}_{P_r}(a) \quad (10.37)$$

$$P(U|\theta) = C(\theta)^{-1}C(U,v_0,v_1,\lambda)\prod_{i<j}\pi^{U_{ij}}(1-\pi)^{(1-U_{ij})} \quad (10.38)$$

where $C(\theta)$ and $C(U,v_0,v_1,\lambda)$ are intractable constants of normalization. The aim of the paper is then to build a block Gibbs sampler to sample from the space of (A,U). The graph selected is the median probability graph. The general methodology is adapted to both the precision graph and covariance graph models. The approach is from now on similar to that in [47]. Let us first consider the precision graph model with $A = K$. Let $V = (v^2_{z_{ij}})$ be an $r\times r$ symmetric matrix with zeros on the diagonal and $v_{ij} = v_{z_{ij}}$ for off-diagonal entries. Splitting K, S and V along the last column as in [47] and making the same change of variable as in (10.34), we obtain that the conditional distribution of $(K_{2\cdot 1}, K_{12})$ given $U, z^1,\dots,z^n$ is a product of independent Gamma and Normal distributions. As for the conditional of U given $K, z^1,\dots,z^n$, we obtain independent Bernoulli($p(U_{ij}) = 1$) distributions with

$$p(U_{ij}=1|K,z^1,\dots,z^n) = \frac{N(K_{ij}|0,v_1^2)\pi}{N(K_{ij}|0,v_1^2)\pi + N(K_{ij}|0,v_0^2)(1-\pi)}\,.$$

For the covariance graph model, $A = \Sigma$. We partition Σ, U, V along the last column as above and make the change of variable (10.34) but with Σ replacing K. Because of the expression of the likelihood in terms of Σ and the same form of the prior on Σ as on K for the precision graph model, we have that the conditional distribution $p(\Sigma_{2\cdot 1}|\Sigma_{11}, U, z^1,\dots,z^n)$ is now generalized inverse Gaussian, not Gamma, and that $p(\Sigma_{12}|\Sigma_{11},U,z^1,\dots,z^n,\Sigma_{2\cdot 1})$ is normal. Moreover, $\Sigma_{2\cdot 1}$ and Σ_{12} are, of course, no longer independent. The block Gibbs sampler is obtained by successively sampling from the conditional distributions of $\Sigma_{2\cdot 1}, \Sigma_{12}$ and U, rotating between the columns of Σ. Through numerical experiments, it is shown that this block Gibbs sampler with a continuous spike and slab prior scales up to high-dimensional problems.

10.5 Matrix Variate Graphical Gaussian Models

A random $p_R \times p_C$ matrix $\mathbf{X}$ is said to follow a matrix normal $N_{p_R,p_C}(0,(K_C\otimes K_R)^{-1})$ distribution if it has density

$$p(\mathbf{X}|K_R,K_C) = (2\pi)^{-p_Rp_C/2}|K_R|^{p_C/2}|K_C|^{p_R/2}\exp\{-\frac{1}{2}\,\mathrm{tr}\,(K_R\mathbf{X}K_C\mathbf{X}^t)\}. \quad (10.39)$$

Let $\mathbf{X}_{1*},\dots,\mathbf{X}_{p_R*}$ denote the rows of $\mathbf{X}$ and let $\mathbf{X}_{*1},\dots,\mathbf{X}_{*p_C}$ denote the column of $\mathbf{X}$. Then it can easily be shown that $\mathbf{X}^t_{i*}\sim N_{p_C}(0,(K_R)_{ii}K_C^{-1})$, $i=1,\dots,p_R$ and that $\mathbf{X}_{*j}\sim N_{p_C}(0,(K_C)_{jj}K_R^{-1})$, $j=1,\dots,p_C$. If we expand $\mathbf{X}$ as $vec(\mathbf{X})$ it is clear that the covariance matrix of $vec(\mathbf{X})$ is the same whether we have $K_R\otimes K_C$ as the precision matrix or $zK_R\otimes z^{-1}K_C$ where z is a positive scalar. We therefore have to put one constraint on the entries of K_C or K_R. We will assume here, as in [51], that $(K_C)_{11} = 1$.

We will define two graphical models for the rows and columns of $\mathbf{X}$. Let $G_R = (V_R, E_R)$

and $G_C = (V_C, E_C)$ be two undirected graphs with vertex set $V_R = \{1, \ldots, p_R\}$ and $V_C = \{1, \ldots, p_C\}$ respectively. We will assume that the rows of $\mathbf{X}$ follow a graphical model (in the regular sense) Markov with respect to G_R and the columns of $\mathbf{X}$ follow a graphical model with respect to G_C. It can be readily verified that in this model, if $s \neq t, s \in V_R, t \in V_R, (s,t) \notin E_R$, then the rows X_{s*} and X_{t*} are such that

$$X_{s*} \perp\!\!\!\perp X_{t*} | X_{V_R \setminus \{s,t\}*} \Leftrightarrow (K_R)_{st} = (K_R)_{ts} = 0.$$

In a similar fashion, if $(s,t) \notin E_C$, the columns X_{*s} and X_{*t} are such that

$$X_{*s} \perp\!\!\!\perp X_{*t} | X_{*V_R \setminus \{s,t\}} \Leftrightarrow (K_C)_{st} = (K_C)_{ts} = 0.$$

To work in a Bayesian framework, we now need to define a prior on $K_C \otimes K_R$. The conjugate prior for K_R and K_C are the G_R- and G_C-Wishart distribution respectively. However, we must remember that we have adopted the constraint $(K_C)_{11} = 1$ and therefore the hyper inverse Wishart must be on zK_C rather than K_C. The Jacobian of the change of variable from $zK_C \to (z, K_C)$ is $z^{|V_C|+|E_C|-1}$ and thus the prior on (z, K_C) has density

$$p(z, K_C | G_C, \delta_C, \theta_C) = \frac{1}{I_{G_C}(\delta_C, \theta_C)} |K_C|^{\frac{\delta_C - 2}{2}} \exp(-\frac{1}{2}\langle K_C, \theta_C \rangle) z^{p_C \frac{\delta_C - 2}{2} + |V_C| + |E_C| - 1}. \tag{10.40}$$

Let $\mathcal{D} = (\mathbf{X}_1, \ldots, \mathbf{X}_n)$ be a sample of size n from the graphical Gaussian model defined above. Then the posterior distribution of the parameters given the data is

$$\begin{aligned} p(K_R, z, K_C, G_R, G_C | \mathcal{D}) &= z^{p_C \frac{\delta_C - 2}{2} + |V_C| + |E_C| - 1} |K_R|^{\frac{n p_C + \delta_R - 2}{2}} |K_C|^{\frac{n p_R + \delta_C - 2}{2}} \\ &\quad \times \exp\{-\frac{1}{2} \operatorname{tr} (\sum_{j=1}^{n} K_R \mathbf{x}_j K_C \mathbf{x}_j^t + K_R \theta_R + z K_C \theta_C)\} \\ &\quad \times \pi(G_C)\pi(G_R) \end{aligned} \tag{10.41}$$

where $\pi(G_C), \pi(G_R)$ are priors on the space of column and row graphs respectively.

Carvalho and West [10] worked with matrix Gaussian variates for multivariate time series in a Bayesian context but then the row covariance matrix was fixed. Wang and West [51] were the first to consider matrix variate graphical Gaussian models in a full Bayesian framework but only for decomposable graphs G_R and G_C. Also, choosing G required the computation of marginal likelihoods of the form

$$\int p(\mathcal{D} | K_R, K_C) p(K_R | G_R, \delta_R, \theta_R) p(z, K_C | G_C, \delta_C, D_C) dK_R dK_C dz$$

which cannot be done explicitly even for decomposable graphs because of the constraint $(K_C)_{11} = 1$. Dobra et al. [18] developed a search algorithm which is a reversible jump consisting of successively sampling from the row graph and row precision matrix and then the column graph and column precision matrix following the method given in the same paper and described in Section 10.3.3 above. Sampling from z involves sampling from its Gamma full conditional distribution (see (10.41) above). Both Wang and West [51] and Dobra et al. [18] use the G-Wishart as prior on the precision matrices K_C and K_R as we did above. More recently Ni et al. [36] consider graphical Gaussian array variate model where G_C and G_R are restricted to be either DAGs or undirected decomposable graphs. Following the ordering of the vertices, they write the modified Cholesky decomposition of K_R and K_C as $L_R^t D_R^{-1} L_R$ and $L_C^t D_C^{-1} L_C$ respectively where L_R and L_C are upper triangular matrices as in (10.24) but in reverse order. For the prior distributions on the entries of D, they take the usual inverse Gamma found in the decomposition of the hyper inverse Wishart but depart from the hyper inverse Wishart or simply the Wishart by taking independent mixtures of Normals centered at 0 and point mass at 0, for the entries of L_R. For the particular case of a matrix-variate model, the set of priors is as follows

$$\begin{aligned} L_R^{k,i} \mid D_R^{k,i}, \pi_R^{k,i} &\sim \pi_R^{k,i} N(0, \tau_R D_R^k) + (1 - \pi_R^{k,i})\delta_0, \;\; k = 1, \ldots, p_R, i = k+1, \ldots, p_R \\ D^k &\sim IG(\alpha_R^k, \beta_R^k), \;\; \pi_R^{k,i} \mid \rho_R \sim Bernoulli(\rho_R), \;\; \rho_R \sim Beta(a_{\rho_R}, b_{\rho_R}) \end{aligned}$$

for α_R^k and β_R^k defined in terms of $\delta_R, \pi_R^{k,i}, \tau_R$. Similar priors are of course defined for L_C, D_C. These can be generalized to array-variate data. The priors are such that the partial marginal likelihoods $l(L_R, T_r, \pi_R, \pi_L)$ can be computed analytically and similarly for the column parameters. They develop two algorithms depending on whether the dimension p_R, p_C are of the same order or not. These algorithms use a combination of Metropolis-Hasting and Gibbs moves.

10.6 Fractional Bayes Factors

In general, if we have a collection of graphical models $\mathcal{M}_G, G \in \mathcal{G}$ with density $f(y|\theta, G)$ and prior distributions $\pi(\theta|G)$ and $\pi(G)$ respectively for θ and G, to compare two models $\mathcal{M}_G$ and $\mathcal{M}_{G'}$, we use the Bayes factor which consists of the ratio of the posterior distributions of G and G'. Integrating out θ and assuming that $\pi(G) = \pi(G')$, this yields

$$B_{G,G'} = \frac{\int f(y|\theta, G)\pi(\theta|G)d\theta}{\int f(y|\theta, G')\pi(\theta|G')d\theta}.$$

Sometimes one wishes to use an objective default prior for θ. In such cases, the prior might be improper and the normalizing constant of $\pi(\theta|G)$ simply does not exist. One way to get around this problem is to use fractional Bayes factor (see [37]) which can be defined as follows. Let $\pi_N(\theta \mid G)$ be an improper prior. We assume that $\int f(y|\theta, G)\pi_N(\theta \mid G)d\theta$ exists and choose $b \in (0, 1)$ such that $\int f(y|\theta, G)^b \pi_N(\theta \mid G)d\theta$ also exists. Define

$$m_G(y, b) = \frac{\int f(y|\theta, G)\pi_N(\theta|G)d\theta}{\int f(y|\theta, G)^b \pi_N(\theta|G)d\theta}.$$

Then the fractional Bayes factor is

$$FBF_{G,G'} = \frac{m_G(y, b)}{m_{G'}(y, b)}.$$

Using the fractional Bayes factor, Carvalho and Scott [9] propose to do model selection in the space of Gaussian models Markov with respect to decomposable graphs. A natural objective default prior is a measure inspired from $\mu_G(\Sigma_G)$ defined in (10.20). They actually use a slight variation on it. They take

$$\pi_N(\Sigma_G|G) \propto \frac{\prod_{C \in \mathcal{C}} \det(\Sigma_C)^{-c}}{\prod_{S \in \mathcal{S}} \det(\Sigma_S)^{-s}}. \tag{10.42}$$

Then for a sample $\mathcal{D}$ from the normal Markov with respect to G, we obtain

$$\begin{aligned} f(z^1, \ldots, z^n|\Sigma, G)^b \pi_N(\Sigma|G) &\propto \left(\frac{\prod_{C \in \mathcal{C}} \det(\Sigma_C)^c}{\prod_{S \in \mathcal{S}} \det(\Sigma_S)^s}\right)^{-nb/2} \frac{\prod_{C \in \mathcal{C}} \det(\Sigma_C)^{-c}}{\prod_{S \in \mathcal{S}} \det(\Sigma_S)^{-s}} \times \frac{e^{-\frac{b}{2}\langle \Sigma_C^{-1}, u_C \rangle}}{e^{-\frac{b}{2}\langle \Sigma_S^{-1}, u_S \rangle}} \\ &\propto \frac{\prod_{C \in \mathcal{C}} \det(\Sigma_C)^{-\frac{nb+2c}{2}}}{\prod_{S \in \mathcal{S}} \det(\Sigma_S)^{-\frac{nb+2s}{2}}} \times \frac{e^{-\frac{b}{2}\langle \Sigma_C^{-1}, u_C \rangle}}{e^{-\frac{b}{2}\langle \Sigma_S^{-1}, u_S \rangle}}. \end{aligned}$$

From the expression of (10.16), it is clear that the prior for Σ_G, after "borrowing" from the sampling distribution, is the hyper inverse Wishart $HIW(ng, gu)$ with parameter shape parameter nb and scale parameter $\theta = bu$. Then, using the notation of (10.19), we have

$$m_G(z, b) = \frac{I_G(bn, bu)}{I_G(n, u)}.$$

The fractional Bayes factor can be computed explicitly. Carvalho and Scott also discuss the choice of b. They show consistency of the fractional Bayes factor with the maximum likelihood ratio statistic or the F statistic in a regression framework.

Consonni and La Rocca [11] also use fractional Bayes to do model search in the class of Gaussian models Markov with respect to a DAG or a decomposable model. Following Geiger and Heckerman [21], they use the fact that the marginal likelihood for a model Markov with respect a DAG $\mathcal{G}$ with vertices $i = 1, \ldots, r$ and parent set $pa(i), i = 1, \ldots, r$ can be written as the Markov ratio of marginal likelihoods for the complete models on $Z_{i \cup pa(i)}$ and $Z_{pa(i)}$. If $\mathcal{G}$ is a DAG, we let $\mathcal{G}_c$ be a complete DAG on the same vertices. Then, for a sample $Z^1, \ldots, Z^n$, we have

$$m_{\mathcal{G}}(Z^1, \ldots, Z^n) = \prod_{i=1}^r \frac{m_{\mathcal{G}_c}(Z^\ell_{i \cup pa(i)}, \ell = 1, \ldots n)}{m_{\mathcal{G}_c}(Z^\ell_{pa(i)}, \ell = 1, \ldots n)} = \prod_{i=1}^r m_{\mathcal{G}_c}(Z^\ell_i \mid Z^\ell_{pa(i)}, \ell = 1, \ldots n). \tag{10.43}$$

Each fractional marginal likelihood $m_{\mathcal{G}_c}(Z^\ell_{i \cup pa(i)}, \ell = 1, \ldots n)$ is obtained explicitly for $b = \frac{n_0}{n}$, where $n_0 > 0$ is small, and with objective prior for each $\Omega = \Sigma^{-1}_{i \cup pa(i)}, i = 1, \ldots, r$ equal to

$$p(\Omega) = |\Omega|^{\frac{a_\Omega - r - 1}{2}}$$

thus extending the more familiar $|\Omega|^{-\frac{r+1}{2}}$ by the parameter a_Ω. For $a_\Omega = r + 1$ and G complete, this prior becomes the transformation of (10.42) by $\Sigma \to K = \Sigma^{-1}$. We note that, in $p(\Omega)$, r is common to all $i = 1, \ldots, r$. The expression of $m_{\mathcal{G}_c}(Z^\ell_{i \cup pa(i)}, \ell = 1, \ldots n)$ is given in equation (25) of [11]. Using the Markov equivalence of decomposable and corresponding DAG graphical Gaussian models, a similar expression for decomposable graphs is given in equation (29) of [11]. Model selection is not discussed in [11] but a recent paper by Leppä-aho et al. [30] uses a combination of fractional Bayes factors and pseudo-likelihood to do model selection in the space of Gaussian models Markov with respect to an arbitrary undirected graph G. For a given vector $\mathbf{X} = (X_1, \ldots, X_r) \in R^r$ with density $p(\mathbf{X}|\theta)$, they replace the usual likelihood by the pseudolikelihood

$$\hat{p}(\mathbf{X}|\theta) = \prod_{i=1}^r p(X_i|\mathbf{X}_{-i}, \theta) = \prod_{i=1}^r p(X_i|\mathbf{X}_{ne(i)}, \theta)$$

where $\mathbf{X}_{-j} = \mathbf{X}_{V \setminus \{j\}}$ and $ne(i)$ denotes the set of neighbors of i, and the usual marginal likelihood by the marginal pseudolikelihood

$$\hat{p}(\mathbf{X}|G) = \int_{\Theta_G} p(\theta|G) \prod_{i=1}^r p(X_i|\mathbf{X}_{ne(i)}, \theta) d\theta.$$

Assuming global independence of $\theta_i, i = 1, \ldots, r$, $\hat{p}(\mathbf{X}|G)$ becomes

$$\hat{p}(\mathbf{X}|G) = \prod_{i=1}^r \int_{\Theta_i} p(X_i|\mathbf{X}_{ne(i)}, \theta_i) p(\theta_i) d\theta_i = \prod_{i=1}^r m(X_i|\mathbf{X}_{ne(i)}), \tag{10.44}$$

the product of the marginal likelihoods for X_i given $\mathbf{X}_{ne(i)}$. Leppä-aho et al. view (10.44) as the fractional marginal likelihood for a DAG model as in (10.43) with $i \cup pa(i) \equiv i \cup ne(i)$ and thus, using the results of [11] obtain explicit formulas for the fractional marginal likelihood which they use as a score function for their search algorithm.

10.7 Two Interesting Questions

In this section, we talk about two papers that address important questions but do not quite fit in the narrative we have followed so far. The first paper by Finegold and Drton [19] considers inference in graphical models Markov with respect to a decomposable graph, when the sampling distribution is not quite Gaussian but rather t-distributed. The key to dealing with a multivariate t, rather than Gaussian, response is to write the $t_{r,\nu}(\mu, \Psi)$ random variable in R^r with density $f_\nu(y \mid \mu, \Psi) = \frac{\Gamma(\frac{\nu+r}{2})|\Psi|^{-1/2}}{(\pi\nu)^{r/2}\Gamma(\frac{\nu}{2})[1+\frac{(y-\mu)^t\Psi^{-1}(y-\mu)}{\nu}]^{(\nu+r)/2}}$ as

$$Y = \mu + \frac{X}{\sqrt{\tau}}$$

where $X \sim N_r(0, \Psi)$ and $\tau \sim \gamma(\nu/2, \nu/2)$ are independent of each other. A flexible approach is to introduce a τ_i for each component sample point $Y_i, i = 1, \dots, n$. The joint distribution of $(Y_1, \dots, Y_n, \tau, G, \Psi, \mu)$ is obtained by using the following priors

$$(Y_i|\tau_i, \Psi) \sim N_r(\mu, \Psi/\tau_i),\ (\tau_i|\nu) \sim \Gamma(\nu/2, \nu/2),\ (\Psi|G, \delta, \Phi) \sim HIW(\delta, \Phi),\ \mu \sim N_r(0, \sigma_\mu I_r).$$

They first derive an algorithm using ratios of the type (10.25), sampling the HIW using the direct sampler of [49] and formula (10.14) to obtain K. They then give a second algorithm with an alternative t-distribution where $\tau_i = (\tau_{ij}, j = 1, \dots, r)$ is now a vector with independent $\tau_{ij}|\nu \sim \Gamma(\nu/2, \nu/2)$. They present yet a third algorithm where each $Y_i \in R^r, i = 1, \dots, n$ is such that

$$(Y_i|\tau_i, \Psi, \mu) \sim N_r\Big(\mu, \text{diag}(\frac{1}{\sqrt{\tau_i}})\Psi\text{diag}(\frac{1}{\sqrt{\tau_i}})\Big),\ \tau_{ij} \overset{\text{i.i.d}}{\sim} P_i, j = 1, \dots, r$$

and the P_i are independent draws from a Dirichlet process $DP(\alpha, P_0)$ with $P_0 \sim \Gamma(\nu/2, \nu/2)$.

Fitch et al. [20] consider the behavior of Bayesian procedures that perform model selection for decomposable Gaussian graphical models when the true model is in fact non-decomposable. Comparing the asymptotic behavior of the Bayes factor for two decomposable graphs differing by one edge, Fitch et al. show that the Bayes factor will asymptotically choose minimal triangulations of the true graph. The choice between different minimal triangulations of the true graph is data dependent. This result is most interesting because it tells us that searching the space of decomposable graphs when the true graph might not be decomposable is not such a bad thing after all. The algorithm used to explore the space of decomposable models is the feature-inclusion stochastic search of Scott and Carvalho [44].

Bibliography

[1] D. Altomare, G. Consonni, and L. La Rocca. Objective Bayesian search for Gaussian directed acyclic graphical models for ordered variables with non-local priors. *Biometrics*, 69:478–487, 2013.

[2] C. Asci and M. Piccioni. Functionally compatible local characteristics for the local specification of priors in graphical models. *Scandinavian Journal of Statistics*, 34:829–840, 2007.

[3] A. Atay-Kayis and H. Massam. A Monte Carlo method for computing the marginal likelihood in nondecomposable Gaussian graphical models. *Biometrika*, 92:317–335, 2005.

[4] M. Barbieri, and J. O. Berger. Optimal predictive model selection. *Annals of Statistics*, 32:870–897, 2004.

[5] A.M. Bjerg and T.H. Nielsen. Modelselektion i kovariansselektionsmodeller. *M.Sc. thesis*, Department of Mathematics and Computer Science, Aalborg University, Denmark, 1993.

[6] S. Byrne and A.P. Dawid. Structural Markov graph laws for Bayesian model uncertainty. *Annals of Statistics*, 43:1647–1681, 2015.

[7] C. Carvalho, H. Massam, and M. West. Simulation of hyper-inverse Wishart distributions in graphical models. *Biometrika*, 94:647–659, 2004.

[8] C. M. Carvalho, N. G. Polson, and J. G. Scott. The horseshoe estimator for sparse signals. *Biometrika*, 97:465–480, 2010.

[9] C. M. Carvalho and J. G. Scott. Objective Bayesian model selection in Gaussian graphical models. *Biometrika*, 96:497–512, 2009.

[10] C. M. Carvalho and M. West. Dynamic matrix-variate graphical models. *Bayesian Analysis*, 2:465–480, 2007.

[11] G. Consonni and L. La Rocca. Objective Bayes Factors for Gaussian directed acyclic graphical models. *Scandinavian Journal of Statistics*, 39:743–756, 2012.

[12] D. R. Cox and M. Wermuth. Linear dependencies represented by chain graphs (with discussion). *Statistical Science*, 8:204–218, 1993.

[13] A.P. Dawid. Some matrix-variate distribution theory: notational considerations and a Bayesian application. *Biometrika*, 68:265–274, 1981.

[14] A.P. Dawid and S.L. Lauritzen. Hyper Markov laws in the statistical analysis of decomposable graphical models. *Annals of Statistics*, 21:1275–1317, 1993.

[15] P. Dellaportas, P. Giudici, and G. Roberts. Bayesian inference for nondecomposable graphical Gaussian models. *Sankhya*, 65:43–55, 2003.

[16] A.P. Dempster. Covariance selection. *Biometrics*, 28:157–175, 1972.

[17] P. Diaconis and D. Ylvisaker. Conjugate priors for exponential families. *Annals of Statistics*, 7:269–281, 1979.

[18] A. Dobra, A. Lenkoski, and A. Rodriguez. Bayesian inference for general Gaussian graphical models with application to multivariate lattice data. *Journal of the American Statistical Association*, 106:1418–1432, 2011.

[19] M. Finegold and M. Drton. Robust Bayesian graphical modelling using Dirichlet t-distributions. *Bayesian Analysis*, 9:521–550, 2014.

[20] M. Fitch, B. Jones, and H. Massam. The performance of covariance selection methods that consider decomposable graphs only. *Bayesian Analysis*, 9:659–684, 2014.

[21] D. Geiger and D. Heckerman. Parameter priors for directed acyclic graphical models and the characterization of several probability distributions. *Annals of Statistics*, 30:1412–1440, 2002.

[22] P. Giudici and P. J. Green. Decomposable graphical Gaussian model determination. *Biometrika*, 86:785–801, 1999.

[23] R. Gröne, C. R. Johnson, E. Sá, and H. Wolkowicz. Positive definite completions of partial Hermitian matrices. *Linear Algebra and Its Applications*, 58:109–128, 1984.

[24] H. Armstrong, C.K. Carter, K.K.F. Wong, and Kohn R. Bayesian covariance matrix estimation using a mixture of decomposable graphical models. *Arxiv:0706.1287*, 2007.

[25] B. Jones, C. Carvalho, A. Dobra, C. Carter, and M. West. Experiments in stochastic computation for high-dimensional graphical models. *Statistical Science*, 20:388–400, 2005.

[26] K. Khare and B. Rajaratnam. Wishart distributions for decomposable covariance graph models. *Annals of Statistics*, 39:514–555, 2011.

[27] S.L. Lauritzen. *Graphical Models.* Oxford University Press, Oxford, 1996.

[28] A. Lenkoski. A direct sampler for G-Wishart variates. *Stat*, 2:119–128, 2013.

[29] A. Lenkoski and A. Dobra. Computational aspects related to inference in gaussian graphical models with the G-Wishart prior. *Journal of Computational and Graphical Statistics*, 20:140–157, 2010.

[30] J. Leppä-aho, J. Pensar, T. Ross, and J. Corander. Learning Gaussian graphical models with fractional marginal pseudo-likelihood. *Arxiv 1602.07863*, 2016.

[31] G. Letac and H. Massam. Wishart distributions for decomposable graphs. *Annals of Statistics*, 35:1278–1323, 2007.

[32] D. Madigan and J. York. Bayesian graphical models for discrete data. *International Statistical Review*, 63:215–232, 1995.

[33] N. Mitsakakis, H. Massam and M. D. Escobar. A Metropolis-Hastings based method for sampling from the G-Wishart distribution in Gaussian graphical models. *Electronic Journal of Statistics*, 5:18–30, 2011.

[34] A. Mohammadi and E. C. Witt. Bayesian structure learning in sparse Gaussian graphical models. *Bayesian Analysis*, 10:109–138, 2015.

[35] R. Muirhead. *Aspects of Multivariate Statistical Theory.* Wiley, New York, 1982.

[36] Y. Ni, F.C. Stingo, and V. Baladandayuthapani. Sparse multi-dimensional graphical models: a unified Bayesian framework. *Journal of the American Statistical Association*, 2017.

[37] A. O'Hagan. Fractional Bayes factors for model comparison. *Journal of the Royal Statistical Society, Series B*, 99–138, 1995.

[38] T. Park and G. Casella. The Bayesian Lasso. *Journal of the American Statistical Association*, 103:681–686, 2008.

[39] V.I. Paulsen, S.C. Power, and R.R. Smith. Schur products and matrix completions. *Journal of Functional Analysis*, 36:151–178, 2008.

[40] B. Rajaratnam, H. Massam, and C. Carvalho. Flexible covariance estimation in graphical Gaussian models. *Annals of Statistics*, 36:2818–2849, 2008.

[41] A. Roverato. Cholesky decomposition of a hyper inverse Wishart matrix. *Biometrika*, 87:99–112, 2000.

[42] A. Roverato. Hyper inverse Wishart distribution for non-decomposable graphs and its application to Bayesian inference for Gaussian graphical models. *Scandinavian Journal of Statistics*, 29:391–411, 2002.

[43] J. Scott and J.O. Berger. An exploration of aspects of Bayesian multiple testing. *Journal of Statistical Planning and Inference*, 136:2144–2162, 2006.

[44] J. Scott and C. Carvalho. Feature-inclusion stochastic search for Gaussian graphical models. *Journal of Computational and Graphical Statistics*, 17:790–808, 2008.

[45] R. Silva and Z. Ghahramani. The hidden life of latent variables: Bayesian learning with mixed graph models. *Journal of Machine Learning Research*, 10:1187–1238, 2009.

[46] R. Tibshirani. Regression shrinkage and selection via the Lasso. *Journal of the Royal Statistical Society, Series B*, 58:267–288, 1996.

[47] H. Wang. The Bayesian graphical lasso and efficient posterior computation. *Bayesian Analysis*, 7:771–790, 2012.

[48] H. Wang. Scaling it up: Stochastic search structure learning in graphical models. *Bayesian Analysis*, 10:351–377, 2015.

[49] H. Wang and C.M. Carvalho. Simulation of hyper-inverse Wishart distributions for non-decomposable graph. *Electronic Journal of Statistics*, 4:1467–1470, 2010.

[50] H. Wang and S. Z. Li. Efficient Gaussian graphical model determination under G-Wishart prior distributions. *Electronic Journal of Statistics*, 6:168–198, 2012.

[51] H. Wang and M. West. Bayesian analysis of matrix normal graphical models. *Biometrika*, 96:821–834, 2009.

[52] F. Wong, C.K. Carter, and R. Kohn. Efficient estimation of covariance selection model. *Biometrika*, 90:809–830, 2003.

11

Latent Tree Models

Piotr Zwiernik

Department of Economics and Business, Universitat Pompeu Fabra

CONTENTS

Latent tree models are graphical models defined on trees, in which only a subset of variables is observed. They were first discussed by Judea Pearl [63] as tree-decomposable distributions to generalize star-decomposable distributions such as the *latent class model.* Latent tree models, or their submodels, are widely used in: phylogenetic analysis, network tomography, computer vision, causal modeling, and data clustering. They also contain other well-known classes of models like hidden Markov models, Brownian motion tree model, the Ising model on a tree, and many popular models used in phylogenetics. This chapter offers a concise introduction to the theory of latent tree models. We emphasize the role of *tree metrics* in the structural description of this model class, in designing learning algorithms, and in understanding fundamental limits of what and when can be learned.

11.1 Basics

In this section we define latent tree models and provide motivation to work with this model class. We present Gaussian and general Markov models as subclasses of latent tree models that admits tractable and rigorous analysis.

11.1.1 Definitions

A *tree* is an undirected graph without cycles. A *leaf* of T is a vertex of degree one, an *internal vertex* is a vertex which is not a leaf, and an *inner edge* is an edge whose both ends are internal vertices. Given a tree T define a *rooted tree* as a directed graph obtained from T by picking one of its vertices r and directing all edges away from r. The vertex r is called the *root*. Trees in this chapter will be leaf-labeled with the labelling set $\{1, \ldots, m\}$, where m is the number of leaves. An undirected tree is *trivalent* if each internal vertex has degree precisely three. A rooted tree is a *binary rooted tree* if each internal vertex has precisely two children. In many applications rooted trees are depicted without using arrows, where direction is made implicit by drawing the root on the top and the leaves on the bottom; see Figure 11.1(c). Two special types of undirected trees are: a *star tree* with one internal vertex and a trivalent tree on four leaves called a *quartet tree*; see Figure 11.1(a) and (b). A *forest* is a collection of trees. Forests here are also leaf-labeled with the labelling set is $\{1, \ldots, m\}$, which means that each tree in this collection is leaf-labeled and the corresponding collection of labelling sets forms a set partition of $\{1, \ldots, m\}$. We define three graph operations on trees (forests). *Removing an edge* means removing that edge from the edge set. *Contracting an edge* $u - v$ means removing u, v from the vertex set, adding a new vertex w and edges such that w is adjacent to all vertices which were adjacent to u or v. *Suppressing a vertex of degree two* means removing that vertex and replacing the two edges incident to that vertex by a single edge.

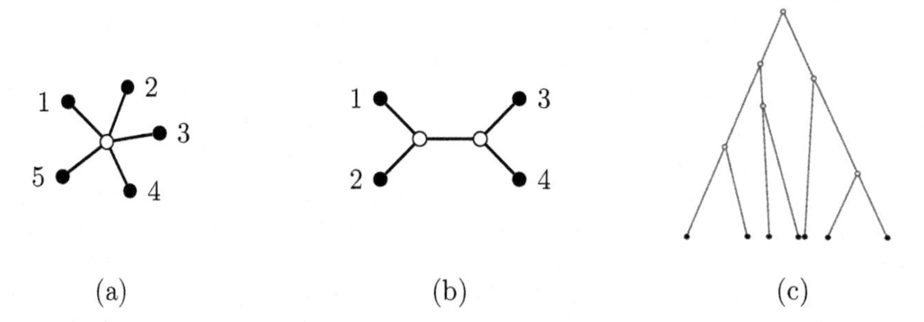

FIGURE 11.1: (a) An undirected star tree with five leaves, (b) a quartet tree, (c) a binary rooted tree.

A latent tree model is defined as follows. Let $Y = (Y_v)_{v \in V}$ be a random vector with coordinates indexed by the vertices V of a tree T and with values in the state space $\mathcal{Y} = \prod_{v \in V} \mathcal{Y}_v$. Suppose that the density of Y, denoted by $p(y)$, lies in the graphical model over T; see Section 4.2 of Chapter 4. We call such a model a *fully-observed* tree model. Denote the set of leaves by $W \subset V$ and let $m := |W|$. Write $X := Y_W$, $\mathcal{X} := \mathcal{Y}_W$, $H := Y_{V \setminus W}$, and $\mathcal{H} := \mathcal{Y}_{V \setminus W}$. The latent tree model $M = M(T, \mathcal{Y})$ is the family of marginal distributions of $p(y)$ over the leaves W:

$$p_W(x) \quad := \quad \int_{\mathcal{H}} p(x, h) \, \mathrm{d}h. \tag{11.1}$$

In other words, the internal vertices represent unobserved random variables. The above definition extends to situations where some internal vertices are also observed. However, this seemingly more general situation does not lead to any new family of distributions; for Gaussian and general Markov models this is shown in Theorem 11.1.3.

Consider now a Bayesian network on a rooted tree obtained from T; see Section 1.7 in Chapter 1. By standard results on Markov equivalence of directed acyclic graphs (see Section 1.8.5 in Chapter 1), this model coincides with the latent tree model on T for any choice of the root location. The model is then fully characterized by the root distribution $p_r(y_r)$ and by the conditional distributions $p_{v|u}(y_v|y_u)$ for all edges $u \to v$ in the rooted tree. The Bayesian network over T rooted at r is parameterized by

$$p(y) = p_r(y_r) \prod_{u \to v} p_{v|u}(y_v|y_u) \quad \text{for all } y \in \mathcal{Y} \tag{11.2}$$

and so the parameterization of the corresponding latent tree model is induced by taking the margin over $\mathcal{X}$ as in (11.1).

Although the definition of latent tree models is fairly general, in this chapter we make additional assumptions on the state space $\mathcal{Y}$ and possible distributions. If $\mathcal{Y}$ is a finite set then we call the corresponding latent tree model *discrete*. *Gaussian latent tree models* are latent tree models for which the vector Y is jointly Gaussian. In the Gaussian case we typically assume that the mean of Y is known and equal to zero. Mixed cases when $\mathcal{H}$ is finite and X is Gaussian conditionally on H are also popular.

The parameters of latent tree models are of two kinds. The underlying tree T is the *discrete parameter*. The *continuous parameter* θ is the parameter specifying the root distribution and the conditional distributions for each edge. To make the parameters explicit, we write $p(y) = p(y; T, \theta)$ and $p_W(x) = p_W(x; T, \theta)$. Formally, the state space $\mathcal{H}$ of the unobserved part of the vector Y is also a parameter of a latent tree model. In certain applications finding $\mathcal{H}$ can be important; see [60, Section 3.2.4]. Here we always assume that $\mathcal{H}$ is fixed. With this convention, there are three main learning tasks related to latent tree models:

(L1) Given a sample of size n from a latent tree model estimate the underlying tree T.

(L2) Given a sample of size n from a latent tree model on a fixed tree T estimate the continuous parameter θ.

(L3) Given a fully specified latent tree model and a single sample at the observed vertices, infer the states at the unobserved vertices.

There are several fundamental questions related to the first two learning tasks that we are going to address in this chapter. We will not discuss here the learning task (L3). If the model is completely specified then, by (11.2), we have access to the full distribution over $\mathcal{Y}$. In this case the learning task (L3) reduces to the marginalization procedure described in Chapter 4.

11.1.2 Motivation and applications

Latent tree models form the most tractable family of Bayesian networks with unobserved variables, which can be used to model dependence structures when unobserved confounders are expected; see, for example, [65, Section 2] or [35]. However, there are several other reasons why latent tree models become popular across sciences. We distinguish three main types of applications.

First, latent tree models represent a larger family of probability distributions than fully-observed tree models but retain some of their computational advantages, which is particularly important in high-dimensional settings. From (11.2) it is clear that having an estimator

$\hat{\theta}$ of model parameters we obtain an estimator of the fully observed distribution $p(y;\hat{\theta})$ and so we can very efficiently compute various marginal distributions in the model as discussed by Bilmes in Chapter 4. Using the max-product algorithm it is also possible to efficiently infer the unobserved states.

Second, a rooted tree can represent evolutionary processes with the root representing the common ancestor and the leaves representing extant species. This makes latent tree models useful in phylogenetic analysis [36, 74]. In this context, the data typically consist of m aligned DNA sequences of length n, where each site in the sequence is treated as an independent realization of the vector X. In this case all state spaces are of the form $\{A, C, T, G\}$ or binary. These applications are not restricted to discrete data. The early evolutionary trees were all built based on morphological characters such as body size. Moreover, with the burst of new genomic data, such as gene expression, phylogenetic models for continuous traits are again becoming important; see [42] and references therein. Latent tree models in this evolutionary context are also popular in linguistics [68, 76], where a tree represents evolution of languages with modern languages represented by the leaves. The data here typically consists of the acoustic structure of spoken words and are continuous although there are also approaches using syntatic (discrete) data; see, for example, [40, 78]. A related application of latent tree models is in network tomography, where it is used to determine the structure of the connections in the Internet [13, 33]. In this application messages are transmitted by sending packets of bits from a source vertex to different destinations and the correlation in arrival times is used in order to infer the underlying network structure. A common assumption is that the underlying network has a tree structure. Then Gaussian latent tree models form a natural correlation model for arrival times because the correlations diminish with the distance on a tree; see Section 11.2.1. Typically in this context a special submodel called the *Brownian motion tree model* is used.

Third, a tree can represent hierarchical structure in complex data sets. This viewpoint was behind the definition of latent tree models that emerged in the machine learning community [10, 53, 96]. In a rooted tree every internal vertex represents a cluster given by all the leaves that descend from it. We refer to [60] for an overview of potential applications. We emphasize that in those applications, it is often not realistic to assume that the true data-generating distribution lies in the latent tree model, which leads to some subtleties in inference; see also Section 11.3.3. Another application in this vein is in computer vision and image processing; see [94] and reference therein. One promising application along these lines is the use of context in computer vision; see, for example, [20].

Latent tree models can be also used as a generalization of hidden Markov models. A hidden Markov model (HMM) is a latent tree model on the caterpillar tree in Figure 11.2(a). Typically hidden Markov models are *homogeneous* in the sense that the conditional distributions $p(H_i|H_{i-1})$ and $p(X_i|H_i)$ do not depend on i; see [67]. The unobserved part in HMMs follows a Markov chain and so respects a very simple dependency structure. This may be a good approximation of the real dependency structure in the context of time series but is often too restrictive in other applications. Hidden Markov Tree models (HMTMs) [24] relax this restriction allowing for any tree structure on the unobserved vector. Still however, like for HMMs, we assume that each internal vertex has a leaf as a child; see Figure 11.2(b). Models of this type were proposed for wavelet-based statistical signal processing but other applications emerged recently: image processing [19, 71], biomedicine [56, 66], computational linguistics [95]. In these settings the unobserved vector typically is binary and the observed part is Gaussian.

This leads to another important reason to study latent tree models: many popular models are submodels of the latent tree models. Examples are given by HMMs, HMTMs, all phylogenetic tree models, but also one factor analysis model, latent class models, and Brownian motion tree models. Models of these types are used virtually everywhere: in biostatistics,

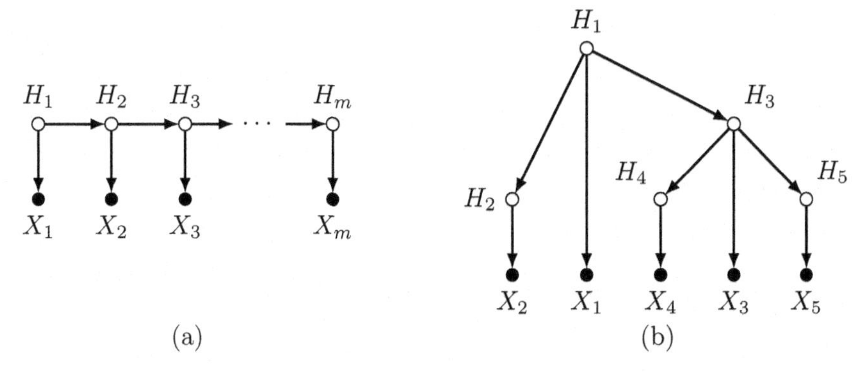

FIGURE 11.2: (a) The caterpillar tree defining the hidden Markov model. (b) An example of a tree defining a hidden Markov tree model.

machine learning, and social sciences. As noted by Wainwright and Jordan in [90], a more general viewpoint gives a unifying framework for existing algorithms used for these different model classes.

11.1.3 Parsimonious latent tree models

In this section we briefly discuss certain redundancy in *discrete* latent tree models. We start by discussing the latent class model, which is a special latent tree model where the state space $\mathcal{Y}$ is finite and the underlying tree is a star as in Figure 11.1(a). In this case the tree is typically rooted at the internal vertex and the parameter consists of the root distribution together with the conditional distributions of the leaves given the internal vertex. The number of latent classes corresponds to the number of states $|\mathcal{H}|$ of the unobserved variable H. This relatively simple class of models gives us first insights into general latent tree models. The following result shows that, if the number of unobserved classes is high enough the model becomes *saturated*, that is, it contains all probability distributions over $\mathcal{X}$.

Proposition 11.1.1. *If a given latent class model* is not *saturated then* $|\mathcal{H}| < |\mathcal{X}|/\max_i |\mathcal{X}_i|$.

The proof of this result can be recovered from the proof of [96, Theorem 3].

A latent tree model is *parsimonious* if there is no other latent tree model with a smaller number of parameters that gives the same family of probability distributions over the observed variables. A *discrete* latent class model is *not* parsimonious if $|\mathcal{H}| \geq |\mathcal{X}|/\max_i |\mathcal{X}_i|$ because, by Proposition 11.1.1, every such model is saturated, and the only parsimonious latent tree model over $\mathcal{X}$ that is saturated is a tree with a single vertex representing the whole vector (X_1, X_2, X_3) as a single variable with $|\mathcal{X}|$ states.

A discrete latent tree model is *regular* if the inequality in Proposition 11.1.1 holds for any unobserved vertex v with neighbors $N_v \subset V \setminus \{v\}$, that is, when

$$|\mathcal{Y}_v| < \prod_{u \in N_v} |\mathcal{Y}_u| / \max_u |\mathcal{Y}_u|. \tag{11.3}$$

The argument for the latent class model can be generalized to conclude the following; see [60]:

Proposition 11.1.2. *Any parsimonious discrete latent tree model is regular.*

Proposition 11.1.2 substantially reduces the space of possible discrete latent tree models to consider. As demonstrated by [96] the size of the space of regular models is bounded by 2^{3m^2}. There are two problems with that in practice. First, this space is still very big with no clear structure. Second, no necessary and sufficient conditions to assure parsimony are known in general and so we do not know how good this reduction is. This leads us to a more tractable subclass of discrete latent tree models called general Markov models, which we discuss in the next section.

11.1.4 Gaussian and general Markov models

A *general Markov model* is a latent tree model for which all $\mathcal{Y}_v$ are equal and $d = |\mathcal{Y}_v| < \infty$. Models of this type, for $d = 2, 4, 20$ or 61 appeared in phylogenetics half a century ago [14], they were formulated in the most general form over 30 years ago [9], but only recently they are becoming increasingly popular; see for example [3, 48]. In statistics, general Markov models appeared in the context of causal inference [64], or simply as the simplest interesting family of graphical models with unobserved variables [75]. As we present in this chapter, general Markov models stand out as the tractable class of discrete latent tree models, as much as Gaussian models stand out as the tractable class in the continuous case.

In the previous section we argued that the space of parsimonious latent tree models is not easy to handle. On the other hand, for general Markov models, inequality (11.3) is satisfied as long as all internal vertices have degree at least three. For general Markov models and Gaussian latent tree models the necessary and sufficient condition for model parsimony is that:

(A1) each unobserved variable has degree at least three,

(A2) any two neighboring variables are neither functionally related nor independent.

It is standard to assume that the underlying tree has no degree two vertices. In fact, for general Markov models and Gaussian latent tree models we can always suppress degree two vertices without changing the model, and so (A1) is always satisfied; see [100, Section 5.3.4]. In particular, the model defined over a binary rooted tree is equal to the model over the corresponding undirected trivalent tree obtained by suppressing the root. We also always assume the following:

(A3) all Y_v are *nondegenerate* meaning that the distribution of Y_v has the full support $\mathcal{Y}_v$.

Working without assuming (A2) is sometimes convenient because of the following result, which allows us to focus on learning latent tree models over trivalent trees.

Theorem 11.1.3. *Every discrete latent tree model satisfying (A1) and (A2) is a submodel of a latent tree model over a trivalent tree that satisfies only (A1). The same applies to Gaussian latent tree models.*

A formal proof can be based on the following two observations; see [100, Section 5.2.2] for more details. First, a tree with no degree two vertices can be obtained from a trivalent tree by edge contraction. Second, if $Y_u = Y_v$ in a tree model, there exists a simpler model on a tree obtained from T by contracting the edge (u, v). This means that every latent tree model can be realized as a submodel over a trivalent tree with some of the vertices identified. For example, consider the discrete latent tree model for the tree on the left of Figure 11.3. Here one of the internal vertices, labeled with 3, represents an observed random variable. We can alternatively consider a discrete latent tree model on the right of Figure 11.3. Here the double edge represent equality between adjacent random variables, and so, the

FIGURE 11.3: Two equivalent latent tree models. In the model on the right extra restrictions are put on parameters so that the double edges represent equality of random variables.

corresponding conditional distribution is degenerate. Directly from the way these models are parameterized, we see that both models are equivalent.

There are numerous advantages of general Markov models. The link to statistical physics, phylogenetics, and Markov processes allows to develop efficient algorithms with strong theoretical guarantees. General Markov models retain very good performance on real-world data [21]. The space of parsimonious general Markov models is also much more tractable than the space of all regular latent tree models over $\mathcal{X}$. In some applications we cannot assume that the state spaces of observed variables are equal but we can assume that the unobserved variables have the same state spaces. This holds for HMMs and HMTMs. Many techniques discussed later in this chapter can be generalized to that case.

For any edge $u \to v$ denote by M^{uv} the $d \times d$ stochastic matrix representing the conditional distribution $p_{v|u}$. In analogy to Markov chains, M^{uv} is called a *transition matrix* and each of its *rows* represents a conditional distribution of Y_v given a particular value of Y_u. In the general Markov model, each M^{uv} is an arbitrary stochastic matrix. Models used in phylogenetics are usually more constrained. They are generated through a continuous time Markov process on T with a given *rate matrix* Q, that is a real $d \times d$ matrix with row elements summing to zero and all off-diagonal elements nonnegative. In this setting the transition matrices are given by

$$M^{uv} = \exp(t_{uv}Q), \tag{11.4}$$

where $t_{uv} > 0$ is an edge parameter and $\exp(\cdot)$ is the matrix exponential function. The rate matrix Q typically has some further symmetries. For example, in the Jukes-Cantor model [49], all off-diagonal entries of Q are equal; see [36] for a review of most popular phylogenetic models.

11.2 Second-Order Moment Structure

Many learning algorithms for latent tree models use the second-order structure of the observed distribution, that is, correlations between the observed variables, mutual information, or other aspects of the pairwise marginal distributions. These algorithms typically exploit links to tree metrics. In this section we start by explaining this link and how it can be used to design robust learning algorithms.

11.2.1 Gaussian latent tree model

The earliest example of Gaussian latent tree models is the factor analysis model with a single unobserved factor [8, 88]. More general Gaussian tree models were not studied until more recently [21, 77]. In the fully observed Gaussian tree model the *inverse covariance* matrix of Y is very sparse. However, the inverse covariance matrix of the observed subvector X has no zeros and it is not convenient to work with. On the other hand, the *covariance* matrix of X provides a great insight into the structure of Gaussian latent tree models and latent tree models in general.

Consider any two vertices i, j of T. Using the Markov properties implied by the tree model, the correlation $\rho_{ij} = \text{corr}(Y_i, Y_j)$ can be written as the product (see, e.g., [21])

$$\rho_{ij} = \prod_{(u,v)\in\overline{ij}} \rho_{uv}, \tag{11.5}$$

where $\overline{ij}$ denotes the unique path between i and j in T. Restricting (11.5) only to pairs of leaves i, j gives the parameterization of the correlations of the Gaussian latent tree model. In particular, vertices that are far from each other in the tree tend to be less correlated. This appealing property makes this model useful for hierarchical data clustering and network tomography problems as described in Section 11.1.2.

The constraints on the correlations induced by (11.5) are the only nontrivial constraints on Gaussian latent tree models. The variances of the observed variables can be arbitrary and together with the *edge correlations*, ρ_{uv} for all edges (u, v), they provide a set of parameters for the latent tree model. The variances of the unobserved variables do not affect the observed distribution and so typically are set to 1. The model is parsimonious as long as it satisfies (A1) and (A2). The condition (A2) is satisfied if the edge correlations satisfy $|\rho_{uv}| \in (0, 1)$.

Example 11.2.1. Consider the quartet tree in Figure 11.1(b) and denote the internal vertices by u, v, where u is closer to 1 and 2. If correlations between the four leaves come from a latent tree model on this quartet tree, then there is a collection of edge correlations ρ_{1u}, ρ_{2u}, ρ_{uv}, ρ_{3v}, and ρ_{4v} all with values in $[-1, 1]$ such that $\rho_{12} = \rho_{1u}\rho_{2u}$, $\rho_{13} = \rho_{1u}\rho_{uv}\rho_{3v}$, $\rho_{14} = \rho_{1u}\rho_{uv}\rho_{4v}$, $\rho_{23} = \rho_{2u}\rho_{uv}\rho_{3v}$, $\rho_{24} = \rho_{2u}\rho_{uv}\rho_{4v}$, and $\rho_{34} = \rho_{3v}\rho_{4v}$. In particular

$$\rho_{13}\rho_{24} = \rho_{14}\rho_{23} = \rho_{12}\rho_{34}\rho_{uv}^2. \tag{11.6}$$

The first equality is an example of a nontrivial relation between the observed correlations. The second equality can be used to recover the value of the edge parameter ρ_{uv} given only the observed correlations, or to establish the inequality $\rho_{14}\rho_{23} \geq \rho_{12}\rho_{34}$.

A systematic study of polynomial constraints defining statistical models is part of algebraic statistics [28]. For Gaussian latent tree models these constraints were studied in [85, Section 6] and [77].

Correlation matrices of latent Gaussian tree models are closely linked to tree metrics. Given a tree T with m leaves assign to each of each edges $u - v$ a nonnegative length d_{uv}. With this choice we can now compute the distance between any two leaves i, j summing the lengths of the edges on the unique path between i and j, that is,

$$d_{ij} = \sum_{(u,v)\in\overline{ij}} d_{uv}. \tag{11.7}$$

Consider an $m \times m$ symmetric matrix $D = [d_{ij}]$ with zeros on the diagonal. Call D a *tree metric* if its entries satisfy (11.7) for some tree T and edge lengths.

Let $\Sigma = [\rho_{ij}]$ be a correlation matrix in a Gaussian latent tree model. Assume first that $\rho_{ij} \neq 0$ for all $1 \leq i < j \leq m$. Consider a symmetric matrix $D = [d_{ij}]$ with $d_{ij} = -\log|\rho_{ij}|$. Then (11.5) translates into (11.7). Since $|\rho_{uv}| \in (0, 1]$ for all $(u, v) \in \overline{ij}$ also $d_{uv} \geq 0$, and so D is a tree metric. If Σ contains zero entries (11.5) implies that zeros cannot appear arbitrarily. For every three indices i, j, k if $\rho_{ij} \neq 0$ and $\rho_{jk} \neq 0$ then also $\rho_{ik} \neq 0$. It follows that the correlation matrix in a Gaussian latent tree model has a block diagonal structure and within each block all entries are non-zero. Each block can be transformed to a tree metric on the corresponding subtree.

The connection to tree metrics will be exploited in many ways. For example, it is a standard result in phylogenetics [11] that for any tree metric D the underlying tree and positive edge lengths d_{uv} can be recovered uniquely. Because $|\rho_{uv}| = \exp(-d_{uv})$, we obtain the following corollary.

Theorem 11.2.2. *If $\Sigma = [\rho_{ij}]$ is a correlation matrix in a Gaussian latent tree model and has non-zero entries, then the underlying tree and the edge correlations are identified uniquely (up to sign).*

To have a concrete example of how it works consider again the quartet tree model in Example 11.2.1. Suppose that we are given a correlation matrix from the Gaussian latent tree model but we do not know the underlying tree. By Theorem 11.1.3 we can first constrain ourselves to trivalent trees, that is, the three possible quartets: 12/34, 13/24, and 14/23, where this notation indicates how leaves group together. In the first case, by (11.6), we have

$$|\rho_{12}\rho_{34}| \;\geq\; |\rho_{13}\rho_{24}| \;=\; |\rho_{14}\rho_{23}| \tag{11.8}$$

and by symmetry we obtain similar relations for the other two trees. Therefore, we can find the underlying tree by computing quantities of the form $|\rho_{ij}\rho_{kl}|$ and choosing the largest. If they all happen to be equal, the underlying tree is the star tree.

We can extend Theorem 11.2.2 to arbitrary correlation matrices but here identifiability of the edge correlations is more involved; see [26] for details. For instance, consider again the model in Example 11.2.1. Suppose that $\rho_{ij} = 0$ for all $i \in \{1,2\}$ and $j \in \{3,4\}$ and $\rho_{12}, \rho_{34} \neq 0$. Then we can easily identify the underlying tree 12/34. Indeed, for no other tree with four leaves there is a choice of edge correlations giving precisely this pattern of zeros. Identifying the parameters is harder. We check that ρ_{uv} must be zero. The other edge correlations must be nonzero but they are identified only up to the relations $\rho_{12} = \rho_{1u}\rho_{2u}$ and $\rho_{34} = \rho_{3v}\rho_{4v}$.

11.2.2 General Markov models

Tree metrics appear also in the description of discrete latent tree models [9, 51, 55, 82]. For any edge $u \to v$ denote by M^{uv} the $d \times d$ transition matrix representing the conditional distribution $p_{v|u}$. Denote by P^{uv} the $d \times d$ matrix of the marginal distribution of (Y_u, Y_v), and by P^{uu} a diagonal matrix with the marginal distribution of Y_u on the diagonal. For any two vertices u, v let

$$\tau_{uv} \;:=\; \frac{\det(P^{uv})}{\sqrt{\det(P^{uu}P^{vv})}}, \tag{11.9}$$

where the denominator is non-zero by assumption (A3). By essentially the same argument as in [74, Theorem 8.4.3] we obtain the following path-product formula

$$\tau_{ij} \;=\; \prod_{(u,v)\in\overline{ij}} \tau_{uv} \qquad \text{for all } i,j \in V. \tag{11.10}$$

In the case of binary variables, $\det P^{ij} = \mathrm{cov}(X_i, X_j)$, $\det(P^{ii}) = \mathrm{var}(X_i)$ and so τ_{ij} is the correlation, which implies that (11.10) reduces to (11.5).

In general the interpretation of the edge parameters τ_{uv} is more complicated. Using the identity $P^{uu}M^{uv} = P^{uv}$ we can write

$$\tau_{uv} \;=\; \det M^{uv}\sqrt{\frac{\det(P^{uu})}{\det(P^{vv})}}. \tag{11.11}$$

Therefore, we have $\tau_{uv} = 0$ if and only if $\det M^{uv} = 0$. If $d = 2$ this is equivalent to independence of Y_u and Y_v but in general it is a strictly weaker condition. For example, if $d = 3$, and the first two rows of M^{uv} are equal but not equal to the third one, then $\det M^{uv} = 0$ but Y_u and Y_v are not independent. Exactly like in the Gaussian case the edge parameters satisfy $|\tau_{uv}| \leq 1$ and the border values ± 1 consider to functional dependence of Y_u and Y_v. Indeed, by the Bayes' theorem, we have

$$M^{vu} = (P^{vv})^{-1}(M^{uv})^T P^{uu}.$$

With this notation, (11.11) gives

$$\tau_{uv}^2 = \det(M^{uv}M^{uv})\det(P^{uu}(P^{vv})^{-1}) = \det M^{uv}\det M^{vu}.$$

Because both M^{uv} and M^{vu} are stochastic matrices, all their eigenvalues lie in the unit circle. In particular, $\tau_{uv} \in [-1, 1]$ and it is equal to ± 1 precisely when M^{uv} is a permutation matrix, or in other words, if Y_u and Y_v are functionally related.

This again gives a direct link to tree metrics and the following result; see [16, 84].

Theorem 11.2.3. *For any distribution in a general Markov model the underlying tree can be uniquely recovered given only its 2-way margins.*

We implicitly assume that τ_{ij} are nonzero but the theorem can be extended to forests.

11.2.3 Linear models

The fact that in the Gaussian and in the binary case the correlation between observed variables decompose as in (11.5) can be proved using the following fact: if X, Y are binary (or jointly Gaussian) then the conditional expectation $\mathbb{E}[X|Y]$ is a linear function of Y. As we see in this section the discrete case can be also interpreted in this way, which gives a unifying framework to understand (11.5) and (11.10). Moreover, it gives much larger families of potential models to consider that also admit a path-product formula linking it to tree metrics.

Let each variable Y_u in the system be modeled as a random vector in $\mathbb{R}^k$ for a fixed k. A ternary variable, for example, will take values $(0, 0)$, $(1, 0)$, $(0, 1)$ in $\mathbb{R}^2$ instead of the typical $1, 2, 3$ in $\mathbb{R}$. Each variable can be either discrete or continuous but we add a minor requirement that the matrix $\Sigma_{vv} = \mathbb{E}Y_vY_v^T - \mathbb{E}Y_v(\mathbb{E}Y_v)^T$ is positive definite. A *linear latent tree model* is a latent tree model in which for every edge $u - w$ in the tree, the conditional expectation $\mathbb{E}[Y_u|Y_w]$ is an affine function of Y_w. Linear latent tree models include (multivariate) Gaussian latent tree models, Kalman filters, Gaussian mixtures, Poisson mixtures and general Markov models. Models of this type were first discussed in [6]. Here we propose a slightly different exposition using ideas from [100, Section 4.3].

We define the normalized version $\bar{Y}_v$ of Y_v as $\bar{Y}_v := (\Sigma_{vv})^{-1/2}(Y_v - \mathbb{E}Y_v)$. Denoting $\Sigma_{uv} = \mathbb{E}Y_uY_v^T - \mathbb{E}Y_u(\mathbb{E}Y_v)^T$ we obtain

$$\mathbb{E}[\bar{Y}_u|\bar{Y}_v] = \Sigma_{uu}^{-1/2}\Sigma_{uv}\Sigma_{vv}^{-1/2}\bar{Y}_v. \tag{11.12}$$

Define $\tau_{uv} := \det(\Sigma_{uu}^{-1/2}\Sigma_{uv}\Sigma_{vv}^{-1/2})$. By the law of total expectation, it follows from (11.12) that $\tau_{uv} = \det(\mathbb{E}[\bar{Y}_u\bar{Y}_v^T])$. Let Y_u, Y_v, Y_w be three random variables with values in $\mathbb{R}^k$ such that $Y_u \perp\!\!\!\perp Y_w | Y_v$. Then

$$\mathbb{E}[\bar{Y}_u\bar{Y}_w^T] = \mathbb{E}\left[\mathbb{E}[\bar{Y}_u|Y_v](\mathbb{E}[\bar{Y}_w^T|Y_v])^T\right] = \Sigma_{uu}^{-1/2}\Sigma_{uv}\Sigma_{vv}^{-1}\Sigma_{vw}\Sigma_{ww}^{-1/2},$$

which implies that $\tau_{uw} = \det(\mathbb{E}[\bar{Y}_u\bar{Y}_w^T]) = \tau_{uv}\tau_{vw}$. Applying this argument recursively we conclude that the path-product decomposition of τ_{ij} given in (11.10) holds for any linear latent tree model.

This clearly generalizes the Gaussian case. To see that this also generalizes (11.10), for each discrete random variable with d states take $k = d - 1$ and set the state-space to be $\{0, e_1, \ldots, e_{d-1}\}$, where 0 is the origin and e_i are the elements of the standard basis of $\mathbb{R}^{d-1}$.

Proposition 11.2.4. *With the above convention* $\det(\Sigma_{uu}) = \det(P^{uu})$ *and* $\det(\Sigma_{uv}) = \det P^{uv}$.

The proposition implies that τ_{uv} as defined in this section is equal to τ_{uv} as defined in Section 11.2.2.

Proof of the proposition. Consider the matrix A obtained from P^{uv} by elementary row and column operations: add all rows to the first row and all columns to the first column. Basic linear algebra implies $\det P^{uv} = \det A$. Matrix A has the following block structure. The top-left 1×1-block is equal to 1. The bottom-right $(d-1) \times (d-1)$-block is equal to $\mathbb{E}Y_u Y_v^T$ and the remaining two blocks are $\mathbb{E}Y_u$ and $\mathbb{E}Y_v^T$. The formula for the determinant of a block matrix implies that $\det A = \det(\mathbb{E}Y_u Y_v^T - \mathbb{E}Y_u \mathbb{E}Y_v^T) = \det \Sigma_{uv}$. The proof of the other equality is analogous. □

11.2.4 Distance based methods

The maximum likelihood tree topology recovery is NP hard [69]. This has motivated a number of investigations of other tractable methods for learning trees as well as theoretical guarantees on performance. The link between tree metrics and latent tree models described in the previous sections makes it possible to come up with consistent methods to learn a tree that work in polynomial time. This approach dates back to [9]; see [44] and references therein.

For a concrete example consider a sample from the Gaussian latent tree model. Given the sample correlation matrix with elements $\hat{\rho}_{ij}$ we compute distances $\hat{d}_{ij} = -\log|\hat{\rho}_{ij}|$. Now use any of the methods to learn a tree metric from observed distances. This gives a tree $\hat{T}$ and edge distances $\hat{d}_{uv}$, or equivalently absolute values of the edge correlations $\hat{\rho}_{uv}$. Such a method will be (statistically) consistent given the original tree distance method is *consistent*, which in this context means that the method outputs the correct tree given a tree metric.

There are many methods that try to recover the underlying tree from noisy distances. The most popular are the Neighbor-Joining (NJ) algorithm and the least-squares method but many other algorithms are available; see Section 7.3 in [74] for an overview. All popular methods are both well studied and widely implemented, for example, in R; see Section 5.1 in [61]. Most of the methods, including NJ and the least squares, are consistent. We also note in passing that these methods output an undirected tree but rooting is also possible by finding an appropriate outgroup; see [29, Section 7.3].

An appealing property of this method, as applied in the Gaussian case, is that there is a one-to-one correspondence between edge lengths and model parameters given by edge correlations (up to sign). In the discrete case the situation is more complicated. Here we obtain noisy distances by defining $\hat{\tau}_{ij}$ like in (11.9) with P^{ij}, P^{ii}, and P^{jj} replaced by their sample versions. This gives $\hat{d}_{ij} = -\log|\hat{\tau}_{ij}|$. Using the NJ algorithm we obtain an estimate of the underlying tree and parameters τ_{uv}. However, this is in general not enough to recover all model parameters. A special case when it is possible is so-called *symmetric discrete distributions*. In these submodels the matrix M^{uv} has all off-diagonal entries equal and the root distribution is assumed to be uniform. In statistical mechanics this corresponds to the Potts model, which in the binary case gives the Ising model; see Example 3.2 in [90].

Although using the NJ algorithm as a tree learning subroutine results in a computationally efficient and consistent method, consistency is only a minor requirement, and other tree learning procedures can be preferred in order to allow for a more sophisticated statistical analysis. An early example is the Dyadic Closure Tree Construction method (DCTC) [31] and witness-antiwitness method (WAM) [32] both focusing on learning the underlying tree by learning certain quartets that hold; see the latter paper for more details and references. In [21] the authors propose two other algorithms: recursive grouping (RG) and CLGrouping. Recursive grouping builds the latent tree recursively by identifying sibling groups using distances d_{ij}. CLGrouping starts with a pre-processing procedure in which a tree over the observed variables is constructed, or more precisely, the Chow–Liu tree. This global step groups the observed vertices that are likely to be close to each other in the true latent tree, thereby guiding subsequent recursive grouping (or equivalent procedures such as neighbor-joining) on much smaller subsets of variables. This results in more accurate and efficient learning of latent trees. This can be further improved by using distance information to learn locally small trees and then glue them together. An example of such a divide-and-conquer algorithm is given in [45].

The distance based methods to learn the underlying tree are based predominantly on second order margins and so typically are very robust with respect to the sampling error and model misspecifications. In the next section we present other methods to estimate model parameters that use much more information about the underlying distribution.

11.3 Selected Theoretical Results

Latent tree models, like all models with unobserved variables, suffer from various problems and learning is generally complicated. The complex geometry of models with unobserved variables usually leads to difficulties in establishing the identifiability of their parameters, and the likelihood function has many local maxima, which lie on the boundary of the parameter space; see, for example, [22, 101]. In consequence, standard inference and model selection procedures are not fully justified in this setting. In this section we discuss parameter identifiability, sample complexity, and model selection methods. These three seemingly unrelated topics all deliver one important message: the class of latent tree models is well behaved from the statistical point of view as long as all the edge correlations are sufficiently large in the absolute value. Otherwise, it should be used with caution. Some further issues with the likelihood function for this model class will be discussed in Section 11.4.1.

11.3.1 Identifiability

A parametric model (P_θ) is *identifiable* if $P_\theta = P_{\theta'}$ implies $\theta = \theta'$. In other words, the parameterization map $\theta \mapsto P_\theta$ is a bijection between the parameter space and the model. Latent tree models are never identifiable. This can be seen for latent class models, where permuting labels of the unobserved variable makes no difference in the observed distribution; see, for example, [98]. This is known as the *label swapping problem.* The label swapping is not a serious problem in practice. We can always take account of it by restricting the parameter space. However, this still does not make the model identifiable because there are special subspaces in the parameter space that map to the same observed distribution. We illustrate this issue in the simplest possible example.

Example 11.3.1. Consider the Gaussian latent tree model on the star tree with three leaves, see Figure 11.1(a). Denote $\rho_{ij} = \mathrm{corr}(X_i, X_j)$ and $\rho_i = \mathrm{corr}(X_i, H)$ for $i = 1, 2, 3$. By (11.5), this model is parameterized by $\rho_{12} = \rho_1\rho_2$, $\rho_{13} = \rho_1\rho_3$, and $\rho_{23} = \rho_2\rho_3$. If the observed correlations are non-zero we can identify ρ_1 up to the sign and then the other parameters as follows

$$\rho_1^2 = \frac{\rho_{12}\rho_{13}}{\rho_{23}}, \qquad \rho_2 = \frac{\rho_{12}}{\rho_1}, \qquad \rho_3 = \frac{\rho_{13}}{\rho_1}.$$

Suppose now that some observed correlations vanish. The form of the parameterization implies that it is impossible for only one of them to vanish. If two correlations are zero, say $\rho_{12} = \rho_{13} = 0$, then the set of all triples (ρ_1, ρ_2, ρ_3) mapping to $(0, 0, \rho_{23})$ is a smooth one-dimensional subset given by $\rho_1 = 0$ and $\rho_2\rho_3 = \rho_{23}$. Suppose now that *all* observed correlations are zero. Then the corresponding parameters form the union of three intervals

$$\{\rho_1 = \rho_2 = 0\} \cup \{\rho_1 = \rho_3 = 0\} \cup \{\rho_2 = \rho_3 = 0\}.$$

This example motivates the following definition.

Definition 11.3.2. *A model is* generically identifiable, *if the parameterization map is finite-to-one everywhere outside of a measure zero set.*

Geometrically, generic identifiability means that for a typical point in the model, its preimage under the parameterization map, also called a *fiber*, is a finite collection of points. Showing that the Gaussian latent tree model satisfying (A1) is generically identifiable can be done by arguments as in Example 11.3.1. For general Markov models generic identifiability is more subtle and, in general, the second-order moments contain not enough information about the underlying parameters. It turns out that three-way margins are already enough. The following is the main result of [15].

Theorem 11.3.3. *Every general Markov model satisfying (A1) is generically identifiable. In fact, to identify the parameters it is enough to know the 3-way marginal distributions.*

Theorem 11.3.3 does not cover the case when the state-spaces $\mathcal{Y}_v$ are allowed to vary. In this case techniques of [1] may be useful to establish identifiability.

The analysis of when exactly identifiability fails can be in general complicated. In the Gaussian and the binary case these special points correspond to some correlations being zero [102]. As illustrated by Example 11.3.1, depending on the situation the corresponding fiber can be either a smooth subset of the parameter space or it can be singular. These theoretical results and examples provide the following insight. If the true data-generating distribution is characterized by high correlations between variables, it is also far from any of the special singular points. However, if some variables have a low degree of dependence, then estimation may become difficult and standard asymptotic theory breaks down.

11.3.2 Guarantees for tree reconstruction

A basic question regarding learning of latent tree models is that of the *sample complexity* of the tree reconstruction problem. Given an estimator $\hat{T}$ of the true tree T we want to assure that $\mathbb{P}(\hat{T} = T) > 1 - \delta$ for some fixed small δ. This is only possible if the sample size is big enough. It is known that irrespective of the method $n = O(\log m)$ is necessary [31, 57] but it is typically not enough. The first systematic study of when the logarithmic bound is sufficient was offered in [31] for the binary symmetric model and in [32] for general Markov models. The link to tree metrics shows that, in order to recover the underlying tree with high probability, the edge lengths cannot be too small or too large. For linear latent tree models this means that the edge parameters τ_{uv} must satisfy $c \leq |\tau_{uv}| \leq C$ for some

$0 < c < C < 1$. Sample bounds we discuss will always necessarily depend on the number of observed variables m and the parameters c, C.

One of the important concepts developed in [31, 32] is that of the tree depth: the *depth* of a tree T with vertices V and leaves $W \subset V$ is $\max_{i \in W \setminus V} \min_{j \in V} |\overline{ij}|$. For example, the depth of the tree in Figure 11.1(c) is 2 and of both trees in Figure 11.2 is 1. Latent tree models for trees with few unobserved variables or small depth are generally easier to learn. This leads to one group of results assuming that the depth of the underlying tree is $O(1)$, that is, constant in the number of leaves of T. It was shown in [32, Theorem 14] that, under constant tree depth, there is an algorithm for general Markov models whose sample complexity is logarithmic in m. This has been generalized to the Gaussian case in [21].

Similarly strong results are possible without assuming constant depth. It was first conjectured by Mike Steel [83] that for the binary symmetric latent tree model the sample complexity is logarithmic in m as long as $c \leq |\tau_{uv}| \leq C$ for some $\frac{\sqrt{2}}{2} < c < C < 1$, that is, when the parameters lie in the *Kesten–Stigum (KS) regime*. This conjecture was proved in [25, 58]. In the subsequent work, it has been shown in multiple scenarios, including the Gaussian case, that in the KS regime, high probability tree topology reconstruction may be achieved with $n = O(\log m)$ samples; see [59] and references therein. The KS regime plays also an important role in the sample complexity analysis of the maximum likelihood estimator. In general the sample complexity is polynomial in m under very general assumptions. However, for symmetric models and under a discretization assumption, it becomes logarithmic in m if the true distribution lies in the KS regime [70].

11.3.3 Model selection

As presented in the previous section, a lot of effort in the literature is put to obtain performance guarantees for the proposed learning algorithms. This analysis is done under assumption that the data come from a latent tree model. Although this assumption seems to be reasonable for phylogenetics and several other applications mentioned here, it is certainly not so for many other kinds of data (e.g. survey data from medicine and social sciences) for which these models are used, c.f. Section 11.1.2. In all these cases model selection techniques tend to outperform algorithms that aim at finding the "true" tree [97].

An example of a model selection algorithm is EAST (Expansion, Adjustment and Simplification until Termination) [18], which aims to find the model with the highest Bayesian Information Criterion (BIC) score. Given a random sample $x^{(1)}, \ldots, x^{(n)}$ the BIC is

$$\mathrm{BIC}(M(T, \mathcal{Y})) \;=\; \ell(\hat{\theta}, T) - \frac{d}{2} \log n,$$

where $\hat{\theta}$ is the maximum likelihood estimator, and d is the dimension of the model $M(T, \mathcal{Y})$. Computing the dimension of a model with unobserved variables is not always easy because it need not correspond to the number of parameters in the model. In that case a more detailed study of the generic rank of the Jacobian of the parameterization is needed. However, for general Markov models, Theorem 11.3.3 can be used to show that simple parameter count does suffice to compute the dimension of the model whenever (A1) holds. For Gaussian latent tree models the analysis in Section 11.2.1 showed that the dimension is $m + |E|$, where E is the set of edges of the underlying tree and m is the number of its leaves.

Theoretical importance of the BIC criterion [73] is that it provides an asymptotic approximation to the marginal likelihood of the model, and so it can guide model selection in the Bayesian setting. However, latent tree models lead to a new difficulty in that the Fisher information matrix of a latent tree model is singular along certain submodels. Such singularities invalidate the mathematical arguments that lead to the Bayesian information criterion; see, for example, [27, 54, 92]. This again shows that BIC must be used with caution in the case of weak correlations. In that case the BIC may be too conservative and select too small models. Correcting BIC to account for singularities involves an indepth study of the model geometry. For binary latent tree models this has been done in [99] and for Gaussian latent tree models in [26]. Watanabe's WAIC [93] is an alternative that can be implemented generally through Monte Carlo integration but involves specific choices of priors.

11.4 Estimation and Inference

In Section 11.2.4 we described some natural tree-metric based approaches to learn latent tree models. In this section we briefly describe other popular learning algorithms.

11.4.1 Fixed tree structure

The maximum likelihood estimator is one of the most popular estimators of the continuous parameter. It cannot be computed in a closed form and the likelihood function is very complicated to analyze directly; see, for example, [22, 101]. However, there are various numerical methods to maximize the likelihood function that were developed in the context of latent tree models or, more generally, Bayesian networks with unobserved variables. The most natural choice is the EM algorithm, which is easy to set up; see, for example, Section 3.1.1 in [60]. There are also several methods building upon the idea of the EM algorithm. For example the progressive EM algorithm of [17] uses a method of moments estimator as a subroutine, which leads to a more computationally efficient estimator.

As with all other EM-based methods, these approaches depend on the initialization and suffer from the possibility of being trapped in local optima. This algorithm seems to work well in the case when the observed variables are highly correlated [91]. The situation becomes much more complicated in the presence of weak correlations as the numerical procedures become unstable. Another problem is that the maximum likelihood estimator often lies on the boundary of the parameter space [101], which has many important consequences. First, the gradient of the likelihood function does not vanish at such a point and so the standard asymptotics does not apply. Second, there may be relatively distant points in the parameter space that give a similar value of the likelihood function as the maximum likelihood estimator. Finally, the boundary points typically correspond to situations when the distribution of the unobserved vector is degenerate. In many applications this may be problematic as a natural interpretation for the unobserved process may be lost.

11.4.2 The structural EM algorithm

Suppose that given a random sample of size n, we are interested in finding the tree that maximizes the likelihood function over all *fully-observed* tree models. For every tree the maximum likelihood estimate is easily obtained; see [52, Section 4.4.2 and Section 5.2.1]. In the naive approach we can search over all possible trees with a given number of vertices and find the one that gives the highest value of the likelihood function. The Chow–Liu algorithm [23] is a remarkably simple algorithm, which gives an efficient way to find the best tree approximation that maximizes the likelihood.

The original approach was proposed for discrete data but it can be easily extended to the Gaussian case [87]. It only uses the fact that the MLE decomposes according to a tree so it can be also used in other similar scenarios to find best BIC and AIC trees [30]; see [43] for further references and implementations. This approach boils down to using the Kruskal's algorithm to find the maximum cost spanning tree of a complete graph with edge weights given by sample mutual informations [50]:

1. Compute all mutual informations of the sample distribution and order them from the largest to the smallest.
2. Move along this ordered sequence adding subsequently the corresponding edges unless adding an edge introduces a cycle. Stop when no more edges can be added.

If all mutual informations are different, then there is a unique best solution. If some of the weights are equal, then multiple solutions are possible, but they will all give the same value of the likelihood function. In the Gaussian case the mutual information is a simple monotone function of the corresponding correlations squared. Therefore, the same tree will be obtained after replacing mutual informations in step 1 above with squares of sample correlations.

A natural idea is to use the Chow–Liu algorithm to learn latent tree models by using an EM-type algorithm. The *structural EM algorithm* [38] is a numerical procedure to maximize the likelihood simultaneously over the continuous parameter θ and the discrete parameter T. It starts from a given parameter value and moves at each step strictly increasing the likelihood unless it is already in a local optimum. The E-step of the algorithm is the standard E-step of the EM algorithm. The M-step follows essentially the Chow–Liu algorithm with several modifications. A minor problem with the Chow–Liu algorithm, when used in the EM algorithm, is that it does not get any information about which vertices represent unobserved variables and so it often outputs a tree whose leaves can be potentially unobserved or internal vertices that can represent observed random variables. In this case it is easy to provide a tree, with leaves precisely corresponding to observed vertices, that gives the same observed likelihood. This procedure is described in more detail for the discrete case in [38, Section 5]. We describe the general idea in the Gaussian case. Suppose that the Chow–Liu algorithm outputs a tree T' with edge correlations ρ_{uv}, then:

- Remove all degree one vertices that represent unobserved vertices.
- If there is an induced chain $\overset{i_1}{*} - \overset{i_2}{\circ} - \overset{i_3}{\circ} - \cdots - \overset{i_{k-1}}{\circ} - \overset{i_k}{*}$, where $*$ stands for any vertex of degree at least three, then replace this chain with a single edge between i_1 and i_k. Set the correlation $\rho_{i_1 i_k}$ equal to the product $\rho_{i_1 i_2} \cdots \rho_{i_{k-1} i_k}$.
- If there is an internal vertex v representing an observed random variable then add an auxiliary copy v' of v and an edge (v, v'). Set $\rho_{vv'} = 1$.

This operation gives a leaf-labeled tree that leads to the same observed likelihood as T' and we take it as the output of the M-step. If we are interested only in trivalent trees, the above procedure can be easily modified so that the output is a trivalent tree; c.f. Theorem 11.1.3.

11.4.3 Phylogenetic invariants

Given latent tree model over T, we associate to it the set $\mathcal{I}_T$ of all polynomials vanishing on $M(T, \mathcal{Y})$. The polynomials are expressed in terms of correlations in the Gaussian case and in terms of the raw probabilities in the discrete case. Every polynomial $f \in \mathcal{I}_T$ is called a *phylogenetic invariant*. Equation (11.6) provides an example of an equation that must hold for a Gaussian latent tree model over a quartet tree. It is a basic result in algebraic geometry that $\mathcal{I}_T$ is finitely generated, that is, it admits a finite basis of polynomials $\{f_1, \ldots, f_r\}$ such that every polynomial in $\mathcal{I}_T$ is a polynomial combination of the f_i.

The basic idea behind application of phylogenetic invariants is as follows. Given n independent observations of X we compute the sample distribution $\hat{p}$, which, by the law of large numbers, converges almost surely to the true data generating distribution p^*. Because $f(p^*) = 0$ for every $f \in \mathcal{I}_T$, for large n also $f(\hat{p}) \approx 0$. The methods proposed in the phylogenetic literature are mainly simple diagnostic tests that work with a given fixed finite set of invariants in $\mathcal{I}_T$, which do not necessarily generate $\mathcal{I}_T$. There is now considerable literature on the method of phylogenetic invariants and for many models all defining polynomials are understood. We refer to [2, 86] for an overview.

The advantage of this approach is that the method based on phylogenetic invariants does not require parameter estimation. The disadvantage is that to a large extent statistical theory behind their use is lacking. The method of phylogenetic invariants gives a way to select the best tree under a given criterion. It does not, however, give a way of quantifying how well the chosen tree fits the data because, in general, the distribution of phylogenetic invariants is too hard to analyze.

Recently there has been some effort aimed at organizing the statistical theory behind phylogenetic invariants. For example, it has been observed by many authors that not all invariants are equally important and from the statistical point of view there is no sense in working with a full generating set of $\mathcal{I}_T$. For example, invariants linking directly to tree metrics, called the *edge invariants*, tend to be more robust with respect to the sampling error and are enough to distinguish between different models [12]. To test the edge invariants, [34] uses the singular value decomposition and the Frobenius norm to compute the distance of a matrix to the set of matrices of certain rank. Recently this method has been further improved by [37]. In its current form the method is robust and simulations show that it outperforms most of the commonly used methods.

Focusing only on quadratic invariants allows us to generalize directly asymptotic chi-square tests for independence in a contingency table; see [72]. In the Gaussian setting we can proceed as follows. For any four leaves of T, ij/kl *forms a quartet* in T if the paths $\overline{ij}$ and $\overline{kl}$ have no vertices in common. It is clear from (11.5) and (11.10) that for any quartet ij/kl we have that $\rho_{ik}\rho_{jl} - \rho_{il}\rho_{jk} = 0$ in the Gaussian case and $\tau_{ik}\tau_{jl} - \tau_{il}\tau_{jk} = 0$ in the discrete case. For example (11.6) gives the equation that holds for the quartet tree in Example 11.2.1. Equations of this form are called tetrads [79]. In the context of latent tree models using tetrads was suggested already by Judea Pearl [64, Section 8.3.5] but with no statistical guidance. An example of a statistically guided quartet-based analysis was given recently for Gaussian latent tree models [77].

Finally, the geometric description of latent tree models involves not only polynomial equalities but also polynomial inequalities. These inequalities cut out a large portion of the space described solely by equalities and therefore they should not be neglected [101]. The inequalities are harder to study than equalities but they are also understood well for general Markov models [5, 75, 101]. Recently [4] showed how basic inequalities can be easily tested within the Bayesian framework to obtain a preliminary assessment of whether the data come from a Gaussian latent tree model.

11.5 Discussion

In this chapter we gave a concise overview of the theory of latent tree models. We argued that, from the theoretical and the practical point of view, the general Markov models, and more generally, linear latent tree models, form the most important subfamily. We showed that for linear latent tree models the link to tree metrics provides a wide variety of learning procedures. The geometric viewpoint gives important insights but was not covered here in much detail. We refer to [100] for further details. A more algorithmic overview of the latent tree model class is provided in [60]. We also skipped other research directions that we believe will become increasingly important in the coming years. Tensor representations of discrete latent tree models help to design learning procedures for latent tree models and more general graphical models with unobserved variables [7, 46, 47, 62]. In the numerical analysis community a closely related concept of hierarchical tensors has become popular [41]. There are also several recent approaches that introduce the nonparametric setting for latent tree models [80, 81].

Acknowledgments

I would like to thank Anima Anandkumar, Sebastien Roch, and Nevin Zhang for helpful comments, clarifications, and literature suggestions. Comments from the anonymous referees allowed to substantially improve the manuscript.

Bibliography

[1] Elizabeth S. Allman, Catherine Matias, and John A. Rhodes. Identifiability of parameters in latent structure models with many observed variables. *Ann. Statist.*, 37(6A):3099–3132, 2009.

[2] Elizabeth S. Allman and John A. Rhodes. Phylogenetic invariants. In *Reconstructing Evolution: New Mathematical and Computational Advances*, 108–146. Oxford Univ. Press, Oxford, 2007.

[3] Elizabeth S. Allman and John A. Rhodes. Phylogenetic ideals and varieties for the general Markov model. *Adv. in Appl. Math.*, 40(2):127–148, 2008.

[4] Elizabeth S. Allman, John A. Rhodes, Bernd Sturmfels, and Piotr Zwiernik. Tensors of nonnegative rank two. *Linear Algebra Appl.*, 473:37–53, 2015.

[5] Elizabeth S. Allman, John A. Rhodes, and Amelia Taylor. A semialgebraic description of the general Markov model on phylogenetic trees. *SIAM J. Discrete Math.*, 28(2):736–755, 2014.

[6] Animashree Anandkumar, Kamalika Chaudhuri, Daniel J Hsu, Sham M Kakade, Le Song, and Tong Zhang. Spectral methods for learning multivariate latent tree structure. In *Advances in Neural Information Processing Systems*, 2025–2033, 2011.

[7] Animashree Anandkumar, Rong Ge, Daniel Hsu, Sham M. Kakade, and Matus Telgarsky. Tensor decompositions for learning latent variable models. *J. Mach. Learn. Res.*, 15:2773–2832, 2014.

[8] T. W. Anderson and Herman Rubin. Statistical inference in factor analysis. In *Proceedings of the Third Berkeley Symposium on Mathematical Statistics and Probability, 1954–1955, volume V*, 111–150. University of California Press, Berkeley and Los Angeles, 1956.

[9] Daniel Barry and John A. Hartigan. Statistical analysis of hominoid molecular evolution. *Statist. Sci.*, 2(2):191–210, 1987. With comments by Stephen Portnoy and Joseph Felsenstein and a reply by the authors.

[10] Christopher M. Bishop and Michael E. Tipping. A hierarchical latent variable model for data visualization. *IEEE Transactions on Pattern Analysis and Machine Intelligence*, 20(3):281–293, 1998.

[11] Peter Buneman. The recovery of trees from measures of dissimilarity. In F. Hodson et al., editor, *Mathematics in the Archaeological and Historical Sciences*, 387–395. Edinburgh University Press, 1971.

[12] Marta Casanellas and Jesús Fernández-Sánchez. Relevant phylogenetic invariants of evolutionary models. *J. Math. Pures Appl. (9)*, 96(3):207–229, 2011.

[13] Rui Castro, Mark Coates, Gang Liang, Robert Nowak, and Bin Yu. Network tomography: recent developments. *Statistical Science*, 499–517, 2004.

[14] L.L. Cavalli-Sforza and AWF Edwards. Phylogenetic analysis: models and estimation procedures. *Evolution*, 21(3):550–570, 1967.

[15] Joseph T. Chang. Full reconstruction of Markov models on evolutionary trees: Identifiability and consistency. *Mathematical Biosciences*, 137(1):51–73, 1996.

[16] Joseph T Chang and John A Hartigan. Reconstruction of evolutionary trees from pairwise distributions on current species. In *Computing science and statistics: Proceedings of the 23rd symposium on the interface*, volume 254, page 257. Interface Foundation, Fairfax Station, VA, 1991.

[17] Peixian Chen, Nevin L. Zhang, Leonard Poon, and Zhourong Chen. Progressive EM for latent tree models and hierarchical topic detection. In *AAAI*, 1498–1504, 2016.

[18] Tao Chen, Nevin L Zhang, Tengfei Liu, Kin Man Poon, and Yi Wang. Model-based multidimensional clustering of categorical data. *Artificial Intelligence*, 176(1):2246–2269, 2012.

[19] Hyeokho Choi and Richard G. Baraniuk. Multiscale image segmentation using wavelet-domain hidden Markov models. *IEEE Trans. Image Process.*, 10(9):1309–1321, 2001.

[20] Myung Jin Choi, Joseph J Lim, Antonio Torralba, and Alan S. Willsky. Exploiting hierarchical context on a large database of object categories. In *Computer Vision and Pattern Recognition (CVPR)*, 129–136. IEEE, 2010.

[21] Myung Jin Choi, Vincent Y. F. Tan, Animashree Anandkumar, and Alan S. Willsky. Learning latent tree graphical models. *J. Mach. Learn. Res.*, 12:1771–1812, 2011.

[22] Benny Chor, Michael D. Hendy, Barbara R. Holland, and David Penny. Multiple maxima of likelihood in phylogenetic trees: An analytic approach. *Molecular Biology and Evolution*, 17(10):1529–1541, 2000.

[23] C. K. Chow and C. N. Liu. Approximating discrete probability distributions with dependence trees. *IEEE Trans. Inform. Theory*, 14:462–467, 1968.

[24] Matthew S. Crouse, Robert D. Nowak, and Richard G. Baraniuk. Wavelet-based statistical signal processing using hidden Markov models. *IEEE Trans. Signal Process.*, 46(4):886–902, 1998.

[25] Constantinos Daskalakis, Elchanan Mossel, and Sebastien Roch. Phylogenies without branch bounds: Contracting the short, pruning the deep. *SIAM Journal on Discrete Mathematics*, 25(2):872–893, 2011.

[26] Mathias Drton, Shaowei Lin, Luca Weihs, and Piotr Zwiernik. Marginal likelihood and model selection for Gaussian latent tree and forest models. *Bernoulli*, 23(2):1202–1232, 2017.

[27] Mathias Drton and Martyn Plummer. A Bayesian information criterion for singular models. *Journal of the Royal Statistical Society: Series B (Statistical Methodology)*, 79(2):323–380, 2017.

[28] Mathias Drton, Bernd Sturmfels, and Seth Sullivant. *Lectures on Algebraic Statistics*. Oberwolfach Seminars Series. Birkhauser Verlag AG, 2009.

[29] Richard Durbin, Anders Krogh, Graeme Mitchison, and Sean R. Eddy. *Biological Sequence Analysis: Probabilistic Models of Proteins and Nucleic Acids*. Cambridge University Press, 1998.

[30] David Edwards, Gabriel CG De Abreu, and Rodrigo Labouriau. Selecting high-dimensional mixed graphical models using minimal AIC or BIC forests. *BMC Bioinformatics*, 11(1):1, 2010.

[31] Péter L. Erdős, Michael A. Steel, László A. Székely, and Tandy J. Warnow. A few logs suffice to build (almost) all trees (part 1). Random Structures and Algorithms, 14(2):153–184, 1999.

[32] Péter L. Erdős, Michael A. Steel, László A. Székely, and Tandy J. Warnow. A few logs suffice to build (almost) all trees (part 2). *Theoretical Computer Science*, 221(1):77–118, 1999.

[33] Brian Eriksson, Gautam Dasarathy, Paul Barford, and Robert Nowak. Toward the practical use of network tomography for internet topology discovery. In *INFOCOM, 2010 Proceedings IEEE*, 1–9.

[34] Nicholas Eriksson. Using invariants for phylogenetic tree construction. In M. Putinar and S. Sullivant, editors, *Emerging Applications of Algebraic Geometry*. Springer, New York, 2009.

[35] Robin J. Evans. Graphs for margins of Bayesian networks. *Scand. J. Statist.*, 43(3):625–648, 2016. 10.1111/sjos.12194.

[36] Joseph Felsenstein. *Inferring phylogenies*. Sinauer Associates Sunderland, 2004.

[37] Jesús Fernández-Sánchez and Marta Casanellas. Invariant versus classical approaches when evolution is heterogeneous across sites and lineages. *arXiv:1405.6546*, 2014.

[38] Nir Friedman, Matan Ninio, Itsik Pe'er, and Tal Pupko. A structural EM algorithm for phylogenetic inference. Journal of Computational Biology, 9(2):331–353, 2002.

[39] Morten Frydenberg. The chain graph Markov property. *Scand. J. Statist.*, 17(4):333–353, 1990.

[40] Cristina Guardiano and Giuseppe Longobardi. Parametric comparison and language taxonomy. In Montserrat Batllori, Maria-Lluïsa Hernanz, Carme Picallo, and Francesc Roca, editors, *Grammaticalization and Parametric Variation*, 149–174. Oxford University Press, Aug 2005.

[41] Wolfgang Hackbusch and Stefan Kühn. A new scheme for the tensor representation. Journal of Fourier Analysis and Applications, 15(5):706–722, 2009.

[42] Gordon Hiscott, Colin Fox, Matthew Parry, and David Bryant. Efficient recycled algorithms for quantitative trait models on phylogenies. *Genome Biology and Evolution*, 8(5):1338–1350, 2016.

[43] Søren Højsgaard, David Edwards, and Steffen Lauritzen. *Graphical Models with R*. Springer Science & Business Media, 2012.

[44] Barbara R. Holland, Peter D. Jarvis, and Jeremy G. Sumner. Low-parameter phylogenetic inference under the general Markov model. *Systematic Biology*, 62(1):78–92, 2013.

[45] Furong Huang, U.N. Niranjan, Ioakeim Perros, Robert Chen, Jimeng Sun, and Anima Anandkumar. Scalable latent tree model and its application to health analytics. *arXiv:1406.4566*, 2014.

[46] Mariya Ishteva, Haesun Park, and Le Song. Unfolding latent tree structures using 4th order tensors. In *ICML (3)*, 316–324, 2013.

[47] Mariya Ishteva, L. Song, H. Park, A. Parikh, and E. Xing. Hierarchical tensor decomposition of latent tree graphical models. In *The 30th International Conference on Machine Learning (ICML 2013)*, 2013.

[48] Vivek Jayaswal, John Robinson, and Lars S. Jermiin. Estimation of phylogeny and invariant sites under the general Markov model of nucleotide sequence evolution. *Systematic Biology*, 56(2):155–162, 2007.

[49] Thomas H. Jukes and Charles R. Cantor. Evolution of protein molecules. *Mammalian Protein Metabolism*, 3:21–132, 1969.

[50] Joseph B. Kruskal. On the shortest spanning subtree of a graph and the traveling salesman problem. *Proceedings of the American Mathematical Society*, 7(1):48–50, 1956.

[51] James A. Lake. Reconstructing evolutionary trees from DNA and protein sequences: Paralinear distances. *Proceedings of the National Academy of Sciences of the United States of America*, 91(4):1455–1459, 1994.

[52] Steffen L. Lauritzen. *Graphical Models*, volume 17 of *Oxford Statistical Science Series*. Oxford University Press, 1996. Oxford Science Publications.

[53] Neil D. Lawrence. Gaussian process latent variable models for visualisation of high dimensional data. *Advances in Neural Information Processing Systems*, 16(3):329–336, 2004.

[54] Shaowei Lin. Ideal-theoretic strategies for asymptotic approximation of marginal likelihood integrals. *Journal of Algebraic Statistics*, 8(1), 2017.

[55] Peter J. Lockhart, Michael A. Steel, Michael D. Hendy, and David Penny. Recovering evolutionary trees under a more realistic model of sequence evolution. *Molecular Biology and Evolution*, 11(4):605–612, 1994.

[56] Mahender K. Makhijani, Niranjan Balu, Kiyofumi Yamada, Chun Yuan, and Krishna S. Nayak. Accelerated 3d merge carotid imaging using compressed sensing with a hidden Markov tree model. *Journal of Magnetic Resonance Imaging*, 36(5):1194–1202, 2012.

[57] Elchanan Mossel. On the impossibility of reconstructing ancestral data and phylogenies. *Journal of Computational Biology*, 10(5):669–676, 2003.

[58] Elchanan Mossel. Phase transitions in phylogeny. *Transactions of the American Mathematical Society*, 356(6):2379–2404, 2004.

[59] Elchanan Mossel, Sébastien Roch, and Allan Sly. Robust estimation of latent tree graphical models: Inferring hidden states with inexact parameters. *IEEE Trans. Inform. Theory*, 59(7):4357–4373, 2013.

[60] Raphaël Mourad, Christine Sinoquet, Nevin Lianwen Zhang, Tengfei Liu, Philippe Leray, et al. A survey on latent tree models and applications. *J. Artif. Intell. Res.(JAIR)*, 47:157–203, 2013.

[61] Emmanuel Paradis. *Analysis of Phylogenetics and Evolution with R*. Springer Science & Business Media, 2011.

[62] Ankur P. Parikh, Le Song, and Eric P. Xing. A spectral algorithm for latent tree graphical models. In *Proceedings of the 28th International Conference on Machine Learning (ICML-11)*, 1065–1072, 2011.

[63] Judea Pearl. Fusion, propagation, and structuring in belief networks* 1. *Artificial Intelligence*, 29(3):241–288, 1986.

[64] Judea Pearl. *Probabilistic Reasoning in Intelligent Systems: Networks of Plausible Inference*. The Morgan Kaufmann Series in Representation and Reasoning. Morgan Kaufmann, San Mateo, CA, 1988.

[65] Judea Pearl. *Causality: Models, Reasoning, and Inference*. Cambridge University Press, New York, 2000.

[66] Michael Pfeiffer, Marion Betizeau, Julie Waltispurger, Sabina Sara Pfister, Rodney J. Douglas, Henry Kennedy, and Colette Dehay. Unsupervised lineage-based characterization of primate precursors reveals high proliferative and morphological diversity in the OSVZ. *Journal of Comparative Neurology*, 524(3):535–563, 2016.

[67] Lawrence R. Rabiner. A tutorial on hidden Markov models and selected applications in speech recognition. *Proceedings of the IEEE*, 77(2):257–286, 1989.

[68] Don Ringe, Tandy Warnow, and Ann Taylor. Indo-european and computational cladistics. *Transactions of the Philological Society*, 100(1):59–129, 2002.

[69] Sebastien Roch. A short proof that phylogenetic tree reconstruction by maximum likelihood is hard. *IEEE/ACM Trans. Comput. Biol. Bioinformatics*, 3(1):92–94, January 2006.

[70] Sebastien Roch and Allan Sly. Phase transition in the sample complexity of likelihood-based phylogeny inference. *arXiv:1508.01964*, 2015.

[71] Justin K. Romberg, Hyeokho Choi, and Richard G. Baraniuk. Bayesian tree-structured image modeling using wavelet-domain hidden Markov models. *IEEE Transactions on image processing*, 10(7):1056–1068, 2001.

[72] David Sankoff. Designer invariants for large phylogenies. *Molecular Biology and Evolution*, 7(3):255, 1990.

[73] Gideon Schwarz. Estimating the dimension of a model. *Annals of Statistics*, 6(2):461–464, 1978.

[74] Charles Semple and Mike Steel. *Phylogenetics*, volume 24 of *Oxford Lecture Series in Mathematics and Its Applications.* Oxford University Press, Oxford, 2003.

[75] Raffaella Settimi and Jim Q. Smith. Geometry, moments and conditional independence trees with hidden variables. *Ann. Statist.*, 28(4):1179–1205, 2000.

[76] Nathaniel Shiers, John A.D. Aston, Jim Q. Smith, and John S. Coleman. Gaussian tree constraints applied to acoustic linguistic functional data. Journal of Multivariate Analysis, 154:199–215, 2017.

[77] Nathaniel Shiers, Piotr Zwiernik, John A. Aston, and James Q. Smith. The correlation space of Gaussian latent tree models and model selection without fitting. *Biometrika*, 2016.

[78] Kevin Shu, Sharjeel Aziz, Vy-Luan Huynh, David Warrick, and Matilde Marcolli. Syntactic phylogenetic trees. *arXiv:1607.02791*, 2016.

[79] Ricardo Silva, Richard Scheine, Clark Glymour, and Peter Spirtes. Learning the structure of linear latent variable models. Journal of Machine Learning Research, 7(Feb):191–246, 2006.

[80] Le Song, Animashree Anandkumar, Bo Dai, and Bo Xie. Nonparametric estimation of multi-view latent variable models. In *Proceedings of the 31st International Conference on Machine Learning (ICML-14)*, 640–648, 2014.

[81] Le Song, Eric P. Xing, and Ankur P. Parikh. Kernel embeddings of latent tree graphical models. In J. Shawe-Taylor, R. S. Zemel, P. L. Bartlett, F. Pereira, and K. Q. Weinberger, editors, *Advances in Neural Information Processing Systems 24*, 2708–2716. Curran Associates, Inc., 2011.

[82] M. Steel. Recovering a tree from the leaf colourations it generates under a Markov model. *Applied Mathematics Letters*, 7(2):19–23, 1994.

[83] M. Steel. My favourite conjecture. *Preprint*, 2001.

[84] M. Steel, M.D. Hendy, and D. Penny. Invertible models of sequence evolution. Mathematical and Information Science report 93/02, Massey University, 1993.

[85] Seth Sullivant. Algebraic geometry of Gaussian Bayesian networks. Advances in Applied Mathematics, 40(4):482–513, 2008.

[86] Jeremy G. Sumner, Amelia Taylor, Barbara R. Holland, and Peter D. Jarvis. Developing a statistically powerful measure for quartet tree inference using phylogenetic identities and Markov invariants. *Journal of Mathematical Biology*, 1–36, 2017.

[87] Vincent Y. F. Tan, Animashree Anandkumar, and Alan S. Willsky. Learning high-dimensional Markov forest distributions: analysis of error rates. *J. Mach. Learn. Res.*, 12:1617–1653, 2011.

[88] Louis L. Thurstone. The vectors of mind. *Psychological Review*, 41(1):1, 1934.

[89] Thomas S. Verma and Judea Pearl. Equivalence and synthesis of causal models. In Piero P. Bonissone, Max Henrion, Laveen N. Kanal, and John F. Lemmer, editors, *UAI '90: Proceedings of the Sixth Annual Conference on Uncertainty in Artificial Intelligence, MIT, Cambridge, MA, USA, July 27-29, 1990.* Elsevier, October 1991.

[90] Martin J. Wainwright and Michael I. Jordan. Graphical models, exponential families, and variational inference. *Foundations and Trends in Machine Learning*, 1(1-2):1–305, 2008.

[91] Yi Wang and Nevin L. Zhang. Severity of local maxima for the EM algorithm: Experiences with hierarchical latent class models. In *Probabilistic Graphical Models*, 301–308. Citeseer, 2006.

[92] Sumio Watanabe. *Algebraic Geometry and Statistical Learning Theory.* Number 25 in Cambridge Monographs on Applied and Computational Mathematics. Cambridge University Press, 2009. ISBN-13: 9780521864671.

[93] Sumio Watanabe. A widely applicable Bayesian information criterion. Journal of Machine Learning Research, 14(Mar):867–897, 2013.

[94] Alan S. Willsky. Multiresolution Markov models for signal and image processing. *Proceedings of the* IEEE, 90(8):1396–1458, 2002.

[95] Zdeněk Žabokrtský and Martin Popel. Hidden Markov tree model in dependency-based machine translation. In *Proceedings of the ACL-IJCNLP 2009 Conference Short Papers*, 145–148. Association for Computational Linguistics, 2009.

[96] Nevin L. Zhang. Hierarchical latent class models for cluster analysis. *J. Mach. Learn. Res.*, 5:697–723, 2003/04.

[97] Nevin L. Zhang and Leonard K.M. Poon. Latent tree analysis. In *AAAI*, 4891–4898, 2017.

[98] Liwen Zou, Edward Susko, Chris Field, and Andrew J. Roger. The parameters of the Barry and Hartigan general Markov model are statistically nonidentifiable. *Systematic Biology*, 2011.

[99] Piotr Zwiernik. Asymptotic behaviour of the marginal likelihood for general Markov models. *J. Mach. Learn. Res.*, 12:3283–3310, 2011.

[100] Piotr Zwiernik. *Semialgebraic statistics and latent tree models*, volume 146 of *Monographs on Statistics and Applied Probability.* Chapman & Hall/CRC, 2016.

[101] Piotr Zwiernik and Jim Q. Smith. Implicit inequality constraints in a binary tree model. *Electron. J. Statist.*, 5:1276–1312, 2011.

[102] Piotr Zwiernik and Jim Q. Smith. Tree-cumulants and the geometry of binary tree models. *Bernoulli*, 18(1):290–321, January 2012.

12

Neighborhood Selection Methods

Po-Ling Loh

Department of Electrical and Computer Engineering, University of Wisconsin – Madison

CONTENTS

12.1 Introduction

In this chapter, we discuss the problem of edge estimation in undirected graphical models, also known as Markov networks. Given observations from a joint distribution, the goal is to construct an estimate of the edge set of the underlying graph. Since the edges encode conditional independence relationships between individual random variables given all other variables, it is natural to expect that jointly observed vectors reveal information about the unknown edge structure. Applications are widespread in numerous scientific fields, including computer vision, political science, epidemiology, neuroscience, and genetics, where it may be desirable to infer the connectivity between individual pixels, people, organisms, neurons, or genes. We are particularly interested in situations where the number of nodes exceeds the number of observations, and sparse edge structure of the underlying graph may be leveraged to perform edge recovery based on a relatively small sample size.

The chapter is organized as follows: We first establish some notation on graphical models to be used throughout the chapter. We then present algorithms for edge estimation for multivariate Gaussian and Ising models, focusing first on population-level results and then discussing statistical theory. The algorithms we cover generally fall into one of two categories, either involving estimating the adjacency matrix of the graph based on the support of an appropriate matrix, or estimating individual node neighborhoods via penalized regression. We then discuss generalizations of these methods to other classes of distributions, as well as adaptations for contaminated or incomplete data. We close by highlighting a few interesting methods for edge recovery based on conditional independence testing. All mentions of "graphical models" in this chapter refer specifically to undirected graphical models which are also known as Markov networks; recall the terminology introduced in Chapter 1.

12.2 Notation

Throughout this chapter, we will use $(X_1, \ldots, X_p)$ to denote a random vector sampled from the underlying distribution. For a subset $A \subseteq \{1, \ldots, p\}$, we write X_A to denote the random vector indexed only by elements of A. We also write $\backslash A$ to denote the set $A^c = \{1, \ldots, p\} \backslash A$.

Let $G = (V, E)$ denote the undirected graphical model associated with the joint distribution, where $V = \{1, \ldots, p\}$ and $E \subseteq \{0, 1\}^{\binom{p}{2}}$, and $(j, k) \notin E$ if and only if $X_j \perp\!\!\!\perp X_k \mid X_{V \backslash \{j,k\}}$. The graph G is also known as the *conditional independence graph.* For each $j \in V$, let $N(j) = \{k \in V : (j, k) \in E\}$ denote the neighborhood set of j. Let $d = \deg(G)$ denote the degree of G.

For a vector $v \in \mathbb{R}^p$, we write $\|v\|_1$ and $\|v\|_2$ to denote the ℓ_1- and ℓ_2-norms of v, respectively. For a matrix $M \in \mathbb{R}^p$, we write $\det(M)$ and $\text{tr}\,(M)$ to denote the determinant and trace of M, respectively. We write $\|M\|_{\max} = \max_{j,k} |M_{jk}|$ to denote the elementwise ℓ_∞-norm of M, and $|\!|\!|M|\!|\!|_1$ to denote the ℓ_1-operator norm of M. We use $\text{supp}(M)$ to denote the support (i.e., set of indices of nonzero entries) of M. For subsets $A, B \subseteq \{1, \ldots, p\}$, we write M_{AB} to denote the $|A| \times |B|$ submatrix of M with rows indexed by A and columns indexed by B. We write $M \succ 0$ when M is positive definite and $M \succeq 0$ when M is positive semidefinite.

Finally, for results concerning statistical theory, we let n denote the number of i.i.d. samples drawn from the joint distribution.

12.3 Gaussian Graphical Models

We begin by discussing edge recovery methods for Gaussian graphical models, which were also discussed in Chapter 9. Consider the case where $(X_1, \ldots, X_p) \sim N(0, \Sigma)$ are joint observations from a multivariate normal distribution, and let $\Theta = \Sigma^{-1}$ denote the inverse covariance matrix. The edge recovery algorithms presented in this section are largely based on a critical observation regarding the relationship between entries of Θ and edges of the conditional independence graph G.

12.3.1 Inverse covariance matrix and edge structure

Recall that the probability density function of the multivariate Gaussian distribution is given by

$$q(x_1, \dots, x_p) = \frac{1}{(2\pi)^{p/2} \det(\Sigma)^{1/2}} \exp\left(-\frac{1}{2} x^T \Theta x\right)$$
$$\propto \exp\left(-\frac{1}{2} \sum_{j,k} \Theta_{jk} x_j x_k\right).$$

It is easy to see that for any $j \neq k$ with $\Theta_{jk} = 0$, we may write

$$q(x_1, \dots, x_p) = q_1(x_j, x_{\setminus\{j,k\}}) q_2(x_k, x_{\setminus\{j,k\}}),$$

for functions $q_1, q_2 > 0$, from which we may deduce that $X_j \perp\!\!\!\perp X_k \mid X_{\setminus\{j,k\}}$. Conversely, if $\Theta_{jk} \neq 0$, such a decomposition is impossible, implying that $X_j \not\perp\!\!\!\perp X_k \mid X_{\setminus\{j,k\}}$. It follows that for all $j \neq k$,

$$(j, k) \in E \iff \Theta_{jk} \neq 0. \tag{12.1}$$

In other words, the support set $\text{supp}(\Theta)$, discounting diagonals, corresponds precisely to the edge structure of the graph. This is the mainstay of many neighborhood selection algorithms for multivariate Gaussian graphical models.

12.3.2 Edge recovery via matrix estimation

Based on the observations of the previous section, it suffices to devise an estimate of the inverse covariance matrix Θ when we are given a data matrix $X \in \mathbb{R}^{n \times p}$ of n i.i.d. observations from the joint distribution. A simple calculation shows that the maximum likelihood estimator

$$\widehat{\Theta}_{MLE} = \arg\min_{\Theta \succ 0} \left\{ \text{tr}\,(\widehat{\Sigma}\Theta) - \log\det(\Theta) \right\} \tag{12.2}$$

is given by $\widehat{\Theta}_{MLE} = \left(\widehat{\Sigma}\right)^{-1}$, provided the sample covariance matrix $\widehat{\Sigma} = \frac{1}{n}\sum_{i=1}^n x_i x_i^T$ is invertible. However, when the number of nodes p exceeds the number of dimensions n, the matrix $\widehat{\Sigma}$ is low-rank, hence not invertible. Various alternative estimators have consequently been proposed that are applicable in high dimensions; we will discuss two such methods below. Both algorithms produce an estimate $\widehat{\Theta}$ of Θ, from which we may in turn estimate the nonzero pattern $\text{supp}(\Theta)$. We will discuss statistical guarantees, as a function of the problem dimensions, in Section 12.3.4.

Graphical Lasso. The graphical Lasso, first proposed in the literature by Yuan and Lin [47], adds a penalty term to the maximum likelihood expression (12.2):

$$\widehat{\Theta}_{GLASSO} = \arg\min_{\Theta \succ 0} \left\{ \text{tr}\,(\widehat{\Sigma}\Theta) - \log\det(\Theta) + \lambda \sum_{j \neq k} |\Theta_{jk}| \right\}. \tag{12.3}$$

The penalty term consists of the regularization parameter λ, multiplied by the ℓ_1-norm of off-diagonal entries of Θ, and encourages a sparse matrix solution. We then define our estimate of the edge set to be $\widehat{E} = \text{supp}(\widehat{\Theta})$, where we abuse notation slightly and disregard the diagonal entries of $\widehat{\Theta}$.

Note that the graphical Lasso is a convex program. Hence, the solution $\widehat{\Theta}_{GLASSO}$ may be obtained efficiently using standard interior point methods [8]. However, generic optimization algorithms may be fairly slow when applied to extremely large data sets, and various authors have proposed more efficient methods specifically designed for solving the graphical Lasso program (12.3) (e.g., [15, 16, 31, 19]). We include further comments regarding computation of the graphical Lasso program at the end of Section 12.3.4.

CLIME. The constrained ℓ_1-minimization estimator (CLIME) was introduced by Cai et al. [11] as an alternative to the graphical Lasso, and solves the following optimization problem:

$$\begin{aligned} \min \quad & \sum_{j,k} |\Theta_{jk}| \\ \text{s.t.} \quad & \|\widehat{\Sigma}\Theta - I_p\|_{\max} \leq \lambda. \end{aligned} \tag{12.4}$$

Again, the elementwise ℓ_1-norm encourages sparsity in the matrix solution, whereas the constraint requires Θ to be a near-inverse of $\widehat{\Sigma}$ in the entrywise ℓ_∞-norm. The statistical results described in Section 12.3.4 guarantee that under appropriate conditions, $\widehat{\Theta}$ and Θ are close in elementwise ℓ_∞-norm. Hence, the estimate of E is obtained by first thresholding the entries of $\widehat{\Theta} = \widehat{\Theta}_{CLIME}$ at a prespecified level τ, and then determining the support:

$$\widehat{\Theta}^{\tau}_{jk} = \widehat{\Theta}_{jk} 1\{|\widehat{\Theta}_{jk}| \geq \tau\}, \quad \forall 1 \leq j, k \leq p, \qquad \text{and} \qquad \widehat{E} = \text{supp}(\widehat{\Theta}^{\tau}).$$

Finally, note that the CLIME program (12.4) is convex and the solution $\widehat{\Theta}_{CLIME}$ may be obtained efficiently, e.g., via linear programming. In fact, the program (12.4) may be decoupled into p optimization problems on each of the column vectors of Θ, which may then be optimized in parallel.

12.3.3 Edge recovery via linear regression

We now outline a fundamental relationship between linear regression and estimation of Θ. Note that for each $1 \leq j \leq p$, by properties of Gaussian random variables, we may write

$$X_j = \theta_j^T X_{\backslash\{j\}} + W_j,$$

where

$$\theta_j = \left(\Sigma_{\backslash\{j\},\backslash\{j\}}\right)^{-1} \Sigma_{\backslash\{j\},j} \in \mathbb{R}^{p-1}, \tag{12.5}$$

$W_j \in \mathbb{R}$ is normally distributed with mean 0, and $W_j \perp\!\!\!\perp X_{\backslash\{j\}}$.

On the other hand, note that by block matrix inversion [18], we have

$$\begin{aligned} \Theta &= \begin{pmatrix} \Sigma_{1,1} & \Sigma_{1,\backslash\{1\}} \\ \Sigma_{\backslash\{1\},1} & \Sigma_{\backslash\{1\},\backslash\{1\}} \end{pmatrix}^{-1} \\ &= \begin{pmatrix} a_1 & -a_1 \Sigma_{1,\backslash\{1\}} \left(\Sigma_{\backslash\{1\},\backslash\{1\}}\right)^{-1} \\ -a_1 \left(\Sigma_{\backslash\{1\},\backslash\{1\}}\right)^{-1} \Sigma_{\backslash\{1\},1} & \left(\Sigma_{\backslash\{1\},\backslash\{1\}} - \Sigma_{\backslash\{1\},1}\Sigma_{1,1}^{-1}\Sigma_{1,\backslash\{1\}}\right)^{-1} \end{pmatrix}, \end{aligned} \tag{12.6}$$

where $a_1 = \left(\Sigma_{1,1} - \Sigma_{1,\backslash\{1\}}\Sigma^{-1}_{\backslash\{1\},\backslash\{1\}}\Sigma_{\backslash\{1\},1}\right)^{-1}$. Hence, the first column of Θ is a constant multiple of the vector $\begin{pmatrix} -1 \\ \theta_1 \end{pmatrix}$, and an analogous statement may be made for each value of j. In particular, recovering $\text{supp}(\theta_j)$, for each $1 \leq j \leq p$, allows us to recover $\text{supp}(\Theta)$; using the equivalence (12.1), the set $\text{supp}(\theta_j)$ corresponds precisely to the neighborhood set $N(j)$.

Nodewise Lasso. Motivated by the relationship described above, Meinshausen and Bühlmann [33] proposed a method for estimating the neighborhood sets $\{N(j) : 1 \leq j \leq p\}$ via successive linear regressions: In particular, the relation (12.5) implies that the random variable $\theta_j^T X_{\setminus\{j\}}$ is the best linear predictor of X_j in terms of $X_{\setminus\{j\}}$, so performing a linear regression of the observations corresponding to X_j upon the vector-valued observations $X_{\setminus\{j\}}$ converges to θ_j as $n \to \infty$. In the setting of interest where $p > n$, the estimate of θ_j is given by the solution of the Lasso program

$$\widehat{\theta}_j = \arg\min_{\theta} \left\{ \|X^j - X^{\setminus j}\theta\|_2^2 + \lambda_j \|\theta\|_1 \right\}, \tag{12.7}$$

where X^j denotes the jth column vector of the $n \times p$ data matrix and $X^{\setminus j}$ denotes the $n \times (p-1)$ block containing the remaining columns. We then define our neighborhood estimates to be

$$\widehat{N(j)} = \text{supp}(\widehat{\theta}_j), \qquad \forall 1 \leq j \leq p,$$

and combine our neighborhood estimates into a global edge estimate $\widehat{E}$ using an AND or OR rule:

Edge recovery from neighborhood estimates

Input: Neighborhood estimates $\{N(j)\}_{1 \leq j \leq p}$

Output: Edge set estimate $\widehat{E}$

AND rule: For each $j \neq k$,

$$(j,k) \in \widehat{E} \iff j \in \widehat{N(k)} \text{ AND } k \in \widehat{N(j)}$$

OR rule: For each $j \neq k$,

$$(j,k) \in \widehat{E} \iff j \in \widehat{N(k)} \text{ OR } k \in \widehat{N(j)}.$$

The theoretical results described in the next subsection guarantee that with a sufficiently large sample size and under certain regularity conditions, we have $\widehat{N(j)} = N(j)$, with high probability, for all j. Hence, the estimated edge set $\widehat{E}$ will be a consistent estimate of E, regardless of which rule is applied.

12.3.4 Statistical theory

We now highlight some theoretical results concerning the success of the algorithms described above. At a high level, all the results guarantee that when the number of samples n scales as a power of d times $\log p$, we have $\widehat{E} = E$, with high probability. However, the results differ in the types of conditions imposed on the underlying distribution. For the results of this section, we will use Σ^* and Θ^* to denote the true covariance and inverse covariance matrices, respectively, so our data matrix $X \in \mathbb{R}^{n \times p}$ consists of n i.i.d. draws from the distribution $N(0, \Sigma^*)$. The constants c_i appearing in the statistical results refer to universal constants, the values of which may vary between theorems.

Theory for graphical Lasso. We begin by discussing the performance of the graphical Lasso (12.3). Numerous theoretical results have been derived concerning the convergence of $\widehat{\Theta}_{GLASSO}$ to Θ^*, where convergence is measured in various norms (e.g, Frobenius norm [37], spectral norm, and elementwise ℓ_∞-norm [35]). Since the topic of this chapter is neighborhood selection, we will focus on edge recovery guarantees; i.e., conditions under which $\widehat{E} = E$, with high probability. In the following theorem, we require the *α-incoherence* condition:

$$\max_{e \in S^c} \|\Gamma^*_{eS} (\Gamma^*_{SS})^{-1}\|_1 \leq 1 - \alpha \tag{12.8}$$

where $\Gamma^* = \Sigma^* \otimes \Sigma^* \in \mathbb{R}^{p^2 \times p^2}$ is the tensor product of true covariance matrices, the augmented edge set S is defined according to $S := E \cup \{(j,j) : j \in V\}$, and $S^c := (V \times V) \backslash S$.

Theorem 12.3.1 (Ravikumar et al. [35]). *Suppose we have the α-incoherence condition (12.8) for some $\alpha \in (0, 1]$. Also suppose the regularization parameter is chosen to be $\lambda = \frac{c_1}{\alpha}\sqrt{\frac{\log p}{n}} + \delta$, for some $\delta \in [0, 1]$, and suppose the sample size satisfies $n \geq c_2 \left(1 + \frac{8}{\alpha}\right)^2 d^2 \log p$. Then with probability at least $1 - c_3 \exp(-c_4 n \delta^2)$, the estimated edge set $\widehat{E}$ based on the graphical Lasso satisfies $\widehat{E} \subseteq E$. Furthermore, $\widehat{\Theta}_{GLASSO}$ satisfies the elementwise ℓ_∞-norm bound*

$$\|\widehat{\Theta}_{GLASSO} - \Theta^*\|_{\max} \leq c_5 \left(\lambda + \left(1 + \frac{8}{\alpha}\right)\sqrt{\frac{\log p}{n}}\right),$$

so if Θ^* also satisfies the minimum signal strength condition

$$\min_{(j,k) \in E} |\Theta^*_{jk}| > c_5 \left(\lambda + \left(1 + \frac{8}{\alpha}\right)\sqrt{\frac{\log p}{n}}\right),$$

we are guaranteed that $\widehat{E} = E$.

Note that the first part of Theorem 12.3.1, which stipulates that $\widehat{E} \subseteq E$ with high probability, guarantees that the edge set generated by the graphical Lasso algorithm does not include any false edges.

Theory for CLIME. We now present theory for the output of the CLIME program (12.4). The following theorem provides a high-probability guarantee for variable selection consistency:

Theorem 12.3.2 (Cai et al. [11]). *Suppose the regularization parameter satisfies $\lambda \geq c_1 \left\|\Theta^*\right\|_1 \sqrt{\frac{\log p}{n}}$, and the sample size satisfies $n \geq c_2 \log p$. Then with probability at least $1 - c_3 \exp(-c_4 \log p)$, we have the elementwise bound*

$$\|\widehat{\Theta}_{CLIME} - \Theta^*\|_{\max} \leq c_5 \lambda \left\|\Theta^*\right\|_1.$$

In particular, if Θ^* satisfies the minimum signal strength condition

$$\min_{(j,k) \in E} |\Theta^*_{jk}| > 2 c_5 \lambda \left\|\Theta^*\right\|_1,$$

the estimated edge set $\widehat{E}$ based on the support of $\widehat{\Theta}^\tau$ with threshold parameter $\tau = c_5 \lambda \left\|\Theta^*\right\|_1$ satisfies $\widehat{E} = E$.

For Frobenius and spectral norm error bounds on $\widehat{\Theta}_{CLIME}$, we refer the reader to Cai et al. [11].

Note that $\widehat{\Theta}_{CLIME}$ may not be positive semidefinite or even symmetric. If a positive semidefinite estimate is desired, the output of the CLIME program (12.4) may be symmetrized by assigning the off-diagonal entries to be the minimum of the (i, j) and (j, i) entries of $\widehat{\Theta}_{CLIME}$. It may be shown that the resulting estimator is also positive definite with high probability.

Theory for nodewise Lasso. The theory for the nodewise regression method may be derived from variable selection guarantees for the Lasso algorithm [48, 40]. Under an incoherence assumption on functions of subblocks of the data matrix X, as well as a minimum signal strength assumption on the true regression coefficients

$$\theta_j^* = \left(\Sigma^*_{\backslash\{j\},\backslash\{j\}}\right)^{-1} \Sigma^*_{\backslash\{j\},j},$$

we may guarantee that the solutions $\{\widehat{\theta}_j\}_{1\leq j\leq p}$ to the nodewise regression programs (12.7) satisfy

$$\text{supp}(\widehat{\theta}_j) = \text{supp}(\theta_j^*), \qquad \forall 1 \leq j \leq p,$$

with high probability. This in turn implies that $\widehat{N(j)} = N(j)$ for all j, so the nodewise regression method succeeds. We summarize in the following theorem. To declutter the theorem statement, we assume that the columns of X have been renormalized so that $\frac{1}{\sqrt{n}} \max_{1\leq j\leq p} \|X^j\|_2 \leq 1$.

Theorem 12.3.3. *Suppose there exists a parameter $\alpha \in (0, 1]$ such that*

$$\max_{1\leq j\leq p} \left\{ \max_{k\in(N(j)\cup j)^c} \left\| \Sigma^*_{k,N(j)} \left(\Sigma^*_{N(j),N(j)}\right)^{-1} \right\|_1 \right\} \leq 1 - \alpha. \tag{12.9}$$

Suppose the regularization parameters for the nodewise Lasso are chosen such that $\lambda_j = \frac{c_1}{\alpha}\sqrt{\frac{\log p}{n}} + \delta$, for some $\delta \in \left[0, \frac{1}{\alpha^2 d}\right]$, and suppose the sample size satisfies $n \geq c_2 d \log p$. Then with probability at least $1 - c_3 \exp(-c_4 \alpha^2 n \delta^2)$, the estimated edge set $\widehat{E}$ based on nodewise regression satisfies $\widehat{E} \subseteq E$. Furthermore,

$$\max_{1\leq j\leq p} \|\widehat{\theta}_j - \theta_j^*\|_\infty \leq c_5 \left(\lambda + \sqrt{\frac{\log p}{n}}\right),$$

so if we also have the minimum signal strength condition

$$\min_{1\leq j\leq p} \min_{j,k\in N(j)} |(\theta_j^*)_k| > c_5 \left(\lambda + \sqrt{\frac{\log p}{n}}\right), \tag{12.10}$$

we are guaranteed that $\widehat{E} = E$.

Note that the condition (12.9) is the linear regression analog of the α-incoherence condition (12.8). Furthermore, the minimum signal strength condition (12.10) may be translated into a minimum signal strength condition on Θ^* via the relation (12.6).

Additional remarks. We have discussed several practical algorithms for edge recovery in Gaussian graphical models. A natural question concerns the optimality of the algorithms: Is it possible that an entirely different approach might also be guaranteed to produce a consistent estimate of the true edge set, say, with a smaller number of samples? Wang et al. [41] partially address this question, proving that when the elements of Θ^* exhibit some level of homogeneity, a sample size requirement scaling as $\log p$ times a polynomial in d is indeed necessary. Minimax optimal rates of estimation for the inverse covariance matrix Θ^* are discussed in detail in Cai et al. [12].

We also remark briefly on the variety of assumptions appearing in the statements of the theorems above. Incoherence conditions such as the condition (12.9), though rather restrictive, are known to be both necessary and sufficient for variable selection consistency of the Lasso [34, 40]. As noted in Meinshausen [32], the incoherence conditions required when the graphical Lasso is used for edge recovery are stronger in comparison to those required for the nodewise Lasso. An illustration of the α-incoherence condition (12.8) for several classes of graphs is included in Ravikumar et al. [35]. Cai et al. [11] also discusses the assumptions imposed in the statistical theory for the CLIME algorithm in comparison to those imposed for the graphical Lasso, noting that the CLIME algorithm does not require restrictive incoherence conditions. On the other hand, the minimum signal strength condition imposed in Theorem 12.3.2 is somewhat stronger than the minimum signal strength condition in Theorem 12.3.1, since it involves an ℓ_1-operator norm of Θ^*. Finally, note that the guarantees for the CLIME algorithm are somewhat weaker than those delivered by Lasso-type algorithms; in particular, whereas the Lasso output is guaranteed to be sparse and satisfy $\widehat{E} \subseteq E$ even without a minimum signal strength requirement, the CLIME algorithm requires both a minimum signal strength assumption and a thresholding step in order to ensure correct edge recovery.

Lastly, we comment on several aspects regarding computational complexity. Although the nodewise Lasso involves performing p separate linear regressions, in contrast to the graphical Lasso, which produces a single matrix estimator, each neighborhood regression problem (12.7) is a simple regularized quadratic program. Such programs are substantially easier to solve than the log-determinant program (12.3); furthermore, the p nodewise regressions may be run in parallel on the data set. Note, however, that if the desired output is not simply the edge set, but also an estimate of Θ^*, it may be advantageous to use a more computationally intensive matrix estimation method rather than nodewise regression. Much work has therefore been done to obtain more efficient algorithms for optimizing the graphical Lasso, including methods that exploit known graph structure [42]. State-of-the-art optimization algorithms are able to handle tens of thousands of variables in a matter of seconds [19].

12.4 Ising Models

We now shift our focus from Gaussian to discrete random variables as discussed in Chapter 8. Recall from Section 5.4.1 that the probability mass function of an Ising model, parametrized by node potentials $\{\theta_j\}_{1 \leq j \leq p}$ and edge potentials $\{\theta_{jk}\}_{(j,k) \in E}$, with $E \subseteq V \times V$, is given by

$$q(x_1, \ldots, x_p) = \frac{1}{Z} \exp\left(\sum_{j=1}^{p} \theta_j x_j + \sum_{(j,k) \in E} \theta_{jk} x_j x_k \right), \qquad \forall x \in \{-1, 1\}^p,$$

where

$$Z = Z(\theta) = \sum_{x \in \{-1,1\}^p} \exp\left(\sum_{j=1}^{p} \theta_j x_j + \sum_{(j,k) \in E} \theta_{jk} x_j x_k \right)$$

is the normalizing constant or *partition function* [21, 4]. Using similar reasoning as in the Gaussian setting, we have the relation

$$(j,k) \in E \iff \Theta_{jk} \neq 0, \qquad \forall j \neq k,$$

where $\Theta \in \mathbb{R}^{p \times p}$ is the symmetric matrix with diagonal entries equal to $\{\theta_j\}$ and off-diagonals equal to

$$\Theta_{jk} = \begin{cases} \theta_{jk} & \text{if } (j,k) \in E \\ \theta_{kj} & \text{if } (k,j) \in E \\ 0 & \text{if } (j,k),(k,j) \notin E. \end{cases}$$

Importantly, although the matrix Θ encodes the edges of the graphical model, it no longer corresponds to the inverse covariance matrix of the joint distribution.

12.4.1 Logistic regression

A straightforward calculation shows that the conditional distributions in the Ising model take the form

$$\log q(x_j \mid x_{\setminus\{j\}}) = -f\left(2\theta_j x_j + 2\sum_{k \in N(j)} \theta_{jk} x_j x_k\right), \tag{12.11}$$

where $f(t) = \frac{1}{1+\exp(t)}$ is the logistic function. This motivates a nodewise neighborhood selection method based on logistic regression. In particular, for high-dimensional graphical models, we optimize the ℓ_1-penalized logistic regression programs

$$\widehat{\theta^j} = \arg\min_{\theta \in \mathbb{R}^p} \left\{ \underbrace{\frac{1}{n}\sum_{i=1}^{n} f\left(2\theta_j x_{ij} + 2\sum_{k \in V\setminus\{j\}} \theta_{jk} x_{ij} x_{ik}\right)}_{\mathcal{L}_n(\theta)} + \lambda_j \sum_{k \in V\setminus\{j\}} |\theta_{jk}| \right\}. \tag{12.12}$$

Here, the minimization is over vectors θ with one coordinate denoted by θ_j and $p-1$ coordinates denoted by $\{\theta_{jk} : k \in V\setminus\{j\}\}$, and the ℓ_1-penalty encourages sparsity. The estimated neighborhood sets are given by

$$\widehat{N(j)} = \text{supp}(\widehat{\theta^j}), \qquad \forall 1 \leq j \leq p,$$

and we combine our neighborhood estimates into an estimate $\widehat{E}$ of the edge set via an AND or OR rule, just as in the case of the nodewise Lasso for Gaussian graphical models. Note that the programs (12.12) are all convex, and may be solved efficiently (e.g., see Koh et al. [23] for an interior-point implementation).

Statistical theory. We now describe a result providing statistical guarantees for the nodewise logistic regression algorithm. Let $\{(\theta^j)^\star\}_{1 \leq j \leq p}$ denote the true parameter vectors, where

$$(\theta^j)^\star = (\theta_j^*; \theta_{jk}^* : k \in V\setminus\{j\}) \in \mathbb{R}^p.$$

The result below assumes an incoherence condition on the Fisher information matrix $J = \nabla^2 \mathcal{L}_n(\theta^*)$:

$$\max_{1 \leq j \leq p} \left\{ \max_{k \notin N(j)} \left\| J_{k,N(j)} \left(J_{N(j),N(j)}\right)^{-1} \right\|_1 \right\} \leq 1 - \alpha, \quad \text{for some } \alpha \in (0,1]. \tag{12.13}$$

This is the logistic regression analog to the earlier incoherence condition (12.9) for linear regression. We then have the following result:

Theorem 12.4.1 (Ravikumar et al. [36]). *Suppose the true parameters of the Ising model satisfy the incoherence condition* (12.13). *If the regularization parameters of the nodewise logistic regression program are chosen such that* $\lambda_j = \frac{c_1}{\alpha}\sqrt{\frac{\log p}{n}} + \delta$*, for some* $\delta \in [0,1]$*, and the sample size satisfies* $n \geq c_2 d^3 \log p$*, then with probability at least* $1 - c_3 \exp(-c_4 n \delta^2)$*, the estimated edge set* $\widehat{E}$ *based on nodewise logistic regression satisfies* $\widehat{E} \subseteq E$*. Furthermore,*

$$\max_{1 \leq j \leq p} \|\widehat{\theta^j} - (\theta^j)^\star\|_\infty \leq c_5 \lambda \sqrt{d},$$

so if we also have the minimum signal strength condition

$$\min_{1 \leq j \leq p} \left\{ \min_{k \in N(j)} |(\theta^j)^\star_k| \right\} > c_5 \lambda \sqrt{d},$$

we are guaranteed that $\widehat{E} = E$.

Santhanam and Wainwright [38] establish lower bounds for edge recovery in Ising models, showing that when $n \leq d^2 \log p$, no method for edge recovery can succeed with probability greater than $\frac{1}{2}$. Bento and Montanari [5] also study necessary conditions for edge recovery in Ising models in terms of the behavior of long-range correlations.

12.4.2 Other methods

Several alternative methods have been proposed for edge recovery in Ising models. We mention two such methods here.

Approximate sparse maximum likelihood. Banerjee et al. [3] suggest using a convex relaxation of the ℓ_1-penalized maximum likelihood objective to estimate the parameter matrix Θ of an Ising model. In particular, rather than solving the exact penalized log likelihood problem

$$\arg\min_{\Theta \in \mathbb{R}^{p \times p}} \left\{ \frac{1}{n} \sum_{i=1}^n \left(\sum_{j=1}^p \Theta_{jj} x_{ij} + \sum_{j<k} \Theta_{jk} x_{ij} x_{ik} \right) - \log Z(\Theta) + \lambda \sum_{j<k} |\Theta_{jk}| \right\}, \quad (12.14)$$

they propose transforming the program (12.14) into a log-determinant program akin to the graphical Lasso formulation (12.3) using a variational upper bound on $\log Z(\Theta)$. Since the specific form of the convex relaxation is rather complicated, we refer the reader to the paper [3] for further details.

Pseudolikelihood. Höfling and Tibshirani [17] propose a different algorithm for estimating all entries of Θ simultaneously. They solve the optimization program

$$\widehat{\Theta} = \arg\min_{\Theta \in \mathbb{R}^{p \times p} : \Theta = \Theta^T} \left\{ \mathcal{L}_n^{\text{pseudo}}(\Theta) + \sum_{j,k} \lambda_{jk} |\Theta_{jk}| \right\}, \quad (12.15)$$

where the pseudolikelihood loss (cf. Besag [7]) is given by the sum of log conditional probabilities across all nodes:

$$\mathcal{L}_n^{\text{pseudo}}(\Theta) = \frac{1}{n} \sum_{i=1}^n \left(\sum_{j=1}^p \log q(x_{ij} \mid x_{i,\setminus\{j\}}) \right).$$

Due to the simple form of individual conditional distributions (12.11), the composite program (12.15) may be solved efficiently. Speed and accuracy comparisons for the pseudolikelihood method and other proposed edge recovery algorithms are provided in Höfling and Tibshirani [17].

12.5 Generalizations and Extensions

Having presented the core theory on neighborhood selection in Gaussian and Ising models, we now discuss how the aforementioned techniques have been extended to edge recovery in other classes of graphical models, as well.

12.5.1 Nonparanormal distributions

In a series of papers [27, 26], Liu et al. presented the notion of a *nonparanormal* distribution, which generalizes the multivariate Gaussian distribution. We outline the ideas here—more details will be given in Chapter 13. A random vector $(X_1, \dots, X_p)$ is said to have a nonparanormal distribution if a set of monotone, univariate functions $f = \{f_1, \dots, f_p\}$ exists, such that $f(X) := (f_1(X_1), \dots, f_p(X_p)) \sim N(0, \Sigma)$, for some $\Sigma \in \mathbb{R}^{p \times p}$. Without loss of generality, we may rescale f so that Σ is a correlation matrix (i.e., $\operatorname{diag}(\Sigma) = 1$).

The key insight is that for $j \neq k$, we have the relation

$$X_j \perp\!\!\!\perp X_k \mid X_{\setminus\{j,k\}} \iff \Theta_{jk} \neq 0,$$

where $\Theta = \Sigma^{-1}$. This may be derived, as in the Gaussian case, from factorizing the nonparanormal density:

$$q(x_1, \dots, x_p) = \frac{1}{(2\pi)^{p/2} \det(\Sigma)^{1/2}} \exp\left(-\frac{1}{2} f(x)^T \Theta f(x)\right) \prod_{j=1}^{p} |f_j'(x_j)|.$$

Hence, in order to estimate the edge set of the graphical model of X, it suffices to estimate $\operatorname{supp}(\Theta)$. Liu et al. [26] and Xue and Zou [43] propose to estimate the latter quantity using a modification of the graphical Lasso (12.3), where the covariance matrix $\widehat{\Sigma}$ is replaced by an estimate of the correlation matrix Σ of $f(X)$. Furthermore, they observe that it is possible to estimate Σ without estimating the individual transformation functions $\{f_j\}$. Indeed, rank-based correlation estimators such as Spearman's rho and Kendall's tau are invariant to monotone transformations.

Illustration. To illustrate the estimation method, we will consider the case of a nonparanormal graphical model estimator based on Spearman's rho estimator. For each pair $j \neq k$, we compute

$$\widehat{\rho}_{jk} = \frac{\sum_{i=1}^n (r_j^i - \bar{r}_j)(r_k^i - \bar{r}_k)}{\sqrt{\sum_{i=1}^n (r_j^i - \bar{r}_j)^2 \sum_{i=1}^n (r_k^i - \bar{r}_k)^2}},$$

where r_j^i denotes the rank of x_{ij} among $\{x_{1j}, \dots, x_{nj}\}$. We then define the correlation matrix estimator $\widehat{\Sigma}^S \in \mathbb{R}^{p \times p}$ according to

$$\widehat{\Sigma}_{jk}^S = \begin{cases} 2\sin\left(\frac{\pi}{6}\widehat{\rho}_{jk}\right) & \text{if } j \neq k \\ 1 & \text{if } j = k, \end{cases}$$

and optimize the program

$$\widehat{\Theta} = \arg\min_{\Theta \succ 0} \left\{ \operatorname{tr}(\widehat{\Sigma}^S \Theta) - \log\det(\Theta) + \lambda \sum_{j \neq k} |\Theta_{jk}| \right\}. \tag{12.16}$$

Statistical theory for the nonparanormal estimator (12.16), analogous to the theory for the graphical Lasso (cf. Theorem 12.3.1), may be found in Liu et al. [26] and Xue and Zou [43].

12.5.2 Augmented inverse covariance matrices

Next, we discuss a useful generalization of the matrix estimation and nodewise regression methods that applies to edge recovery in graphical models when nodes take values in a finite, discrete set. Note that this setting is completely disjoint from the nonparanormal graphical model setting described in the previous section, since no monotone, univariate transformations can transform a discrete random variable into a Gaussian. The main result is to show that although the convenient relationship (12.1) between the support of the inverse covariance matrix and the edge structure of the graphical model does *not* hold in general, the inverse covariance matrix of a suitably augmented random vector nonetheless reveals information about the underlying edge structure.

Throughout this subsection, we assume that each variable in the joint distribution $X = (X_1, \ldots, X_p)$ takes on a value in the finite set $\{0, 1, \ldots, m-1\}$. For the case where the random variables exhibit at most pairwise interactions, we have the following exponential family representation:

$$q(x_1, \ldots, x_p) \propto \exp\left(\sum_{j=1}^{p} \sum_{a=1}^{m-1} \theta_j^a 1\{x_j = a\} + \sum_{j \neq k} \sum_{a=1}^{m-1} \sum_{b=1}^{m-1} \theta_{jk}^{ab} 1\{x_j = a, x_k = b\} \right), \tag{12.17}$$

where the parameters of the model are the $\theta := \{\theta_j^a\} \cup \{\theta_{jk}^{ab}\}$, and the sufficient statistics are the corresponding indicator variables. The class of distributions described in equation (12.17) is also known as the Potts model, and includes the Ising model ($m = 2$) as a special case. In the more general case, which allows for higher-order interactions, we may express the joint distribution as

$$q(x_1, \ldots, x_p) \propto \exp\left(\sum_{C \in \mathcal{C}} \sum_{a \in a_C} \theta_C^a 1\{x_C = a\} \right), \tag{12.18}$$

where $\mathcal{C}$ corresponds to the set of (maximal and non-maximal) cliques in the graphical model, and for each $C \in \mathcal{C}$, we write $a_C = \{1, \ldots, m-1\}^{|C|}$ to denote the set of all possible nonzero configurations of variables in C. We use the shorthand $\phi_C(X)$ to denote the collection of sufficient statistics $\{1\{X_C = a\} : a \in a_C\}$ corresponding to vertices in C. For a subset $\mathcal{S} \subseteq \mathcal{C}$, we write

$$\phi_{\mathcal{S}}(X) := \bigcup_{C \in \mathcal{S}} \phi_C(X).$$

Loh and Wainwright [29] derive the following result:

Theorem 12.5.1 (Loh and Wainwright [29]). *Let $\widetilde{G} = (V, \widetilde{E})$ be a triangulation of the graphical model G corresponding to the distribution of X. Let $\widetilde{C}$ denote the set of all cliques in $\widetilde{G}$. Let $\Gamma = \left(\operatorname{cov}(\phi_{\widetilde{C}}(X))\right)^{-1}$. Then Γ is graph-structured, in the following sense:*

(a) *For any two subsets* $A, B \in \widetilde{C}$ *that are not subsets of the same maximal clique, the block* Γ_{AB} *is identically zero.*

(b) *For almost all parameters* θ *(i.e., all parameters apart from a set of Lebesgue measure zero), the entire block* Γ_{AB} *is nonzero whenever* A *and* B *belong to a common maximal clique.*

The matrix Γ above is the inverse covariance matrix of the vector $\phi_{\widetilde{C}}(X)$ of all sufficient statistics corresponding to cliques of $\widetilde{G}$. Furthermore, since the singleton vertex sets are certainly contained in $\widetilde{C}$, the vector $\phi_{\widetilde{C}}$ contains the $m-1$ indicator variables corresponding to each of these singleton vertex sets, as well as additional indicator sets. For this reason, we call Γ an augmented inverse covariance matrix.

Various algebraic manipulations involving block matrix inversion allow us to obtain corollaries of Theorem 12.5.1 that are more easily interpretable. For instance, a similar result regarding the support of the inverse covariance matrix holds if we only include sufficient statistics corresponding to singleton vertices and subsets of the separator sets of a junction tree representation of G (for more background on triangulations and junction tree representations, see Koller and Friedman [24]). Note that if $m = 2$, corresponding to the binary case, the indicator variables corresponding to singleton vertices are simply equal to the variables $\{X_j\}_{1 \le j \le p}$ themselves. Hence, we have the following corollary:

Corollary 12.5.2. *Suppose* $m = 2$ *and a junction tree representation of* G *exists with only singleton separator sets. Then the ordinary inverse covariance matrix* $\Theta = (\mathrm{cov}(X_1, \dots, X_p))^{-1}$ *satisfies the property that*

$$(j,k) \in E \iff \Theta_{jk} \neq 0, \qquad \forall j \neq k. \tag{12.19}$$

In particular, the relation (12.19) *holds when* G *is a tree.*

The population-level results mentioned in this subsection, together with the observations in Section 12.3.3 relating inverse covariance matrices to linear regression coefficients, lead to natural algorithms for estimating the edges in a discrete-valued graphical model, and analogous statistical theory may subsequently be derived. Note that the block-graph-structure of the augmented inverse covariance matrix requires using multivariate linear regression with a group Lasso penalty when performing nodewise regression. For more details on the edge recovery algorithms and corresponding statistical guarantees, we refer the reader to Loh and Wainwright [29].

12.5.3 Other exponential families

Lastly, we describe several recent results applicable to situations where the conditional distributions of individual nodes adhere to particular classes of exponential families.

Generalized linear models. Yang et al. [44] study the case where node conditional distributions take the general form

$$\begin{aligned} q(x_j \mid x_{\setminus\{j\}}) \propto \exp\Bigg(B(x_j)\Bigg(\theta_j + \sum_{k_1 \in N(j)} \theta_{jk} B(x_{k_1}) \\ + \sum_{k_1,k_2 \in N(j)} \theta_{j,k_1,k_2} B(x_{k_1}) B(x_{k_2}) + \cdots \\ + \sum_{k_1,\dots,k_m \in N(j)} \theta_{j,k_1,\dots,k_m} \prod_{\ell=2}^{m} B(x_{k_\ell}) \Bigg) + C(X_j) \Bigg), \end{aligned} \tag{12.20}$$

for some univariate functions B and C. Thus, the conditional distribution of X_j is specified by an exponential family with sufficient statistics of products of univariate functions $\{B(X_k)\}_{k\in N(j)}$, of order at most m. The class of joint distributions with corresponding conditional densities (12.20) clearly includes the Gaussian and Ising graphical models, but also includes multinomial, Poisson, and exponential graphical models.

The parameters of the exponential family distributions (12.20) may be estimated via nodewise regression, where the loss function is defined to be the negative conditional log likelihood and a (group) ℓ_1-penalty is included to encourage sparsity in the solution. The support of the estimated parameter vectors is then used to recover an estimate of the edge set. For more details, including statistical theory, we refer the reader to Yang et al. [44]. A detailed treatment of multinomial graphical models is provided in Jalali et al. [22].

Poisson graphical models. A special case of the aforementioned setting involves conditional Poisson distributions. This is explored in depth in Yang et al. [46], and is used to model multivariate count data. Specifically, Yang et al. [46] study the setting where the joint density is defined with respect to the conditional distributions (12.20), with $B(z) = z$, $C(z) = -\log(z!)$, and $m = 2$. However, such a distribution is only normalizable if $\theta_{jk} \leq 0$ for all $(j,k) \in E$ [6]. In order to capture positive conditional relationships between variables, Yang et al. [46] propose to use truncated Poisson conditionals. Hence, for a prespecified truncation level R, the conditional distributions are given by

$$q(x_j \mid x_{\setminus\{j\}}) \propto \exp\left(\theta_j x_j + \sum_{k\in N(j)} \theta_{jk} x_j x_k - \log(x_j!)\right), \quad \forall x \in \{0, 1, \ldots, R\}^p.$$

Other proposals for multivariate conditional Poisson modeling that allow variables to take on all integer values include the quadratic Poisson graphical model (QPGM), which contains a quadratic term inside the exponential in each conditional distribution [46], the sub-linear Poisson graphical model (SPGM) [46], and the square-root Poisson graphical model (SQR-PGM) [20]. As before, the edge set may then be estimated via nodewise regression with the appropriate conditional log likelihoods and an ℓ_1-penalty.

Mixed graphical models. Finally, note that the framework developed in this subsection applies equally well to cases where conditional distributions of different nodes belong to different classes of exponential families. Lee and Hastie [25] propose to optimize the pseudolikelihood of such models with a group Lasso penalty, and develop statistical and optimization theory when the conditional distribution of each node is either Gaussian or multinomial logistic, and interactions are pairwise; Cheng et al. [14] extend the analysis to third-order interactions, and study a nodewise regression algorithm. Yang et al. [45] further generalize the theoretical framework to settings where the conditional density of each node may be distinct.

12.6 Robustness

In many practical applications, cleanly-observed i.i.d. data may be unobtainable. We now turn our attention to adaptations of the neighborhood selection techniques that allow for consistent estimation of the edge set, even in the presence of contamination.

12.6.1 Noisy and missing data

Loh and Wainwright [30] develop a framework for high-dimensional errors-in-variables linear regression. Translated into the setting of graphical models, the data matrix X may be observed subject to independent additive noise, so we observe $Z = X + W$, where $W \in \mathbb{R}^{n \times p}$ is a matrix of i.i.d. noise such that $w_i \perp\!\!\!\perp x_i$. We assume the additive noise covariance $\Sigma_w = \text{cov}(w_i)$ is known. In another setting of interest, each entry of X is missing independently with a (possibly unknown) probability α.

The key insight of Loh and Wainwright [30] is that Lasso-based linear regression may be performed in the presence of noisy or missing observations by replacing particular terms in the regularized quadratic program by appropriate surrogates that take into account the sources of systematic error. Indeed, for observations from the linear model

$$y_i = x_i^T \beta^* + \epsilon_i,$$

the Lasso program is equivalent to

$$\widehat{\beta} \in \arg\min \left\{ \frac{1}{2} \beta^T \widehat{\Gamma} \beta - \widehat{\gamma}^T \beta + \lambda \|\beta\|_1 \right\},$$

where $(\widehat{\Gamma}, \widehat{\gamma}) = \left(\frac{X^T X}{n}, \frac{X^T y}{n} \right)$. In the case of additive errors-in-variables, with $Z = X + W$, we may use the alternative pair of estimators $(\widehat{\Gamma}, \widehat{\gamma}) = \left(\frac{Z^T Z}{n} - \Sigma_w, \frac{Z^T y}{n} \right)$; in the case of missing data, we may obtain appropriate estimators by rescaling $\widehat{\Gamma}$ and $\widehat{\gamma}$ by multiples of $1 - \alpha$. For details and statistical guarantees for edge recovery in Gaussian graphical models via nodewise linear regression with the corrected Lasso, see Loh and Wainwright [30].

12.6.2 Corrected graphical Lasso

The graphical Lasso program (12.3) is also readily amenable to correction in the presence of contaminated data. Indeed, the only data-dependent term in the objective function is the sample covariance matrix $\widehat{\Sigma}$; as discussed in the previous subsection, if variables are observed subject to independent additive perturbations with covariance matrix Σ_w, we may replace $\widehat{\Sigma}$ by the natural estimator $\widetilde{\Sigma} = \frac{Z^T Z}{n} - \Sigma_w$. Furthermore, the corrected graphical Lasso is still a convex program of essentially the same form as before, and analogous theoretical results may be derived concerning the consistency of the overall edge recovery algorithm.

In fact, the adjusted form of the graphical Lasso mentioned previously in our discussion of nonparanormal graphical models (12.16) also has attractive robustness properties. As demonstrated empirically by Liu et al. [26] and rigorized by Loh and Tan [28], the graphical Lasso estimator with a rank-based correlation matrix estimator is robust to cellwise contamination. Rather than applying only to a systematic contamination mechanism of a specific form (e.g., additive noise or missing data), the adjusted graphical Lasso (12.16) is guaranteed to produce an output with a bounded influence function, when measured in terms of the elementwise ℓ_∞-norm.

12.6.3 Latent variables

A final type of nonideality arises when joint observations are only available for a subset of variables. Chandrasekaran et al. [13] analyze the setting where $X = (X_1, \ldots, X_p) \sim N(0, \Sigma)$, and the set of vertices $V = \{1, \ldots, p\}$ is partitioned into subsets O and H of observed and hidden variables, respectively. The goal is to recover the edge structure of the graphical model on p nodes, restricted to the subset of observed nodes. In other words, we wish to estimate

$$E_O = \{(j,k) : j,k \in O, (j,k) \in E\}.$$

Although the observed variables X_O are also jointly Gaussian, simply performing nodewise regression or optimizing the graphical Lasso on the observed data matrix would be inconsistent for E_O, since we would instead obtain an estimate of the edge set of the *marginalized* distribution.

Writing the inverse covariance matrix $\Theta = \Sigma^{-1}$ in block form, we have

$$\Theta = \begin{pmatrix} \Theta_{OO} & \Theta_{OH} \\ \Theta_{HO} & \Theta_{HH} \end{pmatrix}.$$

By block matrix inversion, we then have the relation

$$(\Sigma_{OO})^{-1} = \underbrace{\Theta_{OO}}_{S} - \underbrace{\Theta_{OH}\,(\Theta_{HH})^{-1}\,\Theta_{HO}}_{L}. \tag{12.21}$$

Assuming the underlying graphical model is sparse, the submatrix $S = \Theta_{OO}$ is also sparse. Furthermore, the rank of the matrix product $L = \Theta_{OH}\,(\Theta_{HH})^{-1}\,\Theta_{HO}$ is at most $|H|$. Hence, if we assume that the number of hidden variables is relatively small, equation (12.21) provides a decomposition of the matrix $(\Sigma_{OO})^{-1}$ into a sum of sparse and low-rank components.

Using these ideas, Chandrasekaran et al. [13] propose the following estimator, which outputs a pair of estimates $(\widehat{S}, \widehat{L})$ for the matrices (S, L):

$$(\widehat{S}, \widehat{L}) = \arg\min_{S-L\succ 0, L\succeq 0} \left\{ \operatorname{tr}\left((S-L)\widehat{\Sigma}_{OO}\right) - \log\det(S-L) \right.$$
$$\left. + \lambda\left(\gamma \sum_{j\neq k} |S_{jk}| + \operatorname{tr}(L)\right)\right\}. \tag{12.22}$$

The objective function appearing in the program (12.22) is a penalized version of the Gaussian log likelihood over the observed variables, where the form of the penalty term simultaneously encourages S to be sparse and L to be low-rank. The sparsity pattern of S is then used to estimate the edge set E_O, as in the case of the usual graphical Lasso. For additional details and statistical theory, see Chandrasekaran et al. [13].

12.7 Further Reading

Due to space limitations, we have necessarily been selective in our choice of topics concerning neighborhood estimation. Another fascinating branch of work, which is mostly orthogonal to the preceding exposition but may be of interest to the reader, involves conditional independence testing. We remark briefly here.

A natural (but potentially computationally expensive) proposal for recovering the edge set of a graph is to determine whether each of the $\binom{p}{2}$ possible edges is present in the graph. Note that by the Markov property, we could check whether $X_j \perp\!\!\!\perp X_k \mid X_{\backslash\{j,k\}}$. However, if the graph is known to be d-sparse, it suffices to test whether $X_j \perp\!\!\!\perp X_k \mid X_S$ for every subset $S \subseteq V \backslash \{j,k\}$ such that $|S| \le d$; the edge (j,k) is included if and only if X_j and X_k are not found to be conditionally independent with respect to any subset of cardinality at most d. This avoids the large sample complexity requirement of a statistical test involving conditioning on all $p-2$ variables. (Note the similarity between this algorithm and the PC algorithm for estimation in directed graphical models [39], and an analogous argument may be used to show that this procedure outputs the correct undirected graphical model under a faithfulness assumption.)

More efficient algorithms have been devised to speed up the process of conditional independence testing to be faster than $\Theta(p^d)$ time, when the graph possesses certain structural properties or the distribution exhibits correlation decay [10]. The interested reader should refer to the papers by Anandkumar et al. [2] and Bresler [9] for efficient estimation in Ising models, and Anandkumar et al. [1] for efficient estimation in Gaussian graphical models via conditional covariance thresholding.

Bibliography

[1] A. Anandkumar, V. Y. F. Tan, F. Huang, and A. S. Willsky. High-dimensional Gaussian graphical model selection: Walk summability and local separation criterion. *Journal of Machine Learning Research*, 13(Aug):2293–2337, 2012.

[2] A. Anandkumar, V. Y. F. Tan, F. Huang, and A. S. Willsky. High-dimensional structure estimation in Ising models: Local separation criterion. *The Annals of Statistics*, 1346–1375, 2012.

[3] O. Banerjee, L. El Ghaoui, and A. d'Aspremont. Model selection through sparse maximum likelihood estimation for multivariate Gaussian or binary data. *Journal of Machine Learning Research*, 9(Mar):485–516, 2008.

[4] R. J. Baxter. *Exactly Solved Models in Statistical Mechanics.* Dover Books on Physics. Dover Publications, 2007.

[5] J. Bento and A. Montanari. Which graphical models are difficult to learn? In *Advances in Neural Information Processing Systems*, 1303–1311, 2009.

[6] J. Besag. Spatial interaction and the statistical analysis of lattice systems. *Journal of the Royal Statistical Society. Series B (Methodological)*, 192–236, 1974.

[7] J. Besag. Statistical analysis of non-lattice data. *The Statistician*, 179–195, 1975.

[8] S. Boyd and L. Vandenberghe. *Convex Optimization.* Cambridge University Press, New York, 2004.

[9] G. Bresler. Efficiently learning Ising models on arbitrary graphs. In *Proceedings of the Forty-Seventh Annual ACM Symposium on Theory of Computing*, 771–782. ACM, 2015.

[10] G. Bresler, E. Mossel, and A. Sly. Reconstruction of Markov random fields from samples: Some observations and algorithms. In *Approximation, Randomization and Combinatorial Optimization. Algorithms and Techniques*, 343–356. Springer, 2008.

[11] T. Cai, W. Liu, and X. Luo. A constrained ℓ_1 minimization approach to sparse precision matrix estimation. *Journal of the American Statistical Association*, 106:594–607, 2011.

[12] T. T. Cai, Z. Ren, and H. H. Zhou. Estimating structured high-dimensional covariance and precision matrices: Optimal rates and adaptive estimation. *Electronic Journal of Statistics*, 10(1):1–59, 2016.

[13] V. Chandrasekaran, P. A. Parrilo, and A. S. Willsky. Latent variable graphical model selection via convex optimization. *The Annals of Statistics*, 40(4):1935–1967, 2012.

[14] J. Cheng, T. Li, E. Levina, and J. Zhu. High-dimensional mixed graphical models. *arXiv:1304.2810*, 2013.

[15] A. d'Aspremont, O. Banerjee, and L. El Ghaoui. First order methods for sparse covariance selection. *SIAM Journal on Matrix Analysis and Its Applications*, 30(1):55–66, 2008.

[16] J. Friedman, T. Hastie, and R. Tibshirani. Sparse inverse covariance estimation with the graphical Lasso. *Biostatistics*, 9(3):432–441, July 2008.

[17] H. Höfling and R. Tibshirani. Estimation of sparse binary pairwise Markov networks using pseudo-likelihoods. *Journal of Machine Learning Research*, 10(Apr):883–906, 2009.

[18] R. A. Horn and C. R. Johnson. *Matrix Analysis.* Cambridge University Press, 1990.

[19] C.-J. Hsieh, M. A. Sustik, I. S. Dhillon, and P. Ravikumar. QUIC: Quadratic approximation for sparse inverse covariance estimation. *Journal of Machine Learning Research*, 15:2911–2947, 2014.

[20] D. Inouye, P. Ravikumar, and I. Dhillon. Square root graphical models: Multivariate generalizations of univariate exponential families that permit positive dependencies. In *Proceedings of The 33rd International Conference on Machine Learning*, 2445–2453, 2016.

[21] E. Ising. Beitrag zur theorie des ferromagnetismus. *Zeitschrift für Physik*, 31(1):253–258, 1925.

[22] A. Jalali, P. K. Ravikumar, V. Vasuki, and S. Sanghavi. On learning discrete graphical models using group-sparse regularization. In *AISTATS*, 378–387, 2011.

[23] K. Koh, S.-J. Kim, and S. Boyd. An interior-point method for large-scale ℓ_1-regularized logistic regression. *Journal of Machine Learning Research*, 8(Jul):1519–1555, 2007.

[24] D. Koller and N. Friedman. *Probabilistic Graphical Models: Principles and Techniques.* MIT Press, 2009.

[25] J. D. Lee and T. J. Hastie. Learning the structure of mixed graphical models. *Journal of Computational and Graphical Statistics*, 24(1):230–253, 2015.

[26] H. Liu, F. Han, M. Yuan, J. Lafferty, and L. Wasserman. High-dimensional semiparametric Gaussian copula graphical models. *The Annals of Statistics*, 2293–2326, 2012.

[27] H. Liu, J.D. Lafferty, and L.A. Wasserman. The nonparanormal: Semiparametric estimation of high dimensional undirected graphs. *Journal of Machine Learning Research*, 10:2295–2328, 2009.

[28] P. Loh and X. L. Tan. High-dimensional robust precision matrix estimation: Cellwise corruption under ϵ-contamination. *arXiv:1509.07229*, 2015.

[29] P. Loh and M. J. Wainwright. Structure estimation for discrete graphical models: Generalized covariance matrices and their inverses. *The Annals of Statistics*, 41(6):3022–3049, 2013.

[30] P. Loh and M.J. Wainwright. High-dimensional regression with noisy and missing data: Provable guarantees with non-convexity. *Annals of Statistics*, 40(3):1637–1664, 2012.

[31] R. Mazumder and T. Hastie. The graphical Lasso: New insights and alternatives. *Electronic Journal of Statistics*, 6:2125–2149, 2012.

[32] N. Meinshausen. A note on the Lasso for Gaussian graphical model selection. Statistics & Probability Letters, 78(7):880–884, 2008.

[33] N. Meinshausen and P. Bühlmann. High-dimensional graphs and variable selection with the Lasso. *Annals of Statistics*, 34:1436–1462, 2006.

[34] N. Meinshausen and B. Yu. Lasso-type recovery of sparse representations for high-dimensional data. *Annals of Statistics*, 37(1):246–270, 2009.

[35] P. Ravikumar, M. J. Wainwright, G. Raskutti, and B. Yu. High-dimensional covariance estimation by minimizing ℓ_1-penalized log-determinant divergence. *Electronic Journal of Statistics*, 4:935–980, 2011.

[36] P. Ravikumar, M.J. Wainwright, and J.D. Lafferty. High-dimensional Ising model selection using ℓ_1-regularized logistic regression. *Annals of Statistics*, 38:1287, 2010.

[37] A. J. Rothman, P. J. Bickel, E. Levina, and J. Zhu. Sparse permutation invariant covariance estimation. *Electronic Journal of Statistics*, 2:494–515, 2008.

[38] N. P. Santhanam and M. J. Wainwright. Information-theoretic limits of selecting binary graphical models in high dimensions. *IEEE Transactions on Information Theory*, 58(7):4117–4134, 2012.

[39] P. Spirtes, C. N. Glymour, and R. Scheines. *Causation, Prediction, and Search.* MIT press, 2000.

[40] M. J. Wainwright. Sharp thresholds for high-dimensional and noisy sparsity recovery using ℓ_1-constrained quadratic programming (Lasso). *IEEE Transactions on Information Theory*, 55(5):2183–2202, May 2009.

[41] W. Wang, M. J. Wainwright, and K. Ramchandran. Information-theoretic bounds on model selection for Gaussian Markov random fields. In *ISIT*, 1373–1377, 2010.

[42] D. M. Witten, J. H. Friedman, and N. Simon. New insights and faster computations for the graphical Lasso. *Journal of Computational and Graphical Statistics*, 20(4):892–900, 2011.

[43] L. Xue and H. Zou. Regularized rank-based estimation of high-dimensional nonparanormal graphical models. *Annals of Statistics*, 40(5):2541–2571, 2012.

[44] E. Yang, G. Allen, Z. Liu, and P. K. Ravikumar. Graphical models via generalized linear models. In *Advances in Neural Information Processing Systems*, 1358–1366, 2012.

[45] E. Yang, Y. Baker, P. Ravikumar, G. I. Allen, and Z. Liu. Mixed graphical models via exponential families. In *AISTATS*, 1042–1050, 2014.

[46] E. Yang, P. K. Ravikumar, G. I. Allen, and Z. Liu. On Poisson graphical models. In *Advances in Neural Information Processing Systems*, 1718–1726, 2013.

[47] M. Yuan and Y. Lin. Model selection and estimation in the Gaussian graphical model. *Biometrika*, 94(1):19–35, 2007.

[48] P. Zhao and B. Yu. On model selection consistency of Lasso. *Journal of Machine Learning Research*, 7:2541–2567, 2006.

13

Nonparametric Graphical Models

Han Liu

Department of Electrical Engineering and Computer Science, Northwestern University

John Lafferty

Department of Statistics and Data Science, Yale University

CONTENTS

13.1 Introduction

This chapter introduces different approaches to graphical modeling for continuous and mixed data, using semiparametric techniques that make weak assumptions compared with the default Gaussian graphical model that was reviewed in Chapter 9.

Recalling the standard definition of an undirected graphical model from Section 1.7, if $X = (X_1, \ldots, X_d)$ is a random vector with distribution P, the *undirected graph* $G = (V, E)$ corresponding to P consists of a vertex set V and an edge set E where V has d elements, one for each variable X_i. The edge between (i, j) is excluded from E if and only if X_i is independent of X_j given the other variables $X_{\backslash\{i,j\}} \equiv (X_s : \ 1 \leq s \leq d, \ \ s \neq i, j)$, written

$$X_i \amalg X_j \,\Big|\, X_{\backslash\{i,j\}}. \tag{13.1}$$

The graph G is also called the *conditional independence graph*.

The general form for a strictly positive probability density encoded by an undirected graph is

$$p(x) = \frac{1}{Z(\psi)} \exp\left(\sum_{C \in \mathcal{C}(G)} \psi_C(x_C) \right) = \prod_{C \in \mathcal{C}(G)} \phi_C(X_C) \tag{13.2}$$

where the sum is over all cliques, or fully connected subsets of vertices of the graph; the factors $\phi_C(X_C) > 0$ are sometimes referred to as *potential functions.* This is the graphical model analogue of the general nonparametric regression model. However, without further assumptions, it is too general to be practical, since it evidently requires access to the *normalizing constant* $Z(\psi)$; which cannot in general be efficiently computed or approximated.

In this chapter we outline four different ways of making restrictions on the model that lead to computationally tractable models with favorable statistical properties. The models we present are expressed in terms of *edge potentials* and *vertex potentials* as

$$p(x) = \frac{1}{Z(\psi)} \exp\Big(\underbrace{\sum_{(i,j) \in E(G)} \psi_{ij}(x_i, x_j)}_{\text{interaction terms}} + \underbrace{\sum_{i \in V(G)} \psi_i(x_i)}_{\text{main effects}} \Big) = \prod_{(i,j) \in E(G)} \phi_{i,j}(x_i, x_j) \prod_{i \in V(G)} \phi_i(x_i) \tag{13.3}$$

where $Z(\psi)$ is the partition function that ensures $p(x)$ is a valid density. The univariate terms $\psi_i(x_i)$ can be thought of as representing the *main effects*, while the bivariate terms $\psi_{ij}(x_i, x_j)$ represent *interaction terms.* In Section 13.2 we describe a family of semiparametric graphical models, called *exponential family graphical models.* In terms of the potential function representation (13.3) this family uses linear edge potentials and general vertex potentials. That is, $\psi_{ij} = \beta_{ij} x_i x_j$ while $\psi_i(x_i)$ is a function subject only to regularity assumptions. Section 13.3 describes the case where both the edge and vertex potentials are nonparametric, but where tractability is achieved by assuming that the conditional independence graph has no cycles, leading to tree-structured graphical models. Section 13.4 presents variants of *copula graphical models.* In the case of the Gaussian copula or *nonparanormal*, the interaction terms take the form $\psi_{ij}(x_i, x_j) = \beta_{ij} f_i(x_i) f_j(x_j)$, where the functions f_i are strictly increasing. The main effects are then given by $\psi_i(x_i) = \beta_i f_i(x_i)^2 + \log f_i'(x_i)$ where f_i' is the derivative. As described in Section 13.4, it's also possible to combine trees and copulas in natural ways. Section 13.5 presents an approach based on pairwise tensor products of smoothing splines, a type of log-density ANOVA model where the computational bottleneck of computing the normalizing constant is circumvented by using a surrogate loss function called *score matching.*

As summarized in Figure 13.1, these approaches make different assumptions and employ different regularization and computational mechanisms to make the pairwise Markov random field model tractable. Specifically, the nonparanormal model allows arbitrary graphs, but makes a distributional restriction through the use of Gaussian copula. This can be seen as imposing a form of *marginal regularization*, requiring a set of univariate strictly increasing (possibly nonsmooth) functions $\{f_1, \ldots, f_d\}$ such that the main effects are $\psi_j(x_j) = c_j f_j^2(x_j) + \log f_j'(x_j)$ and the pairwise interactions are $\psi_{k\ell}(x_k, x_\ell) = c_{k\ell} f_k(x_k) f_\ell(x_\ell)$. The semiparametric exponential family graphical model imposes a type of *interaction regularization* that allows arbitrary (even nonsmooth) main effects ψ_j while constraining the pairwise interactions to be $\psi_{k\ell}(x_k, x_\ell) = \beta_{k\ell} x_k x_\ell$. The forest graphical model imposes a type of *graph regularization* that permits arbitrary Hölder smooth ψ_j and $\psi_{k\ell}$ but restricts the graphs to trees and forests, with no cycles. Finally, the tensor-product smoothing spline ANOVA model exploits a form of *smoothness regularization* that allows arbitrary graphs, while restricting ψ_j and $\psi_{k\ell}$ to lie in tensor-product smoothing spline spaces. Each of these approaches enables the construction of graphical models having nearly the same computa-

	Nonparanormal	*Forests*	*Exponential family*	*Log-density ANOVA*
type	semiparametric	nonparametric	semiparametric	nonparametric
regularization	marginal	graph structure	interaction	smoothness
computation	glasso	greedy	pseudo-likelihood	score matching
statistical tools	rank correlation	entropy concentration	GLMs	smoothing splines

FIGURE 13.1: Overview of the different nonparametric graphical models described here. Four basic models are presented, each making different assumptions and employing different techniques to gain computational and statistical advantages.

tional and statistical efficiency as the classical Gaussian graphical model, but under much weaker and more flexible assumptions.

13.2 Semiparametric Exponential Family Graphical Models

The semiparametric exponential family graphical model specifies the joint distribution of $X = (X_1, \ldots, X_d)^T$ in terms of the conditional distribution of X_j given

$$X_{\setminus j} := (X_1, \ldots, X_{j-1}, X_{j+1}, \ldots, X_d)^T \tag{13.4}$$

for each $j \in V$. Specifically, the conditional density takes an exponential form

$$p(x_j \mid x_{\setminus j}) = \exp\bigl(\eta_j(x_{\setminus j}) \cdot x_j + f_j(x_j) - b_j(\eta_j, f_j)\bigr), \tag{13.5}$$

where $\eta_j(x_{\setminus j}) = \alpha_j + \sum_{k \neq j} \beta_{jk} x_k$ is the canonical parameter, $f_j(\cdot)$ is an unknown base measure function and $b_j(\cdot, \cdot)$ is the log-partition function, with $x_{\setminus j} = (x_1, \ldots, x_{j-1}, x_{j+1}, \ldots, x_d)$. To make the model identifiable, we set $\alpha_j = 0$ and absorb the term $\alpha_j x_j$ into $f_j(x_j)$. By the Hammersley-Clifford theorem (Observation 1.7.2 in Chapter 1), it is seen that $\beta_{jk} = 0$ if and only if X_j and X_k are conditionally independent given $\{X_\ell \colon \ell \neq j, k\}$. Therefore, we set $(j, k) \in E$ if and only if $\beta_{jk} \neq 0$. The model in (13.5) is semiparametric, treating both $\beta_j = (\beta_{j1}, \ldots, \beta_{jj-1}, \beta_{jj+1}, \ldots, \beta_{jd})^T \in \mathbb{R}^{d-1}$ and the univariate function $f_j(\cdot)$ as parameters. This model can also be used for mixed data, for which X contains both continuous and discrete random variables.

Because the model in (13.5) is only specified by the conditional distribution of each variable, it is important to understand the conditions under which a valid joint distribution of X exists. As shown by [2], sufficient conditions for the existence of joint distribution of X are that $\beta_{jk} = \beta_{kj}$ for $1 \leq j, k \leq d$ and that $g(x) := \exp\{\sum_{j<k} \beta_{jk} x_j x_k + \sum_{j=1}^d f_j(x_j)\}$ is integrable. Under these conditions, there exists a joint probability distribution for the model defined in (13.5), with density of the form

$$p(x) = \exp\Bigl(\sum_{k<\ell} \beta_{k\ell} x_k x_\ell + \sum_{j=1}^{d} f_j(x_j) - A\bigl(\{\beta_i, f_i\}_{i \in [d]}\bigr)\Bigr), \tag{13.6}$$

where $\beta_{k\ell} \neq 0$ if and only if $(k, \ell) \in E$. The log-partition function $A(\cdot)$ is defined as

$$A\big(\{\beta_i, f_i\}_{i\in[d]}\big) := \log\Big(\int_{\mathbb{R}^d} \exp\Big(\sum_{k<\ell} \beta_{k\ell} x_k x_\ell + \sum_{j=1}^{d} f_j(x_j)\Big)\nu(\mathrm{d}x)\Big), \tag{13.7}$$

where $\nu(\cdot)$ is the corresponding probability measure. See also [26].

13.2.1 Examples

Many widely used parametric graphical model families are special cases of this model.

Gaussian graphical models: Under the Gaussian graphical model $X \in \mathbb{R}^d$ follows a multivariate Gaussian distribution $N(\mathbf{0}, \Theta^{-1})$ where $\Theta \in \mathbb{R}^{d\times d}$ is the precision matrix, satisfying $\Theta_{jj} = 1$ for $j \in [d]$. The conditional distribution of X_j given $X_{\backslash j}$ satisfies

$$X_j \mid X_{\backslash j} = \alpha_j^T X_{\backslash j} + \epsilon_j \quad \text{with} \quad \epsilon_j \sim N(0, 1),$$

where $\alpha_j = \Theta_{\backslash j,j}$. The conditional density is given by

$$p(x_j \,|\, x_{\backslash j}) = \sqrt{1/(2\pi)}\exp\big[-x_j(\Theta_{\backslash j,j}^T x_{\backslash j}) - 1/2 \cdot x_j^2 - 1/2 \cdot (\Theta_{\backslash j,j}^T x_{\backslash j})^2\big].$$

Considering (13.5), we obtain $\beta_j = -\Theta_{\backslash j,j}$, $f_j(x) = -x^2/2$ and $b_j(\beta_j, f_j) = (\Theta_{\backslash j,j}^T x_{\backslash j})^2/2 + \log(2\pi)/2$.

Ising models: In an Ising model with no external field, X takes values in $\{0, 1\}^d$ and the joint probability mass function is given by $p(x) \propto \exp\{\sum_{j<k} \theta_{jk} x_j x_k\}$. The conditional distribution of X_j given $X_{\backslash j}$ is of the form

$$p(x_j \,|\, x_{\backslash j}) = \frac{\exp\{\sum_{k<\ell} \theta_{k\ell} x_k x_\ell\}}{\sum_{x_j\in\{0,1\}} \exp\{\sum_{k<\ell} \theta_{k\ell} x_k x_\ell\}} = \exp\Big\{x_j(\theta_j^T x_{\backslash j}) - \log\big[1 + \exp(\theta_j^T x_{\backslash j})\big]\Big\}$$

where $\theta_j = (\theta_{j1}, \ldots, \theta_{j,j-1}, \theta_{j,j+1}, \ldots, \theta_{jd})^T$. In this case we have $\beta_j = \theta_j$, $f_j(x) = 0$ and $b_j(\beta_j, f) = \log[1 + \exp(\beta_j^T x_{\backslash j})]$.

Exponential graphical models: For exponential graphical models, X takes values in $[0, +\infty)^d$ and the joint probability density satisfies $p(x) \propto \exp\{-\sum_{i=1}^d \phi_i x_i - \sum_{k<\ell} \theta_{k\ell} x_k x_\ell\}$. In order to ensure this probability distribution is normalizable, we require that $\phi_j > 0, \theta_{jk} \geq 0$ for all $j, k \in [d]$. The conditional probability density of X_j given $X_{\backslash j}$ is

$$\begin{aligned} p(x_j \,|\, x_{\backslash j}) &= \exp\Big\{-\sum_{i=1}^{d} \phi_i x_i - \sum_{k<\ell} \theta_{k\ell} x_k x_\ell\Big\} \Big/ \int_{x_j\geq 0} \exp\Big\{-\sum_{i=1}^{d} \phi_i x_i - \sum_{k<\ell} \theta_{k\ell} x_k x_\ell\Big\} \mathrm{d}x_j \\ &= \exp\Big\{-x_j(\phi_j + \theta_j^T x_{\backslash j}) - \log(\phi_j + \theta_j^T x_{\backslash j})\Big\}. \end{aligned}$$

In this case $\beta_j = -\theta_j$, $f_j(x) = -\phi_j x$ and $b_j(\beta_j, f_j) = \log(\beta_j^T x_{\backslash j} + \phi_j)$.

Poisson graphical models: In a Poisson graphical model, every node X_j is a discrete random variable taking values in $\mathbb{N} = \{0, 1, 2, \ldots\}$. The joint probability mass function is given by

$$p(x) \propto \exp\Big\{\sum_{j=1}^{d} \phi_j x_j - \sum_{j=1}^{d} \log(x_j!) + \sum_{k<\ell} \theta_{k\ell} x_k x_\ell\Big\}.$$

As for exponential graphical models, restrictions on the parameters are needed to ensure that the probability mass function is normalizable. Here we require that $\theta_{jk} \leq 0$ for all $j, k \in [d]$. The conditional probability mass function of X_j given $X_{\setminus j}$ is then

$$p(x_j \,|\, x_{\setminus j}) = \exp\Big\{x_j(\theta_j^T x_{\setminus j}) + \phi_j x_j - \log(x_j!) - b_j(\theta_j, f_j)\Big\}.$$

Thus, $\beta_j = \theta_j$, $f_j(x) = \phi_j x - \log(x!)$ and $b_j(\beta_j, f_j) = \log\big\{\sum_{y=0}^{\infty} \exp\big[y(\beta_j^T x_{\setminus j}) + f_j(y)\big]\big\}$.

13.2.2 A nuisance-free loss function

Let $X_1, \ldots, X_n$ be n i.i.d. random samples from the distribution of X. To fit the model (13.5) to the data, we treat β_j as the parameter of interest and the base measures $f_j(\cdot)$ as nuisance parameters. Due to the presence of $f_j(\cdot)$, finding the conditional maximum likelihood estimator of β_j is intractable. [18] address this problem by using a pseudo-likelihood loss function based on pairwise local order statistics that is invariant to changes in the nuisance parameters.

Let $x_1, x_2, \ldots, x_n$ be n data points that are realizations of $X_1, X_2, \ldots, X_n$. For any pair of data indices i and i', let $\mathcal{A}_{ii'}^j := \big\{(X_{ij}, X_{i'j}) = (x_{ij}, x_{i'j}), X_{i\setminus j} = x_{i\setminus j}, X_{i'\setminus j} = x_{i'\setminus j}\big\}$ be the event where we observe $X_{i\setminus j}$ and $X_{i'\setminus j}$ and the order statistics of X_{ij} and $X_{i'j}$, but not the relative ranks of X_{ij} and $X_{i'j}$. In other words, we know the values of a pair of observations x_i and $x_{i'}$, but we do not know whether $x_{ij} > x_{i'j}$ or $x_{ij} < x_{i'j}$. [18] show that

$$\mathbb{P}\big(X_i = x_i, X_{i'} = x_{i'} \big| \mathcal{A}_{ii'}^j\big) = \Big\{1 + \exp\big[-(x_{ij} - x_{i'j})\beta_j{}^T(x_{i\setminus j} - x_{i'\setminus j})\big]\Big\}^{-1},$$

which is free of the nuisance function $f_j(\cdot)$. Defining $R_{ii'}^j(\beta_j) := \exp[-(x_{ij} - x_{i'j})\beta_j{}^T(x_{i\setminus j} - x_{i'\setminus j})]$, the pseudo-likelihood loss function

$$L_j(\beta_j) := \frac{2}{n(n-1)} \sum_{1 \leq i < i' \leq n} \log\big[1 + R_{ii'}^j(\beta_j)\big]. \tag{13.8}$$

only involves β_j and resembles a logistic loss function.

To obtain sparse estimates, a penalty is added to the loss function, leading to the optimization

$$\widehat{\beta}_j = \underset{\mathbb{R}^{d-1}}{\arg\min}\, L_j(\beta_j) + \sum_{k \neq j} p_\lambda(|\beta_{jk}|), \tag{13.9}$$

where λ is a tuning parameter and $p_\lambda(\cdot) : [0, +\infty) \to [0, +\infty)$ is a penalty function, required to satisfy some regularity conditions; see for example [4]. The optimization problem in (13.9) is in general nonconvex and may have multiple local solutions. Several strategies have been studied to approach this, including local linear approximation [30, 4] and multi-stage convex relaxations [28, 29]. The loss function $L_j(\beta_j)$ is a U-statistic based logistic loss, which is well suited to statistical analysis [27].

13.3 Tree and Forest Graphical Models

Graphs without cycles generally lead to models that are simpler to work with computationally; recall Chapters 4 and 11. This is because the joint probability can be factored into conditional probabilities involving just two variables. These probabilities can be estimated nonparametrically using bivariate density estimates.

Let $p^*(x)$ be a probability density with respect to Lebesgue measure $\mu(\cdot)$ on $\mathbb{R}^d$ and let $X_1, \ldots, X_n$ be n independent identically distributed $\mathbb{R}^d$-valued data vectors sampled from $p^*(x)$ where $X_i = (X_{i1}, \ldots, X_{id})^T$. Let $\mathcal{X}_j \subset \mathbb{R}$ denote the domain of the jth variable and let $\mathcal{X} = \mathcal{X}_1 \times \cdots \times \mathcal{X}_d$. A graph is a forest if it is acyclic. We say that a probability density function $p(x)$ is *supported by a forest* F if the density can be written as

$$p_F(x) = \prod_{(j,k)\in E_F} \frac{p(x_j, x_k)}{p(x_j)\,p(x_k)} \prod_{\ell \in V_F} p(x_\ell), \tag{13.10}$$

where each $p(x_j, x_k)$ is a bivariate density on $\mathcal{X}_j \times X_k$, and each $p(x_\ell)$ is a univariate density on $\mathcal{X}_\ell$.

For any density $q(x)$, the negative log-likelihood risk $R(q)$ is defined as

$$R(q) = -\mathbb{E} \log q(X) = -\int_{\mathcal{X}} p^*(x) \log q(x)\, dx. \tag{13.11}$$

Let $\mathcal{F}_d$ be the family of forests with d nodes, and $\mathcal{P}_d$ be the corresponding family of densities, then it can be shown [1] that the *oracle forest*

$$q^* = \arg\min_{q\in\mathcal{P}_d} D(p^* \| q) = \arg\min_{q\in\mathcal{P}_d} R(q) \tag{13.12}$$

is given by

$$q^* = p^*_{F^*} = \prod_{(j,k)\in E_{F^*}} \frac{p^*(x_j, x_k)}{p^*(x_j)\,p^*(x_k)} \prod_{\ell\in V_{F^*}} p^*(x_\ell) \tag{13.13}$$

for some forest $F^* \in \mathcal{F}_d$, where $p^*(x_j, x_k)$ and $p^*(x_\ell)$ are the bivariate and univariate marginal densities of p^*.

13.3.1 Tree estimation

The risk of a forest density p^*_F is given by

$$\begin{aligned} R(p^*_F) &= -\int_{\mathcal{X}} p^*(x) \Bigg(\sum_{(j,k)\in E_F} \log \frac{p^*(x_j, x_k)}{p^*(x_j)p^*(x_k)} + \sum_{\ell\in V_F} \log\left(p^*(x_\ell)\right) \Bigg) dx \\ &= -\sum_{(j,k)\in E_F} I(X_j; X_k) + \sum_{\ell\in V_F} H(X_\ell), \end{aligned} \tag{13.14}$$

where

$$I(X_j; X_k) = \int_{\mathcal{X}_j\times\mathcal{X}_k} p^*(x_j, x_k) \log \frac{p^*(x_j, x_k)}{p^*(x_j)\,p^*(x_k)}\, dx_j dx_k \tag{13.15}$$

is the mutual information between the pair of variables X_i, X_j and

$$H(X_k) = -\int_{\mathcal{X}_k} p^*(x_k) \log p^*(x_k)\, dx_k \tag{13.16}$$

is the entropy. Thus, it can be seen that the oracle forest is in fact a maximum weight spanning tree of the complete graph on d nodes with edge weight $I(X_j; X_k)$ on edge (j, k).

Methods for estimating a forest density can be based on plug-in estimates of the mutual information and greedy procedures such as Kruskal's algorithm for finding maximum weight spanning trees. The details of such estimators are studied by [11].

13.3.2 Oracle properties

The statistical properties of forest density estimator can be analyzed under the same type of assumptions that are made for classical kernel density estimation. In particular, assume that the univariate and bivariate densities lie in a Hölder class with exponent β. Under this assumption the minimax rate of convergence in the squared error loss is $O(n^{\beta/(\beta+1)})$ for bivariate densities and $O(n^{2\beta/(2\beta+1)})$ for univariate densities. Technical assumptions on the kernel yield L_∞ concentration results on kernel density estimation [5].

Let $\mathcal{P}_d^{(k)}$ be the family of d-dimensional densities that are supported by forests with at most k edges. Then

$$\mathcal{P}_d^{(0)} \subset \mathcal{P}_d^{(1)} \subset \cdots \subset \mathcal{P}_d^{(d-1)}. \tag{13.17}$$

Due to this nesting property,

$$\inf_{q_F \in \mathcal{P}_d^{(0)}} R(q_F) \geq \inf_{q_F \in \mathcal{P}_d^{(1)}} R(q_F) \geq \cdots \geq \inf_{q_F \in \mathcal{P}_d^{(d-1)}} R(q_F). \tag{13.18}$$

This means that a full spanning tree would generally be selected if we had access to the true distribution. However, with access to finite data to estimate the densities $(\widehat{p}_{n_1})$ the optimal procedure is to use fewer than $d-1$ edges. When the best k-edge forest is selected using plug-in estimators and variants of Kruskal's algorithm, it can be shown that the risk satisfies the following oracle type inequality:

$$R(\widehat{p}_{\widehat{F}_d^{(k)}}) - \inf_{q_F \in \mathcal{P}_d^{(k)}} R(q_F) = O_P\left(k\sqrt{\frac{\log n + \log d}{n^{\beta/(1+\beta)}}} + d\sqrt{\frac{\log n + \log d}{n^{2\beta/(1+2\beta)}}}\right) \tag{13.19}$$

We refer to [11] for details.

Stronger statistical guarantees come from a more detailed analysis of the mutual information estimates. Note that for a second-order Hölder class, the optimal bandwidth for a two-dimensional density estimator scales as $h \asymp n^{-1/6}$. By choosing a smaller bandwidth $h \asymp n^{-1/4}$, the density is undersmoothed. Under suitable regularity assumptions, the resulting plugin estimates of entropy and mutual information are exponentially concentrated around their means [14]. Using this concentration result, one shows that the optimal k-node forest can be estimated at the optimal rate:

$$\sup_{p \in \mathcal{P}_d} \mathbb{P}\left(\widehat{F} \neq F_k^*\right) = O\left(\sqrt{\frac{k}{n}}\right) \tag{13.20}$$

where the supremum is over all d-dimensional densities with bivariate marginals that are in a second-order Hölder space. If, after estimating the forest, the bandwidths are adjusted and the edge densities are then reestimated, the optimal minimax rate of estimation is achieved:

$$\sup_{p \in \mathcal{P}_d} \mathbb{E} \int_{\mathcal{X}} |\widehat{p}_{\widehat{F}}(x) - p_{F^*}(x)| dx = O\left(\sqrt{\frac{k}{n^{2/3}} + \frac{d-k}{n^{4/5}}}\right). \tag{13.21}$$

See [14] for details.

13.4 Gaussian Copulas and Variants

We say that a random vector $X = (X_1, \ldots, X_d)^T$ has a *nonparanormal* distribution and write

$$X \sim NPN(\mu, \Sigma, f)$$

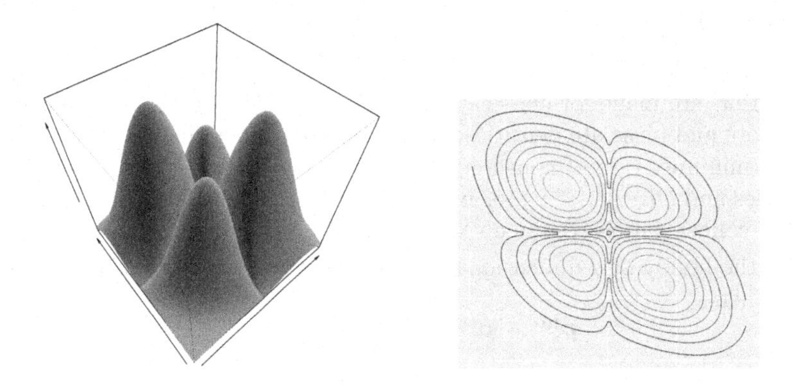

FIGURE 13.2: An example density for a 2-dimensional nonparanormal (copula). The component functions of the form $f_\alpha(x) = \text{sign}(x)|x|^\alpha$, with $\alpha_1 = 1.3$, and $\alpha_2 = 1.4$, with mean $\mu = (0,0)$ and covariance $\Sigma = \left(\begin{smallmatrix} 1 & -0.5 \\ -0.5 & 1 \end{smallmatrix}\right)$. The monotonic transformations have the effect of creating multiple modes, making the density resemble a mixture.

in case there exist functions $\{f_j\}_{j=1}^d$ such that $Z \equiv f(X) \sim N(\mu, \Sigma)$, where

$$f(X) = (f_1(X_1), \ldots, f_d(X_d)).$$

When the f_j's are monotone and differentiable, the joint probability density function of X is given by

$$p_X(x) = \frac{1}{(2\pi)^{d/2}|\Sigma|^{1/2}} \exp\left\{-\frac{1}{2}\left(f(x)-\mu\right)^T \Sigma^{-1}\left(f(x)-\mu\right)\right\} \prod_{j=1}^d |f_j'(x_j)|, \tag{13.22}$$

where the product term is a Jacobian. To make the family identifiable we demand that f_j preserves marginal means and variances:

$$\mu_j = \mathbb{E}(Z_j) = \mathbb{E}(X_j) \quad \text{and} \quad \sigma_j^2 \equiv \Sigma_{jj} = \text{Var}\left(Z_j\right) = \text{Var}\left(X_j\right). \tag{13.23}$$

These conditions only depend on $\text{diag}(\Sigma)$ but not the full covariance matrix.

Let $F_j(x)$ denote the marginal distribution function of X_j. Since the component $f_j(X_j)$ is Gaussian, it follows that

$$F_j(x) = \mathbb{P}\left(X_j \le x\right) = \mathbb{P}\left(Z_j \le f_j(x)\right) = \Phi\left(\frac{f_j(x) - \mu_j}{\sigma_j}\right)$$

so that the transform f_j is given by

$$f_j(x) = \mu_j + \sigma_j \Phi^{-1}\left(F_j(x)\right). \tag{13.24}$$

The form of the density in (13.22) implies that the conditional independence graph of the nonparanormal is encoded in $\Omega = \Sigma^{-1}$, as for the parametric Normal, since the density factors with respect to the graph of Ω, and therefore obeys the global Markov property of the graph. Figure 13.2 shows an example of a 2-dimensional nonparanormal density, exhibiting properties of a mixture.

The assumption that $f(X) = (f_1(X_1), \ldots, f_d(X_d)$ is Normal leads to a semiparametric model where only one dimensional functions need to be estimated. But the monotonicity of the functions f_j, which map onto $\mathbb{R}$, enables computational tractability of the nonparanormal. For more general functions f, the normalizing constant for the density

$$p_X(x) \propto \exp\left\{-\frac{1}{2}\left(f(x)-\mu\right)^T \Sigma^{-1}\left(f(x)-\mu\right)\right\}$$

cannot be computed in closed form.

To make the connection with the Gaussian copula, recall that a copula density takes the form

$$p(x_1,\ldots,x_d) = c(F_1(x_1),\ldots,F_d(x_d))\prod_{j=1}^{d} p(x_j)$$

where $p(x_j)$ is the marginal density of X_j and F_j is the distribution function. For the nonparanormal we have

$$F(x_1,\ldots,x_d) = \Phi_{\mu,\Sigma}(\Phi^{-1}(F_1(x_1)),\ldots,\Phi^{-1}(F_d(x_d))) \tag{13.25}$$

where $\Phi_{\mu,\Sigma}$ is the multivariate Gaussian cdf and Φ is the univariate standard Gaussian cdf. The Gaussian copula is usually expressed in terms of the correlation matrix, which is given by $R = \operatorname{diag}(\sigma)^{-1}\Sigma\operatorname{diag}(\sigma)^{-1}$. Note that the univariate marginal density for a Normal can be written as $p(x_j) = \frac{1}{\sigma_j}\phi(u_j)$ where $u_j = (x_j-\mu_j)/\sigma_j$. The multivariate Normal density can thus be expressed as

$$\begin{aligned} p_{\mu,\Sigma}(x_1,\ldots,x_d) &= \frac{1}{(2\pi)^{d/2}|R|^{1/2}\prod_{j=1}^{d}\sigma_j}\exp\left(-\frac{1}{2}u^T R^{-1}u\right) && (13.26)\\ &= \frac{1}{|R|^{1/2}}\exp\left(-\frac{1}{2}u^T(R^{-1}-I)u\right)\prod_{j=1}^{d}\frac{\phi(u_j)}{\sigma_j}. && (13.27)\end{aligned}$$

Since the distribution F_j of the jth variable satisfies $F_j(x_j) = \Phi((x_j-\mu_j)/\sigma_j) = \Phi(u_j)$, we have that $(X_j-\mu_j)/\sigma_j \overset{d}{=} \Phi^{-1}(F_j(X_j))$. The Gaussian copula density is thus

$$c(F_1(x_1),\ldots,F_d(x_d)) = \frac{1}{|R|^{1/2}}\exp\left\{-\frac{1}{2}\Phi^{-1}(F(x))^T(R^{-1}-I)\Phi^{-1}(F(x))\right\} \tag{13.28}$$

where $\Phi^{-1}(F(x)) = (\Phi^{-1}(F_1(x_1)),\ldots,\Phi^{-1}(F_d(x_d)))$. This is seen to be equivalent to (13.22) using the chain rule and the identity

$$(\Phi^{-1})'(\eta) = \frac{1}{\phi\left(\Phi^{-1}(\eta)\right)}. \tag{13.29}$$

13.4.1 Estimation

To estimate the model one can define

$$\widetilde{h}_j(x) = \Phi^{-1}(\widetilde{F}_j(x))$$

where $\widetilde{F}_j$ is an estimator of F_j, and then estimate the marginal transformations using (13.24) as

$$\widetilde{f}_j(x) = \widehat{\mu}_j + \widehat{\sigma}_j\widetilde{h}_j(x) \tag{13.30}$$

where $\widehat{\mu}_j$ and $\widehat{\sigma}_j$ are the sample mean and standard deviation:

$$\widehat{\mu}_j \equiv \frac{1}{n}\sum_{i=1}^{n} X_j^{(i)} \quad \text{and} \quad \widehat{\sigma}_j = \sqrt{\frac{1}{n}\sum_{i=1}^{n}\left(X_j^{(i)} - \widehat{\mu}_j\right)^2}.$$

Now, let $S_n(\widetilde{f})$ be the sample covariance matrix of the transformed variables $\widetilde{f}(X^{(1)}),\ldots,\widetilde{f}(X^{(n)})$; that is,

$$S_n(\widetilde{f}) \equiv \frac{1}{n}\sum_{i=1}^{n}\left(\widetilde{f}(X^{(i)}) - \mu_n(\widetilde{f})\right)\left(\widetilde{f}(X^{(i)}) - \mu_n(\widetilde{f})\right)^T \tag{13.31}$$

$$\mu_n(\widetilde{f}) \equiv \frac{1}{n}\sum_{i=1}^{n}\widetilde{f}(X^{(i)}).$$

We then estimate Ω using $S_n(\widetilde{f})$. For instance, the maximum likelihood estimator is $\widehat{\Omega}_n^{\text{MLE}} = S_n(\widetilde{f})^{-1}$. The ℓ_1-regularized estimator is

$$\widehat{\Omega}_n = \arg\min_{\Omega}\left\{\operatorname{tr}\left(\Omega S_n(\widetilde{f})\right) - \log|\Omega| + \lambda\|\Omega\|_1\right\} \tag{13.32}$$

where λ is a regularization parameter, and $\|\Omega\|_1 = \sum_{j=1}^{d}\sum_{k=1}^{d}|\Omega_{jk}|$. The estimated graph is then $\widehat{E}_n = \{(j,k) : \widehat{\Omega}_{jk} \neq 0\}$.

13.4.2 Rank correlation

The above is a two-step procedure to estimate the graph. Greater statistical efficiency is obtained by treating the nonparametric transformation functions f_j as nuisance parameters, and estimating the correlation matrix R directly using rank correlation statistics. This is the approach developed in [13].

Specifically, if $\tau_{jk} = \operatorname{Cor}(\operatorname{sign}(X_j - X_j'), \operatorname{sign}(X_k - X_k'))$ is Kendall's tau statistic [9], then the covariance is given by

$$\Sigma_{jk}^0 = \sin\left(\frac{\pi}{2}\tau_{jk}\right). \tag{13.33}$$

Concentration analysis based on U-statistics shows that

$$\max_{jk}\left|\widehat{S}_{jk}^{\tau} - \Sigma_{jk}^0\right| = O_P\left(\sqrt{\frac{\log(pn)}{n}}\right) \tag{13.34}$$

where $\widehat{S}_{jk}^{\tau}$ is the plug-in estimator based on (13.33) for the sample correlation

$$\widehat{\tau}_{jk} = \frac{1}{n(n-1)}\sum_{1\le i<i'\le n}\operatorname{sign}\big((X_j^{(i)} - X_j^{(i')})(X_k^{(i)} - X_k^{(i')})\big). \tag{13.35}$$

This leads to parametric rates of graph estimation for the Gaussian copula under high-dimensional scaling [13, 25].

13.4.3 Trees and copulas

As an example of how new families of graphical models are formed using existing families as building blocks, we briefly mention "blossom tree graphical models," which combine forests with the nonparanormal [15]. The model consists of a forest of trees, and a collection of "blossom" subgraphs containing many cycles. Nonparanormal graphical models are associated with each of the blossoms, and nonparametric bivariate densities are used over the branches (edges) of the trees.

In more detail, the graph $G = (V, E)$ has edge set decomposed as $E = F \cup \{\cup_{B\in\mathcal{B}} B\}$ so that:

(1) F is acyclic

(2) $V(B) \cap V(B') = \emptyset$, for $B, B' \in \mathcal{B}$ with $B \neq B'$, where $V(B)$ denotes the vertex set of B

(3) $|V(B) \cap V(F)| \leq 1$ for each $B \in \mathcal{B}$

(4) $V(F) \cup \bigcup_{\mathcal{B}} V(B) = V$

The subgraphs $B \in \mathcal{B}$ are called *blossoms*. The unique node $\rho(B) \in V(B) \cap V(F)$, which may be empty, is called the *pedicel* of the blossom. The set of pedicels is denoted $\mathcal{P}(F) \subset V(F)$.

With these definitions, suppose $p(x)$ is a density given by a blossom tree with edge set $F \cup \{\cup_{\mathcal{B}} B\}$. Then it is straightforward to show that

$$p(x) = \prod_{(s,t)\in F} \frac{p(X_s, X_t)}{p(X_s)p(X_t)} \prod_{s\in V(F)\setminus\mathcal{P}(F)} p(X_s) \prod_{B\in\mathcal{B}} p(X_{V(B)}).$$

The single node marginal probabilities $p(X_s)$ are arbitrary and each blossom distribution satisfies $X_{V(B)} \sim NPN(\mu_B, \Sigma_B, f_B)$, while enforcing that the single node marginal of the pedicel $\rho(B)$ agrees with the marginals of this node defined by the forest.

To model bivariate non-Gaussianity on the edges, the forest edges can be constructed with Kruskal's algorithm with *negentropy* as a measure of distance to normality, defined as

$$J(X_i; X_j) = H(\phi(x_i, x_j)) - H(p^*(x_i, x_j)), \tag{13.36}$$

where $H(\cdot)$ denotes the differential entropy of a density, and $\phi(x_i, x_j)$ is an Gaussian density with the same mean and covariance matrix as $p^*(x_i, x_j)$. To group the non-forest nodes into blossoms, a nonparametric partial correlation measure is used to associate each node with a pedicel. The forest structure is modeled with kernel density estimates, while each blossom is modeled by a nonparanormal; see [15] for details.

13.5 Tensor Product Smoothing Spline ANOVA Models

Thus far, all of the models have been of the form (13.3), where the potential function can be decomposed as the summation of the main effects and bivariate interaction effects. In this section, we consider graphical models with possibly higher-order interactions, from the general density function of the form (13.2). The method is based on iterative screening to estimate the component functions in a tensor product smoothing spline framework [23]. To fit the model, a surrogate loss function is used, which avoids the partition function $Z(\psi)$. In related work, [8] studies the use of score matching for both Gaussian and nonparametric graphical models using exponential series.

13.5.1 Tensor product smoothing splines

We first introduce the tensor product space, referring the reader to Lin [10] for a detailed review. For any $m \geq \frac{1}{2}$, let H_j^m be the mth order Sobolev Hilbert space of univariate functions of x_j over $[0, 1]$. Let $J \subseteq [d]$ be an index set with cardinality $|J| = r$, and let $\mathcal{H}_J = \otimes_{j \in J} H_j^m$ be the completed tensor product space of H_j^m for $j \in J$. We denote the multivariate function $\psi_J = \psi_{j_1 \cdots j_r}$ and assume that $\psi_J \in \mathcal{H}_J$. The tensor product space is

$$\{1\} \oplus \sum_{j=1}^{d} H_j^m \oplus \sum_{J \subseteq [d], |J|=2} \mathcal{H}_J \oplus \sum_{J \subseteq [d], |J|=3} \mathcal{H}_J \oplus \cdots,$$

where $|J|$ is the cardinality of J; a function $\psi(x)$ in this space can be decomposed as

$$\psi(x) = \mu + \sum_{j=1}^{d} \psi_j(x_j) + \sum_{j<k} \psi_{jk}(x_j, x_k) + \sum_{j<k<\ell} \psi_{jk\ell}(x_j, x_k, x_\ell) + \cdots. \tag{13.37}$$

Without loss of generality, we assume that $\mu = 0$. This allows for higher-order interaction terms corresponding to cliques in a graph. For instance, $\psi_{jk\ell}(x_j, x_k, x_\ell)$ne0 if and only if the variables X_j, X_k, and X_ℓ form a triangle.

13.5.2 Fisher-Hyvärinen scoring

To avoid computing the normalizing constant $Z(\psi)$, we use a surrogate loss function in place of the traditional log-likelihood function. In what follows, we briefly introduce the score matching loss function and refer the reader to Hyvärinen [6] and Hyvärinen [7] for more details. Let X be a random vector with distribution $\mathcal{P}$ and joint density function $p(\cdot)$. For a distribution $\mathcal{Q}$ with density $q(\cdot)$, we define the *Fisher-Hyvärinen score matching* loss of $\mathcal{Q}$ with respect to $\mathcal{P}$ as

$$R(p, q) = \frac{1}{2} \int_{[0,1]^d} p(x) \left\| \mathbf{r}(x) \circ [\nabla \log p(x) - \nabla \log q(x)] \right\|_2^2 dx, \tag{13.38}$$

where $\circ$ is the Hadamard product between two vectors and $\mathbf{r}(x)$ is a d-dimensional vector with elements $r_j(x) = x_j(1 - x_j)$. Equation (13.38) is minimized as a function of $\mathcal{Q}$ when $\mathcal{Q} = \mathcal{P}$. Under some regularity conditions, an asymptotically consistent estimate of $\psi(x)$ for the nonparametric graphical model (13.2) can be obtained by minimizing the score matching loss, which does not depend on the partition function $Z(\psi)$:

$$S(x, \psi) = -2 \sum_{j=1}^{d} r_j(x_j) r_j'(x_j) \psi^{(j)}(x) - \sum_{j=1}^{d} r_j^2(x_j) \psi^{(jj)}(x) + \frac{1}{2} \sum_{j=1}^{d} r_j^2(x_j) \left(\psi^{(j)}(x)\right)^2, \tag{13.39}$$

where $\psi^{(j)}$ and $\psi^{(jj)}$ are the first and second order derivative of $\psi(x)$ with respect to x_j; see [23, 8].

Tan et al. [23] propose a layerwise screening algorithm to estimate $\psi(x)$ by minimizing the score matching loss. Let $X_1, \ldots, X_n$ be n independent observations drawn from the density $p_\psi(x)$. The strategy is to approximate the function sequentially, starting from models with only main effects. At the first step of the algorithm, the main effects are selected according to

$$\widehat{\psi}^{(1)} = \arg\min_{\psi \in \mathcal{H}^{(1)}} \frac{1}{n} \sum_{i=1}^{n} S(X_i, \psi) \quad \text{subject to} \quad \sum_{j=1}^{d} \|P_j(\psi)\|_2 \leq L_1, \tag{13.40}$$

where $P_j(\psi)$ is the orthogonal projection of ψ onto H_j^m, L_1 a positive tuning parameter, which encourages the estimated main effects to be zero, and $\mathcal{H}^{(1)}$ is the tensor product of the function spaces for the main effects.

Conditioned on the support of the estimated main effects, the bivariate interaction terms are estimated only when the main effects are estimated to be non-zero. Thus, the fitted model is hierarchical. Note that this will miss interaction terms that do not have main effects. Given the support $\widehat{\mathcal{S}}^{(1)}$ of $\widehat{\psi}^{(1)}$, the second step of the layerwise screening algorithm selects $\psi^{(2)}$ according to

$$\begin{aligned}\widehat{\psi}^{(2)} = & \underset{\psi \in \mathcal{H}^{(2)}, \mathcal{S}(\psi)=\{\sigma(\widehat{\mathcal{S}}^{(1)}) \cup \widehat{\mathcal{S}}^{(1)}\}}{\arg\min} \frac{1}{n}\sum_{i=1}^{n} S(X_i, \psi) \\ & \text{subject to} \sum_{j\in\widehat{\mathcal{S}}^{(1)}} \|P_j(\psi)\|_2 + \sum_{\{j_1,j_2\}\in\sigma(\widehat{\mathcal{S}}^{(1)})} \|P_{j_1 j_2}(\psi)\|_2 \leq L_2,\end{aligned} \tag{13.41}$$

where $P_{j_1 j_2}(\psi)$ is the orthogonal projection of ψ onto $H_{j_1}^m \otimes H_{j_2}^m$ and $\sigma(\widehat{\mathcal{S}}^{(1)})$ is an index set consisting of all possible pairwise combinations of the main effects that are estimated to be non-zero. For instance, if $\widehat{\mathcal{S}}^{(1)} = \{\{1\},\{2\},\{3\}\}$, then $\sigma(\widehat{\mathcal{S}}^{(1)}) = \{\{1,2\},\{1,3\},\{2,3\}\}$. This process is continued until no more higher-order interaction effects are identified. In practice, (13.40) and (13.41) can be solved using basis expansion, and efficient algorithm can be developed. We refer the reader to Tan et al. [23] for more details.

It can be shown that the layerwise screening estimator is risk consistent under the score matching risk function in (13.38). Let $\mathcal{F}^{(r)}$ be the function class for functions of the form $\psi(x) = \sum \beta_J g_J(x_J)$ with support restricted to $\widehat{\mathcal{S}}^{(r-1)} \cup \sigma(\widehat{\mathcal{S}}^{(r)})$ and that $\sum |\beta_J| \leq L_r$. Let $\mathcal{Q}^{(r)} = \{q \mid q \propto \exp(\psi(x)), \psi \in \mathcal{F}^{(r)}\}$. Then, conditioned on the support $\widehat{\mathcal{S}}^{(r-1)}$, the estimator $\widehat{p}^{(r)} \propto \exp(\widehat{\psi}^{(r)})$ is post-selection risk consistent under the score matching risk function

$$R(p, \widehat{p}^{(r)}) - \inf_{q\in\mathcal{Q}^{(r)}} R(p,q) = O_P\left(L_r^2 \sqrt{\frac{r^3 \log d}{n}}\right)$$

if $L_r = o\big((n/(r^3 \log d))^{1/4}\big)$; see Tan et al. [23] for details. [8] approaches regularized score matching using algorithms based on consensus alternating direction method of multipliers (ADMM).

We note that the score matching loss can be derived from the perspective of Stein's method, which defines a discrepancy measure

$$d_{\mathcal{G}}(\theta_0, \theta) = \sup_{g\in\mathcal{G}} |\mathbb{E}_{\theta_0}(g(X)) - \mathbb{E}_{\theta}(g(X))| \tag{13.42}$$

for a suitable class of functions $\mathcal{G}$ [19]. In particular, for exponential family densities specified by

$$\nabla \log p_\theta(x) = \theta^T \psi(x) + b(x) \tag{13.43}$$

the score matching method can be seen as arising from the Stein discrepancy applied to the class of functions of the form

$$g(x) = \langle h(x), \nabla \log p_\theta(x)\rangle + \langle \nabla, h(x)\rangle. \tag{13.44}$$

Exploration of this connection could be an interesting direction for future work.

13.6 Summary and Extensions

In this chapter, we have introduced several basic approaches to nonparametric and semiparametric graphical modeling that make different assumptions for computational and statistical efficiency. Extensions to these models have included dynamic and mixed Gaussian copula models [17, 3]. The nonparanormal has also been generalized to transelliptical graphical models [12], forest graphical models have been extended to triangle-free graphs [16], and latent variable versions have been introduced [20, 24]. In an approach we did not discuss here, [21] and [22] use reproducing kernel Hilbert spaces for flexible modeling of pairwise interactions. Many further useful extensions are surely possible.

Bibliography

[1] Francis R. Bach and Michael I. Jordan. Beyond independent components: Trees and clusters. *Journal of Machine Learning Research*, 4:1205–1233, 2003.

[2] Shizhe Chen, Daniela M Witten, and Ali Shojaie. Selection and estimation for mixed graphical models. *Biometrika*, 102(1):47–64, 2015.

[3] Jianqing Fan, Han Liu, Yang Ning, and Hui Zou. High dimensional semiparametric latent graphical model for mixed data. *arXiv:1404.7236*, 2014.

[4] Jianqing Fan, Lingzhou Xue, Hui Zou et al. Strong oracle optimality of folded concave penalized estimation. *The Annals of Statistics*, 42(3):819–849, 2014.

[5] E. Giné and A. Guillou. Rates of strong uniform consistency for multivariate kernel density estimators. *Annales de l'institut Henri Poincaré (B), Probabilités et Statistiques*, 38:907–921, 2002.

[6] Aapo Hyvärinen. Estimation of non-normalized statistical models by score matching. *Journal of Machine Learning Research*, 6:695–708, 2005.

[7] Aapo Hyvärinen. Some extensions of score matching. *Computational Statistics & Data Analysis*, 51(5):2499–2512, 2007.

[8] Eric Janofsky. *Exponential Series Approaches for Nonparametric Graphical Models.* PhD thesis, University of Chicago, Department of Statistics, 2015. arXiv:1506.03537.

[9] M. G. Kendall. *Rank correlation methods.* London: Griffin, 1962.

[10] Yi Lin. Tensor product space ANOVA models. *The Annals of Statistics*, 28(3):734–755, 2000.

[11] Han Liu, Min Xu, Haijie Gu, Anupam Gupta, John Lafferty, and Larry Wasserman. Forest density estimation. *Journal of Machine Learning Research*, 12:907–951, March 2011.

[12] Han Liu, Fang Han, and Cun hui Zhang. Transelliptical graphical models. *Advances in Neural Information Processing Systems*, 800–808, 2012.

[13] Han Liu, Fang Han, Ming Yuan, John Lafferty, Larry Wasserman et al. High-dimensional semiparametric Gaussian copula graphical models. *The Annals of Statistics*, 40(4):2293–2326, 2012.

[14] Han Liu, John D. Lafferty, and Larry A. Wasserman. Exponential concentration for mutual information estimation with application to forests. In *Advances in Neural Information Processing Systems 25: 26th Annual Conference on Neural Information Processing Systems 2012. Proceedings of a meeting held December 3-6, 2012, Lake Tahoe, Nevada*, 2546–2554, 2012.

[15] Zhe Liu and John Lafferty. Blossom tree graphical models. In Z. Ghahramani, M. Welling, C. Cortes, N. D. Lawrence, and K. Q. Weinberger, editors, *Advances in Neural Information Processing Systems 27*, 1458–1465, 2014.

[16] Junwei Lu and Han Liu. Graphical fermat's principle and triangle-free graph estimation. *arXiv:1504.06026*, 2015.

[17] Junwei Lu, Mladen Kolar, and Han Liu. Post-regularization inference for dynamic nonparanormal graphical models. *arXiv:1512.08298*, 2015.

[18] Yang Ning, Tianqi Zhao, and Han Liu. A likelihood ratio framework for high dimensional semiparametric regression. *arXiv:1412.2295*, December 2014.

[19] Nathan Ross. Fundamentals of Stein's method. *Probability Surveys*, 2011:210–293, 2011.

[20] Le Song, Han Liu, A. Parikh, and Eric P Xing. Nonparametric latent tree graphical models: Inference, estimation, and learning. *Journal of Machine Learning Research*, 2016.

[21] Bharath Sriperumbudur, Kenji Fukumizu, Arthur Gretton, Aapo Hyvärinen, and Revant Kumar. Density estimation in infinite dimensional exponential families. *Journal of Machine Learning Research*, 18(57):1–59, 2017.

[22] Siqi Sun, Mladen Kolar, and Jinbo Xu. Learning structured densities via infinite dimensional exponential families. In C. Cortes, N. D. Lawrence, D. D. Lee, M. Sugiyama, and R. Garnett, editors, *Advances in Neural Information Processing Systems 28*, 2287–2295. Curran Associates, Inc., 2015.

[23] Kean Ming Tan, Junwei Lu, and Han Liu. Layer-wise learning strategy for nonparametric tensor-product smoothing spline regression and graphical models. Princeton Technical Report, 2016.

[24] Kean Ming Tan, Yang Ning, Daniela M Witten, and Han Liu. Replicates in high dimensions, with applications to latent variable graphical models. *Biometrika*, 2017.

[25] Lingzhou Xue, Hui Zou, et al. Regularized rank-based estimation of high-dimensional nonparanormal graphical models. *The Annals of Statistics*, 40(5):2541–2571, 2012.

[26] Eunho Yang, Pradeep Ravikumar, Genevera I. Allen, and Zhandong Liu. Graphical models via univariate exponential family distributions. *Journal of Machine Learning Research*, 16:3813–3847, 2015.

[27] Zhuoran Yang, Yang Ning, and Han Liu. On semiparametric exponential family graphical models. *arXiv:1412.8697*, 2014.

[28] Tong Zhang. Analysis of multi-stage convex relaxation for sparse regularization. *The Journal of Machine Learning Research*, 11:1081–1107, 2010.

[29] Tong Zhang et al. Multi-stage convex relaxation for feature selection. *Bernoulli*, 19 (5B):2277–2293, 2013.

[30] Hui Zou and Runze Li. One-step sparse estimates in nonconcave penalized likelihood models. *Annals of Statistics*, 36(4):1509, 2008.

14

Inference in High-Dimensional Graphical Models

Jana Janková

Statistical Laboratory, University of Cambridge

Sara van de Geer

Seminar for Statistics, ETH Zurich

CONTENTS

14.1 Undirected Graphical Models

14.1.1 Introduction

Undirected graphical models, also known as Markov random fields, have become a popular tool for representing network structure of high-dimensional data in a large variety of areas including genetics, brain network analysis, social networks and climate studies. Let $G = (\mathcal{V}, \mathcal{E})$ be an undirected graph with a vertex set $\mathcal{V} = \{1, 2, \ldots, p\}$ and an edge set $\mathcal{E} \subset \mathcal{V} \times \mathcal{V}$. Let $X^0 = (X_1, X_2, ..., X_p)$ be a random vector indexed by graph's vertices. The joint distribution of X^0 belongs to the graphical model determined by G if X_j and X_k are conditionally independent given all other variables whenever j and k are not adjacent in G. The graph then encodes conditional independence structure among the entries of X^0. A detailed overview of the basic concepts relating to undirected graphical models is given in Chapter 1 and for further in-depth study of the structure in Gaussian graphical models, we refer to Chapter 9.

If we assume that X^0 is normally-distributed with a covariance matrix Σ_0, one can show that the edge structure of the graph is encoded by the precision matrix $\Theta_0 := \Sigma_0^{-1}$ (assumed to exist). If Θ_{ij}^0 denotes the (i,j)-th entry of the matrix Θ_0, it is well known that $\Theta_{ij}^0 = 0 \Leftrightarrow (i,j) \notin \mathcal{E}$; recall Corollary 9.1.2. The non-zero entries in the precision matrix correspond to edges in the associated graphical model and the absolute values of these entries correspond to edge weights.

Therefore to estimate the edge structure of a Gaussian graphical model, we consider the problem of estimating the precision matrix, based on a sample of n independent instances $X^1, \ldots, X^n$, distributed as X^0. We are not only interested in point estimation, but in quantifying the uncertainty of estimation such as constructing confidence intervals and tests for edge weights. Confidence intervals and tests can be used for identifying significant variables or testing whether networks corresponding to different populations are identical.

Our focus is on the challenges that arise in regimes of high dimensionality, that is, when the number of unknown parameters can be much larger than the number of observations n. However, it is instructive to first inspect the low-dimensional setting. In the regime when p is fixed and the observations are normally distributed with $\mathbb{E}X^i = 0$, $i = 1, \ldots, n$, the sample covariance matrix $\widehat{\Sigma} := X^T X/n$ (where X is the $n \times p$ matrix of observations $X^1, \ldots, X^n$) is the maximum likelihood estimator of the covariance matrix. The inverse of the sample covariance matrix $\widehat{\Theta} := \widehat{\Sigma}^{-1}$ is the maximum likelihood estimator of the precision matrix. Asymptotic linearity of $\widehat{\Theta}$ follows by the decomposition

$$\widehat{\Theta} - \Theta_0 = -\Theta_0(\widehat{\Sigma} - \Sigma_0)\Theta_0 + \mathrm{rem}_0, \tag{14.1}$$

where $\mathrm{rem}_0 := -\Theta_0(\widehat{\Sigma} - \Sigma_0)(\widehat{\Theta} - \Theta_0)$ is the remainder term. The term rem_0 can be bounded by Hölder's inequality to obtain

$$\|\mathrm{rem}_0\|_\infty \leq \|\Theta_0(\widehat{\Sigma} - \Sigma_0)\|_\infty \left|\left|\left| \widehat{\Theta} - \Theta_0 \right|\right|\right|_1,$$

where we used the notation $\|A\|_\infty = \max_{1\leq i,j\leq p} |A_{ij}|$ for the supremum norm of a matrix A and $|||A|||_1 := \max_{1\leq j\leq p} \sum_{i=1}^p |A_{ij}|$ for the ℓ_1-operator norm. If the fourth moments of X^i's are bounded, the decomposition (14.1) implies rates of convergence $\|\widehat{\Theta} - \Theta_0\|_\infty = \mathcal{O}_P(1/\sqrt{n})$, where $\mathcal{O}_P(1)$ denotes boundedness in probability. The remainder term then satisfies $\|\mathrm{rem}_0\|_\infty = o_P(1/\sqrt{n})$, $o_P(1)$ denoting convergence in probability to zero. Therefore, $\widehat{\Theta}$ is indeed an asymptotically linear estimator of Θ_0 and in this sense, we can say it is asymptotically unbiased. Moreover, $\widehat{\Theta}$ is asymptotically normal with a limiting normal distribution.

In high-dimensional settings, the sample covariance matrix does not perform well (see [20] and [19]) and if $p > n$, it is singular with probability one. Various methods have been proposed that try to reduce the variance of the sample covariance matrix at the price of introducing some bias. The idea of banding or thresholding the sample covariance matrix was studied in [2], [3] and [12]. Methods inducing sparsity through Lasso regularization were studied by another stream of works. These can be divided into two categories: global methods and local (nodewise) methods. Global methods estimate the precision matrix typically via a regularized log-likelihood, while nodewise methods split the problem into a series of linear regressions by estimating neighborhood of each node in the underlying graph. A popular global method is the ℓ_1-penalized maximum likelihood estimator, known as the graphical Lasso. Its theoretical properties were studied in a number of papers, including [40], [13], [29] and [26]. The local approach on estimation of precision matrices in particular includes the regression approach [23],[39],[5],[30] which uses a Lasso-type algorithm or Dantzig selector [6] to estimate each column or a smaller part of the precision matrix individually. The major theoretical results that were developed for these methods fall into two main categories: "oracle" error bounds and variable selection properties. In this chapter, we will rely on the oracle bounds; for an overview of variable selection properties of these methods, we refer to Chapter 12.

Inference for parameters in high-dimensional undirected graphical models was studied in several papers. Multiple testing for conditional dependence in Gaussian graphical models with asymptotic control of false discovery rates was considered in [21]. The work [37] proposes methodology for inference about edge weights based on Berry-Esseen bounds and the bootstrap for certain special classes of high-dimensional graphs. Another line of work ([28], [15] and [16]) proposes asymptotically normal estimators for edge weights in Gaussian graphical models based on different modifications of initial Lasso-regularized estimators. In particular, the paper [28] proposes nodewise regression where each pair of variables, (X_i, X_j), is regressed on all the remaining variables; this yields estimates for the parameters of the joint conditional distribution of (X_i, X_j) given all the other variables. The papers [15] and [16] propose methodology inspired by the de-biasing approach in high-dimensional linear regression that was studied in [41], [35] and [18]. This chapter discusses and unifies the ideas from the papers [15] and [16].

A different approach to structure learning in undirected graphical models is the Hyvärinen score matching (see e.g. [10] for a discussion of this approach). Methodology for asymptotically normal estimation of edge parameters in pairwise (not necessarily Gaussian) graphical models based on Hyvärinen scoring was proposed in [38].

14.1.2 De-biasing regularized estimators

The idea of using regularized estimators for construction of asymptotically normal estimators is based on removing the bias associated with the penalty. Consider a real-valued loss function ρ_Θ and let $R_n(\Theta) := \sum_{i=1}^n \rho_\Theta(X^i)/n$ denote the average risk, given an independent sample $X^1, \ldots, X^n$. Under differentiability conditions, a regularized M-estimator $\widehat{\Theta}$ based on the risk function R_n can often be characterized by estimating equations

$$\dot{R}_n(\widehat{\Theta}) + \xi(\widehat{\Theta}) = 0, \tag{14.2}$$

where $\dot{R}_n$ is the gradient of R_n and $\xi(\widehat{\Theta})$ is a (sub-)gradient corresponding to the regularization term, evaluated at $\widehat{\Theta}$. The idea is to improve on the initial estimator by finding a root $\widehat{\Theta}_{\text{de-bias}}$ closer to the solution of estimating equations without the bias term $\xi(\widehat{\Theta})$, i.e. a new estimator $\widehat{\Theta}_{\text{de-bias}}$ such that $\dot{R}_n(\widehat{\Theta}_{\text{de-bias}}) \approx 0$. A natural way is to define a corrected estimator $\widehat{\Theta}_{\text{de-bias}}$ from a linear approximation to $\dot{R}_n$

$$\dot{R}_n(\widehat{\Theta}) + \ddot{R}_n(\widehat{\Theta})(\widehat{\Theta}_{\text{de-bias}} - \widehat{\Theta}) = 0. \tag{14.3}$$

In high-dimensional settings, the matrix $\ddot{R}_n(\widehat{\Theta})$ is typically rank deficient and thus not invertible. Suppose that we have an approximate inverse denoted by $\ddot{R}_n(\widehat{\Theta})^{\text{inv}}$. Then we can approximately solve (14.3) for $\widehat{\Theta}_{\text{de-bias}}$ to obtain

$$\widehat{\Theta}_{\text{de-bias}} \approx \widehat{\Theta} - \ddot{R}_n(\widehat{\Theta})^{\text{inv}} \dot{R}_n(\widehat{\Theta}), \tag{14.4}$$

provided that the remainder term is small. We refer to the step (14.4) as the de-biasing step since the correction term is proportional to the bias term. Generally speaking, if the initial estimator $\widehat{\Theta}$ and the approximate inverse of $\ddot{R}_n(\widehat{\Theta})$ are consistent in a strong-enough sense, then the new estimator $\widehat{\Theta}_{\text{de-bias}}$ will be a consistent estimator of its population version Θ_0 per entry at the parametric rate. The de-biasing step (14.4) may be viewed as one step of the Newton-Raphson scheme for numerical optimization.

In consecutive sections, we will look in detail at the bias of several particular examples of regularized estimators, including the graphical Lasso ([40]) and nodewise Lasso ([23]). We now provide a unified de-biasing scheme which covers both special cases treated below (see also [32], Chapter 14). Suppose that a (possibly non-symmetric) estimator $\widehat{\Theta}$ is available which is an approximate inverse of $\widehat{\Sigma}$ in the sense that the following condition is satisfied

$$\widehat{\Sigma}\widehat{\Theta} - I + \eta(\widehat{\Theta}) = 0, \tag{14.5}$$

where $\eta(\widehat{\Theta})$ is a bias term. This condition in some sense corresponds to the estimating equations (14.2). We can express $\widehat{\Theta}$ from (14.5) by straightforward algebra which leads to the decomposition

$$\widehat{\Theta} + \widehat{\Theta}^T\eta(\widehat{\Theta}) - \Theta_0 = -\Theta_0(\widehat{\Sigma} - \Sigma_0)\Theta_0 + \text{rem}_0 + \text{rem}_{\text{reg}}, \tag{14.6}$$

where

$$\text{rem}_{\text{reg}} = (\widehat{\Theta} - \Theta_0)^T\eta(\widehat{\Theta}).$$

Compared to the regime with p fixed, there is an additional remainder rem_{reg} corresponding to the bias term. Provided that the remainder terms rem_0 and rem_{reg} are small enough, we can take as a new, de-biased estimator, $\widehat{T} := \widehat{\Theta} + \widehat{\Theta}^T\eta(\widehat{\Theta})$. The bias term $\eta(\widehat{\Theta})$ can be expressed from (14.5) as $\eta(\widehat{\Theta}) = -(\widehat{\Sigma}\widehat{\Theta} - I)$. Hence we obtain

$$\widehat{T} = \widehat{\Theta} + \widehat{\Theta}^T - \widehat{\Theta}^T\widehat{\Sigma}\widehat{\Theta}. \tag{14.7}$$

Bounding the remainders rem_0 and rem_{reg} in high-dimensional settings requires more refined arguments than when p is fixed. Looking at the remainder rem_{reg}, we can again invoke Hölder's inequality to obtain

$$\|\text{rem}_{\text{reg}}\|_\infty = \|(\widehat{\Theta} - \Theta_0)^T\eta(\widehat{\Theta})\|_\infty \le \left|\left|\left|\widehat{\Theta} - \Theta_0\right|\right|\right|_1 \|\eta(\widehat{\Theta})\|_\infty.$$

Thus it suffices to control the rates of convergence of $\widehat{\Theta}$ in $|||\cdot|||_1$-norm and control the absolute size of entries of the bias term.

Provided that the remainders are of small order $1/\sqrt{n}$ in probability, asymptotic normality per elements of $\widehat{T}$ is a consequence of asymptotic linearity and can be established under tail conditions on X^i's, by applying the Lindeberg's central limit theorem.

14.1.3 Graphical Lasso

If the observations are independent $\mathcal{N}(0, \Sigma_0)$-distributed, the log-likelihood function is proportional to

$$\ell(\Theta) := \text{tr}(\widehat{\Sigma}\Theta) - \log\det(\Theta).$$

The graphical Lasso (see [40], [9], [13]) is based on the Gaussian log-likelihood function but regularizes it via an ℓ_1-norm penalty on the off-diagonal elements of the precision matrix. The diagonal elements of the precision matrix correspond to certain partial variances and thus should not be penalized. The graphical Lasso is defined by

$$\widehat{\Theta} = \text{argmin}_{\Theta=\Theta^T,\Theta\succ 0}\text{tr}(\widehat{\Sigma}\Theta) - \log\det(\Theta) + \lambda\|\Theta^-\|_1, \tag{14.8}$$

where λ is non-negative tuning parameter and we optimize over symmetric positive definite matrices, denoted by $\succ$. Here Θ^- represents the matrix obtained by setting the diagonal elements of Θ to zero and $\|\Theta^-\|_1$ is the ℓ_1-norm of the vectorized version of Θ^-. The usefulness of the graphical Lasso is not limited only to Gaussian settings; the theoretical results below show that it performs well as an estimator of the precision matrix in a large class of non-Gaussian settings.

A disadvantage of the graphical Lasso (14.8) is that the penalization does not take into account that the variables have in general a different scaling. To take these differences in the variances into account, we define a modified graphical Lasso with a weighted penalty. To this end, let $\widehat{W}^2 := \text{diag}(\widehat{\Sigma})$ be the diagonal matrix obtained from the diagonal of $\widehat{\Sigma}$. We let

$$\widehat{\Theta}_{\text{w}} = \text{argmin}_{\Theta=\Theta^T,\Theta\succ 0}\text{tr}(\widehat{\Sigma}\Theta) - \log\det(\Theta) + \sum_{i\neq j}\widehat{W}_{ii}\widehat{W}_{jj}|\Theta_{ij}|. \tag{14.9}$$

The weighted graphical Lasso $\widehat{\Theta}_{\text{w}}$ is related to a graphical Lasso based on the sample correlation matrix $\widehat{R} := \widehat{W}^{-1}\widehat{\Sigma}\widehat{W}^{-1}$. To clarify the connection, we define

$$\widehat{\Theta}_{\text{norm}} = \text{argmin}_{\Theta=\Theta^T,\Theta\succ 0}\text{tr}(\widehat{R}\Theta) - \log\det(\Theta) + \|\Theta^-\|_1. \tag{14.10}$$

Then it holds that $\widehat{\Theta}_{\text{w}} = \widehat{W}^{-1}\widehat{\Theta}_{\text{norm}}\widehat{W}^{-1}$. The estimator $\widehat{\Theta}_{\text{norm}}$ is of independent interest, if the parameter of interest is the inverse correlation matrix rather than the precision matrix. The estimators $\widehat{\Theta}_{\text{w}}$ and $\widehat{\Theta}_{\text{norm}}$ are also useful from a theoretical perspective as will be shown in the sequel.

We now apply the de-biasing ideas of Section 14.1.2 to the graphical Lasso estimators defined above, demonstrating the procedure on $\widehat{\Theta}$. By definition, the graphical Lasso is invertible, and the Karush-Kuhn-Tucker (KKT) conditions yield

$$\widehat{\Sigma} - \widehat{\Theta}^{-1} + \lambda\widehat{Z} = 0,$$

where

$$\widehat{Z}_{ij} = \text{sign}(\widehat{\Theta}_{ij}) \text{ if } \widehat{\Theta}_{ij} \neq 0, \quad \text{and} \quad \|\widehat{Z}\|_\infty \leq 1.$$

Multiplying by $\widehat{\Theta}$, we obtain

$$\widehat{\Sigma}\widehat{\Theta} - I + \lambda\widehat{Z}\widehat{\Theta} = 0.$$

In line with Section 14.1.2 above, this implies the decomposition

$$\widehat{\Theta} + \widehat{\Theta}^T\eta(\widehat{\Theta}) - \Theta_0 = -\Theta_0(\widehat{\Sigma} - \Sigma_0)\Theta_0 + \text{rem}_0 + \text{rem}_{\text{reg}},$$

with the bias term $\eta(\widehat{\Theta}) = \lambda\widehat{Z}\widehat{\Theta}$. To control the remainder terms rem_0 and rem_{reg}, we need bounds for the ℓ_1-error of $\widehat{\Theta}$ and to control the bias term, it is sufficient to control the upper bound $\|\eta(\widehat{\Theta})\|_\infty = \|\lambda\widehat{Z}\widehat{\Theta}\|_\infty \leq \lambda \left|\left|\left|\widehat{\Theta}\right|\right|\right|_1$.

Oracle bounds

Oracle results for the graphical Lasso were studied in [29] under sparsity conditions and mild regularity conditions. In [26], stronger results were derived under stronger regularity conditions (and weaker sparsity conditions). Here we revisit these results and provide several extensions.

We summarize the main theoretical conditions which require boundedness of the eigenvalues of the true precision matrix and certain tail conditions.

Condition A1 (Bounded spectrum). The precision matrix $\Theta_0 := \Sigma_0^{-1}$ exists and there exists a universal constant $L \geq 1$ such that

$$1/L \leq \Lambda_{\min}(\Theta_0) \leq \Lambda_{\max}(\Theta_0) \leq L.$$

Condition A2 (Sub-Gaussianity). The observations $X^i, i = 1, \dots, n$, are uniformly sub-Gaussian vectors, i.e. there exists a universal constant $K > 0$ such that for every $\alpha \in \mathbb{R}^p$, $\|\alpha\|_2 = 1$ it holds

$$\mathbb{E}\exp\left(|\alpha^T X^i|^2/K^2\right) \le 2 \quad (i = 1, \dots, n).$$

Under Condition A2, the Bernstein inequality implies concentration results for $\widehat{\Sigma}$ as formulated in Lemma 14.1.1 below. The proof is omitted and may be found in [4] (Lemma 14.13). We denote the Euclidean norm by $\|\cdot\|_2$ and the i-th column of a matrix A by A_i.

Lemma 14.1.1. *Assume Condition A2 and that for non-random matrices $A, B \in \mathbb{R}^{p\times p}$ it holds that $\|A_i\|_2 \le M$ and $\|B_i\|_2 \le M$ for all $i = 1, \dots, p$. Then for all $t > 0$, with probability at least $1 - e^{-nt}$ it holds that*

$$\|A^T(\widehat{\Sigma} - \Sigma_0)B\|_\infty/(2M^2K^2) \le t + \sqrt{2t} + \sqrt{\frac{2\log(2p^2)}{n}} + \frac{\log(2p^2)}{n}.$$

To derive oracle bounds for the graphical Lasso, we rely on certain sparsity conditions on the entries of the true precision matrix. To this end, we define for $j = 1, \dots, p$,

$$D_j := \{(i,j) : \Theta^0_{ij} \ne 0, i \ne j\}, \quad d_j := \mathrm{card}(D_j), \quad d := \max_{j=1,\dots,p} |d_j|.$$

The quantity d_j is then the degree of a node X_j and d corresponds to the maximum vertex degree in the associated graphical model (excluding vertex self-loops). Furthermore, we define

$$S := \bigcup_{j=1}^{p} D_j, \quad s := \sum_{j=1}^{p} d_j,$$

thus S denotes the overall off-diagonal sparsity pattern and s is the overall number of edges (excluding self-loops).

The following theorem is an extension of the result for the graphical Lasso in [29] to the ℓ_1-norm. The paper [29] derives rates in Frobenius norm $\|\cdot\|_F$, which is defined as $\|A\|_F^2 := \sum_{i,j} |A_{ij}^2|$ for a matrix A.

Theorem 14.1.2 (Regime $p \ll n$). *Let $\widehat{\Theta}$ be the minimizer defined by (14.8). Assume Conditions A1 and A2. Then for λ satisfying $2\lambda_0 \le \lambda \le 1/(8Lc_L)$ and $8c_L^2 s\lambda^2 + 8c_L p\lambda_0^2 \le \lambda_0/(2L)$, on the set $\|\widehat{\Sigma} - \Sigma_0\|_\infty \le \lambda_0$, it holds that*

$$\|\widehat{\Theta} - \Theta_0\|_F^2/c_L + \lambda\|\widehat{\Theta}^- - \Theta_0^-\|_1 \le 8c_L^2 s\lambda^2 + 8c_L p\lambda_0^2,$$

and

$$\left|\left|\left|\widehat{\Theta} - \Theta_0\right|\right|\right|_1 \le 16c_L^2(p+s)\lambda,$$

where $c_L = 8L^2$.

The slow rate in the result above arises from the part of the estimation error $\mathrm{tr}[(\widehat{\Sigma} - \Sigma_0)(\widehat{\Theta} - \Theta_0)]$ which is related to the diagonal elements of the precision matrix. However, proper normalizing removes this part of the estimation error.

The following theorem derives an extension of [29] for the normalized graphical Lasso $\widehat{\Theta}_{\text{norm}}$. Denote the true correlation matrix by R_0 and let $K_0 := R_0^{-1}$ denote the inverse correlation matrix.

Theorem 14.1.3 (Regime $p \gg n$). *Assume that Conditions A1 and A2 hold. Then for λ satisfying $2\lambda_0 \leq \lambda \leq 1/(8L^2)$ and $8c_L^2 s\lambda^2 \leq \lambda_0/(2L)$, on the set $\|\widehat{R} - R_0\|_\infty \leq \lambda_0$ it holds for some constant $C_L > 0$ that*

$$\|\widehat{\Theta}_{\text{norm}} - K_0\|_F^2 + \lambda \|\widehat{\Theta}_{\text{norm}}^- - K_0^-\|_1 \leq 8c_L^2 s\lambda^2,$$

$$\left|\left|\left|\widehat{\Theta}_{\text{norm}} - K_0\right|\right|\right|_1 \leq 8c_L s\lambda^2 + 8c_L^2 s\lambda.$$

$$\left|\left|\left|\widehat{\Theta}_{\text{w}} - \Theta_0\right|\right|\right|_1 \leq C_L s\lambda,$$

where $c_L = 8L^2$.

Using the normalized graphical Lasso leads to faster rates in Frobenius norm and ℓ_1-norm as shown above. The rates for $\widehat{\Theta}_{\text{w}}$ in $|||\cdot|||_1$-norm can be then established immediately. To derive a high-probability bound for $\|\widehat{R} - R_0\|_\infty$, we may apply Lemma 14.1.1 together with Hölder's inequality to obtain $\|\widehat{R} - R_0\|_\infty = \mathcal{O}_P(\sqrt{\log p/n})$. Hence, $\left|\left|\left|\widehat{\Theta}_{\text{w}} - \Theta_0\right|\right|\right|_1 = \mathcal{O}_P(s\sqrt{\log p/n})$.

Remark 14.1.4. The above result requires a strong condition on the sparsity in Θ_0, i.e. there can only be very few non-zero coefficients due to the restriction $8c_L^2 s\lambda^2 \leq \lambda_0/(2L)$. This condition guarantees that a margin condition is satisfied. An example of a graph satisfying the condition is a star graph with order $\sqrt{n}$ edges.

Asymptotic normality

Once oracle results in ℓ_1-norm are available, we can easily obtain results on asymptotic normality of the de-biased estimator $2\widehat{\Theta} - \widehat{\Theta}\widehat{\Sigma}\widehat{\Theta}$ for the graphical Lasso and its weighted version. We denote the asymptotic variance of the de-biased estimator by $\sigma_{ij}^2 := n\text{var}((\Theta_i^0)^T \widehat{\Sigma}\Theta_j^0)$. The arrow $\rightsquigarrow$ denotes convergence in distribution and for a matrix A we denote $(A)_{ij}$ its (i,j)-entry.

Theorem 14.1.5 (Regime $p \ll n$). *Assume Conditions A1, A2, $\lambda \asymp \sqrt{\log p/n}$ and that $(p+s)\sqrt{d} = o(\sqrt{n}/\log p)$. Then it holds that*

$$2\widehat{\Theta} - \widehat{\Theta}\widehat{\Sigma}\widehat{\Theta} - \Theta_0 = -\Theta_0(\widehat{\Sigma} - \Sigma_0)\Theta_0 + \text{rem}, \tag{14.11}$$

where

$$\|\text{rem}\|_\infty = \mathcal{O}_P\left(24(8L^2)^2 L(p+s)\sqrt{d+1}\lambda^2\right) = o_P(1/\sqrt{n}).$$

Moreover, for $i, j = 1, \ldots, p$,

$$\sqrt{n}(2\widehat{\Theta} - \widehat{\Theta}\widehat{\Sigma}\widehat{\Theta} - \Theta_0)_{ij}/\sigma_{ij} \rightsquigarrow \mathcal{N}(0,1).$$

The result of Theorem 14.1.5 gives us tools to construct approximate confidence intervals and tests for individual entries of Θ_0. However, we need a consistent estimator of the asymptotic variance σ_{ij}. For the Gaussian case, one may take $\widehat{\sigma}_{ij}^2 := \widehat{\Theta}_{ii}\widehat{\Theta}_{jj} + \widehat{\Theta}_{ij}^2$. We omit the proof of consistency of $\widehat{\sigma}_{ij}$ and point the reader to [16], where other distributions are treated as well. Moreover, Theorem 14.1.5 implies parametric rates of convergence for estimation of individual entries and a rate of order $\sqrt{\log p/n}$ for the error in supremum norm. Theorem 14.1.5 requires a stronger sparsity condition than the corresponding oracle-type inequality in Theorem 14.1.2. This is to be expected as will be argued in Section 14.1.7.

Using the weighted graphical Lasso, the results of Theorem 14.1.5 can be established under weaker conditions as shown in the following theorem.

Theorem 14.1.6 (Regime $p \gg n$). *Assume Conditions A1, A2 and* $s\sqrt{d} = o(\sqrt{n}/\log p)$. *Then for* $\lambda \asymp \sqrt{\log p/n}$, *the asymptotic linearity* (14.11) *holds with* $\widehat{\Theta}_{\mathrm{w}}$, *where*

$$\|\mathrm{rem}\|_\infty = \mathcal{O}_P\left(12(8L^2)^2 s\sqrt{d+1}\lambda^2\right) = o_P(1/\sqrt{n}).$$

Moreover, for $i,j = 1,\dots,p$, $\sqrt{n}(2\widehat{\Theta}_{\mathrm{w}} + \widehat{\Theta}_{\mathrm{w}}\widehat{\Sigma}\widehat{\Theta}_{\mathrm{w}} - \Theta_0)_{ij}/\sigma_{ij} \rightsquigarrow \mathcal{N}(0,1)$.

If the parameter of interest is instead the inverse correlation matrix, we formulate a partial result below.

Proposition 14.1.7 (Regime $p \gg n$). *Assume Conditions A1, A2,* $\lambda \asymp \sqrt{\log p/n}$ *and that* $s\sqrt{d} = o(\sqrt{n}/\log p)$. *Then it holds*

$$2\widehat{\Theta}_{\mathrm{norm}} - \widehat{\Theta}_{\mathrm{norm}}\widehat{R}\widehat{\Theta}_{\mathrm{norm}} - K_0 = -K_0(\widehat{R} - R_0)K_0 + \mathrm{rem},$$

$$\|\mathrm{rem}\|_\infty = \mathcal{O}_P\left(12(8L^2)^2 Ls\sqrt{d+1}\lambda^2\right) = o_P(1/\sqrt{n}).$$

To claim asymptotic normality of $K_0(\widehat{R} - R_0)K_0$ per entry would require extensions of central limit theorems to high-dimensional settings (see [7]) and an extension of the δ-method. We do not study these extensions in the present work. To give a glimpse, in the regime when p is fixed, by the central limit theorem it follows that $\sqrt{n}(\widehat{\Sigma} - \Sigma_0) \rightsquigarrow \mathcal{N}_{p^2}(0, C)$, where C is the asymptotic covariance matrix. Then we may apply the δ-method with $h_{ij}(\Sigma) := (K_i^0)^T \mathrm{diag}(\Sigma)^{-1/2}\Sigma\,\mathrm{diag}(\Sigma)^{-1/2}K_j^0$ to obtain $\sqrt{n}(h_{ij}(\widehat{\Sigma}) - h_{ij}(\Sigma_0)) \rightsquigarrow \mathcal{N}(0, \dot{h}(\Sigma_0)^T C \dot{h}(\Sigma_0))$.

Finally, we show that the sparsity conditions in the above results may be further relaxed under a stronger regularity condition on the true precision matrix. We provide here a simplified version of the result in [15] which assumes an irrepresentability condition on the true precision matrix. Let κ_{Σ_0} be the ℓ_∞-operator norm of the true covariance matrix Σ_0, i.e. $\kappa_{\Sigma_0} = |||\Sigma_0|||_1$. Let H_0 be the Hessian of the expected Gaussian log-likelihood evaluated at Θ_0, i.e. $H_0 = \Sigma_0 \otimes \Sigma_0$. When calculating the Hessian matrix, we treat the precision matrix as non-symmetric; this will allow us to accommodate non-symmetric estimators as well. For any two subsets T and T' of $\mathcal{V} \times \mathcal{V}$, we use $H^0_{TT'}$ to denote the $|T| \times |T'|$ matrix with rows and columns of H_0 indexed by T and T' respectively. Define $\kappa_{H_0} = |||(H^0_{SS})^{-1}|||_1$.

Condition A3. (Irrepresentability condition) There exists $\alpha \in (0,1]$ such that $\max_{e\in S^c} \|H^0_{eS}(H^0_{SS})^{-1}\|_1 \le 1 - \alpha$, where S^c is the complement of S.

Condition A3 is an analogy of the irrepresentable condition for variable selection in linear regression (see [33]). If we define the zero-mean edge random variables as $Y_{(i,j)} := X_iX_j - \mathbb{E}(X_iX_j)$, then the matrix H_0 corresponds to covariances of the edge variables, in particular $H^0_{(i,j),(k,l)} + H^0_{(j,i),(k,l)} = \mathrm{cov}(Y_{(i,j)}, Y_{(k,l)})$. Condition A3 means that we require that no edge variable $Y_{(j,k)}$, which is not included in the edge set S, is highly correlated with variables in the edge set (see [26]). The parameter α then is a measure of this correlation with the correlation growing when $\alpha \to 0$. Some examples of matrices satisfying the irrepresentable condition may be found in [16].

Theorem 14.1.8 (Regime $p \gg n$). *Assume that Conditions A1, A2 and A3 are satisfied,* $d = o(\sqrt{n}/\log p)$, $\kappa_{\Sigma_0} = \mathcal{O}(1)$ *and* $\kappa_{H_0} = \mathcal{O}(1)$. *Then for* $\lambda \asymp \sqrt{\log p/n}$, *the asymptotic linearity* (14.11) *holds with* $\widehat{\Theta}$, *where* $\|\mathrm{rem}\|_\infty = \mathcal{O}_P(d\log p/n) = o_P(1/\sqrt{n})$. *Moreover,*

$$\sqrt{n}(2\widehat{\Theta} - \widehat{\Theta}\widehat{\Sigma}\widehat{\Theta} - \Theta_0)_{ij}/\sigma_{ij} \rightsquigarrow \mathcal{N}(0,1).$$

The proof of Theorem 14.1.8 may be found in [15]. We remark that under the irrepresentability condition, one can show that $|\widehat{\Theta}_{ij} - \Theta^0_{ij}| = \mathcal{O}_P(1/\sqrt{n})$ (see [26]). This means that one could use the graphical Lasso itself to construct confidence intervals of asymptotically optimal (parametric) size. However, this holds under the strong irrepresentability condition which is often violated in practice.

Comparing the results obtained for the (weighted) graphical Lasso, we see that the strongest result was attained under the irrepresentable condition and under the sparsity condition $d = o(\sqrt{n}/\log p)$. An analogous result has not yet been obtained for the graphical Lasso without the irrepresentable condition (under the same sparsity condition). However, without the irrepresentable condition, we showed the same results for the weighted graphical Lasso under the sparsity condition $s\sqrt{d} = o(\sqrt{n}/\log p)$. In the next section, we will consider a procedure based on pseudo-likelihood, for which we can derive identical results under weaker conditions, namely under the sparsity condition $d = o(\sqrt{n}/\log p)$ and under the Conditions A1 and A2.

14.1.4 Nodewise square-root Lasso

An alternative approach to estimate the precision matrix is based on linear regression. The idea of nodewise Lasso is to estimate each column of the precision matrix by doing a projection of every column of the design matrix on all the remaining columns. While this is a pseudo-likelihood method, the decoupling into linear regressions gains more flexibility in estimating the individual scaling levels compared to the graphical Lasso which aims to estimate all the parameters simultaneously. Moreover, by splitting the problem up into a series of linear regressions, the computational burden is reduced compared to the graphical Lasso.

In low-dimensional settings, regressing each column of the design matrix on all the other columns would simply recover the inverse of the sample covariance matrix $\widehat{\Sigma}$. However, due to the high-dimensionality of our setting, the matrix $\widehat{\Sigma}$ is not invertible and we can only do approximate projections. If we assume sparsity in the precision matrix (and thus also in the partial correlations), this idea can be effectively carried out using the square-root Lasso ([1]).

The theoretical motivation can be understood in greater detail from the population version of the method. For each $j = 1, \ldots, p$, we define the vector of partial correlations $\gamma^0_j = \{\gamma^0_{j,k}, k \neq j\}$ as follows

$$\gamma^0_j := \text{argmin}_{\gamma \in \mathbb{R}^{p-1}} \mathbb{E}\|X_j - X_{-j}\gamma\|_2^2/n, \tag{14.12}$$

and we denote the noise level by $\tau_j^2 = \mathbb{E}\|X_j - X_{-j}\gamma^0_j\|_2^2/n$. Then one may show that the j-th column of Θ_0 can be recovered from the partial correlations γ^0_j and the noise level τ_j using the following identity: $\Theta^0_j = (-\gamma_{j,1}, \ldots, -\gamma_{j,j-1}, 1, -\gamma_{j,j+1}, \ldots, -\gamma_{j,p})^T/\tau_j^2$.

Hence, the idea is to define for each $j = 1, \ldots, p$ the estimators of the partial correlations, $\widehat{\gamma}_j = \{\widehat{\gamma}_{j,k}, k = 1, \ldots, p, j \neq k\} \in \mathbb{R}^{p-1}$ using, for instance, the square-root Lasso with weighted penalty,

$$\widehat{\gamma}_j := \text{argmin}_{\gamma \in \mathbb{R}^{p-1}} \|X_j - X_{-j}\gamma\|_2/n + 2\lambda\|\widehat{W}_{-j}\gamma\|_1, \tag{14.13}$$

where by A_{-j} we denote the matrix A without its j-th column. We further define estimators of the noise level

$$\widehat{\tau}_j^2 := \|X_j - X_{-j}\widehat{\gamma}_j\|_2^2/n, \quad \widetilde{\tau}_j^2 := \widehat{\tau}_j^2 + \lambda\widehat{\tau}_j\|\widehat{\gamma}_j\|_1,$$

for $j = 1, \ldots, p$. Finally, we define the nodewise square-root Lasso estimator

$$\widehat{\Theta} := \begin{pmatrix} 1/\widetilde{\tau}_1^2 & -\widetilde{\gamma}_{1,2}/\widetilde{\tau}_1^2 & \cdots & -\widetilde{\gamma}_{1,p}/\widetilde{\tau}_1^2 \\ -\widetilde{\gamma}_{2,1}/\widetilde{\tau}_2^2 & 1/\widetilde{\tau}_2^2 & \cdots & -\widetilde{\gamma}_{2,p}/\widetilde{\tau}_2^2 \\ \vdots & \vdots & \ddots & \vdots \\ -\widetilde{\gamma}_{p,1}/\widetilde{\tau}_p^2 & \cdots & -\widetilde{\gamma}_{p,p-1}/\widetilde{\tau}_p^2 & 1/\widetilde{\tau}_p^2 \end{pmatrix} \tag{14.14}$$

An equivalent way of formulating the definitions above is

$$(\widehat{\gamma}_j, \widehat{\tau}_j) = \text{argmin}_{\gamma \in \mathbb{R}^{p-1}, \tau \in \mathbb{R}} \|X_j - X_{-j}\gamma\|_2^2/n/(2\tau) + \tau/2 + 2\lambda\|\widehat{W}_{-j}\gamma\|_1. \tag{14.15}$$

Alternative versions of the above estimator were considered in the literature. One may use the Lasso ([31]) instead of the square-root Lasso (as in [16]) or the Dantzig selector (see [36]). Furthermore, one may define the nodewise square-root Lasso with $\widehat{\tau}_j$ in place of $\widetilde{\tau}_j$.

The properties of the column estimator $\widehat{\Theta}_j$ were studied in several papers (following [23]) and it has been shown to enjoy oracle properties under the Conditions A1, A2 and under the sparsity condition $d = \mathcal{O}(n/\log p)$ (see [35], where a similar version was considered).

In line with Section 14.1.2, we consider a de-biased version of the nodewise square-root Lasso estimator. The KKT conditions for the optimization problem (14.13) give

$$-\widehat{\tau}_j X_{-j}^T(X_j - X_{-j}\widehat{\gamma}_j)/n + \lambda\widehat{\kappa}_j = 0, \tag{14.16}$$

for $j = 1, \ldots, p$, where $\widehat{\kappa}_j$ is an element of the sub-differential of the function $\gamma_j \mapsto \|\gamma_j\|_1$ at $\widehat{\gamma}_j$, i.e. for $k \in \{1, \ldots, p\} \setminus \{j\}$,

$$\widehat{\kappa}_{j,k} = \text{sign}(\widehat{\gamma}_{j,k}) \text{ if } \widehat{\gamma}_{j,k} \neq 0, \quad \text{and} \quad \|\widehat{\kappa}_j\|_\infty \leq 1.$$

If we define $\widehat{Z}_j$ to be a $p \times 1$ vector

$$\widehat{Z}_j := (\widehat{\kappa}_{j,1}, \ldots, \widehat{\kappa}_{j,j-1}, 0, \widehat{\kappa}_{j,j+1}, \ldots, \widehat{\kappa}_{j,p}),$$

then the KKT conditions may be equivalently stated as follows

$$\widehat{\Sigma}\widehat{\Theta}_j - e_j - \lambda\frac{\widehat{\tau}_j}{\widetilde{\tau}_j^2}\widehat{Z}_j = 0.$$

Let $\widehat{Z}$ be a matrix with columns $\widehat{Z}_j$ for $j = 1, \ldots, p$, $\widehat{\tau}$ be a diagonal matrix with elements $(\widehat{\tau}_1, \ldots, \widehat{\tau}_p)$ and similarly $\widetilde{\tau} := \text{diag}(\widetilde{\tau}_1, \ldots, \widetilde{\tau}_p)$. As in Section 14.1.2, this yields the decomposition (14.6) with a bias term $\eta(\widehat{\Theta}) := \widehat{Z}\Lambda\widehat{\tau}\widetilde{\tau}^{-2}$. The bias term can then be controlled as

$$\|\eta(\widehat{\Theta})\|_\infty \leq \lambda\|\widehat{\tau}\|_\infty\|\widetilde{\tau}^{-2}\|_\infty\|\widehat{Z}\|_\infty \leq \lambda \max_{1\leq j\leq p} \widehat{\tau}_j/\widetilde{\tau}_j^2.$$

Theorem 14.1.9 (Regime $p \gg n$). *Suppose that Conditions A1, A2 are satisfied and $d = o(\sqrt{n}/\log p)$. Let $\widehat{\Theta}_{\text{node}}$ be the estimator defined in* (14.14) *and let $\lambda \asymp \sqrt{\log p/n}$. Then it holds*

$$\widehat{\Theta} + \widehat{\Theta}^T - \widehat{\Theta}^T\widehat{\Sigma}\widehat{\Theta} - \Theta_0 = -\Theta_0(\widehat{\Sigma} - \Sigma_0)\Theta_0 + \text{rem},$$

where $\|\text{rem}\|_\infty = \mathcal{O}_P(d\lambda^2) = o_P(1/\sqrt{n})$. *Moreover,*

$$\sqrt{n}(\widehat{\Theta} + \widehat{\Theta}^T - \widehat{\Theta}^T\widehat{\Sigma}\widehat{\Theta} - \Theta_0)_{ij}/\sigma_{ij} \rightsquigarrow \mathcal{N}(0, 1).$$

When the parameter of interest is the inverse correlation matrix, we can use the normalized version of the nodewise square-root Lasso and we obtain an analogous result.

Proposition 14.1.10 (Regime $p \gg n$). *Suppose that Conditions A1, A2 are satisfied, let $\lambda \asymp \sqrt{\log p/n}$ and $d = o(\sqrt{n}/\log p)$. Then*

$$\widehat{\Theta}_{\text{norm}} + \widehat{\Theta}_{\text{norm}}^T - \widehat{\Theta}_{\text{norm}}^T\widehat{R}\widehat{\Theta}_{\text{norm}} - \Theta_0 = -K_0(\widehat{R} - R_0)K_0 + \text{rem},$$

where $\|\text{rem}\|_\infty = o_P(1/\sqrt{n})$.

14.1.5 Computational view

For the nodewise square-root Lasso, we need to solve p square-root Lasso regressions, which can be efficiently handled using interior-point methods with polynomial computational time or first-order methods (see [1]). Alternatively to nodewise square-root Lasso, the nodewise Lasso studied in [16] may be used, which requires selection of a tuning parameter for each of the p regressions. This can be achieved e.g. by cross-validation and can be implemented efficiently using parallel methods ([11]). The graphical Lasso presents a more computationally challenging problem; we refer the reader to e.g. [22]. The computation of the de-biased estimator itself only involves simple matrix addition and multiplication.

14.1.6 Simulation results

We consider a setting with n independent observations generated from $\mathcal{N}_p(0, \Theta_0^{-1})$, where the precision matrix Θ_0 follows one of the three models:

1. Model 1: Θ_0 has two blocks of equal size and each block is a five-diagonal matrix with elements $(1, 0.5, 0.4)$ and $(2, 1, 0.6)$, respectively.
2. Model 2: Θ_0 is a sparse precision matrix generated using the R package GGMselect (using the function simulateGraph() with parameter 0.07). The matrix was converted to a correlation matrix with the function cov2cor().
3. Model 3: $\Theta^0_{ij} = 0.5^{|i-j|}$, $i, j = 1, \dots, p$.

We consider 6 different methods: the de-biased estimator based on the

(1) graphical Lasso (glasso),

(2) weighted graphical Lasso (glasso-weigh),

(3) nodewise square-root Lasso (node-sqrt) as defined above,

(4) nodewise square-root Lasso with alternative $\tilde{\tau}$ as in [30] (node-sqrt-tau),

(5) nodewise Lasso as in [16] (node)

and we also consider

(6) the maximum likelihood estimator (MLE).

Furthermore, as a benchmark we report the oracle estimator (oracle) which applies maximum likelihood using the knowledge of true zeros in the precision matrix. We also report the target coverage and the efficient asymptotic variance of asymptotically regular estimators (see [17]) (perfect).

For the graphical Lasso (1) and the weighted graphical Lasso (2) we choose the tuning parameter by maximizing the likelihood on a validation data set (a new data of the size n). For methods (3), (4) and (5), the universal choice $\sqrt{\log p/n}$ is used.

We display results on confidence intervals for nominal coverage 95% and for normally distributed observations in Tables 14.1, 14.2, 14.3 and 14.4. For other nominal coverages, we obtain similar performance (these results are not reported). For other than Gaussian distributions, we refer the reader to the simulation results in [16]. Firstly, the results of the simulations suggest that the de-biased estimators perform significantly better than the maximum likelihood estimator even though $p < n$ and secondly, the nodewise methods seem to outperform the graphical Lasso methods in our settings.

Model 1: Block 1: $(1, 0.5, 0.4)$, Block 2: $(2, 1, 0.6)$

$p = 100, n = 200$

	Method	Coverage S_0	Coverage S_0^c	Length S_0	Length S_0^c	Average λ
1	glasso	77.19	98.07	0.36	0.32	0.088
2	glasso-weigh	35.02	98.65	0.31	0.27	0.088
3	node-sqrt	89.92	94.02	0.48	0.42	0.152
4	node-sqrt-tau	83.48	97.40	0.38	0.33	0.152
5	node	90.58	96.77	0.41	0.35	0.152
6	MLE	20.92	84.27	0.97	0.81	-
7	oracle*	94.95	-	0.49	0.40	-
8	perfect*	95.00	95.00	0.48	0.40	-

$p = 100, n = 400$

	Method	Coverage S_0	Coverage S_0^c	Length S_0	Length S_0^c	Average λ
1	glasso	84.28	97.53	0.27	0.23	0.049
2	glasso-weigh	46.22	98.41	0.24	0.20	0.049
3	node-sqrt	91.57	94.40	0.34	0.29	0.107
4	node-sqrt-tau	87.11	97.13	0.28	0.24	0.107
5	node	91.48	96.40	0.30	0.25	0.107
6	MLE	41.42	91.29	0.41	0.38	-
7	oracle*	94.87	-	0.34	0.29	-
8	perfect*	95.00	95.00	0.34	0.29	-

Model 1		Block 1						Block 2					
		Θ_{11}		Θ_{12}		Θ_{13}		$\Theta_{50,50}$		$\Theta_{50,51}$		$\Theta_{50,52}$	
	Method	cov	len	cov	len	cov	len	cov	len	cov	len	cov	len
1	glasso	81	0.26	92	0.20	88	0.19	80	0.26	99	0.25	99	0.24
2	glasso-weigh	92	0.24	65	0.17	55	0.16	85	0.23	99	0.24	99	0.23
3	node-sqrt	95	0.29	93	0.22	82	0.21	92	0.29	95	0.29	95	0.28
4	node-sqrt-tau	96	0.26	84	0.20	69	0.19	86	0.26	96	0.26	96	0.25
5	node	90	0.27	94	0.21	85	0.20	86	0.27	99	0.27	97	0.26
6	oracle*	95	0.28	96	0.22	92	0.21	90	0.28	98	0.28	96	0.28
7	perfect*	95	0.28	95	0.22	95	0.21	95	0.28	95	0.28	95	0.28

FIGURE 14.1: Visualization of graphical models used in simulations. Models 1,2 and 3 from left to right. For Model 3, we only plot edges with a weight greater than 0.1.

Model 2

$p = 100, n = 400$

		Coverage		Length		Average λ
	Method	S_0	S_0^c	S_0	S_0^c	
1	glasso	64.17	98.65	0.16	0.15	0.067
2	glasso-weigh	16.80	98.56	0.05	0.05	0.067
3	node-sqrt	87.23	94.43	0.24	0.21	0.107
4	node-sqrt-tau	89.81	97.23	0.20	0.18	0.107
5	node	38.19	99.07	0.10	0.10	0.107
6	MLE	50.98	91.22	0.30	0.26	-
7	oracle*	98.51	-	0.23	0.20	-
8	perfect*	95.00	95.00	0.22	0.20	-

14.1.7 Discussion

We have shown several constructions of asymptotically linear estimators of the precision matrix (and inverse correlation matrix) based on regularized estimators, which immediately lead to inference in Gaussian graphical models. Efficient algorithms are available for both methods as discussed in Section 14.1.5. The constructed estimators achieve entrywise estimation at the parametric rate and a rate of convergence of order $\sqrt{\log p/n}$ in supremum norm.

To provide a brief comparison of the two methods analyzed above, both theoretical and computational results seem in favor of the de-sparsified nodewise Lasso. Theoretical results for nodewise Lasso in the regime $p \gg n$ only need the mild Conditions A1, A2 and

Model 3: $\Theta_{ij} = 0.5^{i-j}$

$p = 100, n = 200$

	Method	Coverage	Length	Average λ
1	glasso	90.43	0.19	0.138
2	glasso-weigh	75.81	0.33	0.138
3	node-sqrt	93.36	0.28	0.152
4	node-sqrt-tau	92.91	0.22	0.152
5	node	89.88	0.20	0.152
6	MLE	80.41	0.56	-
7	perfect*	95.00	0.28	-

$d = o(\sqrt{n}/\log p)$ and are uniform over the considered model. Moreover, the de-sparsified nodewise Lasso may be thresholded again to yield recovery of the set S with no false positives, and under a beta-min type condition, exact recovery of the set S, asymptotically, with high probability. The graphical Lasso requires that we impose the strong irrepresentability condition in the high-dimensional regime. However, the graphical Lasso might be preferred on the grounds that it does not decouple the likelihood. Moreover, the graphical Lasso estimator is always strictly positive definite and thus yields an estimator of the covariance matrix as well. The invertibility of the nodewise Lasso has not yet been explored.

We remark that the sparsity condition $d = o(\sqrt{n}/\log p)$ implied by our analysis is stronger than the condition needed for oracle inequalities and recovery, namely $d = o(n/\log p)$. However, one can show that this sparsity condition is essentially necessary for asymptotically normal estimation. This follows by inspection of the minimax rates (see [28]).

14.2 Directed Acyclic Graphs

In this section, we use the de-biasing ideas to construct confidence intervals for edge weights in directed acyclic graphs (abbreviated as DAGs). A directed acyclic graph is a directed graph (we distinguish between edges (j, k) and (k, j)) without directed cycles. For an overview of fundamental concepts on DAGs, we refer to Section 1.8 in Chapter 1 and to Chapter 15. We consider the Gaussian DAG model, where the DAG represents the probability distribution of a random vector $(X_1, \dots, X_p)$ with a Gaussian distribution $\mathcal{N}(0, \Sigma_0)$, where $\Sigma_0 \in \mathbb{R}^{p \times p}$ is an unknown covariance matrix. A Gaussian DAG may be represented by the linear structural equations model

$$X_j = \sum_{k \in \mathrm{pa}(j)} \beta^0_{k,j} X_k + \epsilon_j, \qquad j = 1, \dots, p,$$

where $\epsilon_1, \dots, \epsilon_p$ are independent and $\epsilon_j \sim \mathcal{N}(0, (\omega^0_j)^2)$. The set $\mathrm{pa}(j)$ is called the set of parents of a node j and it contains all nodes $k \in \{1, \dots, p\}$ such that there exists a directed edge $k \to j$.

Our aim is to construct confidence intervals for edge weights $\beta^0_{k,j}$. However, without further conditions, the DAG and the $\beta^0_{k,j}$'s may not be identifiable from the structural equations model. To ensure identifiability, we assume that the error variances are equal: $\omega^0_j = \omega_0$ for all $j = 1, \dots, p$. We remark that one might equivalently assume that the error variances are all known up to a multiplicative constant. In this setting, the DAG is identifiable as shown in [25]. Our strategy is to use a two-step procedure: in the first step we use the estimator proposed in [34] to estimate the ordering of the variables and in the second step, we use a de-biased version of nodewise regression to construct the confidence intervals.

Given an $n \times p$ matrix $X = [X_1, \dots, X_p]$, with rows being n independent observations from the structural equations model, one may rewrite the above model in a matrix form

$$X = XB_0 + E,$$

where $B_0 := (\beta^0_{k,j})$ is a $p \times p$ matrix with $\beta^0_{j,j} = 0$ for all j, and E is an $n \times p$ matrix of noise vectors $E := (\epsilon_1, \dots, \epsilon_p)$ with columns ϵ_j independent of X_k whenever $\beta^0_{k,j} \neq 0$. The rows of E are independent $\mathcal{N}(0, \omega_0^2 I)$-distributed random vectors. The model then implies that X has covariance matrix

$$\Sigma_0 = \omega_0^2((I - B_0)^{-1})^T (I - B_0)^{-1}.$$

We define the precision matrix (assumed to exist) by $\Theta_0 := \Sigma_0^{-1}$. Notice that

$$\Theta_0 = \frac{1}{\omega_0^2}(I - B_0)(I - B_0)^T.$$

We further consider the class of precision matrices corresponding to DAGs. That is, we let

$$\Theta := \Theta(B, \omega) = \frac{1}{\omega^2}(I - B)(I - B)^T,$$

where (B, ω) is such that there exists a DAG representing the distribution $\mathcal{N}(0, \Sigma)$ with $\Sigma = \omega^2((I - B)^{-1})^T(I - B)^{-1}$. This means that $\omega > 0$ and B can be written as a lower-diagonal matrix, up to permutation of rows. Further we let s_B denote the number of nonzero entries in B, which corresponds to the number of edges in the DAG. Moreover, we denote by $\mathcal{B}$ the set of all edge weights B of DAGs with parameters (B, ω) which have at most $\alpha n / \log p$ incoming edges (parents) at each node, where $\alpha > 0$ is given.

14.2.1 Maximum likelihood estimator with ℓ_0-penalization

In the first step, we use an ℓ_0-penalized maximum likelihood estimator to estimate the DAG. Let $\widehat{\Sigma} = X^T X/n$ be the Gram matrix based on the design matrix X. The minus log-likelihood is proportional to $\ell(\Theta) = \text{trace}(\Theta\widehat{\Sigma}) - \log\det(\Theta)$. Consider the penalized maximum likelihood estimator proposed in [34],

$$\begin{aligned}(\widehat{B}, \widehat{\omega}) \quad := \quad & \text{argmin}_{B,\omega}\{\, \ell(B, \omega) + \lambda^2 s_B : \Theta = \Theta(B, \omega), \text{ for some DAG} \\ & \text{with parameters } (B, \omega) \text{ where } B \in \mathcal{B}\}, \end{aligned} \qquad (14.17)$$

where $\lambda \geq 0$ is a tuning parameter. The estimator is denoted by $\widehat{\Theta} = \Theta(\widehat{B}, \widehat{\omega})$ and it has $\widehat{s} := s_{\widehat{B}}$ edges. Calculating the ℓ_0-penalized maximum likelihood estimator over the class of DAGs is a computationally intensive task, especially because it involves a search through a class of DAGs under a non-convex constraint of acyclicity of the graph and due to the ℓ_0-penalty. For large scale problems, greedy algorithms may be used, see e.g. [8, 14]. The reason for using the ℓ_0-penalty instead of ℓ_1-penalization in the definition of (14.17) was discussed in [34]. The ℓ_1-penalty leads to an objective function which is not constant over equivalent DAGs encoding the same distribution. The ℓ_0-penalization leads to invariant scores over equivalent DAGs. The theoretical properties of $\widehat{\Theta}$ were studied in [34] under the conditions summarized below. We remark that the paper [34] primarily studies the estimator (14.17) with unequal variances and shows that the estimator converges to some member of the Markov equivalence class (cf. [24]) of a DAG with a minimal number of edges, under certain conditions.

To make their result precise, we define some further notions. For any vector $\beta \in \mathbb{R}^p$, let $\|X\beta\| := (\beta^T\Sigma_0\beta)^{1/2}$. By an ordering of variables we mean any permutation of the set $\{1, \ldots, p\}$. For any ordering of the variables, π, we let $\widetilde{B}(\pi)$ be the matrix obtained by doing a Gram-Schmidt orthogonalization of the columns of X in the ordering given by π, with respect to the norm $\|\cdot\|$. Moreover, let $\widetilde{\Omega}_0(\pi) = (I - \widetilde{B}_0(\pi))^T\Sigma_0(I - \widetilde{B}_0(\pi)) = \text{diag}((\widetilde{\omega}_1^0(\pi))^2, \ldots, (\widetilde{\omega}_p^0(\pi))^2)$. We restate the conditions assumed in [34].

Condition B1. There exists a universal constant $L \geq 1$ such that

$$1/L \leq \Lambda_{\min}(\Sigma_0) \leq \Lambda_{\max}(\Sigma_0) \leq L.$$

Condition B2. There exists a constant $\eta_\omega > 0$ such that for all π such that $\widetilde{\Omega}_0(\pi) \neq \omega_0^2 I$ it holds

$$\frac{1}{p}\sum_{i=1}^{p}(|\widetilde{\omega}_j^0(\pi)|^2 - \omega_0^2)^2 > 1/\eta_\omega.$$

Condition B3. There exists a sufficiently small constant α_* such that $p \le \alpha_* n / \log p$.

Condition B2 is an "omega-min" condition: it imposes that if one uses the wrong permutation then the error variances are far enough from being equal.

Under the above conditions, the ℓ_0-penalized maximum likelihood estimator with high-probability correctly estimates the ordering of the variables as shown in [34]. Let π_0 be an ordering of the variables such that a Gram-Schmidt orthogonalization of the columns of X in the order given by π_0 with respect to the norm $\|\cdot\|$ yields B_0. Denote the ordering of variables estimated by the ℓ_0-penalized maximum likelihood estimator by $\widehat{\pi}$. Then the result in [34] states that under Conditions B1, B2 and B3, with high-probability it holds that $\widehat{\pi} = \pi_0$.

14.2.2 Inference for edge weights

Given that we have recovered the true ordering π_0, the problem reduces to estimation of regression coefficients in a nodewise regression model, where each variable is a function of a known set of its "predecessors". Therefore to construct asymptotically normal estimators for the $\beta^0_{k,j}$'s, we may use a nodewise regression approach.

An estimated ordering $\widehat{\pi}$ yields estimates $\widehat{\mathrm{p}}(j)$ of the predecessor sets for each node $j = 1, \dots, p$. If we have recovered the true ordering, that is $\widehat{\pi} = \pi_0$, the estimated predecessor sets $\widehat{\mathrm{p}}(j)$ are equal to the true predecessor sets, which are supersets of the parent sets $\mathrm{pa}(j)$ for each $j = 1, \dots, p$. Consequently, given the predecessor sets, we may obtain a new estimator for the edge weights by regressing the j-th variable X_j on all its predecessors. The predecessor sets $\widehat{\mathrm{p}}(j)$ might be as large as $p-1$, therefore it is necessary to use regularization. Then we use the de-biasing technique in a similar spirit as in Section 14.1.2. We remark that in the initial step, one may use any estimator which guarantees exact recovery of the ordering π_0.

For any non-empty subset $T \subseteq \{1, \dots, p\}$, we denote by X_T the $n \times |T|$ matrix formed by taking the columns X_k of X such that $k \in T$. We define the nodewise regression estimator as proposed in [16] (altenatively, one may use the nodewise square-root Lasso studied in Section 14.1.4)

$$\widehat{\beta}_j = \operatorname*{argmin}_{\beta \in \mathbb{R}^{|\widehat{\mathrm{p}}(j)|}} \|X_j - X_{\widehat{\mathrm{p}}(j)}\beta\|_2^2/n + 2\lambda_j \|\beta\|_1. \tag{14.18}$$

The Karush-Kuhn-Tucker conditions for the above optimization problem give

$$-X^T_{\widehat{\mathrm{p}}(j)}(X_j - X_{\widehat{\mathrm{p}}(j)}\widehat{\beta}_j)/n + \lambda_j \widehat{Z}_j = 0,$$

where the entries of $\widehat{Z}_j$ satisfy $\widehat{Z}_{k,j} = \mathrm{sign}(\widehat{\beta}_{k,j})$ if $\widehat{\beta}_{k,j} \neq 0$, and $\|\widehat{Z}_j\|_\infty \le 1$ ($\widehat{\beta}_{k,j}$ denotes the k-th entry of $\widehat{\beta}_j$). Similarly as in the case of undirected graphical models, we can define a de-biased estimator. The Hessian matrix of the risk function in (14.18) is given by

$$\widehat{\Sigma}_{\widehat{\mathrm{p}}(j)} := X^T_{\widehat{\mathrm{p}}(j)} X_{\widehat{\mathrm{p}}(j)}/n.$$

To find a surrogate inverse for $\widehat{\Sigma}_{\widehat{\mathrm{p}}(j)}$, we construct $\widehat{\Theta}_{\widehat{\mathrm{p}}(j)}$ using the nodewise Lasso with tuning parameters $\lambda_{k,j}$ for $k \in \widehat{\mathrm{p}}(j)$. Using $\widehat{\Theta}_{\widehat{\mathrm{p}}(j)}$, we define the de-biased estimator

$$\widehat{b}_j := \widehat{\beta}_j + \widehat{\Theta}^T_{\widehat{\mathrm{p}}(j)} X^T_{\widehat{\mathrm{p}}(j)}(X_j - X_{\widehat{\mathrm{p}}(j)}\widehat{\beta}_j)/n. \tag{14.19}$$

Theorem 14.2.1 below shows that the entries of the de-biased estimator are asymptotically normal. To formulate the result, we define $\Theta^0_{\mathrm{p}(j)}$ to be the matrix obtained by taking the rows and columns of Θ_0 contained in the true predecessor set $\mathrm{p}(j)$. Denote the k-th column of $\Theta^0_{\mathrm{p}(j)}$ by $\Theta^0_{\mathrm{p}(j),k}$. To provide asymptotically normal estimators for the parameters $\beta^0_{\mathrm{p}(j)} = (\beta^0_{k,j} : k \in \mathrm{p}(j))$, we need to impose a sparsity condition on the sizes of the parent sets, which will be denoted by $d_j = |\mathrm{pa}(j)|$.

Theorem 14.2.1 (Regime $p \leq n$). *Let $\widehat{B}$ be the estimator defined by* (14.17) *with $\lambda \asymp \sqrt{\log p/n}$ and denote the predecessor sets estimated based on $\widehat{B}$ by $\widehat{\mathrm{p}}(j)$ for $j = 1, \ldots, p$. Let $\widehat{b}_j$ be defined in* (14.19) *with sufficiently large tuning parameters $\lambda_j \asymp \lambda_{k,j} \asymp \sqrt{\log p/n}$, uniformly in j, k, where $k \in \widehat{\mathrm{p}}(j)$. Assume Conditions B1, B2 and B3 are satisfied with $1/(|\alpha_*| + |\eta_\omega|) = \mathcal{O}(1)$ and assume that $d_j = o(\sqrt{n}/\log p)$. Then it holds*

$$\widehat{b}_j - \beta^0_{\mathrm{p}(j)} = (\Theta^0_{p(j)})^T X^T_{p(j)} \epsilon_j / n + \mathrm{rem}, \tag{14.20}$$

where

$$\|\mathrm{rem}\|_\infty = o_P(1/\sqrt{n}).$$

Furthermore, for every $k \in p(j)$,

$$\sqrt{n}(\widehat{b}_{k,j} - \beta^0_{k,j})/\sigma_{k,j} \rightsquigarrow \mathcal{N}(0,1),$$

where the asymptotic variance of the de-sparsified estimator is given by

$$\sigma^2_{k,j} := n\mathrm{var}((\Theta^0_{p(j),k})^T X^T_{p(j)} \epsilon_j) = \omega^2_0 (\Theta^0_{p(j)})_{kk}.$$

The result of Theorem 14.2.1 can be used to construct confidence intervals for the edge weights $\beta^0_{k,j}$. To estimate the asymptotic variance, we may define $\widehat{\omega}^2_j := \|X_j - X_{\widehat{\mathrm{p}}(j)} \widehat{\beta}_j\|^2_2/n$ and $\widehat{\sigma}^2_{k,j} := \widehat{\omega}^2_j (\widehat{\Theta}_{\widehat{\mathrm{p}}(j)})_{kk}$. The consistency of this estimator may be easily checked.

14.3 Conclusion

We have provided a unified approach to construct asymptotically linear and normal estimators of low-dimensional parameters of the precision matrix based on regularized estimators. These estimators allow us to construct confidence intervals for edge weights in high-dimensional Gaussian graphical models and, under an identifiability condition, for edge weights in the high-dimensional Gaussian DAG model.

For Gaussian graphical models, we provided two explicit simple constructions: one based on a global method using the graphical Lasso and the second based on a local method using nodewise Lasso regressions. Efficient computational methods are available for both methods as discussed in Section 14.1.5. The constructed estimators are asymptotically normal per entry, achieving the efficient asymptotic variance from the parametric setting. For a detailed analysis of semi-parametric efficiency bounds in Gaussian graphical models, we refer to [17]. For testing hypothesis about a set of edges, the usual multiple testing corrections may be used although in practical applications, these might turn out to be too conservative. More efficient methods for multiple testing in this setting are yet to be developed. While throughout the presented results we have imposed "exact" sparsity constraints on the underlying parameters, we remark that the results might as well be extended to models which are only approximately sparse (see e.g. [4]).

Our main interest lied in developing methodology for graphical models representing continuous random vectors. However, many applications involve *discrete* graphical models, where random variables X_j at each vertex $j \in \mathcal{V}$ take values in a discrete space. A popular family of distributions for the binary case where $X_j \in \{-1, 1\}$ is the Ising model. This model finds applications in statistical physics, neuroscience or modeling of social networks. The Ising model can be efficiently estimated via a nodewise method: the individual neighborhoods can be estimated with ℓ_1-penalized logistic regression as proposed in [27]. Logistic regression falls into the framework of generalized linear models for which the de-biasing methodology was proposed in [35]. Consequently, one may compute the neighborhood estimator via ℓ_1-penalized logistic regression and then compute the de-biased estimator along the lines of [35].

For directed acyclic graphs, we showed that confidence intervals for edge weights may be constructed for the Gaussian DAG when it is identifiable and $p \leq \alpha_* n / \log p$. To this end, we require that the error variances in the structural equations model are equal, or known up to a multiplicative constant. If the variance are not equal, the model may not be identifiable and work on inference in this setting is yet to be developed.

14.4 Proofs

14.4.1 Proofs for undirected graphical models

Lemma 14.4.1. *Assume that* $1/L \leq \Lambda_{\min}(\Theta_0) \leq \Lambda_{\max}(\Theta_0) \leq L$ *for some constant* $L \geq 1$. *Let* $\mathcal{E}(\Delta) := \mathrm{tr}[\Delta\Sigma_0] - [\log\det(\Delta + \Theta_0) - \log\det(\Theta_0)]$. *Then for all* Δ *such that* $\|\Delta\|_F \leq 1/(2L)$, $\mathcal{E}(\Delta)$ *is well defined and*

$$\mathcal{E}(\Delta) \geq \frac{1}{2(L + 1/(2L))^2}\|\Delta\|_F^2. \tag{14.21}$$

Proof of Lemma 14.4.1. First we show that $\mathcal{E}(\Delta)$ is well defined for all Δ such that $\|\Delta\|_F \leq 1/(2L)$. To this end, we need to check that $\Lambda_{\min}(\Theta_0 + \Delta) \geq c_1$ for some $c_1 > 0$. Denote the spectral norm of a matrix M by $\|M\| := \sqrt{\Lambda_{\max}(MM^T)}$. We have

$$\Lambda_{\min}(\Theta_0 + \Delta) = \min_{\|x\|_2 = 1} x^T(\Theta_0 + \Delta)x \geq \Lambda_{\min}(\Theta_0) - \|\Delta\|_F \geq 1/(2L),$$

where we used that $|x^T\Delta x| \leq \|\Delta\| x^T x$ and that $\|\Delta\| \leq \|\Delta\|_F$.
A second order Taylor expansion with remainder in integral form yields

$$\begin{aligned}&\log\det(\Delta + \Theta_0) - \log\det(\Theta_0)\\ &= \mathrm{tr}(\Delta\Sigma_0) - \mathrm{vec}(\Delta)^T\left(\int_0^1 (1-v)(\Theta_0 + v\Delta)^{-1} \otimes (\Theta_0 + v\Delta)^{-1} dv\right)\mathrm{vec}(\Delta).\end{aligned}$$

Then for all Δ such that $\|\Delta\|_F \leq 1/(2L)$, it holds

$$\mathcal{E}(\Delta) = \mathrm{vec}(\Delta)^T\left(\int_0^1 (1-v)(\Theta_0 + v\Delta)^{-1} \otimes (\Theta_0 + v\Delta)^{-1} dv\right)\mathrm{vec}(\Delta),$$

where $\otimes$ denotes the Kronecker product and the remainder in the Taylor expansion is in the integral form. Using the fact that the eigenvalues of Kronecker product of symmetric matrices is the product of eigenvalues of the factors, it follows for all Δ such that $\|\Delta\|_F \leq$

$1/(2L)$

$$\begin{aligned} & \Lambda_{\min}\left(\int_0^1 (1-v)(\Theta_0+v\Delta)^{-1}\otimes(\Theta_0+v\Delta)^{-1}dv\right) \\ \geq & \int_0^1 (1-v)\Lambda^2_{\min}((\Theta_0+v\Delta)^{-1})dv \\ \geq & \frac{1}{2}\min_{0\leq v\leq 1}\Lambda^2_{\min}((\Theta_0+v\Delta)^{-1}) \\ \geq & \frac{1}{2}\min_{\Delta:\|\Delta\|_F\leq 1/(2L)}\Lambda^2_{\min}((\Theta_0+\Delta)^{-1}). \end{aligned}$$

Next we obtain

$$\Lambda^2_{\min}((\Theta_0+\Delta)^{-1}) = \Lambda^{-2}_{\max}(\Theta_0+\Delta) \geq (\|\Theta_0\|+\|\Delta\|)^{-2} \geq \frac{1}{(L+1/(2L))^2} > 0,$$

where we used $\|\Delta\| \leq \|\Delta\|_F \leq 1/(2L)$. Finally this yields that $\mathcal{E}(\Delta) \geq \frac{1}{2(L+1/(2L))^2}\|\Delta\|_F^2$ for all $\|\Delta\|_F \leq 1/(2L)$, as required. □

Proof of Theorems 14.1.2 and 14.1.3. We will prove both Theorem 14.1.2 and Theorem 14.1.3 at the same time, since the proofs only differ slightly. For the proof of Theorem 14.1.3, one has to replace $\widehat{\Sigma}$, Σ_0, $\widehat{\Theta}$, Θ_0 in the proof below by $\widehat{\Gamma}$, Γ_0, $\widehat{\Theta}_{\text{norm}}$, K_0, respectively.

Let $\widetilde{\Theta} := \alpha\widehat{\Theta}+(1-\alpha)\Theta_0$, where $\alpha := \frac{M}{M+\|\widehat{\Theta}-\Theta_0\|_F}$, for some $M>0$ to be specified later. The definition of $\widetilde{\Theta}$ implies that $\|\widetilde{\Theta}-\Theta_0\|_F \leq M$. By the convexity of the loss function and by the definition of $\widehat{\Theta}$, we have

$$\text{tr}(\widetilde{\Theta}\widehat{\Sigma}) - \log\det(\widetilde{\Theta}) + \lambda\|\widetilde{\Theta}^-\|_1 \leq \text{tr}(\Theta_0\widehat{\Sigma}) - \log\det(\Theta_0) + \lambda\|\Theta_0^-\|_1. \tag{14.22}$$

Denote $\Delta = \widetilde{\Theta} - \Theta_0$ and let

$$\mathcal{E}(\Delta) := \text{tr}(\Delta\Sigma_0) - [\log\det(\Delta+\Theta_0) - \log\det(\Theta_0)].$$

The inequality (14.22) implies the basic inequality

$$\mathcal{E}(\Delta) + \lambda\|\widetilde{\Theta}^-\|_1 \leq -\text{tr}[\Delta(\widehat{\Sigma}-\Sigma_0)] + \lambda\|\Theta_0^-\|_1.$$

On the set $\{\|\widehat{\Sigma}-\Sigma_0\|_\infty \leq \lambda_0\}$, we can bound the empirical process term by

$$\begin{aligned} |\text{tr}[\Delta(\widehat{\Sigma}-\Sigma_0)]| &\leq \|\widehat{\Sigma}^- - \Sigma_0^-\|_\infty\|\Delta^-\|_1 + \|\widehat{\Sigma}^+ - \Sigma_0^+\|_2\|\Delta^+\|_F \\ &\leq \lambda_0\|\Delta^-\|_1 + \|\widehat{\Sigma}^+ - \Sigma_0^+\|_2\|\Delta^+\|_F. \end{aligned} \tag{14.23}$$

In what follows, we work on the set $\{\|\widehat{\Sigma}-\Sigma_0\|_\infty \leq \lambda_0\}$.
We now choose M such that $M \leq 1/(2L)$, this then implies $\|\widetilde{\Theta}-\Theta_0\|_F \leq 1/(2L)$. But then Lemma 14.4.1 implies that $\mathcal{E}(\widetilde{\Theta}-\Theta_0)$ is well defined and

$$\mathcal{E}(\widetilde{\Theta}-\Theta_0) \geq c\|\widetilde{\Theta}-\Theta_0\|_F^2, \tag{14.24}$$

where one can take $c := 1/(8L^2)$. Using bounds (14.23) and (14.24), we obtain from the basic inequality

$$c\|\Delta\|_F^2 + \lambda\|\widetilde{\Theta}^-\|_1 \leq \lambda_0\|\Delta^-\|_1 + \lambda\|\Theta_0^-\|_1 + \|\widehat{\Sigma}^+ - \Sigma_0^+\|_2\|\Delta^+\|_F$$

By the triangle inequality and taking $\lambda \geq 2\lambda_0$, we obtain

$$2c\|\Delta\|_F^2 + \lambda\|\widetilde{\Theta}_{S^c}^-\|_1 \leq 3\lambda\|\Delta_S^-\|_1 + 2\|\widehat{\Sigma}^+ - \Sigma_0^+\|_2\|\Delta^+\|_F$$

Consequently,

$$\begin{aligned} 2c\|\Delta\|_F^2 + \lambda\|\Delta^-\|_1 &\leq 4\lambda\|\Delta_S^-\|_1 + 2\|\widehat{\Sigma}^+ - \Sigma_0^+\|_2\|\Delta^+\|_F \\ &\leq 4\lambda\sqrt{s}\|\Delta_S^-\|_F + 2\|\widehat{\Sigma}^+ - \Sigma_0^+\|_2\|\Delta^+\|_F \\ &\leq 8s\lambda^2/c^2 + c\|\Delta_S^-\|_F^2/2 + 8\|\widehat{\Sigma}^+ - \Sigma_0^+\|_2^2/c + c\|\Delta^+\|_F^2/2. \end{aligned}$$

Taking M such that $\lambda_0 M \geq 8s\lambda^2/c^2 + 8\|\widehat{\Sigma}^+ - \Sigma_0^+\|_2^2/c$,

$$c\|\Delta\|_F^2 + \lambda\|\Delta^-\|_1 \leq 8s\lambda^2/c^2 + 8\|\widehat{\Sigma}^+ - \Sigma_0^+\|_2^2/c \leq \lambda_0 M.$$

Taking $M \geq 4\lambda_0/c$,

$$\|\Delta\|_F^2 \leq \lambda_0 M/c \leq M^2/4.$$

But then $\|\Delta\|_F \leq M/2$. The definition of $\widetilde{\Theta}$ in turn implies that $\|\widehat{\Theta} - \Theta_0\|_F \leq M$, and we can repeat all the arguments with $\widehat{\Theta}$ in place of $\widetilde{\Theta}$. Repetition of the arguments leads to the oracle inequality

$$c\|\widehat{\Theta} - \Theta_0\|_F^2 + \lambda\|\widehat{\Theta}^- - \Theta_0^-\|_1 \leq 8s\lambda^2/c^2 + 8\|\widehat{\Sigma}^+ - \Sigma_0^+\|_2^2/c.$$

Finally we distinguish the case of non-normalized graphical Lasso (based on the covariance matrix) and the normalized graphical Lasso (based on the correlation matrix). We have for the case of

a) normalized graphical Lasso: $\widehat{\Sigma}^+ - \Sigma_0^+ = 0$ (recall here that $\widehat{\Sigma} \equiv \widehat{R}, \Sigma_0 \equiv R_0$) and the oracle inequality gives

$$c\|\widehat{\Theta} - \Theta_0\|_F^2 + \lambda\|\widehat{\Theta}^- - \Theta_0^-\|_1 \leq 8s\lambda^2/c^2.$$

b) non-normalized graphical Lasso: we can bound

$$\|\widehat{\Sigma}^+ - \Sigma_0^+\|_2 \leq \sqrt{p}\|\widehat{\Sigma}^+ - \Sigma_0^+\|_\infty \leq \sqrt{p}\lambda_0.$$

Hence the oracle inequality gives

$$c\|\widehat{\Theta} - \Theta_0\|_F^2 + \lambda\|\widehat{\Theta}^- - \Theta_0^-\|_1 \leq 8s\lambda^2/c^2 + 8p\lambda_0^2/c.$$

To show the second statement of the theorems, we use the above oracle inequalities and the following upper bound

$$\begin{aligned} \left|\left|\left|\widehat{\Theta} - \Theta_0\right|\right|\right|_1 &\leq \max_{j=1,\dots,p} |\widehat{\Theta}_{jj} - \Theta_{jj}^0| + \|\widehat{\Theta}_j^- - (\Theta_j^0)^-\|_1 \\ &\leq \|\widehat{\Theta} - \Theta_0\|_F + \|\widehat{\Theta}^- - \Theta_0^-\|_1. \end{aligned}$$

To show the third statement of Theorem 14.1.3, we use the upper bound

$$\begin{aligned} \left|\left|\left|\widehat{\Theta}_{\mathrm{w}} - \Theta_0\right|\right|\right|_1 &= \left|\left|\left|\widehat{W}^{-1}\widehat{\Theta}_{\mathrm{norm}}\widehat{W}^{-1} - W_0^{-1}K_0W_0^{-1}\right|\right|\right|_1 \\ &\leq \|\widehat{W}\|_\infty^2\left|\left|\left|\widehat{\Theta}_{\mathrm{norm}} - K_0\right|\right|\right|_1 + \|\widehat{W} - W_0\|_\infty|||K_0|||_1\|\widehat{W}\|_\infty \\ &\quad + \|W_0\|_\infty|||K_0|||_1\|\widehat{W} - W_0\|_\infty \end{aligned}$$

□

Proof of Theorem 14.1.5. Denote $C_L := 16(8L^2)^2$. Using the results of Theorem 14.1.2, we obtain

$$\begin{aligned}
\|\text{rem}\|_\infty &\leq \|(\widehat{\Theta}-\Theta_0)^T(\widehat{\Sigma}\Theta_0 - I)\|_\infty + \|(\widehat{\Theta}-\Theta_0)^T\lambda\widehat{Z}\widehat{\Theta}\|_\infty \\
&\leq \left|\left|\left|\widehat{\Theta}-\Theta_0\right|\right|\right|_1 \|\widehat{\Sigma}-\Sigma_0\|_\infty |||\Theta_0|||_1 + \left|\left|\left|\widehat{\Theta}-\Theta_0\right|\right|\right|_1 \lambda\|\widehat{Z}\|_\infty \left|\left|\left|\widehat{\Theta}\right|\right|\right|_1 \\
&\leq C_L(p+s)\lambda\sqrt{d+1}\Lambda_{\max}(\Theta_0)\lambda_0 + 2C_L(p+s)\lambda\sqrt{d+1}\Lambda_{\max}(\Theta_0)\lambda \\
&\leq \frac{3}{2}C_L L(p+s)\sqrt{d+1}\lambda^2.
\end{aligned}$$

Taking $\lambda \asymp \sqrt{\log p/n}$, Lemma 14.1.1 implies $\|\widehat{\Sigma}-\Sigma_0\|_\infty = \mathcal{O}_P(\sqrt{\log p/n})$. Then by the sparsity condition, we obtain $\|\text{rem}\|_\infty = \mathcal{O}_P(1/\sqrt{n})$. By Conditions A1 and A2, the random variable $(\Theta_0(\widehat{\Sigma}-\Sigma_0)\Theta_0)_{ij}$ has bounded fourth moments and asymptotic normality per entry follows by application of Lindeberg's central limit theorem for triangular arrays (see [16] for more details). □

Proof of Theorem 14.1.6. For the remainder, we obtain similarly as in the proof of Theorem 14.1.7, $\|\text{rem}\|_\infty \leq \frac{3}{2}C_L Ls\sqrt{d+1}\lambda^2$. Asymptotic normality follows by analogous arguments. □

Proof of Proposition 14.1.7. Denote $C_L := 16(8L^2)^2$. Using the results of Theorem 14.1.3, we obtain

$$\begin{aligned}
\|\text{rem}\|_\infty &\leq \left|\left|\left|\widehat{\Theta}_{\text{norm}}-K_0\right|\right|\right|_1 \|\widehat{\Gamma}K_0 - I\|_\infty + \left|\left|\left|\widehat{\Theta}_{\text{norm}}-K_0\right|\right|\right|_1 \lambda\|\widehat{Z}\|_\infty \left|\left|\left|\widehat{\Theta}_{\text{norm}}\right|\right|\right|_1 \\
&\leq C_L s\lambda\sqrt{d+1}\Lambda_{\max}(\Theta_0)\lambda_0 + C_L s\lambda\sqrt{d+1}\Lambda_{\max}(\Theta_0)\lambda \\
&\leq \frac{3}{2}C_L Ls\sqrt{d+1}\lambda^2.
\end{aligned}$$

The sparsity condition implies the result. □

Proof of Theorem 14.1.9. The proof follows along the same lines as the proof of Theorem 1 in [16]. The only difference is that here we consider a weighted Lasso to estimate the partial correlations and the estimator $\widehat{\tau}_j$ is defined slightly differently. But for the weighted Lasso (with weights bounded away from zero and bounded from above with high probability), oracle inequalities of the same order can be obtained, see Section 6.9 in [4], i.e.

$$\|X_{-j}(\widehat{\gamma}_j - \gamma_j^0)\|_2^2/n + \lambda\|\widehat{\gamma}_j - \gamma_j^0\|_1 = \mathcal{O}_P(d_j \log p/n).$$

For the estimator of variance we have

$$\begin{aligned}
|\widehat{\tau}_j^2 - \tau_j^2| &\leq \|X_{-j}(\widehat{\gamma}_j - \gamma_j^0)\|_2^2/n + 2|(X_j - X_{-j}\gamma_j^0)^T X_{-j}(\widehat{\gamma}_j - \gamma_j^0)/n| \\
&= \|X_{-j}(\widehat{\gamma}_j - \gamma_j^0)\|_2^2/n + 2\|(X_j - X_{-j}\gamma_j^0)^T X_{-j}/n\|_\infty \|\widehat{\gamma}_j - \gamma_j^0\|_1 \\
&= \mathcal{O}_P(1/\sqrt{n}).
\end{aligned}$$

The rest of the proof follows as in [16].

□

14.4.2 Proofs for directed acyclic graphs

Proof of Theorem 14.2.1. By Theorem 5.1 in [34], we have under the conditions of the theorem that $\widehat{\pi} = \pi_0$ with high probability. Then also $\widehat{\mathrm{p}}(j) = \mathrm{p}(j)$ for all j, with high probability. Therefore, the estimated $\widehat{\mathrm{p}}(j)$ in the definitions of $\widehat{\beta}_j$ and $\widehat{\Theta}_{\widehat{\mathrm{p}}(j),k}, k \in \widehat{\mathrm{p}}(j)$ (and elsewhere) can be replaced by $\mathrm{p}(j)$. The nodewise Lasso then yields oracle estimators $\widehat{\beta}_j$ and $\widehat{\Theta}_k, k \in \mathrm{p}(j)$ under the Condition B1 and under the sparsity $d_j = o(\sqrt{n}/\log p)$ (see [16]). This gives in particular that for all $j = 1, \dots, p$

$$\max_k \|\widehat{\Theta}_{\mathrm{p}(j),k} - \Theta^0_{\mathrm{p}(j),k}\|_1 = \mathcal{O}_P(\max_k d_j \lambda_j), \quad \max_k \|\widehat{\Sigma}_{\mathrm{p}(j)} \widehat{\Theta}_{\mathrm{p}(j),k} - e_k\|_\infty = \mathcal{O}_P(\max_k \lambda_j),$$

$$\|\widehat{\beta}_j - \beta^0_j\|_1 = \mathcal{O}_P(\max_{j=1,\dots,p} d_j \lambda_j).$$

We can write the decomposition

$$\widehat{b}_{k,j} - \beta^0_{k,j} = (\Theta^0_{\mathrm{p}(j),k})^T X^T_{\mathrm{p}(j)} \epsilon_j / n + \mathrm{rem}_{k,j}, \tag{14.25}$$

where $\mathrm{rem}_{k,j} = (\widehat{\Theta}_{\mathrm{p}(j),k} - \Theta^0_{\mathrm{p}(j),k})^T X^T_{\mathrm{p}(j)} \epsilon_j / n - (\widehat{\Sigma}_{\mathrm{p}(j)} \widehat{\Theta}_{\mathrm{p}(j),k} - e_k)^T (\widehat{\beta}_j - \beta^0_j)$. First note that by normality and by the independence of $X_{\mathrm{p}(j)}$ and ϵ_j (which follows by the independence of ϵ_j's and acyclicity of the graph), it holds $\|X^T_{\mathrm{p}(j)} \epsilon_j / n\|_\infty = \mathcal{O}_P(\sqrt{\log p / n})$. By Hölder's inequality

$$\begin{aligned} \max_k |\mathrm{rem}_{k,j}| &\leq \max_k \|\widehat{\Theta}_{\mathrm{p}(j),k} - \Theta^0_{\mathrm{p}(j),k}\|_1 \|X^T_{\mathrm{p}(j)} \epsilon_j / n\|_\infty \\ &+ \max_k \|\widehat{\Theta}^T_{\mathrm{p}(j),k} \widehat{\Sigma}_{\mathrm{p}(j)} - e_k\|_\infty \|\widehat{\beta}_j - \beta^0_j\|_1 \\ &= \mathcal{O}_P(d_j \log p / n) = o_P(1/\sqrt{n}), \end{aligned}$$

where we used the sparsity assumption $d_j = o(\sqrt{n}/\log p)$. Thus we have shown that the remainder in (14.25) is of small order $1/\sqrt{n}$. Then applying Lindeberg's central limit theorem for triangular arrays and by Conditions A1 and A2,

$$(\Theta^0_{\mathrm{p}(j),k})^T X^T_{\mathrm{p}(j)} \epsilon_j / (\sigma_{k,j} \sqrt{n}) \rightsquigarrow \mathcal{N}(0, 1),$$

which shows the claim. □

Bibliography

[1] A. Belloni, V. Chernozhukov, and L. Wang. Square-root Lasso: Pivotal recovery of sparse signals via conic programming. *Biometrika*, 98(4):791–806, 2011.

[2] P. J. Bickel and E. Levina. Covariance regularization by thresholding. *Annals of Statistics*, 36(6):2577–2604, 2008.

[3] P. J. Bickel and E. Levina. Regularized estimation of large covariance matrices. *Annals of Statistics*, 36(1):199–227, 2008.

[4] P. Bühlmann and S. van de Geer. *Statistics for High-Dimensional Data.* Springer, 2011.

[5] T. Cai, W. Liu, and X. Luo. A constrained l1 minimization approach to sparse precision matrix estimation. *Journal of the American Statistical Association*, 106, 2011.

[6] E. Candes and T. Tao. The dantzig selector: Statistical estimation when p is much larger than n. *Annals of Statistics*, 35(6):2313–2351, 12 2007.

[7] V. Chernozhukov, D. Chetverikov, and K. Kato. Central limit theorems and bootstrap in high dimensions. *arXiv: 1412.3661*, 2014.

[8] D.M. Chickering. Optimal structure identification with greedy search. *Journal of Machine Learning Research*, 3:507–554, 2002.

[9] A. d'Aspremont, O. Banerjee, and L. El Ghaoui. First-order methods for sparse covariance selection. *SIAM Journal on Matrix Analysis and Applications*, 30(1):56–66, 2008.

[10] M. Drton and M. H. Maathuis. Structure learning in graphical modeling. *Annual Review of Statistics and Its Application*, 4:365–393, 2017.

[11] B. Efron, T. Hastie, I. Johnstone, and R. Tibshirani. Least angle regression. *Ann. Statist.*, 32(2):407–451, June 2004.

[12] N. El Karoui. Operator norm consistent estimation of large dimensional sparse covariance matrices. *Annals of Statistics*, 36(6):2717–2756, 2008.

[13] J. Friedman, T. Hastie, and R. Tibshirani. Sparse inverse covariance estimation with the graphical lasso. *Biostatistics*, 9:432–441, 2008.

[14] A. Hauser and P. Bühlmann. Characterization and greedy learning of interventional Markov equivalence classes of directed acyclic graphs. *Journal of Machine Learning Research*, 13:2409–2464, 2012.

[15] J. Janková and S. van de Geer. Confidence intervals for high-dimensional inverse covariance estimation. *Electronic Journal of Statistics*, 9(1):1205–1229, 2014.

[16] J. Janková and S. van de Geer. Honest confidence regions and optimality for high-dimensional precision matrix estimation. *TEST*, 26(1):143–162, 2016.

[17] J. Janková and S. van de Geer. Semi-parametric efficiency bounds and efficient estimation for high-dimensional models. *ArXiv:1601.00815*, 2016.

[18] A. Javanmard and A. Montanari. Confidence intervals and hypothesis testing for high-dimensional regression. *Journal of Machine Learning Research*, 15(1):2869–2909, 2014.

[19] I. M. Johnstone and A. Y. Lu. On consistency and sparsity for principal components analysis in high dimensions. *Journal of the American Statistical Association*, 104(486):682–693, 2009.

[20] I. M. Johnstone. On the distribution of the largest eigenvalue in principal components analysis. *Annals of Statistics*, 29(2):295–327, 04 2001.

[21] W. Liu et al. Gaussian graphical model estimation with false discovery rate control. *Annals of Statistics*, 41(6):2948–2978, 2013.

[22] R. Mazumder and T. Hastie. The graphical Lasso: New insights and alternatives. *Electronic Journal of Statistics*, 6:2125–2149, 2012.

[23] N. Meinshausen and P. Bühlmann. High-dimensional graphs and variable selection with the Lasso. *Annals of Statistics*, 34(3):1436–1462, 06 2006.

[24] J. Pearl. *Causality: Models, Reasoning and Inference.* Cambridge University Press, 2016.

[25] J. Peters and P. Bühlmann. Identifiability of Gaussian structural equation models with equal error variances. *Biometrika*, 101:219–228, 2014.

[26] P. Ravikumar, G. Raskutti, M. J. Wainwright, and B. Yu. High-dimensional covariance estimation by minimizing l1-penalized log-determinant divergence. *Electronic Journal of Statistics*, 5:935–980, 2008.

[27] P. Ravikumar, M. J. Wainwright, J. D. Lafferty, et al. High-dimensional Ising model selection using ℓ_1-regularized logistic regression. *Annals of Statistics*, 38(3):1287–1319, 2010.

[28] Z. Ren, T. Sun, C.-H. Zhang, H. H. Zhou, et al. Asymptotic normality and optimalities in estimation of large Gaussian graphical models. *Annals of Statistics*, 43(3):991–1026, 2015.

[29] A. J. Rothman, P. J. Bickel, E. Levina, and J. Zhu. Sparse permutation invariant covariance estimation. *Electronic Journal of Statistics*, 2:494–515, 2008.

[30] T. Sun and C.-H. Zhang. Sparse matrix inversion with scaled Lasso. *Journal of Machine Learning Research*, 14:3385–3418, 2012.

[31] R. Tibshirani. Regression shrinkage and selection via the lasso. *Journal of The Royal Statistical Society: Series B*, 58:267–288, 1996.

[32] S. van de Geer. *Estimation and Testing under Sparsity: École d'Été de Saint-Flour XLV.* Springer, 2016.

[33] S. van de Geer and P. Bühlmann. On the conditions used to prove oracle results for the lasso. *Electronic Journal of Statistics*, 3:1360–1392, 2009.

[34] S. van de Geer and P. Bühlmann. ℓ_0-penalized maximum likelihood for sparse directed acyclic graphs. *Annals of Statistics*, 41(2):536–567, 2013.

[35] S. van de Geer, P. Bühlmann, Y. Ritov, and R. Dezeure. On asymptotically optimal confidence regions and tests for high-dimensional models. *Annals of Statistics*, 42(3):1166–1202, 2014.

[36] S. van de Geer. Worst possible sub-directions in high-dimensional models. *Journal of Multivariate Analysis*, 146:248–260, 2016.

[37] L. Wasserman, M. Kolar, A. Rinaldo, et al. Berry-Esseen bounds for estimating undirected graphs. *Electronic Journal of Statistics*, 8(1):1188–1224, 2014.

[38] M. Yu, M. Kolar, and V. Gupta. Statistical inference for pairwise graphical models using score matching. In D. D. Lee, M. Sugiyama, U. V. Luxburg, I. Guyon, and R. Garnett, editors, *Advances in Neural Information Processing Systems 29*, 2829–2837. Curran Associates, Inc., 2016.

[39] M. Yuan. High dimensional inverse covariance matrix estimation via linear programming. *The Journal of Machine Learning Research*, 11:2261–2286, 2010.

[40] M. Yuan and Y. Lin. Model selection and estimation in the Gaussian graphical model. *Biometrika*, 94(1):19–35, 2007.

[41] C.-H. Zhang and S. S. Zhang. Confidence intervals for low-dimensional parameters in high-dimensional linear models. *Journal of the Royal Statistical Society: Series B*, 76:217–242, 2014.

Part IV

Causal inference

15

Causal Concepts and Graphical Models

Vanessa Didelez
Department of Biometry and Data Management, Leibniz Institute for Prevention Research and Epidemiology – BIPS

CONTENTS

15.1 Introduction

The notion of causality has been much examined, discussed and debated, in science and philosophy, over many centuries (see accounts in [67, 49, 36]). In the social discourse, it often refers to aspects of responsibility, such as moral questions ('whose fault is it?') or historical questions ('what caused the second world war?'). However, for the purpose of the present chapter, we focus mainly on statistical contexts where causal inference uses data to inform *decisions* about *actions*, for example with a view to public health policies. In particular we consider different approaches to formalizing causal inquiries which, despite subtle differences, all build on a probabilistic graphical representation of the problem at hand. We have deliberately chosen to present different approaches in order to illustrate the flexibility and representational power of graphical models.

Overview

We start by clarifying the notation and terminology used in this chapter as well as the role of graphical models, recalling how they relate to conditional independence structures. In Section 15.2, a brief tour through different causal notations and frameworks introduces common key concepts as well as differences between formal approaches. Two ways of combining these with graphical models are addressed and compared in Section 15.3. A central question of causal inference is whether a desired causal effect is *identified* from observational data. Early ideas of how to obtain graph-based answers are discussed in Section 15.4.

The usefulness of graphs becomes especially clear when investigating structural sources of bias. They allow to detect, for example, different ways how confounding may be introduced or removed from an analysis; see Section 15.5. We conclude with an outlook onto the many further topics of causal inference where graphs play a facilitating role.

Notation and terminology

We use $\mathbf{X}_V = \mathbf{X} = (X_1, \ldots, X_K)$, $V = \{1, \ldots, K\}$, to refer to a random vector in connection with a graph $\mathcal{G} = (V, E)$. Within specific examples we alternatively use X, Y, Z, C, T, W, U etc. as random variables. The domains of random variables are given by $\mathcal{X}_k = \{x_k \in \mathbb{R} : p(x_k) > 0\}$, $k \in V$, and similar for other variables. We loosely refer to p as distribution or probability, but let it also stand for either the probability density or probability mass function of a distribution. Occasionally we will write $\mathbf{X}^i$ or Y^i etc. when referring to an individual in the sample or population, but mostly we suppress this additional index. For $A \subset V$, subvectors and induced subgraphs are denoted by $\mathbf{X}_A$ and $\mathcal{G}_A$. Graphical parents, children, descendants and non-descendants in a directed acyclic graph (DAG) are denoted by $\text{pa}(A), \text{ch}(A), \text{de}(A), \text{nd}(A)$ respectively (see formal definitions in Section 1.6 of Chapter 1).

The role of graphs

The standard problem of causal inference in statistics is concerned with analyzing observational data in order to infer, and often quantify, causal relations. We would intuitively expect that the validity of such inference requires specialized methods. In particular it relies on assumptions that are somewhat different in their nature, and arguably stronger, than those for traditional statistical inference. Only a thorough understanding of these assumptions, and indeed of the causal target of inference itself, enables us to come up with ways of checking them either empirically or based on subject matter plausibility. Such an understanding is therefore a prerequisite for assessing the soundness of a proposed causal conclusion in any given situation. This is of great importance as causal findings are designed to inform decisions and policy interventions with practical consequences. For example, it can have very different implications if we claim that dietary fat intake causes coronary heart disease than if we say that dietary fat intake is only associated with coronary heart disease.

Graphical approaches assist us with formalizing causal targets of inference. Moreover, they have proved especially useful for being transparent and general about the assumptions required for causal conclusions. Hence they facilitate the detection and elimination of possible sources of bias, or suggest specific sensitivity analyses. Also, we often gain a better insight into the logic behind methods of causal inference when using graphs. Finally, graphs are sometimes a causal target of inference in their own right.

Before going into the details of how to combine causal concepts and graphical models, let us first revisit some key aspects of the latter and establish that the *Markov properties do not automatically supply directed edges with any causal meaning.* We mainly consider DAGs, but see Section 15.6 for references to approaches that use other types of graphs.

Recall that the distribution for a random vector $\mathbf{X} = (X_1, \ldots, X_K)$ factorizes according to a DAG $\mathcal{G} = (V, E)$ if the joint distribution satisfies

$$p(\mathbf{x}) = \prod_{k \in V} p(x_k | \mathbf{x}_{\text{pa}(k)}). \tag{15.1}$$

Conditional independencies can be read off the DAG using various criteria. For example, the above factorization is equivalent to $X_k \perp\!\!\!\perp \mathbf{X}_{\mathrm{nd}(k)\setminus \mathrm{pa}(k)} \,|\, \mathbf{X}_{\mathrm{pa}(k)}$. Most commonly, the d-separation criterion based on blocking of paths can be used (or equivalently the moralization criterion); see Section 1.8 in Chapter 1 or also [40]. For example, the two DAGs $1 \longleftarrow 2 \longleftarrow 3$ and $1 \longrightarrow 2 \longrightarrow 3$ imply the exact same single conditional independence $X_1 \perp\!\!\!\perp X_3 \,|\, X_2$ because node 2 d-separates nodes 1 and 3 in both DAGs. When graphs with different orientations of edges represent the same set of conditional independencies they are called *Markov equivalent.* But even if a DAG is uniquely determined by its Markov properties it still represents no more than certain conditional independencies.

Hence, if we want a graph to encode some notion of causality an extra ingredient is required. Graphical and causal modeling can, in fact, be combined in many ways resulting in differing representational power. We aim to provide the reader with a broad overview and will focus on essentially two approaches: (i) augmenting traditional conditional independence DAGs with a separate type of non-random nodes representing interventions; (ii) retaining the original set of nodes pertaining to the domain variables but modifying the meaning of edges, hence supplying a causal interpretation on top of the graphical Markov properties. We return to this distinction in more detail in Section 15.3.

15.2 Association versus Causation: Seeing versus Doing

Traditional statistical regression models for an outcome Y and an exposure X specify some aspect of the conditional distribution $p(y|x)$, or in words: the distribution of Y when we *observe* $X = x$. This describes an association in the sense of how *seeing* different values of X helps us to *predict* possibly different (expected) values of Y. A regression-based analysis would typically be accompanied by a warning that 'association is not causation'. The warning seems necessary so that people do not expect the predicted change in Y to actually materialize when they *intervene* in a system to *manipulate* the value of X. In other words, the warning is given so that individuals or policy makers do not base *decisions* about their *actions* on a finding that is 'just associational'. Intervening can be understood as actively *doing* something to the system instead of passively observing it. To make this explicit, Pearl [49] introduced the so-called do($\cdot$)-notation defined below.

As a toy example, consider the positive association between the amount of books in a household and the household income. This presumably reflects that higher education is accompanied by more books as well as better paid jobs and that with a higher income one can afford more books. It would be misguided to think that one's income can be raised purely by increasing the number of books in one's home.

If we want to address these issues formally we require a notation allowing us to express for instance that, while there can be an association between X and Y, a manipulation of X may not necessarily result in a corresponding change in Y. We can regard this as making an explicit distinction between concepts of *association* on the one hand and *causation* on the other. Note, however, that none of the approaches covered below actually define causality. They are simply all based on the premise that causation, in contrast to association, is relevant to decisions about actions, such as interventions or manipulations. Pragmatically, without going into philosophical subtleties, we say that X is causal for Y if some manipulation of X has an effect on Y.

The following brief overview of some notations will give an idea of the key concepts employed in the causal literature, each providing a slightly different angle on the problem (see also overviews [21, 55, 13]). Section 15.3 revisits these in more detail and combines them with graphical models.

Do–calculus

An intuitive notation distinguishing conditioning on observing versus intervening has been introduced by Pearl [49]. The former is denoted by $p(y; \mathrm{see}(X = \widetilde{x}))$ and can be equated with ordinary conditioning $p(y; \mathrm{see}(X = \widetilde{x})) = p(y|\widetilde{x})$. Here, 'seeing' refers to X taking its natural value, where 'natural' is always relative to a given context. For example, 'natural' could be the situation where in an observational study we simply ask individuals of a specific population how many books they have at home. In contrast, the notation $p(y; \mathrm{do}(X = \widetilde{x}))$[1] refers to the distribution of Y under the situation where X has been *forced* to take the value $\widetilde{x}$ by some *intervention*. For example, the state could provide every household with $\widetilde{x}$ books. With Y = 'income' and X = '# books', it is intuitively plausible that $p(y|\widetilde{x}) \neq p(y; \mathrm{do}(X = \widetilde{x}))$. This inequality is a formal way to state that association is not causation. Note that several different notations have been used to indicate conditioning on intervention, see [46, 47, 67, 49, 40, 11].

Pearl [47, 49] further develops a do-calculus which relates the corresponding intervention to a DAG and then uses graphical rules to convert conditioning on 'doing' into conditioning on 'seeing'. The defining property of a *causal DAG* demands that a $\mathrm{do}(X_j = \widetilde{x}_j)$-intervention modularly replaces the factor $p(x_j|\mathbf{x}_{\mathrm{pa}(j)})$ by $\mathbf{1}\{x_j = \widetilde{x}_j\}$ in the factorized joint distribution (15.1) [33]. This do-intervention is called an *atomic* intervention as it affects a single variable that is being set to a specific value [46].

Regime indicator

We can consider the above 'seeing' and 'doing' as two types of *regimes*, a natural one and a set of interventional ones. Here, regimes refer to external circumstances under which we expect some aspects of the joint distribution of $\mathbf{X}$ to differ [13]. In many situations a do-intervention is somewhat idealized or hypothetical. How would one for instance fix the dietary fat intake or BMI of a person exactly at a given value? It is also implicit in the notation that the manner in which a variable is manipulated is irrelevant. However, in practice it may matter whether a medical treatment is, for example, given orally or as an injection. For greater generality, we may therefore want to consider a possibly larger and more detailed set $\mathcal{S}$ of different regimes describing different circumstances under which a system might be observed and manipulated. Each regime then induces a different probability measure for the joint distribution of $\mathbf{X}$. Hence, let σ be an indicator for the regime taking values in $\mathcal{S}$ and index the distribution accordingly, so that $p(\mathbf{x}; \sigma = s) = p(\mathbf{x}; s)$, $s \in \mathcal{S}$ denotes the joint distribution of $\mathbf{X}$ under regime s[2]. As addressed in Section 15.3.1, suitable factorizations of the joint density under each regime express assumptions about interventions, for example, what type of interventions in which variables take place and how these then affect further variables.

Returning to the example of an outcome Y and exposure X, the joint distributions under different regimes are given by $p(y, x; s), s \in \mathcal{S}$. Under any regime we always have $p(y, x; s) = p(x; s)p(y|x; s)$. If a particular regime $s \in \mathcal{S}$ indicates an intervention that physically manipulates X we can express this exactly by specifying $p(x; \sigma = s)$. For instance,

---

[1] Pearl [49] uses the notation $p(y|\,\mathrm{do}(X = \widetilde{x}))$; but here we want to avoid confusion with the measure theoretic definition of probabilistic conditioning on random variables. Instead we regard $\mathrm{do}(X = \widetilde{x})$ as an index to the distribution, $p(\cdot; \mathrm{do}(X = \widetilde{x}))$.

[2] Several authors have used the notation F_X to indicate an atomic intervention in X in a similar way to our use of σ here, see [46, 47, 40, 11, 13]. We use σ for greater generality, alluding to intervention strategies as in [14].

let the above 'seeing' be denoted by $\sigma = \emptyset$, also called the 'idle' or 'observational' regime[3]. In contrast, forcing X to be $\tilde{x}$ is denoted $\sigma = \tilde{x}$, $\tilde{x} \in \mathcal{X}$, such that in this case $\mathcal{S} =$ '$\emptyset$'$\cup\mathcal{X}$, where $\mathcal{X}$ is the domain of X. Hence $p(x; \sigma = \emptyset)$ is the distribution of the exposure we observe naturally, while $p(x; \sigma = \tilde{x}) = \mathbf{1}\{x = \tilde{x}\}$ is an atomic intervention. With these choices of regimes, $p(y; \sigma = \emptyset)$ denotes the distribution of Y when X arises naturally, and $p(y|\tilde{x}; \sigma = \emptyset)$ is the corresponding conditional distribution of Y when we passively observe $X = \tilde{x}$. In contrast, $p(y; \sigma = \tilde{x})$ is the distribution of Y when X is forced to take on the value $\tilde{x}$; now $p(y|x; \sigma = \tilde{x})$ is undefined unless $x = \tilde{x}$, and $p(y|\tilde{x}; \sigma = \tilde{x}) = p(y; \sigma = \tilde{x})$ due to the particular choice of $p(x; \sigma = \tilde{x})$ as a point-distribution.

As with the do($\cdot$)-notation, we can express that 'association is not causation' by allowing $p(y|\tilde{x}; \sigma = \emptyset)$ to differ from $p(y; \sigma = \tilde{x})$. But we can now also formulate different types of interventions, for example experimental settings where X is randomly generated from a distribution $\tilde{p}_X$. To represent this, let $\mathcal{S}$ denote the set of interventions defined by demanding that $p(x; \sigma = \tilde{p}_X) = \tilde{p}_X(x), \tilde{p}_X \in \mathcal{S}$. Now, $p(y; \sigma = \tilde{p}_X)$ describes the behavior of the outcome Y when X is drawn from the distribution $\tilde{p}_X$. Interventions that do not fix a variable at a value, but just 'nudge' it, adding a random error, or somehow shift its distribution are sometimes called *soft* interventions [22].

Another important type of regimes is given by conditional or dynamic interventions [56]. Here, we may want to force X to take on a value that is a specified function g_X of pre-exposure covariates C. The purpose is to reflect, for example, a treatment strategy that is adapted to the patient's history. Under such a regime we have that $p(x|C = c; \sigma = g_X) = \mathbf{1}\{x = g_X(c)\}$. More generally even, we may also include situations where potentially more aspects of the system than just a single variable X are affected by a manipulation, possibly even in a partially unknown manner.

Aspects that remain the same under different regimes are called *stable* or *invariant.* These properties can formally be expressed by statements such as $Y \perp\!\!\!\perp \sigma \,|\, X$, meaning $p(y|x; \sigma = s) = p(y|x; \sigma = s'), s \neq s' \in \mathcal{S}$. Such invariances can be read off suitable graphs via d-separations if these graphs include a node for σ. These graphs are called *intervention graphs* and are addressed in Section 15.3.1.

Potential outcomes

A third notation in the context of causal inference uses *potential outcomes* [61, 62]. If, as above, we want to consider some causal effect of X on an outcome Y, we define the potential outcome $Y(\tilde{x})$ to be the value of Y that we would observe if X were set (forced) to $\tilde{x}$. Hence this approach is essentially based on atomic interventions. Similar to before, the possibility that $p(Y(x) = y)$ does not equal $p(y|x)$ allows us to express that causation is not association. But note that potential outcomes are formulated at the level of variables instead of distributions. They can hence be used to express individual causal effects as functions of $Y^i(\tilde{x})$ for different $\tilde{x}$. Potential outcomes also allow us to express 'cross-world' independencies such as $Y(\tilde{x}) \perp\!\!\!\perp W(x') \,|\, (Z, X)$ for two different values $\tilde{x}, x'$.

These examples illustrate why potential outcomes are also referred to as *counterfactuals.* As X^i can only ever take one value at a time for a given individual i, only one of $Y^i(\tilde{x}), \tilde{x} \in \mathcal{X}$, can ever be observed and the others are 'counter to the fact'. One may therefore want to be careful with implicit or explicit assumptions about the joint distribution of counterfactuals as these can never be observed together [10]. Richardson and Robins [55] advocate a restriction to models and assumptions allowing only 'single-world' interventions.

It may not be immediately obvious how potential outcomes can be combined with graphs,

[3]We use the terms 'seeing', observational / natural / unmanipulated / idle regime interchangeably — they all refer to the distribution from which data is actually available which can sometimes be a randomized controlled trial.

and often they are not. An overview of different approaches to combine them is given in [55]. One approach uses structural equation models (SEMs). These assume that all variables are functionally related at an individual level in a way that is invariant to interventions. The functional relations correspond to the equations of a SEM. The equations, in turn, relate each variable to its parents in an associated DAG. Finally, the functional relations induce a joint distribution on all counterfactuals. We return to this in more detail in Section 15.3.2.

Causal effects

As alluded to earlier, we say that X has a causal effect on Y if an intervention in the former affects the distribution of the latter. For example, the presence of a causal effect means that for two values $\widetilde{x} \neq \widetilde{x}'$ we have $p(y; \mathrm{do}(X = \widetilde{x})) \neq p(y; \mathrm{do}(X = \widetilde{x}'))$. Similarly, if $s, s' \in \mathcal{S}$ denote two different, not necessarily atomic, interventions in X, then a causal effect means that $p(y; \sigma = s) \neq p(y; \sigma = s')$. Particular causal parameters can be formalized as functions of intervention distributions. For instance, the average causal effect (ACE) is given by $E(Y; \mathrm{do}(X = \widetilde{x})) - E(Y; \mathrm{do}(X = \widetilde{x}'))$, but other contrasts such as odds ratios or risk ratios are popular. Using potential outcomes the individual causal effect (ICE) is defined as $Y^i(\widetilde{x}) - Y^i(\widetilde{x}')$ while the ACE is the expectation over the individuals in the population $E(Y(\widetilde{x}) - Y(\widetilde{x}'))$. In the following we typically refer to the whole intervention distributions as causal effects unless stated otherwise.

15.3 Extending Graphical Models for Causal Reasoning

As announced earlier, we now consider two different approaches to extending DAGs with a view to a causal interpretation. The first adds a particular type of node representing experimental manipulation or intervention. The second approach modifies the semantics of a DAG to obtain a *causal DAG*.

15.3.1 Intervention graphs

Intervention graphs are conditional independence graphs augmented by a separate type of node for σ as an indicator for the regime[4]. As statistical models they retain the interpretation in terms of Markov properties. However, this raises a question as the regime indicator is not a random variable. It may not make sense to specify a distribution over the possible regimes, especially if some of them correspond to 'what if' scenarios such as: 'banning smoking' or 'what if doctors followed strategy 1 versus strategy 2 when treating HIV patients' etc. Nevertheless we can make conditional independence statements such as $\mathbf{X}_B \perp\!\!\!\perp \sigma \mid \mathbf{X}_A$ to indicate that the conditional distribution of $\mathbf{X}_B$ given $\mathbf{X}_A$ is identical across the particular choices of regimes $\mathcal{S}$ (for a formal treatment see Constantinou and Dawid [6]). In fact, this *stability* [14] or *invariance* [33, 53] of conditional distributions across regimes is central to causal inference where a key aim is to predict effects of interventions from observational data. Clearly, in order to be able to infer anything about the former based on the latter at least some aspects of the two situations must be assumed 'similar'. Graphs are useful to make such assumptions explicit. We explain this with an example.

[4]Unfortunately, the terminology is not in agreement here. Lauritzen [40] calls these 'intervention graphs', while Dawid [11] calls them 'augmented' and uses 'intervention' graphs for causal DAGs; they can also be regarded as 'decision' or 'influence diagrams' [39]. As graphs can be augmented in several different ways, we revert to 'intervention graphs'.

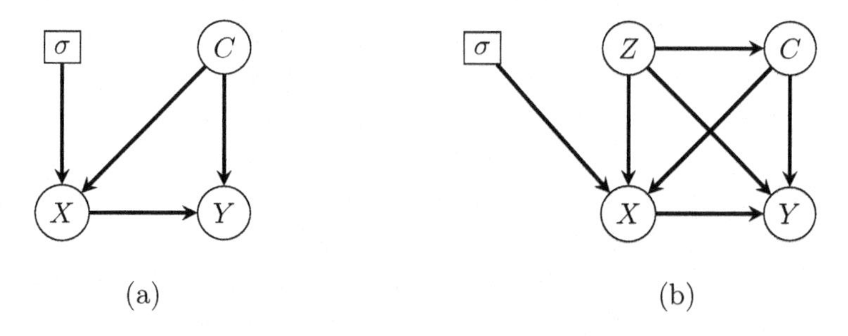

FIGURE 15.1: Examples of intervention DAGs.

Example: Consider the case where a patient suffers from some ailment and may or may not take a treatment. Let X be a treatment indicator, C a blood-measurement and Y a health outcome. The different regimes are: (P) the patient decides what treatment to take, (O) the doctor decides the treatment based on old guidelines, or (N) the doctor decides the treatment based on new guidelines. A typical question would be how to compare regime O versus N based on data gathered only under regime P and whether this is possible at all. (We return to this as a question of *identifiability* later.) As the circumstances are not necessarily the same in the three situations we may want our model to allow three different joint distributions $p(\cdot; \sigma = s), s =$ P, O, N. For each we have the factorization

$$p(x, c, y; s) = p(y|x, c; s)p(x|c; s)p(c; s).$$

Some of $p(x|c; s)$ may be unknown, for example $p(x|c; \sigma = \text{P})$, while others are determined by background knowledge, for example the old guidelines may specify a decision rule according to which treatment is given if C is larger than a threshold τ, yielding $p(X = 1|c; \sigma = \text{O}) = \mathbf{1}\{c > \tau\}$. Now assume that there is reason to believe that, across the three regimes, only the conditional distribution of treatment $p(x|c; s)$ is possibly different, but the conditional distribution of Y given (X, C) as well as the marginal distribution of C remain the same. Formally, this is implied by the equalities $p(y|x, c; s) = p(y|x, c; s')$ and similarly $p(c; s) = p(c; s'), s \neq s'$. The assumptions are represented in Figure 15.1(a), where the box around σ reminds us that this node is not a random variable but a regime indicator. It is easy to justify the invariance of the marginal distribution of C if the covariates in C are measured before 'deciding' the value of X. Its distribution will then obviously be the same under all regimes. A sufficient condition for the invariance of Y given (X, C) across regimes, in this example, would be that the only information available to the decision maker, in any regime, is the blood measurement. However, under regime P this may be implausible.

The ideas sketched above can be employed very flexibly to represent a wide range of situations and assumptions. An intervention graph, as defined next, describes the probabilistic dependence structure of a system that factorizes according to a given DAG, where certain factors are determined by interventions.

Definition 15.3.1 (Intervention DAG and Model). *Consider the random vector* $\mathbf{X} = (X_1, \ldots, X_K)$*, on vertices* $V = \{1, \ldots, K\}$ *of a DAG* $\mathcal{G}$*. Let* $\mathcal{S}$ *denote a set of regimes and let* $p(\cdot; \sigma = s), s \in \mathcal{S}$*, be the joint distributions under the respective regimes. The augmented DAG* $\mathcal{G}^\sigma = (V \cup \{\sigma\}, E^\sigma)$ *is called the* intervention DAG *for* $\mathbf{X}$ *under regimes* $\mathcal{S}$ *if it has the following properties:*
(i) the node σ *is a source node (has no incoming directed edges),*
(ii) each distribution $p(\mathbf{x}; s), s \in \mathcal{S}$*, factorizes according to* $\mathcal{G}$*,*

(iii) for disjoint $A, B \subset V$, whenever B is d-separated from σ by A in $\mathcal{G}^\sigma$ we have: $p(\mathbf{x}_B|\mathbf{x}_A; s) = p(\mathbf{x}_B|\mathbf{x}_A; s')$, for all $s \neq s'$. This is denoted by $\mathbf{X}_B \perp\!\!\!\perp \sigma|\mathbf{X}_A$.

A key aspect of intervention DAGs is that the factors $p(x_k|\mathbf{x}_{\mathrm{pa}(k)}; \sigma = s)$ for $k \in \mathrm{ch}(\sigma)$ characterize the particular regime s, while the remaining factors are assumed invariant. Moreover, the graph itself is assumed invariant; with the above definition it is, for instance, not possible to describe a situation where under one regime we have $j \longleftarrow k$ and under the other regime we have $j \longrightarrow k$.

Example ctd.: This example illustrates that a given intervention DAG can be implausible if important information is ignored. First, note that the sub-DAG on $\{C, X, Y\}$ in Figure 15.1(a) is Markov equivalent to any other orientation of the edges. This means that as a conditional independence DAG it imposes no restrictions at all on the joint distribution of (C, X, Y). However, adding $\sigma \longrightarrow X$, as shown, means that there is no other Markov equivalent DAG on $\{C, X, Y, \sigma\}$. The stability assumptions therefore make this a unique model that can in principle be *verified with data from more than one regime.* For instance if we have three data sets, one each under P, O, and N, respectively, then it can be checked empirically that the marginal of C and the conditional of Y given (C, X) are the same in all data sets.

In practice, we rarely have data sets from more than one regime (but see [53] for examples with data from different regimes). Assumptions such as in Figure 15.1(a) then need to be justified with subject matter knowledge about, for example, the physical mechanisms linking (C, X, Y) and the contemplated intervention(s). In many situations it will be plausible that, if the system is specified in sufficient detail, certain aspects of it will remain invariant across a range of specified situations, and the edges out of σ then correspond to specific physical manipulations. As a counterexample assume that an important factor Z has been omitted, for example the patient's temperature which could be influencing the treatment decision of the patient or doctor. Then the DAG of Figure 15.1(a) becomes implausible and we may want to consider the DAG in (b) instead, where $Y \perp\!\!\!\perp \sigma \,|\, (Z, C, X)$ but $Y \not\!\perp\!\!\!\perp \sigma \,|\, (C, X)$, hence the incorrectness of (a).

The appropriateness of a particular intervention DAG also depends on the set of regimes $\mathcal{S}$ considered. Figure 15.1(a) may be suitable to represent $\sigma \in \mathcal{S} = \{\text{O, N}\}$ where X is assigned based on the old or new guidelines if these use only the blood measurement as information. But it may not be suitable when including the patient as decision maker, so when $\sigma \in \mathcal{S}' = \{\text{P, O, N}\}$, as the patient may have measured her temperature before deciding on treatment.

15.3.2 Causal DAGs

Causal DAGs could be defined as intervention DAGs where the set of regimes consist of the natural (observational) regime and atomic interventions in *all* nodes [13]. Under these two conventions the indicator σ and the set $\mathcal{S}$ can be dropped, as σ would simply have edges into all nodes and all interventions are of the same type. Pearl's do($\cdot$)-notation is often combined with causal DAGs. However, the notion of a causal DAG is in fact more in the spirit of Spirtes et al. [67] who refer to *unmanipulated* and *manipulated* populations. Pearl [47], in contrast, links causal DAGs, or *causal diagrams*, very much with SEMs and potential outcomes.

The following defines causal DAGs such that atomic interventions in arbitrary nodes are adequately represented by a modular modification to the factorized joint distribution.

Definition 15.3.2 (Causal DAG). *Consider a DAG $\mathcal{G} = (V, E)$ and a random vector $\mathbf{X} = (X_1, \ldots, X_K)$ with distribution p. Then $\mathcal{G}$ is called a* causal DAG *for $\mathbf{X}$ if p satisfies*

the following:
(i) p factorizes, and thus is Markov, according to $\mathcal{G}$, and
(ii) for any $A \subset V$ and any $\widetilde{\mathbf{x}}_A, \mathbf{x}_B$ in the domains of $\mathbf{X}_A, \mathbf{X}_B$, where $B = V \backslash A$,

$$p(\mathbf{x}; \text{do}(\widetilde{\mathbf{x}}_A)) = \prod_{k \in B} p(x_k | x_{\text{pa}(k)}) \prod_{j \in A} \mathbf{1}\{x_j = \widetilde{x}_j\}. \tag{15.2}$$

The second factor of (15.2) makes explicit that $\text{do}(\widetilde{\mathbf{x}}_A)$ refers to interventions that fix $\mathbf{X}_A$ at the given value $\widetilde{\mathbf{x}}_A$. In other words, the difference between the observational distribution $p(\mathbf{x})$ and the intervention distribution $p(\mathbf{x}; \text{do}(\widetilde{\mathbf{x}}_A))$ is that all factors $p(x_j|x_{\text{pa}(j)}), j \in A$, are removed and replaced by degenerate probabilities $\mathbf{1}\{x_j = \widetilde{x}_j\}$, while all remaining factors $p(x_k|x_{\text{pa}(k)}), k \in B$, stay the same. Equation (15.2) is therefore known as the *truncated factorization* and the substitution of individual factors $p(x_j|x_{\text{pa}(j)})$ as *causal modularity* [33].

The validity of (15.2) in a given context could in principle be decided by carrying out all possible atomic interventions. It is more common to employ subject matter knowledge, and include unobservables, to justify a causal DAG. Similar to the points made in connection with intervention graphs, the plausibility of causal modularity will crucially depend on the set of nodes (measured or unmeasured) being sufficiently rich. This is often expressed in the recommendation that when a causal DAG is posited, the absence of possible further edges and further nodes needs to be justified. In other words, for a DAG to be causal it needs to include all 'common causes' of its nodes [36].

Example ctd.: To compare intervention graphs with causal DAGs consider again Figure 15.1. The corresponding causal DAGs would simply omit the node σ, but would additionally assume that atomic interventions in C and Z are also correctly represented. This makes a difference in (b), where the direction of the edge between Z and C is not relevant to the assumptions expressed in the intervention DAG because a reversal of the edge is Markov equivalent to the original intervention DAG. However, it is relevant in the causal DAG as it implies a different causal ordering between Z and C.

Randomization

It can easily be seen that (15.2) would be obtained as the ordinary conditional distribution of $\mathbf{X}_B$ given $\mathbf{X}_A = \widetilde{\mathbf{x}}_A$ in a graph where all incoming edges into $\mathbf{X}_A$ have been deleted. Such a truncated DAG corresponds to the situation where $\mathbf{X}_A$ was drawn randomly and independently of any predecessors while all other relations remain the same. In other words, it mimics the ideal randomized experiment reflecting the special status of randomized trials for causal inference.

Similarly it can be seen that (15.2) is obtained from the original joint distribution upon dividing by $p(x_j|x_{\text{pa}(j)})$, $j \in A$. This motivates the principle of 'inverse probability of treatment weighting' (IPTW). With IPTW a dataset is reweighted to recreate empirically the situation of a randomly assigned treatment or unconfounded exposure [59, 36]. Hence, IPTW empirically 'removes' the arrows into the variable targeted by an intervention.

Faithfulness

With Definition 15.3.2, the absence of a directed edge from X_j to X_k in a causal DAG can be interpreted as 'X_j has no direct causal effect on X_k relative to the given set of nodes'. More precisely, if we were to fix the parent nodes of X_k, and then manipulate X_j, $j \notin \text{pa}(j)$, we would see no change in the distribution of X_k. Conversely, one might say that a directed edge always represents a direct causal effect. However, this is not implied by the Definition 15.3.2. It requires the additional assumption of *causal faithfulness* [67].

To make this more formal, recall that the Markov properties imply that every d-separation in a DAG $\mathcal{G}$ induces a conditional independence in the corresponding distribution p. If, in addition, the converse is true and every conditional independence under p corresponds to a d-separation in the DAG, then we have the probabilistic version of *faithfulness*. If the DAG is supplemented with an interpretation in terms of atomic intervention as in Definition 15.3.2, probabilistic faithfulness implies *causal faithfulness*. Hence, every directed edge corresponds to an intervention effect and every absence of a directed edge to the absence of a direct causal effect. Moreover, under causal faithfulness it follows that if, and only if, there is a directed path from X_j to X_k in $\mathcal{G}$ then some manipulation of the former will affect the distribution of the latter.

Faithfulness is often invoked when we want to empirically construct or check parts of the causal model. *Causal search* (or *causal discovery*) refers to methods that reconstruct the whole causal DAG, up to Markov equivalence, based on observational data [67]. Chapter 18 in this book deals with these methods in more detail. We return to faithfulness in Section 15.5.1.

Structural equation models

Causal DAGs are often combined with the potential outcomes framework. One way of doing this relies on (recursive) structural equation models (SEMs) [26]. To keep it general we only consider non-parametric ones (NPSEMs) where the shape of the functional relations is left unspecified. These models assume a set of variables to be functionally related at an individual level such that these functional relations are invariant to how the input comes about. Consider $\mathbf{X} = (X_1, \ldots, X_K)$ and a DAG $\mathcal{G}$. A structural equation model for $\mathbf{X}$ on $\mathcal{G}$ assumes for each node $k \in V$ that X_k is a function of its graphical parents and possibly a random variable ϵ_k [47]

$$X_k := f_k(\mathbf{X}_{\text{pa}(k)}, \epsilon_k),$$

where in the simplest case all ϵ_k are assumed mutually independent. A violation of this independence assumption can graphically be depicted by corresponding bi-directed edges resulting in a so-called *semi-Markovian* graph [49]. We write ':=' to remind the reader that the above should be regarded as an asymmetric *assignment* of the value of f_k to X_k [47]. The joint distribution of $(\epsilon_1, \ldots, \epsilon_K)$ together with the above functional relations induces a distribution on $(X_1, \ldots, X_K)$.

The *structural* nature of such a system means that potential outcomes can be constructed as follows. Consider an intervention forcing X_j to have the value $\widetilde{x}_j$. Under an NPSEM such an intervention corresponds to replacing the function f_j with $\mathbf{1}\{x_j = \widetilde{x}_j\}$, and x_j is replaced by $\widetilde{x}_j$ in the argument of f_k whenever $j \in \text{pa}(k)$. The latter yields equations for the potential outcomes $X_k(\widetilde{x}_j)$. However, instead of *replacing* equations, NPSEMs allow to simply *add* equations for each intervention. The result is a system of equations that simultaneously describes what would happen if X_j was fixed at value $\widetilde{x}_j$ as well as at value $\widetilde{x}'_j$ etc. Hence, the joint distribution of $(\epsilon_1, \ldots, \epsilon_K)$ also induces a joint distribution on all potential outcomes $\{X_k(\widetilde{x}_j); k \in V, \widetilde{x}_j \in \mathcal{X}_j\}$.

Note that the above combination of DAGs with NPSEMs agrees with Definition 15.3.2 in the sense that with the above construction the joint distribution $p(\mathbf{X}_B(\widetilde{\mathbf{x}}_A))$ will satisfy the definition of $p(\mathbf{x}_B; \text{do}(\widetilde{\mathbf{x}}_A))$. However, a causal DAG as in Definition 15.3.2 does not encompass joint distributions under interventional settings of a variable simultaneously to different values.

Example: Consider the DAG in Figure 15.4(b) to which we return later in the context of bias amplification. Linked with a NPSEM it induces variables C, U, X, Y as follows:

$$C := \epsilon_C, \quad U := \epsilon_U, \quad X := f_X(C, U, \epsilon_X), \quad Y := f_Y(X, U, \epsilon_Y).$$

A set of potential outcomes is defined for any fixed values $\tilde{c} \in \mathcal{C}, \tilde{x} \in \mathcal{X}$ as

$$X(\tilde{c}) := f_X(\tilde{c}, U, \epsilon_X), \quad Y(\tilde{x}) := f_Y(\tilde{x}, U, \epsilon_Y),$$

yielding $4+|\mathcal{C}|+|\mathcal{X}|$ equations / variables in our system if C and X are discrete, and infinitely many if they are continuous. Note that we could extend this to interventions in U yielding even more potential outcomes. Any joint distribution of $(\epsilon_C, \epsilon_U, \epsilon_X, \epsilon_Y)$ induces a joint distribution on all of these, that is on $(C, U, X, Y, \{X(\tilde{c}) : \tilde{c} \in \mathcal{C}\}, \{Y(\tilde{x}) : \tilde{x} \in \mathcal{X}\})$, where we use that $Y(\tilde{x}, \tilde{c}) = Y(\tilde{x})$. Within this set-up, certain counterfactual quantities are therefore well-defined, such as for instance (for binary C and X) $E(Y(1) - Y(0)|X(1) > X(0))$. In the context of an RCT with imperfect compliance, let C be the randomized treatment and X the actual treatment taken. Then, $E(Y(1) - Y(0)|X(1) > X(0))$ can be regarded as the effect of treatment on the group of people who 'comply' with their assignment as these are characterized exactly by $X(1) > X(0)$. As we can never observe $X(1), X(0)$ together, this is a latent subgroup of the population. Moreover, a common assumption in this context is the one of monotonicity which demands that f_X is such that $X(1) \geq X(0)$ always. Such counterfactual assumptions cannot be read off the graph and are an additional specification of the NPSEM.

The above gives an idea of how extensive a latent structure is imposed by NPSEMs. A less restrictive approach for a graph-based construction of potential outcomes is discussed in Richardson and Robins [55].

15.3.3 Comparison

The differences between the above approaches are subtle and lie partly in the different graphical displays and partly in the different semantics. Intervention DAGs use the additional intervention nodes to supplement the graph with causal meaning, while causal DAGs modify the meaning of edges and separations via the causal Markov properties. With the above definitions, intervention DAGs are the more general class of models. Causal DAGs are a special case where there is an (invisible) intervention node into every variable and all interventions are atomic ones. The NPSEM interpretation of causal DAGs further imposes a counterfactual structure as discussed.

Causal DAGs can and have been generalized to cases where the interventions are other than atomic, for example random, and where not all nodes need to be manipulable. For instance, Definition 15.3.2 can be relaxed by demanding in (ii) that the property holds *locally*, that is only for a *specific* set $A \subset V$ [40, 49]. Such a specific set could for instance be the set of all treatment-like variables that can be manipulated in practice. However, such a relaxation is rarely used in the literature, even though many results that have first been derived for causal DAGs can be shown to hold more generally for locally causal DAGs. The latter is more explicit in intervention DAGs due to the corresponding intervention nodes. Arguably, in practice, it is often not plausible nor required to assume that all variables are manipulable and we find that using intervention DAGs is a helpful tool to focus on the essential assumptions. For instance, in the next section we give two results, the back-door and the front-door criterion. The former was proposed by Pearl [47] using causal DAGs, but a close look at his proof shows that it uses a simpler intervention DAG as also shown by Dawid [11]. The same is true for the front-door criterion as we demonstrate in Section 15.4.2.

15.4 Graphical Rules for the Identification of Causal Effects

One of the most prominent uses of DAGs in causal inference is to help decide whether and how the available data identifies a desired causal target of inference under the assumed causal model. Chapter 16 in this book deals with this topic in more detail. Here we consider some classic fundamental results. Let us first clarify what is meant by 'identifiability' [56, 67, 49].

Definition 15.4.1 (Identifiability). *Consider $\mathbf{X} = \mathbf{X}_V$ and let (O, U) be a partition of V. Let $\mathcal{P}$ be a class of distributions for $\mathbf{X}$, including the relevant observational and interventional distributions, for instance, but not necessarily, as induced by a causal DAG. Finally, let $A \subset O$ and $D \subset O \backslash A$.*
We say that the causal effect of $\mathbf{X}_A$ on $\mathbf{X}_D$ is identified by $\mathbf{X}_O$ under $\mathcal{P}$ if for any two distributions $p', p'' \in \mathcal{P}$

$$p'(\mathbf{x}_O) = p''(\mathbf{x}_O) \quad \Rightarrow \quad p'(\mathbf{x}_D; \mathrm{do}(\widetilde{\mathbf{x}}_A)) = p''(\mathbf{x}_D; \mathrm{do}(\widetilde{\mathbf{x}}_A)) \quad \forall \widetilde{\mathbf{x}}_A \in \mathcal{X}_A. \tag{15.3}$$

More generally consider a situation where we are interested in comparing a set of regimes $s \in \widetilde{\mathcal{S}}$ based on data from an observational regime $\sigma = \emptyset$. We say that the consequences of regimes $\widetilde{\mathcal{S}}$ for $\mathbf{X}_D$ are identified if for any two distributions $p', p'' \in \mathcal{P}$

$$p'(\mathbf{x}_O; \sigma = \emptyset) = p''(\mathbf{x}_O; \sigma = \emptyset) \quad \Rightarrow \quad p'(\mathbf{x}_D; \sigma = s) = p''(\mathbf{x}_D; \sigma = s) \quad \forall s \in \widetilde{\mathcal{S}}. \tag{15.4}$$

The above says that the effect of atomic interventions, or the consequence of a regime s, is not a function of the unobserved part $\mathbf{X}_U$. Once $\mathbf{X}_O$ is observed, $p(\mathbf{x}_D; \mathrm{do}(\widetilde{\mathbf{x}}_A))$ can uniquely be determined. Typically, identifiability is demonstrated by finding an expression for $p(\mathbf{x}_D; \mathrm{do}(\widetilde{\mathbf{x}}_A))$ (or for $p(\mathbf{x}_D; \sigma = s)$) in terms of the observational $p(\mathbf{x}_O; \sigma = \emptyset)$. Note that identification as defined above concerns the whole interventional distribution $p(\mathbf{x}_D; \mathrm{do}(\widetilde{\mathbf{x}}_A))$ within the whole model class $\mathcal{P}$. In practice, interest may be restricted to certain parameters such as the ACE or causal odds ratio, possibly within a parametric subclass such as linear or logistic models. There are situations where specific causal parameters can be identified without the whole intervention distribution being identified, for example odds ratios in a case-control study [19]. However, we do not go into these details here. Moreover, we ignore any issues due to finite sample size. In Section 15.5.2 we briefly address situations where identification can be achieved for almost the whole model class $\mathcal{P}$ except a lower dimensional subset.

Positivity

A general requirement for identifiability is that there is an empirical basis for estimating the consequences of the contemplated interventions. This means that combinations of values possible under the interventional regimes must also be possible under the observational regime. A violation could be if we wanted to compare 'treatment' with 'no treatment' in a patient group where some patients are so ill that they are never left untreated in practice. The exact formulation of this assumption depends on the context and the causal target of inference, but usually requires a positive probability for the relevant combinations under the observational regime — hence known as *positivity*. It will generally be assumed to hold in the following.

15.4.1 The Back-Door Theorem

A first identification result based on causal DAGs is known as the 'Back-Door Theorem' [46, 47]. It is closely related to the assumption of 'no unobserved confounding' as addressed in Section 15.5.1. The name stems from the following terminology: a *back-door path* from j to k in a DAG $\mathcal{G}$ is a path that does not have an edge emanating from j. In other words, it is of the shape $j \leftarrow \cdots k$, where $j, k \in V, j \neq k$.

Theorem 15.4.2 (Back-Door Theorem). *Let $\mathcal{G}$ be a causal DAG for $\mathbf{X}$. The causal effect of X_j on X_k is identified by $\mathbf{X}_{\{j,k\}\cup C}$, $C \subset V\backslash\{j,k\}$ in the sense of (15.3) if*
(i) $C \cap \mathrm{de}(j) = \emptyset$,
(ii) C blocks every back-door path from j to k in $\mathcal{G}$.
The intervention distribution is then identified as

$$p(x_k; \mathrm{do}(\widetilde{x}_j)) = \int p(x_k|\widetilde{x}_j, \mathbf{x}_C)p(\mathbf{x}_C)\, \mathrm{d}\,\mathbf{x}_C. \tag{15.5}$$

Equation (15.5) can be recognized as the adjustment or standardization formula, which makes many appearances in the literature with different motivations, see [9, 56, 30, 28, 38]. Note that the type of positivity required here demands that for all $\widetilde{x}_j \in \mathcal{X}_j, \mathbf{x}_C \in \mathcal{X}_C$, we have $p(\widetilde{x}_j, \mathbf{x}_C) > 0$.

The Back-Door Theorem can also be derived from Theorem 7.1 of Spirtes et al. [67]. Using intervention DAGs, Lauritzen [40] and Dawid [11] provide slightly different versions and more general proofs of the above theorem than Pearl [49]. These alternatives demonstrate that the same criterion can be applied without assuming a causal DAG as long as the model is valid (the assumed invariances hold) with regard to an intervention in X_j, or when a corresponding intervention DAG is used for representing an atomic intervention in X_j.

Example ctd.: Returning to the examples in Figure 15.1 (omit the node σ to read these as causal DAGs), we find that in (a) C blocks the only back-door path, while in (b) it needs to be supplemented by Z in order to block all three back-door paths. As alluded to earlier, this is an example where it is actually not necessary to decide the causal direction between Z and C, or to assume a causal relation between them at all, for inference to be valid. In fact, Figure 15.1(a) expresses exactly the assumptions of Dawid's [11] version of the Back-Door Theorem, where the interventional regimes target X, and C is characterized by the invariances in (a).

Use of the Back-Door Theorem in practice

A way to use Theorem 15.4.2 is to construct an assumed causal DAG $\mathcal{G}$, for example based on subject matter background knowledge. *To be plausible this typically involves unobserved quantities.* With the assumed causal DAG, it can be checked whether a set of *measured* covariates $\mathbf{X}_C$ can be found such that C satisfies the back-door criterion relative to (j, k) in this DAG. In that case, no further measurements are required to identify any causal effect that is a function of $p(x_k; \mathrm{do}(\widetilde{x}_j))$.

Example: Staplin et al. [68] aim to investigate the effect of the binary indicator $X =$ 'smoker at entry into the study' on the binary indicator $Y =$ 'End Stage Renal Disease' (ESRD), where a number of covariates have been observed, but smoking status in early adulthood is unobserved. First we note that condition (i) $C \cap \mathrm{de}(X) = \emptyset$ of the back-door criterion implies that C consists only of *pre-exposure covariates.* This means that they must be known to not be affected by an intervention in the exposure, for instance because they are prior in time. Condition (i) mirrors the recommendation that for estimating a total causal effect we should not adjust for post-exposure covariates [7]. Without referring to

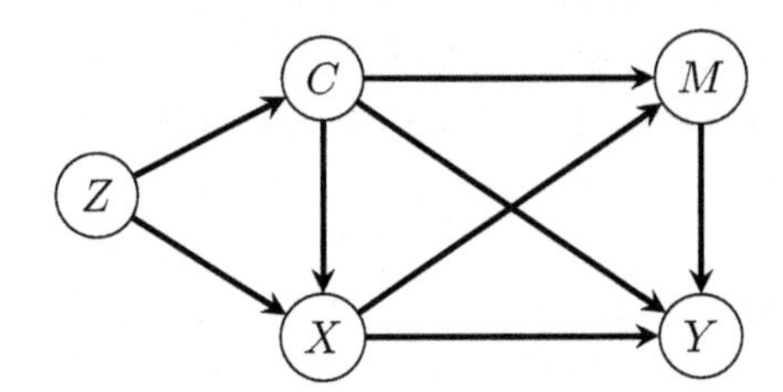

FIGURE 15.2: Assumed causal DAG for end stage renal disease example [68]. Here, X = 'smoker' (at entry in study), Y = 'ESRD', Z = 'early smoker' (unobserved). Further C are observed pre-exposure covariates such as age, sex, ethnicity, prior diseases, and M are post-exposure covariates such as the urinary albumin-to-creatinine ratio, for example.

a graph this requirement can be stated as $C \perp\!\!\!\perp \sigma$ or $p(c; \mathrm{do}(X = \widetilde{x})) = p(c)$.[5] Hence, for the elicitation of the assumed causal DAG, subject matter knowledge should be used to determine, first of all, which covariates are pre-exposure. Graphically, these cannot be descendants of the exposure node. In the example, early smoking status, age, sex, ethnicity, and prior diseases cannot possibly be affected by current smoking and are hence considered pre-exposure covariates. In contrast, covariates such as the urinary albumin-to-creatinine ratio and renal status could be affected by current smoking status and are therefore post-exposure. These and further subject matter assumptions are depicted in Figure 15.2. The authors argue that smoking status in early adulthood carries no further information for ESRD once we account for current smoking status and prior diseases, hence there is no edge from Z to Y in Figure 15.2. The assumed causal DAG can now be queried as to whether the measured pre-exposure covariates C satisfy (ii) so that the unmeasured earlier smoking status Z can safely be ignored. This is indeed the case with C including covariates age, sex, ethnicity as well as information on prior diseases [68]. The example illustrates that subject matter knowledge combined with the back-door criterion sometimes leads to the selection of a subset of the covariates as it turns out that not all of them are required for valid causal inference.

Using a DAG in this way strengthens any analysis as it makes explicit and transparent the key assumptions. It can also help to identify weaknesses, for example important unmeasured covariates, and can thus potentially lead to improvements of design and data-collection for future studies.

An apparent limitation of using causal DAGs in the above way is that researchers often find it difficult to specify the whole DAG, including all pairwise relations between any two variables, observed as well as relevant unobserved variables. For instance, it may not be clear what the causal direction between alcohol consumption and smoking status are as these are processes over time and both reflect a general life-style. In such cases it may be helpful to elaborate the DAG, for example to include 'life-style choice' as an underlying unobservable quantity that encompasses several issues.

Remarks and generalizations

Theorem 15.4.2 provides a sufficient (graphical) criterion for the identification of $p(x_k; \mathrm{do}(\widetilde{x}_j))$; necessary conditions are given for instance in [64] and Chapter 16 in the present part of this book. However, these are necessary only in the context of causal DAGs.

[5] Section 15.4.2 gives a different graphical criterion showing how post-exposure covariates can be used. Section 15.5.1 gives an example for so-called M-bias demonstrating that adjusting for pre-exposure covariates does not protect from introducing bias.

As has been discussed earlier, we may not want to assume causal validity with regard to *all* nodes, but only locally, or we may want to work with intervention graphs. In these cases it is unclear whether necessary conditions can be given.

Perković et al. [52] provide a complete investigation of graphical criteria for identification specifically through the adjustment formula (15.5). They generalize the existing results in a number of ways, for instance showing that sets C exist that do not satisfy (i) but for which (15.5) is still valid. Further, they consider larger classes of graphs than DAGs so as to cover latent variables and relevant Markov equivalence classes.

Sequential treatments

A further generalization concerns multiple, possibly sequential, interventions. In fact, the Back-Door Theorem can immediately be extended to sets of intervention nodes [46, 47, 52]. Here we consider the additional challenge when variables that are descendants of earlier intervention nodes are needed to identify later interventions. To give a brief idea, consider a time-ordered sequence $(\mathbf{X}_{C_0}, X_i, \mathbf{X}_{C_1}, X_j, X_k)$ and assume we want to know the effect of a joint intervention in (X_i, X_j) on X_k, denoted by $p(x_k; \text{do}(\widetilde{x}_i, \widetilde{x}_j))$. Such situations occur for example in the treatment of chronically ill patients. Let the causal DAG be such that $C_0 \cap \text{de}(i,j) = \emptyset$ and $C_1 \cap \text{de}(j) = \emptyset$ but $C_1 \cap \text{de}(i) \neq \emptyset$ meaning that $\mathbf{X}_{C_1}$ can be affected by interventions in X_i but not X_j. In other words, $\mathbf{X}_{C_1}$ are post-X_i but pre-X_j covariates. It is not obvious how to apply Theorem 15.4.2 when C_1 is needed to block some back-door paths from X_j to X_k but contains at the same time descendants of X_i. Extending the back-door criterion [51, 14], causal DAGs or influence diagrams can again be used to characterize when $(\mathbf{X}_{C_0}, \mathbf{X}_{C_1})$ identify the desired causal effect such that

$$p(x_k; \text{do}(\widetilde{x}_i, \widetilde{x}_j)) = \int p(x_k | \widetilde{x}_i, \widetilde{x}_j, \mathbf{x}_{C_0}, \mathbf{x}_{C_1}) p(\mathbf{x}_{C_1} | \widetilde{x}_i, \mathbf{x}_{C_0}) p(\mathbf{x}_{C_0}) \, \text{d}\, \mathbf{x}_{C_0} \mathbf{x}_{C_1}. \tag{15.6}$$

The above has been termed 'g-formula' by Robins [56] and is a generalization of the adjustment formula (15.5). It is straightforward to further generalize the above such that, for example, $\widetilde{x}_i$ is a function of $\mathbf{x}_{C_0}$ and $\widetilde{x}_j$ a function of $\mathbf{x}_{C_1}$, for instance to represent the case where treatment decisions depend on previous measurements taken on the patient.

15.4.2 The Front-Door Theorem

As mentioned above, the Back-Door Theorem provides a sufficient criterion for the identification of a causal effect. In this section we consider a second sufficient graphical criterion, the Front-Door Theorem [47]. The complete identification algorithm of Shpitser and Pearl [64] can be regarded as combining and generalizing these two criteria in the context of causal DAGs.

We use this opportunity to re-state the theorem not in its original version, but in the *framework of intervention DAGs.* This illustrates the use of intervention graphs, and demonstrates that it is not necessary to assume a fully causal DAG to obtain identifiability. The following theorem therefore generalizes the original result. Moreover, we hope that this helps the reader to become more familiar with the intervention DAG framework.

Theorem 15.4.3 (Front-Door Theorem). *Let $\mathcal{G}^\sigma = (V \cup \sigma, E^\sigma\})$ be an intervention DAG for $\mathbf{X}$ under $p(\mathbf{x}; s)$. Let $s = \emptyset$ denote the observational regime and $s = \widetilde{x}_j \in \mathcal{X}_j$ denote an atomic intervention fixing X_j at x_j. Let, further, $C \subset V \backslash \{j,k\}$ and $U \subset V \backslash (\{j,k\} \cup C)$. The consequences of regimes $s = \widetilde{x}_j \in \mathcal{X}_j$ are identified by $\mathbf{X}_{\{j,k\} \cup C}$ in the sense of (15.4) if*

(i) $\mathbf{X}_U \perp\!\!\!\perp \sigma$, and

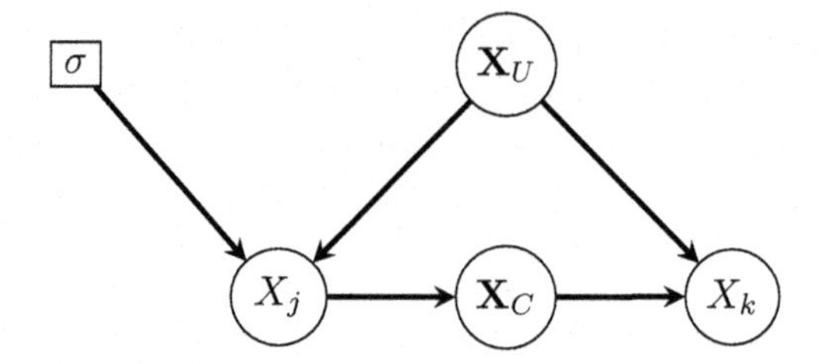

FIGURE 15.3: Intervention graph illustrating the Front-Door Theorem.

(ii) $\mathbf{X}_C \perp\!\!\!\perp (\sigma, \mathbf{X}_U) \,|\, X_j$, *and*
(iii) $X_k \perp\!\!\!\perp (\sigma, X_j) \,|\, (\mathbf{X}_C, \mathbf{X}_U)$.
The intervention distribution is then identified as

$$p(x_k; \sigma = \widetilde{x}_j) = \int p(\mathbf{x}_C|\widetilde{x}_j; \sigma = \emptyset) \int p(x_k|x_j, \mathbf{x}_C; \sigma = \emptyset) p(x_j; \sigma = \emptyset) \, \mathrm{d}\, x_j \, \mathrm{d}\, \mathbf{x}_C. \quad (15.7)$$

Figure 15.3 shows the intervention graph depicting the invariance assumptions of Theorem 15.4.3.

Proof of Theorem 15.4.3: Using all of conditions (i, ii, iii) and that $p(x_j|\mathbf{x}_U; \sigma = \widetilde{x}_j) = \mathbf{1}\{x_j = \widetilde{x}_j\}$ by definition, we have

$$\begin{aligned} p(x_k; \sigma = \widetilde{x}_j) &= \int p(x_k|x_j, \mathbf{x}_C, \mathbf{x}_U; \sigma = \widetilde{x}_j) p(\mathbf{x}_C|x_j, \mathbf{x}_U; \sigma = \widetilde{x}_j) \\ &\quad \times p(x_j|\mathbf{x}_U; \sigma = \widetilde{x}_j) p(\mathbf{x}_U; \sigma = \widetilde{x}_j) \, \mathrm{d}(x_j, \mathbf{x}_C, \mathbf{x}_U) \\ &= \int p(\mathbf{x}_C|\widetilde{x}_j; \sigma = \emptyset) \\ &\quad \times p(x_k|\mathbf{x}_C, \mathbf{x}_U; \sigma = \emptyset) p(\mathbf{x}_U; \sigma = \emptyset) \, \mathrm{d}\, \mathbf{x}_U \, \mathrm{d}\, \mathbf{x}_C. \end{aligned}$$

The claim now follows upon verifying that with the assumed conditional independencies, under the idle regime,

$$\int p(x_k|\mathbf{x}_C, \mathbf{x}_U; \sigma = \emptyset) p(\mathbf{x}_U; \sigma = \emptyset) \, \mathrm{d}\, \mathbf{x}_U = \int p(x_k|x_j, \mathbf{x}_C; \sigma = \emptyset) p(x_j; \sigma = \emptyset) \, \mathrm{d}\, x_j.$$

This last term corresponds to the back-door formula (15.5) for the 'effect' of $\mathbf{X}_C$ on X_k with X_j blocking the only back-door path. In this sense, the front-door formula (15.7) can be interpreted as exploiting the Markov structure so as to combine the effect of X_j on $\mathbf{X}_C$ and the 'effect' of $\mathbf{X}_C$ on X_k. However, as can be seen from the assumptions of the above theorem and its proof, no intervention in $\mathbf{X}_C$ is assumed. The result follows as long as the conditional independencies and invariances of assumptions (i)-(iii) regarding intervention in X_j can be justified.

The Front-Door Theorem appears to be used very little in practice. Its assumptions require that the causal effect of X_j on X_k is known to be 'fully mediated' by $\mathbf{X}_C$ while at the same time this $\mathbf{X}_C$ is known to be conditionally independent of the unobserved $\mathbf{X}_U$ given X_j. With 'fully mediated' we refer to the assumptions of the theorem implying that $p(x_k|\mathbf{x}_C; \sigma = \widetilde{x}_j)$ is in fact not a function of $\widetilde{x}_j$. Pearl [47] gives the following example: let X_j be smoking intensity, X_C the amount of tar deposit in the lungs and X_k an indicator for lung cancer. Note that while it may be feasible to intervene in and modify the smoking intensity, it seems much less practical to intervene in the tar deposit without changing the smoking intensity. Hence an approach only considering an intervention in X_j but not X_C seems appropriate. Further, it appears plausible that while smoking itself is related to many observed and unobserved social and life-style factors under the observational regime, the tar deposit is independent of these given smoking intensity. In turn it is plausible that once we account for the actual tar deposit in the lungs smoking intensity does not further contribute to developing lung cancer. Hence the key assumptions are plausible in this situation but the difficulty is to obtain measurements on X_C.

15.5 Graphical Characterization of Sources of Bias

The main threats to the validity of causal inference are threefold: (i) confounding, (ii) measurement error, and (iii) selection bias. Of these, confounding is the one most specific to causal inference, while measurement error or selection also impede consistent estimation of associational or other statistical targets of inference, such as estimating the prevalence of a disease in a population. We therefore focus on the use of graphs especially in the context of confounding, followed by a brief overview of the issues surrounding selection.

15.5.1 Confounding

One of the most fundamental assumptions underlying causal inference is that of 'no unmeasured confounding' [57], also known as 'ignorability' [62], 'exchangeability' [29, 36], or 'exogeneity' [37]. Dawid [11] speaks of a 'sufficient set of covariates' or 'unconfounders', which allow identification of causal targets if measured. Note that the assumption typically concerns conditions for the *absence* of a source of bias, namely 'confounding', but does not define confounders themselves. The former appears to be easier to formalize than the latter. Nevertheless, one can sometimes find the recommendation that we need to adjust for 'all confounders'. VanderWeele and Shpitser [71] give a detailed account of a number of issues around the notion of confounders. Graphs help clarifying many aspects of the assumption, eliminating ambiguities and lead to a better understanding. To see this, we first discuss some definitions of the absence of confounding that do not refer to graphs.

Sufficient adjustment set

Assume that we are interested in the causal effect of an exposure X on an outcome Y, and consider a set of pre-exposure covariates C. Depending on the chosen framework, we say that there is no confounding of the effect of X on Y given this set of covariates C if

$$
\begin{gathered}
Y(x) \perp\!\!\!\perp X \mid C, \quad \forall x \quad \text{or} \\
p(y|c; \mathrm{do}(\tilde{x})) = p(y|c, \tilde{x}), \quad \forall \tilde{x}, c \text{ or} \\
Y \perp\!\!\!\perp \sigma \mid (X, C),
\end{gathered}
\tag{15.8}
$$

where σ takes values in the set of the idle regime and interventions in X. The set C is then *sufficient to adjust for confounding* or a *sufficient adjustment set*. For such a set C it can easily be seen that equation (15.5) is valid with $X = X_j$, $Y = X_k$ and $C = \mathbf{X}_C$. It follows that we can graphically characterize the sufficiency: whether we use a DAG augmented with an intervention node σ or a causal DAG, any set C blocking all back-door paths from X to Y is a sufficient adjustment set. If no proper subset of C satisfies the condition, we call C

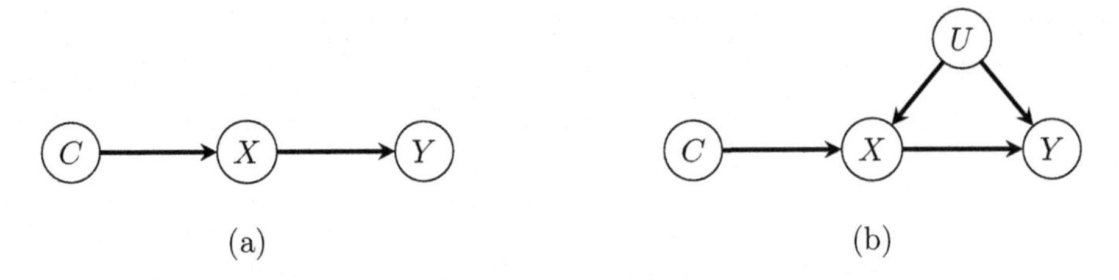

FIGURE 15.4: Causal DAGs: (a) no confounding by C, (b) potential for bias amplification by C.

minimally sufficient adjustment set. For there to be 'no unmeasured confounding' at least one such set needs to be measured. If $C = \emptyset$ satisfies the above we say that there is no confounding.

All the above versions of the assumption essentially express that whether X was generated observationally or by an intervention does not further predict Y once C is taken into account. This ensures that inference about the effect of an intervention can indeed be based on data on X, Y, C. The most popular way of adjusting for confounding is based on equation (15.8) and relies on a regression of Y on X and C to estimate $p(y|c; \text{do}(\widetilde{x}))$. The regression coefficient of X receives a causal interpretation as the effect of an intervention $\text{do}(\widetilde{x})$. This approach is known as *regression adjustment*. Strictly speaking, it results in a conditional, or subgroup, causal effect, namely an effect of X for given values of C. This is not necessarily equal to the marginal causal effect as obtained using equation (15.5), where C is integrated out. The distinction is relevant if (i) the effect of X varies for different values of C (effect modification by C), or (ii) the conditional and marginal effects are not identical, a phenomenon known as non-collapsibility (see [7, 28, 38] for discussions).

Examples and misconceptions

A 'confounder' is sometimes loosely defined as a covariate that predicts both exposure and outcome [71]. This is too imprecise and can lead to overadjustment as seen with the following examples.

Bias amplification

In Figure 15.4(a), C predicts X and Y but while it does not violate condition (15.8), it is not required because $p(y; \text{do}(\widetilde{x})) = p(y|\widetilde{x})$ meaning there is no confounding in the first place. An extension of this example is given in Figure 15.4(b). Here, C is not a sufficient adjustment set and, moreover, adjusting for C in the presence of an unobserved U may be detrimental: the estimator of the coefficient of X in a linear regression of Y on both X, C will be *more biased* than in a regression of Y on X alone under causal DAG (b). This phenomenon is known as bias amplification [50, 44]. It occurs in linear regressions when covariates are used for adjustment based on strongly predicting exposure X without regard to their relation with the outcome Y as is sometimes recommended in the context of propensity score methods [63].

M-bias

In Figure 15.5(a), C is a pre-exposure covariate as it is not a descendant of X. Moreover, it may appear to be a confounder as it predicts (or is associated with) both X and Y but, as in Figure 15.4(a), $p(y; \text{do}(\widetilde{x})) = p(y|\widetilde{x})$ so that C is not required to adjust for confounding.

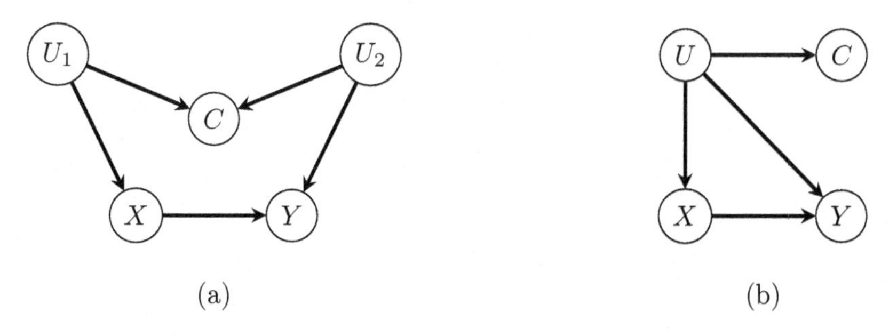

FIGURE 15.5: Causal DAGs: (a) potential for M-bias when adjusting for C, (b) mismeasured confounder.

In fact, it is again detrimental to adjust for C because as a collider it opens a back-door path so that $p(y|c; \mathrm{do}(\widetilde{x})) \neq p(y|c, \widetilde{x})$. This phenomenon is known as *M-bias* [65] and C is in this case a *bias inducer* [44]. It follows from this example that if a set of pre-exposure covariates is a sufficient adjustment set, a superset will not necessarily be sufficient anymore because additionally conditioning on colliders may open back-door paths. Note that while the exact situation of Figure 15.5(a) may not often be encountered in practice, the one where C additionally has directed edges into X and Y could be quite common. In such a case, a sufficient adjustment set would need to include either U_1 or U_2. This is illustrated with the following application.

Williamson et al. [73] consider a population-based longitudinal cohort study of children. Interest lies in estimating the causal effect of personal smoking (X) on adult asthma (Y), and the question is whether to adjust for childhood asthma (C). Relating this to the DAG in Figure 15.5(a), U_1 is parental smoking, while U_2 stands for underlying atopy. Note that the actual DAG used by these authors to represent subject matter background knowledge is more elaborate with more variables. Fortunately, data on parental smoking U_1 was available so that together with a number of further covariates, including childhood asthma, a plausible measured adjustment set could be found.

Mismeasured confounder

The causal DAG in Figure 15.5(b) could represent the situation where only an imperfect measurement C of the underlying quantity of interest U is available. While C is associated with both X and Y, it is not a sufficient adjustment set. In fact, if U was available, C would not be relevant. However, with U unobserved, the conditional independence structure as in Figure 15.5(b) characterizes the measurement error as *non-differential.* Then, with additional parametric assumptions (such as lack of effect modification by U, for example), adjusting for C reduces the bias compared to no adjustment. However, as shown by Ogburn and VanderWeele [45], bias reduction is not guaranteed. It requires specific measurement error assumptions that cannot be represented graphically.

Minimal adjustment sets

Finally, in Figure 15.6(a) none of C_1, C_2 alone is a sufficient adjustment set, only (C_1, C_2) is, while in Figure 15.6(b) each C_1 or C_2 or (C_1, C_2) are sufficient adjustment sets. Moreover, in (b), each C_1 or C_2 alone are minimally sufficient adjustment sets, demonstrating that minimality does not imply uniqueness.

FIGURE 15.6: Causal DAGs: (a) both (C_1, C_2) required for adjustment, (b) C_1 or C_2 each sufficient.

Selection of adjustment sets

We started this section by claiming that the problem of confounding is specific to causal inference. As detailed by VanderWeele and Shpitser [71], no associational definition of confounding is satisfactory. It is evident from the definition of a sufficient adjustment set and from the above examples that the assumption of no unmeasured confounding cannot be tested based on observational data alone as it relates an observational to an interventional distribution. As mentioned in the example after Definition 15.3.1, equation (15.8) could in principle be tested if data from both the observational and interventional regimes were available. We could then compare the conditional distributions to see if they are the same. In practice, the assumption typically needs to be justified based on subject matter knowledge assisted by graphs as demonstrated in the above examples.

Nevertheless, there is clearly a strong motivation to somehow assert empirically that a given set of covariates is sufficient to adjust for confounding, or to empirically *select* a sufficient adjustment set from a possibly very large pool of covariates. This is also an increasingly important topic in the context of high-dimensional data. There are two approaches: (i) First, we can assume that a large set of measured covariates C is a sufficient adjustment set, but we want to determine empirically whether and how this set can be reduced. (ii) We wish to determine empirically whether a large set of measured pre-exposure covariates C contains a sufficient adjustment set.

Reducing a sufficient adjustment set

Assume we are interested in the effect of an intervention in X on Y and let C be a known sufficient adjustment set. Loosely speaking, the reduction of C is based on the idea that variables that do not affect X and (possibly other) variables that do not affect Y can be discarded. In this vein, Robins [58] shows that for disjoint $W_1, W_2 \subset C$ if $Y \perp\!\!\!\perp W_1 \mid X, C \backslash W_1$ and $X \perp\!\!\!\perp W_2 \mid C \backslash (W_1, W_2)$ then $C \backslash (W_1, W_2)$ is also a sufficient adjustment set. De Luna et al. [15] elaborate this procedure and give two versions of a selection algorithm as well as conditions under which the resulting subsets are minimal. In Figure 15.6(b) the two versions of the algorithm result in reduction either to C_1 or to C_2. Guo and Dawid [31] speak of 'treatment' versus 'response' sufficient reduction of covariates. VanderWeele and Shpitser [71] give a forward selection algorithm which requires causal faithfulness. Certain pitfalls in the empirical selection of covariates in this way can be illustrated graphically [15]. For instance in the situation of Figure 15.5(a), starting the above algorithms with C will not find that $\emptyset$ is sufficient, while starting the algorithms with (U_1, U_2, C) will correctly conclude that $\emptyset$ is sufficient.

Establishing a sufficient adjustment set

If C is a large set of pre-exposure covariates, but not known to be sufficient for adjustment, Entner et al. [24] invoke an underlying faithful causal DAG model to prove the following. If we can find a set $W \subset C$ as well as $Z \subset (C \backslash W)$ such that $W \not\perp\!\!\!\perp Y|Z$ but $W \perp\!\!\!\perp Y|(Z, X)$, then we can conclude not only that Z is a sufficient adjustment set, but also that an intervention in X has a causal effect on Y. This follows because the two conditions, together with causal faithfulness, imply that there must be directed paths from W to Y blocked by X. Hence there are directed paths from X to Y implying a causal effect, and all back-door paths from X to Y must be blocked by Z. However, when such a W does not exist the question cannot be decided.

15.5.2 Selection bias

As shown above, confounding can be addressed by conditioning on a suitable set of covariates. Selection bias is the dual problem because it occurs when 'wrongly' conditioning by mistake, design or analysis [34, 7, 19, 5].

When DAGs are used to model and illustrate the problem we are typically alerted to the possibility of selection bias by opening paths due to conditioning on a collider, similar to the phenomenon of M-bias. The problem is therefore also known as *collider-stratification bias* [27]. We refrain from a full treatment here due to space limitation, but give some examples.

Example — sequential treatments ctd.: Consider the example in Figure 15.7(a), where (X_1, X_2) could be two sequentially administered treatments similar to the situation around equation (15.6). Assume X_1 is randomized but not X_2 so that the latter may be affected by unobserved confounding U. We are interested in the joint effect of (X_1, X_2) on Y, or formally in $p(y; \text{do}(\tilde{x}_1, \tilde{x}_2))$. Naively, we might carry out a regression of Y on (X_1, X_2). Unbeknown to us, the true causal structure is as shown. In particular, there is no direct or indirect effect neither of X_1 nor X_2 on Y. Obviously, due to unobserved confounding by U, any effect estimate of X_2 will typically be biased. However, the analysis will also typically find a non-zero effect of X_1 because a joint regression conditions also on X_2. As X_2 is a collider this conditioning opens a path between X_1 and Y. In other words, even without a causal effect of X_1 on Y we have $Y \not\perp\!\!\!\perp X_1|X_2$ as verified via d-separation in Figure 15.7(a). In contrast, a regression of Y on X_1 alone would correctly reveal the absence of any causal effect. Hence, it is the unfortunate choice of a joint regression including X_2 that impedes inference regarding the effect of X_1, a phenomenon pointed out by Robins [56]. In the particular situation of sequential treatments a joint regression is therefore usually not appropriate, and methods such as IPTW or g-computation should be used [59]. Note that an analogous phenomenon applies to the analysis of direct and indirect causal effects. For instance one may attempt to justify conditioning on X_2 with targeting a direct effect of X_1 on Y, but this conditioning introduces selection bias for the same reason as in the sequential treatment example [48, 4].

Sampling selection

While the above problem occurs due to an unfortunate choice of method for the data analysis, a structurally similar type of problem is due to *sampling selection*. Let S be a binary indicator for being sampled, so that our data is drawn from $p(\mathbf{x}|S = 1)$. It is common to assume random sampling $\mathbf{X} \perp\!\!\!\perp S$ and hence $p(\mathbf{x}|S = 1) = p(\mathbf{x})$. However, many designs are based on non-random sampling, for example case-control studies. Hence, we may ask when valid inference about $p(\mathbf{x})$ is possible given data from $p(\mathbf{x}|S = 1)$. For instance, in a case-control study it is well-known that the odds ratio can often still be identified as a

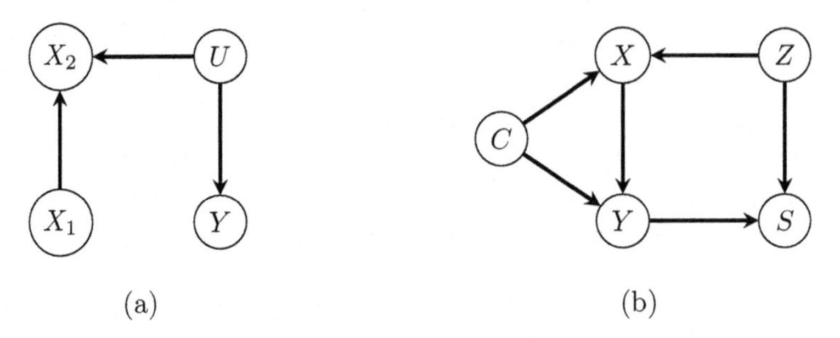

FIGURE 15.7: Causal DAGs with potential for selection bias, (a) sequential treatments; (b) time-to-pregnancy problem.

measure of association between outcome and exposure. With suitable covariate data, the corresponding conditional odds ratio will also be identified and have a causal interpretation, see overview of Didelez and Evans [18].

Didelez et al. [19] give graphical conditions for valid causal inference under sampling selection. They consider conditional causal odds ratios as well as the testability of the causal null-hypothesis as their targets of inference. A key issue in this context is that of collapsibility: A quantity being collapsible over S means that inference within the sampled is the same as in the whole population. For instance, the conditional odds ratio between X and Y given C is collapsible over S, $OR_{YX}(C,S) \equiv OR_{YX}(C)$, if, and only if, either $Y \perp\!\!\!\perp S \mid (X,C)$ or $X \perp\!\!\!\perp S \mid (Y,C)$. Moreover, under these conditions we can also test the null-hypothesis $Y \perp\!\!\!\perp X \mid C$ by testing within the sampled, that is conditional on $S = 1$. As these are conditional independence properties, they can easily be checked via separations in a graph. Generalizations of the conditions take more available information into account and hence apply in a wider range of situations. We illustrate this with an example, for a detailed treatment see [19, 2, 3].

Example — sampling selection: Let X be exposure to a toxic substance and Y time-to-pregnancy (TTP). This is a common measure of fertility and defined as the time, from initiation, that it takes a couple to become pregnant. As discussed by Weinberg et al. [72], there is a fundamental problem with certain retrospective designs: If couples who are trying to become pregnant or who just gave birth are interviewed during a given time-window, then those with long TTP must have started earlier. Hence, within the sampled (that is, given $S = 1$) initiation time and Y = TTP are necessarily associated. This could induce a non-causal (X,Y)-association if exposure to the substance has changed over time due to legislation, say. The problem is illustrated in Figure 15.7(b), where Z is the initiation time and C is a set of covariates. Key assumptions in the DAG are that $Y \perp\!\!\!\perp Z \mid (X,C)$ in the whole population, while $Y \not\perp\!\!\!\perp Z \mid (X,C,S)$. It is also assumed that sampling does not depend directly on the exposure so that $S \perp\!\!\!\perp X \mid (Y,Z,C)$. With these conditional independencies, it follows from results of [19, 2] that by taking Z into account, we can first collapse over S and subsequently over Z. Hence, the conditional odds ratio in the sample corresponds to the population conditional odds ratio, $OR_{YX}(C,Z,S = 1) = OR_{YX}(C)$. The former can be estimated based on the sampled data, while the latter is the actual target of inference. Similarly, testing for a conditional independence between exposure and outcome given Z and C in the sample, $Y \perp\!\!\!\perp X \mid (C,Z,S = 1)$, is equivalent to testing for conditional independence given only C for the whole population, $Y \perp\!\!\!\perp X \mid C$. A causal interpretation of these odds ratios and tests hinges further on C being a sufficient adjustment set.

The results referred to above concern odds ratios and testing of a relevant null-hypothesis, but they do not achieve full identification of intervention distributions. In the context of sampling selection this is indeed usually not possible [3], but some information can still be extracted from the available data. Evans and Didelez [25] consider *generic identification* of the marginal distribution $p(\mathbf{x})$ from data on $p(\mathbf{x}|S = 1)$, meaning that identification is possible almost everywhere except on a lower dimensional subset of the model parameter space. We do not go into details here, but note that graphical criteria can again be given to characterize when generic identification is possible.

15.6 Discussion and Outlook

In this chapter, we have reviewed key causal concepts in the form of distributional invariances under different regimes, effects of atomic interventions or stable mechanisms. We showed how these can be combined with graphs in different ways. The resulting DAGs can be queried with respect to inferential questions, for instance regarding identifiability of causal quantities or possible sources of bias. There are many further examples where a graphical approach adds clarity by making crucial assumptions explicit and guiding the analysis. A large body of work has developed for instance on the topic of *instrumental variables* which allow some causal inference even when there is unobserved confounding but require very specific and somewhat subtle conditional independence assumptions which need to be carefully justified in any given case [35, 20].

We have only briefly alluded to the converse task, that of causal discovery, where the question is how to determine empirically the causal relations with no or only partial subject matter knowledge (see also Chapter 18 in this book). This is relevant to the estimation of intervention effects especially in high-dimensional settings without prior assumptions [42], or when the whole causal structure is unknown and itself of interest. As an example consider the investigation of genetic regulatory or cellular signaling systems where data from some limited experimental conditions are available and researchers want to determine the most promising future experiments [43]. Numerous causal search algorithms with innumerable variations have been put forward [67], and the field is more active than ever. Naturally this endeavor relies on strong assumptions, such as causal faithfulness in addition to specific parametric assumptions. For critical discussions of causal discovery see Robins et al. [60] and Dawid [12]. Some of the recent developments are concerned with integrating different sources of information, such as data sets that have been obtained under different experimental conditions [22, 32, 70]. For example Peters et al. [53] reverse our earlier reasoning in the following way: If data sets are available from different regimes, such as a number of different experimental settings, their method searches for invariances among conditional distributions to partially reconstruct causal structures. In this context, intervention graphs under general regimes may turn out more useful than causal DAGs [17].

Finally let us point out some extensions to other types of graphs and to dynamic systems. Some generalizations of causal interpretations to a wider class of graphs are motivated by the causal search problem. For instance, so-called maximal ancestral graphs (MAGs) allow for the possibility of latent variables [54, 52]. Other generalizations address the fact that interventions in dynamic systems cannot necessarily be modeled using DAGs due to the continuous nature of time [8, 1]. Also, when the dynamics are in an equilibrium one may want to employ either partly undirected or cyclic graphs. Hence, an alternative is based on chain graphs [41], where the relevant intervention corresponds to holding fixed a process while the system returns to a new equilibrium. Those alternatives based on cyclic graphs address for example non-recursive structural equation models [66], interventions in time-series [23], or the causal relations among events as modeled by a multi-state process [16]. Here, the cyclicity is usually a shortcut for expressing feedback, where the present of one variable depends on the past of another and vice versa. Furthermore, certain dynamic aspects of causal reasoning can alternatively be represented and analyzed with chain event graphs [69]. The fundamental causal concepts and uses of graphs presented in this chapter essentially carry over to all these more complex models.

Bibliography

[1] O. O. Aalen, K. Røysland, J. M. Gran, R. Kouyos, and T. Lange. Can we believe the DAGs? A comment on the relationship between causal DAGs and mechanisms. *Statistical Methods in Medical Research*, 25(5):2294–2314, 2016.

[2] E. Bareinboim and J. Pearl. Controlling selection bias in causal inference. In *Proceedings of the 15th International Conference on Artificial Intelligence and Statistics*, 100–108. *Journal of Machine Learning Research*, 2012.

[3] E. Bareinboim and J. Tian. Recovering causal effects from selection bias. In *Proceedings of the 29th National Conference on Artificial Intelligence*, 3475–3481. AAAI Press, Menlo Park, CA, 2015.

[4] S. R. Cole and M. A. Hernán. Fallibility in estimating direct effects (with discussion). *International Journal of Epidemiology*, 31:163–165, 2002.

[5] S. R. Cole, R. W. Platt, E. F. Schisterman, H. Chu, D. Westreich, D. Richardson, and C. Poole. Illustrating bias due to conditioning on a collider. *International Journal of Epidemiology*, 39:417–420, 2010.

[6] P. Constantinou and A. P. Dawid. Extended conditional independence and applications in causal inference. Accepted by *Annals of Statistics*. 2016.

[7] D. R. Cox and N. Wermuth. Causality: a statistical view. *International Statistical Review*, 72(3):285–305, 2004.

[8] D. Dash and M. Druzdzel. Caveats for causal reasoning with equilibrium models. In *European Conference on Symbolic and Quantitative Approaches to Reasoning and Uncertainty*, 192–203. Springer, 2001.

[9] J. A. Davis. Extending Rosenberg's technique for standardizing percentage tables. *Social Forces*, 62:679–708, 1984.

[10] A. P. Dawid. Causal inference without counterfactuals (with discussion). *Journal of the American Statistical Association*, 95:407–448, 2000.

[11] A. P. Dawid. Influence diagrams for causal modelling and inference. *International Statistical Review*, 70:161–89, 2002.

[12] A. P. Dawid. Beware of the DAG! *NIPS Causality: Objectives and Assessment*, 6:59–86, 2010.

[13] A. P. Dawid. Statistical causality from a decision-theoretic perspective. *Annual Review of Statistics and Its Application*, 2(1):273–303, 2015.

[14] A. P. Dawid and V. Didelez. Identifying the consequences of dynamic treatment strategies: a decision-theoretic overview. *Statististical Surveys*, 4:184–231, 2010.

[15] X. De Luna, I. Waernbaum, and T. S. Richardson. Covariate selection for the nonparametric estimation of an average treatment effect. *Biometrika*, 98(4):861–875, 2011.

[16] V. Didelez. Causal reasoning for events in continuous time: A decision–theoretic approach. In *Proceedings of the 31st Annual Conference on Uncertainty in Artifical Intelligence—Causality Workshop*, 40–45, 2015.

[17] V. Didelez. Discussion of: 'Causal inference by using invariant prediction: identification and confidence intervals'. *Journal of the Royal Statistical Society, Series B*, 78(5):990–991, 2016.

[18] V. Didelez and R. J. Evans. Causal inference from case-control studies. In N. Breslow, O. Borgan, N. Chatterjee, A. Scott, and G. Mitchell, editors, *Handbook of Case-Control Studies*, Handbooks of Modern Statistical Methods. Chapman & Hall/CRC, 2017.

[19] V. Didelez, S. Kreiner, and N. Keiding. Graphical models for inference under outcome-dependent sampling. *Statistical Science*, 25(3):368–387, 2010.

[20] V. Didelez and N. A. Sheehan. Mendelian randomisation as an instrumental variable approach to causal inference. *Statistical Methods in Medical Research*, 16(4):309–330, 2007.

[21] V. Didelez and N. A. Sheehan. Mendelian randomisation: why epidemiology needs a formal language for causality. In F. Russo and J. Williamson, editors, *Causality and Probability in the Sciences*, volume 5 of *Texts in Philosophy*, 263–292. College Publications, London, 2007.

[22] F. Eberhardt and R. Scheines. Interventions and causal inference. *Philosophy of Science*, 74(5):981–995, 2007.

[23] M. Eichler and V. Didelez. On Granger causality and the effect of interventions in time series. *Lifetime Data Analysis*, 16(1):3–32, 2010.

[24] D. Entner, P. Hoyer, and P. Spirtes. Data-driven covariate selection for nonparametric estimation of causal effects. In *Proceedings of the 16th International Conference on Artificial Intelligence and Statistics*, 256–264. JMLR W&CP, 2013.

[25] R. J. Evans and V. Didelez. Recovering from selection bias using marginal structure in discrete models. In *Proceedings of the 31st Annual Conference on Uncertainty in Artifical Intelligence—Causality Workshop*, 46–55, 2015.

[26] A. Goldberger. Structural equation methods in the social sciences. *Econometrica*, 40(6):979–1001, 1972.

[27] S. Greenland. Quantifying biases in causal models: Classical confounding vs. collider–stratification bias. *Epidemiology*, 14(3):300–306, 2003.

[28] S. Greenland and J. Pearl. Adjustments and their consequences — collapsibility analysis using graphical models. *International Statistical Review*, 79(3):401–426, 2011.

[29] S. Greenland and J. M. Robins. Identifiability, exchangeability, and epidemiological confounding. *International Journal of Epidemiology*, 15(3):413–419, 1986.

[30] S. Greenland, J. M. Robins, and J. Pearl. Confounding and collapsibility in causal inference. *Statistical Science*, 14(1):29–46, 1999.

[31] H. Guo and A. P. Dawid. Sufficient covariates and linear propensity analysis. In *Proceedings of the 13th International Conference on Artificial Intelligence and Statistics*, 281–288. JMLR W&CP, 2010.

[32] A. Hauser and P. Bühlmann. Jointly interventional and observational data: estimation of interventional Markov equivalence classes of directed acyclic graphs. *Journal of the Royal Statistical Society: Series B*, 77(1):291–318, 2015.

[33] D. M. Hausman and J. Woodward. Independence, invariance and the causal Markov condition. *The British Journal for the Philosophy of Science*, 50(4):521–583, 1999.

[34] M. A. Hernán, S. Hernández-Díaz, and J. M. Robins. A structural approach to selection bias. *Epidemiology*, 15(5):615–625, 2004.

[35] M. A. Hernán and J. M. Robins. Instruments for causal inference: an epidemiologist's dream? *Epidemiology*, 17(4):360–372, 2006.

[36] M. A. Hernán and J. M. Robins. *Causal Inference*. Chapman & Hall/CRC, 2018. Forthcoming.

[37] G. W. Imbens. Nonparametric estimation of average treatment effects under exogeneity: A review. *Review of Economics and Statistics*, 86(1):4–29, 2004.

[38] N. Keiding and D. Clayton. Standardization and control for confounding in observational studies: A historical perspective. *Statistical Science*, 29(4):529–558, 11 2014.

[39] U. B. Kjaerulff and A. L. Madsen. *Bayesian Networks and Influence Diagrams: A Guide to Construction and Analysis*. Springer. 2010.

[40] S. L. Lauritzen. Causal inference from graphical models. In O. E. Barndorff-Nielsen, D. R. Cox, and C. Klüppelberg, editors, *Complex Stochastic Systems*, 63–107. CRC Press, London, 2000.

[41] S. L. Lauritzen and T. S. Richardson. Chain graph models and their causal interpretations. *Journal of the Royal Statistical Society: Series B*, 64(3):321–348, 2002.

[42] M. H. Maathuis, M. Kalisch, and P. Bühlmann. Estimating high-dimensional intervention effects from observational data. *The Annals of Statistics*, 37(6A):3133–3164, 2009.

[43] F. Markowetz and R. Spang. Inferring cellular networks – a review. *BMC Bioinformatics*, 8(6):22–43, 2007.

[44] J. A. Middleton, M. A. Scott, R. Diakow, and J. L. Hill. Bias amplification and bias unmasking. *Political Analysis*, 24(3):307–323, 2016.

[45] E. L. Ogburn and T. J. VanderWeele. Bias attenuation results for nondifferentially mismeasured ordinal and coarsened confounders. *Biometrika*, 100(1):241–248, 2013.

[46] J. Pearl. Aspects of graphical models connected with causality. In *Proceedings of the 49th Session of the International Statistical Institute*, 391–401, 1993.

[47] J. Pearl. Causal diagrams for empirical research. *Biometrika*, 82(4):669–688, 1995.

[48] J. Pearl. Direct and indirect effects. In *Proceedings of the 7th Conference on Uncertainty in Artificial Intelligence (UAI-01)*, 411–420. Morgan Kaufmann, 2001.

[49] J. Pearl. *Causality.* Cambridge University Press, second edition, 2009.

[50] J. Pearl. On a class of bias-amplifying variables that endanger effect estimates. In *Proceedings of the 26th Conference on Uncertainty in Artificial Intelligence (UAI-10)*, 417–424. Morgan Kaufmann, 2010.

[51] J. Pearl and J. Robins. Probabilistic evaluation of sequential plans from causal models with hidden variables. In *Proceedings of the Eleventh Annual Conference on Uncertainty in Artificial Intelligence (UAI-95)*, 444–453, San Francisco, CA, 1995. Morgan Kaufmann.

[52] E. Perković, J. Textor, M. Kalisch, and M. H. Maathuis. Complete graphical characterization and construction of adjustment sets in Markov equivalence classes of ancestral graphs. *Journal of Machine Learning Research*, 18(220): 1–62.

[53] J. Peters, P. Bühlmann, and N. Meinshausen. Causal inference by using invariant prediction: identification and confidence intervals. *Journal of the Royal Statistical Society, Series B*, 78(5):947–1012, 2016.

[54] T. Richardson and P. Spirtes. Ancestral graph Markov models. *Annals of Statistics*, 30(4):962–1030, 2002.

[55] T. S. Richardson and J. M. Robins. Single world intervention graphs (SWIGs): A unification of the counterfactual and graphical approaches to causality. *Working Paper No.128*, 2013. Center for Statistics and the Social Sciences of the University of Washington.

[56] J. M. Robins. A new approach to causal inference in mortality studies with sustained exposure periods — application to control for the healthy worker survivor effect. *Mathematical Modelling*, 7:1393–1512, 1986.

[57] J. M. Robins. Estimation of the time-dependent accelerated failure time model in the presence of confounding factors. *Biometrika*, 79:321–334, 1992.

[58] J. M. Robins and S. Greenland. Causal inference from complex longitudinal data. *Latent Variable Modeling and Applications to Causality*, Lecture notes in Statistics Vol. 120, ed. M. Berkane, Springer, NY120:69–117, 1997.

[59] J. M. Robins, M. A. Hernán, and B. Brumback. Marginal structural models and causal inference in epidemiology. *Epidemiology*, 11(5):550–560, 2000.

[60] J. M. Robins, R. Scheines, P. Spirtes, and L. Wasserman. Uniform consistency in causal inference. *Biometrika*, 90(3):491–515, 2003.

[61] D. B. Rubin. Estimating causal effects of treatments in randomized and nonrandomized studies. *Journal of Educational Psychology*, 66(5):688–701, 1974.

[62] D. B. Rubin. Bayesian inference for causal effects: The role of randomization. *Annals of Statistics*, 6:34–58, 1978.

[63] D. B. Rubin. Should observational studies be designed to allow lack of balance in covariate distributions across treatment groups. *Statistics in Medicine*, 28:1420–1423, 2009.

[64] I. Shpitser and J. Pearl. Identification of conditional interventional distributions. In *Proceedings of the Twenty-Second Annual Conference on Uncertainty in Artificial Intelligence (UAI-06)*, 437–444, Arlington, Virginia, 2006. AUAI Press.

[65] A. Sjölander. Propensity scores and M-structures. *Statistics in Medicine*, 28(9):1416–1420, 2009.

[66] P. Spirtes. Directed cyclic graphical representations of feedback models. In *Proceedings of the 11th Conference on Uncertainty in Artificial Intelligence (UAI-95)*, 491–498, San Francisco, CA, USA, 1995. Morgan Kaufmann.

[67] P. Spirtes, C. Glymour, and R. Scheines. *Causation, Prediction and Search.* MIT Press, second edition, 2000.

[68] N. Staplin, W. G. Herrington, P. K. Judge, C. A. Reith, R. Haynes, M. J. Landray, C. Baigent, and J. Emberson. Use of causal diagrams to inform the design and interpretation of observational studies: An example from the study of heart and renal protection (SHARP). *Clinical Journal of the American Society of Nephrology*, 2016.

[69] Peter Thwaites, Jim Q. Smith, and Eva Riccomagno. Causal analysis with chain event graphs. *Artificial Intelligence*, 174(12):889–909, 2010.

[70] I. Tsamardinos, S. Triantafillou, and V. Lagani. Towards integrative causal analysis of heterogeneous data sets and studies. *Journal of Machine Learning Research*, 13:1097–1157, 2012.

[71] T. J. VanderWeele and I. Shpitser. On the definition of a confounder. *The Annals of Statistics*, 41(1):196–220, 2013.

[72] C. R. Weinberg, D. D. Baird, and A. S. Rowland. Pitfalls inherent in retrospective time-to-event studies: The example of time to pregnancy. *Statistics in Medicine*, 12(9):867–879, 1993.

[73] E. J. Williamson, Z. Aitken, J. Lawrie, S. C. Dharmage, J. A. Burgess, and A. B. Forbes. Introduction to causal diagrams for confounder selection. *Respirology*, 19(3):303–311, 2014.

16

Identification in Graphical Causal Models

Ilya Shpitser

Department of Computer Science, Johns Hopkins University

CONTENTS

16.1 Introduction

Previous chapters introduced statistical graphical models: sets of distributions defined by conditional independences linked, via Markov properties, to absences of edges in directed and mixed graphs.

Directed acyclic graphs (DAGs) have also been extended to represent causal models—a viewpoint developed in the preceding Chapter 15. Like statistical models, causal models can be viewed as sets of distributions defined by certain restrictions. However, unlike statistical models, a causal model also includes distributions that are counterfactual, rather than factually observed. Random variables following such counterfactual distributions, called potential outcomes, correspond to results of hypothetical experiments by means of which causality is represented. Restrictions given by causal models provide a link between factual and counterfactual distributions. This link allows identification of certain distributions of potential outcomes from the observed data distribution, and ultimately allows causal inferences to be made from data.

In this chapter, we will describe a simple formulation of identification theory for common targets of inference that arise in causal inference, developed in the context of non-parametric graphical causal models. In addition, we will describe extensions of this theory to an important type of causal model where counterfactual random variables are determined via linear causal mechanisms and Gaussian noise. These models are known as linear structural equation models with correlated errors.

16.2 Causal Models of a DAG

Throughout this section, fix an arbitrary DAG $\mathcal{G}$ with a vertex set $\mathbf{V}$ whose elements have a one to one correspondence to the random variables under consideration. When necessary, we will denote $\mathcal{G}$ with an explicit vertex set as $\mathcal{G}(\mathbf{V})$. The statistical model of $\mathcal{G}$, or a Bayesian network, is the set of all joint distributions $p(\mathbf{V})$ obeying the following factorization restriction

$$p(\mathbf{V}) = \prod_{V \in \mathbf{V}} p(V \mid \text{pa}_{\mathcal{G}}(V)), \tag{16.1}$$

where $\text{pa}_{\mathcal{G}}(V)$, also known as the set of *parents* of V, is defined as $\{W \in \mathbf{V} \mid W \to V \text{ exists in } \mathcal{G}\}$. In cases where $\mathcal{G}$ is clear, we will omit $\mathcal{G}$ to yield $\text{pa}(V)$. Such distributions are also said to be Markov relative to $\mathcal{G}$.

Causal models of a DAG $\mathcal{G}$ are also defined via restrictions on sets of joint distributions. Unlike statistical models, however, causal models consist of joint distributions over *potential outcome* random variables, defined via an intervention operation. Let $\mathbf{A} \subset \mathbf{V}$ be a set of random variables, and let $\mathbf{a}$ be a set of possible values for the variables in $\mathbf{A}$. Then $\text{do}(\mathbf{a})$ denotes an intervention in which an experimenter controls the variables in $\mathbf{A}$ and sets them to the values determined by $\mathbf{a}$. If $Y \in \mathbf{V}$ is another variable then $Y(\mathbf{a})$ is used to denote the response of Y to the intervention $\text{do}(\mathbf{a})$. These types of random variables are called potential outcomes, because Y is often an outcome of interest, and the intervention is often hypothetical, rather than actually occurring. To represent causality on a DAG $\mathcal{G}$, we will encode the outcomes of every possible intervention operation, starting with interventions on sets $\text{pa}(V)$ for every $V \in \mathbf{V}$. For all value assignments $\mathbf{a}$ to $\text{pa}(V)$, we assume the existence of the potential outcome $V(\mathbf{a})$, and a well-defined joint distribution over these random variables. One way to justify the existence of $V(\mathbf{a})$ is by postulating a set of *structural equations* $\{f_V(\mathbf{a}, \epsilon_V) \mid V \in \mathbf{V}\}$. Each function f_V maps a set of values $\mathbf{a}$ of parents of V in $\mathcal{G}$, and a random variable ϵ_V representing exogenous factors not captured by the model, to values of V. These functions act as causal mechanisms, and are assumed to be invariant to intervention operations. For any value set $\mathbf{a}$, the function f_V, and the exogenous variable ϵ_V induce the random variable $V(\mathbf{a})$.

We use these potential outcomes, and the associated joint distributions, to define other potential outcomes using *recursive substitution*. For any $\mathbf{A} \subseteq \mathbf{V}$, and any values $\mathbf{a}$ of $\mathbf{A}$, we define for every $Y \in \mathbf{V}$

$$Y(\mathbf{a}) \equiv Y(\mathbf{a}_{\text{pa}(Y) \cap \mathbf{A}}, \{W(\mathbf{a}) \mid W \in \text{pa}(Y) \setminus \mathbf{A}\}) \tag{16.2}$$

In words, this states that the response of Y to $\text{do}(\mathbf{a})$ is defined as the potential outcome

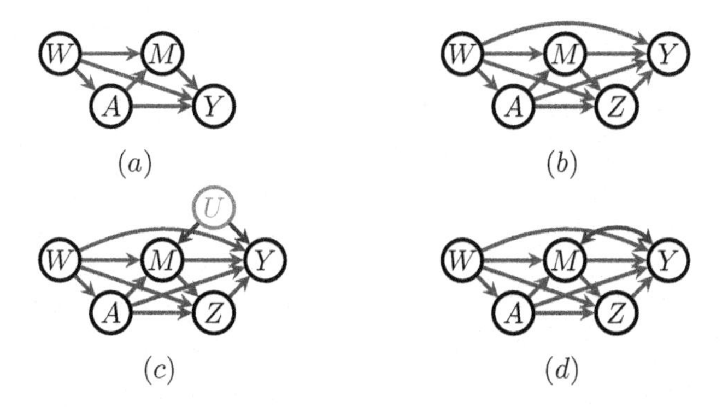

FIGURE 16.1: (a) A causal DAG with a single treatment A, a single outcome Y, a vector W of baseline variables, and a single mediator M. (b) A more complex causal DAG with two mediators M and Z. (c) A version of the DAG in (b) with an unobserved confounder U. (d) The ADMG obtained from the DAG in (c) via the latent projection operation collapsing over the unobserved variable U.

where all parents of Y which are in $\mathbf{A}$ are assigned an appropriate value from $\mathbf{a}$, and all other parents $W \in \text{pa}(Y) \setminus \mathbf{A}$ are assigned whatever value they would have attained under $\text{do}(\mathbf{a})$. These are defined recursively, and the definition terminates because of the lack of directed cycles in $\mathcal{G}$. For example, in the graph in Fig. 16.1 (a), $Y(a) = Y(a, M(a, W), W)$. The distribution $p(\{Y(\mathbf{a}) \mid Y \in \mathbf{Y}\})$ is sometimes written as $p(\mathbf{Y} \mid \text{do}(\mathbf{a}))$, as in [16].

Just as a statistical model of $\mathcal{G}$ is a set of distributions over $\mathbf{V}$ defined by (16.1), a causal model of $\mathcal{G}$ is a set of distributions over random variables in the set

$$\{V(\mathbf{v}_{\text{pa}(V)}) \mid V \in \mathbf{V}, \text{ any set of values } \mathbf{v} \text{ of } \mathbf{V}\} \tag{16.3}$$

defined by some restrictions. We consider two such models, described in [21]. The *finest fully randomized causally interpretable structured tree graph model (FFRCISTGM)*, or the *single world model* [29], is the set of all distributions such that variables in

$$\{V(\mathbf{v}_{\text{pa}(V)}) \mid V \in \mathbf{V}\}$$

are mutually independent for every set of values $\mathbf{v}$ of $\mathbf{V}$. The *non-parametric structural equation model with independent errors (NPSEM-IE)*, also known as the *functional model*, or the *multiple worlds model* [29], is the set of all distributions such that variables in

$$\left\{\{V(\mathbf{a}_{\text{pa}(V)}) \mid \mathbf{a}_{\text{pa}(V)} \text{ any set of values of } \text{pa}(V)\} \middle| V \in \mathbf{V}\right\} \tag{16.4}$$

are mutually independent. The multiple worlds model associated with $\mathcal{G}$ is a submodel of the single world model associated with $\mathcal{G}$, because it always places at least as many restrictions on potential outcome responses, and in most cases many more. Specifically, the single world model only imposes independence restrictions within single hypothetical worlds, as specified by consistent interventions $\text{do}(\mathbf{v})$, for any set of values $\mathbf{v}$ of $\mathbf{V}$. By contrast, the multiple worlds model, in addition, imposes independence restrictions across multiple hypothetical worlds.

For example, the binary single world model associated with the DAG in Fig. 16.1 (a) asserts that variables W, $A(w)$, $M(a, w)$, $Y(a, m, w)$ are mutually independent for any $a, m, w \in \{0, 1\}$, while the binary multiple worlds model associated with the same DAG asserts that sets $\{W\}, \{A(w) \mid w \in \{0, 1\}\}, \{M(a, w) \mid a \in \{0, 1\}, w \in \{0, 1\}\}, \{Y(a, m, w) \mid$

$a \in \{0,1\}, m \in \{0,1\}, w \in \{0,1\}\}$ are mutually independent. As their names imply, the single world model only imposes restrictions on a set of variables under a single consistent set of interventions, while the multiple worlds model may also impose restrictions on variables across multiple conflicting sets of interventions simultaneously. If the set of random variables in (16.3) is viewed as defined by a set of structural equations

$$\{f_V(\mathbf{v}_{\text{pa}(V)}, \epsilon_V) \mid V \in \mathbf{V}, \text{ any set of values } \mathbf{v} \text{ of } \mathbf{V}\},$$

then the assumption (16.4) corresponding to the multiple worlds model can be interpreted to mean that the distribution $p(\{\epsilon_V \mid V \in \mathbf{V}\})$ factorizes as $\prod_{V \in \mathbf{V}} p(\epsilon_V)$.

16.2.1 Causal, direct, indirect, and path-specific effects

Targets of inference in causal inference are functions of potential outcomes. Chapter 15 has already described counterfactual mean contrasts such as the average causal effect (ACE):

$$\mathbb{E}[Y(\mathbf{a})] - \mathbb{E}[Y(\mathbf{a}')],$$

which quantify the overall causal effect of a set of treatments $\mathbf{A}$ on the outcome Y, and the direct and indirect effects on the difference scale [23, 15]:

$$\begin{aligned}\mathbb{E}[Y(a)] - \mathbb{E}[Y(a')] &= \underbrace{(\mathbb{E}[Y(a)] - \mathbb{E}[Y(a, M(a'))])}_{\text{total indirect effect}} + \underbrace{(\mathbb{E}[Y(a, M(a'))] - \mathbb{E}[Y(a')])}_{\text{pure direct effect}} \\ &= \underbrace{(\mathbb{E}[Y(a)] - \mathbb{E}[Y(a', M(a))])}_{\text{total direct effect}} + \underbrace{(\mathbb{E}[Y(a', M(a))] - \mathbb{E}[Y(a')])}_{\text{pure indirect effect}}\end{aligned}$$

which quantify the extent to which the causal effect of A on Y is mediated by a third variable M.

Direct and indirect effects are defined using nested potential outcomes such as $Y(a, M(a'))$. These variables represent hypothetical situations where A is set to one value for the purposes of the direct causal path $A \to Y$ from treatment to outcome, and to another value for the purposes of the indirect causal path $A \to M \to Y$ mediated by M. These different treatment settings are crucial for defining effects which isolate the causal influence along one but not the other path.

The intuition behind direct and indirect effects can be generalized to settings where an effect along a particular causal path that is neither direct nor through a single mediator M is of interest. Effects of exposure on outcome along a predefined set of paths are called path-specific effects [15]. To define such effects we need to define a potential outcome Y where the treatment is set to one value a with respect to the set of paths of interest, and to another value a' with respect to all other paths. This is a natural generalization of the way direct effects were defined above, with the direct path $A \to Y$ being a path of interest, and the indirect path $A \to M \to Y$ being "all other paths." We can modify the recursive substitution definition (16.2) to define such potential outcomes as follows. Given a set of directed paths π from A to Y, and values a, a', define the π-specific potential outcome Y as

$$\begin{aligned}&Y(\pi, a, a') \equiv a \text{ if } Y = A \\ &Y(\pi, a, a') \equiv Y(\{W(\pi, a, a') \mid W \in \text{pa}^{\pi}(Y)\}, \{W(a') \mid W \in \text{pa}^{\overline{\pi}}(Y)\})\end{aligned} \tag{16.5}$$

where $W(a') \equiv a'$ if $W = A$, $\text{pa}^{\pi}(Y)$ is the set of parents of Y along an edge which is a part of a path in π, and $\text{pa}^{\overline{\pi}}(Y)$ is the set of all other parents of Y. Applying this definition to π consisting of a single path $A \to Z \to Y$ in Fig. 16.1 (b) yields

$Y(a', Z(a, M(a', W)), M(a', W), W)$. By analogy with direct and indirect effects, we can use this counterfactual to define the total effect not through π as

$$\mathbb{E}[Y(a)] - \mathbb{E}[Y(\{A \to Z \to Y\}, a, a')] =$$
$$\mathbb{E}[Y(a)] - \mathbb{E}[Y(a', Z(a, M(a', W)), M(a', W), W)].$$

and the pure π-specific effect as

$$\mathbb{E}[Y(\{A \to Z \to Y\}, a, a')] - \mathbb{E}[Y(a')] =$$
$$\mathbb{E}[Y(a', Z(a, M(a', W)), M(a', W), W)] - \mathbb{E}[Y(a')].$$

Equation (16.5) generalizes in a natural way to a set of treatments $\mathbf{A}$, and multiple outcomes $\mathbf{Y}$. In such cases, attention is restricted to *proper causal paths* for $\mathbf{A}$ and $\mathbf{Y}$, which are directed paths from an element in $\mathbf{A}$ to an element in $\mathbf{Y}$ that otherwise do not intersect $\mathbf{A}$.

16.2.2 Responses to dynamic treatment regimes

In settings such as precision medicine, the primary goal is obtaining good outcomes for every individual, rather than assessing average treatment effects in a population, of the kind defined above. In simple versions of this setting, shown in Fig. 16.1 (a), the goal is to use data generated from the observed data distribution

$$p(W)\, p(A|W)\, p(M|A, W)\, p(Y|M, A, W),$$

where the treatment A can be viewed as assigned based on a (likely suboptimal, possibly stochastic) policy represented by $p(A|W)$, to infer a "better" hypothetical policy $g_A(W)$. Here $g_A(W)$ is a mapping, taken from some class, from values of W to treatment values a. Such policies are sometimes known as *dynamic treatment regimes* [4]. Policy quality can be defined in a number of ways, but is often defined using expected outcomes given the policy: $\mathbb{E}[Y(A = g_A(W))]$, where the expectation is taken with respect to the distribution $p(W)$. Effective policies like g_A may assign a known primary treatment a for patients unless W suggests severe side effects for a, in which case an alternative a' is given.

More generally, given a set of treatments $\mathbf{A}$ in a causal model represented by a DAG $\mathcal{G}(\mathbf{V})$, fix a topological ordering $\prec$ on $\mathbf{V}$ consistent with $\mathcal{G}$, and consider for every $A \in \mathbf{A}$ a set of variables $\mathbf{W}_A$ earlier in the ordering $\prec$ than A. Fix a set of policies $\mathbf{g_A} \equiv \{g_A(\mathbf{W}_A) \mid A \in \mathbf{A}\}$ that determine the value of each $A \in \mathbf{A}$ using values of $\mathbf{W}_A$. Policies in $\mathbf{g_A}$ could be either deterministic or stochastic, depending on the application. For any $Y \in \mathbf{V} \setminus \mathbf{A}$, the counterfactual response Y had every $A \in \mathbf{A}$ been determined by $\mathbf{g_A}$ is defined using the natural generalization of the recursive substitution definition (16.2):

$$Y(\mathbf{g_A}) = Y(\{A = g_A(\mathbf{W}_A(\mathbf{g_A})) | A \in \text{pa}(Y) \cap \mathbf{A}\}, \{W(\mathbf{g_A}) | W \in \text{pa}(Y) \setminus \mathbf{A}\}). \quad (16.6)$$

As an example, in Fig. 16.1 (b), if we are interested in the potential outcome Y, had A and Z been set using policies $g_A(W)$ and $g_Z(M, W)$, this outcome would be

$$Y(\mathbf{g_A}) \equiv Y(A = g_A(W), Z = g_Z(M = M(\mathbf{g_A}), W), W)$$
$$M(\mathbf{g_A}) \equiv M(A = g_A(W), W). \quad (16.7)$$

These kinds of examples arise in management of cancer, HIV, or other chronic diseases where choices of primary therapy (such as induction chemotherapy for cancer patients) A depend on baseline patient characteristics such as age, and the choice to switch to a second line therapy Z (such as salvage chemotherapy) depends on intermediate outcomes M which themselves depend on the choice of primary therapy.

16.2.3 Identifiability

Causal models described so far are sets of joint distributions over random variables in the set $\{V(\mathbf{a}_{\text{pa}(V)}) \mid V \in \mathbf{V}\}$, and other random variables defined from this set via (16.2), (16.5), and (16.6). In particular, the *observed distribution* $p(\mathbf{V})$ is always a marginal of any joint distribution that is an element of a causal model. This distribution is important in causal inference applications since data drawn from this distribution is what is typically available.

A distribution $\widetilde{p}$ is said to be globally identified from $p(\mathbf{V})$ in a causal model, if there exists a function g such that $\widetilde{p} = g(p(\mathbf{V}))$ in every element of the model. Weaker notions of identifiability, such as generic identifiability where $\widetilde{p}$ is a function $g(p(\mathbf{V}))$ in most but not all elements of a model, are discussed in Section 16.4. Identification is important to establish if there is to be any hope of estimating $\widetilde{p}$ from observed data. A distribution $\widetilde{p}$ is said to be non-identified from $p(\mathbf{V})$ in a causal model if there exist two elements in the model which share $p(\mathbf{V})$ but differ in $\widetilde{p}$. No estimation strategy coherent for the entire model is possible for non-identified distributions.

In this chapter, rather than considering identification of counterfactual expectations, which were used to define causal effects and evaluate the quality of dynamic treatment regimes, we will concentrate on identification of distributions of potential outcomes, such as $p(Y(\mathbf{a}))$. Generally, causal effects and effects of dynamic treatment regimes are identified under the weaker single world model, while direct, indirect, and path-specific effects require the stronger multiple worlds model to yield identification, except in very simple cases.

16.2.4 Identification of causal effects

Under the single world model, for any value set $\mathbf{a}$ of $\mathbf{A} \subseteq \mathbf{V}$, the interventional distribution $p(\mathbf{V} \setminus \mathbf{A} \mid \text{do}(\mathbf{a}))$ over counterfactuals $\{V(\mathbf{a}) \mid V \in \mathbf{V} \setminus \mathbf{A}\}$ is identified by

$$p(\mathbf{V} \setminus \mathbf{A} \mid \text{do}(\mathbf{a})) = \prod_{V \in \mathbf{V} \setminus \mathbf{A}} p(V \mid \text{pa}(V))|_{\mathbf{A}=\mathbf{a}} \, . \tag{16.8}$$

This equation is known as the *g-formula* [22], the *manipulated distribution* [31], or the *truncated factorization* [16].

An intuitive interpretation of (16.8) is as follows. In causal models of a DAG $\mathcal{G}$, the value of each variable V is determined by values of $\text{pa}(V)$, and an exogenous source of noise ϵ_V, via a structural equation f_V. The conditional distribution induced by $\text{pa}(V)$, ϵ_V and f_V that captures the variation of V as a function of values $\mathbf{w}$ of $\text{pa}(V)$ is simply $p(V \mid \mathbf{w}_{\text{pa}(V)}) = p(V \mid \text{do}(\mathbf{w}))$. An intervention operation that sets variables $\mathbf{A}$ to $\mathbf{a}$ in particular implies that the value of any variable $A \in \mathbf{A}$ is now a constant, and no longer determined by either $\text{pa}(A)$ or ϵ_A via f_A. Thus the overall distribution of the remaining variables can be obtained from the DAG factorization (16.1) by simply dropping all terms from the factorization that are no longer relevant post-intervention, namely terms of the form $p(A \mid \text{pa}(A))$, and setting all remaining occurrences of elements in $\mathbf{A}$ to appropriate values in $\mathbf{a}$.

The g-formula has a number of well-known special cases. For instance, in Fig. 16.1 (a),

$$\begin{aligned} p(Y \mid \text{do}(a)) &= \sum_{M,W} p(Y, M, W \mid \text{do}(a)) \\ &= \sum_{M,W} p(Y \mid a, M, W) p(M \mid a, W) p(W) \\ &= \sum_{W} p(Y \mid a, W) p(W), \end{aligned}$$

which recovers the well-known adjustment or backdoor formula [16, 30]. In Fig. 16.1 (b),

$$\begin{aligned} p(Y \mid \text{do}(a,z)) &= \sum_{M,W,U} p(Y, M, W, U \mid \text{do}(a,z)) \\ &= \sum_{M,W,U} p(Y \mid a, z, M, W, U) p(M \mid a, W, U) p(W) \\ &= \sum_{W,M} p(Y \mid a, z, M, W) p(M \mid a, W) p(W), \end{aligned}$$

which recovers the g-computation algorithm [22] for inferring causal effects in longitudinal studies (in this example for two time points, as that is the length shown in Fig. 16.1 (b)). Here and elsewhere, we use the summation notation for the sake of clarity of exposition, although summations only apply for discrete random variables. For continuous random variables similar expressions arise with an appropriately defined set of integrals.

16.2.5 Identification of path-specific effects

Distributions of nested counterfactuals involved in defining path-specific effects are more general objects than interventional distributions involved in defining average causal effects. As a consequence, not every such distribution is identified even in causal DAG models, and those that are identified require the stronger multiple worlds model. Identification for path-specific effects in a DAG is governed by a simple criterion known as the *recanting witness criterion* [1].

Let $\text{ch}_{\mathcal{G}}(V)$ be the set $\{W \in \mathbf{V} \mid V \to W \text{ exists in } \mathcal{G}\}$. We omit $\mathcal{G}$ from the notation in cases where the graph is clear, yielding $\text{ch}(V)$. If we are interested in the path-specific effect of $\mathbf{A}$ on $\mathbf{Y}$ along a set π of proper causal paths for $\mathbf{A}$ and $\mathbf{Y}$, we say a variable $W \in \text{ch}(A)$ for some $A \in \mathbf{A}$ is a recanting witness for π if there exists a path in π with the first edge $A \to W$ and another proper causal path with the same first edge (but not necessarily the same final vertex in $\mathbf{Y}$) which is not in π. As an example, in Fig. 16.1 (b), if we are interested in the path-specific effect of A on Y along a single path $A \to M \to Y$, that is if $\pi = \{A \to M \to Y\}$, then M is a recanting witness for π, since the path $A \to M \to Z \to Y$ is a proper causal path for A and Y, is not an element of π, and has as its first edge $A \to M$ which is also the first edge of $A \to M \to Y$.

A well known result [1, 24] states that in a causal DAG $\mathcal{G}$, a path-specific effect from $\mathbf{A}$ to $\mathbf{Y}$ along a set of paths π is identified if and only if there does *not* exist a recanting witness for π. If the recanting witness does not exist, then the joint counterfactual distribution over variables $\{Y(\pi, \mathbf{a}, \mathbf{a}') \mid Y \in \mathbf{Y}\}$ is identified via a generalization of equation (16.8) called the *edge g-formula* [29]:

$$p(\{Y(\pi, \mathbf{a}, \mathbf{a}') | Y \in \mathbf{Y}\}) = \sum_{\mathbf{V} \setminus (\mathbf{A} \cup \mathbf{Y})} \prod_{V \in \mathbf{V} \setminus \mathbf{A}} p(V | \mathbf{a}_{\text{pa}^{\pi}(V) \cap \mathbf{A}}, \mathbf{a}'_{\text{pa}^{\overline{\pi}}(V) \cap \mathbf{A}}, \text{pa}(V) \setminus \mathbf{A}). \quad (16.9)$$

Just as the ordinary g-formula, the edge g-formula can be viewed as a truncated DAG factorization. However, in the edge g-formula a variable $A \in \mathbf{A} \cap \text{pa}(V)$ in a Markov factor $p(V \mid \text{pa}(V))$ can be set to either its value in $\mathbf{a}$ or its value in $\mathbf{a}'$, depending on whether the edge $A \to V$ is a part of a path in π or not.

As an example, a recanting witness does not exist for the path-specific effect of A on Y along the path set $\pi \equiv \{A \to M \to Y; A \to M \to Z \to Y\}$ in Fig. 16.1 (b). The counterfactual distribution $p(Y(\pi, a, a'))$ corresponding to this set of paths is then identified as

$$\sum_{W,M,Z} p(Y \mid M, W, Z, a') p(Z \mid M, W, a') p(M \mid W, a) p(W).$$

In the simple, but practically important case of Fig. 16.1 (a), where we are interested in the direct effect of A on Y, in other words in the path set $\pi \equiv \{A \to Y\}$, the counterfactual distribution $p(Y(\{A \to Y\}, a, a')) = p(Y(a, M(a')))$ is identified by the edge g-formula as

$$\sum_{W,M} p(Y \mid M, W, a)p(M \mid a', W)p(W).$$

If we are interested in using this counterfactual distribution to obtain the pure direct effect $\mathbb{E}[Y(a, M(a'))] - \mathbb{E}[Y(a')]$, we recover the well-known *mediation formula* [17]:

$$\sum_{W,M} \{\mathbb{E}[Y \mid M, W, a] - \mathbb{E}[Y \mid M, W, a']\}p(M \mid a', W)p(W).$$

16.2.6 Identification of responses to dynamic treatment regimes

Given a set of treatments $\mathbf{A}$ in a causal model represented by a DAG $\mathcal{G}$, the distribution $p(\{V(\mathbf{g_A})|V \in \mathbf{V} \setminus \mathbf{A}\})$ over responses in $\mathbf{V} \setminus \mathbf{A}$ to a treatment regime $\mathbf{g_A}$ is identified via the following generalization of (16.8):

$$\prod_{V \in \mathbf{V} \setminus \mathbf{A}} p(V|\text{pa}(V) \setminus \mathbf{A}, \{A = g_A(\mathbf{W}_A)|A \in \text{pa}(V) \cap \mathbf{A}\}). \tag{16.10}$$

As an example, in Fig. 16.1 (b) the potential outcome $Y(\mathbf{g_A}) = Y(\{g_A, g_Z\})$, where g_A is a mapping from values of W to values of A, and g_Z is a mapping from values of M, W to values of Z, is identified as

$$\sum_{W,M} p(Y|A = g_A(W), Z = g_Z(M, W), M, W)p(M|A = g_A(W), W)p(W)$$

Equation (16.10) can be viewed as a version of (16.8) where values of $\mathbf{A}$ are not set to constants $\mathbf{a}$, but are instead set using appropriate functions in $\mathbf{g_A}$, which potentially creates dependence on other variables.

16.3 Causal Models of a DAG with Hidden Variables

Identification results in the previous section assumed all variables in a DAG are observed. While this assumption allows an elegant characterization of identifiability via variations of the g-formula (16.8), (16.9), and (16.10), it is extremely unrealistic in practical applications. This motivates the study of causal models of a DAG where some variables are hidden. Unfortunately, the presence of hidden variables introduces a number of complications, both because hidden variables may prevent identification, and because functions $g(p(\mathbf{V}))$ of the observed distribution that correspond to counterfactual distributions that are identified can potentially become much more complex than (16.8), (16.9), and (16.10). In this section, we will describe a characterization of identifiable targets of causal inference in hidden variable causal DAGs, and identification algorithms that yield appropriate generalizations of the g-formula.

16.3.1 Latent projections and targets of inference

Identification in a causal model described by a hidden variable DAG $\mathcal{G}(\mathbf{V} \cup \mathbf{H})$, where $\mathbf{V}$ are observed variables and $\mathbf{H}$ are hidden variables, is often described in terms of an acyclic directed mixed graph (ADMG) $\mathcal{G}(\mathbf{V})$ constructed from $\mathcal{G}(\mathbf{V} \cup \mathbf{H})$ via the latent projection operation described in Definition 2.3.2 from Chapter 2.

An ADMG is a mixed graph containing directed ($\rightarrow$) and bidirected ($\leftrightarrow$) edges that contains no directed cycles. In an ADMG, a pair of vertices can share at most two edges, and if the pair does share two edges, one of them is directed and one is bidirected. The latent projection ADMG represents an infinite class of hidden variable DAGs that, as we will describe later, all share identification theory.

Notation for a latent projection $\mathcal{G}(\mathbf{V})$ of a DAG $\mathcal{G}(\mathbf{V} \cup \mathbf{H})$ with hidden variables $\mathbf{H}$ intentionally resembles the notation for marginalization in distributions. The latent projection operation can be viewed as a graphical analogue of the marginalization operation. A detailed exploration of this connection, and in particular a way of using mixed graphs derived from hidden variable DAGs to define statistical models that are supersets of marginals of distributions in DAG models is found in [9, 20].

Definitions of $Y(\mathbf{a})$ via (16.2), $Y(\pi, a, a')$ via (16.5), and $Y(\mathbf{g_A})$ via (16.6) in a fully observed DAG $\mathcal{G}(\mathbf{V})$ carry over to a hidden variable DAG $\mathcal{G}(\mathbf{V} \cup \mathbf{H})$ without change. Furthermore, if certain additional properties hold, these counterfactuals can be defined directly on $\mathcal{G}(\mathbf{V})_{\rightarrow}$, the edge subgraph of the latent projection $\mathcal{G}(\mathbf{V})$ of $\mathcal{G}(\mathbf{V} \cup \mathbf{H})$ that only contains directed edges.

These properties are listed below.

- For $Y(\mathbf{a})$ in $\mathcal{G}(\mathbf{V} \cup \mathbf{H})$, $Y \in \mathbf{V}$, and $\mathbf{A} \subseteq \mathbf{V}$.
- For $Y(\mathbf{g_A})$ in $\mathcal{G}(\mathbf{V} \cup \mathbf{H})$, $Y \in \mathbf{V}$, $\mathbf{A} \subseteq \mathbf{V}$, and $\mathbf{W}_A \subseteq \mathbf{V}$ for every $A \in \mathbf{A}$.
- For $Y(\pi, \mathbf{a}', \mathbf{a})$ in $\mathcal{G}(\mathbf{V} \cup \mathbf{H})$, $Y \in \mathbf{V}$, $\mathbf{A} \subseteq \mathbf{V}$, and for any two directed paths π_i, π_j in $\mathcal{G}(\mathbf{V} \cup \mathbf{H})$ such that the largest subsets of vertices in π_i, π_j that are in $\mathbf{V}$ form the same directed path $\widetilde{\pi}$ in $\mathcal{G}(\mathbf{V})$, either $\pi_i, \pi_j \in \pi$, or $\pi_i, \pi_j \notin \pi$.

If these properties hold, in any causal model of $\mathcal{G}(\mathbf{V} \cup \mathbf{H})$,

- Every $Y(\mathbf{a})$ defined using (16.2) in $\mathcal{G}(\mathbf{V} \cup \mathbf{H})$ is equivalent to $Y(\mathbf{a})$ defined using (16.2) in $\mathcal{G}(\mathbf{V})_{\rightarrow}$.
- Every $Y(\mathbf{g_A})$ defined using (16.6) in $\mathcal{G}(\mathbf{V} \cup \mathbf{H})$ is equivalent to $Y(\mathbf{g_A})$ defined using (16.2) in $\mathcal{G}(\mathbf{V})_{\rightarrow}$.
- Every $Y(\pi, \mathbf{a}', \mathbf{a})$ defined using (16.5) in $\mathcal{G}(\mathbf{V} \cup \mathbf{H})$ is equivalent to $Y(\widetilde{\pi}, \mathbf{a}', \mathbf{a})$ defined using (16.2) in $\mathcal{G}(\mathbf{V})_{\rightarrow}$, where $\widetilde{\pi}$ is the set of directed paths formed from largest subsets of vertices in every element of π that intersect $\mathbf{V}$.

We will consider targets of inference defined on $\mathcal{G}(\mathbf{V})_{\rightarrow}$ directly, with the understanding that these are equivalent to targets of inference in any underlying hidden variable DAG model in the class represented by the latent projection $\mathcal{G}(\mathbf{V})$.

16.3.2 Conditional mixed graphs and kernels

The identifying formulas described in Section 16.2 can be viewed as obtained by an algorithm that, in a single step, drops terms corresponding to elements in $\mathbf{A}$ from a DAG factorization, and possibly relabels remaining occurrences of $\mathbf{A}$ in remaining terms. In hidden variable DAGs represented by ADMGs, many counterfactual distributions of interest are not identifiable. In addition, even for distributions that are identifiable, their identifying formulas are obtained from more complicated algorithms that perform sequences of operations that drop terms in a particular order. To effectively describe how these algorithms operate, we introduce a special type of graph and a special type of distribution that represent intermediate outputs these algorithms produce.

A *conditional ADMG (CADMG)* $\mathcal{G}(\mathbf{V}, \mathbf{W})$ is an ADMG with a vertex set $\mathbf{V} \cup \mathbf{W}$, where the set of vertices $\mathbf{W}$ are marked as *fixed*, and no edges of the form $\circ \to W$, or $\circ \leftrightarrow W$ exist. Vertices in $\mathbf{V}$ in a CADMG are meant to represent random variables as in an ordinary DAG, while vertices in $\mathbf{W}$ are meant to represent variables that were previously random, but which are currently set to a constant. Note that there are no restrictions on vertices in $\mathbf{V}$. In particular there may exist $V \in \mathbf{V}$ such that edges $\circ \to V$, or $\circ \leftrightarrow V$ also do not exist.

A bidirected path in a CADMG (or ADMG) is a path consisting entirely of bidirected edges. Given $V \in \mathbf{V}$, the maximum subset of $\mathbf{V}$ connected to V by a bidirected path (and V itself) is called the district [19] (or a c-component [34]) of V, abbreviated as $\text{dis}_{\mathcal{G}}(V)$. The set of districts in a CADMG $\mathcal{G}(\mathbf{V}, \mathbf{W})$ is denoted by $\mathcal{D}(\mathcal{G}(\mathbf{V}, \mathbf{W}))$ and forms a partition of elements in $\mathbf{V}$.

A *kernel* $q_{\mathbf{V}}(\mathbf{V} \mid \mathbf{W})$ is a mapping from values $\mathbf{w}$ of $\mathbf{W}$ to normalized densities $q_{\mathbf{V}}(\mathbf{V} \mid \mathbf{w})$ over $\mathbf{V}$. A conditional distribution is a type of kernel, though other types of kernels are possible. Given a subset $\mathbf{A} \subseteq \mathbf{V}$, marginalization and conditioning for kernels are defined in the usual way:

$$q_{\mathbf{V}}(\mathbf{A} \mid \mathbf{W}) \equiv \sum_{\mathbf{V} \setminus \mathbf{A}} q_{\mathbf{V}}(\mathbf{V} \mid \mathbf{W}); \qquad q_{\mathbf{V}}(\mathbf{V} \setminus \mathbf{A} \mid \mathbf{A} \cup \mathbf{W}) \equiv \frac{q_{\mathbf{V}}(\mathbf{V} \mid \mathbf{W})}{q_{\mathbf{V}}(\mathbf{A} \mid \mathbf{W})}.$$

Kernels were called Q-factors in [34, 32]. The dependence of kernels on values of $\mathbf{W}$ was kept implicit in the Q-factor notation.

An ADMG $\mathcal{G}(\mathbf{V})$ representing a hidden variable DAG $\mathcal{G}(\mathbf{H} \cup \mathbf{V})$, and the corresponding observed data distribution $p(\mathbf{V})$ are special cases of a CADMG and a kernel, respectively. CADMGs and kernels involved in identification are derived from $\mathcal{G}(\mathbf{V})$ and $p(\mathbf{V})$ by sequential applications of the fixing operation, defined in [20].

16.3.3 The fixing operation

Given a CADMG $\mathcal{G}(\mathbf{V}, \mathbf{W})$, a vertex $V \in \mathbf{V}$ is said to be fixable if $\text{de}_{\mathcal{G}}(V) \cap \text{dis}_{\mathcal{G}}(V) = \{V\}$, where $\text{de}_{\mathcal{G}}(V)$ (descendants of V) is the set of all vertices Z in $\mathcal{G}(\mathbf{V}, \mathbf{W})$, including V itself, with a directed path from V to Z. Given a fixable vertex V in $\mathcal{G}$ define the fixing operator on graphs $\phi_V(\mathcal{G}(\mathbf{V}, \mathbf{W}))$ to be an operator that produces a CADMG $\widetilde{\mathcal{G}}(\mathbf{V} \setminus \{V\}, \mathbf{W} \cup \{V\})$ which is obtained from $\mathcal{G}(\mathbf{V}, \mathbf{W})$ by removing all edges pointing to V, and marking V as fixed.

Given a CADMG $\mathcal{G}(\mathbf{V}, \mathbf{W})$, a kernel $q_{\mathbf{V}}(\mathbf{V} \mid \mathbf{W})$, and any fixable V in $\mathcal{G}$, define the fixing operator on kernels $\phi_V(q_{\mathbf{V}}(\mathbf{V} \mid \mathbf{W}); \mathcal{G}(\mathbf{V}, \mathbf{W}))$ to be one that produces the kernel

$$\widetilde{q}_{\mathbf{V} \setminus \{V\}}(\mathbf{V} \setminus \{V\} \mid \mathbf{W} \cup \{V\}) \equiv \frac{q_{\mathbf{V}}(\mathbf{V} \mid \mathbf{W})}{q_{\mathbf{V}}(V \mid \text{nd}_{\mathcal{G}}(V) \cup \mathbf{W})},$$

where $\text{nd}_{\mathcal{G}}(V) \equiv (\mathbf{V} \cup \mathbf{W}) \setminus \text{de}_{\mathcal{G}}(V)$ is the set of non-descendants of V in $\mathcal{G}(\mathbf{V}, \mathbf{W})$.

A sequence $\langle V_1, \ldots, V_k \rangle$ is said to be fixable in $\mathcal{G}$ if V_1 is fixable in $\mathcal{G}$, V_2 is fixable in $\phi_{V_1}(\mathcal{G})$, V_3 is fixable in $\phi_{V_2}(\phi_{V_1}(\mathcal{G}))$, and so on. We extend the fixing operators for graphs and kernels in the natural way via function composition for any fixable sequence $\langle V_1, \ldots, V_k \rangle$:

$$\phi_{\langle V_1, \ldots, V_k \rangle}(\mathcal{G}) \equiv \phi_{V_k}(\ldots \phi_{V_2}(\phi_{V_1}(\mathcal{G})) \ldots)$$
$$\phi_{\langle V_1, \ldots, V_k \rangle}(q_{\mathbf{V}}; \mathcal{G}) \equiv \phi_{V_k}(\ldots \phi_{V_2}(\phi_{V_1}(q_{\mathbf{V}}; \mathcal{G}); \phi_{V_1}(\mathcal{G})) \ldots; \phi_{\langle V_1, \ldots, V_{k-1} \rangle}(\mathcal{G}))$$

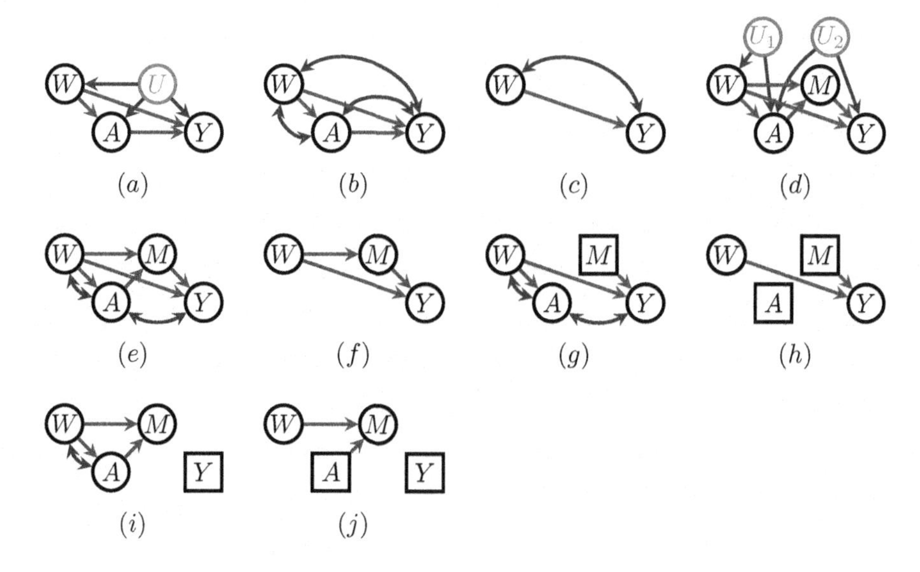

FIGURE 16.2: (a) A causal DAG with an unobserved common cause of the treatment A and the outcome Y which prevents identification of $p(Y(a))$. (b) The ADMG obtained from the DAG in (a) via the latent projection operation collapsing over the unobserved variable U. (c) A subgraph of the ADMG in (b) relevant for identification of $p(Y \mid \text{do}(a))$. (d) A causal DAG with an unobserved common cause of the baseline variables W, the treatment A and the outcome Y. This DAG also contains a mediator M that "captures" all the causal influence of A on Y that is also not confounded by U_2. (e) The ADMG $\mathcal{G}$ obtained from the DAG in (e) via the latent projection operation collapsing over the unobserved variables U_1, U_2. (f) A subgraph of the ADMG in (e) relevant for identification of $p(Y \mid \text{do}(a))$. (g) The CADMG $\phi_M(\mathcal{G})$ obtained from the ADMG in (e). (h) The CADMG $\phi_{\langle M,A\rangle}(\mathcal{G}) = \phi_{\{M,A\}}(\mathcal{G})$ obtained from the ADMG in (e). (i) The CADMG $\phi_Y(\mathcal{G})$ obtained from the ADMG in (e). (j) The CADMG $\phi_{\langle Y,A\rangle}(\mathcal{G}) = \phi_{\{Y,A\}}(\mathcal{G})$ obtained from the ADMG in (e).

It is known [20] that any two distinct fixable sequences of the same set of variables applied to both graphs and kernels yield the same object. This invariance property justifies considering fixable sets, which are any subsets of $\mathbf{V}$ with vertices that can be arranged in a fixable sequence, and fixing operators $\phi_{\mathbf{W}}(\mathcal{G})$, $\phi_{\mathbf{W}}(q_{\mathbf{V}}; \mathcal{G})$. These operators are defined on any fixable subset $\mathbf{W}$ of $\mathbf{V}$, and obtained via any fixable sequence on elements in $\mathbf{W}$.

16.3.4 The ID algorithm

Given a hidden variable DAG $\mathcal{G}(\mathbf{V} \cup \mathbf{H})$ representing a causal model, fix disjoint subsets $\mathbf{A}, \mathbf{Y}$ of $\mathbf{V}$ representing the set of treatments and outcomes, respectively, where we are interested in identification of the interventional distribution $p(\mathbf{Y} \mid \text{do}(\mathbf{a}))$. Identification of these types of distributions in hidden variable DAGs was phrased in terms of a recursive algorithm called the ID algorithm [34, 28]. Here we give a simple reformulation of the ID algorithm in terms of the fixing operator ϕ, CADMGs and kernels, where the recursion is folded into the iterative application of the fixing operator.

Let $\mathcal{G}(\mathbf{V})$ be the latent projection of $\mathcal{G}(\mathbf{V} \cup \mathbf{H})$ onto observable vertices $\mathbf{V}$, and let $\mathbf{Y}^* \equiv \text{an}_{\mathcal{G}(\mathbf{V})_{\mathbf{V}\setminus\mathbf{A}}}(\mathbf{Y})$, where for any subset $\mathbf{S}$ of $\mathbf{V}$, $\mathcal{G}(\mathbf{V})_{\mathbf{S}}$ is the subgraph of $\mathcal{G}(\mathbf{V})$ containing only vertices in $\mathbf{S}$ and edges in $\mathcal{G}(\mathbf{V})$ between pairs of vertices in $\mathbf{S}$. Then if for every

$\mathbf{D} \in \mathcal{D}(\mathcal{G}(\mathbf{V})_{\mathbf{Y}^*})$, $\mathbf{V} \setminus \mathbf{D}$ is a fixable set,

$$p(\mathbf{Y} \mid \text{do}(\mathbf{a})) = \sum_{\mathbf{Y}^* \setminus \mathbf{Y}} \prod_{\mathbf{D} \in \mathcal{D}(\mathcal{G}(\mathbf{V})_{\mathbf{Y}^*})} \phi_{\mathbf{V} \setminus \mathbf{D}}(p(\mathbf{V}); \mathcal{G}(\mathbf{V}))|_{\mathbf{A}=\mathbf{a}}. \tag{16.11}$$

If some $\mathbf{V} \setminus \mathbf{D}$ is not fixable, then $p(\mathbf{Y} \mid \text{do}(\mathbf{a}))$ is not identifiable, meaning that the ID algorithm is complete for identification of interventional distributions in non-parametric hidden variable DAG causal models. The proof of soundness of the original formulation of the ID algorithm appears in [34], and of completeness in [28]. The proof of soundness of the simplified version shown in (16.11) appears in [20]. Since preconditions for the application of (16.11), and all expressions within (16.11) itself were phrased in terms of the latent projection $\mathcal{G}(\mathbf{V})$, all causal models associated with a hidden variable DAG $\mathcal{G}(\mathbf{V} \cup \mathbf{H})$ that yields the latent projection $\mathcal{G}(\mathbf{V})$ agree on which distributions $p(\mathbf{Y} \mid \text{do}(\mathbf{a}))$ are identifiable and which are not, and further agree on all identifying functionals. It is for this reason that identification theory for counterfactuals in the presence of hidden variables is phrased in terms of the latent projection graph.

If $\mathbf{A} = \emptyset$, and $\mathbf{Y} = \mathbf{V}$, (16.11) can be viewed as a factorization of $p(\mathbf{V})$ known as the district or c-component factorization:

$$p(\mathbf{V}) = \prod_{\mathbf{D} \in \mathcal{D}(\mathcal{G})} q_{\mathbf{D}}(\mathbf{D} \mid \mathbf{V} \setminus \mathbf{D}) = \prod_{\mathbf{D} \in \mathcal{D}(\mathcal{G})} \phi_{\mathbf{V} \setminus \mathbf{D}}(p(\mathbf{V}); \mathcal{G}(\mathbf{V})). \tag{16.12}$$

This type of factorization has been used to define conditional independence supermodels of sets of marginals of hidden variable DAGs [10, 26, 20].

We now illustrate how (16.11) is applied with two examples. The first example is the graph in Fig. 16.2 (a), which contains an unobserved common cause of A and Y. The latent projection of this graph is the ADMG $\mathcal{G}(\{W, A, Y\})$ shown in Fig. 16.2 (b). Here $\mathbf{Y}^* = \text{an}_{\mathcal{G}_{\{Y,W\}}}(Y) = \{Y, W\}$, with $\mathcal{G}_{\mathbf{Y}^*}$ shown in Fig. 16.2 (c). It's easy to verify that $\mathcal{D}(\mathcal{G}_{\mathbf{Y}^*}) = \{\{W, Y\}\}$. Unfortunately, the set $\{A\}$ is not fixable, since Y is a descendant of A and lies in the district of A. From this we conclude that $p(Y \mid \text{do}(a)) = p(Y(a))$ is not identified in the causal model represented by Fig. 16.2 (a).

Consider now the graph in Fig. 16.2 (d). Like Fig. 16.2 (a), there is an unobserved common cause of A and Y, namely U_2. However, in Fig. 16.2 (d) there is, in addition, a mediator variable M which lies on a causal pathway from A to Y, indicated by the presence of a directed path $A \to M \to Y$. In fact, this is the only directed path from A to Y, meaning that M captures or mediates all of the causal influence of A on Y. Finally, M is not a child of either U_1 or U_2, meaning it is determined entirely by W and A and remains unconfounded by hidden variables, unlike A and Y. The presence of a mediator of this type allows non-parametric identification of $p(Y \mid \text{do}(a))$ in Fig. 16.2 (d), as was first noticed by [16]. We now give the identifying formula for $p(Y \mid \text{do}(a))$ using (16.11).

The latent projection of Fig. 16.2 (d) is the ADMG $\mathcal{G}(\{W, A, M, Y\})$ shown in Fig. 16.2 (e). Here $\mathbf{Y}^* = \text{an}_{\mathcal{G}_{\{Y,W,M\}}}(Y) = \{Y, M, W\}$, with $\mathcal{G}_{\mathbf{Y}^*}$ shown in Fig. 16.2 (f). It's easy to verify that $\mathcal{D}(\mathcal{G}_{\mathbf{Y}^*}) = \{\{Y\}, \{W\}, \{M\}\}$. In this case, the sets $\mathbf{V} \setminus \{Y\} = \{A, M, W\}$, $\mathbf{V} \setminus \{W\} = \{Y, M, A\}$, and $\mathbf{V} \setminus \{M\} = \{Y, W, A\}$ are fixable. We first consider $\{A, M, W\}$. M is fixable in Fig. 16.2 (e), which yields the CADMG in Fig. 16.2 (g), with the corresponding kernel

$$q_{\{W,A,Y\}}(W, A, Y|M) = \frac{p(W, A, M, Y)}{p(M|A, W)} = p(Y|M, A, W)p(A, W).$$

In this CADMG, A becomes fixable (it was not fixable in Fig. 16.2 (e)), yielding the CADMG in Fig. 16.2 (h), with the corresponding kernel

$$q_{\{W,Y\}}(W, Y|A, M) = \frac{q_{\{W,A,Y\}}(W, A, Y|M)}{q_{\{W,A,Y\}}(A|W, Y, M)} = \sum_{A} p(Y|M, A, W,)p(A, W).$$

Finally, in Fig. 16.2 (h), W is fixable, yielding a CADMG obtained from Fig. 16.2 (h) by drawing W as a square, with the corresponding kernel

$$q_{\{Y\}}(Y|W,A,M) = \frac{\sum_A p(Y|M,A,W,)p(A,W)}{\sum_{Y,A} p(Y|M,A,W,)p(A,W)} = \sum_A p(Y|M,A,W)p(A|W).$$

Similarly, the set $\{Y, W, A\}$ is also fixable. First, Y is fixable in Fig. 16.2 (e), yielding the CADMG in Fig. 16.2 (i), with the corresponding kernel

$$q_{\{W,A,M\}}(W,A,M|Y) = \frac{p(W,A,M,Y)}{p(Y|W,A,M)} = p(W,A,M).$$

Next, A becomes fixable in Fig. 16.2 (i), yielding the CADMG in Fig. 16.2 (j), with the corresponding kernel

$$q_{\{W,M\}}(W,M|A,Y) = \frac{q_{\{W,A,M\}}(W,A,M|Y)}{q_{\{W,A,M\}}(A|W,Y)} = p(M|A,W)p(W).$$

Finally, W is fixable in Fig. 16.2 (j), yielding a CADMG obtained from Fig. 16.2 (j) by drawing W as a square, with the corresponding kernel

$$q_{\{M\}}(M|W,A,Y) = \frac{q_{\{W,M\}}(W,M|A,Y)}{q_{\{W,M\}}(W|A,Y)} = p(M|A,W).$$

The set $\{Y, M, A\}$ is also fixable. We have already shown above that we can fix Y in Fig. 16.2 (e), and then fix A in Fig. 16.2 (i), yielding the graph in Fig. 16.2 (j), and the kernel $p(M|A,W)p(W)$. But M is fixable in Fig. 16.2 (j), yielding the kernel

$$q_{\{W\}}(W|M,A,Y) = \frac{p(M|A,W)p(W)}{p(M|A,W)} = p(W).$$

Combining these three kernels into the expression (16.11), where $\mathbf{Y} \setminus \mathbf{Y}^* = \{W, M\}$, and evaluating $q_{\{M\}}(M \mid W, A, Y)$ at $A = a$ yields

$$\begin{aligned} p(Y \mid \text{do}(a)) &= \sum_{W,M} \phi^a_{\{Y,W,A\}}(p;\mathcal{G})\phi_{\{A,M,W\}}(p;\mathcal{G})\phi_{\{Y,M,A\}}(p;\mathcal{G}) \\ &= \sum_{W,M} q_{\{M\}}(M|Y,W,a)q_{\{Y\}}(Y|A,M,W)q_{\{W\}}(W|Y,M,A) \\ &= \sum_{W,M} p(M|a,W)\left(\sum_A p(Y|M,A,W)p(A|W)\right)p(W). \end{aligned}$$

This is known as the front-door formula [16]. A connection between formulas of the above type, and the edge g-formula is given in [29].

In general, expressions obtained from (16.11) may become quite complicated. Nevertheless, (16.11) may be viewed as a generalization of the g-formula (16.8) appropriate to the hidden variable DAG setting. In particular, just as (16.8) is a truncated version of the DAG factorization (16.1), (16.11) is a truncated version of the district factorization in (16.12), in the sense that no kernel terms in (16.11) are densities over elements of $\mathbf{A}$. This is accomplished for each term in (16.11) by repeated applications of the fixing operation, which sequentially drop terms associated with variables in $\mathbf{V} \setminus \mathbf{D}$, which include the set $\mathbf{A}$ for every $\mathbf{D}$. Some fixing operations resemble conditioning, some resemble marginalization, and some resemble neither. Finally, just like in the g-formula (16.8), remaining terms in (16.11) that are functions of $\mathbf{A}$ are evaluated at $\mathbf{A} = \mathbf{a}$.

16.3.5 Controlled direct effects

An interesting special case of the identification problem described in the previous section occurs when $\mathbf{Y} = \{Y\}$, and $\mathbf{A} = \text{pa}_{\mathcal{G}(\mathbf{V})}(Y)$ in a hidden variable DAG $\mathcal{G}(\mathbf{V} \cup \mathbf{H})$. The resulting counterfactual distribution $p(Y(\mathbf{a}))$ is used to define *controlled direct effects* of the form

$$\mathbb{E}[Y \mid a, \mathbf{a}_{\text{pa}_{\mathcal{G}(\mathbf{V})}(Y)\setminus\{A\}}] - \mathbb{E}[Y \mid a', \mathbf{a}_{\text{pa}_{\mathcal{G}(\mathbf{V})}(Y)\setminus\{A\}}].$$

These types of effects are relevant in settings where we are interested in understanding the direct effect of A on Y within particular levels of all other observed direct causes of Y.

An ADMG $\mathcal{G}(\mathbf{V})$ is said to be an arborescence converging at Y if $\text{an}_{\mathcal{G}}(Y) = \mathbf{V}$, and $\mathcal{D}(\mathcal{G}) = \{\mathbf{V}\}$. The distribution $p(Y(\mathbf{a}))$ of this type is identified if and only if the largest subset $\mathbf{V}'$ of $\mathbf{V}$ such that $\mathcal{G}_{\mathbf{V}'}$ is an arborescence converging at Y is $\{Y\}$. If $p(Y(\mathbf{a}))$ is identified, we have, as a special case of (16.11),

$$p(Y(\mathbf{a})) = q_Y(Y \mid \mathbf{V} \setminus \{Y\})|_{\mathbf{A}=\mathbf{a}} = \phi_{\mathbf{V}\setminus\{Y\}}(p(\mathbf{V}); \mathcal{G}(\mathbf{V}))|_{\mathbf{A}=\mathbf{a}}.$$

As an example, $p(Y(a))$ in Fig. 16.2 (b) is not identified since the graph in Fig. 16.2 (b) is an arborescence converging at Y. However, $p(Y(m, w))$ in $\mathcal{G}^{(e)}$ shown in Fig. 16.2 (e) is identified:

$$\begin{aligned}
p(Y(m, w)) &= \phi_{\{W,M,A\}}(p(W, A, M, Y); \mathcal{G}^{(e)}) \\
&= \phi_{\{W,A\}}(p(Y|M, A, W)p(A, W); \phi_M(\mathcal{G}^{(e)}))|_{M=m} \\
&= \phi_{\{W\}}\left(\sum_A p(Y|M, A, W)p(A, W); \phi_{\{M,A\}}(\mathcal{G}^{(e)})\right)\Bigg|_{M=m} \\
&= \sum_A p(Y|M, A, W)p(A|W)|_{M=m,W=w} = \sum_A p(Y|m, A, w)p(A|w).
\end{aligned}$$

16.3.6 Conditional causal effects

A common variation of the problem of identification of joint interventional distributions considers instead identification of conditional interventional distributions of the form $p(Y_1(\mathbf{a}), \ldots, Y_k(\mathbf{a})|Z_1(\mathbf{a}), \ldots, Z_m(\mathbf{a}))$, where $\mathbf{Y} = \{Y_1, \ldots, Y_k\}$, $\mathbf{Z} = \{Z_1, \ldots, Z_m\}$. This distribution is sometimes written as $p(\mathbf{Y}|\mathbf{Z}, \text{do}(\mathbf{a}))$. These types of distributions arise in causal inference applications where causal effects in particular subpopulations are of interest.

A simple modification of (16.11) gives a characterization of identifiability of these distributions. Let $\mathbf{Z}$ be partitioned into $\mathbf{W}$,$\mathbf{Z}'$ such that $\mathbf{W}$ is any maximal set with the property that for any $\mathbf{z}$, $p(\mathbf{Y} \mid \mathbf{z}, \text{do}(\mathbf{a})) = p(\mathbf{Y} \mid \mathbf{z}', \text{do}(\mathbf{w} \cup \mathbf{a}))$, and $\mathbf{z}', \mathbf{w}$ are the appropriate partition of values in $\mathbf{z}$. Such a set $\mathbf{W}$, which is known to be unique and thus also maximum, can be obtained by application of the rule 2 of do-calculus [16]. Then $p(\mathbf{Y} \mid \mathbf{z}, \text{do}(\mathbf{a}))$ is identified from $p(\mathbf{V})$ if and only if $p(\mathbf{Y} \cup \mathbf{Z}' \mid \text{do}(\mathbf{w} \cup \mathbf{a}))$ is identified from $p(\mathbf{V})$, and is equal to

$$p(\mathbf{Y}|\mathbf{z}, \text{do}(\mathbf{a})) = \left.\frac{p(\mathbf{Y}, \mathbf{Z}'|\text{do}(\mathbf{w} \cup \mathbf{a}))}{p(\mathbf{Z}'|\text{do}(\mathbf{w} \cup \mathbf{a}))}\right|_{\mathbf{Z}'=\mathbf{z}'},$$

where $p(\mathbf{Y} \cup \mathbf{Z}'|\text{do}(\mathbf{w} \cup \mathbf{a}))$ is identified as usual via (16.11), and $p(\mathbf{Z}'|\text{do}(\mathbf{w} \cup \mathbf{a}))$ is obtained from $p(\mathbf{Y} \cup \mathbf{Z}'|\text{do}(\mathbf{w} \cup \mathbf{a}))$ via marginalization [27].

16.3.7 Path-specific effects

Path-specific effects in DAGs with all variables observed were not always identified due to the presence of recanting witnesses. Unsurprisingly, additional complications arise in hidden variable DAGs.

Fix $\mathbf{A}, \mathbf{Y}$ and a set of proper causal paths π for $\mathbf{A}$ and $\mathbf{Y}$, where each $A \in \mathbf{A}$ is the origin of at least one path in π. Let $\mathbf{Y}^* \equiv \text{an}_{\mathcal{G}_{\mathbf{V}\setminus\mathbf{A}}}(\mathbf{Y})$. Then $\mathbf{D} \in \mathcal{D}(\mathcal{G}(\mathbf{V})_{\mathbf{Y}^*})$ is said to be a *recanting district* for π if there exists a path in π with the first edge of the form $A \to D$, where $A \in \mathbf{A}, D \in \mathbf{D}$, and another proper causal path not in π, with the first edge of the form $A \to D'$, $D' \in \mathbf{D}$.

In the absence of a recanting district in $\mathcal{G}(\mathbf{V})$, and if $p(\mathbf{Y} \mid \text{do}(\mathbf{a}))$ is identified from $p(\mathbf{V})$ in $\mathcal{G}(\mathbf{V})$, the joint counterfactual distribution over variables $\{Y(\pi, \mathbf{a}, \mathbf{a}') \mid Y \in \mathbf{Y}\}$ is identified as

$$p(\{Y(\pi, \mathbf{a}, \mathbf{a}') \mid Y \in \mathbf{Y}\}) = \sum_{\mathbf{Y}^* \setminus \mathbf{Y}} \prod_{\mathbf{D} \in \mathcal{D}(\mathcal{G}_{\mathbf{Y}^*})} \phi_{\mathbf{V}\setminus\mathbf{D}}(p(\mathbf{V}); \mathcal{G}(\mathbf{V}))|_{\mathbf{A}\cap\text{pa}_{\mathcal{G}}(\mathbf{D})=\tilde{\mathbf{a}}_{\mathbf{D}}}, \tag{16.13}$$

where $\tilde{\mathbf{a}}_{\mathbf{D}}$ is defined to be the subset of values of $\mathbf{a}$ corresponding to $\text{pa}_{\mathcal{G}}(\mathbf{D}) \cap \mathbf{A}$ if all elements in $\text{pa}_{\mathcal{G}}(\mathbf{D}) \cap \mathbf{A}$ are connected to elements in $\mathbf{D}$ via edges in π, defined to be the subset of values of $\mathbf{a}'$ corresponding to $\text{pa}_{\mathcal{G}}(\mathbf{D}) \cap \mathbf{A}$ if all elements in $\text{pa}_{\mathcal{G}}(\mathbf{D}) \cap \mathbf{A}$ are connected to elements in $\mathbf{D}$ via edges not in π, and defined to be the empty set if $\text{pa}_{\mathcal{G}}(\mathbf{D}) \cap \mathbf{A} = \emptyset$. The absence of a recanting district guarantees these three possibilities are exhaustive.

If a recanting district is present, or $p(\mathbf{Y} \mid \text{do}(\mathbf{a}))$ is not identified, then $p(\{Y(\pi, \mathbf{a}, \mathbf{a}') \mid Y \in \mathbf{Y}\})$ is also not identified for some values $\mathbf{a}, \mathbf{a}'$. Just as (16.11) was an appropriate generalization of the g-formula (16.8) to hidden variable DAGs, so is (16.13) an appropriate generalization of the edge g-formula (16.9) to hidden variable DAGs.

As an example, consider Fig. 16.1 (c), where we are interested in the path-specific effect of A on Y via the path $A \to Z \to Y$. The latent projection of this graph is shown as a graph $\mathcal{G}^{(d)}$ in Fig. 16.1 (d). Here $\mathbf{Y}^* = \{W, M, Z, Y\}$, and $\mathcal{D}(\mathcal{G}_{\mathbf{Y}^*}) = \{\{W\}, \{Z\}, \{M, Y\}\}$. Note that there is no recanting district – the district containing the first post-exposure variable on the only path of interest is $\{Z\}$, and no path other than $A \to Z \to Y$ has the first post-exposure variable in this district. Furthermore, $p(Y \mid \text{do}(a))$ is identifiable. Thus, the counterfactual corresponding to the path-specific effect is identified:

$$\begin{aligned} p(Y(\pi, a, a')) &= \sum_{W,Z,M} \left(\phi_{\{W,A,Z\}}(p; \mathcal{G}^{(d)})|_{A=a'}\right) \cdot \left(\phi_{\{W,A,M,Y\}}(p; \mathcal{G}^{(d)})|_{A=a}\right. \\ &\qquad\qquad \left.\phi_{\{A,M,Z,Y\}}(p; \mathcal{G}^{(d)})\right) \\ &= \sum_{W,Z,M} (p(Y \mid a', M, Z, W)p(M \mid a', W))\, p(Z \mid a, M, W)p(W). \end{aligned}$$

On the other hand, if we were interested in the path-specific effect of A on Y along paths $\pi = \{A \to Z \to Y; A \to Y\}$, this path-specific effect is not identified. This is because the path $A \to M \to Y$ is not in π but has $A \to M$ as the first edge, while $A \to Y$ is a path in π. M and Y share a district in $\mathcal{G}(\mathbf{V})_{\mathbf{Y}^*}$, where $\mathbf{Y}^* = \{W, M, Z, Y\}$. This implies $\{M, Y\}$ is a recanting district, and will prevent identification of $Y(\{A \to Z \to Y; A \to Y\}, a, a')$.

16.3.8 Responses to dynamic treatment regimes

A general algorithm for identification of distributions $p(\{Y(\mathbf{g_A}) \mid Y \in \mathbf{Y}\})$ in causal models represented by a hidden variable DAG $\mathcal{G}(\mathbf{H} \cup \mathbf{V})$, and the corresponding latent projection $\mathcal{G}(\mathbf{V})$ was given in [32]. Here we reformulate this algorithm in terms of the fixing operator. Define the graph $\mathcal{G}(\mathbf{V})_{\mathbf{g_A}}$ to be an ADMG obtained from $\mathcal{G}(\mathbf{V})$ by removing all edges pointing into $\mathbf{A}$ and adding a directed edge $W \to A$ for any $W \in \mathbf{W}_A$, and the set $\mathbf{Y}^* \equiv \text{an}_{\mathcal{G}(\mathbf{V})_{\mathbf{g_A}}}(\mathbf{Y}) \setminus \mathbf{A}$. Then $p(\{Y(\mathbf{g_A}) \mid Y \in \mathbf{Y}\})$ is identified if $p(\mathbf{Y}^* \mid \text{do}(\mathbf{a}))$ is identified.

Moreover, the identification formula is

$$p(\{Y(\mathbf{g_A}) \mid Y \in \mathbf{Y}\}) = \sum_{(\mathbf{Y}^* \cup \mathbf{A}) \setminus \mathbf{Y}} \prod_{\mathbf{D} \in \mathcal{D}(\mathcal{G}(\mathbf{V})_{\mathbf{Y}^*})} \phi_{\mathbf{V} \setminus \mathbf{D}}(p(\mathbf{V}); \mathcal{G}(\mathbf{V}))|_{\{A = g_A(\mathbf{W}_A) | A \in \mathbf{A}\}}.$$

The sum over $\mathbf{A}$ is vacuous if $\mathbf{g_A}$ is a set of deterministic policies, since in this case there is no variation in values of $\mathbf{A}$ in any $Y(\mathbf{g_A})$. This algorithm is conjectured, but currently not known, to be complete for identification of responses to dynamic treatment regimes in hidden variable DAGs models.

As an example, in Fig. 16.1 (c), if $\mathbf{g_A} = \{g_A(W), g_Z(M, W)\}$, $\mathcal{G}(\mathbf{V})$ is shown in Fig. 16.1 (d), and $\mathcal{G}(\mathbf{V})_{\mathbf{g_A}}$ is the same graph as $\mathcal{G}(\mathbf{V})$. Since $p(Y, M, W \mid \text{do}(a, z))$ is identified as

$$\phi_{\{A,M,Z,Y\}}(p; \mathcal{G}(\mathbf{V}))\phi_{\{W,A,Z\}}(p; \mathcal{G}(\mathbf{V}))_{A=a,Z=z},$$

which is equal to $p(W)p(Y \mid z, M, a, W)p(M \mid a, W)$, $p(Y(\mathbf{g_A}))$ is also identified as

$$\begin{aligned} &\sum_{W,M,A,Z} \phi_{\{A,M,Z,Y\}}(p; \mathcal{G}(\mathbf{V}))\phi_{\{W,A,Z\}}(p; \mathcal{G}(\mathbf{V}))|_{\{A=g_A(W), Z=g_Z(M,W)\}} \\ = &\sum_{W,M,A,Z} p(W)p(Y|Z = g_Z(M, W), M, A = g_A(W), W)p(M|A = g_A(W), W). \end{aligned}$$

16.4 Linear Structural Equation Models

An important parametric causal model representable directly by an ADMG (and not necessarily a DAG) is the linear structural equation model (SEM). The linear SEM associated with an ADMG $\mathcal{G}(\mathbf{V})$ represents potential outcomes $V_i(\mathbf{v}_{\text{pa}_\mathcal{G}(V_i)})$ via linear functions $f_{V_i}(\mathbf{v}_{\text{pa}_\mathcal{G}(V_i)}, \epsilon_{V_i})) \equiv \sum_{M \in \text{pa}_\mathcal{G}(V)} w_M M + \epsilon_V$, where $(\epsilon_{V_1}, \ldots, \epsilon_{V_k})$ is a random vector following the multivariate normal distribution $\mathcal{N}(0, \Omega)$, where Ω is a positive-definite matrix with the property that it contains a 0 entry in the ij cell for any V_i, V_j such that $V_i \neq V_j$ and $V_i \leftrightarrow V_j$ is absent in $\mathcal{G}(\mathbf{V})$. A well known result states that $p(\mathbf{V})$ specified via a linear SEM associated with an ADMG is multivariate normal. The value of the first moment of this distribution does not affect identification considerations, and is often assumed without loss of generality to be the 0 vector of the appropriate size.

Linear SEMs have a long pedigree in applied data analysis problems, particularly in the social sciences [35, 13]. Aside from its non-parametric generalizations given by the single world model and the multiple worlds model, current literature has generalized linear SEMs in a number of useful directions, such as factor analysis models [14] which allow inclusion of meaningful unobserved variables, and non-Gaussian additive noise models [18] which permit inferences about causal directionality of edges from observed data. However, the classical linear SEM remains a very important and useful model class. An attractive property of this model is its relative simplicity for computation and statistical inference, and its additional algebraic structure layered on top of the non-parametric structure of the graphical causal model. This additional structure permits identification in a number of cases where identification is impossible non-parametrically.

Parameters of interest in linear SEMs are the coefficients which suffice to specify linear functions f_{V_i} for every $V_i \in \mathbf{V}$. Most interesting causal targets of inference can be phrased in terms of these coefficients. For example, if $A \in \text{pa}_\mathcal{G}(Y)$, and the linear structural equation

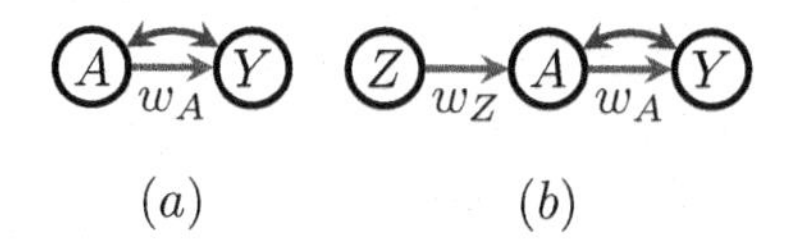

FIGURE 16.3: (a) A simple ADMG representing a linear SEM where the coefficient w_A is not identified from $p(A, Y)$. (b) A more complex ADMG, the instrumental variable graph, representing a linear SEM where the coefficient w_A is generically, but not globally identified from $p(A, Y)$.

f_Y for Y is

$$Y = w_A A + \sum_{V \in \mathrm{pa}_{\mathcal{G}}(V) \setminus \{A\}} w_V V + \epsilon_Y,$$

then a simple substitution argument, and linearity of f_Y implies that w_A corresponds to the controlled direct effect:

$$\mathbb{E}[Y(A=1, \mathbf{a})] - \mathbb{E}[Y(A=0, \mathbf{a})] = w_A, \tag{16.14}$$

for any assignment $\mathbf{a}$ to $\mathrm{pa}_{\mathcal{G}}(Y) \setminus \{A\}$.

Identification results in linear SEMs take two forms, *global* and *generic*. A parameter w_V is said to be globally identified in a graph $\mathcal{G}(\mathbf{V})$, if it is a function of $p(\mathbf{V})$ in any linear SEM corresponding to $\mathcal{G}(\mathbf{V})$. A parameter is said to be generically identified in $\mathcal{G}(\mathbf{V})$ if it is a function of $p(\mathbf{V})$ in almost every element of the linear SEM corresponding to $\mathcal{G}(\mathbf{V})$, except perhaps a set of elements of measure zero. If $\mathcal{G}(\mathbf{V})$ is a DAG, then every parameter is globally identified due to (16.8) and (16.14).

In linear SEMs associated with ADMGs, some parameters are generically, but not globally identified, and some are not identified at all. For example, in the linear SEM corresponding to Fig. 16.3 (a), given by

$$\begin{aligned} A &= \epsilon_A \\ Y &= w_A A + \epsilon_Y, \end{aligned}$$

where $\mathrm{cov}[\epsilon_A, \epsilon_Y] \neq 0$, the coefficient w_A, representing the average causal effect of A on Y is not identified from $p(\mathbf{V})$.

On the other hand, in Fig. 16.3 (b) known as the *instrumental variable graph*, corresponding to the model

$$\begin{aligned} Z &= \epsilon_Z \\ A &= w_Z Z + \epsilon_A \\ Y &= w_A A + \epsilon_Y, \end{aligned} \tag{16.15}$$

where $\mathrm{cov}[\epsilon_A, \epsilon_Y] \neq 0$, the parameter w_A is not globally identified (since setting $w_Z = 0$ recovers the model in Fig. 16.3 (a)), but is generically identified as $w_A = \mathrm{cov}[Z, Y]/\mathrm{cov}[Z, A]$.

16.4.1 Global identification of linear SEMs

An elegant characterization of global identifiability of the linear SEM model (that is global identifiability of every coefficient of every structural equation) was given in [8]. Specifically the linear SEM model associated with an ADMG $\mathcal{G}(\mathbf{V})$ is globally identified if and only if for every $V \in \mathbf{V}$, the largest subgraph of $\mathcal{G}(\mathbf{V})$ that forms an arborescence converging at V contains a single vertex.

There is a connection between this result and the earlier results on identifiability of controlled direct effects described in Section 16.3.5. Identification of controlled direct effects of A on Y, for any A and Y in an ADMG $\mathcal{G}(\mathbf{V})$ representing a causal model of a hidden variable DAG $\mathcal{G}(\mathbf{V} \cup \mathbf{H})$ is characterized by the absence of an arborescence of size 2 or more converging on Y. This implies that if the largest convergent arborescence in $\mathcal{G}(\mathbf{V})$ has size 1, then every controlled direct effect is identified non-parametrically. In fact, this identification result applies, without change, to a linear SEM defined directly on an ADMG $\mathcal{G}(\mathbf{V})$, and implies the entire model is identified due to (16.14). The converse is also true, although this is considerably more difficult to show [8].

16.4.2 Generic identification of linear SEMs

The characterization of generic identifiability of a linear SEM remains an open problem, despite decades of work [3, 2, 33, 12, 11, 7, 5]. It is known via computer algebra methods [12] that showing whether a particular parameter is identified in a linear SEM is a decidable problem, although its computational complexity is not currently known. We describe a general method based on features of the graph described in [11], although this method is known to not be complete, and further extensions are possible.

In an ADMG $\mathcal{G}(\mathbf{V})$ a half-trek between vertices V_i, V_j is a path of one of the following two types:

$$\pi \equiv \underbrace{V_i \to \circ \to \ldots \to \circ \to V_j}_{\text{Right}(\pi)}$$

$$\pi \equiv \underbrace{V_i}_{\text{Left}(\pi)} \leftrightarrow \underbrace{\circ \to \ldots \to \circ \to V_j}_{\text{Right}(\pi)}$$

with subset Left(π) defined to be $\emptyset$ for the first type of half-trek, and Left(π) and Right(π) defined as shown otherwise. A half-trek can be an empty path (a path with no edges and $V_i = V_j$). A set of half-treks is called a *system* if no two half-treks in the set share initial or final vertices. A system of half-treks $\{\pi_1, \ldots, \pi_k\}$ is said to have *no sided intersection* if for every $i, j \in \{1, \ldots, k\}$, and $i \neq j$,

$$\text{Left}(\pi_i) \cap \text{Left}(\pi_j) = \emptyset = \text{Right}(\pi_i) \cap \text{Right}(\pi_j).$$

In Fig. 16.4 (a), $A \to B \to C$ and $A \leftrightarrow D \to E$ are half-treks while $A \leftrightarrow E \leftrightarrow D$ and $A \to B \leftrightarrow C$ are not. The pair of half-treks $\pi_1 \equiv C \to D \to E$ and $\pi_2 \equiv D \leftrightarrow A$ form a system of half-treks $\{\pi_1, \pi_2\}$ from $\{C, D\}$ to $\{A, E\}$. This system has no sided intersection, since $D \in \text{Right}(\pi_1) \cap \text{Left}(\pi_2)$. On the other hand, π_1 and $\pi_3 \equiv A \leftrightarrow D$ form a system of half-treks $\{\pi_1, \pi_3\}$ from $\{A, C\}$ to $\{D, E\}$ with a sided intersection since $D \in \text{Right}(\pi_1) \cap \text{Right}(\pi_3)$. This illustrates that initial and final vertices in a half-trek matter for the purposes of determining the existence of sided intersections.

For any $V \in \mathbf{V}$ in an ADMG $\mathcal{G}(\mathbf{V})$ denote the set of siblings of V, $\text{sib}_{\mathcal{G}(\mathbf{V})}(V) \equiv \{W \leftrightarrow V \mid W \in \mathbf{V}\}$. For every $V \in \mathbf{V}$ define the following subset of vertices reachable from V by a half-trek:

$$\text{htr}_{\mathcal{G}(\mathbf{V})}(V) \equiv \left\{ W \in \mathbf{V} \setminus (\{V\} \cup \text{sib}_{\mathcal{G}(\mathbf{V})}(V)) \;\middle|\; \begin{array}{c} V \leftrightarrow \circ \to \ldots \to \circ \to W \\ \text{or} \\ V \to \circ \to \ldots \to \circ \to W \end{array} \right\}.$$

A (possibly empty) vertex set $\mathbf{Y} \subseteq \mathbf{V}$ satisfies the *half-trek criterion* with respect to $V \in \mathbf{V}$ if $|\mathbf{Y}| = |\text{pa}_{\mathcal{G}(\mathbf{V})}(V)|$, $\mathbf{Y} \cap (\{V\} \cup \text{sib}_{\mathcal{G}(\mathbf{V})}(V)) = \emptyset$, and there is a system of half-treks with no sided intersection from $\mathbf{Y}$ to $\text{pa}_{\mathcal{G}(\mathbf{V})}(V)$. If $\text{pa}_{\mathcal{G}(\mathbf{V})}(V) = \emptyset$, then $\mathbf{Y} = \emptyset$ satisfies the half-trek criterion with respect to V.

The half-trek criterion gives a very general condition for generic identifiability of the linear SEM called **HTC-identifiability**. A linear SEM model represented by an ADMG $\mathcal{G}(\mathbf{V})$ is HTC-identifiable if there is a set

$$\{\mathbf{Y}_V \subseteq \mathbf{V} \mid \mathbf{Y}_V \text{ satisfies the half-trek criterion with respect to } V\}$$

in $\mathcal{G}(\mathbf{V})$, and there is a total ordering $\prec$ on $\mathbf{V}$ in $\mathcal{G}(\mathbf{V})$ such that $W \prec V$ whenever $W \in \mathbf{Y}_V \cap \text{htr}_{\mathcal{G}(\mathbf{V}}(V)$. HTC-identifiability guarantees that, except in a small set of elements of the model, every coefficient of every structural equation is a rational function of the observed distribution $p(\mathbf{V})$. These functions are obtained by solving systems of linear equations following the ordering $\prec$, with the details given in [11].

A closely related condition called **HTC-nonidentifiability** strongly precludes identification of the model. A linear SEM model represented by an ADMG $\mathcal{G}(\mathbf{V})$ is HTC-nonidentifiable if for every set of sets $\{\mathbf{Y}_V \subseteq \mathbf{V} \mid V \in \mathbf{V}\}$ in $\mathcal{G}(\mathbf{V})$, either some $\mathbf{Y}_V$ does not satisfy the half-trek criterion with respect to V or there exist $\mathbf{Y}_V$, $\mathbf{Y}_W$ in this set such that $V \in \mathbf{Y}_W$ and $W \in \mathbf{Y}_V$. HTC-nonidentifiability guarantees the existence of an infinite set of elements of a linear SEM model corresponding to $\mathcal{G}(\mathbf{V})$ that agree on the observed distribution $p(\mathbf{V})$.

The graph $\mathcal{G}^{(a)}$ in Fig. 16.4 (a) is HTC-identifiable. Let

$$Y_A = \emptyset; \quad Y_B = \{E\}; \quad Y_C = \{B\}; \quad Y_D = \{B\}; \quad Y_E = \{C\}.$$

Then each set above satisfies the half-trek criterion with respect to its vertex:

- Y_A: since $\text{pa}_{\mathcal{G}^{(a)}}(A) = \emptyset$;
- Y_B: since $Y_B \cap (\{B\} \cup \text{sib}_{\mathcal{G}^{(a)}}(B)) = \{E\} \cap \{B, A\} = \emptyset$, and the single half-trek system $\{E \leftrightarrow A\}$ has no sided intersection;
- Y_C: since $Y_C \cap (\{C\} \cup \text{sib}_{\mathcal{G}^{(a)}}(C)) = \{B\} \cap \{C, D\} = \emptyset$, and the single half-trek system consisting of the empty half-trek $\{\emptyset\}$ has no sided intersection;
- Y_D: since $Y_D \cap (\{D\} \cup \text{sib}_{\mathcal{G}^{(a)}}(D)) = \{B\} \cap \{A, C, D, E\} = \emptyset$, and the single half-trek system $\{B \to C\}$ has no sided intersection;
- Y_E: since $Y_E \cap (\{E\} \cup \text{sib}_{\mathcal{G}^{(a)}}(E)) = \{C\} \cap \{A, D, E\} = \emptyset$, and the single half-trek system $\{C \to D\}$ has no sided intersection.

Finally, we have

$$\begin{array}{cc}
\text{htr}_{\mathcal{G}^{(a)}}(A) = \{C\}; & \text{htr}_{\mathcal{G}^{(a)}}(A) \cap Y_A = \emptyset \\
\text{htr}_{\mathcal{G}^{(a)}}(B) = \{C, D, E\}; & \text{htr}_{\mathcal{G}^{(a)}}(B) \cap Y_B = \{E\} \\
\text{htr}_{\mathcal{G}^{(a)}}(C) = \{E\}; & \text{htr}_{\mathcal{G}^{(a)}}(C) \cap Y_C = \emptyset \\
\text{htr}_{\mathcal{G}^{(a)}}(D) = \{B\}; & \text{htr}_{\mathcal{G}^{(a)}}(D) \cap Y_D = \{B\} \\
\text{htr}_{\mathcal{G}^{(a)}}(E) = \{B, C\}; & \text{htr}_{\mathcal{G}^{(a)}}(E) \cap Y_E = \{C\}
\end{array}$$

Thus, any ordering $\prec$ which asserts $C \prec E \prec B \prec D$ will satisfy the criterion for HTC-identifiability. Both HTC-identifiability and HTC-nonidentifiability are properties that can be checked in polynomial time using algorithms for determining maximum flow in a graph. The details are found in [11].

HTC-identifiability and HTC-nonidentifiability are not complete methods in the sense that there exist identifiable and non-identifiable models that are neither HTC-identifiable

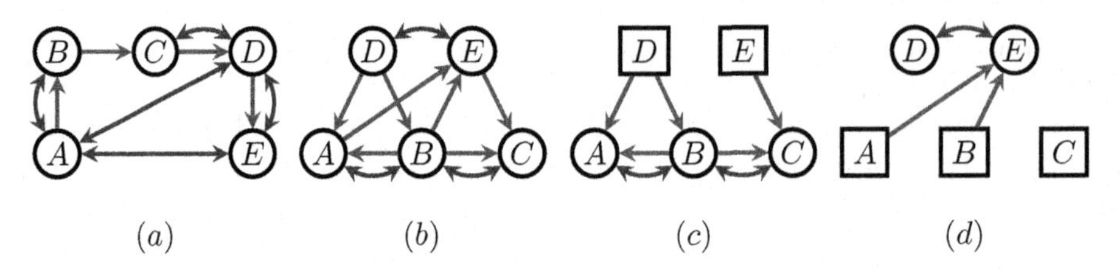

FIGURE 16.4: (a) An ADMG. (b) An ADMG representing a linear SEM that is not generically identified according to the half-trek criterion. (c) (d) CADMGs corresponding to the district factorization of the observed distribution corresponding to the linear SEM in (a), where each district factor is identified according to the half-trek criterion.

nor HTC-nonidentifiable. An interesting example occurs in Fig. 16.4 (b). The linear SEM corresponding to this graph is neither HTC-identifiable, nor HTC-nonidentifiable. However, it is possible to show that the linear SEM corresponding to this graph is identified by first applying (16.12), which yields

$$\begin{aligned} p(A,B,C,D,E) &= q_{\{A,B,C\}}(A,B,C \mid D,E)q_{\{D,E\}}(D,E \mid A,B,C) \\ &= [p(A,C \mid B,D,E)p(B \mid D)] \cdot [p(E \mid B,D)p(D)], \end{aligned}$$

with CADMGs corresponding to $q_{\{A,B,C\}}$ and $q_{\{D,E\}}$ shown in Fig. 16.4 (c) and (d), respectively.

The linear SEM corresponding to Fig. 16.4 (d) satisfies the HTC-identifiability criterion because Fig. 16.4 (d) is a simple graph (every vertex pair shares at most one edge). In any simple graph $\mathcal{G}$, for every vertex V, $\text{pa}_{\mathcal{G}}(V)$ satisfies the half-trek criterion for V and any ordering topological for $\mathcal{G}$ will then satisfy HTC-identifiability. See also proposition 1 in [11].

The linear SEM corresponding to Fig. 16.4 (c) is HTC-identifiable with

$$\mathbf{Y}_D = \mathbf{Y}_E = \emptyset; \quad \mathbf{Y}_B = \{A\}; \quad \mathbf{Y}_C = \{D,E\}; \quad \mathbf{Y}_A = \{C,D\},$$

and any ordering $\prec$ where $C \prec A$. This implies that every parameter of $q_{\{A,B,C\}}$ and $q_{\{D,E\}}$ is generically identified, and thus so is every parameter in $p(A,B,C,D,E)$ in a linear SEM for the graph in Fig. 16.4 (b). General methods that combine the half-trek criterion and district decomposition appear in [6, 5, 7].

16.5 Summary

In classical statistical inference parameter identification is often trivial, and thus not usually discussed. In causal inference problems, parameter identification is in general a subtle problem.

In causal models represented by directed acyclic graphs (DAGs) where all relevant variables are observed, counterfactual responses to interventions that set variables to constants or via a known function of other variables are identified by versions of the g-formula (16.8), (16.10). Counterfactual responses corresponding to path-specific effects are identified by the edge g-formula (16.9) if and only if no recanting witness variables exist in the DAG.

In hidden variable DAGs, counterfactual responses to interventions may no longer be identified. However, complete identification algorithms exist for counterfactual distributions used to define causal effects, conditional causal effects, controlled direct effects, and path-specific effects. These algorithms, originally given a recursive formulation in [34, 28, 27], have been given a simple formulation using the fixing operator defined in [20], and latent projection mixed graphs representing classes of hidden variable DAGs that share identification theory. In addition, very general identification results are known for counterfactual responses to variables being set according to a known policy, and in parametric causal models defined on acyclic directed mixed graphs (ADMGs) termed linear structural equation models.

Some definitions, examples and existing identification results that appear in this chapter also appear in [25].

Bibliography

[1] Chen Avin, Ilya Shpitser, and Judea Pearl. Identifiability of path-specific effects. In *Proceedings of the Nineteenth International Joint Conference on Artificial Intelligence (IJCAI-05)*, volume 19, 357–363. Morgan Kaufmann, San Francisco, 2005.

[2] Carlos Brito and Judea Pearl. Graphical condition for identification in recursive SEM. In *Proceedings of the Twenty-Second Conference on Uncertainty in Artificial Intelligence*, 47–54. AUAI Press, Arlington, VA, 2006.

[3] Carlos Brito and Judea Pearl. A graphical criterion for the identification of causal effects in linear models. In *Eighteenth National Conference on Artificial Intelligence*, 533–538. American Association for Artificial Intelligence, 2002.

[4] Bibhas Chakraborty and Erica E. M. Moodie. *Statistical Methods for Dynamic Treatment Regimes (Reinforcement Learning, Causal Inference, and Personalized Medicine)*. Springer, New York, 2013.

[5] Bryant Chen. Identification and overidentification of linear structural equation models. In *Advances in Neural Information Processing Systems*, volume 29, 1579–1587. Curran Associates, Inc., 2016.

[6] Bryant Chen, Jin Tian, and Judea Pearl. Testable implications of linear structural equation models. In *Proceedings of the Twenty-Eighth AAAI Conference on Artificial Intelligence*, 2424–2430, 2014.

[7] Mathias Drton and Luca Weihs. Generic identifiability of linear structural equation models by ancestor decomposition. *Scandinavian Journal of Statistics*, 2016.

[8] Mathias Drton, Rina Foygel, and Seth Sullivant. Global identifiability of linear structural equation models. *Annals of Statistics*, 39(2):865–886, 2011.

[9] Robin J. Evans and Thomas S. Richardson. Markovian acyclic directed mixed graphs for discrete data. *Annals of Statistics*, 1–30, 2014.

[10] Robin J. Evans and Thomas S. Richardson. Smooth, identifiable supermodels of discrete DAG models with latent variables. Unpublished, 2015.

[11] Rina Foygel, Jan Draisma, and Mathias Drton. Half-trek criterion for generic identifiability of linear structural equation models. *Annals of Statistics*, 40(3):1682–1713, 2012.

[12] Luis David Garcia-Puente, Sarah Spielvogel, and Seth Sullivant. Identifying causal effects with computer algebra. In *Proceedings of the Twenty-sixth Conference on Uncertainty in Artificial Intelligence (UAI)*. AUAI Press, 2010.

[13] Trygve Haavelmo. The statistical implications of a system of simultaneous equations. *Econometrica*, 11:1–12, 1943.

[14] Rex B. Kline. *Principles and Practice of Structural Equation Modeling*. The Guilford Press, 2005.

[15] Judea Pearl. Direct and indirect effects. In *Proceedings of the Seventeenth Conference on Uncertainty in Artificial Intelligence (UAI-01)*, 411–420. Morgan Kaufmann, San Francisco, 2001.

[16] Judea Pearl. *Causality: Models, Reasoning, and Inference*. Cambridge University Press, second edition, 2009.

[17] Judea Pearl. The causal mediation formula – a guide to the assessment of pathways and mechanisms. Technical Report R-379, Cognitive Systems Laboratory, University of California, Los Angeles, 2011.

[18] Jonas Peters, Joris M. Mooij, Dominik Janzing, and Bernhard Scholkopf. Causal discovery with continuous additive noise models. *Journal of Machine Learning Research*, 2009–2053, 2014.

[19] Thomas Richardson and Peter Spirtes. Ancestral graph Markov models. *Annals of Statistics*, 30:962–1030, 2002.

[20] Thomas S. Richardson, Robin J. Evans, James M. Robins, and Ilya Shpitser. Nested Markov properties for acyclic directed mixed graphs. Working paper, 2017.

[21] Thomas S. Richardson and Jamie M. Robins. Single world intervention graphs (SWIGs): A unification of the counterfactual and graphical approaches to causality. *preprint*: `http://www.csss.washington.edu/Papers/wp128.pdf`, 2013.

[22] James M. Robins. A new approach to causal inference in mortality studies with sustained exposure periods – application to control of the healthy worker survivor effect. *Mathematical Modeling*, 7:1393–1512, 1986.

[23] James M. Robins and Sander Greenland. Identifiability and exchangeability of direct and indirect effects. *Epidemiology*, 3:143–155, 1992.

[24] Ilya Shpitser. Counterfactual graphical models for longitudinal mediation analysis with unobserved confounding. *Cognitive Science (Rumelhart special issue)*, 37:1011–1035, 2013.

[25] Ilya Shpitser. Identification in causal models with hidden variables. *Journal of the French Statistical Society (to appear)*, 2017.

[26] Ilya Shpitser, Robin J. Evans, Thomas S. Richardson, and James M. Robins. Introduction to nested Markov models. *Behaviormetrika*, 41(1):3–39, 2014.

[27] Ilya Shpitser and Judea Pearl. Identification of conditional interventional distributions. In *Proceedings of the Twenty-Second Conference on Uncertainty in Artificial Intelligence (UAI-06)*, 437–444. AUAI Press, Corvallis, Oregon, 2006.

[28] Ilya Shpitser and Judea Pearl. Identification of joint interventional distributions in recursive semi-Markovian causal models. In *Proceedings of the Twenty-First National Conference on Artificial Intelligence (AAAI-06)*. AAAI Press, Palo Alto, 2006.

[29] Ilya Shpitser and Eric Tchetgen Tchetgen. Causal inference with a graphical hierarchy of interventions. *Annals of Statistics*, 44(6):2433–2466, 2016.

[30] Ilya Shpitser, Tyler VanderWeele, and James M. Robins. On the validity of covariate adjustment for estimating causal effects. In *Proceedings of the Twenty-Sixth Conference on Uncertainty in Artificial Intelligence (UAI-10)*, 527–536. AUAI Press, 2010.

[31] Peter Spirtes, Clark Glymour, and Richard Scheines. *Causation, Prediction, and Search*. Springer Verlag, New York, 2 edition, 2001.

[32] Jin Tian. Identifying dynamic sequential plans. In *Proceedings of the Twenty-Fourth Annual Conference on Uncertainty in Artificial Intelligence (UAI-08)*, 554–561, Corvallis, Oregon, 2008. AUAI Press.

[33] Jin Tian. Parameter identification in a class of linear structural equation models. In *Proceedings of the Twenty-First International Joint Conference on Artificial Intelligence (IJCAI)*, 1970–1975. AAAI Press, Palo Alto, CA, 2009.

[34] Jin Tian and Judea Pearl. On the testable implications of causal models with hidden variables. In *Proceedings of the Eighteenth Conference on Uncertainty in Artificial Intelligence (UAI-02)*, volume 18, 519–527. AUAI Press, Corvallis, Oregon, 2002.

[35] Sewall Wright. Correlation and causation. *Journal of Agricultural Research*, 20:557–585, 1921.

17

Mediation Analysis

Johan Steen

Department of Intensive Care, Ghent University Hospital

Stijn Vansteelandt

Department of Applied Mathematics, Computer Science and Statistics, Ghent University

CONTENTS

17.1 Introduction

In many applications across a wide range of scientific disciplines, scholars aim to understand the mechanisms behind established cause-effect relationships, as witnessed by the widespread usage of mediation analyses. Such understanding may not only be of pure scientific or etiologic interest, but may also aid policymakers in making informed decisions about public health interventions or reforms.

The Job Search Intervention Study (JOBS II), for instance, was designed to assess the effectiveness of a job training intervention to facilitate re-employment and reduce depressive symptoms in unemployed job seekers [47, 48]. 1,249 randomly assigned job seekers were invited to participate in several sessions of job search skills workshops (the treatment group), whereas the remaining 552 unemployed workers received a booklet with job search tips (the control group). Vinokur and Schul [48] hypothesized that the treatment group would benefit from the workshops, assuming workshop attendance improves one's sense of self-efficacy and increases chances of getting re-employed, which, in turn, leads to a reduction in depressive symptoms. Researchers thus believed an enhanced sense of mastery and re-employment to be active ingredients of the intervention's beneficial effect on mental health.

More generally, interventions or exposures essentially always realize their effects via a combination of causal chains or mechanisms. Mediation analysis seeks to unravel and to quantify specific bundles of these pathways. To fix ideas, consider the causal DAG in Figure 17.1, which may represent hypothesized causal mechanisms underlying the effect of the job search intervention A on mental health Y. Suppose that workshop attendance increases participants' sense of self-efficacy L, which may, in turn, exert beneficial effects on mental health, either by increasing chances of getting re-employed M (along pathway $A \to L \to M \to Y$) or by other subsequent (unspecified) mechanisms (along pathway $A \to L \to Y$). The intervention may also positively affect re-employment through other mechanisms before finally exerting its effect on mental health (along pathway $A \to M \to Y$), or it may reduce depressive symptoms through none of the putative mediators (along pathway $A \to Y$). Mediation analysis then aims to answer questions such as *"How much of the intervention's effect on mental health is mediated by increased chances of re-employment?"* It does so by disentangling the indirect effect that captures all pathways along which re-employment status M, the mediator of interest, transmits the intervention effect ($A \to M \to Y$ and $A \to L \to M \to Y$) from the direct effect that captures all remaining pathways ($A \to Y$ and $A \to L \to Y$). More generally, it aims to assess what effect the exposure realizes along one or multiple pathways. We will informally refer to this as a *path-specific effect* and give a precise definition later.

Bias-free estimation of path-specific effects crucially relies on certain structural assumptions and may often be compromised due to the subtle interplay between causal mechanisms. In this chapter, we therefore aim to further elaborate on *identifiability* of path-specific effects — that is, whether or not a certain set of causal assumptions suffices (or may even be deemed

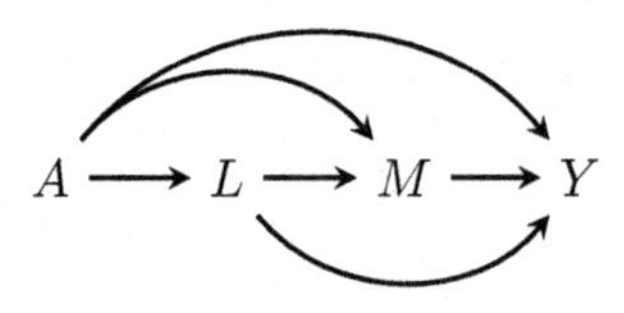

FIGURE 17.1: The treatment effect conceived as a combination of the effects along multiple causal chains.

necessary) to identify such components from observed data. In contrast to Chapter 16, we wish to provide a more comprehensive overview of the literature on causal mediation analysis. We first briefly review definitions of path-specific effects, in particular of natural direct and indirect effects [18, 24], which are the standard targets of inference in causal mediation analysis. Through various worked-out examples we next aim to develop intuition into non-parametric identification[1] of this class of path-specific effects and, in particular, the technical nature of certain assumptions on which mediation analysis relies.

17.2 Definitions and Notation

To enable clear and formal definitions of the target causal estimands, let A denote the exposure or treatment (e.g. workshop participation) and Y the outcome of interest (e.g. presence of depressive symptoms). Throughout, we will use counterfactual notation where, for instance, $Y(a)$ denotes the value of the outcome that would have been observed had A (possibly contrary to the fact) been set to level a. The (population-)*average causal effect* can then be defined as $E\{Y(a)-Y(a')\}$, where a and a' correspond to meaningful levels of treatment. This is essentially identical to the interventional contrast $E(Y|\text{do}(a)) - E(Y|\text{do}(a'))$ in terms of Pearl's do-operator. For expositional simplicity, we will restrict our presentation to binary treatments (with $a = 1$ and $a' = 0$), although definitions and results extend to multicategorical or continuous treatments. For instance, in our motivating example, $E\{Y(1) - Y(0)\}$ expresses the difference in prevalence of depressive symptoms if all unemployed workers were invited to participate in the job search skills workshop versus all received a booklet with job search tips.

17.2.1 Natural direct and indirect effects

Robins and Greenland [24] laid the foundations for effect decomposition by invoking *nested counterfactuals* to conceptualize the intuitive notion of changing treatment assignment along specific pathways but not others. For instance, the nested counterfactual $Y(a, M(a'))$ denotes the outcome that would have been observed had (possibly contrary to the fact) A been set to level a and M to $M(a')$, the mediator value that had been observed had A been set to a'. Consequently, nested counterfactual expressions enable us to isolate and quantify part of the intervention effect that is transmitted through the mediator M by leaving treatment unchanged at $A = 1$, but changing the counterfactual intermediate outcome $M(1)$ to $M(0)$. This then leads to the definition of the so-called *natural indirect effect*

$$E\{Y(1) - Y(1, M(0))\} = E\{Y(1, M(1)) - Y(1, M(0))\}.$$

Its complement, the *natural direct effect*

$$\begin{aligned} &E\{Y(1) - Y(0)\} - E\{Y(1) - Y(1, M(0))\} \\ = &E\{Y(1, M(0)) - Y(0)\} = E\{Y(1, M(0)) - Y(0, M(0))\} \end{aligned}$$

captures the notion of blocking the intervention's effect on the mediator by keeping the latter fixed at whatever value it would have attained under no intervention.

In our motivating example, the natural direct effect expresses by how much the prevalence of depressive symptoms would change if all unemployed workers' treatment assignment

[1]For brevity, we will loosely use terms such as 'identification', 'identify', 'recover from observed data' to refer to *non-parametric* identification.

status were to be changed, but their employment status were to be fixed to whatever status would be observed if they had originally received a booklet with job search tips. In contrast, the natural indirect effect expresses the change in prevalence of depressive symptoms if all unemployed workers were to be invited to participate in the workshop, but their employment status were changed to whatever status would be observed if they had received a booklet with job search tips.

The main appeal of effect definitions that utilize nested counterfactuals, as opposed to equivalent formulations in the linear structural equation modeling tradition, is that they are model-free. That is, they combine to produce the total effect, irrespective of the scale of interest or presence of interactions or nonlinearities, under the composition assumption that $Y(a, M(a)) = Y(a)$. For instance, although the above effects are expressed in terms of mean (or risk) differences, the causal risk ratio of a binary outcome can similarly be expressed as the product of the natural direct effect risk ratio and the natural indirect effect risk ratio

$$\frac{E\{Y(1)\}}{E\{Y(0)\}} = \frac{E\{Y(1, M(0))\}}{E\{Y(0, M(0))\}} \frac{E\{Y(1, M(1))\}}{E\{Y(1, M(0))\}}.$$

Consequently, mean nested counterfactuals can be parameterized using a class of marginal structural models [25] for mediation analysis, so-called *natural effect models* [10, 11, 12, 34, 45], for instance

$$E\{Y(a, M(a'))\} = g^{-1}(\beta_0 + \beta_1 a + \beta_2 a' + \beta_3 aa'), \tag{17.1}$$

for all a and a' and where $g(\cdot)$ is a known link function. If $g(\cdot)$ is chosen to be the identity link, β_1 captures the natural direct effect and $\beta_2 + \beta_3$ captures the natural indirect effect on an additive scale. Similarly, effects can be expressed on multiplicative scales, such as risk or odds ratios, by choosing $g(\cdot)$ to represent the log or logit link function. Robins and Greenland [24] originally termed these parameters the *pure direct effect* and *total indirect effect*, respectively. By differently apportioning the interaction term β_3, an alternative decomposition is obtained in terms of the *total direct effect* $E\{Y(1, M(1)) - Y(0, M(1))\}$, captured by $\beta_1 + \beta_3$, and the *pure indirect effect* $E\{Y(0, M(1)) - Y(0, M(0))\}$, captured by β_2. In accordance with VanderWeele [40], any of these two decompositions can thus be further refined, leading to the same unique three-way decomposition into the pure direct effect β_1, the pure indirect effect β_2, and a mediated interactive effect β_3, which can be interpreted to capture the extent to which direct and indirect pathways interact in their effect on the outcome.

Pearl [18] later adopted the same definitions but named these parameters *natural* (rather than pure) direct and indirect effects to emphasize that pure direct effects, as opposed to *controlled direct effects* $E\{Y(1, m) - Y(0, m)\}$, allow for *natural* variation in the mediator. That is, natural direct effects reflect the effect of treatment upon fixing the mediator at values that would, for each individual, have *naturally* occurred under no treatment, rather than at some predetermined level m (uniformly across the population). In the remainder of this chapter, we will adopt Pearl's terminology of natural effects.

17.2.2 Path-specific effects

In graphical terms, the natural indirect effect quantifies the contribution along all pathways through which a single mediator transmits the treatment's effect on the outcome. Its counterpart, the natural direct effect, quantifies the contribution along all remaining pathways from treatment to outcome. Both of their counterfactual definitions refer to specific instances of nested counterfactuals of the form $Y(a, M(a'))$, with a possibly different from a'. Contributions along other predefined sets of directed paths π from treatment A

to outcome Y can similarly be defined in terms of contrasts of path-specific nested counterfactuals, which we will denote $Y(\pi, a, a')$, in accordance with notation in the previous Chapter 16.[2] As for natural effects, these π-specific counterfactuals represent two (possibly incompatible) hypothetical interventions which, for instance, set A to a for the purpose of all directed paths in π, or to a' for the purpose of directed paths not in π. For notational convenience, we denote $\overline{\pi}$ to be the set of directed pathways from A to Y not in π.

Suppose that, in our motivating example, interest lies in the effect of the job search intervention mediated by re-employment (M in Figure 17.1) but not by possible prior changes in perceived self-efficacy (L in Figure 17.1), as captured by $\pi = \{A \to M \to Y\}$. This π-specific effect has been referred to as the *partial* [6] or *semi-natural* [19] indirect effect with respect to M. Its corresponding π-specific nested counterfactual

$$Y(\{A \to M \to Y\}, a, a') = Y(a', L(a'), M(a, L(a')))$$

can be obtained by recursive substitution, as discussed more formally in Chapter 16, Equation (16.5). Just as counterfactuals of the form $Y(a, M(a'))$ give rise to definitions for natural effects, recursively nested counterfactuals of the above form enable us to define, for instance, the *pure* π-specific effect as

$$E\{Y(0, L(0), M(1, L(0))) - Y(0)\}$$

and the *total* path-specific effect along pathways not in π (or in $\overline{\pi}$) as

$$E\{Y(1) - Y(0, L(0), M(1, L(0)))\}$$

By symmetry and the composition assumption, these components again combine to produce the total effect of treatment. The natural effect model

$$E\{Y(a, L(a), M(a', L(a)))\} = \gamma_0 + \gamma_1 a + \gamma_2 a' + \gamma_3 aa',$$

for all a and a', maps these path-specific effects to γ_1 and $\gamma_2 + \gamma_3$, respectively. Natural effect models that parameterize more fine-grained decompositions of the total causal effect into $k > 2$ path-specific effects (along $k - 1$ ordered mediators) have been discussed in [34].

17.3 Cross-World Quantities Call for Cross-World Assumptions

Despite the formal and intuitive appeal of path-specific effects, their non-parametric identification is subtle and a source of much controversy. The reason is that the usual consistency assumptions alone — for instance, that $M(a) = M$ when $A = a$ and that $Y(a, m) = Y$ when $A = a$ and $M = m$ — do not suffice to link all counterfactuals to observed data. In particular, nested counterfactual outcomes such as $Y(a, M(a'))$ are unobservable whenever $anea'$. Data, whether experimental or observational, thus never carry information about the distribution of these counterfactuals as they imply a union of two incompatible states a and a' that may only seem to coexist 'across multiple worlds'. Because of their 'cross-world' nature, path-specific effects cannot in general be expressed in terms of interventional contrasts, which typically refer to ideal interventions in a single hypothetical world. Mediation analyses based on natural or path-specific effects are thus bound to rely on assumptions that cannot be empirically verified or guaranteed by any study design [9, 24, 26].

[2]For expositional simplicity, we will restrict settings to those with A and Y being singletons.

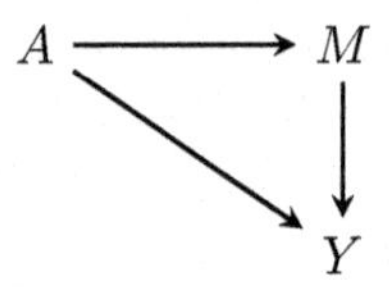

FIGURE 17.2: A causal DAG representing a simple yet overly simplistic mediation setting.

To gradually develop intuition into non-parametric identification of natural and path-specific effects we will work through a number of simple but typical illustrative examples, spanning the next few sections of this chapter. Unless stated otherwise, we shall assume throughout that treatment A is randomized, in order to exclusively focus on assumptions characteristic to mediation analysis. In this section, we particularly highlight that the distinct nature of nested counterfactuals calls for a type of assumption that cannot be verified empirically but that is, nonetheless, naturally encoded in so-called non-parametric structural equation models (NPSEMs).

17.3.1 Imposing cross-world independence

Identification of natural effects in the causal DAG $\mathcal{G}(\mathbf{V})$ with $\mathbf{V} = \{A, M, Y\}$, in Figure 17.2, can be obtained if we recover the distribution $p(Y(a, M(a')) = y)$ of nested counterfactuals. This requires summing (or integrating) the joint counterfactual distribution $p(Y(a,m) = y, M(a') = m)$ over m. When $anea'$, observed data carry no information about the dependence of $Y(a,m)$ on $M(a')$. This articulates why natural effects cannot, in general, be identified from experimental data without further, untestable assumptions. One such assumption is that of *cross-world independence*

$$Y(a,m) \perp\!\!\!\perp M(a'). \tag{i}$$

Under this assumption, we can factorize $p(Y(a,m) = y, M(a') = m)$ as a product of interventional distributions, each of which is identified from observed data under the assumptions encoded in $\mathcal{G}(\mathbf{V})$, as follows

$$\begin{aligned} p(Y(a, M(a')) = y) &= \sum_m p(Y(a,m) = y, M(a') = m) \\ &= \sum_m p(Y(a,m) = y) p(M(a') = m) = \sum_m p(y|a,m)p(m|a'), \end{aligned}$$

where for arbitrary variables V and W, $p(v|w)$ is shorthand notation for $p(V = v|W = w)$.[3] Pearl [18] claimed cross-world assumption (i) to be key to 'experimental' identification of natural effects. With this, he indicated that, if interventional distributions $p(Y(a,m) = y)$ and $p(M(a') = m)$ were known from previous randomized interventions $\text{do}(a,m)$ and $\text{do}(a')$, this assumption could be considered the missing link required to piece together these distributions in order to recover $p(Y(a, M(a')) = y)$.

[3] In this chapter, we will mostly use counterfactual notation instead of Pearl's do-notation, especially when cross-world counterfactuals cannot be expressed using do-notation. However, we will refer to counterfactual distributions as interventional distributions, whenever applicable.

17.3.2 Cross-world independence and NPSEMs

Subtleties surrounding cross-world assumptions such as (i) have long been obscured to practitioners because of reliance on stringent parametric constraints or, more recently, on representations of causal DAGs as NPSEMs. In fact, as discussed in more detail in Chapters 15 and 16, NPSEMs impose restrictions on the joint distribution of all counterfactual outcomes, including those that inhabit different worlds, that is, worlds under conflicting hypothetical interventions or treatment assignments such as do(a) and do(a'). As a result, cross-world independencies are naturally encoded by the set of (recursive) structural equations that defines a particular NPSEM. For instance, the NPSEM representation of $\mathcal{G}(\mathbf{V})$ in Figure 17.2 is characterized by the following set of structural equations:

$$\begin{aligned} A &:= f_A(\epsilon_A) \\ M &:= f_M(A, \epsilon_M) \\ Y &:= f_Y(A, M, \epsilon_Y) \end{aligned}$$

where f_A, f_M and f_Y are unknown deterministic functions and ϵ_A, ϵ_M and ϵ_Y are mutually independent random error terms (representing unobserved background variables). The assumed invariance of these equations endows them with a causal interpretation and permits us to deduce the counterfactual independencies they encode. For example, under the interventions do(a, m) and do(a'), the structural equations can respectively be written as

$$\begin{aligned} A &:= a & \qquad A &:= a' \\ M(a) &:= m & \qquad M(a') &:= f_M(a', \epsilon_M) \\ Y(a,m) &:= f_Y(a, m, \epsilon_Y) & \qquad Y(a') &:= f_Y(a', M(a'), \epsilon_Y) \end{aligned}$$

Under this representation, the joint distribution of the one-step ahead counterfactuals

$$V(\mathbf{x}_{\text{pa}_{\mathcal{G}}(V)}) := f_V(\mathbf{x}_{\text{pa}_{\mathcal{G}}(V)}, \epsilon_V),$$

where $\text{pa}_{\mathcal{G}}(V)$ denotes the set of parents of V in $\mathcal{G}(\mathbf{V})$ and $\mathbf{x}_{\text{pa}_{\mathcal{G}}(V)}$ the set of values to which these parents are set via the intervention do($\mathbf{x}$), is fully determined by the mutually independent error terms ϵ_V. It thus follows that all such one-step ahead counterfactuals are also mutually independent, irrespective of the choice of hypothetical values $\mathbf{x}_{\text{pa}_{\mathcal{G}}(V)}$ to which we set the parents of V. As a result, independence of the error terms $\epsilon_M \perp\!\!\!\perp \epsilon_Y$ in the above structural equations not only translates into $Y(a,m) \perp\!\!\!\perp M(a)$ but also into cross-world independence (i). This may sound reassuring, but also signals the restrictiveness of NPSEMs, as they inherently seem to encode independence assumptions that can never be verified from randomized interventions.

17.3.3 Single world versus multiple worlds models

Robins and Richardson [26] extensively discuss these restrictions encoded by NPSEMs. They moreover contrast the latter with another class of graphical causal models, Robins' [23] *Finest Fully Randomized Causally Interpretable Structured Tree Graph Model* (FFRCISTGM) representation of causal DAGs, which only enforces restrictions that are (in principle) empirically verifiable. Because this less restrictive class of models only imposes independence restrictions on sets of counterfactuals under a single set of (non-conflicting) interventions such models have been referred to as 'single world models', as opposed to NPSEMs which were termed 'multiple worlds models'. A more formal treatment of NPSEMs, FFRCISTGs and *Single World Intervention Graphs* (SWIGs) [22], which encode counterfactual independencies implied by a 'single world model', is given in Chapters 15 and 16.

17.3.4 Further outline

Because NPSEMs naturally encode cross-world independence assumptions, they have provided a framework for the recent development of a fairly intuitive graphical rule that governs whether and how nested 'cross-world' counterfactual quantities relate to observed variables [1], even in the presence of unobserved or hidden variables [27]. In Section 17.5, we demonstrate that specific cross-world independence assumptions can indeed be relatively easily interrogated from a (hidden variable) causal DAG interpreted as an NPSEM by this graphical rule. As it turns out, this type of assumption forms the extra necessary layer on top of a set of assumptions that *is* subject to experimental verification and serves to identify total treatment effects. When combined with complete identification algorithms for total treatment effects [5, 29, 38], the proposed graphical criterion therefore not only delineates sufficient, but also necessary conditions for identification of path-specific effects.

Essentially, when it comes to natural direct and indirect effects, this sound and complete criterion indicates that identification can be obtained under NPSEMs with the aid of two different types of auxiliary variables, provided that no mediator-outcome confounders are themselves affected by treatment (and the total treatment effect is identifiable). As the connection with earlier sufficient identification conditions for natural effects [18] seems to be somewhat missing from the literature, we choose to review the main (two) graphical identification algorithms in chronological order (in Sections 17.4 and 17.5, respectively) and to revisit earlier assumptions in the light of this recent graphical criterion (in Section 17.6). In doing so, we point out that certain identification strategies have long been concealed because of the initial and exclusive focus on a single type of auxiliary variable that recovers identification by establishing a conditional version of cross-world assumption (i). Finally, in Section 17.7, we provide insights that may help to put a longstanding conceptual discussion regarding the very nature of mediation analysis into perspective.

17.4 Identification 1.0

In this section, we further extend the simple causal DAG in Figure 17.2 to illustrate the logic and reasoning behind sufficient conditions for identification of natural effects.

17.4.1 Unmeasured mediator-outcome confounding

In most settings, the assumptions encoded by Figure 17.2 are unrealistic. Indeed, even if treatment were randomized, as represented by the absence of back-door paths into A, we cannot generally assume the absence of confounding of the mediator-outcome relation (other than by A) because typically M is not randomized. Nonetheless, independence of the error terms $\epsilon_M \perp\!\!\!\perp \epsilon_Y$, as encoded in the NPSEM representation of Figure 17.2, critically hinges on the assumption of *no unmeasured mediator-outcome confounding.* Because the latter assumption can be considered unlikely, the assumption of independent error terms is therefore almost guaranteed to be violated. In this subsection, we will therefore relax assumptions by adding a hidden node U that captures unmeasured confounding of the mediator-outcome relation (and induces dependence between their respective error terms when structural equations are expressed only in terms of observed variables $\mathbf{V}$), as in Figure 17.3A. More generally, we will represent unobserved variables $\mathbf{H}$ on hidden variable causal DAGs $\mathcal{G}(\mathbf{V} \cup \mathbf{H})$ by circled nodes.

By treatment randomization we have that $U \perp\!\!\!\perp A$ such that, not surprisingly, the *g-formula* [23] yields

$$p(Y(a) = y) = \sum_{u,m} p(y|a, m, u)p(m|a, u)p(u) = \sum_{m} p(y|a, m)p(m|a) = p(y|a). \quad (17.2)$$

Unfortunately, U cannot similarly be integrated out 'across worlds', which prevents us from identifying $p(Y(a, M(a')) = y)$, even if conditional cross-world independence $Y(a, m) \perp\!\!\!\perp M(a')|U$ were to hold. Indeed, we obtain

$$\begin{aligned} p(Y(a, M(a')) = y) &= \sum_{u,m} p(Y(a, m) = y|u)p(M(a') = m|u)p(u) \\ &= \sum_{u,m} p(y|a, m, u)p(m|a', u)p(u), \end{aligned} \quad (17.3)$$

an expression that cannot further be reduced to a functional of observed variables (such as Equation 17.2) because of the conflicting treatment assignments in its first two factors.

17.4.2 Adjusting for mediator-outcome confounding

Issues of non-identifiability of $p(Y(a, M(a')) = y)$ may, however, be remedied when one has available a measured set of prognostic covariates $\mathbf{C} \subseteq \mathbf{V} \setminus \{A, M, Y\}$ for mediator and/or outcome that renders the mediator-outcome relationship unconfounded given treatment assignment. This can be understood because the availability of such a set $\mathbf{C}$, as, for instance, in the simplified example DAGs in Figures 17.3B and 17.3C where $\mathbf{C} = \{C\}$, no longer necessitates stratifying on hidden variables such as U to establish cross-world independence.

For example, in Figure 17.3B, conditioning on $\mathbf{C}$ suffices, since the structural equations

$$\begin{aligned} M(a') &:= f_M(a', U, \epsilon_M) \\ Y(a, m) &:= f_Y(a, m, \mathbf{C}, \epsilon_Y), \end{aligned}$$

indicate that cross-world independence holds within strata of $\mathbf{C}$, that is

$$Y(a, m) \perp\!\!\!\perp M(a')|\mathbf{C}. \quad \text{(ii)}$$

This then implies the same functional as expression (17.3) but with unobserved U replaced by the observed adjustment set $\mathbf{C}$

$$p(Y(a, M(a')) = y) = \sum_{\mathbf{c},m} p(Y(a, m) = y|\mathbf{c})p(M(a') = m|\mathbf{c})p(\mathbf{c}) \quad (17.4)$$

$$= \sum_{\mathbf{c},m} p(y|a, m, \mathbf{c})p(m|a', \mathbf{c})p(\mathbf{c}). \quad (17.5)$$

This functional is commonly referred to as Pearl's [18] *mediation formula.*

To appreciate the importance of adjustment for prognostic factors $\mathbf{C}$, reconsider our motivating example. Randomization of the intervention in itself did not suffice to eliminate potential confounding between re-employment M and the outcome. It is therefore essential to adjust for the pretreatment level of depression, a strong prognostic factor of the outcome of interest and most likely also related to re-employment. Measurements on a range of other baseline covariates, including demographics, previous occupation and financial strain, were also collected and adjusted for to strengthen the validity of cross-world assumption (ii).

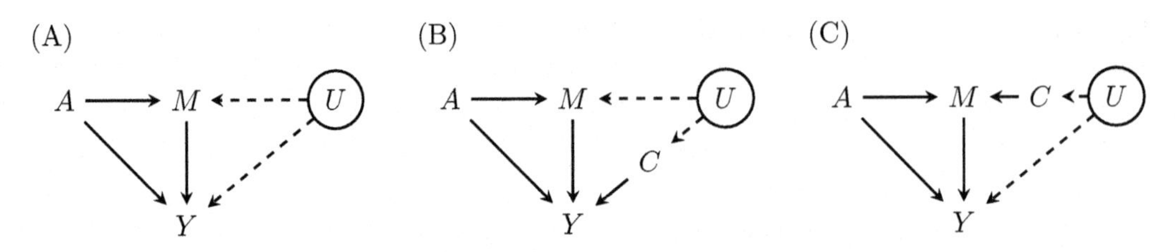

FIGURE 17.3: Causal DAGs that reflect more realistic mediation settings with unmeasured mediator-outcome confounding (A) along with two scenarios where a measured covariate C may deconfound the mediator-outcome relation (B,C).

17.4.3 Treatment-induced mediator-outcome confounding

The previous example may have led the reader to erroneously conclude that, given treatment randomization, adjustment for a measured covariate set $\mathbf{C}$ that deconfounds the mediator-outcome relation within treatment arms suffices to establish cross-world independence (ii) under NPSEMs, thus enabling identification of $p(Y(a, M(a')) = y)$. An important additional requirement is that no prognostic factor $L \in \mathbf{C}$ is affected by treatment.

Intuitively, if L were a common cause of both M and Y, as in the causal DAG $\mathcal{G}(\mathbf{V})$ with $\mathbf{V} = \{A, L, M, Y\}$, in Figure 17.4A, adjustment for L (as in Equation 17.5 with $\mathbf{C} = \{L\}$) would block the pathway $A \to L \to M \to Y$, which makes up part of the natural indirect effect of interest. Lack of identification can formally be understood as follows. According to the NPSEM associated with Figure 17.4A, $Y(a, m) \perp\!\!\!\perp M(a')$ holds conditional on $\{L(a) = l, L(a') = l'\}$ since

$$M(a') := f_M(a', L(a'), \epsilon_M)$$
$$Y(a, m) := f_Y(a, m, L(a), \epsilon_Y).$$

Under the remaining assumptions encoded by this NPSEM, this allows us to express $p(Y(a, M(a')) = y)$ as

$$\sum_{l,l',m} p(Y(a,m) = y|L(a) = l, L(a') = l')p(M(a') = m|L(a) = l, L(a') = l') \times p(L(a) = l, L(a') = l')$$
$$= \sum_{l,l',m} p(y|a, l, m)p(m|a', l')p(L(a) = l, L(a') = l').$$

As in the example of the previous section, this expression cannot further be reduced to a functional of the observed data as it requires the joint cross-world counterfactual distribution $p(L(a) = l, L(a') = l')$. Because this distribution again involves conflicting treatment assignments, strong untestable restrictions (beyond those encoded in NPSEMs) would be needed to enable identification.

In our motivating example, all available covariates were measured prior to randomization. It may thus be safely assumed that none of them was affected by the intervention. However, other mediators of the intervention's effect on mental health, such as an altered sense of self-efficacy, may well have affected re-employment and thus manifest themselves as mediator-outcome confounders that are affected by the intervention. In that case, cross-world independence (ii) is likely violated.

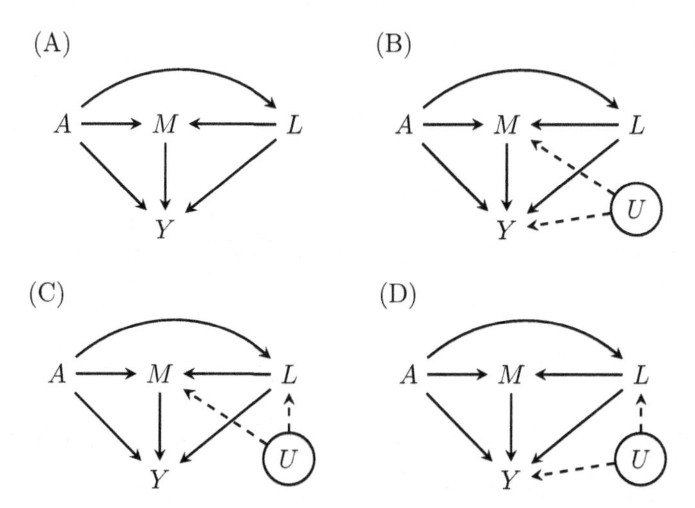

FIGURE 17.4: Causal DAGs that reflect mediation settings with treatment-induced confounding by L.

17.4.4 Pearl's graphical criteria for conditional cross-world independence

Pearl [18] devised two graphical criteria for assessing cross-world independence (ii) under an NPSEM associated with a certain hidden variable causal DAG $\mathcal{G}(\mathbf{V} \cup \mathbf{H})$. The logic for these criteria can be understood from the previous two examples in Sections 17.4.1 and 17.4.3.

The first criterion requires the availability of an adjustment set $\mathbf{C}$ that is sufficient, along with treatment A, to adjust for confounding of the mediator-outcome relation. Such covariate set $\mathbf{C}$ should block all back-door paths between M and Y (except those traversing A) in the sense that

$$(Y \perp\!\!\!\perp M|\mathbf{C})_{\mathcal{G}(\mathbf{V}\cup\mathbf{H})_{\underline{AM}}}. \tag{ii.a}$$

That is, $\mathbf{C}$ d-separates Y from M in $\mathcal{G}(\mathbf{V} \cup \mathbf{H})_{\underline{AM}}$, the subgraph constructed from the original graph $\mathcal{G}(\mathbf{V} \cup \mathbf{H})$ by deleting all arrows emanating from A and M.

The second criterion requires that

$$\text{no element of } \mathbf{C} \text{ is affected by treatment.} \tag{ii.b}$$

We will henceforth refer to this criterion as 'no treatment-induced confounding' or 'no intermediate confounding'.

In the next two subsections, we review sufficient conditions for identifying natural direct and indirect effects from i) experimental data from studies where treatment is randomized or from ii) observational data. In doing so, we highlight that identification from purely observational data typically requires additional assumptions, which (in contrast to cross-world assumption (ii)) are empirically falsifiable. Following Pearl [19], we compare different formulations of these additional assumptions in terms of their identification power.

17.4.5 Sufficient conditions to recover natural effects from experimental data

Equation 17.4 illustrates that cross-world independence (ii) enables expressing the cross-world counterfactual distribution $p(Y(a, M(a')) = y)$ in terms of 'single world' interventional distributions $p(M(a') = m|\mathbf{c})$ and $p(Y(a, m) = y|\mathbf{c})$. It is easily demonstrated that these interventional distributions are identified under an NPSEM if treatment is randomized and cross-world independence (ii) holds. In other words, when combined with the ignorability condition that represents treatment randomization

$$\{Y(a,m), M(a), \mathbf{C}\} \perp\!\!\!\perp A, \tag{iii}$$

conditional cross-world independence (ii) enables identification of $p(Y(a, M(a')) = y)$ under an NPSEM from data obtained from a single randomized intervention do(a) [7]. The latter implication could be considered an extension of Pearl's [18] 'experimental' identification, which formulates that marginal cross-world independence (i) suffices to recover $p(Y(a, M(a')) = y)$ from two sequentially randomized interventions do(a, m) and do(a') (see Section 17.3.1; also see [9]). This extension basically illustrates that, under a single randomized intervention do(a), we need a measured set of baseline covariates $\mathbf{C}$ such that identification can be obtained under conditional cross-world independence (ii), without reliance on additional ignorability assumptions.

17.4.6 Sufficient conditions to recover natural effects from observational data

As opposed to randomized trials, assumption (ii) is not sufficient for identifying natural effects from purely observational data. This is because recovering interventional distributions $p(M(a') = m|\mathbf{c})$ and $p(Y(a,m) = y|\mathbf{c})$ from observational data requires additional assumptions.

17.4.6.1 The adjustment criterion for natural effects

The availability of an adjustment set $\mathbf{C}$ that simultaneously satisfies assumption (ii) and the following conditional ignorability assumption

$$\{Y(a,m), M(a)\} \perp\!\!\!\perp A|\mathbf{C} \tag{iv}$$

restores identifiability of $p(Y(a, M(a')) = y)$ under an NPSEM from observational data by the mediation formula (Equation 17.5) [8, 32]. Moreover, because, under NPSEMs, assumptions (ii) and (iv) are exchangeable (as a set) with the following set of conditional ignorability assumptions[4] [32]

$$M(a) \perp\!\!\!\perp A|\mathbf{C} \tag{v.a}$$
$$Y(a,m) \perp\!\!\!\perp \{A, M\}|\mathbf{C}, \tag{v.b}$$

it follows that the search for such a sufficient adjustment set $\mathbf{C}$ may as well be restricted to covariate sets that simultaneously identify $p(M(a) = m)$ and $p(Y(a,m) = y)$ by the *back-door* [17] or *adjustment formula* [33]. This led Shpitser and VanderWeele [32] to develop a complete graphical criterion for identification of $p(Y(a, M(a')) = y)$ by the mediation formula (under NPSEMs). They instead termed this the *adjustment formula for natural direct and indirect effects* as it generalizes the adjustment criterion for total effects [33] to mediation settings.

Intuitively, this criterion can be thought of as aiming to establish both cross-world independence (ii) and conditions (v.a) and (v.b) solely by means of adjustment for a common measured covariate set $\mathbf{C}$. First, it demands no unmeasured mediator-outcome confounding (as in Figure 17.3A), which, if not met, violates cross-world independence (ii) and,

[4]Assumption (v.b) encodes cross-world independence, be it in a more subtle way. That is, by the consistency assumption, it implies $Y(a,m) \perp\!\!\!\perp M(a')|\{A = a', \mathbf{C}\}$, which is inherently cross-world counterfactual.

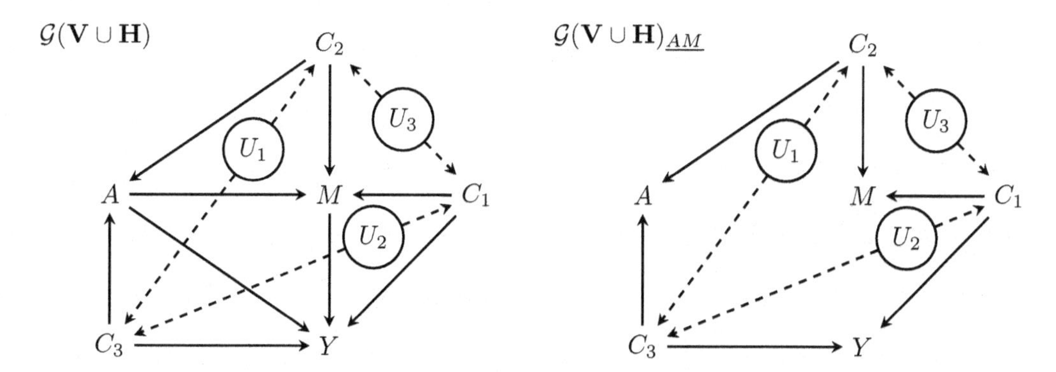

FIGURE 17.5: Hidden variable causal DAG $\mathcal{G}(\mathbf{V} \cup \mathbf{H})$ which permits identification of $p(Y(a, M(a')) = y)$ by the adjustment criterion (under its NPSEM representation). The subgraph $\mathcal{G}(\mathbf{V} \cup \mathbf{H})_{\underline{AM}}$ aids in selecting a candidate covariate set that satisfies cross-world independence (ii) by Pearl's graphical criteria described in Section 17.4.4.

moreover, hampers identification of $p(Y(a, m) = y)$ by the adjustment formula. Second, it demands the absence of treatment-induced mediator-outcome confounders (such as L in Figure 17.4A), because their presence both violates cross-world independence (ii) and hinders the availability of a common set $\mathbf{C}$ that enables identification of both $p(M(a) = m)$ and $p(Y(a, m) = y)$ by the adjustment formula. Crucially, establishing cross-world independence (ii) and conditions (v.a) and (v.b) by means of this generalized adjustment criterion goes hand in hand.

A relatively simple example

Consider the causal DAG $\mathcal{G}(\mathbf{V} \cup \mathbf{H})$ with observed variables $\mathbf{V} = \{C_1, C_2, C_3, A, M, Y\}$ and hidden variables $\mathbf{H} = \{U_1, U_2, U_3\}$ in Figure 17.5 (adopted from [19]; Figure 3B). The search for a candidate covariate set $\mathbf{C}$ that satisfies assumption (ii) may be guided by graphical criteria (ii.a) and (ii.b), as discussed in Section 17.4.4. Because $\mathcal{G}(\mathbf{V} \cup \mathbf{H})$ includes no intermediate confounders, it suffices to search for a set of baseline covariates that d-separates M from Y in the subgraph $\mathcal{G}(\mathbf{V} \cup \mathbf{H})_{\underline{AM}}$. These candidate adjustment sets include $\{C_1, C_2\}$, $\{C_1, C_3\}$ and $\{C_1, C_2, C_3\}$. Assessing whether $p(Y(a, M(a')) = y)$ can be recovered from observed data by the mediation formula now boils down to verifying whether both $p(M(a) = m)$ and $p(Y(a, m) = y)$ are identified by the adjustment formula upon adjustment for one of these candidate sets. It turns out that only the set $\{C_1, C_2, C_3\}$ identifies both of these interventional distributions by the adjustment formula, such that under the NPSEM representation of $\mathcal{G}(\mathbf{V} \cup \mathbf{H})$, we obtain

$$p(Y(a, M(a')) = y) = \sum_{c_1, c_2, c_3, m} p(y|a, m, c_1, c_2, c_3) p(m|a', c_1, c_2, c_3) p(c_1, c_2, c_3).$$

In fact, under NPSEMs, any covariate set $\mathbf{C}$ that suffices to identify both $p(M(a) = m)$ and $p(Y(a, m) = y)$ by the adjustment formula, will also satisfy cross-world independence (ii), such that the initial step can simply be skipped [32].

17.4.6.2 Identification beyond the adjustment criterion

A major appeal of identification via the adjustment criterion for natural effects is that it leads to a standard identifying functional. This, in turn, allows for general modeling and estimation strategies. However, as the following examples illustrate, it may unnecessarily

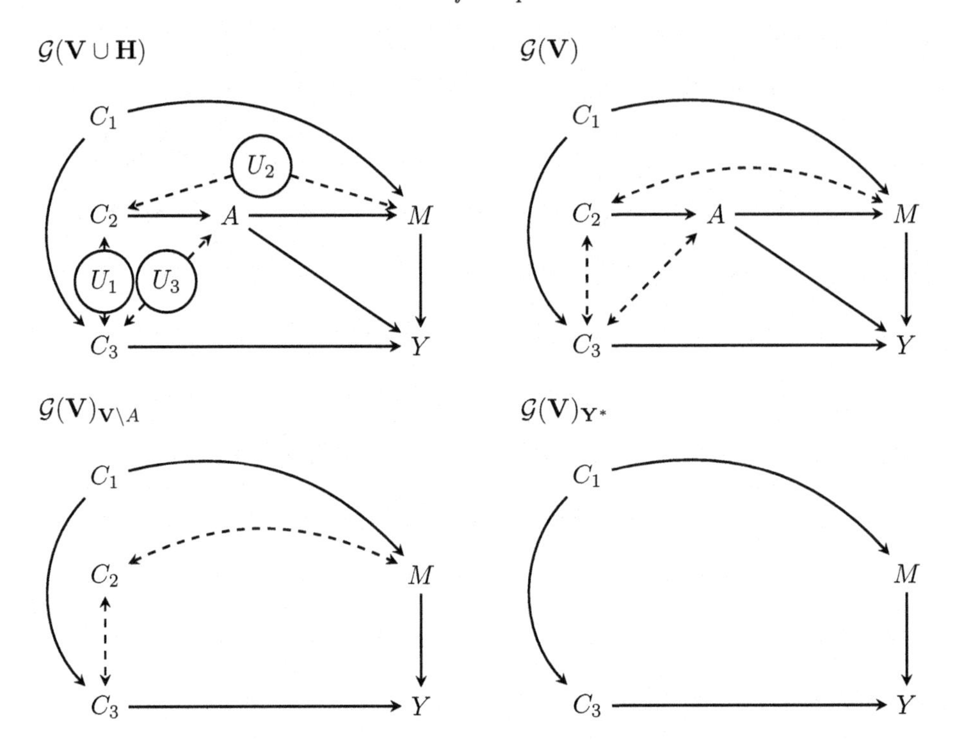

FIGURE 17.6: A somewhat more involved hidden variable causal DAG $\mathcal{G}(\mathbf{V} \cup \mathbf{H})$, its latent projection ADMG $\mathcal{G}(\mathbf{V})$ and subgraphs $\mathcal{G}(\mathbf{V})_{\mathbf{V}\setminus A}$ and $\mathcal{G}(\mathbf{V})_{\mathbf{Y}^*}$.

increase modeling demands and limit the ability to identify $p(Y(a, M(a')) = y)$.

A relatively simple example revisited

By exploiting the following exclusion restrictions encoded in $\mathcal{G}(\mathbf{V} \cup \mathbf{H})$ in Figure 17.5

$$M \perp\!\!\!\perp C_3 | \{A, C_1, C_2\}$$
$$Y \perp\!\!\!\perp C_2 | \{A, M, C_1, C_3\},$$

the identification result can be simplified as

$$\sum_{c_1, c_2, c_3, m} p(y|a, m, c_1, c_3) p(m|a', c_1, c_2) p(c_1, c_2, c_3),$$

thereby reducing modeling demands (although see [7] for a critical discussion on such exclusion restrictions).

A somewhat more involved example

Consider next the causal DAG $\mathcal{G}(\mathbf{V} \cup \mathbf{H})$ with observed variables $\mathbf{V} = \{C_1, C_2, C_3, A, M, Y\}$ and hidden variables $\mathbf{H} = \{U_1, U_2, U_3\}$ in Figure 17.6 (adopted from [19]; Figure 5F). To identify $p(M(a) = m)$ we must, in any case, adjust for C_2. Since C_3 is a collider, adjusting for it opens spurious pathways that cannot be blocked by additionally adjusting for C_1, leaving $\{C_2\}$ and $\{C_1, C_2\}$ as the only viable adjustment sets for identifying $p(M(a) = m)$ by the adjustment formula. However, identification of $p(Y(a, m) = y)$ by the adjustment formula requires that C_3 is included in the adjustment set, because the back-door path from A to Y via U_3 can only be blocked by C_3. As a result, $p(Y(a, M(a')) = y)$ cannot be identified by the adjustment criterion for natural direct and indirect effects.

Nonetheless, identification can be obtained by resorting to an alternative identification strategy. Such strategy may consist of first listing all sufficient adjustment sets that identify $p(M(a) = m)$ and all sufficient adjustment sets that identify $p(Y(a, m) = y)$. Progress can then be made if a subset of the intersection of any two of these respective candidate adjustment sets satisfies assumption (ii). For instance, $p(M(a) = m)$ is identified by adjusting for $\{C_1, C_2\}$, whereas $p(Y(a, m) = y)$ is identified by adjusting for $\{C_1, C_3\}$. Moreover, $\{C_1\}$, the intersection of these separate adjustment sets, satisfies cross-world independence (ii). Relying on the conditional independence $Y \perp\!\!\!\perp C_1 | \{A, M, C_3\}$, $p(Y(a, M(a')) = y)$ may then be identified from observed data by

$$\begin{aligned}
&\sum_{c_1,m} p(Y(a,m) = y|c_1)p(M(a') = m|c_1)p(c_1) \\
&= \sum_{c_1,m} \left(\sum_{c_3} p(y|a,m,c_1,c_3)p(c_3|c_1)\right)\left(\sum_{c_2} p(m|a',c_1,c_2)p(c_2)\right)p(c_1) \\
&= \sum_{c_1,c_2,c_3,m} p(y|a,m,c_3)p(m|a',c_1,c_2)p(c_1)p(c_2)p(c_3|c_1). \qquad (17.6)
\end{aligned}$$

The above examples demonstrate that $p(M(a') = m|\mathbf{c})$ and $p(Y(a, m) = y|\mathbf{c})$ in Equation 17.4 can be identified under a much wider range of scenarios than those that lead to identification by the adjustment criterion for natural effects [19]. That is, identification of $p(Y(a, M(a')) = y)$ can be obtained if, for any candidate covariate set $\mathbf{C}$ that satisfies (ii), $p(M(a') = m|\mathbf{c})$ and $p(Y(a, m) = y|\mathbf{c})$ are identified by Shpitser's complete **IDC** algorithm for conditional treatment effects [28] (see Section 16.3.6 of the previous chapter). This resonates Pearl's original formulations [18] which state that, to recover 'non-experimental' identification, assumption (ii) needs to be complemented with the following two assumptions

$$p(M(a) = m|\mathbf{c}) \text{ is identifiable by some means, and} \qquad \text{(vi.a)}$$

$$p(Y(a, m) = y|\mathbf{c}) \text{ is identifiable by some means.} \qquad \text{(vi.b)}$$

One way to increase identification power is by resorting to the 'divide and conquer' strategy described in the previous example, which Pearl [19] referred to as *piecemeal deconfounding.* However, in certain settings, this strategy may still be overly restrictive and identification may then, instead, sometimes be recovered by exploiting so-called *mediating instruments* [19]. That is, if specific instruments can be found that fully mediate certain crucial but confounded paths (that cannot be deconfounded by observed covariates), further progress can be made by local application of the front-door formula. Specific examples are given in [19].

17.5 Identification 2.0

Widening the scope to also include mediating instruments in our 'identification toolbox' still does not enable us to fully characterize all possible settings that enable non-parametric identification of $p(Y(a, M(a')) = y)$ under NPSEMs. In other words, while the assumption set (ii)-(vi.a)-(vi.b) may be sufficient for recovering natural effects from observational data, it is not necessary and, consequently, not *complete* for identification. This lack of completeness can be demonstrated using a simple illustrating example.

Cross-world independence (ii) is violated in the causal DAG $\mathcal{G}(\mathbf{V} \cup \mathbf{H})$ with $\mathbf{V} = \{A, L, M, Y\}$ and $\mathbf{H} = \{U\}$, in Figure 17.7A, because of unmeasured mediator-outcome confounding. Nonetheless, by exploiting both conditional independencies that are naturally encoded in $\mathcal{G}(\mathbf{V}\cup\mathbf{H})$ and conditional counterfactual independencies implied by the following NPSEM representation of $\mathcal{G}(\mathbf{V}\cup\mathbf{H})$

$$\begin{aligned} L(a') &:= f_L(a', \epsilon_L) \\ M(l) &:= f_M(l, U, \epsilon_M) \\ Y(a,m) &:= f_Y(a, m, U, \epsilon_Y), \end{aligned}$$

$p(Y(a, M(a')) = y)$ can be identified from observed data as follows

$$\begin{aligned} p(Y(a, M(a')) = y) &= p(Y(a, M(L(a'))) = y) \\ &= \sum_{l,m} p(Y(a,m) = y, M(l) = m, L(a') = l) \\ &= \sum_{u,l,m} p(Y(a,m) = y|u)p(M(l) = m|u)p(L(a') = l)p(u) \\ &= \sum_{u,l,m} p(y|a,m,u)p(m|l,u)p(l|a')p(u) \\ &= \sum_{u,l,m} p(y|a,l,m,u)p(m|a,l,u)p(l|a')p(u|a,l) \\ &= \sum_{l} p(y|a,l)p(l|a'). \end{aligned} \tag{17.7}$$

This result should not come as a big surprise: since the effect of treatment on the mediator is (assumed to be) entirely mediated by L, and, in addition, L only affects the outcome via M, L can simply substitute for M. However, the sufficient conditions outlined so far (especially assumption (ii)) do not naturally lead to this simple result.

In the remainder of this section, we therefore take a step back. Armed with the tools and concepts from Chapter 16, we take a closer look at the commonalities that characterize the key problems in the examples in Sections 17.4.1 and 17.4.3. The resulting insights offer a framework that allows to extend complete algorithms for identification of total causal effects (see Sections 16.3.4 to 16.3.6 of the previous chapter) to mediation settings. Moreover, this extension has produced a complete graphical criterion for identification under NPSEMs, not only of natural direct and indirect effects, but of path-specific effects in general [27]. In Sections 17.6 and 17.7, we highlight that this graphical criterion gives rise to complementary identification strategies that may, in addition, help to shed new light on an ongoing debate about the controversial cross-world nature of path-specific effects.

17.5.1 Building blocks for complete graphical identification criteria

Key to violation of assumptions (ii.a) and (ii.b) in the examples in Sections 17.4.1 and 17.4.3, respectively, is the occurrence of conflicting treatment assignments in certain factors of the identifying functional. It can be shown that this problem arises whenever the conflict is situated in factors involving distributions of ancestors of the outcome — which, by convention, include the outcome itself — that belong to a common *confounded component* (abbreviated: *c-component*) [38] or *district* [21]. In this subsection, we first provide the necessary conceptual background on districts, district factorization and complete identification algorithms for total treatment effects, as discussed in more technical detail in the previous chapter.

Confounded paths, districts and district factorizations

Following [30], we first define a *confounded path* to be a path where all directed arrowheads

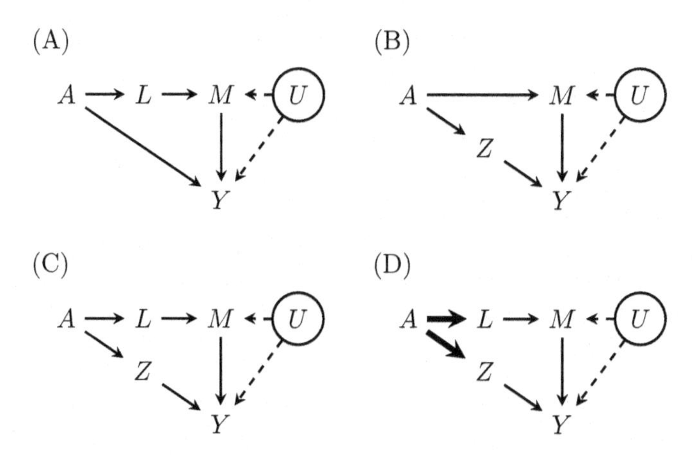

FIGURE 17.7: Hidden variable causal DAGs with mediating instruments L for the path $A \to M$ (A), Z for the path $A \to Y$ (B), and a combination of both (C). The DAG in panel (D) is an extended deterministic graph of the DAG in Figure 17.3A, with thick edges indicating a deterministic relationship.

point at observed nodes, and never away from observed nodes. To avoid cluttering causal DAGs $\mathcal{G}(\mathbf{V} \cup \mathbf{H})$ with large numbers of hidden variables, unobserved common causes of any two observed nodes are often omitted, while their presence is, instead, indicated by bidirected edges ($\leftrightarrow$). This latent projection operation, as discussed in previous chapters, gives rise to *acyclic directed mixed graphs* (ADMGs) [21], which contain only observed nodes $\mathbf{V}$ and both directed and bidirected edges (see Section 2.3 of Chapter 2). These graphs encode conditional independencies between observed variables via m-separation [21], a graphical criterion closely related to d-separation for causal DAGs containing only directed edges. Throughout, we will denote latent projection ADMGs of hidden variable causal DAGs $\mathcal{G}(\mathbf{V} \cup \mathbf{H})$ by $\mathcal{G}(\mathbf{V})$. In ADMGs, confounded paths can similarly be defined as paths that contain only bidirected edges.

A district may now be defined as the maximal set of observed nodes that are pairwise connected by confounded paths. The set of all observed nodes can thus be partitioned into disjoint districts $\mathbf{S} \in \mathcal{D}(\mathcal{G}(\mathbf{V}))$, where $\mathcal{D}(\mathcal{G}(\mathbf{V}))$ denotes the set of districts in the latent projection $\mathcal{G}(\mathbf{V})$. Each of these districts consists of either a single observed node or a set of observed nodes that are pairwise connected by confounded paths. The importance of districts may be appreciated by the fact that their disjointness implies that the marginal distribution of observed variables $p(\mathbf{x_V})$ in $\mathcal{G}(\mathbf{V} \cup \mathbf{H})$ factorizes as the product of their corresponding *kernels* or *c-factors*

$$p(\mathbf{x_V}) = \prod_{\mathbf{S} \in \mathcal{D}(\mathcal{G}(\mathbf{V}))} Q[\mathbf{S}],$$

where each kernel corresponds to

$$Q[\mathbf{S}] = \sum_{\mathbf{x}_{\mathrm{u}_\mathcal{G}(\mathbf{S})}} \prod_{V \in \mathbf{S}} p(x_V | \mathbf{x}_{\mathrm{pa}_\mathcal{G}(V)}, \mathbf{x}_{\mathrm{u}_\mathcal{G}(V)}) p(\mathbf{x}_{\mathrm{u}_\mathcal{G}(\mathbf{S})}),$$

and where $\mathrm{pa}_\mathcal{G}(V)$ and $\mathrm{u}_\mathcal{G}(V)$ denote the set of observed and unobserved parents of V in $\mathcal{G}(\mathbf{V} \cup \mathbf{H})$, respectively. For each $Q[\mathbf{S}]$, the product runs across all observed nodes $V \in \mathbf{S}$ and the summation over all possible realizations of unobserved parents of $V \in \mathbf{S}$.

For instance, in the causal DAG $\mathcal{G}(\mathbf{V} \cup \mathbf{H})$ with $\mathbf{V} = \{A, M, Y\}$ and $\mathbf{H} = \{U\}$, in

Figure 17.3A, M and Y are connected by a confounded path as they share an unmeasured parent U. The set of observed variables can thus be partitioned into two districts: $\{A\}$ and $\{M, Y\}$. Moreover, $p(a, m, y)$ factorizes as the product of the kernels of these districts:

$$p(a, m, y) = Q[\{A\}]Q[\{M, Y\}] = p(a)\sum_{u} p(y|a, m, u)p(m|a, u)p(u).$$

Tian and Pearl [38] pointed out that every $Q[\mathbf{S}]$ can be interpreted as $p(\mathbf{x_S}|\text{do}(\mathbf{x}_{\text{pa}_{\mathcal{G}}(\mathbf{S})}))$, the distribution of $\mathbf{S}$ under an intervention that sets all its observed parents $\text{pa}_{\mathcal{G}}(\mathbf{S})$ to $\mathbf{x}_{\text{pa}_{\mathcal{G}}(\mathbf{S})}$. Moreover, they proved that every $Q[\mathbf{S}]$, for which $\mathbf{S} \in \mathcal{D}(\mathcal{G}(\mathbf{V}))$, is identifiable from observed data. For example, since $A \perp\!\!\!\perp U$ in Figure 17.3A, it is relatively trivial to show that

$$Q[\{M, Y\}] = \sum_{u} p(y|a, m, u)p(m|a, u)p(u) = p(y|a, m)p(m|a).$$

We next illustrate that districts (and their corresponding kernels) form the building blocks of complete graphical identification algorithms for total causal effects [5, 29, 38], since interventional distributions can always be expressed as the marginal of a truncated district factorization.

Truncated district factorizations in hidden variable causal DAGs

In hidden variable causal DAGs $\mathcal{G}(\mathbf{V} \cup \mathbf{H})$, observed nodes $V \in \mathbf{V}$ that are connected by confounded paths group together into districts $\mathbf{S} \in \mathcal{D}(\mathcal{G}(\mathbf{V}))$. Consequently, factorizations of $p(\mathbf{x_V})$ involve kernels that can be interpreted as multivariate interventional distributions. Because district factorizations extend usual Markov factorizations for causally sufficient DAGs to hidden variable causal DAGs with respect to $p(\mathbf{x_V})$, this permits us to express $p(Y(a) = y)$ as the marginal of a truncated version of the district factorization of $p(\mathbf{x_V})$:

$$p(Y(a) = y) = \sum_{\mathbf{x}_{\mathbf{Y}^* \setminus Y}} \prod_{\mathbf{D} \in \mathcal{D}(\mathcal{G}(\mathbf{V})_{\mathbf{Y}^*})} p(\mathbf{x_D}|\text{do}(a_{\text{pa}_{\mathcal{G}}(\mathbf{D}) \cap A}, \mathbf{x}_{\text{pa}_{\mathcal{G}}(\mathbf{D}) \setminus A})), \tag{17.8}$$

where $a_{\text{pa}_{\mathcal{G}}(\mathbf{D}) \cap A} = a$ if there exists a directed path of the form $A \to D \to \ldots \to Y$ (with $D \in \mathbf{D}$), $a_{\text{pa}_{\mathcal{G}}(\mathbf{D}) \cap A} = \emptyset$ if no such path exists, and where $\mathbf{Y}^* \equiv \text{an}_{\mathcal{G}_{\mathbf{V} \setminus A}}(Y)$ denotes the set of ancestors of Y (including Y) in a subgraph $\mathcal{G}(\mathbf{V})_{\mathbf{V} \setminus A}$ of the latent projection $\mathcal{G}(\mathbf{V})$. Here, the product runs across all districts $\mathbf{D} \in \mathcal{D}(\mathcal{G}(\mathbf{V})_{\mathbf{Y}^*})$ in a subgraph of $\mathcal{G}(\mathbf{V})$, and the summation is made over all possible realizations of the nodes in these districts, except for the outcome.

Equation 17.8 indicates that the original problem of identifiability of $p(Y(a) = y)$ can be reduced to a set of smaller identification problems within a subgraph of $\mathcal{G}(\mathbf{V} \cup \mathbf{H})$. Logically, each district $\mathbf{D} \in \mathcal{D}(\mathcal{G}(\mathbf{V})_{\mathbf{Y}^*})$ is a subset of a district $\mathbf{S} \in \mathcal{D}(\mathcal{G}(\mathbf{V}))$. Even though each kernel $Q[\mathbf{S}]$ is identifiable from observed data, identification of some kernels $Q[\mathbf{D}]$ may fail under certain hidden variable causal models. In fact, $p(Y(a) = y)$ is identifiable if and only if every kernel $Q[\mathbf{D}]$ in Equation 17.8 can be recovered from observed data. In the absence of hidden variables, $\mathcal{D}(\mathcal{G}(\mathbf{V})_{\mathbf{Y}^*}) = \mathcal{D}(\mathcal{G}(\mathbf{V}))$, such that $p(Y(a) = y)$ is always identifiable from observed data and Equation 17.8 reduces to the well-known g-formula [23].

For instance, in $\mathcal{G}(\mathbf{V} \cup \mathbf{H})$ in Figure 17.3A, $\mathbf{Y}^* = \mathbf{V} \setminus A = \{M, Y\}$, such that there is only one district $\{M, Y\}$ in $\mathcal{G}(\mathbf{V})_{\mathbf{Y}^*}$. Its corresponding kernel $Q[\{M, Y\}]$ in Equation 17.8 perfectly corresponds to a kernel in the latent projection $\mathcal{G}(\mathbf{V})$, such that it can readily be identified from observed data. Consequently, $p(Y(a) = y)$ is identified by

$$\sum_{m} p(y, m|\text{do}(a)) = \sum_{u,m} p(y|a, m, u)p(m|a, u)p(u) = \sum_{m} p(y|a, m)p(m|a) = p(y|a). \tag{17.9}$$

In contrast to this simple example, truncation of district factorizations can become much more complicated under hidden variable DAGs. That is, whenever a district $\mathbf{D} \in \mathcal{D}(\mathcal{G}(\mathbf{V})_{\mathbf{Y}^*})$ is a *proper* subset of a district $\mathbf{S} \in \mathcal{D}(\mathcal{G}(\mathbf{V}))$, identification of $Q[\mathbf{D}]$ requires (repeated) application of the 'fixing operation' (as described in chapter 16) and may eventually fail.[5] Problematic graphical structures that hinder identification under hidden variable causal models have been discussed in detail in [28].

17.5.2 The central notion of recantation

Having provided the necessary theoretical and conceptual background on complete identification algorithms for total treatment effects, we now pick up where we left off in the beginning of Section 17.5.1 and provide more formality and generality by introducing the central notion of *recantation.* This notion will enable us to map crucial cross-world independencies onto an intuitive graphical criterion under NPSEMs. It can therefore be viewed to serve as a passkey that permits easy translations from cross-world quantities, used to define path-specific effects, to 'single-world' interventional quantities. Identification of the latter can then simply be passed on to well-established algorithms for identifying total causal effects. Ultimately, when combined with such complete algorithms, this graphical criterion does not only delineate complete identification criteria for natural effects, but also for more generally defined path-specific effects.

17.5.2.1 The recanting witness criterion

Cross-world counterfactual $Y(a, M(a'))$ in Figure 17.4A corresponds to $Y(a, L(a), M(a', L(a')))$ and thus represents the response of the outcome to two hypothetical interventions which set to A to a for the purpose of $\pi = \{A \to Y; A \to L \to Y\}$, on the one hand, and to a' for the purpose of $\overline{\pi} = \{A \to M \to Y; A \to L \to M \to Y\}$, on the other hand. However, even though the interventional distribution $p(Y(a) = y)$ is identified by

$$\sum_{l,m} p(y|\text{do}(a,l,m))p(m|\text{do}(a,l))p(l|\text{do}(a)) = \sum_{l,m} p(y|a,l,m)p(m|a,l)p(l|a) = p(y|a),$$

identification of $p(Y(a, M(a')) = y)$ is hampered because of conflicting treatment assignments a and a' in the single node district L. Here, L is called a *recanting witness* [1], for the following reason. Identification of the natural indirect effect via M requires a first statement from L that *blocks* the path $A \to L \to Y$ in order to keep treatment from transmitting its effect on the outcome other than through M (as this path is regarded part of the natural direct effect). However, L subsequently needs to retract this statement in favor of a new statement which *refrains from blocking* the path $A \to L \to M \to Y$ in order to allow treatment to transmit its entire effect on the mediator (as blocking would imply adjusting away part of the natural indirect effect). Clearly, we can't have it both ways.

The *recanting witness criterion* [1] formalizes this requirement of having no such witnesses along π to enable identification of the π-specific effect. More specifically, a child L of treatment A is called a recanting witness[6] for π (and by symmetry, also for $\overline{\pi}$) if there exists a directed path in π of the form $A \to L \to ... \to Y$ and another directed path in $\overline{\pi}$ of the form $A \to L \to ... \to Y$. Avin, Shpitser and Pearl [1] demonstrated that if and only if there

[5] Application of the 'fixing operation' in a conditional ADMG (as described in chapter 16) is essentially equivalent to the systematic removal of certain non-essential nodes in subgraphs of the latent projection ADMG $\mathcal{G}(\mathbf{V})$.

[6] The definition we provide here is restricted to π-specific effects from A to Y, with both A and Y being singletons. A more general definition requires making reference to *proper causal paths*, and is given in chapter 16.

is no recanting witness for π in a causal DAG $\mathcal{G}(\mathbf{V})$ without hidden variables, then the π-specific effect is identified under the NPSEM representation of $\mathcal{G}(\mathbf{V})$. More specifically, the distribution of the corresponding nested counterfactual $Y(\pi, a, a')$ is then identified from observed data as

$$p(Y(\pi, a, a') = y) = \sum_{\mathbf{x}_{\mathbf{V}\setminus(A\cup Y)}} \prod_{V\in\mathbf{V}\setminus A} p(x_V | a_{\mathrm{pa}^{\pi}_{\mathcal{G}}(V)\cap A}, a'_{\mathrm{pa}^{\overline{\pi}}_{\mathcal{G}}(V)\cap A}, \mathbf{x}_{\mathrm{pa}_{\mathcal{G}}(V)\setminus A}), \quad (17.10)$$

where $\mathrm{pa}^{\pi}_{\mathcal{G}}(V)$ denotes the set of parents of V in $\mathcal{G}(\mathbf{V})$ along an edge which is part of a path in π. This result has been referred to as the *edge g-formula* [31] because it generalizes the ordinary g-formula [23] in that it permits different treatment assignments along separate sets of edges $A \to V$. More specifically, in different Markov factors, treatment A is set to either a or a' depending on whether or not $\{A \to V \to ... \to Y\} \in \pi$. The recanting witness criterion thus implicitly imposes the restriction that a single edge $A \to V$ can only be assigned a single treatment value or, in other words, that treatment assignment must be *edge consistent* for $p(Y(\pi, a, a') = y)$ to be identified [31]. The mediation formula (Equation 17.5) can be viewed as a specific case of this more general identifying functional.

17.5.2.2 The recanting district criterion

Even though the recanting witness criterion gives a complete characterization of settings when path-specific effects are identified under DAGs without hidden variables, it does not suffice as a graphical identification criterion in hidden variable DAGs. This can be seen from the illustrating example in Section 17.3, which suffers from unmeasured mediator-outcome confounding. For instance, even though $p(Y(a) = y)$ is identified by Equation 17.9 under the hidden variable causal DAG $\mathcal{G}(\mathbf{V} \cup \mathbf{H})$ in Figure 17.3A, $p(Y(a, M(a')) = y)$ is not identified by $\sum_m p(y|a, m)p(m|a')$, simply because we cannot readily integrate out U under conflicting treatment assignments a and a'. Identification of $p(Y(a, M(a')) = y)$ is thus hindered because of conflicting treatment assignments a and a' in the district $\{M, Y\}$, where A is set to a for the purpose of $\pi = \{A \to Y\}$ and to a' for the purpose of $\overline{\pi} = \{A \to M \to Y\}$.

Inspired by complete algorithms for identifying $p(Y(a) = y)$, based on district factorizations, Shpitser [27] extended the recanting witness criterion to hidden variable DAG settings. This extension is conceptually fairly simple. As districts rather than single nodes are the building blocks of the factorization of $p(\mathbf{x}_{\mathbf{V}})$, the term 'witness' simply needs to be replaced by the term 'district'. Informally, this extended criterion requires there to be no 'conflict of interest' between members of a common district within a particular subgraph $\mathcal{G}(\mathbf{V})_{\mathbf{Y}^*}$. Formally, a district $\mathbf{D} \in \mathcal{D}(\mathcal{G}(\mathbf{V})_{\mathbf{Y}^*})$ is said to be a *recanting district* for π if there exists a directed path in π of the form $A \to D \to ... \to Y$ as well as a directed path in $\overline{\pi}$ of the form $A \to D' \to ... \to Y$, where $D, D' \in \mathbf{D}$ (and possibly $D = D'$).

Shpitser [27], moreover, showed that only in the absence of a recanting district for π, the cross-world counterfactual distribution $p(Y(\pi, a, a') = y)$ can be expressed as a functional of interventional distributions

$$p(Y(\pi, a, a') = y) = \sum_{\mathbf{x}_{\mathbf{Y}^*\setminus Y}} \prod_{\mathbf{D}\in\mathcal{D}(\mathcal{G}_{\mathbf{Y}^*})} p(\mathbf{x}_{\mathbf{D}} | \mathrm{do}(a_{\mathrm{pa}^{\pi}_{\mathcal{G}}(\mathbf{D})\cap A}, a'_{\mathrm{pa}^{\overline{\pi}}_{\mathcal{G}}(\mathbf{D})\cap A}, \mathbf{x}_{\mathrm{pa}_{\mathcal{G}}(\mathbf{D})\setminus A})). \quad (17.11)$$

This functional is a generalization of the truncated district factorization used to identify $p(Y(a) = y)$ in hidden variable causal DAGs (Equation 17.8) via Tian's **ID** algorithm [38]. Equation 17.11 closely matches the formulation in Equation 17.10 in that it also permits different treatment assignments along separate sets of edges, while replacing single nodes V by districts $\mathbf{D} \in \mathcal{D}(\mathcal{G}(\mathbf{V})_{\mathbf{Y}^*})$. It can therefore also be considered an extension of the edge g-formula to hidden variable causal DAGs. However, an additional condition must hold to enable further translation of the involved interventional quantities onto observable quantities. That is, whereas each of the kernels in Equation 17.10 is identifiable from observed data in causal DAGs without hidden variables, identifiability of kernels in Equation 17.11 is not guaranteed because of the assumed presence of hidden variables. Nonetheless, if $p(Y(a) = y)$ is identified via Tian's **ID** algorithm, this logically implies that each kernel in the above functional is expressible in terms of observed data. Importantly, the recanting district criterion thus needs to be complemented by identifiability of the total causal effect $p(Y(a) = y)$ to give a complete characterization of identification conditions for path-specific effects under hidden variable causal DAGs (interpreted as NPSEMs) [27]. Whereas the recanting district criterion enables translations from cross-world counterfactual quantities into 'single world' interventional quantities, the **ID** algorithm then verifies whether these interventional quantities can be expressed as functionals of the observed data.

A somewhat more involved example revisited

Consider again $\mathcal{G}(\mathbf{V} \cup \mathbf{H})$ in Figure 17.6A, along with the subgraphs of interest $\mathcal{G}(\mathbf{V})_{\mathbf{V}\setminus A}$ and $\mathcal{G}(\mathbf{V})_{\mathbf{Y}^*}$. Since there is no recanting district for the set of pathways that capture the natural (in)direct effect in $\mathcal{G}(\mathbf{V})_{\mathbf{Y}^*}$, the identifying functional for $p(Y(a, M(a')) = y)$ can be expressed as

$$\sum_{c_1,c_3,m} p(y|do(a, m, c_3))p(m|do(a', c_1))p(c_3|do(c_1))p(c_1) \tag{17.12}$$

by application of Equation 17.11. Moreover, because $p(Y(a) = y)$ is identified by

$$\sum_{c_1,c_2,c_3,m} p(y|a, m, c_3)p(m|a, c_1, c_2)p(c_1)p(c_2)p(c_3|c_1) \tag{17.13}$$

via Tian's **ID** algorithm, we can re-express Equation 17.12 as Equation 17.6 by simply plugging in appropriate treatment assignments in the respective factors of Equation 17.13.

Note that Pearl's 'divide and conquer' approach, as discussed in Section 17.4.6.2, required searching the space of candidate covariate sets $\mathbf{C}$ that not only satisfy cross-world independence (ii) but also conditions (vi.a) and (vi.b). Shpitser's identification approach is not only (more) complete, but arguably also more insightful as it clarifies that the main difficulty in identifying $p(Y(a, M(a')) = y)$ is identification of the total treatment effect, which involves repeated application of the fixing operator.

17.5.2.3 A new perspective on cross-world independence

From the perspective of the recanting district criterion, the need for an observed covariate set $\mathbf{C}$ that is sufficient to adjust for confounding of the mediator-outcome relation (given treatment) serves to establish that mediator and outcome belong to separate districts so that conflicting treatment assignments causes no further identification problems (provided that no member of $\mathbf{C}$ is affected by treatment). For instance, in Figures 17.3B and C, a sufficient adjustment set $\{C\}$ enables to pull apart the district $\{M, Y\}$ and resolve the conflict in order to ensure the validity of cross-world assumption (ii) that permits factorizing $p(Y(a, m) = y, M(a') = m|\mathbf{c})$ as $p(Y(a, m) = y|\mathbf{c})p(M(a') = m|\mathbf{c})$.

Importantly, the central notion of recantation thus groups Pearl's graphical criteria (ii.a) and (ii.b) for establishing cross-world independence (ii) under NPSEMs by offering a framework that allows their respective violations to be interpreted as distinct instances of essentially the same problem. As will be discussed in the next section, the implications of this graphical criterion reach beyond Pearl's [18] sufficient conditions, as discussed in Section 17.4.

17.6 Complementary Identification Strategies

The completeness of Shpitser's [27] new identification approach reveals that Pearl's [18] conditions (ii)-(vi.a)-(vi.b) may not be necessary to recover natural effects from observed data. That is, simply combining cross-world independence (ii) and identifiability of the total causal effect may suffice to identify $p(Y(a, M(a')) = y)$. When cross-world independence (ii) can thus be established upon adjustment for a covariate set $\mathbf{C}$, one may simply assess identifiability of $p(Y(a) = y)$ via the **ID** algorithm instead of assessing identifiability of both $p(M(a) = m|\mathbf{c})$ and $p(Y(a, m) = y|\mathbf{c})$ via the more complicated **IDC** algorithm.

Interestingly, the completeness of this novel result also highlights that, in some rare cases, cross-world independence (ii) — despite being a sufficient condition for 'experimental' identification [18] — may not be required either. This was already exemplified by the case of Figure 17.7A, which we now revisit.

17.6.1 Interchanging cross-world assumptions

Careful inspection of the hidden variable causal DAG $\mathcal{G}(\mathbf{V} \cup \mathbf{H})$ in Figure 17.7A yields that $\mathbf{Y}^* = \{L, M, Y\}$, such that the subgraph $\mathcal{G}(\mathbf{V})_{\mathbf{Y}^*}$ can be partitioned into districts $\{L\}$ and $\{M, Y\}$. The absence of a recanting district for $\pi = \{A \to Y\}$, which transmits the natural direct effect with respect to M, and identifiability of $p(Y(a) = y)$, by randomization of treatment, then leads to the same identification result for $p(Y(a, M(a')) = y)$ as obtained in Equation 17.7, but derived more elegantly via application of Equation 17.11:

$$\sum_{l,m} p(y, m|\text{do}(a, l))p(l|\text{do}(a')) = \sum_{l,m} p(y|a, l, m)p(m|a, l)p(l|a') = \sum_{l} p(y|a, l)p(l|a').$$

This result can be explained by the fact that, in this case, the recanting district criterion does not serve to establish identifiability via cross-world independence (ii), but, instead, via the alternative cross-world independence assumption $Y(a, m) \perp\!\!\!\perp L(a')$ encoded in the NPSEM representation of $\mathcal{G}(\mathbf{V} \cup \mathbf{H})$. Indeed, the derivations in Equation 17.7 illustrate that the *mediating instrument* L achieves to prevent the conflict between treatment assignments a and a' from taking place *within* the district $\{M, Y\}$ by diverting treatment state a' to itself, thereby fulfilling its mediating role, literally and figuratively. A crucial insight here is that when L is assumed to mediate the entire treatment effect on the mediator M, then the latter is no longer a child of A, and hence, cannot receive any input from A that may conflict with input to other children of A in the same district.

A mediating instrument on the path between treatment and outcome, such as Z in Figure 17.7B, would similarly allow to make progress upon substituting (ii) by cross-world independence $Z(a) \perp\!\!\!\perp M(a')$.

17.6.2 Two types of auxiliary variables

The above examples illustrate that, when $p(Y(a) = y)$ is identifiable, further identification of $p(Y(a, M(a')) = y)$ by the recanting district criterion can be achieved under NPSEMs with the aid of two types of auxiliary variables. Each type can be viewed to have its own distinct strategy for preventing recantation.[7]

[7]Note that this classification is analogous to the one often used for auxiliary variables that aid identification of treatment effects, where identification can be achieved via two main strategies: using either the back-door criterion (i.e. standard adjustment for covariates) or the front-door criterion (i.e. sequential adjustment by means of a mediating instrument).

The first type, such as C in Figures 17.3B and 17.3C, aims to prevent conflicting treatment assignments within districts by separating nodes of a common district, such as $\{M, Y\}$ in Figure 17.3A, which is bound to recant due to unmeasured mediator-outcome confounding, into different districts. Adjustment for this type of covariates specifically aims to strengthen cross-world assumption (ii). The second type, such as L or Z in Figures 17.7A, 17.7B and 17.7C, avoids conflicts in a specific district (such as $\{M, Y\}$) not by separating its nodes into different districts, but instead hosting one potential 'troublemaker' in its own district. Such mediating instruments therefore do not aspire to establish assumption (ii), but instead target identification by means of alternative cross-world assumptions that may substitute for assumption (ii).

This result is important because mediating instruments, while useful to identify $p(M(a) = m|\mathbf{c})$ and/or $p(Y(a, m) = y|\mathbf{c})$ in order to satisfy conditions (vi.a) and (vi.b), cannot aid in (avoiding recantation by) establishing cross-world independence (ii) because this can only be achieved by means of covariate adjustment [19]. The recanting district criterion reveals the extended utility of mediating instruments as auxiliary variables that may, nonetheless, help to avoid recantation by establishing cross-world independencies that substitute for cross-world independence (ii).

17.6.3 Mediating instruments — some reasons for skepticism

Contrary to the long-held belief that identification of $p(Y(a, M(a')) = y)$ hinges on the assumption that no mediator-outcome confounding is left unadjusted, mediating instruments arm us with additional identification power in the presence of such unmeasured confounding. This provides researchers different identification strategies, each relative to a specific set of assumptions. One may use this as a basis for sensitivity analysis, or adopt the strategy that corresponds with the most plausible assumptions given a certain research context. However, some caution is warranted.

First, the recanting district criterion indicates that, when resorting to mediating instruments, the requirement of no unmeasured confounding is simply shifted from the mediator-outcome relation to both the instrument-mediator and instrument-outcome relations. This can be seen upon noting that either type of unmeasured confounding results in the instrument being 'absorbed' into the district $\{M, Y\}$, such that it expands to $\{L, M, Y\}$ in Figure 17.7A or to $\{Z, M, Y\}$ in Figure 17.7B, respectively. Both of these would, however, be recanting with respect to the set of pathways π that transmit either the natural direct or natural indirect effect. As for mediator-outcome confounding, neither types of unmeasured confounding can be avoided by treatment randomization.

Second, the assumption that L or Z is a mediating instrument involves strong and often unrealistic exclusion restrictions.[8] For instance, for L in Figure 17.7A to be a mediating instrument, it would need to mediate the entire effect of A on M. Despite being a strong assumption, it is partially testable from observed data when there is no unmeasured $L - M$ or $A - M$ confounding, as in Figure 17.7A, for then M must be conditionally independent of A, given L. Likewise, for Z in Figure 17.7B to be a mediating instrument for the path $A \to Y$, it would need to mediate the entire direct treatment effect on the outcome that is not mediated by M. However, the requirement that Z and M together mediate the entire treatment effect, which implies $Y \perp\!\!\!\perp A|\{Z, M\}$, is untestable in the presence of unmeasured mediator-outcome confounding. Even though these assumptions cannot directly be verified from observed data, in principle, along with the aforementioned unmeasured confounding assumptions, they lend themselves to experimental verification.

[8]Importantly, L or Z may also correspond to covariate sets (rather than being singletons) that satisfy the stated conditions.

Third, mediating instruments do not resolve the previously considered identification problems in the presence of treatment-induced mediator-outcome confounding by a recanting witness. For instance, the exclusion restriction that L does not directly affect Y in Figure 17.7A, or that Z does not affect M in Figure 17.7B, can be thought of as a constraint that prevents the instrument from turning into a recanting witness.

17.7 From Mediating Instruments to Conceptual Clarity

Even though the practical use of mediating instruments, as an alternative route to identification of natural effects, may be debatable, their added value is more immediate on a conceptual level. Such instruments may help to frame some recent conceptual development that aims to cast mediation analysis into a more strict interventionist paradigm, void of untestable cross-world assumptions [26]. Before going on to discuss this development, we briefly sketch some difficulties that may arise when interpreting natural effects, at least from an interventional point of view.

17.7.1 In search of operational definitions

When it comes to the interpretation of natural direct effects, critics adhering to the slogan 'no causation without manipulation' have repeatedly emphasized the operational question of *how* exactly one may go about blocking the treatment's effect on the mediator, in order to recover $M(0)$ in treated subjects, without affecting the direct path from treatment to outcome [e.g. 4]. Inevitably, any answer to this question invokes a mediating instrument, such as L in Figure 17.7A, that can be intervened on in order to prevent treatment from exerting its effect on the mediator. Likewise, it is difficult to imagine an intervention that would block only the direct path from treatment to outcome, without conceptualizing a mediating instrument such as Z in Figure 17.7B.

17.7.2 Deterministic expanded graphs

It thus seems that mediating instruments provide some sort of necessary extension to the original causal diagram that permits interventionist interpretations of natural effects. The conceptual notion of an expanded graph with two mediating instruments, as depicted in Figure 17.7C, corresponds very closely to what has been described by Robins and Richardson [26]. In settings where L and Z can confidently be considered mediating instruments, the recanting district criterion tells us that, given that $p(Y(a) = y)$ is identifiable, identification of $p(Y(a, M(a')) = y)$ may be obtained if the instruments are in separate districts and if neither of the instruments affects the other. The associated cross-world assumption $Z(a) \perp\!\!\!\perp L(a')$ indeed formalizes the need for no unmeasured confounding between the two instruments. However, this cannot be guaranteed unless both L and Z are deterministic functions of (a randomized) treatment [26]. In that case, both $Z(a)$ and $L(a')$ are constants, and hence trivially independent.[9]

Ironically, this required determinism seems to leave us incapable of pulling apart the pathways that we meant to separate in the first place. However, progress can be made if one can conceive of separate interventions on L and Z that would enable to break their perfect correlation. From this perspective, the deterministic characterization of an expanded

[9] In addition, as shown in [26], independence of $Z(a)$ and $L(a')$ leads to cross-world assumption (ii).

causal DAG, such as Figure 17.7D, gives rise to a specific type of experimental design that requires one to think of L and Z as inherent but distinct properties of the treatment, which may be intervened on separately but, when combined, fully capture all of its active ingredients. The feasibility of such designs primarily mirrors the extent to which different active components of treatment can be conceived of being manipulated in isolation [3]. Moreover, when combined with the aforementioned exclusion restrictions, such designs thus entail separate manipulations of L and Z, which capture distinct but exhaustive features of treatment to which, respectively, solely M or Y are (directly) responsive. Importantly, this characterization enables to interpret natural effects as specific interventional contrasts.

Consider the causal DAG $\mathcal{G}(\mathbf{V} \cup \mathbf{H})$ with observed variables $\mathbf{V} = \{A, M, Y\}$ and hidden variable $\mathbf{H} = \{U\}$ in Figure 17.3A, and its deterministic expansion $\mathcal{G}'(\mathbf{V}' \cup \mathbf{H})$, with $\mathbf{V}' = \mathbf{V} \cup \{Z, L\}$, in Figure 17.7D. Let Z and L be deterministic functions which can be conceived as two complementary components that fully characterize what we will refer to as the 'composite' treatment $\mathbf{A}$ such that $\mathbf{A} \equiv \{L, Z\}$, $\mathbf{a} \equiv \{a_L, a_Z\}$, $\mathbf{a}' \equiv \{a'_L, a'_Z\}$, $p(l|\widetilde{\mathbf{a}}) = \mathbf{1}\{l = \widetilde{a}_L\}$ and $p(z|\widetilde{\mathbf{a}}) = \mathbf{1}\{z = \widetilde{a}_Z\}$. As pointed out by Robins and Richardson [26], $p(Y(a, M(a')) = y)$ then corresponds to the interventional distribution $p(Y(a_Z, a'_L) = y)$ since

$$\begin{aligned} p(Y(a, M(a')) = y) &= \sum_{z,l,m} p(y, m|\text{do}(z, l))p(z|\text{do}(\mathbf{a}))p(l|\text{do}(\mathbf{a}')) \\ &= \sum_{z,l,m} p(y|z, l, m)p(m|l)p(z|\mathbf{a})p(l|\mathbf{a}') \qquad (17.14) \\ &= \sum_{z,l} p(y|z, l)\mathbf{1}\{z = a_Z\}\mathbf{1}\{l = a'_L\} \\ &= p(y|a_Z, a'_L) = p(Y(a_Z, a'_L) = y). \qquad (17.15) \end{aligned}$$

The first equality is obtained by application of Equation 17.11, the third by conditional independence $M \perp\!\!\!\perp Z|L$ and determinism, and the second and last by Tian's **ID** algorithm. Note that, because $Z(a) \perp\!\!\!\perp L(a')$ holds by determinism rather than by independence restrictions implied by NPSEMs, this result can be obtained under the 'single world' model associated with the causal DAG in Figure 17.7D.

The above result implies that, if deterministic mediating instruments like L and Z can be assumed to exist and the aforementioned exclusion restrictions are deemed plausible, it is not necessary to actually conduct any experiment, nor to assume any cross-world independencies to identify the interventional distribution $p(Y(a_Z, a'_L) = y)$. Instead, if $Y(a_Z, a'_L) = Y(a, M(a'))$ under $\mathcal{G}'(\mathbf{V}' \cup \mathbf{H})$ with $\mathbf{V}' = \mathbf{V} \cup \{Z, L\}$, then identification of $p(Y(a_Z, a'_L) = y)$ is tantamount to identification of $p(Y(a, M(a')) = y)$ from observed data on $\mathbf{V}$ under the 'single world' causal model representation of $\mathcal{G}(\mathbf{V} \cup \mathbf{H})$ [26]. For instance, if one merely assumes the existence of some (unidentified) deterministic mediating instruments L and Z in Figure 17.7D, measurements on L and Z are typically missing and $p(a'_L, a_Z) = 0$ (for $a'_L \neq a_Z$) in the observed sample. Consequently, we can only express Equation 17.14 in terms of observable data on $\mathbf{V} = \{A, M, Y\}$ in the absence of unmeasured mediator-outcome confounding by U. To recover identifiability we will thus generally need to complement $\mathbf{V}$ with an additional set of observable auxiliary variables of the types described in Section 17.6.2.

17.7.3 Some examples

Some existing designs, such as double-blind placebo-controlled trials, were in fact devised in the spirit of Robins and Richardson's [26] deterministic extended graphs [3]. Such trials aim to isolate part of the effect of the drug A that may be attributed to active chemical components Z, and is not mediated by the patient's or doctor's expectations about the effectiveness of the drug M. In such designs it is often reasonable to assume that expectations are solely affected by the knowledge of (possibly) being treated L and that the active component itself does not affect expectations. The natural direct effect of the drug, not mediated by expectations, could therefore be interpreted as the interventional contrast comparing drug effectiveness between the treatment and placebo arm. Note that experimental designs that reflect an expanded deterministic graph do not require any measurements on the mediator to identify interventional contrasts that correspond to certain natural effects.

Unfortunately, success is not always guaranteed. Side effects in the treatment arm may, for instance, raise suspicions of being on active treatment, thereby violating the crucial exclusion restriction $M \perp\!\!\!\perp Z|L$ encoded in Figure 17.7D. To accommodate for known side effects, active placebos have been designed that mimic side effects of the active treatment [2, 13], illustrating that the ability to increase the credibility of required exclusion restrictions may often be highly dependent on the creativity of the researcher [26].[10]

In other contexts, experimental designs in the spirit of deterministic extended graphs are more difficult to conceive. For instance, even though the JOBS II study [48] involved a job search skills workshop that targeted specific component processes grounded in psychological theory, it may still be hard to imagine similar interventions or workshops that isolate the distinct triggering elements of separate targeted processes, let alone, to conceive of distinct elements that exclusively affect either re-employment or mental health (via direct pathways). Any attempt to endow natural direct and indirect effects with an interventionist interpretation would thus necessarily rely on strong theoretical assertions about the active components of the job training intervention.

17.8 Path-Specific Effects for Multiple Mediators

The focus of this chapter has hitherto been restricted to identification of natural effects. The recanting district criterion, however, delineates conditions that permit identification of any effect along any bundle of pathways that may be of interest. Its utility may thus be particularly appealing in settings with intertwined pathways along multiple mediators or longitudinal settings where both treatment and/or mediators may be time-varying.

Because of the inherent cross-world nature of path-specific effects, non-parametric identification necessarily always relies on untestable cross-world independence assumptions. A major appeal of the recanting district criterion is that it makes explicit formulations of relevant path-specific cross-world independence assumptions essentially redundant for the purpose of identification under NPSEM representations of hidden variable DAGs, as illustrated below. In general, decompositions of the treatment effect into path-specific effects other than natural direct and indirect effects may be motivated by non-identifiability of natural effects or by the simple fact that the primary mediation hypothesis cannot be expressed in terms of such effects.

Alternative decompositions in the presence of intermediate confounding

In our motivating example, the natural indirect effect with respect to re-employment M is not identified if re-employment and mental health Y are believed to be subject to treatment-induced confounding by changes in perceived self-efficacy, as denoted by L in Figure 17.4A.

[10] In a strict sense, active placebo designs also violate the required exclusion restrictions. Nonetheless, they enable to arrive at a measure of a direct effect that more closely resembles the natural direct effect of primary interest [2].

If, nonetheless, the main interest is in the mediating role of re-employment, we may either calculate partial identification bounds for the natural indirect effect (see [14] and references therein), conduct a sensitivity analysis (see [41] and references therein), or abandon focus on the natural indirect effect altogether. Instead, shifting focus to less ambitious decompositions in terms of either the joint natural indirect effect mediated by both perceived self-efficacy and re-employment (transmitted by $\pi_1 = \{A \to M \to Y; A \to L \to Y; A \to L \to M \to Y\}$), or the partial indirect effect with respect to re-employment (transmitted by $\pi_2 = \{A \to M \to Y\}$), may still to some extent enable us to learn about the mediating role of re-employment.

The *joint natural indirect effect* with respect to $\{L, M\}$ requires recovering the cross-world counterfactual distribution $p(Y(\pi_1, a, a') = y) = p(Y(a', L(a), M(a)) = y)$ and is identifiable even if we relax assumptions to allow for unmeasured confounding between sense of self-efficacy and re-employment, as in Figure 17.4C. This can easily be seen upon noting that the district $\{L, M\}$ is not recanting with respect to π_1. In this case, the recanting district criterion serves to establish cross-world independence

$$Y(a, l, m) \perp\!\!\!\perp \{M(a', l), L(a')\}.$$

When combined with experimentally verifiable identifying assumptions for $p(Y(a) = y)$, as encoded in the 'single-world' representation of the hidden variable DAG, this cross-world assumption renders $p(Y(a', L(a), M(a)) = y)$ identifiable from observed data. In contrast, districts $\{M, Y\}$ and $\{L, Y\}$ are recanting with respect to π_1, as in Figures 17.4B and D respectively, such that $p(Y(a', L(a), M(a)) = y)$ is not identifiable under unmeasured confounding of the relation between the outcome and any of the given intermediate variables along paths in π_1. Nonetheless, identification can be restored in the presence of a mediating instrument, for instance, on the edge $A \to M$ or $A \to Y$ if it is hindered due to unmeasured $M - Y$ confounding.

The *partial indirect effect* with respect to re-employment, on the other hand, requires recovering $p(Y(\pi_2, a, a') = y) = p(Y(a', L(a'), M(a, L(a'))) = y)$. This cross-world counterfactual distribution remains identifiable if we relax assumptions by allowing for unmeasured confounding between sense of self-efficacy and mental health, as in Figure 17.4D, because the district $\{L, Y\}$ is not recanting with respect to π_2. Allowing for unmeasured confounding of the relation between re-employment M and either the outcome Y or intermediate confounder L, in contrast, destroys identification because of recantation of the districts $\{M, Y\}$ and $\{L, M\}$ with respect to π_2, in Figures 17.4B and C, respectively. The violated cross-world independence assumption

$$\{Y(a, l, m), L(a)\} \perp\!\!\!\perp M(a', l)$$

which enables identification of $p(Y(a', L(a'), M(a, L(a')) = y)$, can, however, again be interchanged with another assumption which restores identification in the presence of a mediating instrument on one of the respective edges emanating from A.

Addressing different types of mediation questions

In certain cases, the partial indirect effect with respect to a mediator of interest M that is affected by earlier mediators, may, however, be the primary path-specific effect of interest. For example, Miles and colleagues [15] aimed to assess the extent to which treatment adherence driven by non-toxicity factors mediates the effect of antiretroviral therapy (ART) on virological failure in HIV patients in Nigeria. Ignoring potential baseline confounders, their target of inference corresponds to the path-specific effect along $\pi = \{A \to M \to Y\}$ in Figure 17.4D, where A denotes ART, L drug toxicity, M adherence and Y viral load. Their corresponding mediation analysis thus aimed to answer the question *"How much of the medication's effect is mediated by adherence, if we discard the mediating role of adherence driven by drug toxicity?"* Estimation of this contribution to the total effect of ART on viral load enabled Miles and colleagues to address questions that are not only etiologically relevant but that may also have important policy implications. In fact, the corresponding mediation analysis aimed to assess whether conceivable modifications to the treatment regimen, which may increase adherence (but not through changes in toxicity), may magnify the net treatment effect and hence increase its effectiveness. The interpretation of the partial indirect effect as an interventional contrast, which could be estimated from such a hypothetical experiment, can likewise be represented via a deterministic expanded DAG. A more detailed discussion of deterministic expanded DAGs for path-specific effects, as discussed in [26], is, however, beyond the scope of this chapter.

17.9 Discussion and Further Challenges

Most developments on the identification of natural direct and indirect effects have so far focused on single mediator settings where cross-world independence (ii) holds along with conditional ignorability assumptions (v.a) and (v.b). That is, where the data-generating mechanism can be described by an NPSEM in which a common set of baseline covariates $\mathbf{C}$ suffices to adjust for confounding of the treatment-mediator, treatment-outcome and mediator-outcome associations (within levels of treatment), and where moreover none of the elements of $\mathbf{C}$ is affected by treatment. The latter two requirements have, to a large extent, prohibited extensions to settings with multiple, possibly longitudinal, mediators.

Recently, a complete graphical criterion has been devised for identification of any path-specific effect of a treatment on an outcome under NPSEMs [27]. Briefly, it shows that when the total causal effect is identifiable by some means, as can be verified using Tian's **ID** algorithm [38] (as discussed in chapter 16), then every path-specific effect (along a set of pathways) for which there is no recanting district is also identifiable. Identification then essentially proceeds via Tian's identifying functional (Equation 17.8), which extends Robins' g-functional [23] to (NPSEM representations of) hidden variable causal DAGs, while allowing treatment assignments to be different across districts (Equation 17.11).

Increased identification power

It is not too hard to come up with examples of settings where the adjustment criterion for natural effects fails to identify $p(Y(a, M(a')) = y)$, due to violations of assumption set (v.a)-(v.b), yet the recanting district criterion leads to identifiable natural direct and indirect effects (see Figures 17.6 and 17.7, respectively). Even so, from a practitioner's point of view, it can be argued that the associated increased identification power is of limited practical relevance in single mediator settings. The reason is that prior knowledge in practice is often too limited to justify assumptions that could substitute for failure of assumptions (v.a)-(v.b) [7]. On the other hand, one cannot ignore the potential of causal structure learning algorithms (see Chapter 18), especially in the 'big data' age, in which we have increasing access to massive data sets. In particular, such algorithms may aid researchers to construct a class of DAGs that are compatible with the observed data distribution and under which identification may be obtained under more general assumptions as delineated by the **ID** algorithm and the recanting district criterion.

Importantly, the recanting district criterion has been proposed as a criterion that, given identifiability of $p(Y(a) = y)$, delineates conditions for identifying marginal (or population-averaged) path-specific distributions $p(Y(\pi, a, a') = y)$. Nonetheless, its utility is still unclear (but definitely more subtle) when it comes to identification of conditional (or stratum-specific) path-specific distributions $p(Y(\pi, a, a') = y|\mathbf{c})$. Generalizations for complete identification criteria for conditional path-specific effects are undoubtedly less straightforward, and are left as subject for further research. Consequently, the increased identification power that follows from Shpitser's results is currently only well-documented for marginal natural direct and indirect effects, since the adjustment criterion for natural direct and indirect effects delineates identical conditions for identifying both $p(Y(a, M(a')) = y)$ and $p(Y(a, M(a')) = y|\mathbf{c}^*)$ whenever $\mathbf{C}^*$ is a subset of a set of baseline covariates $\mathbf{C}$ that controls for mediator-outcome confounding.

Estimation

If identification of natural effects is achieved under the above 'traditional' but more stringent set of assumptions, this leads to a standard identifying functional, generally known as the mediation formula, for which a well-established suite of (semi-)parametric estimators has been developed (see [44] and references therein). Accordingly, estimation may then proceed via routine application of these methods as implemented in off-the-shelf statistical software packages (see [35] and references therein). Even though software implementations of complete graphical identification algorithms, such as the **ID** algorithm, are now publicly available [39], estimation of more involved or less standard identifying functionals arguably imposes another barrier to routine application of complete identification algorithms. Inevitably, this may have led to a trade-off between postulating realistic causal structure, on the one hand, and simple and accessible estimation strategies, on the other hand. Future research thus needs to focus on the development of a generic and flexible estimation framework for more generic identifying functionals. Such framework should not only incorporate estimation of natural effects, but also of more generally defined path-specific effects (see [15, 31, 34] for some first promising steps in this direction).

Broadening the scope

Even though its added value for identification of natural effects may be debatable, the recanting district criterion offers a major potential for extensions to settings with multiple, possibly longitudinal, mediators. The identification of the partial indirect effect via a given mediator of interest, as in Figure 17.4D, forms a first step towards such extension, as it allows for possibly high-dimensional post-treatment confounders to confound the mediator-outcome association. In particular, it allows for earlier mediators to be confounders of the association between later mediators and outcome, while at the same time being confounded with the outcome by unmeasured common causes.

Cross-world contemplations

Both the cross-world nature of path-specific effects and the required cross-world independence assumptions for identification have been the subject of an ongoing debate [e.g. 16, 26], roughly dividing the field into NPSEM 'skeptics' and 'advocates'. We have tried to shed some light on this controversy, and illustrated the important role of mediating instruments and deterministic expanded graphs [26] in elucidating and bridging this conceptual and ontological divide. The main objection is that such cross-world independence assumptions, on which modern causal mediation analysis generally relies (although see [36] for a recent exception), cannot be enforced experimentally and hence are not falsifiable. However, whether or not researchers should be encouraged to reformulate their mediational hypotheses in terms of feasible potential interventions on defining features of treatment, remains an open question.

An alternative approach to avoiding cross-world definitions and assumptions has recently gained increasing attention. This approach builds on the claim that, even in the absence of any reference to cross-world quantities or restrictions, certain contrasts based on the mediation formula may still carry empirically meaningful interpretations [4, 20]. This has given rise to the more formal definition of so-called *randomized intervention analogs* of natural effects [43, 42, 46], which conceive of setting the mediator at some level that is randomly assigned from the conditional counterfactual mediator distribution $p(M(a') = m|\mathbf{c})$ rather than at the individual counterfactual level (see [13] for a related approach). Importantly, because their definitions do not employ cross-world counterfactuals strong and unfalsifiable assumptions, such as cross-world independence (but also 'no intermediate confounding') may be avoided. These estimands also tend to correspond more closely to relevant policy measures that can be estimated from actual interventions.

Even if one is willing to make untestable cross-world assumptions, identification of natural effects is typically hindered in the presence of treatment-induced confounding. Accordingly, several contributions to the field have articulated alternative assumptions that may allow us to recover natural effects despite treatment-induced confounding. However, such assumptions generally impose additional structure on the joint distribution of counterfactuals, such as rank preservation [26], monotonicity [37] or parametric constraints [24, 20, 37].

Acknowledgments

The authors would like to thank Vanessa Didelez, Robin Evans, Yves Rosseel and Karel Vermeulen for helpful suggestions and feedback on an earlier draft of this chapter, and Ilya Shpitser for valuable discussions that have led to some improved insights presented in this chapter. This work was supported by Research Foundation Flanders (FWO Grant G.0111.12).

Bibliography

[1] Chen Avin, Ilya Shpitser, and Judea Pearl. Identifiability of Path-Specific Effects. In *Proceedings of the 19th International Joint Conference on Artificial Intelligence*, IJCAI'05, 357–363, San Francisco, 2005. Morgan Kaufmann Publishers Inc.

[2] Vanessa Didelez. Basic concepts of causal mediation analysis and some extensions. Talk at Symposium on Causal Mediation Analysis, Ghent, 2013.

[3] Vanessa Didelez. Discussion on the paper by Imai, Tingley and Yamamoto. *Journal of the Royal Statistical Society. Series A (Statistics in Society)*, 176(1):39, 2013.

[4] Vanessa Didelez, A. Philip Dawid, and Sara Geneletti. Direct and Indirect Effects of Sequential Treatments. In Rina Dechter and Thomas Richardson, editors, *Proceedings of the Twenty-Second Conference on Uncertainty in Artificial Intelligence (UAI-06)*, 138–146, Arlington, Virginia, 2006. AUAI Press.

[5] Yimin Huang and Marco Valtorta. Identifiability in causal Bayesian networks: A sound and complete algorithm. *Proceedings of the National Conference on Artificial Intelligence*, 21(2):1149, 2006.

[6] Martin Huber. Identifying causal mechanisms (primarily) based on inverse probability weighting. *Journal of Applied Econometrics*, 29(6):920–943, Sep 2014.

[7] Kosuke Imai, Luke Keele, Dustin Tingley, and Teppei Yamamoto. Comment on Pearl: Practical implications of theoretical results for causal mediation analysis. *Psychological Methods*, 19(4):482–487, 2014.

[8] Kosuke Imai, Luke Keele, and Teppei Yamamoto. Identification, inference and sensitivity analysis for causal mediation effects. *Statistical Science*, 25(1):51–71, Feb 2010.

[9] Kosuke Imai, Dustin Tingley, and Teppei Yamamoto. Experimental designs for identifying causal mechanisms. *Journal of the Royal Statistical Society A*, 176(1):5–51, Jan 2013.

[10] Theis Lange, Mette Rasmussen, and Lau Caspar Thygesen. Assessing natural direct and indirect effects through multiple pathways. *American Journal of Epidemiology*, 179(4):513–8, Feb 2014.

[11] Theis Lange, Stijn Vansteelandt, and Maarten Bekaert. A simple unified approach for estimating natural direct and indirect effects. *American Journal of Epidemiology*, 176(3):190–195, Jul 2012.

[12] Tom Loeys, Beatrijs Moerkerke, Olivia De Smet, Ann Buysse, Johan Steen, and Stijn Vansteelandt. Flexible mediation analysis in the presence of nonlinear relations: beyond the mediation formula. *Multivariate Behavioral Research*, 48(6):871–894, Nov 2013.

[13] Judith J. Lok. Defining and estimating causal direct and indirect effects when setting the mediator to specific values is not feasible. *Statistics in Medicine*, 35, 2016.

[14] Caleb H. Miles, Phyllis Kanki, Seema Meloni, and Eric J. Tchetgen Tchetgen. On partial identification of the natural indirect effect. *Journal of Causal Inference*, Jan 2017.

[15] Caleb H. Miles, Ilya Shpitser, Phyllis Kanki, Seema Meloni, and Eric J. Tchetgen Tchetgen. Quantifying an adherence path-specific effect of antiretroviral therapy in the nigeria PEPFAR program. *Journal of the American Statistical Association*, 2017.

[16] Ashley I. Naimi, Jay S. Kaufman, and Richard F. MacLehose. Mediation misgivings: Ambiguous clinical and public health interpretations of natural direct and indirect effects. *International Journal of Epidemiology*, 43(5):1656–1661, Oct 2014.

[17] Judea Pearl. Causal diagrams for empirical research. *Biometrika*, 82(4):669–710, Dec 1995.

[18] Judea Pearl. Direct and Indirect Effects. In John Breese and Daphne Koller, editors, *Proceedings of the Seventeenth Conference on Uncertainty in Artificial Intelligence (UAI-01)*, UAI-01, 411–420, San Francisco, 2001. Morgan Kaufmann.

[19] Judea Pearl. Interpretation and identification of causal mediation. *Psychological Methods*, 19(4):459–481, Jun 2014.

[20] Maya L. Petersen, Sandra E. Sinisi, and Mark J. van der Laan. Estimation of direct causal effects. *Epidemiology*, 17(3):276–84, May 2006.

[21] Thomas S. Richardson. Markov properties for acyclic directed mixed graphs. *Scandinavian Journal of Statistics*, 30(1):145–157, Mar 2003.

[22] Thomas S. Richardson and James M. Robins. Single World Intervention Graphs (SWIGs): A Unification of the Counterfactual and Graphical Approaches to Causality. Technical Report 128, University of Washington, 2013.

[23] James M. Robins. A new approach to causal inference in mortality studies with a sustained exposure periodâ–application to control of the healthy worker survivor effect. *Mathematical Modelling*, 7(9–12):1393–1512, 1986.

[24] James M. Robins and Sander Greenland. Identifiability and exchangeability for direct and indirect effects. *Epidemiology*, 3(2):143–155, Mar 1992.

[25] James M. Robins, Miguel A. Hernán, and Babette Brumback. Marginal structural models and causal inference in epidemiology. *Epidemiology*, 11(5):550–60, Sep 2000.

[26] James M. Robins and Thomas S. Richardson. Alternative Graphical Causal Models and the Identification of Direct Effects. In P. Shrout, editor, *Causality and Psychopathology: Finding the Determinants of Disorders and Their Cures*, 103–158. Oxford University Press, Oxford, England, 2010.

[27] Ilya Shpitser. Counterfactual graphical models for longitudinal mediation analysis with unobserved confounding. *Cognitive Science*, 37(6):1011–35, Aug 2013.

[28] Ilya Shpitser and Judea Pearl. Identification of Conditional Interventional Distributions. *Proceedings of the Twenty-Second Conference on Uncertainty in Artificial Intelligence*, 437–444, 2006.

[29] Ilya Shpitser and Judea Pearl. Identification of Joint Interventional Distributions in Recursive semi-Markovian Causal Models. *Proceedings of the Twenty-First AAAI Conference on Artificial Intelligence*, 1219–1226, 2006.

[30] Ilya Shpitser and Judea Pearl. Complete identification methods for the causal hierarchy. *The Journal of Machine Learning Research*, 9:1941–1979, 2008.

[31] Ilya Shpitser and Eric J. Tchetgen Tchetgen. Causal inference with a graphical hierarchy of interventions. *The Annals of Statistics*, 44(6):2433–2466, Dec 2016.

[32] Ilya Shpitser and Tyler J. VanderWeele. A Complete Graphical Criterion for the Adjustment Formula in Mediation Analysis. *The International Journal of Biostatistics*, 7(1):1–24, 2011.

[33] Ilya Shpitser, Tyler J. VanderWeele, and James M. Robins. On the validity of covariate adjustment for estimating causal effects. *Proceedings of the Twenty-Sixth Conference on Uncertainty in Artificial Intelligence (UAI-10)*, 527–536, 2010.

[34] Johan Steen, Tom Loeys, Beatrijs Moerkerke, and Stijn Vansteelandt. Flexible mediation analysis with multiple mediators. *American Journal of Epidemiology*, 186(2):184–193, Jul 2017.

[35] Johan Steen, Tom Loeys, Beatrijs Moerkerke, and Stijn Vansteelandt. Medflex: An R package for flexible mediation analysis using natural effect models. *Journal of Statistical Software*, 76(11), 2017.

[36] Eric J. Tchetgen Tchetgen and Kelesitse Phiri. Evaluation of medication-mediated effects in pharmacoepidemiology. *Epidemiology*, 28(3):439–445, May 2017.

[37] Eric J. Tchetgen Tchetgen and Tyler J. VanderWeele. Identification of natural direct effects when a confounder of the mediator is directly affected by exposure. *Epidemiology*, 25(2):282–91, Mar 2014.

[38] Jin Tian and Judea Pearl. On the identification of causal effects. Technical report, Department of Computer Science, University of California, Los Angeles, 2003.

[39] Santtu Tikka and Juha Karvanen. Identifying causal effects with the R package causal effect. *Journal of Statistical Software*, 76(12), 2017.

[40] Tyler J. VanderWeele. A three-way decomposition of a total effect into direct, indirect, and interactive effects. *Epidemiology*, 24(2):224–232, 2013.

[41] Tyler J. VanderWeele and Yasutaka Chiba. Sensitivity analysis for direct and indirect effects in the presence of exposure-induced mediator-outcome confounders. *Epidemiology, Biostatistics and Public Health*, 11(2):1–16, 2014.

[42] Tyler J. VanderWeele and Eric J. Tchetgen Tchetgen. Mediation analysis with time varying exposures and mediators. *Journal of the Royal Statistical Society: Series B (Statistical Methodology)*, Jun 2016.

[43] Tyler J. VanderWeele, Stijn Vansteelandt, and James M. Robins. Effect decomposition in the presence of an exposure-induced mediator-outcome confounder. *Epidemiology*, 25(2):300–306, 2014.

[44] Stijn Vansteelandt. Understanding counterfactual-based mediation analysis approaches and their differences. *Epidemiology*, 23(6):889–91, Nov 2012.

[45] Stijn Vansteelandt, Maarten Bekaert, and Theis Lange. Imputation strategies for the estimation of natural direct and indirect effects. *Epidemiologic Methods*, 1(1):Article 7, Jan 2012.

[46] Stijn Vansteelandt and Rhian M. Daniel. Interventional effects for mediation analysis with multiple mediators. *Epidemiology*, 28(2):258–265, Nov 2017.

[47] Amiram D. Vinokur, Richard H. Price, and Yaacov Schul. Impact of the JOBS intervention on unemployed workers varying in risk for depression. *American Journal of Community Psychology*, 23(1):39–74, 1995.

[48] Amiram D. Vinokur and Yaacov Schul. Mastery and inoculation against setbacks as active ingredients in the JOBS intervention for the unemployed. *Journal of Consulting and Clinical Psychology*, 65(5):867–877, 1997.

18

Search for Causal Models

Peter Spirtes

Department of Philosophy, Carnegie Mellon University

Kun Zhang

Department of Philosophy, Carnegie Mellon University

CONTENTS

18.1 Introduction

In the last two decades there has been a rapid spread of interest in principled methods of search or estimation of causal relations, that has been driven in part by technological developments. These technological developments include the ability to collect and store "big data" with huge numbers of variables and sample sizes, and increases in the speed of computers. Statistics books from 30 years ago often presented examples with fewer than 10 variables, in domains where some background knowledge was plausible. In contrast, in domains where there are measurements such as satellite images of weather, fMRI brain imaging, microarray measurements of gene expression, or SNP data, the number of variables can range into the millions, and there is often very limited background knowledge to reduce the space of alternative causal hypotheses. In such domains, causal discovery techniques without the aid of automated search appear to be hopeless. At the same time, the availability of faster computers with larger memories and disc space allow for the practical implementation of computationally intensive automated search algorithms over large search spaces.

There are a number of different related problems that can be addressed by causal search algorithms. In this chapter, we will assume that causal relations are represented by a directed graph (see Section 18.3 for more details). In its most general form, we assume that the inputs are a set of data points, either experimental or observational, and possible background knowledge of various kinds. The background knowledge could include information about the existence or non-existence of direct or indirect causal relations, or time order, or a parametric family, and could be in the form of hard constraints (e.g. death does not cause smoking) or soft constraints (e.g. the prior probability that smoking causes cancer is .95). The goal of causal search is to use computationally and statistically feasible algorithms for reliably discovering causal relations that are provably correct in the large sample limit (either uniformly consistent or pointwise consistent) under standard sampling assumptions and various kinds of background knowledge. In its most general form, the output is either a set of graphs (possibly a single graph in some cases) or a probability distribution over graphs. In addition to representing causal relations, the output graphs of causal search algorithms provide an efficient representation of joint probability distributions. Examples of the various kinds of inputs and outputs are given in the discussions of the algorithms.

A causal graph by itself gives qualitative information about causal relations (where a directed edge from A to B means that A is a direct cause of B relative to the variables in the graph) but not quantitative information about the strength of causal effects. The qualitative information about causal relations may be of interest in itself. However, if the quantity of interest is quantitative information about the strength of a causal effect, then one common method of inferring the strength of a causal effect is to first perform a search for a causal graph or a set of causal graphs. Then the causal graph or set of causal graphs, together with the data, are used to estimate parameter values or bounds on parameter values or probability distributions over parameter values. Finally, the strength of the causal effect is inferred using the do-calculus or extensions of the do-calculus (see Chapter 15 in this handbook) applied to the causal graphs and the parameter information. In this way, the output of a causal search algorithm can, depending upon the exact form of the output, be used to estimate the strength of a causal effect, bounds on the strength of a causal effect, or some properties of the causal effect. An alternative strategy for inferring the strength of a causal effect is to simultaneously estimate the causal graph and the parameters, instead of separating the estimation into two distinct stages. Examples of both kinds of algorithms will also be described in this chapter.

There are too many causal search algorithms, and too many different kinds of background knowledge and input data to describe all of the alternatives in this chapter. We will give an overview of the major kinds of input data, background knowledge, and algorithms, along with descriptions of the problems that arise and strategies for dealing with those problems. We will also describe the advantages and disadvantages of different algorithms, and describe open problems that remain. For another useful survey on Bayesian network learning, see [14]. A number of the algorithms described in this chapter are available through R packages *bnlearn* and *pcalg*, and the Center for Causal Discovery (http://www.ccd.pitt.edu/tools/), which also allows causal search algorithms to be employed on big data using large memory servers at the Pittsburgh Supercomputing Center.

18.2 Why Causal Search Is Difficult

The dimensions of the causal search problem include at least the following major difficulties:

1. The number of directed acyclic causal graphs grows super-exponentially with the number of variables, and the number of cyclic graphs grows even faster. (For example, for 10 vertices, there are roughly 4.2×10^{18} directed acyclic graphs). This makes examining each graph, no matter how quickly, impossible in practice.
2. There may be unmeasured confounders (i.e. latent direct common causes) of some pairs of measured variables.
3. There may be selection bias, where values of some of the variables of interest are associated with sample membership.
4. The variables of interest may not be directly measured, e.g. "intelligence" is not directly measured, but estimated by a score constructed from answers to a test.
5. There may be feedback (naturally represented by cyclic graphs).
6. The data may be time series data, which may be non-stationary, and for which the rate of sampling may be lower than the rate of change of the underlying dynamic process.
7. The causal system may be nonlinear in various ways, and contain discrete and continuous variables with various types of distributions.
8. The data may come from multiple data sets that have overlapping sets of variables, some of which may be observational and some of which may be experimental. The data sets may also have different distributions.
9. The data may have been aggregated in various ways (e.g. spatial or temporal coarsening, or averaging).
10. There may be significant amounts of measurement error.
11. There may be significant associations between variables that are due to non-causal relations between variables (e.g. one variable is defined in terms of another).
12. In addition to causal relations between variables in a single unit, there may be causal relations between variables in different "natural" units (e.g., one person's smoking causes another person's lung cancer), leading to "interference" or a violation of the Stable Unit Treatment Value Assumption (SUTVA).
13. There may be deterministic relationships between variables.
14. There may be conditional independence relationships between variables that hold by "coincidence" (as explained in Section 18.3).
15. The data may come from a mixture of sub-populations with different causal graphs, or different parameter values for the causal graphs, or from individuals at different points in a process.

Although each of these problems has been addressed to some extent by different methodologies, there is currently no one method that can deal with all of these problems simultaneously.

18.3 Assumptions and Terminology

We will denote sets of random variables with capital boldfaced names, and individual random variables, as well as probability distributions and directed graphs, with capitalized italicized names. Let $I_P(\mathbf{X}, \mathbf{Y}|\mathbf{Z})$ stand for $\mathbf{X}$ is independent of $\mathbf{Y}$ conditional on $\mathbf{Z}$ in distribution P, where we also write $I_P(\mathbf{X}, \mathbf{Y})$ if $\mathbf{Z}$ is the empty set. A *Bayesian network* is a pair $\langle G, P\rangle$ where G is a directed acyclic graph (DAG), which is a finite directed graph with no directed cycles, over a set of variables $\mathbf{V}$, and P is a joint distribution over $\mathbf{V}$ such that P satisfies the local Markov condition for G, i.e. that in P each variable is independent of its non-descendants in G conditional on its parents in G. In the DAG G, two variable sets $\mathbf{X}$ and $\mathbf{Y}$ are said to be d-separated conditional on variable set $\mathbf{Z}$ if for each of the paths between a variable in $\mathbf{X}$ and a variable in $\mathbf{Y}$, it contains a chain $\cdot \rightarrow Z \rightarrow \cdot$ or a fork $\cdot \leftarrow Z \rightarrow \cdot$ such that the middle node Z is in $\mathbf{Z}$ or contains a collider $\cdot \rightarrow W \leftarrow \cdot$ such that neither the middle node W nor its descendants are in $\mathbf{Z}$. If P satisfies the local Markov condition for G and $\mathbf{X}$ is d-separated from $\mathbf{Y}$ conditional on $\mathbf{Z}$, then $\mathbf{X}$ is independent of $\mathbf{Y}$ conditional on $\mathbf{Z}$ in P, and the joint probability factors into the product of the probabilities of each variable conditional on its parents [44].

A *causal graph* is a directed graph over a set of random variables $\mathbf{V}$ in which there is a directed edge from A to B if and only if A is a direct cause of B relative to $\mathbf{V}$ [77]. Equivalently, in terms of manipulations, in a causal graph there is a directed edge from A to B if and only if there is some pair of experimental manipulations of the set of variables in $\mathbf{V} \setminus \{B\}$ that differ only in how A is manipulated that changes the probability distribution of B. See Chapter 15 for more details. A set of variables $\mathbf{V}$ is *causally sufficient* when every direct (relative to $\mathbf{V}$) cause of two variables in $\mathbf{V}$ is also in $\mathbf{V}$ (i.e. there are no unmeasured confounders.)

In order to relate causal graphs to data, some assumptions need to be made about the relationship between causal graphs and probability distributions [77]. The first assumption that we will make throughout this chapter is the Stable Unit Treatment Value Assumption (SUTVA).

The Stable Unit Treatment Value Assumption (SUTVA): The random variables of a unit in the population do not causally interact with the random variables of a different unit in the population.

If the units in a population are taken to be individual people, then my smoking causing my lung cancer would not be a violation of SUTVA, because that is intra-unit causation, but if my smoking caused your lung cancer that would be a violation of SUTVA, because that is inter-unit causation. Violations of SUTVA can be remedied by redefining what the units are. For instance, in the case of an epidemic, the unit could be redefined to be a village, rather than an individual person (if the villages did not causally interact). This, however, creates the problem of multiplying the number of variables (different sets of variables for each person in the village), and reducing the sample size (to the number of villages, rather than the number of people). SUTVA can also fail in time series data, depending again on how the units are defined.

The second assumption that we will make is the Causal Markov Assumption, which presupposes SUTVA. It is a generalization of two commonly made assumptions: the immediate past screens off the present from the more distant past; and if X does not cause Y and Y does not cause X, then X and Y are independent conditional on their common causes.

(Local) Causal Markov Assumption: If G is a causal graph for a given population with causally sufficient vertex set $\mathbf{V}$ and $P(\mathbf{V})$ is the probability distribution over the variables in $\mathbf{V}$ in the population, for every W in $\mathbf{V}$, in $P(\mathbf{V})$ W is independent of its non-descendants in G (the set of non-effects) conditional on its parents in G (direct causes).

The Causal Markov Assumption states that the local Markov condition holds between the probability distribution over random variables in a population, and a causal DAG representing the causal structure in that same population, as long as the set of variables is casually sufficient in the population.

There are a number of cases where the Causal Markov Assumption is violated, e.g. if one variable is defined in terms of another (e.g. when *Total cholesterol* is defined as the sum of *LDL cholesterol* and *HDL cholesterol*), in which case the variables are associated, even if there is no strictly "causal" relationship between them. In some cases this can be remedied by simply choosing a subset of variables (e.g. *LDL cholesterol* and *HDL cholesterol* but not *Total cholesterol*).

Given the Causal Markov Assumption, if $\mathbf{X}$ is d-separated from $\mathbf{Y}$ conditional on $\mathbf{Z}$ in a causal DAG, then $\mathbf{X}$ is independent of $\mathbf{Y}$ conditional on $\mathbf{Z}$ in the population. However, the Causal Markov Assumption does *not* entail that if $\mathbf{X}$ is d-connected to (i.e. not d-separated from) $\mathbf{Y}$ conditional on $\mathbf{Z}$ that $\mathbf{X}$ is dependent on $\mathbf{Y}$ conditional on $\mathbf{Z}$. Hence, while the Causal Markov Assumption allows for some causal conclusions from sample data, it only supports inferences that some causal connections exist (i.e. dependencies entail d-connections as indicated by the contrapositive of the Causal Markov Assumption). The Causal Markov Assumption does not support inferences that some causal connections do not exist (i.e. that independencies entail d-separation). Given any data distribution, a fully connected graph will trivially satisfy the Causal Markov Assumption. In order to draw inferences from conditional independencies to the non-existence of an edge, an additional assumption must be made. The Causal Faithfulness Assumption that we will describe here is assumed by a large number of constraint-based causal search algorithms, described in more detail in Section 18.5. There are also a number of score-based algorithms that substitute various parametric assumptions in place of the Causal Faithfulness Assumption, described in Section 18.6.

Causal Faithfulness Assumption: If G is a causal graph for a given population with causally sufficient vertex set $\mathbf{V}$ and $P(\mathbf{V})$ is the probability distribution over the variables in $\mathbf{V}$ in the population, if $\mathbf{X}$ is d-connected to $\mathbf{Y}$ conditional on $\mathbf{Z}$ in G then $\mathbf{X}$ is dependent on $\mathbf{Y}$ conditional on $\mathbf{Z}$ in $P(\mathbf{V})$.

Together the Causal Markov and Causal Faithfulness Assumptions entail the following equivalence: $\mathbf{X}$ and $\mathbf{Y}$ are independent conditional on $\mathbf{Z}$ in the population if and only if $\mathbf{X}$ is d-separated from $\mathbf{Y}$ conditional on $\mathbf{Z}$ in the causal DAG (over a set of causally sufficient variables).

If $\mathbf{X}$ and $\mathbf{Y}$ are d-connected conditional on $\mathbf{Z}$ in the DAG G but $\mathbf{X}$ and $\mathbf{Y}$ are independent conditional on $\mathbf{Z}$ in P, then P is *unfaithful* to G. This can happen if two paths in the graph cancel each other. For example, suppose taking birth control pills increases the probability of blood clots directly, but also indirectly decreases the probability of blood clots through decreasing the probability of pregnancy. If the two causal paths ($BirthControlPills \rightarrow BloodClots$ and $BirthControlPills \rightarrow Pregnancy \rightarrow BloodClots$) exactly cancel each other, then $I(BirthControlPills, BloodClots)$. The Causal Faithfulness Assumption entails that such unfaithful distributions do not happen - that there may be approximately canceling causal paths but they do not *exactly* cancel each other. Another kind of violation of the Causal Faithfulness Assumption occurs when A is a cause of B and B is a cause of C, but A and C are nevertheless independent. (Necessary and sufficient conditions for a distribution to be faithful to a variety of different kinds of graphical models have recently been discovered [67]). The Causal Faithfulness Assumption has been partially justified by the fact that for any distribution in which conditional independence constraints can be expressed as an algebraic equality among the parameters, violations of the Causal Faithfulness Assumption are Lebesgue measure zero in the parameter space.

The Causal Faithfulness Assumption plays a number of distinct roles in constraint-based algorithms [79]. It assumes that certain very unusual but problematic cases where the correct identification of a true causal pattern cannot be made from observational data do not occur. For example, if (contra the Causal Faithfulness Assumption) a causal graph of the same structure as the *Birth Control Pills* in which the two paths exactly canceled were Gaussian, then no algorithm can reliably choose the correct causal graph without more information than the joint distribution over the variables. However, Causal Faithfulness is often a problematic assumption, and in Section 18.5.1.2 we will discuss in more detail why it is problematic and examine some less problematic variants of it.

18.4 Types of Search

Typically, causal search algorithms fall into one of three broad classes — constraint-based, score-based, or hybrid algorithms.

Constraint-based algorithms treat causal search as a constraint satisfaction problem. The constraints being satisfied can be of a variety of different types, with the most common one being conditional independence and dependence constraints judged to hold in the population. However, there are also non-parametric non conditional independence constraints (so-called Verma constraints) that can be used for causal search [71], as well as constraints for certain parametric families (e.g., rank constraints on sub-matrices of the correlation matrix for linear latent variable models [42, 76]; see Chapter 3 in this book).

In constraint-based searches, under the Causal Markov and Causal Faithfulness Assumptions, the constraints on the DAG are that all and only the conditional independence relations that hold in the population are entailed (via d-separation) by the true causal DAG. In practice, whether a conditional independence relation holds in the population or not is judged by performing a statistical test of conditional independence. However, the number of conditional independence and dependence relations entailed by a DAG (under the Causal Markov and Causal Faithfulness Assumptions) grows exponentially with the number of variables in the DAG, which presents a computational problem. It also raises a statistical problem since it is generally the case that if all of the conditional independence relations are tested on finite data sets, statistical tests will make some errors. If the constraints on the DAG are taken to be that the true causal DAG entails all and only the conditional independence relations that pass a statistical test on finite data, then given the number of such constraints, it is highly probable that no DAG satisfies all of the constraints. Various solutions to this problem are discussed in more detail in Section 18.5.1.2.

In contrast to constraint-based searches, score-based searches look for a causal model that maximizes or minimizes the score of a causal graph relative to a data set. Common scores that are maximized or minimized include posterior probabilities of a graph, or a penalized maximum likelihood score such as the Bayesian Information Criterion (BIC). Score-based searches have several advantages over constraint-based searches. They are often more reliable and less prone to propagate errors, in some cases they give information about how reliable the output is, and in some cases are more informative (i.e. can narrow the output down to a smaller set of candidate graphs). On the other hand, score-based searches may get stuck in local optima, and constraint-based searches often make fewer parametric assumptions than score-based searches, and easily generalize to handle unmeasured confounders and selection bias even for large sets of variables.

Hybrid searches either have some steps based on scores and other steps based on conditional independence tests, or use the outcomes of conditional independence tests to calculate a score. They are able to combine some of the advantages of each kind of both constraint-based and score-based searches. We will describe instances of constraint-based, hybrid, and score-based searches in more detail.

18.5 Constraint-Based Search and Hybrid Search

We will consider constraint-based searches under a variety of different background assumptions.

18.5.1 Acyclicity, no latent confounders, no selection bias

The PC algorithm [77] is an example of an early constraint-based search that we will use to illustrate some features common to constraint-based searches, to show some of the strengths and weaknesses of constraint-based search algorithms, and to demonstrate how various problems that arise with the simplest versions of constraint-based algorithms can be solved. We will then discuss various extensions of PC to deal with errors in statistical independence tests.

18.5.1.1 The PC algorithm

The input to the PC algorithm is an α for the significance level of the statistical tests that it performs, an independent and identically distributed (i.i.d.) sample, and optional background knowledge, including a list of forbidden edges, a list of required edges, and a partial temporal order of the variables. The output of the PC algorithm is a pattern (also known as a PDAG) [4, 8], that represents a Markov equivalence class (defined below) of DAGs.

The algorithm has two phases, a phase for determining adjacencies described in Algorithm 18.1, and a phase for determining orientations described in Algorithm 18.2. We will illustrate the algorithm with data from a study by Sewell and Shah [68] that included five variables from a sample of 10,318 Wisconsin high school seniors. (At the time, unlike today, male students were heavily over-represented in college.) The variables and their values are:

SEX [male = 0, female = 1]
IQ = Intelligence Quotient [lowest = 0, highest = 2]
CP = college plans [yes = 0, no = 1]
PE = parental encouragement [low = 0, high = 1]
SES = socioeconomic status [lowest = 0, highest = 3]

Sewell and Shah's hypothesis was the model shown in Figure 18.1(i), except that they did not include the edges from SES to CP and from IQ to CP.

For any DAG G, let a triple of sets of random variables $\langle \mathbf{X},\mathbf{Y},\mathbf{Z} \rangle$ be a member of $I(G)$ if and only if $\mathbf{X}$ is d-separated from $\mathbf{Y}$ conditional on $\mathbf{Z}$ in G. G_1 and G_2 are *Markov equivalent* if and only if $I(G_1) = I(G_2)$. A set of graphs that are all Markov equivalent to each other is a *Markov equivalence class.*

A *pattern* E represents a Markov equivalence class $\mathbf{M}$ if and only if: (i) E contains the same adjacencies as each of the DAGs in $\mathbf{M}$; (ii) each edge in E is oriented as $X \rightarrow Z$ if and only if the edge is oriented as $X \rightarrow Z$ in every DAG in $\mathbf{M}$, and oriented as $X - Z$ otherwise. There are simple algorithms for generating patterns from a DAG [49, 4, 6]. The pattern for the Markov equivalence class containing G_1 is shown in Figure 18.1(iii). It contains the same adjacencies as G_1, and the edges are the same except that the edge between SEX and IQ is undirected in the pattern, because it is oriented as $SEX \rightarrow IQ$ in G_1, and oriented as $IQ \rightarrow SEX$ in the Markov equivalent DAG G_2. A triple of variables $\langle X, Y, Z \rangle$ is an *unshielded triple* in a DAG or a pattern if and only if X is adjacent to Y but not Z, and Y

and Z are adjacent. An unshielded triple $\langle X, Y, Z\rangle$ is an *unshielded collider* if and only if X and Z are both into Y; otherwise it is an *unshielded non-collider.*

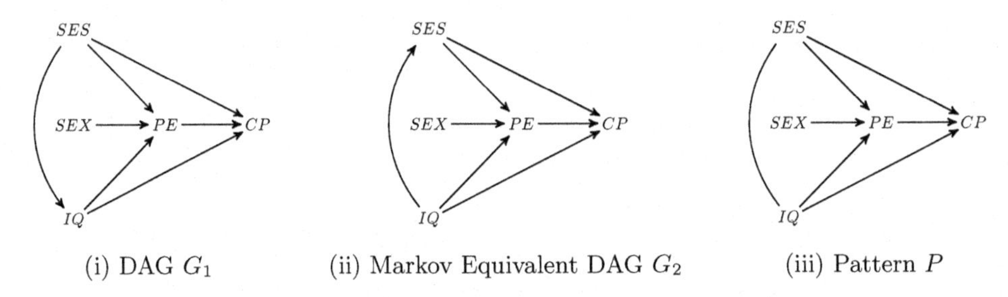

(i) DAG G_1 (ii) Markov Equivalent DAG G_2 (iii) Pattern P

FIGURE 18.1: Markov Equivalence Class and Pattern.

The adjacency phase of the PC algorithm (Algorithm 18.1) outputs a *skeleton* that is an undirected graph that represents the adjacencies (but not the orientations) common to a Markov equivalence class of the causal DAG. It is based on part (i) of the following theorem (which is a simple consequence of Theorem 3.4 and Lemma 5.1.3 in [77]):

Theorem 18.5.1. *Under the Causal Markov and Faithfulness Assumptions, in a causal DAG G over a causally sufficient set of variables with distribution P,*

(i) X and Y are adjacent if and only if they are dependent conditional on every subset of the other variables; and if X and Y are not adjacent, they are independent conditional on some subset of the parents of X or some subset of the parents of Y;

(ii) for all unshielded triples $\langle X, Y, Z\rangle$, either Y is in every subset $\mathbf{W}$ such that $I_P(X, Z|\mathbf{W})$ (an unshielded non-collider), or Y is in no subset $\mathbf{W}$ such that $I_P(X, Z|\mathbf{W})$ (an unshielded collider).

Given an oracle for conditional independence (or d-separation), the adjacency phase never incorrectly removes an adjacency to a parent of X or Y, so at each stage of the adjacency phase the PC algorithm need only search among subsets of variables adjacent to X and variables adjacent to Y for possible d-separating sets of X and Y. The adjacency phase is illustrated in Figure 18.2, where the conditional independence relations leading to the removal of each edge are listed in the captions. In Algorithm 18.1, *Adjacencies*(G, X) refers to the set of vertices adjacent to vertex X in graph G, and *Sepset*(Y,X) is the set of vertices that led to the removal of the edge between X and Y if it is removed, and is NULL if the edge is not removed.

After the adjacency phase of the algorithm, the orientation phase of the algorithm (Algorithm 18.2) is performed. The first part of the orientation phase is based on part (ii) of Theorem 18.5.1, which implies that it is possible to determine whether X and Z are both into Y by checking whether the conditioning set that led to the removal of the edge between X and Z contains Y or not. After the unshielded colliders are found, the second part of the orientation phase draws out the consequences of which triples of variables are unshielded colliders.

The orientation phase of the PC algorithm is illustrated in Figure 18.3, where the rule leading to the orientation of each edge is listed in the captions of the graphs. The last orientation rule in Algorithm 18.2 (Double Triangle) is not used in the example. The orientation rules are sound because if the edges were oriented in ways that violated the rules, there would either be a directed cycle in the pattern or it would introduce an incorrect unshielded collider. The orientation rules are also complete [48], i.e. every edge that has the same orientation in every member of a Markov equivalence class is oriented by these rules.

Algorithm 18.1 Adjacency Phase of PC Algorithm

Input: I.I.D. Sample Data, Optional Background Knowledge, Significance Level for Statistical Tests
Result: Skeleton (Undirected Graph), List of Sepsets
Form an undirected graph G in which every pair of vertices in $\mathbf{V}$ is adjacent.
$n := 0$
repeat
  repeat
    - Select an ordered pair of variables X and Y that are adjacent in G such that $Adjacencies(G, X)\backslash\{Y\}$ has cardinality greater than or equal to n, and a subset $\mathbf{S}$ of $Adjacencies(G, X)\backslash\{Y\}$ of cardinality n.
    - If X and Y are independent conditional on $\mathbf{S}$ delete edge $X - Y$ from G and record $\mathbf{S}$ in $Sepset(X,Y)$ and $Sepset(Y,X)$.
  until all such triples (X,Y,$\mathbf{S}$) have been tested for conditional independence
  $n := n + 1$
until for each ordered pair of adjacent vertices X and Y, the set $Adjacencies(G, X)\backslash\{Y\}$ is of cardinality less than n

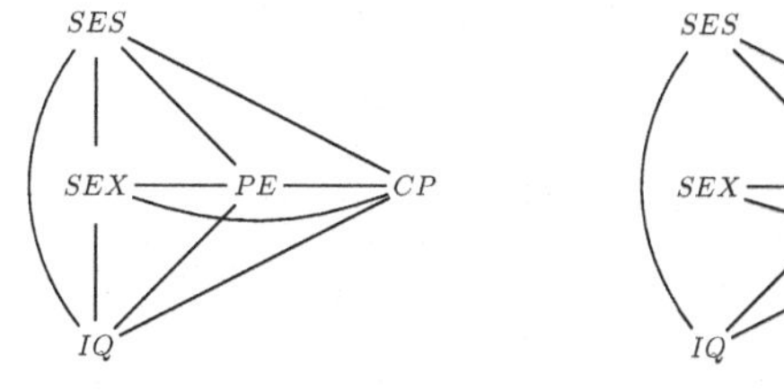

(i) Complete Graph

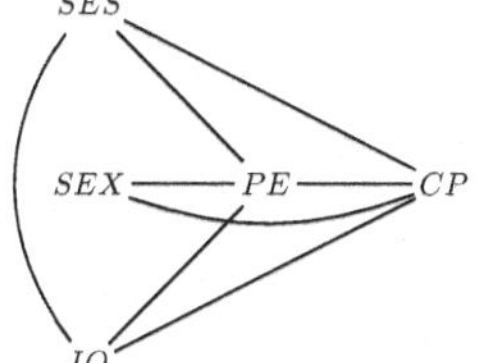

(ii) $I(SEX, SES), I(SEX, IQ)$

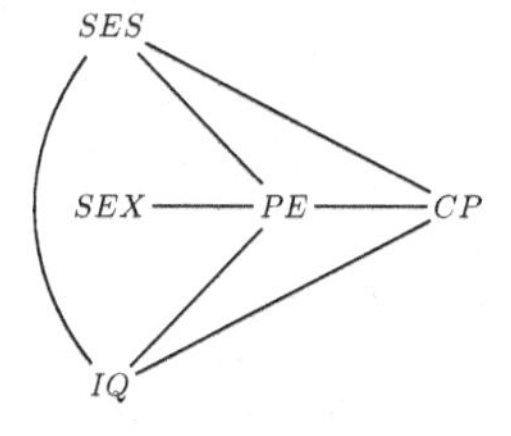

(iii) $I(SEX, CP|\{SES, IQ, PE\})$

FIGURE 18.2: Adjacency Phase of PC.

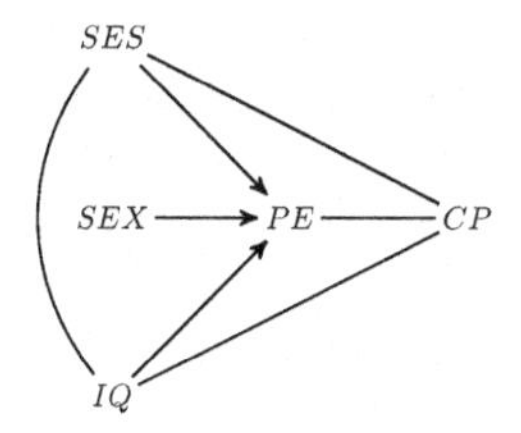

(i) Unshielded Colliders

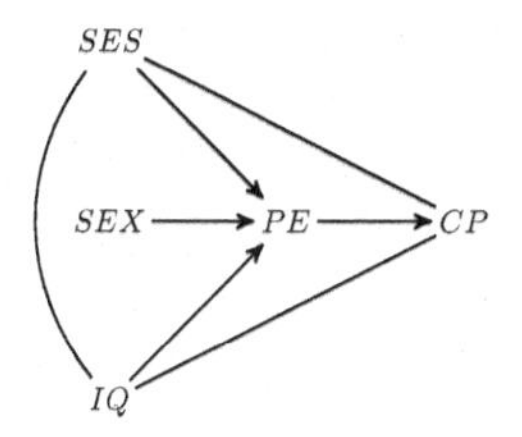

(ii) Away from Collider

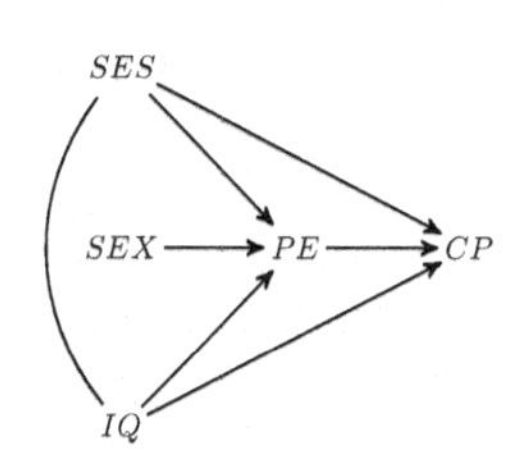

(iii) Away From Cycles

FIGURE 18.3: Orientation Phase.

Algorithm 18.2 Orientation Phase of PC Algorithm

Input: Skeleton, List of Sepsets
Result: Pattern (PDAG)
for all unshielded triple $\langle X, Y, Z \rangle$ **do**
  orient $X - Y - Z$ as $X \to Y \leftarrow Z$ if and only if Y is not in $Sepset(X,Z)$
end for
repeat
  Away from colliders:
    If $A \to B - C$, and A and C are not adjacent, then orient $B - C$ as $B \to C$.
  Away from cycles:
    If $A \to B \to C$ and $A - C$, then orient $A - C$ as $A \to C$.
  Double Triangle:
    If $A \to B \leftarrow C$, A and C are not adjacent, $A - D - C$, and there is an edge $B - D$, orient $B - D$ as $D \to B$
until no more edges can be oriented

Under the Causal Markov and Causal Faithfulness Assumptions, and given as input data from a causal graph that is acyclic and has no latent confounders and no selection bias, in the large sample limit the PC algorithm outputs a pattern that contains the true causal graph, i.e. the algorithm is pointwise consistent. The PC algorithm runs quickly on sparse graphs, since in those cases it only performs independence tests conditional on sets of small size. For sparse graphs, the PC algorithm can easily handle tens of thousands of variables in a few minutes on a laptop.

There are a number of drawbacks to the naive version of the PC algorithm, which will be addressed by various strategies discussed below. First, it does not output any information about the quality of the output. This is in part because it is not uniformly consistent, that is, it is not possible to output even non-trivial probabilistic bounds on the size of the errors in the output on finite samples. As further explained in Section 18.5.1.3, no algorithm is uniformly consistent using just conditional independence facts, given just the Causal Markov and Faithfulness Assumptions. In addition, the output depends upon the α level chosen for the significance levels of the tests of conditional independence and the order in which the tests were performed (because of statistical errors). In practice, this in turn creates a dependence on the order in which the variables were listed in the data. A recent modification of the PC algorithm, the PC-Stable algorithm, removes the order dependence of the output and improves the performance of PC [11].

The output of the PC algorithm in Figure 18.3(iii) is a pattern that contains two DAGs, one with $IQ \to SES$ and one with $SES \to IQ$. It can be used to give point estimates of some of the total effects of one variable on another, and bounds for all of the total effects. Standard techniques for estimating the parameter values for the causal model specified by a given DAG and data cannot be directly applied to the pattern, since it usually represents multiple DAGs. Similarly, the do-calculus cannot be applied directly to the output, since the output is a pattern. The "global" version of the "Intervention-calculus When the DAG is Absent" (IDA) algorithm [46] solves this problem by first forming every possible DAG represented by the output pattern, estimating the parameter values for each vertex and its parent set, and then applying the do-calculus to each estimated DAG to calculate the total effect (or some other effect of interest) of A on B. The results of applying the do-calculus to each DAG can then be recorded in a multiset. From the multiset, the bounds on each effect can be calculated. In the example, the multiset for the total effect of *SEX* on *CP* contains a single value; the multiset for the total effect of manipulating *IQ* on *SES* contains multiple values depending on which way the edge between *SES* and *IQ* is oriented.

One of those values will be no effect (i.e. the probability distribution of *SES* remains the same for all manipulations of *IQ*), as will be the case for every unoriented edge. In more complex patterns, there are cases where there are multiple values for the total effect of a manipulation in the multiset, where none of those values is zero.

However, for large patterns it is not computationally feasible to form every DAG represented by the pattern. For the question of what the total effect of A on B is, the "local" version of the IDA algorithm forms every orientation of the undirected edges containing B that does not create new unshielded colliders or cycles, estimates the total effect of A on B for each orientation, and stores each different value in a set. By selecting pairs of variables whose minimal absolute value of the estimate of the effect of A on B is large, the output can be used to guide experimenters in choosing which experiment to perform next (if the goal is to find strong causes). This algorithm has been successfully applied to find the largest total effects of yeast genes on each other from observational data on gene activity levels, and partially verified with experimental data on yeast [45], and recently been extended to handle estimation of multiple simultaneous interventions [52].

18.5.1.2 Dealing with Statistical Errors

In practice, statistical tests are performed by the PC algorithm to determine if a conditional independence relation is judged to hold. In the case of linear Gaussian models, conditional independence is equivalent to a zero partial correlation, which can be tested using Fisher's z-test. In the multinomial case, a chi-squared test of the G^2 statistic can be used. There exist extensions of the PC algorithm to deal with other cases as well, including Rank PC [23] for Gaussian copula data, and Copula PC [13] for mixed continuous and discrete data. In addition, there are recently discovered non-parametric tests of conditional independence that can be used quite generally [94] (which is one advantage constraint-based algorithms have over score-based algorithms); however, they are often quite slow and require large sample sizes to have useful power.

The use of statistical tests of conditional independence introduces several problems. First, for any given significance level of the tests, there will be a percentage of type I errors (where the null is the hypothesis of conditional independence, and a type I error is a false rejection of a conditional independence). Second, the power of such tests is limited, so there will be some type II errors (i.e. a conditional independence that does not hold in the distribution will be falsely accepted). The former is particularly likely when there are "almost" violations of faithfulness, that is cases where the conditional independence is not entailed, but approximately (and not exactly) holds in the population. In the case of linear Gaussian models, an "almost" violation of the Causal Faithfulness Assumption is one where X and Y are not d-separated conditional on $\mathbf{W}$ in the true causal graph, but $\rho_{X,Y|\mathbf{W}}$ is close to zero. In those cases, only very powerful tests will be able to reject the conditional independence, and they generally require very large sample sizes. The problem is particularly severe for multinomial distributions, where the decrease of the power of the test as the size of the conditioning set increases is much steeper than in the Gaussian case.

There are a number of ways of dealing with type I or type II errors in the output of statistical tests that are performed. The first and simplest method is to simply ignore them. The naive version of the PC algorithm is an example of this. The PC algorithm in effect assumes that there are no errors in the output of the statistical tests applied to finite samples. This allows the PC algorithm to limit the number of conditional independence relations that it tests, which has both a computational advantage (by reducing the number of conditional independence relations that need to be tested) and statistical advantages (by minimizing the size of the conditioning sets in conditional independence tests, leading to greater power of the tests). However, the assumption that the tests are all correct leads the

PC algorithm to ignore some errors that are detectable. All of the following methods for dealing with statistical errors generally improve the performance of the naive version of the PC algorithm.

A second method of dealing with statistical errors is to search for sets of outputs of conditional independence tests that conflict with each other, and not use any of the conflicting members of the set when drawing conclusions. For example, part (ii) of Theorem 18.5.1 entails that if $\langle X, Y, Z \rangle$ is an unshielded triple, then either every set $\mathbf{W}$ that d-separates X and Z contains Y (if $\langle X, Y, Z \rangle$ is a non-collider) or no set that d-separates X and Z contains Y (if $\langle X, Y, Z \rangle$ is a collider). Assume that the output of a test of conditional independence is either "independent" or "dependent" [77]. (This is not the standard interpretation of statistical tests in which the outputs are interpreted as "reject" or "fail to reject", but is necessary for the employment of statistical tests in causal search algorithms.) Hence, if there are two sets $\mathbf{W_1}$ and $\mathbf{W_2}$ such that statistical tests report "independent" for the tests $I(X, Z|\mathbf{W_1})$ and $I(X, Z|\mathbf{W_2})$, where $\mathbf{W_1}$ contains Y and $\mathbf{W_2}$ does not contain Y, under the Causal Markov and Faithfulness Assumptions, at least one of the tests must be in error. We will refer to sets of conditional independencies and dependencies that are not entailed by any DAG given the Causal Markov and Causal Faithfulness assumptions as "conflicted", and the existence of such sets as "conflicts". This is likely to happen if there are "almost" violations of the Causal Faithfulness Assumption, where a conditional dependence is not entailed to be absent, but due to the particular parameter values is quite weak. The Conservative PC algorithm, which is a simple modification of the PC algorithm [64], tests whether X and Z are independent given some subset of the variables adjacent to X or adjacent to Z that contains Y, and also whether X and Z are independent given some subset of the variables adjacent to X or adjacent to Z that does not contain Y, and if both kinds of independencies are judged to hold, the algorithm indicates that it cannot tell whether the correct orientation is a collider or a non-collider. This has both a computational cost (the Conservative PC tests more conditional independence relations than PC), a statistical cost (Conservative PC may test conditional independence relations on larger conditioning sets than PC), and is possibly less informative in its output. However, Conservative PC is provably correct under a weaker assumption than the Causal Faithfulness Assumption: it requires only the Adjacency Faithfulness Assumption [64], i.e. that if two variables are adjacent, then they are not independent conditional on any subset of the other variables, which is strictly weaker than the Causal Faithfulness Assumption. In addition, Conservative PC is not significantly slower than PC, and is generally more accurate. Extending this strategy, the Very Conservative SGS Algorithm [79] performs additional checks, including whether the output of the algorithm is Markov to the distribution. Again this comes at the cost of slowing down the PC algorithm, and increasing the number of DAGs compatible with the output.

A third way of dealing with statistical errors is to use only the test outputs estimated to be most reliable in the algorithm. For example, the COmbINE algorithm [83] and the BCCD algorithm [10] apply heuristics to the p-values of outcomes of conditional independence tests to order them by how reliable they are, and then input the outcomes into the respective algorithms in order of estimated reliability.

A fourth way of dealing with statistical errors is to use bootstrapping. For example, suppose the goal is to select the strongest causes of some variable B. The CStaR algorithm [80] runs IDA on multiple bootstrap samples of size $n/2$ and records what percentage of the runs a given variable appears in the top q variables (estimated strongest causes of B). The ones that appear most often are stably estimated to be among the largest causes of B. The algorithm also estimates a per comparison error rate. In application to the same yeast data as IDA, CStaR outperformed IDA, and when applied to measurements of gene activity in Arabidopsis plants, was also able to successfully select genes that were strong causes of the time to flowering, as confirmed by subsequent knockout experiments on the genes [80]. Stability selection has also been applied to the PC algorithm to reduce order dependence [11].

A fifth way of dealing with statistical errors is to modify the algorithm to control the false discovery rate. This also has the advantage of dealing with two other problems with the PC algorithm - how to choose the significance level α for the tests, and how to evaluate the quality of the output. The naive version of the PC algorithm performs a large number of conditional independence tests at a user-selected α value for the tests, but does no adjustment for multiple testing. Hence the α levels for the individual tests bear at most a very complicated relationship to the probability of errors in the output. It would be possible to use a Bonferroni adjustment, but given the large number of tests that are done, and the fact that it is not known prior to the search how many conditional independence tests will be done, this is generally not practical. It would be highly desirable to control the errors in the output, e.g. the number of false adjacencies in the output. The false discovery rate is the expected ratio of falsely discovered positive hypotheses to all those discovered, and it has been discovered how the α level can be adjusted in order to control the false discovery rate of the adjacencies in the output [38], and of the orientations as well [81].

18.5.1.3 Uniform Consistency

The existence of causal models that contain "almost" violations of the Causal Faithfulness Assumption but are arbitrarily close to models that do not violate the Causal Faithfulness Assumption entails that it is not possible to put non-trivial probabilistic bounds on errors in the output of any constraint-based causal algorithm, that is, there is no uniformly consistent constraint-based algorithm for finding the correct pattern even assuming acyclicity, no latent confounders, no selection bias, and the Causal Markov and Faithfulness Assumptions [66].

However, it has been shown [40] that under a set of assumptions including a strengthened version of the Causal Faithfulness Assumption, the PC algorithm is a uniformly consistent estimator of the pattern that represents the true causal DAG for linear Gaussian models, even in the high-dimensional case where the number of variables and the density of the DAG grows with the sample size. The Strong Causal Faithfulness Assumption states that for Gaussian distributions, the partial correlations that are not entailed to be zero by the graph are all bounded away from zero by a constant (which can depend on the number of variables in the DAG and the sample size). The Strong Causal Faithfulness Assumption has the implausible consequence that there are no very weak but non-zero edges in a DAG. Furthermore, it has been shown that for certain DAGs, the probability of a violation of the Strong Causal Faithfulness Assumption is quite high unless the constant is extremely small (that is, the percentage of the volume of the parameter space taken up by causal models that contain an "almost" violation of the Strong Causal Faithfulness Assumption is large), and grows along with the number of variables [85].

Not every violation of the Strong Causal Faithfulness Assumption would necessarily lead to incorrect output, and in some cases the errors could be quite localized. However, there are also cases where violations of the Strong Causal Faithfulness Assumption do lead to errors, and the errors can propagate throughout the output. (For example, if an unshielded non-collider is mistaken for an unshielded collider, that can lead to other incorrect orientations as well, through use of the "Away from Colliders" orientation rule.)

There is a constraint-based algorithm (the sample version of the Very Conservative SGS algorithm [79]) that is uniformly consistent under a weaker version of the Strong Causal Faithfulness Assumption. The weaker version of the Strong Causal Faithfulness Assumption applies only to pairs of adjacent variables X and Y in a clique of size greater than 2, and bounds partial correlations not entailed to be zero away from zero by a function of the strength of the edge connecting X and Y. However, the Very Conservative SGS algorithm is not computationally feasible on large numbers of variables, nor is it currently known how probable violations of this weaker version of the Strong Causal Faithfulness Assumption are.

18.5.1.4 General strategies

A general framework for causal discovery has been developed [1, 2] that is a generalization and improvement of the strategy used by the PC algorithm, and can be instantiated with constraint-based or hybrid algorithms. At the broadest level, Locally-constrained Global Learning (LGL) has three steps: (i) Find the parents and children of each variable X in the data (denoted $\mathbf{PC}(X)$, where **PC** in this case stands for "the set of parents and children") using an admissible instantiation of Generalized Local Learning - Parents and Children (described below); (ii) piece together the undirected edges; (iii) orient the edges with any desired edge orientation scheme.

The Generalized Local Learning - Parents and Children (GLL-PC) algorithm has the following steps: (i) Begin with an empty set of candidates; (ii) apply a heuristic to rank variables for priority for inclusion in the candidate set, and then include the highest ranked variable or variables; (iii) each variable that is independent of the response variable conditional on any subset of the candidate set is permanently discarded; (iv) elimination and inclusion can be iterated in any fashion as long as iteration stops when no variable not in the candidate set is eligible for inclusion, and no more variables in the candidate set can be removed; (v) filter the candidate set using symmetry criterion after the iteration has stopped; (vi) output the candidate set.

Many algorithms, including two well-known algorithms that have proved successful on real and simulated data, Max-Min Parent Child (MMPC) and Hiton-PC [3] are instances of GLL-PC. The Max-Min Hill Climbing Algorithm (MMHC), which calls MMPC [84] as its first step, is an instance of LGL that can handle up to 10,000 discrete variables on a laptop, and has been shown to be among the most accurate causal search algorithms for discrete data, although it is not sound in the large sample limit due to the constrained search space [51].

18.5.2 Acyclicity, latent confounders, selection bias

Unlike PC-style algorithms, FCI-style algorithms, including the Fast Causal Inference (FCI), Really Fast Causal Inference (RFCI), and FCI+ algorithms, are pointwise consistent without assuming that there are no latent common causes or selection bias [78, 12, 10].

Selection bias introduces complications into drawing causal inferences from conditional independence and dependence relations in the following way. Imagine, for example, *IQ* is independent of *SEX*, but only women with very high *IQ*s attended college, and men of average *IQ*s or higher attended college. It would follow that *SEX* and *IQ* are independent in the population, but dependent in a sample of college students (since anyone with an average *IQ* in college would have to be male). A mechanism that sampled randomly from college students would entail that the population and the sub-population from which the sample was drawn do not exhibit the same conditional independencies and dependencies.

This situation can be represented in a DAG by putting the sampling mechanism into the DAG. The variables $\mathbf{V}$ in a causally sufficient causal DAG G with a given sampling method, can be partitioned into three types: a subset $\mathbf{O}$ of observed variables, a subset $\mathbf{L}$ of latent or unobserved variables, and a set $\mathbf{S}$ of selection variables. For each variable O in $\mathbf{O}$, there can be a corresponding variable S_O in $\mathbf{S}$, which takes the value 1 when the value of O has been measured for a unit in the population and 0 otherwise. (For cases where either all of the variables of a unit are measured or none of them are **S** can be taken to be a single variable indicating whether the unit is in the sample or not, and similarly other sampling mechanisms may lead to combining some of the selection variables.) Since the value of a variable O is measured only when the value of S_O is 1, the sample distribution of O is the population distribution conditional on $S_O = 1$. If missing values for O are missing at random, S_O has no causal connections to any other variables, and the conditional independencies and dependencies in the sample distribution will be the same as in the population distribution. For the *IQ* and *SEX* example, the sampling mechanism would be represented by having directed edges from *SEX* and *IQ* to S (since *SEX* and *IQ* are causes of attending college, which in turn causes whether someone is in the sample or not). In this case, conditioning on S is conditioning on an unshielded collider, which introduces a dependence between *SEX* and *IQ*. Given a causally sufficient DAG G and a partition of the variables in $\mathbf{V}$ into $\mathbf{O}$, $\mathbf{L}$, and $\mathbf{S}$, we will represent the partition as $G(\mathbf{O},\mathbf{L},\mathbf{S})$.

Two DAGs $G_1(\mathbf{O}, \mathbf{L_1}, \mathbf{S})$ and $G_2(\mathbf{O}, \mathbf{L_2}, \mathbf{S})$ are *Markov equivalent over* $\boldsymbol{O}$ *conditional on* S when for all disjoint $\mathbf{X}, \mathbf{Y}, \mathbf{Z} \subseteq \mathbf{O}$, $\mathbf{X}$ and $\mathbf{Y}$ are d-separated conditional on $\mathbf{Z} \cup \mathbf{S}$ in $G_1(\mathbf{O}, \mathbf{L_1}, \mathbf{S})$ if and only if $\mathbf{X}$ and $\mathbf{Y}$ are d-separated conditional on $\mathbf{Z} \cup \mathbf{S}$ in $G_2(\mathbf{O}, \mathbf{L_2}, \mathbf{S})$. The set of all DAGs Markov equivalent to $G(\mathbf{O}, \mathbf{L}, \mathbf{S})$ over $\mathbf{O}$ conditional on $\mathbf{S}$ is denoted by **O**-**S**-$Equiv(G(\mathbf{O}, \mathbf{S}, \mathbf{L}))$.

The output of the FCI and FCI+ algorithms is a complete partial ancestral graph (CPAG), which is a graphical object that represents features common to an **O**-**S**-Markov-equivalence class of DAGs. (A CPAG also represents a Markov equivalence class of Mixed Ancestral graphs: see Chapter 2 for more details about Mixed Ancestral graphs.) Examples of CPAGs are shown in Figures 18.4(i) and 18.4(iii). (The output of the RFCI algorithm is a slight modification of a CPAG.) The CPAG in Figure 18.4(i) represents the DAGs in Figures 18.1(i) and 18.1(ii), as well as an infinite number of other DAGs that may have an arbitrarily large set of unmeasured confounders. Despite the fact that there are important differences between the DAGs in the **O**-**S**-Markov equivalence class of the DAG in Figure 18.1(i), they share a number of important features in common (e.g. in all of the DAGs, PE is a direct cause of CP, there is no unmeasured confounder of PE and CP, and CP is not a cause of PE). The features that all DAGs represented by a CPAG share in common can be read off of the output CPAG according to the following rules.

Let "$*$" be a meta-symbol that stands for any of the three kinds of edge endpoints, tail ("–"), head (">"), and circle ("o"), that can occur as endpoints of edges in a CPAG. H is a complete partial ancestral graph [88] that represents a DAG $G(\mathbf{O}, \mathbf{L}, \mathbf{S})$ or **O**-**S**-$Equiv(G(\mathbf{O}, \mathbf{S}, \mathbf{L}))$ if and only if:

- The set of variables in H is $\mathbf{O}$.
- If there is any edge between A and B in H, it is one of the following kinds: $A \rightarrow B$, $A \circ\!\!\rightarrow B$, $A \leftrightarrow B$, $A \circ\!\!-\!\!\circ B$, $A - B$, or $A \circ\!\!- B$.
- A and B are adjacent in H if and only if for every subset $\mathbf{Z}$ of $\mathbf{O} \setminus \{A, B\}$ A is d-connected to B conditional on $\mathbf{Z} \cup \mathbf{S}$ in every DAG in **O**-**S**-$Equiv(G(\mathbf{O}, \mathbf{S}, \mathbf{L}))$.
- An edge between A and B in H is oriented as $A \,-\!\!* B$ only if A is an ancestor of B or of $\mathbf{S}$ in every DAG in **O**-**S**-$Equiv(G(\mathbf{O}, \mathbf{S}, \mathbf{L}))$.
- An edge between A and B in H is oriented as $A \leftarrow\!\!* B$ only if A is not an ancestor of B or $\mathbf{S}$ in any DAG in **O**-**S**-$Equiv(G(\mathbf{O}, \mathbf{S}, \mathbf{L}))$.
- An edge between A and B in H is oriented as $A \circ\!\!-\!\!* B$ only if A is an ancestor of B or $\mathbf{S}$ in some but not all DAGs in **O**-**S**-$Equiv(G(\mathbf{O}, \mathbf{S}, \mathbf{L}))$.

While a directed edge from X to Y in a CPAG (without selection bias) entails that X is a cause of Y, it does not entail that it is a *direct* cause of Y, or that there is no latent common cause of X and Y as well. However, in addition to the direct meaning of the edge, there are also some other features that all of the DAGs represented by a CPAG may have in common and that can be read off of the CPAG. For example, it is possible to tell for some directed edges $A \to B$ that there is also no unmeasured confounder of A and B, as is the case with *PE* and *CP* [87]. As with patterns, not every causal effect can be calculated from a CPAG. There is an algorithm that outputs an estimate of a causal effect if it can be calculated from a CPAG, but it is not known whether it is complete [87] and there is an algorithm for calculating bounds on causal effects from CPAGs [47] based on the generalized back-door criterion [59].

On the Sewell and Shah data, the output of the FCI algorithm is shown in Figure 18.4(iii). The CPAG that represents the DAG in Figure 18.1(i) (where $\mathbf{L}$ and $\mathbf{S}$ are empty) is shown in Figure 18.4(i). The reason that the two differ is that there is no DAG without latent variables that faithfully represents the set of conditional independencies and dependencies judged to hold in the population, and hence the PC algorithm, which assumed no latent variables, found a pattern that entailed as many of the conditional independence and dependence constraints as possible. The two CPAGs in Figures 18.4(iii) and 18.4(i) give slightly different estimates of the effect of *PE* on *CP*, but are still fairly close to each other. However they disagree strongly about the effect of *SES* and *IQ* on *CP*; the CPAG in Figure 18.4(iii) entails that *SES* and *IQ* have no effect on *CP* (which is not very plausible), whereas the effects of *SES* and *IQ* on *CP* cannot be calculated from the CPAG in Figure 18.4(i).

Although FCI-style algorithms are significantly slower than PC-style algorithms, for sparse CPAGs the algorithm can still handle up to 10,000 variables on a laptop. They tend to be much less accurate than PC-style algorithms, often producing far too many "$\leftrightarrow$" edges. However, a hybrid algorithm, the Greedy Fast Causal Inference (GFCI) algorithm is much more accurate on simulated data [57]. GFCI uses GES (described in Section 18.6.2) as a first step to output a pattern, and the adjacencies and unshielded colliders and non-colliders in the pattern are used as a starting point for an FCI search.

Time series data raises a number of issues, including non-stationarity and the problem of sampling at a rate that is slower than the causal process, in which the unsampled variables act like latent variables. Modifications of FCI [19, 9] as well as other constraint-based algorithms [31] have been applied to time series data.

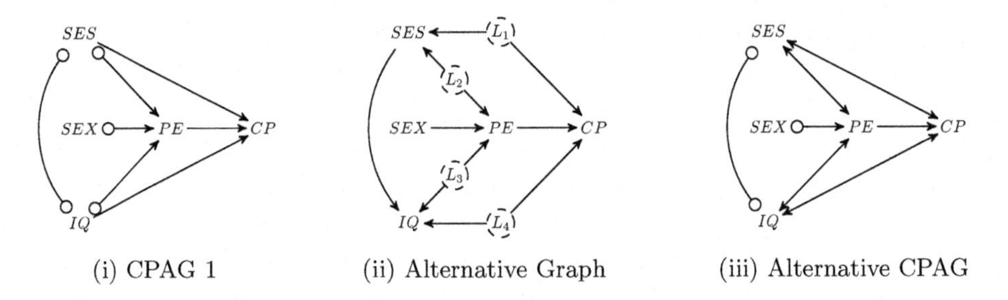

(i) CPAG 1 (ii) Alternative Graph (iii) Alternative CPAG

FIGURE 18.4: Complete Partial Ancestral Graphs.

18.5.3 Cycles, no latent confounders, no selection bias

One way of representing structural equation models of feedback at equilibrium is by allowing directed cycles into a directed graph. If there is feedback between A and B, then clamping A to a fixed value will affect the equilibrium value of B, and alternatively, clamping B at a fixed value will affect the equilibrium value of A, thus leading to a cycle in the directed graph. In the case of linear models [75] and some multinomial models [36, 53, 62] if $\mathbf{X}$ is

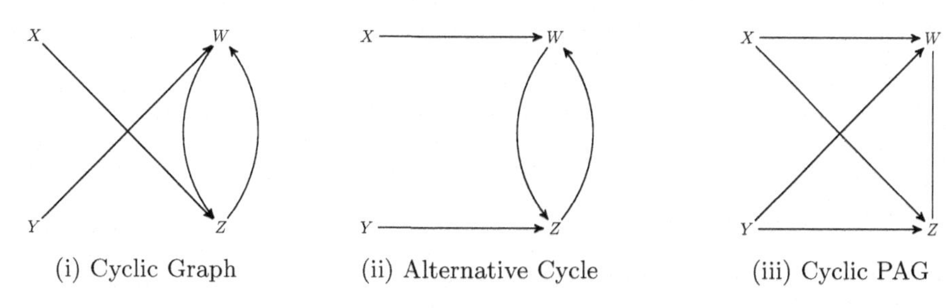

(i) Cyclic Graph (ii) Alternative Cycle (iii) Cyclic PAG

FIGURE 18.5: Cyclic Graphs.

d-separated from $\mathbf{Y}$ conditional on $\mathbf{Z}$ in the directed graph, it is entailed that $I(\mathbf{X}, \mathbf{Y}|\mathbf{Z})$ (note that "d-separation" here refers to the obvious extension of d-separation as defined for acyclic graphs). However, in the general case, only a subset of d-separations in a cyclic directed graph entail the corresponding conditional independence relation [75].

Even for the linear case, the Markov equivalence class of directed graphs with cycles is considerably more complicated than it is in the acyclic case. For example, the two directed graphs shown in Figure 18.5(i) and Figure 18.5(ii) are Markov equivalent to each other, but in contrast to the acyclic case, do not contain the same adjacencies. In addition, it is possible that two variables that are not adjacent in the causal graph are d-connected conditional on every subset of the other variables, e.g. X and W in Figure 18.5(i).

The Cyclic Causal Discovery (CCD) algorithm outputs a graph that represents a Markov equivalence class of cyclic graphs, that is pointwise consistent under the Causal Markov and Faithfulness Assumptions, as long as the variables are causally sufficient and there is no selection bias, and d-separation entails conditional independence (as in the linear Gaussian case) [65].

18.5.4 Cycles, latent confounders, overlapping data sets, experimental and observational data

Unlike the previous cases considered, there is no known efficient test of Markov equivalence for directed graphs with that contain all of cycles, latent confounders, and selection bias, nor is there an analogy of a pattern for representing features common to a Markov equivalence class. Several algorithms have been proposed that use SAT solvers to search for a DAG that most closely matches the conditional independence and dependence constraints judged to hold in the data [83, 29].

The Answer Set Programming Causal Discovery algorithm [29] is a hybrid algorithm that applies the following strategy. First, the class of graphs considered has two kinds of edges: $A \rightarrow B$ represents that A is a direct cause of B, and $A \leftrightarrow B$ represents that there is a latent common cause of A and B. Each pair of variables can have any combination of the possible edge types between them. For each conditional independence or dependence entailed by a graph, the corresponding conditional independence test is performed, and assigned a weight based on its p-value, intended to represent the reliability of the outcome of the test. Each graph G is assigned a score equal to the sum of the weights of the conditional independence tests not compatible with G, that is, it is penalized for predicting a false outcome of a conditional independence test. The weights are assigned in such a way that conflicts among constraints are resolved well by minimizing the sum of the weights. The output G^* of the algorithm is a graph that minimizes the sum of the weights. The constraints can be extended to include data from multiple overlapping data sets that combine both experimental and observational data. There can be other graphs that entail exactly the same conditional independence relations as G^*, but it is possible to determine whether all such graphs orient a particular adjacency in the same direction using another SAT solver algorithm [30]. Because the number of constraints grows exponentially with the number of variables, the Answer Set Programming Causal Discovery algorithm is currently limited to about a dozen variables. Constraint-based causal discovery has also been extended to deal with nonstationary data or multiple, heterogeneous data sets [90], in which the data distribution varies over time or across data sets.

To summarize, constraint-based approaches to causal discovery make use of (conditional) independence relationships between random variables. Given reliable, nonparametric conditional independence tests, such methods are widely applicable because they do not make specific assumptions on causal mechanisms. On the other hand, they involve the multiple testing problem [26], i.e., they require the simultaneous testing of more than one (conditional) independence hypothesis. Constraint-based methods, such as the PC algorithm, also usually use the same significance level for different tests; this may lead to different powers in conditional independence tests with different numbers of conditional variables.

18.6 Score-Based Search

In contrast to constraint-based approaches, score-based methods aim to find the causal structure by optimizing a properly defined score function. The score function may be the Bayesian posterior probability of the causal structure or a penalized likelihood like BIC. We start with score-based Bayesian network learning algorithms, and note that their output DAG structure is not necessarily identical to, but has the same (conditional) independence relations as, the underlying causal structure. We then review classic score-based causal search methods that output the Markov equivalence class of DAGs. These score-based methods usually assume either the case with linear causal relations and Gaussian noise or the case with a multinomial distribution for the data, and the scoring criterion is score equivalent (it has the same score value for Markov equivalent DAGs) [8]. That is, in those two cases we cannot distinguish between different DAGs in the equivalence class; in particular, in the two-variable case, we cannot tell cause from effect. Next, we will discuss causal discovery based on restricted functional causal models; by making use of well-designed functional causal models and certain properties of the data distribution, this set of approaches is able to distinguish between causal structures in the same equivalence class. Roughly speaking, in the continuous case when we drop either the linearity assumption on the causal relations or the Gaussianity assumption on the noise terms, the best solution is usually a single DAG, not an equivalence class any more.

In this section we assume that we have access to i.i.d. data, that the system is causally sufficient and does not have any cycles, and that there is no selection bias in the sampling procedure. At the end of this section we will briefly mention some score-based methods that relax such assumptions.

18.6.1 Score-based DAG search

As stated in Section 18.2, the number of different DAGs grows super-exponentially with the number of variables. In fact, without additional constraints on the Bayesian network structure (or DAG), learning the network structure based on a properly defined score function has been proven to be NP-hard [7]. Therefore, to learn the optimal DAG over a reasonable number of variables, additional constraints are usually assumed. For instance, the method proposed for learning Bayesian networks of bounded treewidth is polynomial both in the size of the graph and the treewidth bound [18]. (The treewidth of a Bayesian network is defined as the treewidth of its moralized (undirected) graph; the treewidth, roughly speaking, indicates how dense the undirected graph is.) Otherwise, learning procedures usually resort to smart tricks or heuristics.

Assuming the scoring function can be decomposed into a sum of local terms, each of which depends on a child node and its parent nodes, a number of well-known exact Bayesian network learning algorithms were developed based on dynamic programming [41], in the way that the traveling salesman problem was solved [5]. In practice, these algorithms are feasible to networks of at most 30 variables. Later different heuristic strategies were used to reduce the search space and improve the scalability of the learning procedure, including the relaxed linear programming approach [37] and the A^*-based shortest path finding approach [86]. The latter approach has been demonstrated to be feasible to be applied to more than 40 variables. Note that these approaches were first developed for discrete data and that they were later extended to handle the linear-Gaussian case.

To take advantage of the merits of both the constraint-based and score-based search procedures, various hybrid search methods have been proposed for DAG learning; examples include the CB algorithm [73] and Max-Min Hill-Climbing (MMHC) [84].

As mentioned before, the DAGs learned by this set of methods do not necessarily have a causal interpretation. For instance, suppose that we are given enough data for two variables which are discrete or Gaussian with a linear causal relation. Such methods will output a directed edge between them, while the direction may be arbitrary. In this case we cannot distinguish different causal structures in the same equivalence class, which share the same (conditional) independence relationships. Generally speaking, this set of methods outputs an arbitrary DAG in this equivalence class, which, nevertheless, give a compact representation of the joint distribution. In order for the output to have a causal interpretation, one may apply some procedures to generate the pattern from the output DAG [49, 4, 6], as mentioned in Section 18.5.1.1, as an additional post-processing step. It is also worth mentioning that in the presence of confounders, there exist score-based methods to learn acyclic directed mixed graphs (ADMGs) [16], "bow-free" ADMGs [56], and nested Markov models [72], from observed data. See Chapter 2.

18.6.2 Score-based equivalence class search

There are a number of algorithms that directly search over the space of equivalence classes. Among them, Greedy Equivalence Search (GES) [8] is a well-known two-phase (step-forward by adding edges, and step-back by removing edges) search procedure. It was shown that if the generative distribution is Markov and faithful with respect to a DAG defined over the measured variables and the scoring function is locally consistent (which is the case for the BIC score), then under the Causal Markov and Faithfulness Assumptions, in the limit of large sample sizes GES outputs the pattern representing the Markov equivalence class of the underlying causal DAG [8]. That is, GES is statistically consistent in classical settings (with a fixed number of variables) even if it involves a greedy search. GES has further been shown to be consistent in certain sparse high-dimensional settings with multivariate Gaussian or nonparanormal distributions [51].

Various extensions of GES have been developed to improve its performance or applicability. The k-greedy equivalence search algorithm [54] allows a trade-off between greediness and randomness, to help escape local optima on finite samples. GES has been extended to learn causal structures from data arising from multiple interventions, yielding an algorithm called Greedy Interventional Equivalence Search (GIES) [25]. Adaptively Restricted GES (ARGES) [51], as a hybrid version of GES, modifies the forward phase of GES by restricting edge additions to only those so-called admissible ones, and still retains statistical consistency. Thanks to modifications that parallelize and reorganize caching in the GES algorithm, a recently proposed search procedure, Fast Greedy Search (FGS), is able to recover sparse causal graphs with one million variables with high precision and good recall [63].

18.6.3 Functional causal discovery

Recently it has been shown that with properly defined structural equation models, or Functional Causal Models (FCMs), the underlying causal structure can be uniquely estimated from observed data, i.e., we are able to distinguish between different DAGs in the same Markov equivalence class. In this section we focus on continuous data. A FCM represents the effect Y as a function of the direct causes X and some noise [58]:

$$Y = f(X, E; \boldsymbol{\theta}_1), \tag{18.1}$$

where E is the noise term that is assumed to be independent from X, the function $f \in \mathcal{F}$ explains how Y is generated from X, $\mathcal{F}$ is an appropriately constrained functional class, and $\boldsymbol{\theta}_1$ is the parameter set involved in f. We assume that the transformation from (X, E) to (X, Y) is invertible, such that E can be uniquely recovered from the observed variables X and Y.

In the FCM (18.1), the noise term is assumed to be independent from the cause. The effectiveness of causal discovery based on functional causal models relies on two properties of the exploited FCMs [92]. On one hand, the FCM is expected to be able to approximate the true cause process from causes to the effect; if very little information about the causal process is known, then the FCM should be as flexible as possible. On the other hand, as shown in Section 18.6.3.1, without suitable constraints on the FCM, one can find independent noise terms for both directions between X and Y. Hence, it is essential for the FCM to be properly constrained, so that the causal direction implied by the FCM is identifiable–this amounts to proving that the model assumptions, especially the independence between the noise and cause, hold only for the true causal direction and are violated for the wrong direction.

In Section 18.6.3.2 we will discuss several FCMs according to which the causal direction is generally identifiable. Generally speaking, for these FCMs if the true causal model is $Y = f(X, E)$, where E is independent of X, then there does not exist a function $g \in \mathcal{F}$ such that $X = g(Y, \widetilde{E})$ with $\widetilde{E}$ independent of Y, thus rendering the causal direction between X and Y identifiable. This identifiability condition can be understood as a special type of faithfulness. As (18.1) is assumed to be invertible, we can write E as a function of Y and X accordingly: $E = h(Y, X)$, where $h \in \mathcal{H}$ and the functional class $\mathcal{H}$ corresponds to the functional class $\mathcal{F}$ for f. Then the causal direction between X and Y is identifiable if and only if for any function $\widetilde{h} \in \mathcal{H}$, $\widetilde{h}(X, Y)$ is always dependent on its input Y.

Here we term causal discovery based on FCMs as "functional causal discovery." Suppose the underlying causal structure is identifiable. Functional causal discovery aims to find the best structure that, while as simple as possible, makes the noise terms independent from their respective causes. It has been shown that for general FCMs, minimizing the mutual information between noise terms and their corresponding hypothetical causes is equivalent to maximizing the data likelihood [95]. (This result was shown for FCMs with additive noise in [91].) Therefore, this set of methods can be put under the umbrella of score-based methods, and the score is a properly penalized likelihood function. Sometimes one further performs model checking by testing for statistical independence between the estimated noise term and hypothetical causes [27, 92].

18.6.3.1 Remark on general functional causal models

If for the reverse of the true causal direction one cannot find a noise term which is independent from the hypothetical cause (which is Y), then we can determine the true causal direction, or distinguish cause from effect, because of this asymmetry. As we shall see in Section 18.6.3.2, in general this is the case for the post-nonlinear (PNL) causal model [92], as well as for the linear and nonlinear models with additive noise. Unfortunately, this is not the case if we do not impose any constraint on the function f.

As discussed in [33], given *any* two random variables V_1 and V_2 with continuous support, no matter how they are related, one can always construct another variable from them, denoted by $E_{1\to 2}$, which is statistically independent from V_1. In [95] the class of functions to produce such an independent variable $E_{1\to 2}$ (or called independent noise term in our causal discovery context) was given, and it was shown that this procedure is invertible: V_2 is a function of V_1 and $E_{1\to 2}$.

This is also the case for the hypothetical causal direction $Y \to X$: if the functional form is not properly constrained, we can always represent X as a function of Y and an independent noise term. That is, any two variables would be symmetric according to the functional causal model, if f is not constrained. Therefore, in order for the functional causal models to be useful to determine the causal direction, we have to introduce certain constraints on the function f such that the independence condition on the noise and hypothetical cause holds for only one direction. Examples of such constraints include the linear model, the nonlinear additive noise model, and the PNL causal model, which we shall discuss below.

18.6.3.2 Examples of well-defined FCMs

Typical FCMs according to which cause and effect are generally asymmetric include the Linear, Non-Gaussian, Acyclic Model (LiNGAM) [69], the PNL causal model [92], and the nonlinear additive noise model (ANM) [27]. Below we review these causal models, ordered from the most to least restrictive. We focus on the two-variable case, and the results are theoretically readily extended to the case with an arbitrary number of variables [60], although computational and statistical considerations limit the number of variables they can be applied to.

Linear, Non-Gaussian, Acyclic Model. The LiNGAM in the two-variable case can be written as

$$Y = bX + E, \tag{18.2}$$

where E is independent of X. Let us first give an illustration with simple examples why it is possible to identify the causal direction between two variables in the linear case. Assume Y is generated from X in a linear form, i.e., $Y = X + E$, where E is independent of X. Figure 18.6 shows the scatterplot of 1000 data points of the two variables X and Y (columns 1 and 3) and that of the predictor and regression residual for two different regression tasks (columns 2 and 4). The three rows correspond to different settings: X and E are both Gaussian (case 1), uniformly distributed (case 2), and distributed according to the Laplace distribution (case 3). In the latter two settings, X and E are non-Gaussian, and one can see clearly that when regressing X on Y (the anti-causal or backward direction), the regression residual is not independent from the predictor anymore–they are always uncorrelated, but the dependence can be seen from higher order statistics. In other words, in those two situations the regression residual is independent from the predictor only for the correct causal direction, giving rise to the causal asymmetry between X and Y.

Rigorously speaking, if at most one of X and E is Gaussian, the causal direction is identifiable, due to the independent component analysis (ICA) theory [32], or more fundamentally, due to the Darmois-Skitovich theorem [39]. Actually, even with more than two

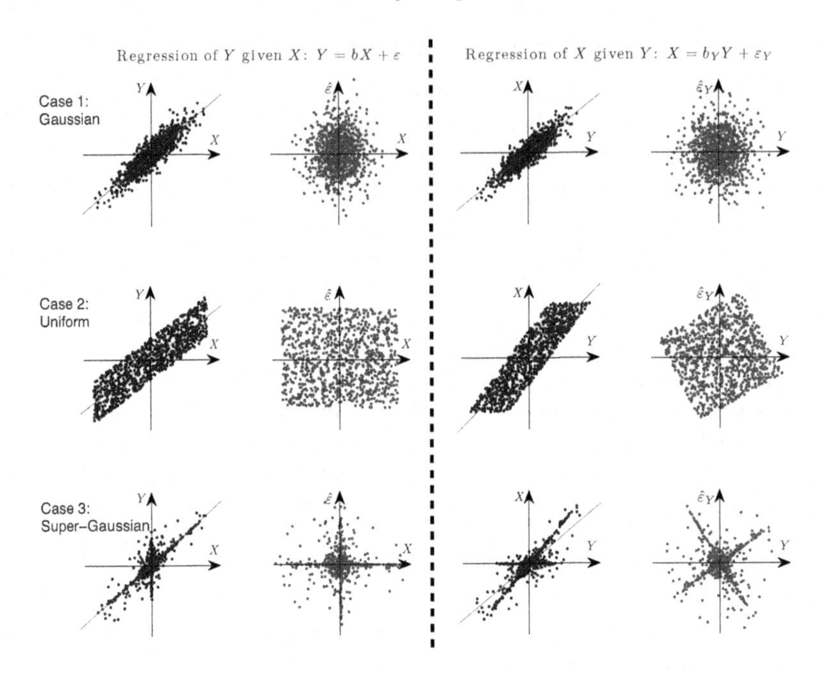

FIGURE 18.6: Illustration of causal asymmetry between two variables with linear relations. The data were generated according to equation 18.2 with E independent of X, i.e., the causal relation is $X \to Y$. From top to bottom: X and E both follow the Gaussian distribution (case 1), uniform distribution (case 2), and Laplace distribution (case 3). The two columns on the left show the scatter plot of X and Y and that of X and the regression residual for regressing Y on X, and the two columns on the right correspond to regressing X on Y. Here we used 1000 data points. One can see that for regression of X given Y, in cases 2 and 3 the residual $\widehat{\epsilon}_Y$ is not independent from the predictor Y because $P(\widehat{\epsilon}_Y \mid Y)$ changes across different values of Y, although they are uncorrelated by construction.

variables, in the linear case the non-Gaussianity assumption of the noise terms enables us to go beyond the equivalence class–we can then uniquely identify the causal structure. ICA provides a tool to estimate this model. Although one may argue that non-Gaussian distributions are actually ubiquitous, practical questions are whether the noise terms are non-Gaussian enough and whether we have enough data to distinguish between Gaussian and non-Gaussian distributions. In the context of ICA, there exist studies on quantifying the identifiability of the ICA mixing matrix when the independent sources (or the independent noise terms in our causal discovery context) are close to Gaussian; for instance, see [74].

Unfortunately, the linearity assumption does not seem to hold in many real-world problems; soon after the development of LiNGAM, some nonlinear extensions of ICA were proposed in order to identify causal models in the presence of various kinds of nonlinearity [89], including the PNL causal model [92]. The ANM [27], which generalizes LiNGAM and is a special case of the PNL model, has received much attention in the causal discovery community.

Nonlinear Additive Noise Model. In practice, a nonlinear transformation is often involved in the data generating process, and should be taken into account in the functional class. As a direct extension of LiNGAM, the nonlinear additive noise model represents the effect as a nonlinear function of the cause plus independent noise [27]:

$$Y = f_{AN}(X) + E. \tag{18.3}$$

It has been shown that the set of all $p(X)$ for which the backward model also admits an independent noise term is contained in a 3-dimensional affine space. Bearing in mind that the space of all possible $p(X)$ is infinite dimensional, one can see that roughly speaking, if the data were generated by the nonlinear additive noise model, the causal direction is identifiable in the generic case. This model is a special case of the PNL causal model discussed below.

Post-Nonlinear Causal Model. The PNL causal model takes into account the nonlinear influence from the cause, the noise effect, and possible sensor or measurement distortion in the observed variables [89, 92]. It assumes that the effect Y is generated by a post-nonlinear transformation on the nonlinear effect of the cause X plus noise term E:

$$Y = f_2(f_1(X) + E), \tag{18.4}$$

where both f_1 and f_2 are nonlinear functions and f_2 is assumed to be invertible. Clearly it contains the linear model and the ANM as special cases. The multiplicative noise model, $Y = X \cdot E$, where all involved variables are positive, is another special case, since it can be written as $Y = \exp(\log X + \log E)$, where $\log E$ is considered as a new noise term, and in (18.4), $f_1(X) = \log(X)$, and $f_2(\cdot) = \exp(\cdot)$.

The identifiability of the causal direction is a crucial issue in FCM-based causal discovery. Since LiNGAM and the nonlinear additive noise model are special cases of the PNL causal model, the identifiability conditions of the causal direction for the PNL causal model also entail those for the former two FCMs. Such identifiability conditions for the PNL causal model were established by a proof by contradiction [92] (when both f_1 and f_2 are assumed to be invertible, the identifiability results were reported in [89]): assuming the causal model holds in both directions, $X \to Y$ and $Y \to X$, one can then show that this implies very strong conditions on the distributions and functions involved in the model. It turns out that under the smoothness assumption on the involved functions and the positivity assumption on the densities of X and E, there are only five specific situations where the causal direction is not identifiable. In other words, the causal direction implied by the PNL causal model is generally identifiable.

Discussion. This set of approaches has been widely used to determine the causal direction between two variables [27, 93, 34]–in this case, comparing the attained maximum likelihood is enough because the model complexity is considered the same for both directions, given the constraints on the functional class. On the cause-effect pairs (which contain about 80 real data sets, available at `http://webdav.tuebingen.mpg.de/cause-effect/`), LiNGAM, the ANM, and the PNL model correctly finds the causal direction for about 62%, 72% [61], and 76% [95] of the data sets.

Given a system with more than two variables, it is straightforward to define a score function, such as BIC, for the linear-Gaussian model or discrete variables, because the likelihood and the number of free parameters in the causal model, both of which are involved in the score function, can be readily calculated. Rigorously speaking, this is usually not the case for the non-Gaussian or nonlinear FCMs. The noise terms in these FCMs are not assumed to be Gaussian. Calculating the likelihood function involves representing their distributions with parametric or nonparametric models, which require additional parameters in the causal model. Furthermore, in nonlinear models it is usually not straightforward to find the number of free parameters. Instead, one might use the estimated effective degrees of freedom, which are defined in various ways [24, Chapter 3.5]. For the reasons given above, to develop the score-based search procedure which assumes the above FCMs, one usually has to approximate the score function in some way or use heuristics; for a score-based learning procedure of nonlinear additive noise models, see [55]. Alternatively, instead of using a single score function to compare candidate structures, one may apply functional causal discovery to find the best DAG by comparing different DAGs in the equivalence class produced by constraint-based methods [92, 82], or iteratively perform (linear or nonlinear) regression and statistical test of independence between the residual and hypothetical causes until finding independent noise for each variable [70, 61]. For LiNGAM-based causal discovery, the DirectLiNGAM approach [70] is feasible to handle up to 100 variables. How to make use of the FCMs to efficiently and reliably estimate moderately large graphs (say, with more than 15 variables) is under investigation.

18.6.3.3 Extensions

In the linear non-Gaussian case it is possible to identify the causal structure in the presence of confounders by exploiting overcomplete ICA (which allows more independent sources than observed variables) [28]. Currently this algorithm is feasible for only several variables; the main reason is that to produce accurate estimates, overcomplete ICA usually requires the underlying independent components to be far from Gaussian and the sample size to be large. Furthermore, when the underlying causal model has cycles or feedbacks, which violates the acyclicity assumption, one may still be able to reveal the causal knowledge under certain assumptions [43]; this idea has been extended to the nonlinear additive noise model [50]. FCMs have also been exploited to handle the selection bias issue [96]; when the causal process follows the PNL model, the causal direction between two variables as well as the FCM is identifiable in the presence of outcome-dependent selection.

Both the constraint-based and score-based approaches to causal discovery are directly applicable to find causal relations over the random variables involved in stochastic processes (or time series). In this scenario, one can further benefit from the temporal constraint that the effect cannot precede the cause, which helps reduce the search space of the causal structure. Eichler provides an overview over various definitions of causation w.r.t. time series and reviews some constraint-based causal discovery methods [17]. Interestingly, with FCMs, especially the LiNGAM, one can discover precise causal information from time series in rather challenging situations. For instance, in the linear non-Gaussian case, one can estimate time-delayed causal relations together with instantaneous effects from moderately high-dimensional time series [35], discover the causal relations at the original causal frequency from subsampled data [22] or temporally aggregated time series [21], and identify the vector autoregressive processes with hidden components [20]. For causal discovery from subsampled data, a completely nonparametric approach has also been proposed [15]. Algorithms to solve these problems usually involve high computational load and do not scale well.

18.7 Conclusion and Discussions

Understanding causal relations can enable making predictions under interventions, be used as input to algorithms for constructing interventions to achieve certain objectives, and provide an understanding of how changes have been brought about. A traditional way to discover causal relations is to use interventions or randomized experiments, which is, however, in many cases of interest too expensive, too time-consuming, unethical, or even impossible. Therefore, inferring the underlying causal structure from purely observational data, or from combinations of observational and experimental data, has drawn much attention in computer science, statistics, philosophy, neuroscience, and other disciplines. With the rapid accumulation of huge volumes of data, it is even more desirable to abstract causal knowledge from data, and it is necessary to develop automatic causal search algorithms that scale well.

Causal search algorithms can be categorized into constraint-based, score-based, and hybrid algorithms, depending on the procedure to find the causal graph. The constraint-based methods aim to find the graph that satisfies certain types of constraints, usually about the (conditional) independence and dependence relations between the variables. Score-based methods find the graph that maximizes or minimizes a score function, which indicates the quality of candidate causal graphs. Hybrid methods try to combine the merits enjoyed by each of them. We used the PC algorithm to demonstrate how the constraint-based methods work and discussed various methods to deal with statistical testing errors, latent confounders, and selection bias in the data. Traditional score-based search methods usually work with data that are assumed to follow the joint Gaussian distribution or the multinomial distribution and, like constraint-based methods, they cannot distinguish between causal graphs in the same Markov equivalence class. We then reviewed some functional causal models which assume non-Gaussianity or nonlinearity in the data–they have been shown to be able to distinguish cause from effect and hence uniquely identify the underlying causal graph, under proper assumptions.

Each method has its own pros and cons. Constraint-based methods do not require specific assumptions on the causal process and can scale well, but they involve the multiple testing problem. Score-based methods avoid multiple testing, and are feasible in the high-dimensional case, given that the data are jointly Gaussian or multinomial and generated by a sparse graph; on the other hand, they have to resort to the functional form of the causal influence, may get stuck in local optima, and cannot handle latent confounders in a straightforward way. The non-Gaussian or nonlinear functional causal models help identify more detailed information of the causal process; they, however, suffer from two drawbacks. First, currently it is computationally demanding to use such models to estimate a large causal graph. Second, nonlinear functional causal models are usually intransitive, limiting their power in causal direction determination–for instance, suppose we have a causal chain in which each causal mechanism follows the nonlinear additive noise model, and then after marginalizing some intermediate variables out, the process does not follow the nonlinear additive noise model anymore. To solve a wider range of real problems, it is essential to develop causal search methods that can either abstract rich causal knowledge with more realistic assumptions on the causal process or solve larger-scale problems more efficiently.

Bibliography

[1] C. Aliferis, A. Statnikov, I. Tsamardinos, S. Mani, and X. Koutsoukos. Local causal and Markov blanket induction for causal discovery and feature selection for classification Part I: Algorithms and empirical evaluation. *Journal of Machine Learning Research*, 11:171–234, 2010.

[2] C. Aliferis, A. Statnikov, I. Tsamardinos, S. Mani, and X. Koutsoukos. Local causal and Markov blanket induction for causal discovery and feature selection for classification Part II: Analysis and extensions. *Journal of Machine Learning Research*, 11:235–284, 2010.

[3] C. Aliferis, I. Tsamardinos, and A. Statnikov. Hiton: A novel Markov blanket algorithm

for optimal variable selection. In *Proceedings of the 2003 American Medical Informatics Association Annual Symposium*, 21–25. Washington, DC, 2003.

[4] S. Andersson, D. Madigan, and M. Perlman. A characterization of Markov equivalence classes for acyclic digraphs. *Annals of Statistics*, 25(2):505–541, 1997.

[5] R. Bellman. Dynamic programming treatment of the traveling salesman problem. *Journal of the ACM*, 9:61–63, 1962.

[6] D. Chickering. A transformational characterization of equivalent Bayesian network structures. *Proceedings of the Eleventh Conference on Uncertainty in Artificial Intelligence*, 87–98, 1995.

[7] D. Chickering. Learning Bayesian networks is NP-complete. In D. Fisher and H. J. Lenz, editors, *Learning from Data: Artificial Intelligence and Statistics V*, 121–130. Lecture Notes in Statistics 112, 1996.

[8] D. Chickering. Optimal structure identification with greedy search. *Journal of Machine Learning Research*, 3:507–554, 2002.

[9] T. Chu and C. Glymour. Search for additive nonlinear time series causal models. *Journal of Machine Learning Research*, 9:967–991, 2008.

[10] T. Claasen, J. Mooij, and T. Heskes. Learning sparse causal models is not NP-hard. *Uncertainty in Artificial Intelligence*, 29:172–181, 2013.

[11] D. Colombo and M. Maathuis. Order-independent constraint-based causal structure learning. *Journal of Machine Learning Research*, 15:3921–3962, 2014.

[12] D. Colombo, M. Maathuis, M. Kalisch, and T. Richardson. Learning high-dimensional directed acyclic graphs with latent and selection variables. *Annals of Statistics*, 40(1):294–321, 2012.

[13] R. Cui, P. Groot, and T. Heskes. Copula PC algorithm for causal discovery from mixed data. *Machine Learning and Knowledge Discovery in Databases. ECML PKDD 2016. Lecture Notes in Computer Science*, 9852:3365–3383, 2016.

[14] R. Daly, Q. Shen, and S. Aitken. Learning Bayesian networks: Approaches and issues. *The Knowledge Engineering Review*, 26:99–157, 2009.

[15] D. Danks and S. Plis. Learning causal structure from undersampled time series. In *NIPS 2014 Workshop on Causality*, 2014.

[16] M. Drton, M. Eichler, and T. S. Richardson. Computing maximum likelihood estimates in recursive linear models with correlated errors. *The Journal of Machine Learning Research*, 10:2329–2348, 2009.

[17] M. Eichler. Causal inference in time series analysis. In C. Berzuini, A. Dawid, and L. Bernardinelli, editors, *Advances in Neural Information Processing Systems 10*, 327–354. John Wiley and Sons, Ltd, 2012.

[18] G. Elidan and S. Gould. Learning bounded treewidth Bayesian networks. *Journal of Machine Learning Research*, 9:2699–2731, 2008.

[19] D. Entner and P. Hoyer. On causal discovery from time series data using FCI. In P. Myllymaki, T. Roos, and T. Jaakkola, editors, *Proceedings of the Fifth European Workshop on Probabilistic Graphical Models*. Helsinki Institute for Information Technology (HIIT), 2010.

[20] P. Geiger, K. Zhang, M. Gong, D. Janzing, and B. Schölkopf. Causal inference by identification of vector autoregressive processes with hidden components. In *Proc. 32nd International Conference on Machine Learning (ICML)*, 2015.

[21] M. Gong, K. Zhang, B. Schölkopf C. Glymour, and D. Tao. Causal discovery from temporally aggregated time series. In *Proceedings of the 33rd Conference on Uncertainty in Artificial Intelligence (UAI 2017)*, 2017.

[22] M. Gong, K. Zhang, D. Tao, P. Geiger, and B. Schölkopf. Discovering temporal causal relations from subsampled data. In *Proc. 32nd International Conference on Machine Learning (ICML 2015)*, 2015.

[23] N. Harris and M. Drton. PC algorithm for nonparanormal graphical models. *Journal of Machine Learning Research*, 14:3365–3383, 2013.

[24] T. Hastie and R. Tibshirani. *Generalized Additive Models.* Chapman & Hall, 1990. Monographs on Statistics and Applied Probability.

[25] A. Hauser and P. Bühlmann. Characterization and greedy learning of interventional Markov equivalence classes of directed acyclic graphs. *Journal of Machine Learning Research*, 13:2409–2464, 2011.

[26] Y. Hochberg and A. Tamhane. *Multiple Comparison Procedures.* Wiley, New York, 1987.

[27] P. Hoyer, D. Janzing, J. Mooji, J. Peters, and B. Schölkopf. Nonlinear causal discovery with additive noise models. In *Advances in Neural Information Processing Systems 21*, Vancouver, B.C., Canada, 2009.

[28] P. Hoyer, S. Shimizu, A. Kerminen, and M. Palviainen. Estimation of causal effects using linear non-Gaussian causal models with hidden variables. *International Journal of Approximate Reasoning*, 49:362–378, 2008.

[29] A. Hyttinen, F. Eberhardt, and J. Jarvisalo. Constraint-based causal discovery: Conflict resolution with answer set programming. *Proceedings of the Thirtieth Annual Conference on Uncertainty in Artificial Intelligence (UAI-14)*, 340–349, 2014.

[30] A. Hyttinen, P. Hoyer, F. Eberhardt, and M. Jarvisalo. Discovering cyclic causal models with latent variables: A general SAT-based procedure. *Proceedings of the Twenty-Ninth Annual Conference on Uncertainty in Artificial Intelligence (UAI-13)*, 301–310, 2013.

[31] A. Hyttinen, S. Plis, M. Järvisalo, F. Eberhardt, and D. Danks. Causal discovery from subsampled time series data by constraint optimization. In A. Antonucci, G. Corani, and G. Campos, editors, *Proceedings of the Eighth International Conference on Probabilistic Graphical Models*, 216–227, 2016.

[32] A. Hyvärinen, J. Karhunen, and E. Oja. *Independent Component Analysis.* John Wiley & Sons, Inc, 2001.

[33] A. Hyvärinen and P. Pajunen. Nonlinear independent component analysis: Existence and uniqueness results. *Neural Networks*, 12(3):429–439, 1999.

[34] A. Hyvärinen and S. Smith. Pairwise likelihood ratios for estimation of non-Gaussian structural equation models. *Journal of Machine Learning Research*, 14:111–152, 2013.

[35] A. Hyvärinen, K. Zhang, S. Shimizu, and P. Hoyer. Estimation of a structural vector autoregression model using non-Gaussianity. *Journal of Machine Learning Research*, 1709–1731, 2010.

[36] J. Pearl and R. Dechter. Identifying independencies in causal graphs with feedback. In *Proceedings of the Twelfth Annual Conference on Uncertainty in Artificial Intelligence (UAI-96)*, 420–426, San Francisco, 1996. Morgan Kaufmann.

[37] T. Jaakkola, D. Sontag, A. Globerson, and M. Meila. Learning Bayesian network structure using lp relaxations. In *Proceedings of the 13th International Conference on Artificial Intelligence and Statistics*, 2010.

[38] L. Junning and Z. Wang. Controlling the false discovery rate of the association/causality structure learned with the PC algorithm. *Journal of Machine Learning Research*, 475–514, 2009.

[39] A. M. Kagan, Y. V. Linnik, and C. R. Rao. *Characterization Problems in Mathematical Statistics.* Wiley, New York, 1973.

[40] M. Kalisch and P. Bühlmann. Estimating high-dimensional directed acyclic graphs with the PC-algorithm. *J Mach Learn Res*, 8:613–636, 2007.

[41] M. Koivisto and K. Sood. Exact Bayesian structure discovery in Bayesian networks. *Journal of Machine Learning Research*, 5:549–573, 2004.

[42] E. Kummerfeld, J. Ramsey, R. Yang, P. Spirtes, and R. Scheines. Causal clustering for 2-factor measurement models. In T. Calders, F. Esposito, E. Hüllermeier, and R. Meo, editors, *Machine Learning and Knowledge Discovery in Databases*, 34–49. Springer, 2014.

[43] G. Lacerda, P. Spirtes, J. Ramsey, and P. O. Hoyer. Discovering cyclic causal models by independent components analysis. In *Proceedings of the 24th Conference on Uncertainty in Artificial Intelligence (UAI)*, Helsinki, Finland, 2008.

[44] S. Lauritzen, A. Dawid, B. Larsen, and H. Leimer. Independence properties of directed Markov fields. *Networks*, 20:491–505, 1990.

[45] M. Maathuis, D. Colombo, M. Kalisch, and P. Bühlmann. Predicting causal effects in large-scale systems from observational data. *Nat Methods*, 7(4):247–248, 2010.

[46] M. Maathuis, M. Kalisch, and P. Bühlmann. Estimating high-dimensional intervention effects from observational data. *Annals of Statistics*, 37(6A):3133–3164, 2009.

[47] D. Malinsky and P. Spirtes. Estimating causal effects with ancestral graph Markov models. *Journal of Machine Learning Research: Workshop and Conference Proceedings (PGM 16)*, 52:299–309, 2016.

[48] C. Meek. Causal inference and causal explanation with background knowledge. *Proceedings of the Eleventh Conference on Uncertainty in Artificial Intelligence*, 403–411, 1995.

[49] C. Meek. Strong completeness and faithfulness in Bayesian networks. *Proceedings of the Eleventh Annual Conference on Uncertainty in Artificial Intelligence (UAI-95)*, 411–419, 1995.

[50] J. Mooij, D. Janzing, T. Heskes, and B. Schölkopf. On causal discovery with cyclic additive noise models. In *Proc. NIPS 2011*, 2011.

[51] P. Nandy, A. Hauser, and M. Maathuis. High-dimensional consistency in score-based and hybrid structure learning. *Annals of Statistics*, 46, 3151–3183, 2015.

[52] P. Nandy, M. Maathuis, and T. Richardson. Estimating the effect of joint interventions from observational data in sparse high-dimensional settings. *Annals of Statistics*, 45, 647–674, 2014.

[53] R. Neal. On deducing conditional independence from d-separation in causal graphs with feedback. *Journal of Artificial Intelligence Research*, 12:87–91, 2000.

[54] J. Nielsen, T. Kočka, and J. Peña. On local optima in learning Bayesian networks. In *Proceedings of the 19th Conference on Uncertainty in Artificial Intelligence*, 2003.

[55] C. Nowzohour and P. Bühlmann. Score-based causal learning in additive noise models. *Statistics*, 50:471–485, 2016.

[56] C. Nowzohour, M. H. Maathuis, R. J. Evans, and P. Bühlmann. Structure learning with bow-free acyclic path diagrams. *Electronic Journal of Statistics*, 11, 5342–5374, 2015.

[57] J. Ogarrio, P. Spirtes, and J. Ramsey. A hybrid causal search algorithm for latent variable models. *Journal of Machine Learning Research: Workshop and Conference Proceedings (PGM 16)*, 52:368–379, 2016.

[58] J. Pearl. *Causality: Models, Reasoning, and Inference.* Cambridge University Press, Cambridge, 2000.

[59] E. Perkovic, J. Textor, M. Kalisch, and M. Maathuis. Complete graphical characterization and construction of adjustment sets in Markov equivalence classes of ancestral graphs. *Journal of Machine Learning Research*, 18(220): 1–62, 2018.

[60] J. Peters, J. Mooij, D. Janzing, and B. Schölkopf. Identifiability of causal graphs using functional models. In *Proc. UAI 2011*, 589–598, 2011.

[61] J. Peters, J. Mooij, D. Janzing, and B. Schölkopf. Causal discovery with continuous additive noise models. *Journal of Machine Learning Research*, 15:2009–2053, 2014.

[62] D. Poole and M. Crowley. Cyclic causal models with discrete variables: Markov chain equilibrium semantics and sample ordering. *IJCAI'13 Proceedings of the Twenty-Third International Joint Conference on Artificial Intelligence*, 1060–1068, 2013.

[63] J. Ramsey, M. Glymour, R. Sanchez-Romero, and C. Glymour. A million variables and more: The fast greedy search algorithm for learning high dimensional graphical causal models, with an application to functional magnetic resonance images. *International Journal of Data Science and Analytics*, 2016. forthcoming.

[64] J. Ramsey, P. Spirtes, and J. Zhang. Adjacency-faithfulness and conservative causal inference. *22nd Conference on Uncertainty in Artificial Intelligence*, 401–408, 2006.

[65] T. Richardson. A discovery algorithm for directed cyclic graphs. *Uncertainty in Artificial Intelligence*, 12:454–461, 1996.

[66] J. Robins, R. Scheines, P. Spirtes, and L. Wasserman. Uniform consistency in causal inference. *Biometrika*, 90(3):491–515, 2003.

[67] K. Sadeghi. Faithfulness of probability distributions and graphs. *eprint arXiv:1701.08366*, 2017.

[68] W. Sewell and V. Shah. Social class, parental encouragement, and educational aspirations. *Am J Sociol*, 73(5):559–572, 1968.

[69] S. Shimizu, P. Hoyer, A. Hyvärinen, and A. Kerminen. A linear non-Gaussian acyclic model for causal discovery. *Journal of Machine Learning Research*, 7:2003–2030, 2006.

[70] S. Shimizu, T. Inazumi, Y. Sogawa, A. Hyvärinen, Y. Kawahara, T. Washio, P. Hoyer, and K. Bollen. Directlingam: A direct method for learning a linear non-Gaussian structural equation model. *Journal of Machine Learning Research*, 12:1225–1248, 2011.

[71] I. Shpitser, R. Evans, T. Richardson, and J. Robins. Sparse nested Markov models with log-linear parameters. *Proceedings of the Twenty-Ninth Annual Conference on Uncertainty in Artificial Intelligence (UAI-13)*, 576–585, 2013.

[72] I. Shpitser, T. S. Richardson, J. M. Robins, and R. Evans. Parameter and structure learning in nested Markov models. *Uncertainty in Artificial Intelligence Workshop on Causal Structure Learning*, 2012.

[73] M. Singh and M. Valtorta. Construction of Bayesian network structures from data: A brief survey and an efficient algorithm. *International Journal of Approximate Reasoning*, 12:111–131, 1995.

[74] A. Sokol, M. H. Maathuis, and B. Falkeborg. Quantifying identifiability in independent component analysis. *Electron. J. Statist.*, 8:1438–1459, 2014.

[75] P. Spirtes. Directed cyclic graphical representations of feedback models. *Eleventh Conference on Uncertainty in Artificial Intelligence*, 491–499, 1995.

[76] P. Spirtes. Calculation of entailed rank constraints in partially non-linear and cyclic models. *Uncertainty in Artificial Intelligence*, 29:606–615, 2013.

[77] P. Spirtes, C. Glymour, and R. Scheines. *Causation, Prediction, and Search, Second Edition (Adaptive Computation and Machine Learning)*. MIT Press, 2001.

[78] P. Spirtes, T. Richardson, and C. Meek. Causal discovery in the presence of latent variables and selection bias. In G. Cooper and C. Glymour, editors, *Computation, Causality, and Discovery*, 211–252. AAAI Press, 1999.

[79] P. Spirtes and J. Zhang. A uniformly consistent estimator of causal estimating and controlling the false discovery rate. *Statistical Science*, 29(4):662–678, 2014.

[80] D. Stekhoven, L. Hennig, M. Izabel, M. Maathuis, and P. Bühlmann. Causal stability ranking. *Bioinformatics.*, 28(21):2819–2823, 2012.

[81] E. Strobl, P. Spirtes, and S. Visweswaran. Estimating and controlling the false discovery rate for the PC algorithm using edge-specific p-values. *arXiv:1607.03975*, 2016.

[82] R. Tillman, A. Gretton, and P. Spirtes. Nonlinear directed acyclic structure learning with weakly additive noise models. In *Advances in Neural Information Processing Systems (NIPS)*, Vancouver, Canada, December 2009.

[83] S. Triantafillou and I. Tsamardinos. Constraint-based causal discovery from multiple interventions over overlapping variable sets. *Journal of Machine Learning Research*, 16:2147–2205, 2015.

[84] I. Tsamardinos, L. Brown, and C. Aliferis. The max-min hill-climbing Bayesian network structure learning algorithm. *Mach Learn*, 65(1):31–78, 2006.

[85] C. Uhler, G. Raskutti, P. Bühlmann, and B. Yu. Geometry of the faithfulness assumption in causal inference. *The Annals of Statistics*, 41:436–463, 2013.

[86] C. Yuan and B. Malone. Learning optimal Bayesian networks: A shortest path perspective. *Journal of Artificial Intelligence Research*, 48:23–65, 2013.

[87] J. Zhang. Causal reasoning with ancestral graphs. *Journal of Machine Learning Research*, 9:1437–1474, 2008.

[88] J. Zhang. On the completeness of orientation rules for causal discovery in the presence of latent confounders and selection bias. *Artificial Intelligence*, volume 172, issues 16–17:1873–1896, 2008.

[89] K. Zhang and L. Chan. Extensions of ICA for causality discovery in the Hong Kong stock market. In *Proc. 13th International Conference on Neural Information Processing (ICONIP 2006)*, 2006.

[90] K. Zhang, B. Huang, J. Zhang, C. Glymour, and B. Schölkopf. Causal discovery from nonstationary/heterogeneous data: Skeleton estimation and orientation determination. In *Proceedings of the International Joint Conference on Artificial Intelligence (IJCAI)*, 2017.

[91] K. Zhang and A. Hyvärinen. Acyclic causality discovery with additive noise: An information-theoretical perspective. In *Proc. European Conference on Machine Learning and Principles and Practice of Knowledge Discovery in Databases (ECML PKDD)*, Bled, Slovenia, 2009.

[92] K. Zhang and A. Hyvärinen. On the identifiability of the post-nonlinear causal model. In *Proceedings of the 25th Conference on Uncertainty in Artificial Intelligence*, Montreal, Canada, 2009.

[93] K. Zhang and A. Hyvärinen. Distinguishing causes from effects using nonlinear acyclic causal models. In *JMLR Workshop and Conference Proceedings, volume 6*, 157–164, 2010.

[94] K. Zhang, J. Peters, D. Janzing, and B. Schölkopf. Kernel-based conditional independence test and application in causal discovery. *Uncertainty in Artificial Intelligence 27*, 804–813, 2011.

[95] K. Zhang, Z. Wang, J. Zhang, and B. Schölkopf. On estimation of functional causal models: General results and application to post-nonlinear causal model. *ACM Transactions on Intelligent Systems and Technologies*, 7, 2016.

[96] K. Zhang, J. Zhang, B. Huang, B. Schölkopf, and C. Glymour. On the identifiability and estimation of functional causal models in the presence of outcome-dependent selection. In *Proceedings of the 32rd Conference on Uncertainty in Artificial Intelligence (UAI 2016)*, 2016.

Part V

Applications

19

Graphical Models for Forensic Analysis

A. Philip Dawid
Statistical Laboratory, University of Cambridge

Julia Mortera
Department of Economics, Universitá Roma Tre

CONTENTS

19.1 Introduction

"Forensic" means relating to or denoting the application of scientific methods and techniques to the investigation of crime. Here we are concerned with systems to assist in the evaluation of evidence presented in a criminal or civil court case. Such a case may have a mixed mass of evidence of many kinds, all of it subject to uncertainty. We describe how such a case can be helpfully represented by means of a *Bayesian Network* (BN), or *Probabilistic Expert System* [2]: a directed graphical model describing the various items of evidence and hypotheses, and the probabilistic relationships between them. Such a representation displays clearly the relevance of the evidence to questions of interest, and supports efficient routines to compute the impact of the evidence presented. In many cases the BN can be constructed as an *object-oriented Bayesian network* (OOBN), a top-down hierarchical structure which hides irrelevant detail and simplifies both construction and interpretation.

In Section 19.2 we describe by means of a fictitious example the way in which different elements in a case (eye-witness, fiber and blood evidence) can be drawn together into a single coherent story structured as a Bayesian network. We use this example to explain how a BN can be used to discover implicit relationships of relevance and irrelevance in the evidence, which in turn can be used to simplify probabilistic calculations. Section 19.3 describes the features of an OOBN, and shows how simple reusable modules can be constructed to represent common features and relationships such as eyewitness testimony and identification. Section 19.4 briefly describes how a BN can be used to simplify the specification and manipulation of probabilities, in particular the use of "evidence propagation" to compute conditional probabilities taking the evidence into account.

The remainder of this chapter focuses on DNA evidence (the appendix gives a very brief glossary of the relevant biological background and terminology). Section 19.5 gives examples of the use of OOBNs to handle cases of criminal identification and simple and complex disputed paternity. These examples deal with cases where "clean" single source DNA profiles are available, whereas Section 19.6 shows how the methods can be extended to deal with more complex cases, where (for example) a crime trace may contain a mixture of DNA from more than one contributor, in varying proportions. Finally, Section 19.7 relaxes some of the simplifying assumptions made so far, to account for such realistic complications as uncertainty about allele frequencies and heterogeneity in the reference population. Network modules that account for these additional features are introduced; these can then be integrated into the variety of identification problems previously described.

The key to the approach in formulating the BNs for forensic DNA analysis is a careful restructuring of the pedigrees as a BN, by appropriate definition of variables and their interrelationships. For example, the family relationships among the individuals form the basis for the graphical network structure and some conditional probability tables are given by Mendelian inheritance laws.

19.2 Bayesian Networks for the Analysis of Evidence

In a legal case, we may have various items of evidence, both lay and scientific, with more or less complex relationships. It can often be helpful to represent such relationships in graphical form, as a BN. As described in Milan Studený's Chapter 1 of this Handbook, a BN is a directed acyclic graph (DAG), with nodes representing relevant variables in the problem, joined by arrows representing probabilistic dependence, and, for each "child" node in the DAG, a specification of its conditional distribution, given the states of its "parents". This can then be used for further analysis, both qualitative and quantitative. We start by considering purely qualitative properties.

Example 19.2.1 (Robbery). We illustrate with a fictional crime story (reproduced with permission from Dawid and Evett [9]).

Eye witness evidence: An unknown number of offenders entered commercial premises late at night through a hole which they cut in a metal grille. Inside, they were confronted by a security guard who was able to set off an alarm before one of the intruders punched him in the face, causing his nose to bleed. The security guard said that there were four men but the light was too poor for him to describe them and he was confused because of the blow he had received. About 10 minutes later the police found the suspect trying to "hot wire" a car in an alley about a quarter of a mile from the incident. The suspect denied having anything to do with it.

Fiber evidence: A tuft of red acrylic fibers was found on the jagged end of one of the cut edges of the grille. The suspect's jumper was red acrylic. The tuft was indistinguishable from the fibers of the jumper by eye, microspectrofluorimetry and thin layer chromatography.

Blood evidence: A spray pattern of blood was found on the front and right sleeve of the suspect's jumper. The blood on the jumper was of a different type from that of the suspect, but the same as that of the security guard.

The DAG of Figure 19.1 contains a number of nodes, corresponding to relevant random events or variables; here a square node corresponds to a variable that has been observed, while a round node indicates an unobserved variable that is required to complete the picture. For example, the actual number of offenders, N, is not information directly available to the court, but is relevant to G_1, the guard's recollection of that number, and (because it embodies alternative possibilities) to C, whether or not the suspect was one of the offenders, and to A and B, the origin of the fibers and the blood. The arrows leading into any node originate from its "parents", the variables on whose value it is supposed to depend (probabilistically). For example, Y_2, the measurement of the blood type of the spray on the

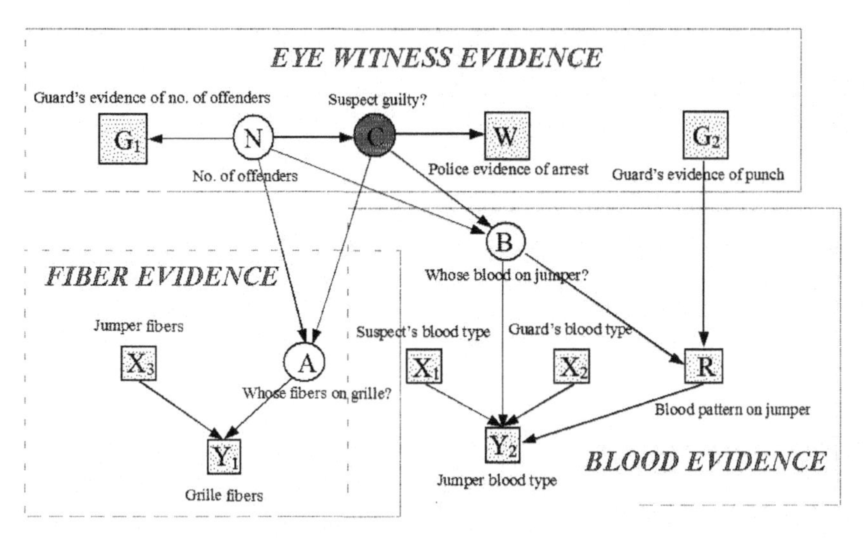

FIGURE 19.1: Directed acyclic graph representing robbery story (adapted with permission from Dawid and Evett [9]). Square nodes represent the available evidence. The arrows encode the probabilistic dependence of a "child" node on its "parents".

jumper is dependent on X_1, the suspect's blood type (because it might be a self stain) and the guard's blood type X_2. But information is also provided by R, describing the shape of the stain, because that sheds light on whether or not it might be a self stain. In turn, the shape of the stain is influenced by the way in which the guard was punched, G_2, and B, the identity of the person who did it. B is in turn influenced by variable C, whether or not the suspect was one of the offenders, and also by N, the number of offenders.

This construction of the DAG utilizes the concept of conditional independence [8]. For example, were we to know N, the number of offenders, and C, whether or not the suspect is one of them, our uncertainty about B, the identity of the person who struck the guard, would (it is supposed) be unaffected by further information about both the eye-witness variables,

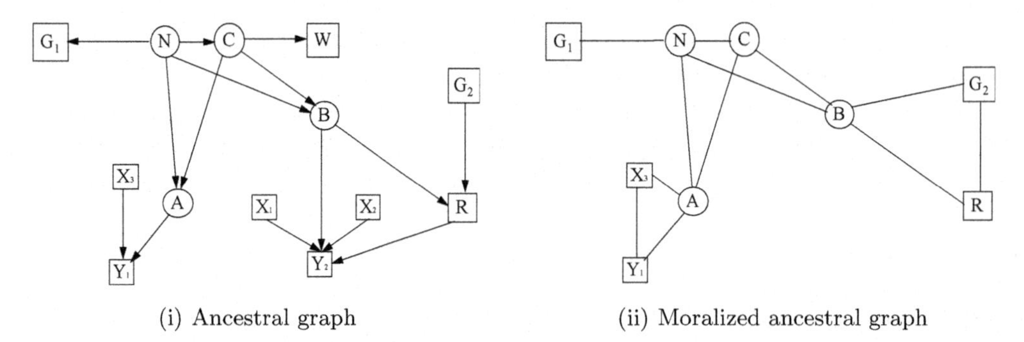

(i) Ancestral graph

(ii) Moralized ancestral graph

FIGURE 19.2: Procedure to query the conditional independence property: $(B,R) \perp\!\!\!\perp (G_1,Y_1) \mid (A,N)$? (adapted with permission from Dawid and Evett [9]). The ancestral graph in Figure 19.2(i) is obtained from Figure 19.1 by retaining only those nodes that are either in the queried sets, or are "ancestors" of these. The moralized graph in Figure 19.2(ii) is then obtained from Figure 19.2(i) by "marrying" unmarried parents (X_1 and A for Y_1, and B and G_2 for R), and then dropping the arrowheads. The queried property is seen to hold since there is no path in Figure 19.2(ii) from a node in (B,R) to one in (G_1,Y_1) that avoids (A,N).

(G_1,G_2,W), and the fiber variables, (A,X_3,Y_1). In conditional independence notation [8]: $B \perp\!\!\!\perp (G_1,G_2,W,A,X_3,Y_1) \mid (N,C)$. This is an example of the general requirement that a variable be independent of its "non-descendants", given its "parents". Similarly, once B is known, then G_1, N, C and W become irrelevant to any variables that are descendants of B in the graph, such as Y_2: $Y_2 \perp\!\!\!\perp (G_1,N,C,W) \mid B$. Note that conditional independence, so interpreted, is a purely qualitative "irrelevance" property, and does not require numerical assessment of any probabilities in the problem. (However, it does impose relationships between these probabilities.)

Further, we have methods for examining a DAG to discover additional, implicit, conditional independence properties. One such method is the "moralization" criterion of Lauritzen et al. [21], which operates as follows. Let S, T, U be sets of nodes. To query the conditional independence $S \perp\!\!\!\perp T \mid U$:

Ancestral graph Form the subgraph containing just the nodes in S, T and U, together with their ancestors.

Moralization "Marry" any unmarried parents, by adding undirected links between any parents of a common child that are not already joined by an arrow; then drop all arrowheads.

Separation Look for a path from S to T avoiding U. If there is none such, deduce $S \perp\!\!\!\perp T \mid U$.

For a description of other, equivalent, graphical criteria, refer again to Chapter 1 of this Handbook.

As an example, suppose we wish to query the conditional independence property $(B,R) \perp\!\!\!\perp (G_1,Y_1) \mid (A,N)$. Figure 19.2 shows the relevant ancestral graph and its moralization. We note that, in the latter, every path from B or R to G_1 or Y_1 passes through either A or N, and so deduce that this conditional independence does indeed hold.

Such properties can be helpful in simplifying algebraic manipulations on probabilities. Thus we can express the likelihood ratio in favor of guilt, given the full evidence

$(G_1, G_2, W, R, X_1, X_2, X_3, Y_1, Y_2) = (g_1, g_2, w, r, x_1, x_2, x_3, y_1, y_2)$, as

$$\frac{\Pr(g_1, g_2, w, r, x_1, x_2, x_3, y_1, y_2 \mid c)}{\Pr(g_1, g_2, w, r, x_1, x_2, x_3, y_1, y_2 \mid \bar{c})} = \frac{\Pr(r, x_1, x_2, x_3, y_1, y_2 \mid c, g_1, g_2, w)}{\Pr(r, x_1, x_2, x_3, y_1, y_2 \mid \bar{c}, g_1, g_2, w)} \times \frac{\Pr(g_1, g_2, w \mid c)}{\Pr(g_1, g_2, w \mid \bar{c})}. \tag{19.1}$$

The second term on the right-hand side of (19.1) is the likelihood ratio based on the eyewitness evidence alone. This term can be simplified using the following conditional independence properties, which follow from application of the moralization criterion:

$$\begin{aligned} G_2 &\perp\!\!\!\perp (G_1, C, W) \\ W &\perp\!\!\!\perp G_1 \mid C. \end{aligned}$$

These allow us to express

$$\frac{\Pr(g_1, g_2, w \mid c)}{\Pr(g_1, g_2, w \mid \bar{c})} = \frac{\Pr(w \mid c)}{\Pr(w \mid \bar{c})} \times \frac{\Pr(g_1 \mid c)}{\Pr(g_1 \mid \bar{c})}.$$

For this term we can thus ignore entirely the guard's evidence of the punch, g_2, and consider independently his evidence g_1 as to the number of offenders and w, the police evidence about the arrest.

The scientific evidence only enters into the first term on the right-hand side of (19.1), which has the form of a conditional likelihood ratio, given the eyewitness evidence. This term can be simplified on applying the following conditional independence properties (again following from application of the moralisation criterion):

$$\begin{aligned} (X_1, X_2, X_3) &\perp\!\!\!\perp (C, G_1, G_2, W) \\ (R, Y_1, Y_2) &\perp\!\!\!\perp W \mid (C, X_1, X_2, X_3, G_1, G_2) \\ Y_1 &\perp\!\!\!\perp (R, Y_2) \mid (C, X_1, X_2, X_3, N, G_2) \\ Y_1 &\perp\!\!\!\perp (X_1, X_2, G_2) \mid (X_3, C, N) \\ (R, Y_2) &\perp\!\!\!\perp X_3 \mid (X_1, X_2, C, N, G_2). \end{aligned}$$

Together with a further simplifying assumption $G_1 = N$ (the guard's evidence of the number of offenders is accurate), the above properties allow us to simplify the conditional likelihood ratio:[1]

$$\frac{\Pr(r, x_1, x_2, x_3, y_1, y_2 \mid c, g_1, g_2, w)}{\Pr(r, x_1, x_2, x_3, y_1, y_2 \mid \bar{c}, g_1, g_2, w)} = \frac{\Pr(y_1 \mid x_3, c, n)}{\Pr(y_1 \mid x_3, \bar{c}, n)} \times \frac{\Pr(r, y_2 \mid x_1, x_2, c, n, g_2)}{\Pr(r, y_2 \mid x_1, x_2, \bar{c}, n, g_2)}. \tag{19.2}$$

We can thus consider entirely separately the fiber evidence and the blood evidence; moreover, in doing so we need not take any account of w, the police evidence of the arrest. The first term on the right-hand side of (19.2), relating to the fiber evidence, requires consideration of the conditional probability of y_1, the observed features of the fibers on the grille, given the information about the fibers on the jumper, x_3, and the number, n, of offenders (as testified by the guard), under each of the two competing hypotheses: that the suspect was (c) or was not ($\bar{c}$), one of the offenders. It does not involve any variables relating to the blood evidence. These appear in the second term, for which we have to consider the conditional probability of (r, y_2), the pattern and type of the blood on the jumper, given (x_1, x_2), the (observed) blood types of the suspect and the guard, g_2, the guard's evidence of the punch, and n, the number of offenders, under each of the competing hypotheses. No variables related to fibers appear here.

---

[1]Still further simplification is possible, using reasonable properties not all of which are represented in the graph: see Dawid and Evett [9] for details.

19.3 Object-Oriented Networks

Many problems have a hierarchical or repetitive structure that is not best represented by a "flat" network such as that of Figure 19.1. An "object-oriented Bayesian network" (OOBN) allows such additional structure to be taken into account, to simplify the construction, display and interpretation of the network. In an OOBN, what looks like a single node in a network can in fact be a network in its own right. This generalization of a BN was first proposed by Laskey and Mahoney [19].

As an example, Figure 19.3 (created using the commercial software package HUGIN) gives a high-level view of the network of Figure 19.1, showing that, conditional on the (unobserved) identification nodes (N, C), the fiber evidence is independent of the blood and eyewitness evidence—whereas the blood evidence remains dependent on the eyewitness evidence (in fact, through the node G_2).

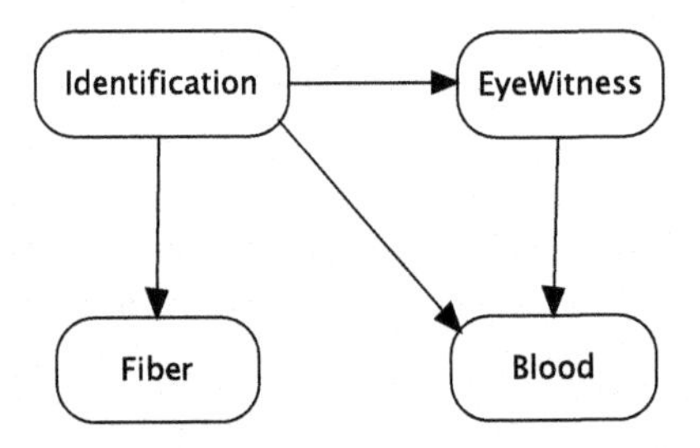

FIGURE 19.3: Object-oriented Bayesian network (OOBN) for Robbery Example 19.2.1. Each block node is an "instance" of one of the generic network modules of Figure 19.4.

The internal structure of the individual submodules is shown in Figure 19.4. A thick grey rim denotes an output node, which can be identified with an input node (dashed grey rim) in another module. This is done as shown in Figure 19.5.

19.3.1 Generic modules

A particularly valuable use of OOBNs is based on generic network modules (also termed fragments, or idioms), that can be reused, both within and across higher-level networks [24, 17, 14]. We indicate such a module by a **boldface** font. Any specific instantiation of a module in a larger network will be set in `teletype` font (like any other node), while a value (state) of a node will by indicated by *italic*.

One generic module, **testimony**, describes features of eyewitness testimony of an event [25]. This is structured into three stages: `sensation`, `objectivity`, and `veracity`, as represented in the network of Figure 19.6, which builds on the submodules shown in Figure 19.7. Here **sensation** models the possibility of mistakes in the witness's perception of the event, due either to his sensory and general physical condition (leading to possible disagreement between the actual and the perceived features of the event), or to the conditions under which the observation is made. The latter aspect is termed "competence". For example, if the witness was hiding under a table, he might not have been in a position to observe what was happening. These two processes are incorporated into Figure 19.7(i), where the node `agreement` is an instance of the generic module **accuracy** of Figure 19.7(ii), which uses a random `Error` to determine whether or not the output node reproduces the input. `objectivity` relates to whether or not the witness's belief is a correct interpretation of

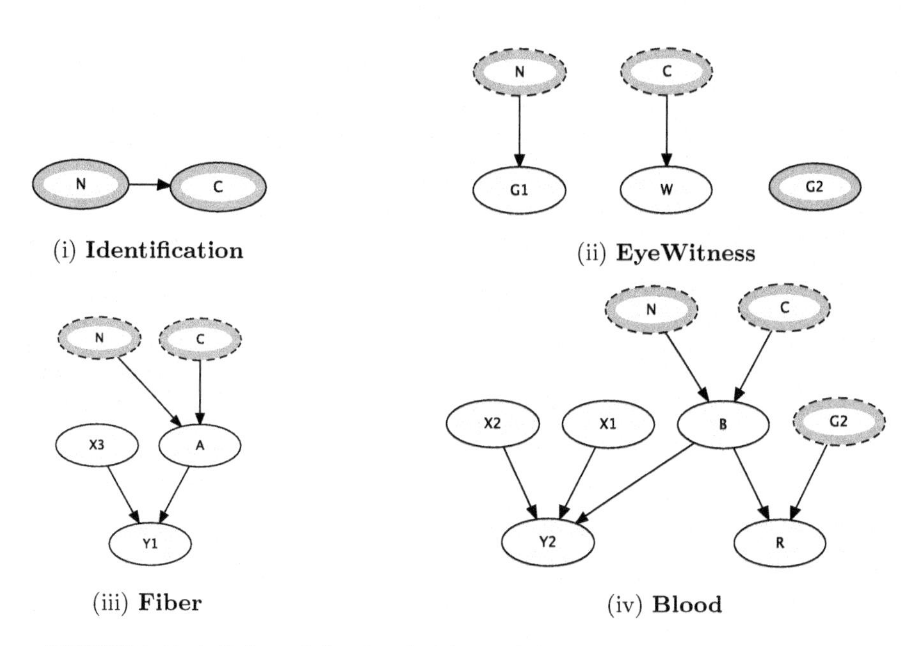

(i) **Identification** (ii) **EyeWitness** (iii) **Fiber** (iv) **Blood**

FIGURE 19.4: Submodules for Robbery OOBN of Figure 19.3.

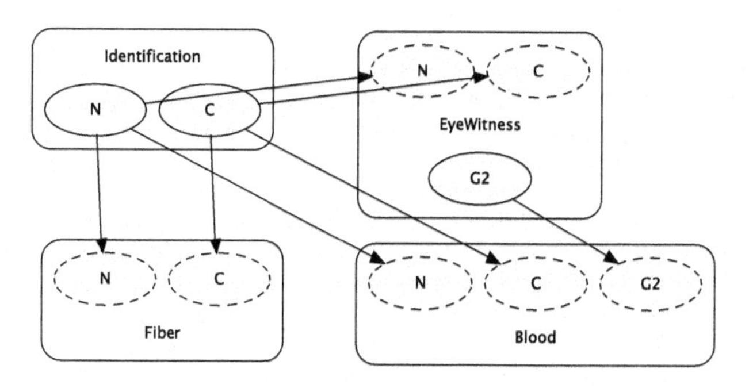

FIGURE 19.5: Expanded view of Robbery OOBN (Figure 19.3) showing identification of output and input nodes across modules.

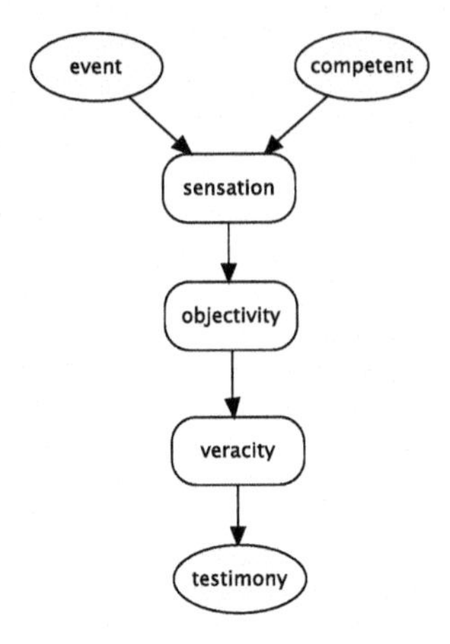

FIGURE 19.6: Generic **testimony** module (adapted with permission from Dawid et al. [12]), showing a series of stages (themselves instances of generic submodules—see Figure 19.7) whereby an actual event influences the testimonial report of a witness.

the evidence of his senses, and `veracity` to whether or not he truthfully reports his belief. The subnetworks `objectivity` and `veracity` in Figure 19.6 are constructed as instances of **accuracy**.

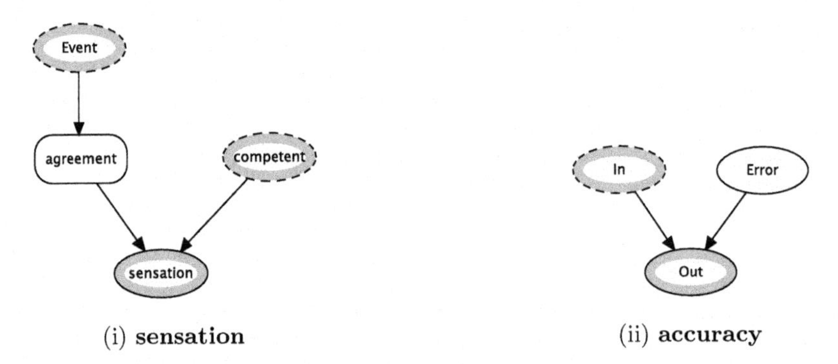

(i) **sensation** (ii) **accuracy**

FIGURE 19.7: Submodules for use in **testimony** module of Figure 19.6 (adapted with permission from Dawid et al. [12]). The subnetworks `agreement` in Figure 19.7(i), and `objectivity` and `veracity` in Figure 19.6, are all instances of **accuracy**.

Other generic evidential modules include **identification** (as in Figure 19.9 below), **contradiction**, **corroboration**, **conflict**, **convergence**, and **explaining away**. See Hepler et al. [17] for details.

19.4 Quantitative Analysis

Our discussion so far has largely concentrated on the qualitative aspects of a BN representation. Such a representation also allows simplification of the tasks of assigning and manipulating probability distributions for the variables. Rather than specify a very large collection of joint probabilities for all the variables in the problem, it is enough (and generally much easier) to specify, for each node, its conditional distribution, given the configuration of states of its "parent" variables. It is then possible, using elegant and efficient computational algorithms, both exact and (for more complex problems) approximate, to extract the marginal distribution of any variable, or ("evidence propagation") its conditional distribution, after taking into account observed values for certain other variables. For details, see Chapters 4 and 5 of this Handbook. There exist a number of software systems that conduct such computations, including HUGIN[2], GENIE[3], NETICA[4], AGENARISK[5],GRAIN[6], GRAPPA[7] and (for approximate inference) WINBUGS[8]. All networks shown in this Chapter were created and analyzed using HUGIN.

19.5 Bayesian Networks for Forensic Genetics

Forensic DNA evidence has special features, principally owing to its pattern of inheritance from parent to child (a very brief introduction to the basic genetics is given in the appendix). These make it possible to address queries such as the following:

Criminal case: Did A leave the trace at the scene of the crime?

Disputed paternity: Is individual A the father of individual B?

Immigration: Is A the mother of B? How is A related to B?

Criminal case: mixed trace: Did A and B both contribute to a stain found at the scene of the crime? Who contributed to the stain?

Disputed inheritance: Is A the daughter of deceased B? Is A the son of a contributor to the mixture?

Disasters: Was A among the individuals involved in a disaster? Who were those involved?

In a simple criminal identification case we have evidence E that a suspect's DNA profile matches that found at the crime scene. The prosecution hypothesis H_p is that the suspect left the DNA trace, while the alternative defense hypothesis, H_d, might be that another individual randomly drawn from some reference population left the trace. In a simple disputed paternity case, the evidence E will comprise DNA profiles from mother, child and putative father. Hypothesis H_p is that the putative father is the true father, while hypothesis H_d might be that the true father is some other individual randomly drawn from the population. We can also entertain other hypotheses, such as that one of one or more other identified individuals is the father, or that the true father is the putative father's brother.

In a complex criminal case, we might find a stain at the scene of the crime having the form of a *mixed trace*, containing DNA from more than one individual. DNA profiles are also taken from the victim and a suspect. We can entertain various hypotheses as to just who—victim?, suspect?, person or persons unknown?—contributed to the mixed stain.

When we are only comparing two hypotheses H_0 and H_1, the impact of the totality of the DNA evidence E available, from all sources, is crystallized in the *likelihood ratio*,

[2]http://www.hugin.com
[3]http://www.bayesfusion.com/genie-modeler
[4]https://www.norsys.com
[5]http://www.agenarisk.com
[6]https://CRAN.R-project.org/package=gRain
[7]https://people.maths.bris.ac.uk/~mapjg/Grappa
[8]http://www.mrc-bsu.cam.ac.uk/software/bugs

$\text{LR} = \Pr(E \mid H_1)/\Pr(E \mid H_0)$. If we wish to compare more than two hypotheses, we require the full *likelihood function*, a function of the various hypotheses H being entertained (and of course the evidence E):

$$\text{lik}(H) \propto \Pr(E \mid H). \tag{19.3}$$

The proportionality sign in (19.3) indicates that we have omitted a factor that does not depend on H, although it can depend on E. Such a factor is of no consequence and need not be specified, since it disappears on forming ratios of likelihoods for different hypotheses on the same evidence. Only such relative likelihoods are required, not absolute values.

We also now need to specify the prior probabilities, $\Pr(H)$, for the full range of hypotheses H. Then posterior probabilities in the light of the evidence are again obtained from Bayes's theorem, which can now be expressed as:

$$\Pr(H \mid E) \propto \Pr(H) \times \text{lik}(H). \tag{19.4}$$

Again the omitted proportionality factor in (19.4) does not depend on H, although it might depend on E. It can be recovered, if desired, as the unique such factor for which the law of total probability, $\sum_H \Pr(H \mid E) = 1$, is satisfied.

19.5.1 Bayesian networks for simple criminal cases

In a simple criminal DNA identification case, the evidence is that the suspect's DNA profile matches a trace found at the scene of the crime. We are interested in comparing two mutually exclusive hypotheses: the *prosecution hypothesis* H_p: "the crime trace belongs to the suspect `s`" (loosely, "the suspect is guilty"), and the *defense hypothesis* H_d: "the crime trace belongs to another actor, `o`, *randomly drawn from the population*". Representation of such problems as BNs was introduced by Dawid et al. [10], and as OOBNs by Dawid et al. [11].

Each genetic marker m is analyzed separately. The relevant OOBN is shown in Figure 19.8, together with its expanded version. Nodes `s` and `o` are each instances of a **founder** network module, with nodes paternal gene `pg`, maternal gene `mg`, and genotype `gt`. Each of the (input) nodes `pg` and `mg` is identified with the output node of a simple module **gene** (not shown), that contains the alleles of that marker, and their frequencies in a relevant reference population, while the (output) node `gt` is constructed as the unordered combination of `pg` and `mg` (since we cannot distinguish the paternal and maternal gene in a genotype). Node `trace` is an instance of the **identification** network module shown in Figure 19.9: its output `trace` is modeled as equal to `sgt` or `ogt`, according as `S guilty?` is *true* or *false*, respectively.

Node `S guilty?` is assigned probability 0.5 for *true*. The observed matching genotype is entered as evidence at `gt` in `s`, and again at `crgt` in `trace`, and propagated through the network. The resulting computed odds on *true* at `S guilty?` can then be interpreted as the likelihood ratio in favor of H_p, based on the evidence of a match at marker m. Finally, under the assumption of independence across markers, multiplying these values across all markers delivers the overall likelihood ratio based on the full DNA evidence.

19.5.2 Bayesian network for simple paternity cases

In a simple case of disputed paternity, a man is alleged to be the father of a child, but disputes this. DNA profiles are obtained from the mother `m`, the child `c`, and the putative father `pf`. On the basis of these data, we wish to assess the likelihood ratio for the hypothesis of *paternity*: H_p: `tf` = `pf`, the true father is the putative father; as against that of *non-paternity*: H_d: `tf` = `af`—where `af` denotes an unspecified alternative father, treated as unrelated to `pf` and randomly drawn from an appropriate reference population.

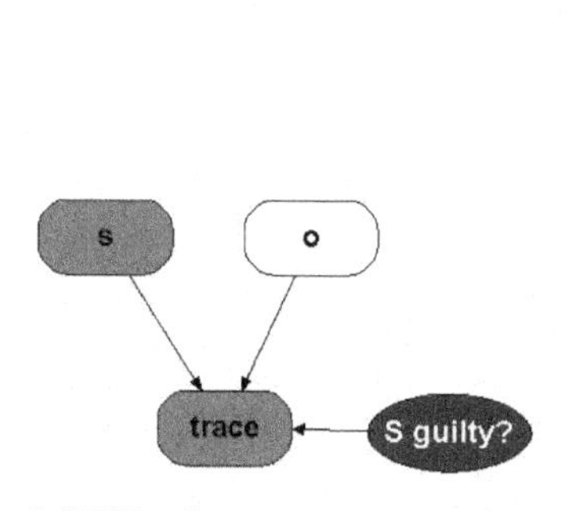

(i) OOBN, showing actors `s` and `o`, crime trace and hypothesis.

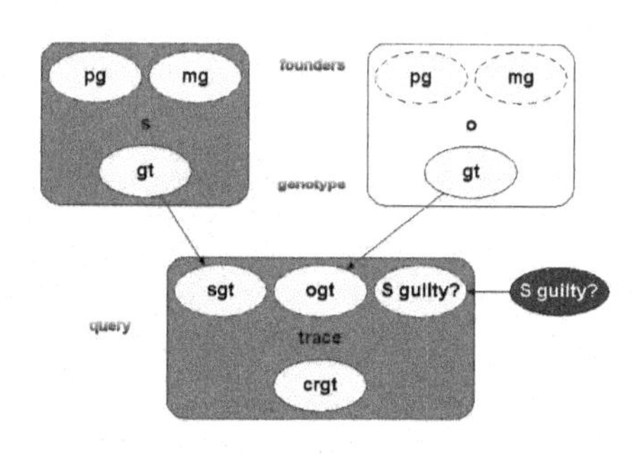

(ii) Expanded network, revealing variables within each block node.

FIGURE 19.8: OOBN for criminal identification (adapted with permission from Green and Mortera [15]). Information is available on a certain feature for both the suspect `s` and a `trace` from the crime scene, and we wish to know if it was `s`, or some other individual `o`, who left the `trace`. The expanded version in Figure 19.8(ii) is specific to DNA profile evidence, where the observed feature is the genotype at a given marker.

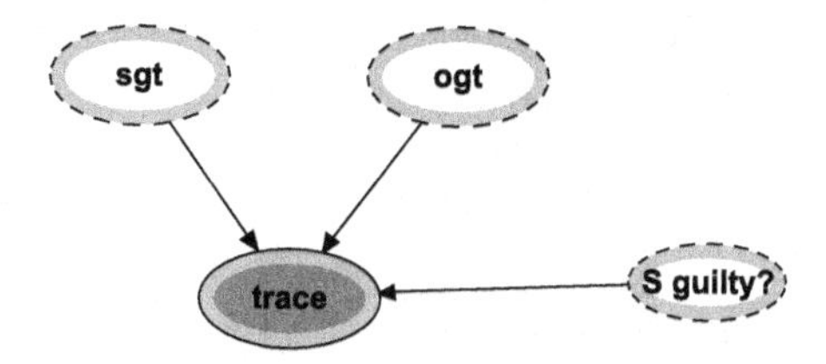

FIGURE 19.9: Module **identification**: `trace` is identical to `sgt` or `ogt` according as `S guilty?` is *true* or *false*.

The disputed pedigree can be represented by the OOBN of Figure 19.10. Nodes `m`, `pf` and `af` are instances of the network module **founder** as in Section 19.5.1, while node `tf` is an instance of **identification**—its output is a genotype copied from that of `pf` or `af`, according as `tf` = `pf`? is *true* (H_p) or *false* (H_d). Node `c` is an instance of a network module **child**, containing two copies (one for each parent) of the module **mendel**, shown in Figure 19.11, whereby, according to Mendel's law, the child `c` inherits its parental gene `cg` by a random draw (represented as a fair coin flip `fcoin`) from the maternal and paternal genes, `mg` and `pg`, of the relevant parent.

As in Section 19.5.1, under the hypothesis of independence across markers, we analyze the markers one at a time. For each, we assign the relevant allele frequency distribution at each founder gene, enter the observed evidence at `m`, `pf` and `c`, and perform probability propagation. This yields a likelihood ratio based on that marker data, and we multiply all these together to obtain the overall likelihood ratio based on the full collection of markers. This can then be combined with the prior odds of paternity, based on external background evidence B, to obtain the posterior odds on paternity.

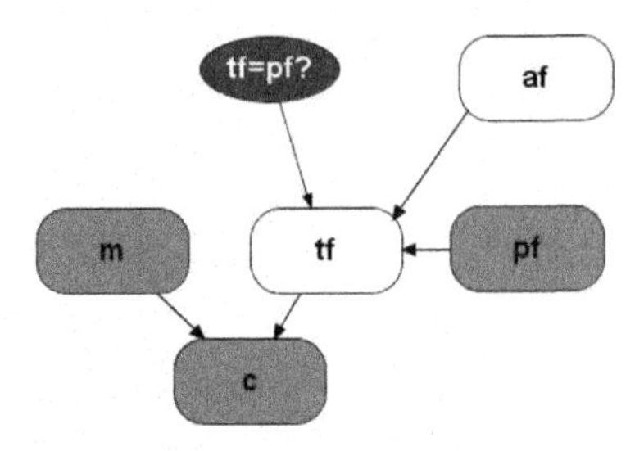

FIGURE 19.10: OOBN representing disputed pedigree for simple paternity (adapted with permission from Green and Mortera [15]).

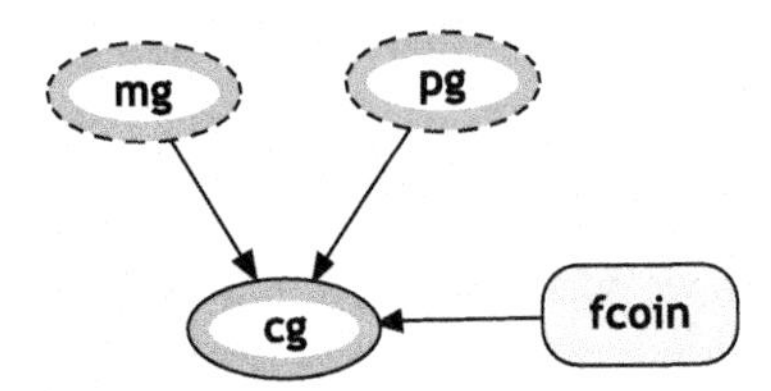

FIGURE 19.11: Module **mendel** representing Mendelian inheritance, whereby a child's gene `cg` is a random draw of either the maternal gene `mg` or the paternal gene `pg`.

19.5.3 Bayesian networks for complex cases

A major advantage of OOBN representations is that they make it easy to elaborate the network with additional features [11]. For example, in the presence of possible mutation, we can modify the network of Figure 19.11 to allow either `mg` or `pg` to mutate, before being possibly selected for transmission to `cg`. Various different mutation models can be constructed and incorporated. Other possible modifications include, for example, allowance for alleles that are not picked up by the instrumentation—a property that can be either inherited (a "silent" allele) or sporadic (a "missed" allele). Such modifications can typically be confined to low-level networks; the other modules, and the overall high-level structure, are unchanged.

Another advantage is the ability to re-use existing network modules in new combinations, to tell different stories. For example, Figure 19.12 puts together instances of **founder** (at `gf`, `gm`, `m1`, `m2` and `af`), of **child** (at `pf`, `b1`, `b2`, `c1`, `c2`) and of **identification** (at `tf`), to analyze a case where it was impossible to collect DNA from `pf`, the putative father of the child `c1` of mother `m1`, but DNA is available from his two full brothers `b1` and `b2` (all children of grandfather `gf` and grandmother `gm`) and his undisputed child `c2` by a different mother `m2`, as well as from `m1`, `m2` and `c1`.

Moreover, the building blocks used in such constructions can themselves be modified, as described above, to incorporate additional features such as mutation.

19.6 Bayesian Networks for DNA Mixtures

When several actors have contributed to a DNA trace found at a crime scene we will have a *mixed* DNA profile. The presence of 3 or more alleles on any marker indicates that the trace is a mixture from more than one contributor. In a two person mixture, one might

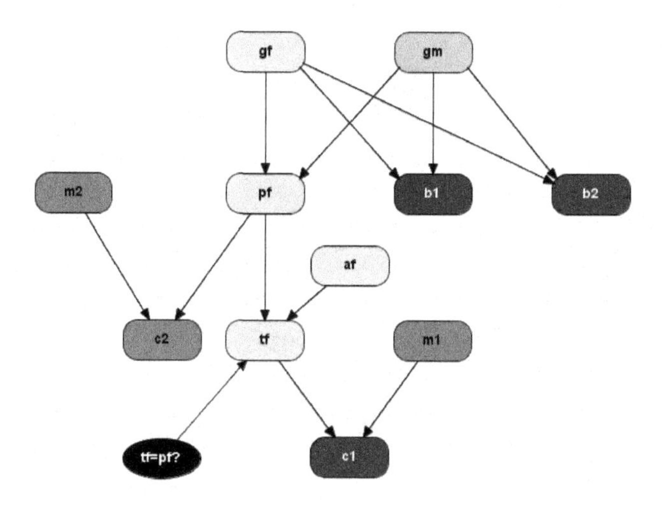

FIGURE 19.12: OOBN representing disputed paternity with absent putative father `pf`. DNA profiles are available for the darker shaded individuals, including two brothers, `b1` and `b2`, of `pf`.

be interested in testing whether the victim and suspect contributed to the mixture, H_p: $v \,\&\, s$, against the hypothesis that the victim and an unknown individual contributed to the mixture, H_d: $v \,\&\, u$. One might alternatively consider an additional unknown individual u_2 instead of the victim, with hypotheses H'_p: $u_2 \,\&\, s$ versus H'_d: $u_2 \,\&\, u_1$.

19.6.1 Qualitative data

We first describe Bayesian networks for analyzing purely *qualitative* data, describing simply which alleles are observed in the trace. Figure 19.13 shows a top-level network which can be used for analyzing a mixture with two contributors, $p1$ and $p2$, and a marker in the trace having three alleles A, B and C (the network can be simply modified to account for different numbers of alleles). Nodes `sgt`, `vgt`, `u1gt` and `u2gt` are instances of the network class **founder**, and represent the suspect's, the victim's and two unknown individuals' genotypes. Node `p1gt`, the genotype of $p1$, is an instance of **identification**, which selects between the two genotypes `sgt` or `u1gt` according to the *true/false* state of the Boolean node `p1=s?`, representing the hypothesis that contributor $p1$ is the suspect s. A similar relationship holds between nodes `p2gt`, `vgt`, `u1gt` and `p1=v?`. The `target` node is the logical combination of the two Boolean nodes `p1=s?` and `p2=v?` and represents the four different hypotheses described above. Node `Ainmix?` determines whether allele A is in the mixture: this will be so if at least one A allele is present in either `p1gt` or `p2gt`. Similarly for `Binmix?`, `Cinmix?` and `Dinmix?` (where `D` refers to all the alleles that are not observed).

For each marker the gene nodes are populated with the relevant allele frequency distribution, and nodes `p1=s?`, `p1=v?` are modeled as coin-flips. Any available genotype information on the suspect and the victim is entered into nodes `sgt` and `vgt`, *true* is entered at `Ainmix?` and `Binmix?`, and `Cinmix?`, and *false* at `Dinmix?`. This evidence is propagated, after which the probability distribution over the four hypotheses at `target` can be interpreted as a likelihood function, based on the data for that marker. Again, an overall likelihood function is obtained on multiplying these across markers.

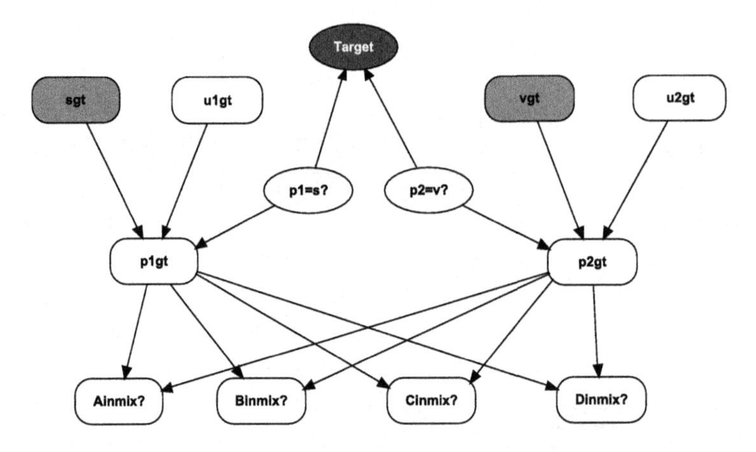

FIGURE 19.13: OOBN for analysing a DNA mixture involving two contributors (qualitative data).

By simple modification of the network, the modular structure of Bayesian networks supports easy extension to mixtures with more contributors, so long as the total number of contributors can be assumed known. Or, if it can be agreed to limit attention to some maximum total number of potential contributors [20], cases where the number of unknown contributors is itself uncertain can be addressed using a Bayesian network, now including nodes for the number of unknown contributors and the total number of contributors [22]. This can be used for computing the posterior distribution of the total number of contributors to the mixture, as well as likelihood ratios for comparing all plausible hypotheses. The modular structure of the Bayesian networks can be used to handle still further complex mixture problems. For example, we can consider together missing individuals, silent alleles and a mixed crime trace simply by piecing together the appropriate modules.

19.6.2 Quantitative data

The networks above only use the qualitative information as to which allele values are present in the mixture and the other profiles. A more sensitive analysis additionally uses measured continuous "peak heights" or "peak areas", which give quantitative information on the amounts of DNA involved. (The DNA is amplified using the polymerase chain reaction (PCR) process and the peak height, or area, is a measure of the amount of the allele in the amplified sample expressed in relative fluorescence units.) This requires much more detailed modeling, but again this can be effected by means of a Bayesian network [5]. Because the mixture proportion `frac` of DNA contributed by one of the parties is a quantity common across all markers, we must now handle the markers all simultaneously within one "super-network". Figure 19.14 shows the top level network for two contributors, involving six markers (D8, vWA, D21, D18, FGA, TH01), each an instance of a module **marker** as shown in Figure 19.15. This network is an extended version of the one shown in Figure 19.13, incorporating additional structure to model the quantitative peak height information. In particular, the nodes `Aweight` etc. in **marker** are instances of a module that models the quantitative information on the peak height.

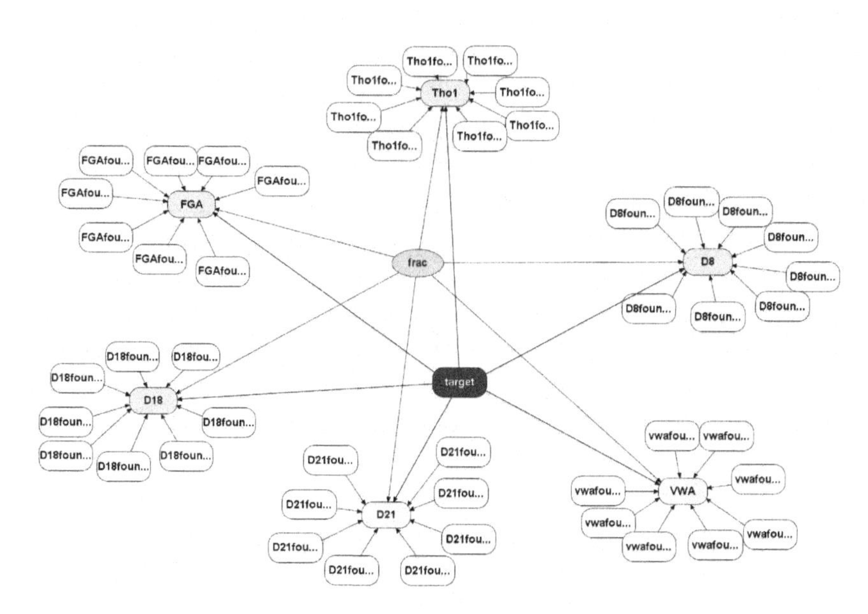

FIGURE 19.14: OOBN for 2-person DNA mixture using quantitative peak area data on six markers, `FGA`, `TH01`, etc. (reproduced from Cowell et al. [3]).

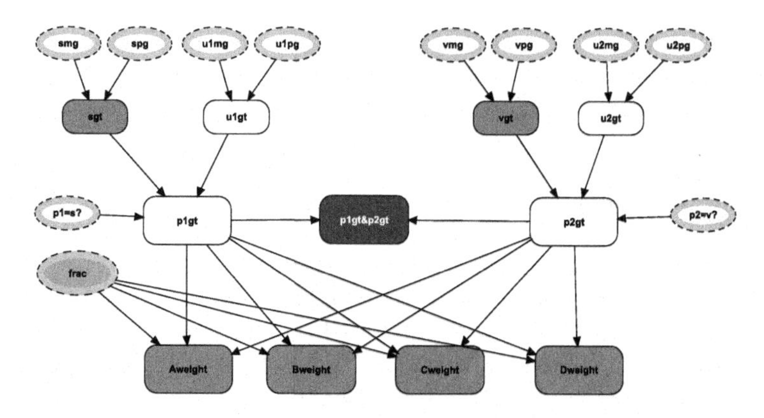

FIGURE 19.15: Module **marker** for use in Figure 19.14, for four observed allele peaks denoted by `Aweight`, etc. (reproduced from Cowell et al. [3]).

Cowell et al. [4, 5] analyze the data shown in Table 19.1, taken from Evett et al. [13], involving a 6-marker mixed profile with between 2 and 4 distinct observed alleles and corresponding peak areas per marker, and a suspect whose profile is contained in these. It is assumed that the crime profile is a mixture either of the suspect and one other unobserved contributor, or of two unknown contributors. Using only the alleles as data, the likelihood ratio for the suspect being a contributor to the mixture is calculated to be around 25,000. On taking account of the peak areas also, this rises 6,800-fold, to about 170,000,000.

Marker	D8			D18			D21			
Alleles	10*	11	14*	13*	16	17	59	65	67*	70*
Peak area	6416	383	5659	38985	1914	1991	1226	1434	8816	8894

Marker	FGA			TH01		vWA			
Alleles	21*	22*	23	8*	9.3*	16*	17	18*	19
Peak area	16099	10538	1014	17441	22368	4669	931	4724	188

TABLE 19.1: Data for mixed trace with two contributors. The starred values are the suspect's alleles.

19.6.3 Further developments on DNA mixtures

Cowell et al. [4, 6, 7] further extend the statistical model in Section 19.6.2 for the quantitative peak information obtained from an electropherogram of a forensic DNA sample. A gamma model is used for the peak heights and the model further develops the modeling of various artefacts that can occur in the DNA amplification process. Thus *dropout* of an allele occurs when its associated peak fails to exceed the detection threshold. Another common artefact is *stutter*, whereby an allele at repeat number a that is present in the sample is mis-copied, and appears as a peak at repeat number $a - 1$. Yet another artefact is *dropin*, referring to the occurrence of small unexpected peaks in the DNA amplification: this can, for example, be due to sporadic contamination of a sample, either at source or in the forensic laboratory. Current technology allows for the amplification of very small amounts of DNA, even as little as contained within one cell; in such a case many of these artefacts can occur. These artefacts are simply represented in a coherent way in this model.

The model can both find likelihood ratios for evidential calculations, and deconvolve a DNA mixture for the purpose of finding likely profiles of one or more unknown contributors to the mixture. Computation from this model rely on an efficient implementation of Bayesian network techniques.

The network in Figure 19.16 shows how a genotype is represented by a vector of allele counts $n_{i,a} = 0, 1$ or 2, for alleles $a = 1, 2, \ldots, 5$, for two individuals $i = 1, 2$ (S_{ia} being the partial sum $S_{ia} = \sum_{b \leq a} n_{ib}$). In this way a genotype is modeled by a Markov structure and computations can be done linearly in the number of alleles.

The exact probability propagation methods of BNs work on discrete or conditional Gaussian variables. In order to model continuous data like the quantitative peak information represented by a gamma model the computation is achieved by introducing auxiliary dummy binary variables O_a into the BN. The quantities of interest are then computed by efficient probability propagation algorithms. For further details see Cowell et al. [7].

The Markov Bayesian network representation allows for ready extension to many unknown contributors and the simultaneous analysis of more than one mixture trace. This modeling of peak height information provides for a very efficient mixture analysis.

Recently Mortera et al. [23] applied this model to analyze a complex disputed paternity case, where the DNA of the putative father was extracted from his corpse, which had been inhumed for over 20 years. This DNA was contaminated and appeared to be a mixture of

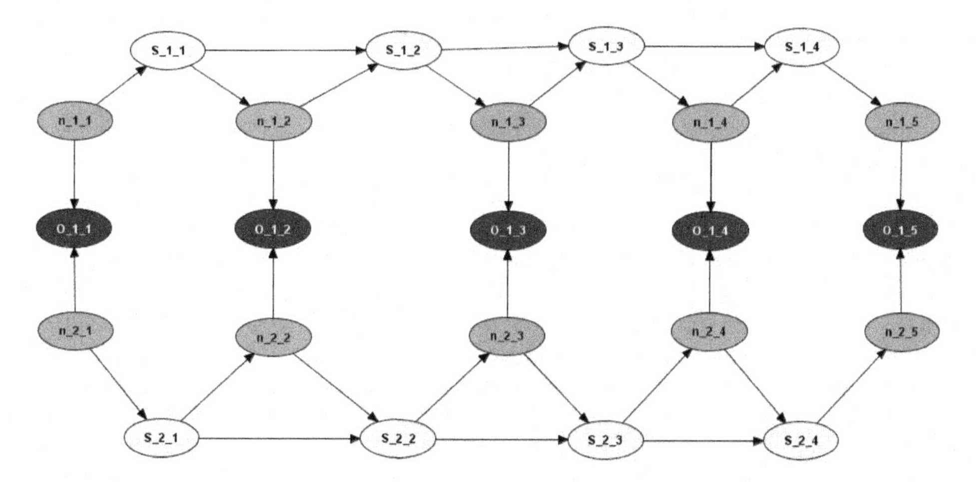

FIGURE 19.16: Markov Bayesian network representation of a genotype for two unknown contributors ($i = 1, 2$) and peak height observations for a marker with five possible allelic types ($a = 1, \ldots, 5$). The observations are described by the allele counts $\{\texttt{n\_i\_a}\}$, having values 0, 1 or 2.

at least two individuals. This case was further analyzed in Green and Mortera [16], which presents general methods for inference about relationships between contributors to a DNA mixture and other individuals of known genotype. The model for relationship inference builds on the approach in Cowell et al. [7], but makes more explicit use of the Bayesian networks in the modeling.

19.7 Analysis of Sensitivity to Assumptions on Founder Genes

Many forensic genetics problems, as we have shown, can be handled using structured systems of variables, for which BNs offer an appealing practical modeling framework, and allow inferences to be computed by probability propagation methods. However, when standard assumptions are violated—for example when allele frequencies are unknown, there is identity by descent or the population is heterogeneous—dependence is generated among founding genes, that makes exact calculation of conditional probabilities by propagation methods less straightforward. The standard assumptions that the allele frequencies are fixed and known, that the individual actors in the model are independent and that the allele frequency database is homogeneous can all be questioned [15]. We now illustrate a couple of these issues and show how they can be represented as a BN module generalizing the networks used for forensic identification.

19.7.1 Uncertainty in allele frequencies

In reality, the allele frequencies assumed when conducting probabilistic forensic inference are not known probabilities, but estimates based on empirical frequencies in a database.

For the criminal case of Section 19.5.1, the joint distribution of the founding genes is

$$\prod_m \{p(\texttt{spg}_m)p(\texttt{smg}_m)p(\texttt{opg}_m)p(\texttt{omg}_m)\}, \tag{19.5}$$

and all questions about sensitivity can be expressed through modifications to (19.5). Some generate dependence between founding genes.

Following Green and Mortera [15], assuming the idealization of a Dirichlet prior and multinomial sampling, the posterior distribution of a set of probabilities is Dirichlet($M\rho(1), M\rho(2), \ldots, M\rho(k)$), where M is the (posterior) sample size and the ρ's are essentially the database allele frequencies (posterior means). The founding genes (`spg`, `smg`, `opg`, `omg`) are drawn from this distribution, (conditionally) independently and identically across alleles. This corresponds to the standard set-up for a Dirichlet process model which, by marginalizing over the Dirichlet distribution, can be represented in a BN using a Pólya urn scheme. This is represented by the network module shown in Figure 19.17: for further details see Green and Mortera [15]. For efficiency of the probability propagation, in order to create smaller clique tables this network is set up so that all choices are binary, following the "divorcing" procedure [18], whereby auxiliary nodes are introduced in order to reduce the number of incoming edges of a selected node. This module can then be incorporated as a building block in a higher level network that computes inference, for example, about a criminal identification case, a simple or complex paternity testing or a DNA mixture problem. Thus Figure 19.18 shows a network for criminal identification that integrates the network of Figure 19.8(i) with that of Figure 19.17. Similarly the module in Figure 19.17, representing uncertain allele frequencies, can be integrated into the networks described in Section 19.5.2, Section 19.5.3, and Section 19.6. In this way, we can introduce uncertain allele frequencies for the reference population into forensic identification problems.

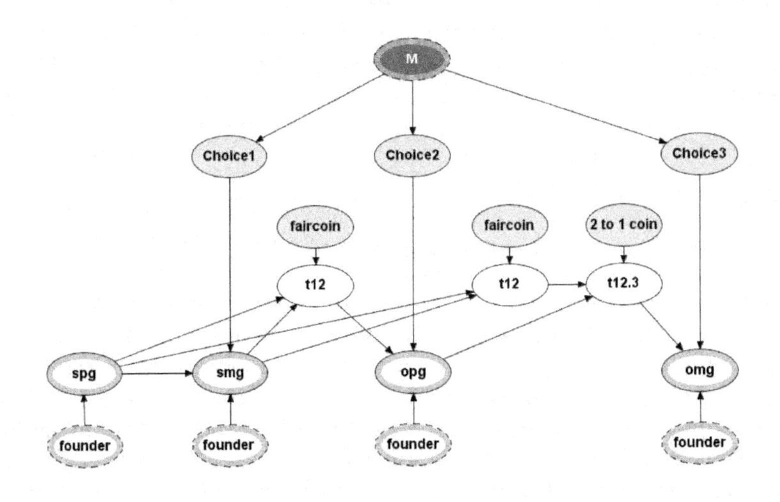

FIGURE 19.17: Network module **UGF** for the Pólya urn scheme (adapted with permission from Green and Mortera [15]).

19.7.2 Heterogeneous reference population

The assumption that the DNA reference population is homogeneous is questionable. The population is typically a mixture of subgroups.

Population heterogeneity raises two kinds of issues in the modeling. First, since unobserved actors are assumed to have genes drawn from a population, results can depend on which population (and corresponding allele frequency database) is used. Secondly, when

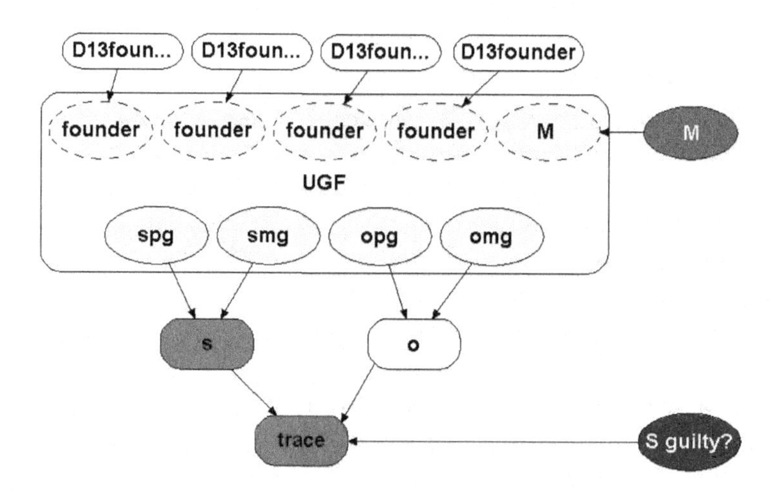

FIGURE 19.18: OOBN for criminal case with uncertain allele frequencies, incorporating an instance of the Pólya urn scheme module **UGF** of Figure 19.17.

there is uncertainty about which population is relevant, this can induce dependence between actors, observed or not. Additionally, when uncertainty about subpopulation relates to untyped actors, dependence between markers is induced.

The upper level network for sensitivity of inferences to population structure for criminal identification, based on a synthetic population (that is a mixture of Afro-Caribbean, Hispanic and Caucasian subpopulations) is shown in Figure 19.19.

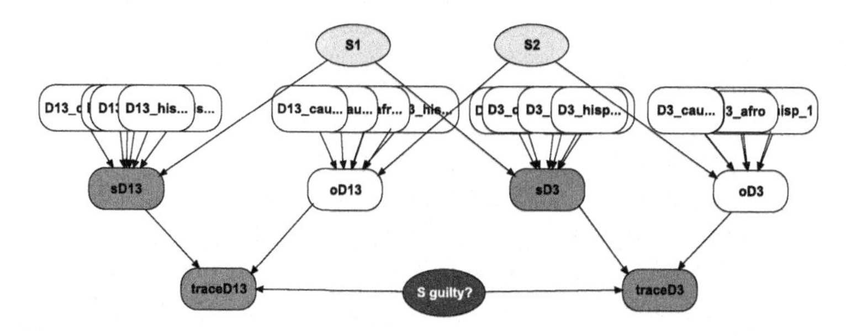

FIGURE 19.19: OOBN for 2 markers in a criminal identification case, allowing for subpopulation effect.

Such problems are easily set up as BNs with the sub-network structure shown in Figure 19.20. The variable S identifies the subpopulation, which may be dependent or independent between actors depending on the scenario of interest. Crucially, for each actor, S is the same for both genes for all markers, so that mixing across subpopulations is not the same as averaging the allele frequencies and assuming an undivided subpopulation. Note that, conditional on subpopulation S, every gene at every marker is drawn independently from the appropriate subpopulation gene pool.

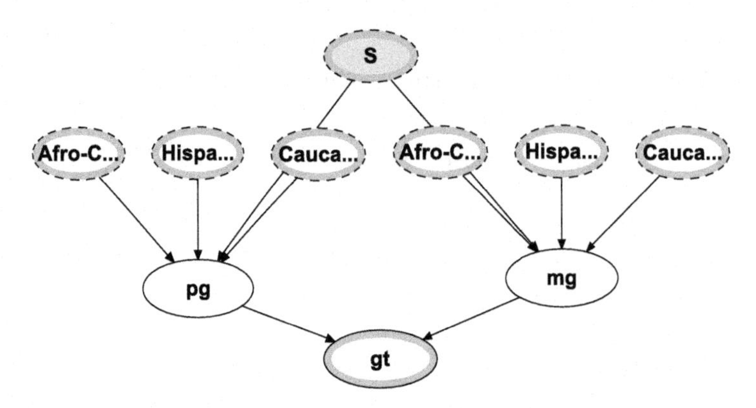

FIGURE 19.20: Network module for a genotype in Figure 19.19, accounting for subpopulation effect, based on a mixture of Afro-Carribean, Hispanic and Caucasian subpopulations (adapted with permission from Green and Mortera [15]).

19.8 Conclusions

We hope we have stimulated the reader's interest in the application of BNs for modeling problems in forensic science.

We have also aimed to show the usefulness of BNs for representing and solving a wide variety of complex forensic problems. Both genetic and non-genetic information can be represented in the same network. A particularly valuable feature is the modular structure of BNs, which allows a complex problem to be broken down into simpler structures that can then be pieced back together in many ways, so allowing us to address a wide range of forensic queries. In particular, using OOBNs we can construct a flexible computational toolkit, and use it to analyze complex cases of DNA profile evidence, accounting appropriately for such features as missing individuals, mutation, silent alleles and mixed DNA traces, accounting for uncertainty in allele frequencies, heterogeneous populations and also inference about relatedness in DNA mixtures [16].

As new technologies for forensic DNA identification are developed, such as single cell sequencing, specific BN modeling will be needed to account for problems like dependence among the measurements (linkage disequilibrium) and appropriate population frequencies. BNs and OOBNs can also be usefully used in many branches of forensic analysis beyond those illustrated here.

Appendix: Genetic Background

We will introduce some basic facts about DNA profiles; for a more detailed explanation see Butler [1].

A *gene* is a particular sequence of the four *bases*, represented by the letters A, C, G and T. A specific position on a chromosome is called a *locus* (hence there are two genes at any locus of a chromosome pair). A *DNA profile* consists of measurements on the genotype at a number of *forensic markers*, which are specially selected *loci* on different chromosomes. In standard forensic identification problems it is customary to assume *Hardy–Weinberg equilibrium*, and that loci are *unlinked*, which corresponds to assuming independence within and across markers.

Current technology uses around 17–23 *short tandem repeat* (STR) markers. At each marker, each gene has a finite number (up to around 20) of possible values, or *alleles*, generally positive integers. For example, an allele value of 5 indicates that a certain word (e.g. *CAGGTG*) in the four letter alphabet is repeated exactly 5 times in the DNA sequence at that locus. In statistical terms, a gene is represented by a random variable, whose realized state is an *allele*.

In a particular forensic context, we will refer to the various human individuals involved in the case as 'actors'. For each marker having two alleles (autosomal marker) a *genotype* consists of an unordered pair of genes, one inherited from the father and one from the mother (though one cannot distinguish which is which). When both alleles are identical the actor is *homozygous* at that marker, and only a single allele value is observed; otherwise the actor is *heterozygous*. An actor's *DNA profile* comprises a collection of genotypes, one for each marker.

Assuming *Mendelian segregation*, at each marker a parent passes a copy of just one of his or her two genes, randomly chosen, to his or her child, independently of the other parent and independently for each child.

Databases have been gathered from which allele frequency distributions, for various populations, can be estimated for each forensic marker.

Acknowledgments

The authors would like to thank the Simons Foundation and the Isaac Newton Institute for Mathematical Sciences for its hospitality and support during the programme *Probability and Statistics in Forensic Science* which was supported by EPSRC Grant Number EP/K032208/1. We also thank Normal Fenton and Nadine Smit for useful comments on a previous version. We also thank two referees for their useful comments.

Bibliography

[1] J. M. Butler. *Forensic DNA Typing.* Elsevier, USA, 2005.

[2] R. G. Cowell, A. P. Dawid, S. L. Lauritzen, and D. J. Spiegelhalter. *Probabilistic Networks and Expert Systems.* Springer, New York, 1999.

[3] R. G. Cowell, S. L. Lauritzen, and J. Mortera. Identification and separation of DNA mixtures using peak area information using a probabilistic expert system. Statistical Research Paper 25, Cass Business School, City University, 2004.

[4] R. G. Cowell, S. L. Lauritzen, and J. Mortera. A gamma model for DNA mixture analyses. *Bayesian Analysis*, 2:333–348, 2007.

[5] R. G. Cowell, S. L. Lauritzen, and J. Mortera. Identification and separation of DNA mixtures using peak area information. *Forensic Science International*, 166:28–34, 2007.

[6] R. G. Cowell, S. L. Lauritzen, and J. Mortera. Probabilistic expert systems for handling artefacts in complex DNA mixtures. *Forensic Science International: Genetics*, 5:202–209, 2011.

[7] R. G. Cowell, T. Graversen, S. L. Lauritzen, and J. Mortera. Analysis of DNA mixtures with artefacts (with Discussion). *Journal of the Royal Statistical Society Series C*, 64: 1–48, 2015.

[8] A. P. Dawid. Conditional independence in statistical theory (with Discussion). *Journal of the Royal Statistical Society Series B*, 41:1–31, 1979.

[9] A. P. Dawid and I. W. Evett. Using a graphical method to assist the evaluation of complicated patterns of evidence. *Journal of Forensic Sciences*, 42:226–231, 1997.

[10] A. P. Dawid, J. Mortera, V. L. Pascali, and D. W. van Boxel. Probabilistic expert systems for forensic inference from genetic markers. *Scandinavian Journal of Statistics*, 29:577–595, 2002.

[11] A. P. Dawid, J. Mortera, and P. Vicard. Object-oriented Bayesian networks for complex forensic DNA profiling problems. *Forensic Science International*, 169:195–205, 2007.

[12] A. P. Dawid, A. B. Hepler, and D. A. Schum. Inference networks: Bayes and Wigmore. In A. P. Dawid, W. L. Twining, and D. Vasilaki, editors, *Evidence, Inference and Enquiry*, volume 171 of *Proceedings of the British Academy*, 119–150. Oxford University Press, Oxford, 2011.

[13] I. W. Evett, P. D. Gill, and J. A. Lambert. Taking account of peak areas when interpreting mixed DNA profiles. *Journal of Forensic Sciences*, 43:62–69, 1998.

[14] N. Fenton, M. Neil, and D. A. Lagnado. A general structure for legal arguments about evidence using Bayesian networks. *Cognitive Science*, 37:61–102, 2013.

[15] P. J. Green and J. Mortera. Sensitivity of inferences in forensic genetics to assumptions about founder genes. *Annals of Applied Statistics*, 3:731–763, 2009.

[16] P. J. Green and J. Mortera. Paternity testing and other inference about relationships from DNA mixtures. *Forensic Science International: Genetics*, 28:128–137, 2016.

[17] A. B. Hepler, A. P. Dawid, and V. Leucari. Object-oriented graphical representations of complex patterns of evidence. *Law, Probability and Risk*, 6:275–293, 2007.

[18] F. V. Jensen. *An Introduction to Bayesian Networks.* UCL Press and Springer Verlag, London, 1996.

[19] K. B. Laskey and S. M. Mahoney. Network fragments: Representing knowledge for constructing probabilistic models. In D. Geiger and P. Shenoy, editors, *Proceedings of the Thirteenth Conference on Uncertainty in Artificial Intelligence*, UAI97, 334–341. Morgan Kaufmann, San Francisco, 1997.

[20] S. L. Lauritzen and J. Mortera. Bounding the number of contributors to mixed DNA stains. *Forensic Science International*, 130:125–126, 2002.

[21] S. L. Lauritzen, A. P. Dawid, B. N. Larsen, and H.-G. Leimer. Independence properties of directed Markov fields. *Networks*, 20:491–505, 1990.

[22] J. Mortera, A. P. Dawid, and S. L. Lauritzen. Probabilistic expert systems for DNA mixture profiling. *Theoretical Population Biology*, 63:191–205, 2003.

[23] J. Mortera, C. Vecchiotti, S. Zoppis, and S. Merigioli. Paternity testing that involves a DNA mixture. *Forensic Science International: Genetics*, 23:50–54, 2016.

[24] M. Neil, N. Fenton, and L. Nielson. Building large-scale Bayesian networks. *The Knowledge Engineering Review*, 15:257–284, 2000.

[25] D. A. Schum and J. R. Morris. Assessing the competence and credibility of human sources of intelligence evidence: Contributions from law and probability. *Law, Probability and Risk*, 6:247–274, 2007.

20

Graphical Models in Molecular Systems Biology

Sach Mukherjee

German Center for Neurodegenerative Diseases

Chris Oates

School of Mathematics, Statistics and Physics, Newcastle University

CONTENTS

Molecular biology has become a quantitative, data-driven science. Biotechnological advances have meant that it is now possible to relatively efficiently measure thousands of molecular variables in biological samples. Such data offer a remarkable window into the processes that underlie events in health and disease. However, the underlying mechanisms are complex and high-level phenomena of interest (such as whether a cell is healthy or diseased) may be rooted in the interplay between large numbers of biochemical species. Graphical models represent a natural statistical framework for linking molecular data to interpretable models of underlying biology. This Chapter surveys graphical models as used in molecular and systems biology, with an emphasis on their use for making inferences about molecular networks, that is the task of estimating molecular networks directly from data on e.g. genes or proteins. The subsequent Chapter 21 covers the use of graphical models in genetics (and in the context of combining genetic and genomic data) as well as in metagenomics.

20.1 Background

20.1.1 DNA, RNA and proteins

The molecular biological revolution of the second half of the 20th century revealed the unity of biological function at the level of information-coding biomolecules [2]. What follows is a very brief summary of some of the key ideas in molecular and systems biology. We focus on the most general concepts and refer the reader to the references for details and special cases.

Living cells store information in a relatively stable form in DNA molecules. These molecules code information in the form of chains of chemical building blocks (DNA bases) whose sequence is encoded using the alphabet $\{A, C, T, G\}$. The information stored in DNA specifies the chemical composition of biologically active molecules including proteins. Due to certain aspects of the chemistry there are a large number of possible proteins that can in principle be produced by a cell at a given time. Cells typically produce relevant proteins in response to specific factors, such as environmental cues. To produce a protein Z, the information needed to specify Z's composition is first copied from the DNA into an intermediate form called messenger RNA (mRNA), which is then used as a template to produce the protein Z from chemical inputs and energy. The process of producing the mRNA from DNA is called transcription and the mRNA itself is called a transcript. The process of producing the protein using the information encoded in the mRNA is called translation. The subsequence of the DNA corresponding to the protein Z is said to be the gene coding for Z. Proteins themselves form parts of the physical structure of the cell and its components, and are active in diverse processes, for example in energy conversion and storage, information transmission and processing and so on.

20.1.2 Biological networks

Molecules in cells, including proteins, transcripts and DNA, influence one another via diverse physical and chemical mechanisms. One kind of influence is clear from the description above, namely that the DNA sequence influences the transcript which in turn influences the protein. However, proteins also influence one another, via various chemical and physical mechanisms, and influence the process of transcription. Thus, there are multiple influences between and within types of biomolecules, that can operate on very different timescales.

The notion of interplay between molecules is fundamental to biology and it is for this reason that networks or graphs are widely used to summarize and reason about biological systems. For the moment we will not try to make the use of graphs in biology mathematically or statistically precise, but simply note that drawing edges between molecules provides an intuitive way to keep track of molecular influences. Such networks, with vertices corresponding to molecules and edges corresponding to direct or indirect influences, are known as biological networks and are a central part of the practical and conceptual toolkit of modern biology.

Although the temporal aspect is not always made explicit, it is important to appreciate the dynamic nature of cellular behavior. It can be useful to think of cells as dynamical systems, in which the current physical and chemical state, including the abundances and locations of a vast number of distinct biomolecules, influences future states. We return to the connection between the biophysical view and graphical models below. It is also important to note that due to the large number of relevant molecules and the diversity of physical and chemical processes, these systems are very high-dimensional in nature.

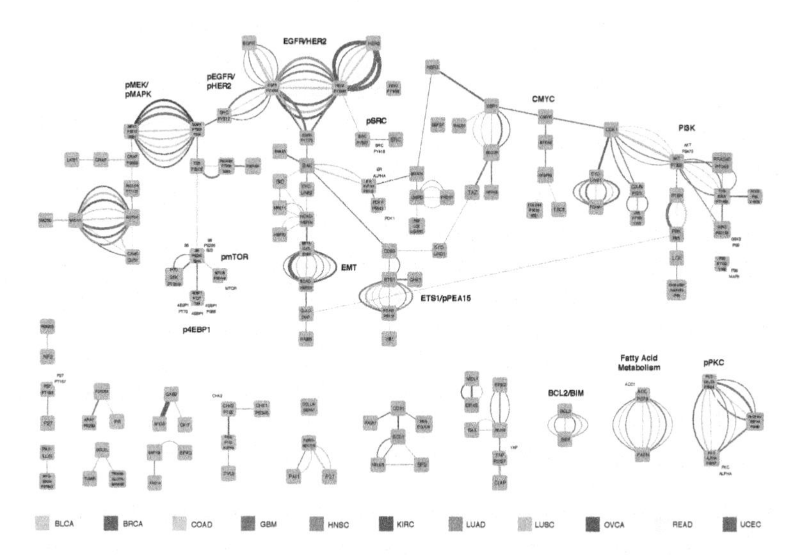

FIGURE 20.1: Biological networks estimated from cancer samples. Each node shown is a protein, and the edges summarize sparse Gaussian graphical models estimated separately for each of 11 cancer types (breast, colorectal etc.). The edges are color coded by cancer type (as shown at the bottom of the figure). Estimation was performed using the graphical lasso. Figure reproduced from [1]. See text for details.

20.1.3 A motivating problem

To fix ideas, we describe a motivating problem from cancer biology. Different cancer types (such as a breast cancer, skin cancer, and so on) have differences in their underlying biology, both in terms of genetics (the aberrant genes that drive the disease) and molecular phenotypes (e.g. the abundances of biomolecules as measured in cancer samples). In recent years cancer biologists have been mounting concerted efforts to exploit modern biotechnologies to survey such variation in a quantitative manner. A recent study (under the aegis of the Cancer Genome Atlas or TCGA) considered protein abundance levels in each of 3,467 cancer samples [1]. Each sample belonged to one of 11 cancer types. The assays focused on nearly 200 proteins that are involved in a process called signaling. The interplay between proteins is central to signaling and signaling is central to cancer biology, so the patterns of influence between signaling molecules in specific cancer types are of biological interest. To shed light on such patterns, sparse Gaussian graphical models (GGMs) were used to model the data with nodes corresponding to proteins. For each cancer type k, the graphical lasso (this is an L_1-penalized estimator for GGMs, see below) was used to estimate an inverse covariance matrix Ω_k. Figure 20.1 shows a visualization of the point estimates, for all 11 cancer types together (here, a missing edge corresponds to a zero entry in the respective inverse covariance matrix). Such estimates can be used to investigate variation across cancer types in terms of patterns of partial correlation. This example is fairly typical of emerging problems in biomedicine (although the dimensionality can be much larger) and we will return to it to illustrate specific issues below.

20.1.4 Notation

We denote by $X = (X_1, \dots, X_p)^{\mathrm{T}}$ a random vector whose components are the random variables X_j and a graph with vertex set $V = \{1, \dots, p\}$ and edge set $E \subseteq V \times V$ by

$G = (V(G), E(G))$. Each vertex in the graph is identified with a random variable. In the example above $j \in \{1, \ldots, p\}$ indexes the proteins under study. The graph may be directed or undirected depending on the specific context. For directed graphs, we use the notation $i \to j \in G$ to indicate the presence of a directed edge from node i to j in G. A $n \times p$ data matrix comprising n samples, each a p-dimensional vector, will be denoted by $\mathbf{X}$. We will use $\widehat{G} = \widehat{G}(\mathbf{X})$ to refer to a graph estimated from data. The specific estimator $\widehat{G}$ and type of graph will be clear from the context. We denote tuning parameters generically by λ where the precise nature of the tuning parameter will be clear from context. A multivariate Normal distribution with mean μ and covariance matrix Σ is denoted $N(\mu, \Sigma)$.

20.2 Methods for Snapshot or Static Data

Biological assays are often used to provide a snapshot of molecular state without an explicit temporal aspect. In such data, the underlying assumption is that even without an explicit temporal aspect, the data can provide some insight into relevant regulatory relationships. Such assays generically give rise to a $n \times p$ data matrix $\mathbf{X}$, comprising n observations for p molecular variables. Although single-cell assays are becoming more common (e.g. [6]), more often measurements are carried out on samples comprised of many cells. This means that the biological material upon which a single observation $i \in \{1, \ldots, n\}$ is based is itself a collection of many individual cells. The assays may be performed for several different groups of samples (as in the motivating example above, where the groups are cancer types), in which case the data would comprise matrices $\mathbf{X}_k$, one for each group k.

Graphical models are widely used to analyze snapshot data to provide an estimate, possibly with uncertainty quantification, of a graph describing relationships between the molecules. In most studies, n is not large relative to p or to the size of the relevant model space and high-dimensional inferential issues come into play. Regularization, whether via penalization or Bayesian formulations, plays a key role. In this Section we survey the most popular model classes used in bioinformatics and systems biology.

20.2.1 Gaussian graphical models

Background. We first recall some basic facts concerning Gaussian graphical models or GGMs, which were treated in depth in Chapter 9. Consider a multivariate Normal vector $X \sim N(\mu, \Sigma)$, $X \in \mathbb{R}^p$. The conditional independence relationships between the variables X_j (the components of X) are given by the zero pattern of the inverse covariance matrix $\Omega = \Sigma^{-1}$, with X_i conditionally dependent on X_j given all other variables if and only if $\Omega_{ij} \neq 0$. Thus, the undirected graph $G = (V, E)$ for a GGM can be defined by setting $V = \{1, \ldots, p\}$ and specifying the edge set as

$$(i, j) \in E \iff \Omega_{ij} \neq 0. \tag{20.1}$$

Note also that the partial correlations ψ_{ij} (between variables X_i and X_j given all other variables) can be obtained directly from Ω as $\psi_{ij} = -\Omega_{ij}/\sqrt{\Omega_{ii}\Omega_{jj}}$.

GGMs are used in molecular biology applications by identifying vertices with molecules (e.g. genes, proteins etc.) and treating the p-dimensional molecular data as the vector X. It is instructive to contrast GGMs and correlation networks. A very simple way to construct a molecular network is to threshold a measure of marginal association, i.e. to define an edge set E_{corr} by

$$(i,j) \in E_{\text{corr}} \iff |\rho_{ij}| \geq \tau, \tag{20.2}$$

where ρ_{ij} is a measure of marginal association between X_i and X_j (e.g. the Pearson correlation coefficient) and τ a threshold that controls sparsity. Such networks are variously referred to as correlation or relevance networks in the bioinformatics literature and as graphical models of marginal independence in the graphical models literature (see e.g. [19]). The GGM is a multivariate model that takes into account all other variables in deciding whether or not to place an edge between a given pair of vertices. Very often in biological systems many variables are correlated with one another, but the correlation can be explained by other variables (for example both variables might be biochemically regulated by a third) such that the variables are marginally correlated but nevertheless conditionally independent given the other variables. In such situations, a GGM may provide estimates that give better insight into underlying mechanisms (see for discussion [34, 1]).

Estimation. In biological settings, n is rarely large enough to allow estimation of a GGM by direct inversion of the sample covariance matrix S (indeed for $p > n$ the matrix is rank deficient). Rather, some form of shrinkage or regularization is usually employed to allow effective estimation in practice (see also Chapter 12).

L_1 penalization has proved effective in high-dimensional regression, offering an attractive combination of sparsity and convexity. In the graphical lasso L_1 penalization is used in the GGM context [11]. Specifically an L_1 penalty is placed on the entries of the inverse covariance matrix Ω. This encourages sparsity in Ω, which in turn corresponds to a sparse GGM with relatively few edges. The penalized criterion used by the graphical lasso is

$$\widehat{\Omega} = \arg\max_{\Omega} \Big\{ \log|\Omega| - \text{Tr}(S\,\Omega) - \lambda\|\Omega\|_1 \Big\} \tag{20.3}$$

where $|\cdot|$ denotes the determinant of its argument, Tr is the trace and $\|\cdot\|_1$ is the sum of the absolute values of the entries of its matrix argument, i.e. the L_1 norm. The parameter λ is a tuning parameter that controls the degree of penalization and thereby the level of sparsity of the estimate. The first two terms on the right-hand side are the multivariate Normal log-likelihood (partially maximized with respect to the mean) and setting $\lambda = 0$ recovers the maximum likelihood estimator. Maximization of the penalized likelihood in equation (20.3) is a convex problem and can be solved in a number of ways. [11] put forward an efficient co-ordinate descent procedure that can be run on problems with hundreds or thousands of vertices. Other related work on penalized GGM estimation includes [4, 37, 23] and [8, 18] from a Bayesian perspective.

Due to the nature and specificity of biochemical interplay the assumption of sparsity is widely thought to be a reasonable one in biology, at least in an approximate sense where only a few effects are large in magnitude. For example, for the signaling proteins in Figure 20.1, it is not expected that a given protein would be strongly influenced by a large number of other proteins, since the interactions are based on quite specific enzyme-substrate relationships.

A popular and convenient alternative to the graphical lasso is to use a shrinkage approach to the covariance estimation step followed by a model selection step. In the context of gene expression data, [30] discuss shrinkage estimators of the form

$$\widehat{\Sigma} = (1-\lambda)S + \lambda T \tag{20.4}$$

where $\lambda \in [0,1]$ is the shrinkage intensity, S denotes the (unbiased) empirical covariance matrix and the $p \times p$ matrix T is a shrinkage target. The latter is often set to a simple form with few estimated parameters (e.g. diagonal with common variance) but there are several possibilities that may be appropriate depending on the setting (see [30] for discussion). Setting $\lambda = 0$ gives $\widehat{\Sigma} = S$ and higher values of λ pull the estimate towards the target T. For further details and discussion of the model selection step see [30].

20.2.2 Directed acyclic graphs

We now turn to graphical models based on directed acyclic graphs or DAGs. The likelihood is given by the recursive factorization

$$p(X \mid G, \theta) = \prod_j p(X_j \mid X_{\pi_G(j)}, \theta_j) \tag{20.5}$$

where $\pi_G(j) = \{i \in V : i \to j \in G\}$ is the set of parents of vertex j in graph G and $\theta = (\theta_1, \ldots, \theta_p)$ are parameters governing the conditional distributions. In a DAG model, each variable is conditionally independent of its non-descendants given its immediate parents (recall Section 1.8 in Chapter 1).

In biological applications, estimation is often challenging due to the vast model space (of DAGs with p vertices) as well as typically limited sample sizes. In many applications the emphasis is on estimation of the structure of the graph, i.e. of the edge set, rather than estimation of the parameters of the model given the structure. We focus on the structure estimation task here.

A good applied example of structure estimation is the work of [29] who use DAG models whose nodes are identified with proteins involved in a biological process called signaling and whose structure is estimated from large-sample data obtained via single-cell assays.

In biological applications, Bayesian approaches have been widely used for estimating the structure of DAG models, as they provide a way to handle small sample sizes with regularization and uncertainty quantification and also give a natural way to incorporate known biology or induce sparsity via suitable prior formulations. The Bayesian marginal likelihood (known as the evidence in the machine learning literature) for a DAG model is

$$p(\mathbf{X} \mid G) = \int p(\mathbf{X} \mid G, \theta)\, p(\theta \mid G)\, d\theta \tag{20.6}$$

where $p(\theta \mid G)$ is a prior over the parameters and the first term in the integral is the DAG model likelihood (20.5). Conjugate formulations allow the marginal likelihood to be obtained in closed form.

One approach to structure estimation is simply to find a graph with a high score $p(\mathbf{X} \mid G)$. This can be done by essentially any discrete optimization method (e.g. greedy search, genetic algorithms etc.). However, the size of the search space (the space of all DAGs with p vertices) and the difficult nature of the search landscape poses challenges for this strategy.

The full posterior distribution over graphs contains additional information, since it gives not only the location of the modes (high scoring graphs) but also their posterior probability. The posterior distribution is given, up to proportionality, by the product of the marginal likelihood and a prior $P(G)$ over graphs, i.e. $P(G \mid \mathbf{X}) \propto p(\mathbf{X} \mid G)\, P(G)$. A number of Markov chain Monte Carlo (MCMC) methods have been proposed to allow this distribution to be approximated. However, for moderate-to-large p MCMC sampling in this setting remains challenging. The simplest sampler is a Metropolis-Hastings sampler with proposals made by making small changes to the current graph (typically single edge changes). This is asymptotically valid, but can converge very slowly. [13] discuss improvements that allow faster mixing. [12] introduce a Gibbs sampler that allows large but efficient moves. They also present an empirical comparison of constraint-based estimation (see Chapter 18) and Bayesian approaches, including applications to biomedical data, which shows that broadly speaking when both p and n are large, the PC algorithm is effective, but that for non-large n state-of-the-art MCMC approaches can outperform PC, also for point estimation.

An appealing feature of Bayesian inference in this setting is that it allows posterior probabilities to be calculated for features of the DAG model, such as the presence or absence of a specific edge or of features involving multiple vertices such as the posterior probability of a specific pathway. Such probabilities are analogous to inclusion probabilities in Bayesian variable selection and can be obtained using the output of MCMC procedures. These can be useful indicators of uncertainty for specific claims and can be used to guide further experiments. Examples of use of such output appear in [16, 36], where Bayesian analyses were followed by experimental validation of specific hypotheses.

20.2.3 Heterogeneous data and biological context

The procedures discussed above are aimed at estimating a single model from a dataset $\mathbf{X}$ whose samples are assumed to be drawn identically from the same model. However, in many biomedical applications data are instead heterogeneous in the sense of arising from multiple related but potentially non-identical data-generating processes. Examples include data from different cell types, tissues or disease subtypes, as in the cancer example above. We refer to these generically as contexts.

Let $k \in \{1, \ldots, K\}$ index the context, n_k denote the context-specific sample size and $\mathbf{X}_k$ denote an $n_k \times p$ context-specific data matrix. Then, collecting the $\mathbf{X}_k$'s together we can form an $n \times p$ data matrix $\mathbf{X}$, with $n = \sum_k n_k$. In this heterogeneous setting, it may be reasonable to assume that each context k has an underlying graphical model G_k. Estimation is challenging because the context-specific estimates $\widehat{G}(\mathbf{X}_k)$, while asymptotically valid for a consistent estimator $\widehat{G}$, must contend with sample sizes still smaller than the total sample size n. On the other hand, applying the estimator to the pooled, heterogeneous data (i.e. the estimate $\widehat{G}(\mathbf{X})$) enjoys a larger sample size but obviously cannot detect context-specific structure (since one model is used for all the contexts). Furthermore, such a pooled estimate may be severely mis-specified if the underlying context-specific models are sufficiently different. For example, conditional independencies at the level of each context are not in general the same as those at the aggregate, pooled level.

However, in many biomedical settings it is reasonable to suppose that the context-specific models G_k are not entirely dissimilar, since at least in the version of the problem sketched above they concern the same variable set V and refer to the same biological processes. Then, a natural approach is to jointly estimate the G_k's, whilst shrinking them towards a common, latent graph, in effect borrowing strength across the K problems.

For GGMs, this strategy is applied via a penalized likelihood approach in [7] using a formulation similar to the graphical lasso, but allowing for multiple models indexed by k, with an additional penalty that encourages similarity between these models. Joint estimation of DAG models in the non-identical setting is discussed in [28] who put forward a hierarchical Bayesian formulation in which the context-specific DAG models are shrunk together based on a undirected graph that describes relationships between the contexts k. Related work for time series data is discussed in [26]. In an applied setting, [14] jointly estimate multiple non-identical graphical models in a study spanning biological contexts defined by combinations of genetics (different cell lines) and growth conditions.

20.3 Biological Dynamics and Models for Time-Varying Data

Biological networks can be thought of as structural summaries of biological processes such as gene regulation or protein signaling. These processes have complicated temporal and spatial aspects that we have so far entirely ignored. Furthermore, we have not detailed what the random vector $X = (X_1, \ldots, X_p)^{\mathrm{T}}$ is intended to model in the underlying physical system. For instance, do the X_j's refer to the concentration of molecules in a single cell or in an aggregate of multiple cells? Such a distinction may be significant, because while the networks that graphical models are intended to represent are usually conceived of at the cellular level, statistical methods are often applied to bulk or aggregate data (i.e. data that are averages over large numbers of cells). In this Section we discuss the temporal and cellular aspects and introduce some graphical models that have been used for time-varying molecular data. For a fuller treatment of dynamical and systems biology perspectives we direct the interested reader to [5] and [3] and for some statistical aspects to [35] and [27].

20.3.1 Cellular dynamics

The dynamics that underlie cellular function are subject to both intrinsic and extrinsic stochastic effects (see e.g. [33, 20]). To retain a simple presentation, consider continuous random variables such that $\mathbf{X}(t) = (X_1(t), \ldots, X_p(t)) \in [0, \infty)^p$ denotes a state vector describing the abundance of molecular quantities of interest, with t indexing time. In general effects due to low molecule numbers will be relevant but here we focus on the case of continuous random variables. Following [27] consider a (time-homogeneous) stochastic differential equation as a working model for cellular dynamics:

$$\mathrm{d}\mathbf{X} \;=\; \mathbf{f}(\mathbf{X})\mathrm{d}t + \mathbf{g}(\mathbf{X})\mathrm{d}\mathbf{B} \tag{20.7}$$

where $\mathbf{f}$ and $\mathbf{g}$ are drift and diffusion functions respectively and $\mathbf{B}$ denotes a p-dimensional standard Brownian motion. One well known example of this model is the Chemical Langevin Equation (CLE).

The drift function $\mathbf{f}$ defines the edge structure $E(G)$ of a biological network G on the variables $V(G) = \{1, \ldots, p\}$, such that $(i, j) \in E \iff f_j(\mathbf{X})$ depends on the ith component X_i. Recent work by [31] formalizes these ideas using a causal interpretation of stochastic differential equations.

20.3.2 Towards linear models

A number of approximations are needed to go from the stochastic differential equation above to a statistical model for data obtained by averaging over many cells. These approximations are discussed in detail in [27]; we refer the interested reader to the reference and emphasize only that a number of steps and assumptions are needed to arrive at a tractable statistical model. In this Section we sketch how starting from the stochastic differential equation one can arrive at a simple linear model. For notational simplicity consider the Markovian regime. A Taylor approximation of the cellular drift $\mathbf{f}$ about the origin gives

$$\mathbf{f}(\mathbf{X}) \;\approx\; \mathbf{f}(\mathbf{0}) + D\mathbf{f}\big|_{\mathbf{x}=\mathbf{0}}\mathbf{X} \tag{20.8}$$

where $D\mathbf{f}$ is the Jacobian matrix of $\mathbf{f}$ (the constant term can be omitted ($\mathbf{f}(\mathbf{0}) = \mathbf{0}$), since absent any molecules there can be no change in expression levels). Then, the Jacobian $D\mathbf{f}$ captures the dynamics approximately and can be used to obtain an autoregressive type likelihood [27].

20.3.3 Dynamic Bayesian networks

A widely used class of models of autoregressive type are dynamic Bayesian networks or DBNs (see e.g. [17, 16]). A DBN is a directed graphical model based on a DAG G' with a vertex for each variable at each (discrete) time point. For T time points, G' has pT vertices. We focus here on the simplest case in which edges are only permitted one step forward in time, there are no edges within a time slice (i.e. no random variables with the same time index are adjacent in the graph) and where the edge structure is fixed in time. Then, the graph G' can be compactly encoded by a graph G with only p vertices, in which an edge between nodes i and j is understood to mean that in the graph G' there is an edge from variable i at each time point to variable j at the next time point. Note that while the "collapsed" graph G need not be acyclic, the corresponding full pT-vertex graph G' is guaranteed to be acyclic.

Using the "collapsed" graph G, the likelihood is

$$p(\mathbf{X} \mid G, \theta) = \prod_{j=1}^{p} \prod_{t=2}^{T} p(X_{j,t} \mid X_{\pi_G(j),t-1}, \theta_j) \tag{20.9}$$

where $X_{j,t}$ is a random variable corresponding to variable j at time t, G is a directed graph with vertex set $V(G) = \{1 \dots p\}$, T is the total number of time points, $\pi_G(j) = \{i \in V : i \to j \in G\}$ is the set of parents of vertex j in graph G and $\theta = (\theta_1, \dots, \theta_p)$ are parameters.

The model outlined above is sometimes called a "feedforward" DBN and allows for particularly tractable estimation [16]. [16] used a feedforward DBN to make inferences regarding a signaling network in a cancer cell line.

In some settings fixing the edge structure and parameters in time is unreasonable, for example in studies over longer time periods or of cells that are changing biological state. Then, it may be useful to consider models that allow the graph and/or parameters to differ between contiguous time segments. Such models have been discussed by [9]. They apply the models to data obtained during the life cycle of the fruit fly *Drosophila melanogaster* and argue that the time-varying models are better able to deal with changes related to passage through clearly distinct life stages. However the greater generality of these time-varying DBNs comes at some computational and statistical cost.

20.3.4 Nonlinear models

Although most commonly used formulations for continuous data are based on linear models, biological systems display nonlinearities in their behavior. If such nonlinearities have major qualitative implications, they may be relevant even when the goal is not to model processes in detail but rather to capture the main features of the system. Furthermore, in some systems biology settings the goal is indeed to model the underlying processes in some detail. These factors motivate an interest in developing nonlinear extensions to the models discussed above. However, as we have seen, many of the inferential issues that arise in biological networks are already challenging for linear models and these issues are exacerbated in the nonlinear case.

A natural way to move towards nonlinear models is to replace the linear approximation in a DBN with a suitable class of nonlinear models. This is the approach taken by [36], who discuss model selection over a small number of candidate models, each defined by a set of coupled nonlinear ordinary differential equations (ODEs) with unknown parameters. They proceed within a Bayesian framework, with the posterior probability of each candidate model calculated by numerically integrating out model parameters.

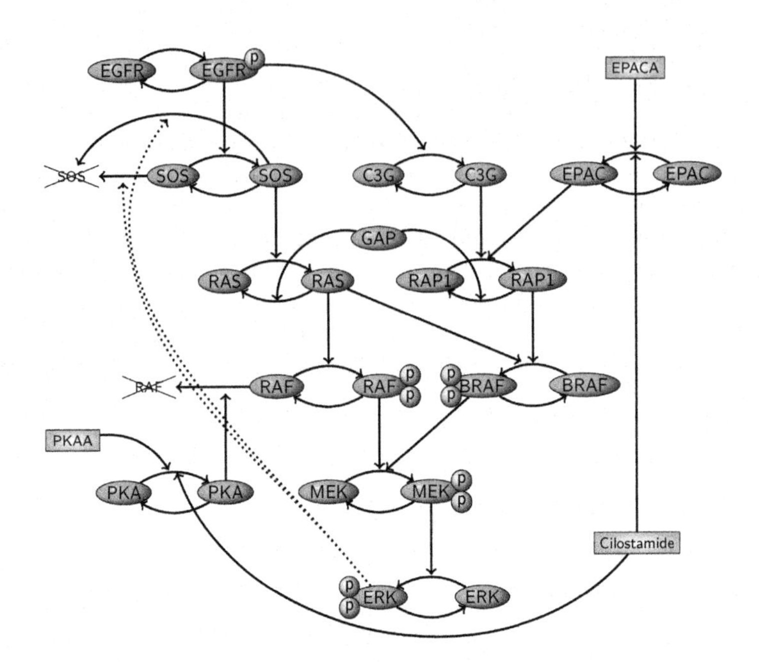

FIGURE 20.2: A chemical reaction graph for a biological network. The graph shown summarizes chemical reactions within a type of biological network called a protein signaling network. The green nodes are proteins that can be modified by a process called phosphorylation (indicated by the adjacent "p" symbol). Edges indicate enzyme-substrate relationships. Reproduced from [24] and based on a model presented in [36]. See text for details.

Moving beyond a small number of candidate models, the more general problem is to learn a network $G \in \mathcal{G}$ (where $\mathcal{G}$ is a suitably defined space of graphs) with the likelihood corresponding to a candidate graph G specified by a nonlinear model with parameters θ_G. In systems biology applications, there is usually uncertainty regarding both the graph and the parameters, because even in the case of well-studied processes the effective reaction rates in samples under study may be quite different from published rates obtained in laboratory experiments.

In this context, [24] propose an approach in which a dynamical model is automatically generated for a candidate graph using an appropriate class of such models. Specifically, for a candidate graph G, they generate a model using a class of nonlinear ODEs based on the relevant biochemistry. The models considered are more detailed than in a standard graphical models formulation; an example of the type of model considered is shown in Figure 20.2. The graph summarizes a set of chemical reactions pertaining to the same class of process (signaling) in our motivating example (see Figure 20.1), but at a greater level of detail. The green nodes are proteins that can be modified by a process called phosphorylation (indicated by the adjacent "p" symbol). Other proteins act to catalyze these phosphorylation reactions and such influences are indicated by directed edges. The model for a graph G is specified up to unknown parameters θ_G. They then proceed in a Bayesian framework using MCMC to approximate the marginal likelihood $p(\mathbf{X} \mid G)$. Using factorization ideas similar to those described above for DBNs, they show that this can be done relatively efficiently, permitting general structure estimation for problems of moderate size. Furthermore, they argue that this can help in causal inference (see below), since the underlying models have appropriate nonlinearities.

We note that nonlinear models can also play a role for "snapshot" data, since for a given class of models of biochemical dynamics the steady-state relationships between the variables are typically not linear and these relationships can be used to define nonlinear likelihoods [25].

Although it is clear that nonlinear models can be closer to the underlying biochemical processes than linear models, it remains unclear in practice whether the additional computational and statistical cost of nonlinear modeling generally leads to substantive gains with respect to specific tasks such as predicting phenotypes or proposing validation experiments. Thus, nonlinear models applied to high-throughput biological data still require careful empirical assessment on a case-by-case basis and ideally benchmarking against simpler linear analogues.

20.4 Causality

In molecular biology networks typically have a causal interpretation, for example a directed edge between molecules A and B would be taken to mean that A influences B either via a direct physical mechanism or via a sequence of such events. Thus, a biologist would typically expect edges in network models to bear some relation to interventional experiments. For this reason the causal aspect of biological networks is of both conceptual and practical importance. We summarize the current state-of-the-art below but note that causal discovery is a challenging and frontier topic in its own right and its interface with molecular biology remains an open and fast moving area.

20.4.1 Causal discovery for biological data

DAG models can be endowed with causal semantics, as discussed in Chapter 15. [29] used causal DAG models to model single-cell protein data, albeit for a small number of variables and using relatively large sample data. The modeling of interventional time-varying data using DAG models is discussed in [32], who build on [10] to propose ways to modify DAG models to model biological interventions. [22] proposed a DAG-based approach called IDA to bound causal effects using only observational data. This is done by estimating an equivalence class of DAGs from the observational data, and then bounding the causal effect of variable i on j by considering the corresponding causal effect estimated from each DAG in the equivalence class. Furthermore, empirical work using data from Yeast knockout experiments [21] has provided empirical evidence of the practical efficacy of IDA. These studies are encouraging, in that they suggest that graphical model based approaches may be able to elucidate causal relationships using molecular data, possibly using even only observational data, but there remain many open questions as discussed below.

20.4.2 Empirical assessment of causal discovery

Despite some positive results in causal discovery using molecular data, in general it remains unclear whether useful inferences can be drawn from a given dataset using a certain method. All causal discovery methods make explicit or implicit assumptions that cannot be easily checked. A common example of such an assumption is that of no hidden common causes: this is hard to verify in practical biological applications. Furthermore, given the complexity of the underlying physical processes, the experimental set-up and measurement processes, a range of issues may arise that make causal discovery difficult in practice. One such example was highlighted above and relates to the fact that although the relevant biochemical events take place in single cells (or subcellular compartments) typical molecular data are from aggregates of many individual cells. Whether this is a major concern for a given application may depend on unknown details of the underlying mechanisms. Another example relates to the fact that individual cells may not be synchronized with respect to the dynamic processes that underlie the network of interest. Again, whether this is critical or not may depend on unknown factors. Indeed, due to a number of such issues it may be that a given dataset simply does not contain signals that would permit good results with respect to causal discovery by any method. Such failure cases cannot be easily ruled out *a priori*.

Thus, simply put, in a given application it is hard to be assured from first principles that a given causal discovery method will work. This motivates a need for suitable empirical diagnostics. Recently, [15] proposed a leave-one-out type scheme for this purpose that is applicable in settings where some interventional data is available. They propose to block data $\mathbf{X}$ into $\mathbf{X} = (\mathbf{X}^{(0)}, \mathbf{X}^{(1)}, \mathbf{X}^{(2)}, \ldots, \mathbf{X}^{(I)})$, where $\mathbf{X}^{(0)}$ denotes all observational data, $\mathbf{X}^{(j)}$ denotes all data that includes an intervention on node j and I is the total number of interventions performed (variables have been reordered for notational convenience). The idea is to apply a candidate causal estimator $\widehat{G}$ to all data except the intervention on node j (denote all data leaving out the intervention on j by $\mathbf{X}^{(-j)}$), and to test the resulting estimate $\widehat{G}(\mathbf{X}^{(-j)})$ using the held-out data $\mathbf{X}^{(j)}$. Comparing $\mathbf{X}^{(j)}$ with $\mathbf{X}^{(0)}$ reveals the effect of intervention on node j so the procedure in effect tests the ability of $\widehat{G}$ to predict the effect of an unseen intervention. [15] used this approach within an open science benchmarking framework using molecular data from human cells and found that several competitors were able to achieve some agreement with held out interventional test data.

20.5 Perspective and Outlook

Graphical models are increasingly becoming a key part of the molecular and systems biology toolkit. They provide a principled statistical framework for network-related questions that arise in systems biology. However, many questions remain open, of which we highlight a few below:

- Applications: Although graphical models have already been used in several applications, networks are ubiquitous in biology and many areas of biomedical inquiry have not yet been explored from the truly multivariate viewpoint that graphical models can allow. Concerted efforts in collaboration with biomedical scientists will be needed to fully explore the potential for these approaches.
- High-dimensional approaches: Biological systems are typically high-dimensional but typically have much sparse and low rank structure. We anticipate that recent and future advances in exploiting such structure will mean that the robustness, stability and scalability of graphical model estimation will continue to improve. As both biotechnologies and estimators continue to develop it remains to be seen whether it will be possible to estimate graphical models that are truly global in the sense of including a large fraction of potentially relevant biomolecules, and if so, whether such large models will prove to be practically useful.
- Integrative models. The majority of graphical model applications in the literature to date have focused on one data type. However, it is increasingly common to obtain multiple complementary data types from the same set of biological samples. Thus, development of methods that can integrate these data in a useful but principled way

will be an important direction for future work. Although the raw dimension of such integrative analyses might be very large, the kinds of low dimensional and sparse structure that make high-dimensional inference possible in general are likely very relevant to analyses spanning multiple data types and exploiting such structure will be a key direction for future work on integrative analysis.

- Causality and semantics. Causal modeling in molecular biology is in its infancy. Further developments in this area will shed light on the feasibility of large-scale causal discovery in biology and have the long-term potential to change the way biological mechanisms are explored. Ongoing developments in gene editing technologies will be relevant as well as integrative data as mentioned above. As the modeling-experimental link becomes closer, we expect that the interface between causal graphical models and experimental research will become an important one, at least in some areas of biology.

Bibliography

[1] R. Akbani, P.K.S. Ng et al. A pan-cancer proteomic perspective on the cancer genome atlas. *Nature Communications*, 5, 2014.

[2] B. Alberts, D. Bray, K. Hopkin, A. Johnson, J. Lewis, M. Raff, K. Roberts, and P. Walter. *Essential Cell Biology.* Garland Science, 2013.

[3] U. Alon. *An Introduction to Systems Biology: Design Principles of Biological Circuits.* CRC Press, 2006.

[4] O. Banerjee, L. El Ghaoui, and A. d'Aspremont. Model selection through sparse maximum likelihood estimation for multivariate Gaussian or binary data. *Journal of Machine Learning Research*, 9:485–516, 2008.

[5] D.A. Beard and H. Qian. *Chemical Biophysics: Quantitative Analysis of Cellular Systems.* Cambridge University Press, 2008.

[6] S.C. Bendall, E.F. Simonds et al. Single-cell mass cytometry of differential immune and drug responses across a human hematopoietic continuum. *Science*, 332(6030):687–696, 2011.

[7] P. Danaher, P. Wang, and D.M. Witten. The joint graphical lasso for inverse covariance estimation across multiple classes. *Journal of the Royal Statistical Society: Series B (Statistical Methodology)*, 76(2):373–397, 2014.

[8] A. Dobra, C. Hans, B. Jones, J.R. Nevins, G. Yao, and M. West. Sparse graphical models for exploring gene expression data. *Journal of Multivariate Analysis*, 90(1):196–212, 2004.

[9] F. Dondelinger, S. Lebre, and D. Husmeier. Non-homogeneous dynamic Bayesian networks with Bayesian regularization for inferring gene regulatory networks with gradually time-varying structure. *Machine Learning*, 90(2):191–230, February 2013.

[10] D. Eaton and K.P. Murphy. Exact Bayesian structure learning from uncertain interventions. In *AISTATS*, 107–114, 2007.

[11] J. Friedman, T. Hastie, and R. Tibshirani. Sparse inverse covariance estimation with the graphical Lasso. *Biostatistics*, 9(3):432–441, 2008.

[12] R.J.B. Goudie and S. Mukherjee. A Gibbs sampler for learning DAGs. *Journal of Machine Learning Research*, 17(30):1–39, 2016.

[13] M. Grzegorczyk and D. Husmeier. Improving the structure MCMC sampler for Bayesian networks by introducing a new edge reversal move. *Machine Learning*, 71(2-3):265–305, 2008.

[14] S. Hill, N.K. Nesser et al. Context-specificity in causal signaling networks revealed by phosphoprotein profiling. *Cell Systems*, 2016.

[15] S.M. Hill, L.M. Heiser et al. Inferring causal molecular networks: empirical assessment through a community-based effort. *Nature Methods*, 13(4):310–318, 2016.

[16] S.M. Hill, Y. Lu, J. Molina, L.M. Heiser, P.T. Spellman, T.P. Speed, J.W. Gray, G.B. Mills, and S. Mukherjee. Bayesian inference of signaling network topology in a cancer cell line. *Bioinformatics*, 2012.

[17] D. Husmeier. Sensitivity and specificity of inferring genetic regulatory interactions from microarray experiments with dynamic Bayesian networks. *Bioinformatics*, 19(17):2271–2282, 2003.

[18] B. Jones, C. Carvalho, A. Dobra, C. Hans, C. Carter, and M. West. Experiments in stochastic computation for high-dimensional graphical models. *Statistical Science*, 388–400, 2005.

[19] G. Kauermann. On a dualization of graphical Gaussian models. *Scandinavian Journal of Statistics*, 105–116, 1996.

[20] S.C. Kou, X. Sunney Xie, and J.S. Liu. Bayesian analysis of single-molecule experimental data. *Journal of the Royal Statistical Society: Series C (Applied Statistics)*, 54(3):469–506, 2005.

[21] M.H. Maathuis, D. Colombo, M. Kalisch, and P. Bühlmann. Predicting causal effects in large-scale systems from observational data. *Nature Methods*, 7(4):247–248, 2010.

[22] M.H. Maathuis, M. Kalisch, and P. Bühlmann. Estimating high-dimensional intervention effects from observational data. *The Annals of Statistics*, 37(6A):3133–3164, 2009.

[23] N. Meinshausen and P. Bühlmann. High-dimensional graphs and variable selection with the Lasso. *The Annals of Statistics*, 34(3):1436–1462, 2006.

[24] C.J. Oates, F. Dondelinger, N. Bayani, J. Korkola, J.W. Gray, and S. Mukherjee. Causal network inference using biochemical kinetics. *Bioinformatics*, 30(17):i468–i474, 2014.

[25] C.J. Oates, B.T. Hennessy, Y. Lu, G.B. Mills, and S. Mukherjee. Network inference using steady-state data and Goldbeter–Koshland kinetics. *Bioinformatics*, 28(18):2342–2348, 2012.

[26] C.J. Oates, J. Korkola, J.W. Gray, and S. Mukherjee. Joint estimation of multiple related biological networks. *The Annals of Applied Statistics*, 8(3):1892–1919, 2014.

[27] C.J. Oates and S. Mukherjee. Network inference and biological dynamics. *The Annals of Applied Statistics*, 6(3):1209, 2012.

[28] C.J. Oates, J.Q. Smith, S. Mukherjee, and J. Cussens. Exact estimation of multiple directed acyclic graphs. *Statistics and Computing*, 1–15, 2015.

[29] K. Sachs, O. Perez, D. Pe'er, D.A. Lauffenburger, and G.P. Nolan. Causal protein-signaling networks derived from multiparameter single-cell data. *Science*, 308(5721):523–529, April 2005.

[30] J. Schäfer and K. Strimmer. A shrinkage approach to large-scale covariance matrix estimation and implications for functional genomics. *Statistical Applications in Genetics and Molecular Biology*, 4(1), 2005.

[31] A. Sokol and N.R. Hansen. Causal interpretation of stochastic differential equations. *Electronic Journal of Probability*, 19(100):1–24, 2014.

[32] S.E.F. Spencer, S.M. Hill, and S. Mukherjee. Inferring network structure from interventional time-course experiments. *The Annals of Applied Statistics*, 9(1):507–524, 2015.

[33] P.S. Swain, M.B. Elowitz, and E.D. Siggia. Intrinsic and extrinsic contributions to stochasticity in gene expression. *Proceedings of the National Academy of Sciences*, 99(20):12795–12800, 2002.

[34] A.V. Werhli, M. Grzegorczyk, and D. Husmeier. Comparative evaluation of reverse engineering gene regulatory networks with relevance networks, graphical Gaussian models and Bayesian networks. *Bioinformatics*, 22(20):2523–2531, 2006.

[35] D.J. Wilkinson. *Stochastic Modelling for Systems Biology*. CRC Press, 2011.

[36] T-R. Xu, V. Vyshemirsky, et al. Inferring signaling pathway topologies from multiple perturbation measurements of specific biochemical species. *Science Signaling*, 2010.

[37] M. Yuan and Y. Lin. Model selection and estimation in the Gaussian graphical model. *Biometrika*, 94:19–35, 2007.

21

Graphical Models in Genetics, Genomics, and Metagenomics

Hongzhe Li

Department of Biostatistics, Epidemiology and Informatics, University of Pennsylvania

Jing Ma

Biostatistics Program, Fred Hutchinson Cancer Research Center

CONTENTS

21.1 Introduction

High-throughput technologies have generated an enormous amount of tissue and cell-type specific genetic, genomic, and metagenomic data. Measurements of gene expression at the single-cell level have also become possible and are promising data sources in studying brains, cancer and immunology [19]. The CRISPR (clustered regularly interspaced short palindromic repeats) screen has recently emerged as a powerful new approach in profiling gene essentiality at the genome scale and in facilitating the dissection of regulatory networks by gene editing [43]. These new data and technologies enable us to experimentally measure and define biomolecular interactions on a large scale. This chapter focuses on graphical models and network-based analysis in genetics, genomics, and metagenomics, with an emphasis on incorporating biomolecular networks in answering fundamental biological questions.

21.1.1 The human interactome

As introduced in the preceding Chapter 20, a biological network consists of a collection of biomolecules and their interactions that correspond to various cellular functional relationships, and is often represented as a graph with directed and/or undirected edges. Throughout the chapter, the word 'interaction' is used to denote the presence of an edge between two nodes, which may be directed or undirected and defined experimentally or statistically depending on the context. Examples of important biological networks include gene regulatory networks, whose directed edges represent activation or repression relationships between genes; protein-protein interaction networks, whose nodes are proteins linked together by physical binding events; metabolic networks, whose nodes are metabolites and edges reflect the chemical reactions of metabolism. Other useful networks are gene co-expression networks [49], which are phenotypic networks in which genes are linked if they share similar co-expression patterns.

Using complex network theory, work of [3, 4] found that the topologies of biomolecular networks are far away from being random, but are in fact scale-free. In addition, these networks are often comprised of physically or functionally connected subnetworks, also called pathways, that work together to achieve certain biological functions. It is worth noting that both the nodes and the interactions described above can be tissue- and context-specific. An important goal of studying the human interactome is to elucidate the functional role of biological networks under selected tissues and contexts, so as to understand the mechanisms of disease onset and progression, and identify previously unknown genes and pathways associated with complex phenotypes.

21.1.2 Publicly available databases

The past few years have seen systematic efforts in collecting and storing biomolecular interactions, that are curated from literature and high-throughput experiments or estimated using statistical methods, in publicly available databases. Some of these databases span a wide range of data types, such as KEGG [25, 26, 27] that has structural information on genes, proteins and pathways, while others contain specific data types, such as iRefIndex [42] and STRING [45] that provide a critical assessment and integration of protein-protein interactions. These well-maintained and regularly updated databases allow us to answer important questions about the factors that control how signals pass through the biological network in response to external stimuli. Efficient and rigorous incorporation of known biological information about pathways and networks into analysis of multiple genomic data is a key component of integrative genomics.

However, an increasing body of evidence suggests that biomolecular interactions and canonical pathways in existing databases are incomplete and largely inaccurate. Complementary to existing knowledge, one can also computationally construct biological networks based on various types of molecular data and use the resulting networks/subnetworks in downstream analysis. Learning biological networks by integrating both perturbation experiments and observational data has been an active area of research. Interested readers are referred to Chapter 20 for an overview of available methods. On the other hand, the validity of biological networks inferred from data-driven approaches may largely depend on the size of the study cohort which is often small compared to the number of genetic features, the quality of the data, and the tissue or context under which the data are collected. One expects that combining existing network information with data-driven approaches may facilitate better understanding of fundamental biological processes.

TABLE 21.1: Definitions and abbreviations of genetic terminologies used in the paper.

SNP	single nucleotide polymorphism; DNA sequence variation occurring in which a single nucleotide differs among individual subjects
eQTL	expression quantitative trait loci; statistical associations between a SNP value and the expression level of a mRNA
Haplotype	alleles across different loci on the same chromosome
GWAS	genome-wide association study; statistical association between a SNP value and a trait (e.g. response to therapy) or disease
LD	linkage disequilibrium; non-random association of alleles (e.g. SNP values) at different genomic locations
Pathway	functionally related set of biomolecules (genes, proteins, metabolites)
Pleiotropy	one gene that influences two or more unrelated phenotypes
Polymorphism	genetic variations between individual subjects
PPI	protein-protein interactions

21.1.3 Genetic terminologies

Table 21.1 lists the key genetic terminologies and their definitions used throughout this chapter. More details about their biological contexts are available in [2] and [35].

21.2 Network-Based Analysis in Genetics

Human genetic research aims to identify the genetic variants that are associated with various complex phenotypes. A genetic variant may refer to (1) a single-nucleotide polymorphism (SNP), which is a common variant that occurs in at least 1% of a population, (2) a mutation, in a case where it is a rare variant, or (3) a copy-number variation/aberration (a CNV is change in copy number in germline cells, whereas a CNA is change in copy number arisen in somatic tissues). Graphical model and network-based methods play an important role in identifying biologically relevant variants by taking into consideration gene-gene interactions.

21.2.1 Network-assisted analysis in genome-wide association studies

Genome-wide association studies (GWAS) attempt to identify commonly occurring genetic variants that contribute to disease risk, and so far have identified thousands of SNPs that are associated with many human traits [5]. In its simplest form, GWAS analysis is formulated as a sequence of logistic regressions where the disease status from all individuals serve as the response and each genotyped SNP is the covariate. The resulting p-value for each SNP is then corrected for multiple comparisons using e.g. the Bonferroni adjustment. Although this standard approach has the power of identifying common SNPs with strong effects on phenotypes, it ignores the possible synergistic effects of genetic variants on disease phenotypes. Therefore network-assisted methods have been proposed to prioritize the GWAS results and to identify subnetwork of genes that are associated with phenotypes. The rationale of such network-based methods is that topologically related genetic variants are more likely to produce similar phenotypic effects.

Among the available databases, the protein-protein interaction (PPI) networks from STRING [45], iRefIndex [42] and Reactome [10] are often used in network-assisted analysis (NAA). In addition, directed graphs such as the protein-DNA regulatory networks [28] have also been used to identify potential causal variants [30]. NAA starts with preprocessing the GWAS data to compute SNP- or gene-based statistical values such as p-values that measure the significance of associations between the tested SNPs and the phenotype. After overlaying these SNP- or gene-based p-values onto the network extracted from public databases, NAA approaches search for subnetworks and assess the combined effects of multiple genes participating in the subnetworks through a gene set analysis. Depending on the null hypothesis tested, one may apply permutation tests that randomly swap case and control labels in the GWAS data or randomization tests that use randomly generated networks to estimate the null distribution, thereby evaluating the significance of the detected subnetworks [24]. Network information has also been explored for identification of the causal variant within a longer haplotype that is associated with the trait and for identification of causal variants among multiple genes within a pathway [30].

Motivated by network-based analysis of gene expression data in [47, 48], [9] and [32] proposed methods that incorporate the biological pathway information via a hidden Markov random field (HMRF) model for GWAS. These types of HMRF models have been further developed and applied to network-based analysis of rare variants (see Section 21.2.2). An important and closely related problem to GWAS is expression quantitative trait locus (eQTL) mapping where the phenotype of interest is gene expression (see detailed discussion in Section 21.3).

21.2.2 Co-expression network-based association analysis of rare variants

Advances in next-generation sequencing technologies have revolutionized biomedical research, including the ability to obtain the exome or whole genome sequencing of a large set of samples. The large number of single nucleotide variants (SNVs) uncovered in each single human genome or exome provide insights into the role of rare genetic variants in the risk of complex diseases, but also create computational challenges for genetic studies. Compared to a SNP which is a common variant, a SNV means a variation in a single nucleotide without any limitations of frequency. Analysis of rare variants from exome or whole genome sequencing has been a very active area of human genetic research. Given the very low minor allele frequencies of such rare variants, grouping the variants based on gene annotation or pathway information is the standard way of testing rare variant associations. However, such approaches have seen limited success because many rare variants are neutral and have no functional relevance.

Combining gene co-expression network with information on rare genetic variants, [34] developed a novel algorithm, DAWN, to model two types of data: rare variations from exome sequencing and gene co-expression in tissues that are related to disease risk. The algorithm is based on a HMRF model [32, 47, 48], whose graph structure is determined by gene co-expression. Specifically, for rare variants, gene-based tests for gene-disease associations are first applied to obtain the p-value for each gene [22]. To construct a more interpretable gene co-expression network, two important screening steps are applied. Step 1 first identifies a set of key genes, which are defined as those with relatively small p-values. In step 2, DAWN trims the set of key genes by excluding those that are not substantially co-expressed with any other measured genes. Further neighborhood selection is used to construct a sparse gene co-expression network.

Let n represent the total number of genes in the final network and Ω the corresponding $n \times n$ adjacency matrix. DAWN converts the gene-based p-values to normal Z-scores, $Z =$

$(Z_1, \cdots, Z_n)$, to obtain a measure of the evidence of disease association for each gene. These Z-scores are assumed to have a Gaussian mixture distribution, where the mixture membership of Z_i is determined by the hidden state I_i indicating whether gene i is a risk gene ($I_i = 1$) or not ($I_i = 0$). The mixture model for Z_i can be formally expressed as

$$Z_i \sim P(I_i = 0)N(0, 1) + P(I_i = 1)N(\mu, \sigma^2),$$

where μ and σ^2 correspond, respectively, to the mean and variance of Z_i under the alternative ($I_i = 1$). An Ising model is used to model the conditional dependence structure of the hidden states $(I_1, \ldots, I_n)$ with the probability mass

$$P(I = \eta) \propto \exp(b^T \eta + c\eta^T \Omega \eta) \text{ for all } \eta \in \{0, 1\}^n.$$

Using an iterative algorithm, one can estimate the parameters (b, c, μ, σ^2) and the posterior probability for each gene $P(I_i = 1 \mid Z)$, which can be used to select disease-associated genes.

It is worth pointing out that although the two screening steps used in DAWN significantly reduce the search space, they may also prematurely remove genes that are in fact risk genes from downstream analysis. This limitation may be addressed with improving quality of the sequencing data and larger study cohort.

21.3 Network-Based eQTL and Integrative Genomic Analysis

Expression quantitative trait locus (eQTL) refers to the genomic regions that carry one or more sequence variants that affect the expression of a gene, typically measured by microarrays or high-throughput RNA sequencing. Such variation may suggest mechanisms under which phenotypic differences arise. Thus eQTL analysis has emerged as a key tool for elucidating the causal effects of regulatory variants on gene expressions and the clinical traits, where tissue-specific gene expressions can serve as possible mediators of the genetic variants.

If the eQTLs are located close to the genes they influence, they are called local eQTLs. Local eQTLs can act in *cis* by directly affecting only the expression of the gene that is on the same physical chromosome with it, as well as in *trans*, owing to changes in the function of a mediator [2]. In contrast, distant eQTLs refer to those that are located further away from the genes they influence and usually act in *trans* [37] (see Figure 21.1). As whole-genome sequencing data from different tissues/cell types become more accessible, there is a growing interest in integrative eQTL studies including joint analysis of eQTL mapping across multiple tissues for improved power [14] and Bayesian methods that combine external functional annotations with genetic association data for prioritizing causal variants in genome-wide association studies [16, 29].

Network-guided methods have also proven useful for identification of regulatory key driver genes associated with coronary heart disease [23], detection of *trans* acting eQTLs by modeling local gene networks [41] and eQTL mapping with mixed graphical Markov models [46]. Below we discuss in detail several novel approaches that incorporate network information for detecting eQTLs, and integration of GWAS and eQTL analysis. In such applications of causal inference in genetics, directed graphs provide an effective way of modeling various causal relationships.

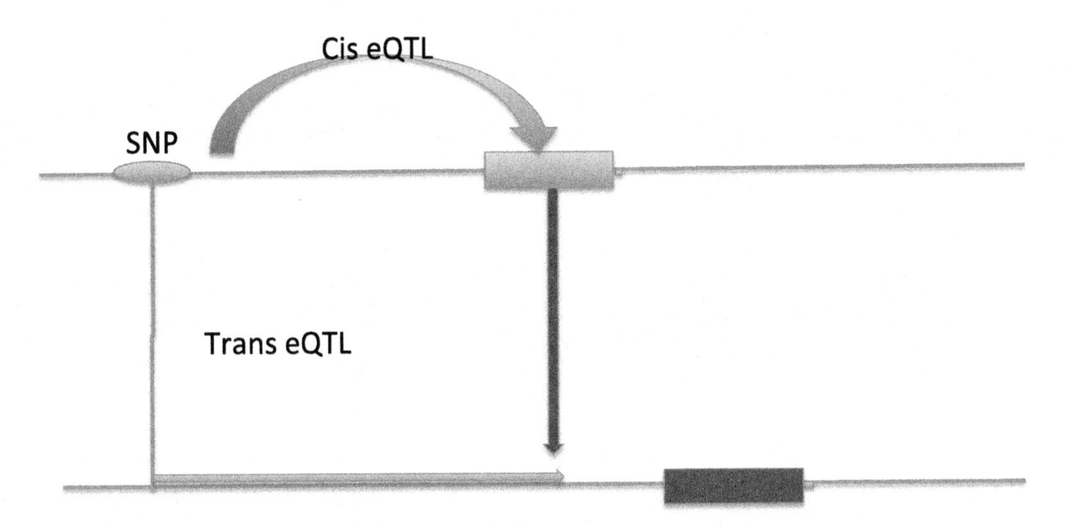

FIGURE 21.1: Schematic of eQTLs, and the two mechanisms *cis* and *trans* that a genetic variant can regulate gene expressions [37]. The boxes represent protein-coding genes influenced by the SNPs, as indicated by the arrows. Local eQTLs that are located close to the genes they regulate can act both in *cis* and in *trans*, whereas distant eQTLs located further away from the genes they regulate usually act in *trans*.

21.3.1 Detection of trans acting genetic effects

Recent studies suggest that a substantial proportion of the heritability in human gene expression cannot be explained by *cis* variants [20, 40], indicating the important contribution from *trans* acting genetic variants. Compared to *cis* eQTLs, detection of trans acting genetic effects remains a major challenge due to the relatively small effects of *trans* eQTLs and their specificity to the particular tissues and contexts [2, 12]. The high-dimensional multivariate nature of gene expression traits imposes a severe multiple testing burden, which further complicates *trans* eQTL mapping.

To enhance the power for *trans* eQTL mapping, a simple but important principle is to account for as many competing sources of variation as possible. Consider the rationale in Figure 21.2. Here SNP A regulates gene A in *cis*. The directed edge between gene B and gene C, when unaccounted for, may reduce the signal and hence the power for detecting the *trans* association between SNP A and gene C. Gene B in Figure 21.2 is also called the *exogenous factor*, which is defined as any gene that (i) has a causal effect on gene C, and (ii) is independent of the genetic variant SNP A. By identifying and conditioning on all exogenous genes B, one is hopeful to increase the power for mapping *trans* eQTLs. This is also the main objective underlying GNet-LMM [41] which detects *trans* acting genetic variants by modeling local gene regulatory networks. To be more specific, for each SNP A - gene C pair to be tested, GNet-LMM evaluates the following statistical dependencies to detect the V structure gene A $\rightarrow$ gene C $\leftarrow$ gene B (Figure 21.2):

$$\begin{array}{lll} \mathrm{dep}(X_A, X_C), & \mathrm{dep}(X_B, X_C), & \mathrm{ind}(X_A, X_B), \\ \mathrm{dep}(X_A, X_B \mid X_C), & \mathrm{dep}(Z_A, X_A), & \mathrm{ind}(Z_A, X_B). \end{array} \tag{21.1}$$

Here $\mathrm{dep}(\cdot,\cdot)$ and $\mathrm{ind}(\cdot,\cdot)$ denote, respectively, a statistical dependence and independence criterion (see [41] for detailed definitions of these criteria). Given the gene expression data, a standard correlation test is employed to assess the dependency between two genes. To

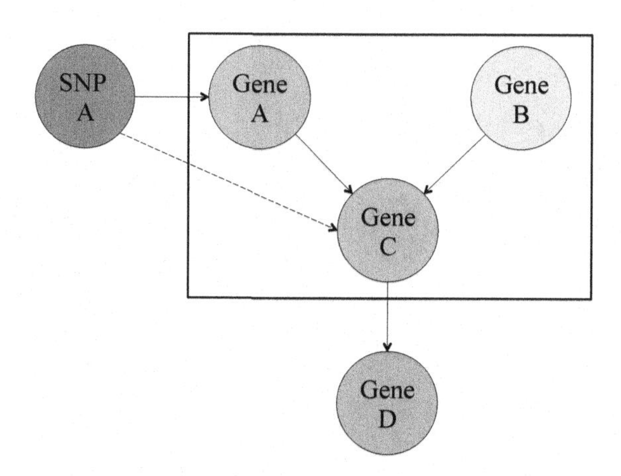

FIGURE 21.2: Graphical model illustration of the GNet-LMM algorithm [41]. The principle to improve power for detecting *trans* association between SNP A and gene C is to identify and condition on all exogenous genes with incoming edges. Exogenous genes represent either additional sources of variation or regulatory effects between genes, and are defined by testing for V-structures gene A $\rightarrow$ gene C $\leftarrow$ gene B (black box) that are linked to SNP A via gene A.

evaluate the dependency between a SNP Z and a gene X, [41] used a linear mixed effects model of the form

$$X \sim N(Z\beta, \sigma_g^2 K_g + \sigma_n^2 \mathbf{I}).$$

Here K_g denotes the random effects covariance defined based on the genotype similarity with σ_g^2 being the variance of the random effects, σ_n^2 is the variance of the noise and $\mathbf{I}$ the identify matrix.

Running the testing procedure in (21.1) over all possible combinations of (SNP A, gene A, gene B, gene C) is a daunting task. To reduce the search space, one possible solution is to require the presence of *cis* or *trans* association between SNP A and gene A. Conditioning on the expressions of all identified exogenous genes B, an extended linear mixed effects model can be applied to detect whether there is a *trans* association between SNP A and gene C [41].

21.3.2 A causal mediation framework for integration of GWAS and eQTL studies

Mapping eQTLs is also frequently used for unraveling the causal mechanism leading from genotype to phenotype, a crucial step for developing effective treatments of diseases. The increasing ability and power to map not only *cis* acting eQTLs but also *trans* acting eQTLs has greatly facilitated investigations into putative causal intermediates between genotype and the phenotype (node A, B in Figure 21.3). It has been recognized that gene expressions, albeit associated with genetic variants, may not be causal for the phenotype as the associations could be the result of responses to the phenotype (node E, F in Figure 21.3) or side effects (node C, G in Figure 21.3). Thus novel statistical methods beyond association analysis are needed to prioritize causal mediating genes.

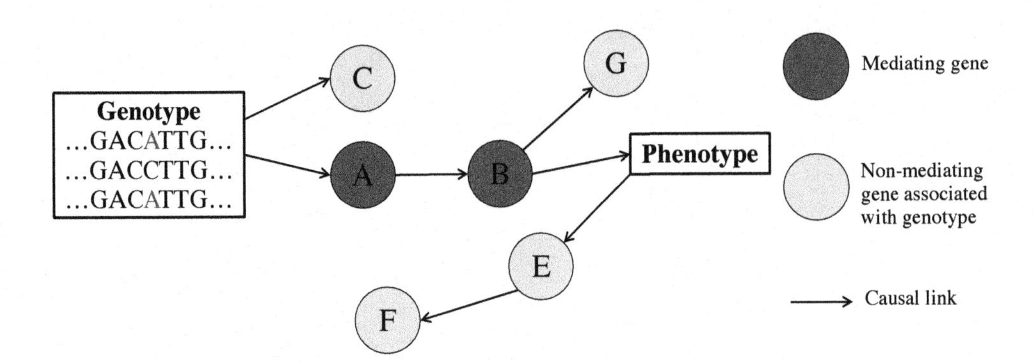

FIGURE 21.3: Distinguishing causal intermediate genes between genetic variation and phenotype (modified based on [17]). Genes can mediate the effect of genetic variation (genotype) on phenotype (nodes A and B), but gene expressions may also be affected by genetic variants irrespective of the phenotype (nodes C and G) or as a consequence of the phenotype (nodes E and F). A major challenge is to go beyond association analysis and identify mediating genes like A and B, which are valuable intervention points for understanding the molecular chain of causality.

When both gene expressions and genetic variants are measured on the same set of individuals with different phenotypes, [33] developed a sparse instrumental variable regression approach in order to identify the phenotype-associated genes whose expressions are controlled by genetic variants, where the genome-wide genetic variants served as instrumental variables. PrediXcan [18] is a gene-based association method that estimates gene expressions determined by an individual's genetic profile and subsequently correlates imputed gene expression with the phenotype under investigation to identify genes involved in the etiology of the phenotype.

There has been a recent interest in integrating summary data from GWAS and eQTL studies in order to identify genes whose expression levels are associated with complex trait because of pleiotropy. One advantage of such approaches is that the summary-level data from GWAS and eQTL studies can come from two completely different sets of individuals, thereby effectively increasing the sample size for association analysis. The rationale for such an integrative analysis is articulated in [50] and is illustrated in Figure 21.4. If the phenotypic difference is caused by a genetic variant mediated by gene expression or transcription, then we should expect simultaneous association between phenotype, gene expression and the genetic variant (see Figure 21.4 (a) and also [50]). Such a simultaneous association can be due to causality with gene expression as mediator, pleiotropy where the same causal variant is associated with both phenotype and gene expression, or due to linkage where the shared association is because of linkage disequilibrium (LD) with two distinct causal variants, one affecting gene expression and another affecting phenotype (see Figure 21.4 (b)).

A summary data-based mediation test was proposed in [50] to identify gene expressions that are associated with complex traits. For each of the GWAS identified SNPs, [50] performed a mediation test with each of the gene expressions that has at least one *cis* eQTL at a p-value $< 5 \times 10^{-8}$. Specifically, let Z be a genetic variant (e.g. a SNP), X the expression level of a gene and Y the trait. Using the mediation framework illustrated in Figure 21.5,

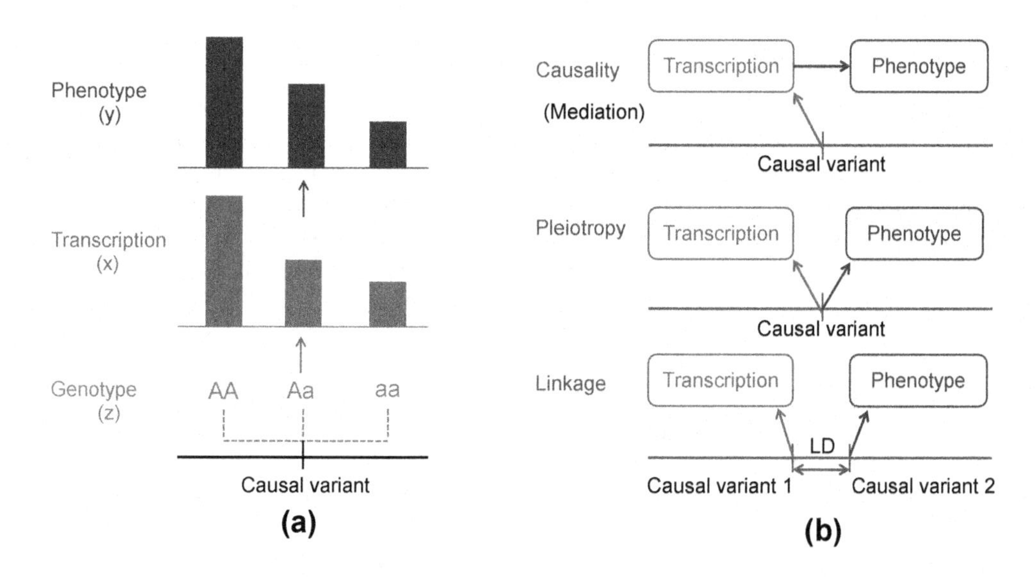

FIGURE 21.4: Association between gene expression and phenotype through shared genotypes (modified based on [50]). (a) Simultaneous associations caused by shared causal variants; (b) three possible causes of simultaneous association, causality, pleiotropy and linkage [21].

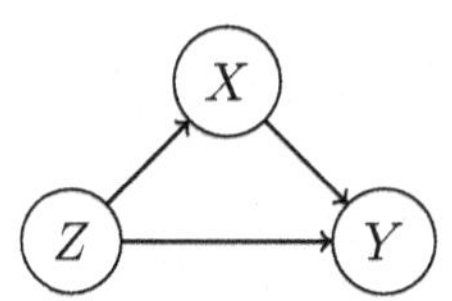

FIGURE 21.5: A simple mediation framework to link genetic variant Z to the gene expression X and the trait of interest Y.

the two-step least squares estimate of the effect of X on Y is

$$\widehat{b}_{XY} = \widehat{b}_{ZY}/\widehat{b}_{ZX},$$

where $\widehat{b}_{ZY}$ and $\widehat{b}_{ZX}$ are the least squares estimates of Y and X on Z, respectively. One can interpret b_{XY} as the effect size of X on Y free of confounding from non-genetic factors. In addition, the variance of $\widehat{b}_{XY}$ can also be estimated from the GWAS and eQTL summary statistics. One can thus use the test statistic $\widehat{b}^2_{XY}/\operatorname{Var}(\widehat{b}_{XY})$ to test whether gene X is significantly associated with the trait Y. To differentiate pleiotropy from the less interesting case of linkage, [50] tested against the null hypothesis that there is a single causal variant, or equivalently the absence of heterogeneity in the b_{XY} values estimated for the SNPs in the *cis* eQTL region.

21.4 Network Models in Metagenomics

Metagenomics has emerged as a powerful tool for learning microbial communities by directly extracting genetic materials from environmental samples. Microorganisms such as bacteria and archaea do not exist in isolation but form complex ecological interaction networks. These microorganisms are naturally assembled into interacting communities, and these community structures are directly linked to microbial processes. Therefore, the identification of key players in a taxonomically complex sample and understanding these complex interdependencies are necessary to understand the ecology of a particular habitat.

21.4.1 Covariance based on compositional data

One challenge in microbial network construction is that we cannot measure the true abundances of the microbes. Instead, the current sequencing technologies such as 16S rRNA sequencing or shotgun metagenomic sequencing only provides information on the relative abundances of the microbial taxa. Such relative abundances are often given in terms of proportions with a unit sum. In other words, the data are compositional. Another feature of the compositional data is the presence of many zeros because many taxa are absent from the sample or their abundances are below the detection level due to insufficient sequencing depths. Recent attempts in microbial network analysis focused on estimating the covariance matrix of compositional data [15]. However, one caveat with learning directly from the compositional data is that the unit sum constraint can lead to large spurious correlations, as illustrated in Figure 6 of [31].

It is instructive to first examine the quantity that is estimable based on compositional data. Let $W = (W_1, \ldots, W_p)^T$ with $W_j > 0$ for all j be a vector of latent variables, called the *basis* counts (e.g., true bacterial counts), that generate the observed compositional data via the normalization

$$X_j = \frac{W_j}{\sum_{i=1}^p W_i}, \quad j = 1, \ldots, p.$$

Estimating the covariance structure of W based on X has traditionally been considered infeasible owing to the apparent lack of identifiability. Nonetheless, [6] showed that the *basis covariance matrix* Ω_0 is approximately identifiable as long as it belongs to a class of large sparse covariance matrices, where $\mathbf{\Omega}_0 = (\omega_{ij}^0)_{p\times p}$ is defined by

$$\omega_{ij}^0 = \mathrm{Cov}(Y_i, Y_j), \quad Y_j = \log W_j.$$

To see this, recall one of the matrix specifications of compositional covariance structures introduced by [1] is the *variation matrix* $\mathbf{T}_0 = (\tau_{ij}^0)_{p\times p}$ defined by

$$\tau_{ij}^0 = \mathrm{Var}(\log(X_i/X_j)) = \mathrm{Var}(\log W_i - \log W_j) = \omega_{ii}^0 + \omega_{jj}^0 - 2\omega_{ij}^0,$$

or in matrix form,

$$\mathbf{T}_0 = \boldsymbol{\omega}_0 \mathbf{1}^T + \mathbf{1}\boldsymbol{\omega}_0^T - 2\mathbf{\Omega}_0, \tag{21.2}$$

where $\boldsymbol{\omega}_0 = (\omega_{11}^0, \ldots, \omega_{pp}^0)^T$ and $\mathbf{1} = (1, \ldots, 1)^T$. One can see from the decomposition (21.2) that $\mathbf{\Omega}_0$ is unidentifiable, since $\boldsymbol{\omega}_0 \mathbf{1}^T + \mathbf{1}\boldsymbol{\omega}_0^T$ and $\mathbf{\Omega}_0$ are in general not orthogonal to each other (with respect to the usual Euclidean inner product).

On the other hand, one can similarly define the *centered log-ratio covariance matrix* $\Gamma_0 = (\gamma_{ij}^0)_{p\times p}$ by

$$\gamma_{ij}^0 = \mathrm{Cov}\{\log(X_i/g(\mathbf{X})), \log(X_j/g(\mathbf{X}))\},$$

where $g(\mathbf{x}) = (\prod_{j=1}^{p} x_j)^{1/p}$ is the geometric mean of a vector $\mathbf{x} = (x_1, \ldots, x_p)^T$. Letting $\gamma_0 = (\gamma_{11}^0, \ldots, \gamma_{pp}^0)^T$, [6] shows that

$$\mathbf{T}_0 = \gamma_0 \mathbf{1}^T + \mathbf{1}\gamma_0^T - 2\mathbf{\Gamma}_0. \tag{21.3}$$

Unlike (21.2), the following proposition shows that (21.3) is an orthogonal decomposition and hence the components $\gamma_0 \mathbf{1}^T + \mathbf{1}\gamma_0^T$ and $\mathbf{\Gamma}_0$ are identifiable. In addition, by comparing the decompositions (21.2) and (21.3), one can bound the difference between $\mathbf{\Omega}_0$ and its identifiable counterpart $\mathbf{\Gamma}_0$ as follows.

Proposition 21.4.1. *The components $\gamma_0 \mathbf{1}^T + \mathbf{1}\gamma_0^T$ and $\mathbf{\Gamma}_0$ in the decomposition* (21.3) *are orthogonal to each other. Moreover, for the covariance parameters $\mathbf{\Omega}_0$ and $\mathbf{\Gamma}_0$ in the decompositions* (21.2) *and* (21.3),

$$\|\mathbf{\Omega}_0 - \mathbf{\Gamma}_0\|_{\max} \leq 3p^{-1}\|\mathbf{\Omega}_0\|_1.$$

Proposition 21.4.1 implies that the covariance parameter $\mathbf{\Omega}_0$ is *approximately* identifiable as long as $\|\mathbf{\Omega}_0\|_1 = o(p)$. Consequently one can use $\mathbf{\Gamma}_0$ as a proxy for $\mathbf{\Omega}_0$, which greatly facilitates the development of new methodology and associated theory. [6] developed a composition-adjusted thresholding (COAT) method under the assumption that the basis covariance matrix is sparse, and showed that the resulting procedure can be viewed as thresholding the sample centered log-ratio covariance matrix and hence is scalable for large covariance matrices.

21.4.2 Microbial community dynamics

The microbial communities are highly dynamic, and are constantly responding to perturbations in the environment. Several large-scale time-series microbiome data have been generated to gain insights into the dynamics of gut microbiome over time. The human microbiota time series study in [7] covers two individuals at four body sites over 396 time points, including gut, tongue, left palm and right palm. [11] reported coupled longitudinal datasets of human lifestyle and microbiota by tracking two healthy male volunteers and their commensal microbial communities each day over the course of a year. These studies not only show overall stability of the microbial communities, but also marked community disturbance following changes in the environment. However, the small number of subjects involved in the above two studies proves to be inadequate for valid inference of time-varying microbial networks.

The most comprehensive study on the progression of the infant gut microbiota thus far examined 58 preterm infants in a neonatal intensive care unit, with repeated measurements taken every few days on all study subjects starting within the first days of life, and ending at approximately one month of age [36]. This set of densely sampled microbiome data provides the relative abundances of microbial taxa measured over time, revealing important information on ecological dynamics. One approach towards inference of time-varying microbial networks is based on multivariate functional data analysis. For each taxon, its abundance trajectory can be treated as functional data. One can then apply techniques developed for estimating the covariance structure of multivariate functional data [39] to reconstruct the microbial network at each time point. However, the compositional nature of the data and also the excessive zeros require appropriate modification to these existing methods.

See [13] for a brief review of dynamic network inference from metagenomic data.

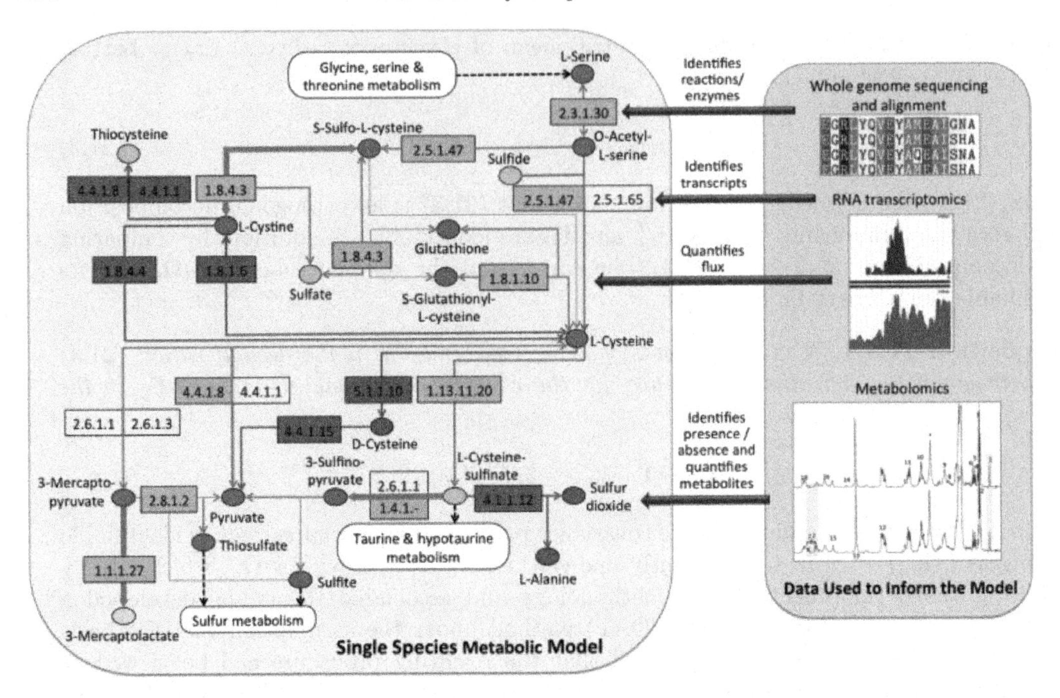

FIGURE 21.6: Subset of a microbial cysteine/methionine metabolic network for one bacterial species. The model is constructed based on the bacterial genome. Each box represents a reaction. The numbers within the boxes are KEGG Enzyme Commission (EC) number and code for specific enzymes present in each reaction. Dark gray boxes represent reactions that occur in this bacteria, as predicted by its genome. Boxes denote reactions that are not predicted by the genome. Circles represent metabolites consumed and produced within the reaction network. Arrows represent reaction pathways that do or do not occur in this bacteria, as predicted by the model. Black dashed arrows indicate input or output from or to other metabolic networks. This figure is reproduced from [44] under a Creative Commons license. doi:10.1016/j.atg.2016.02.001.

21.5 Future Directions and Topics

As our knowledge of biological networks increases, incorporation of such networks in analysis of biological data proves invaluable in genetic association studies, and to some degree in analysis of genetical genomics or eQTL studies. Looking forward, single-cell measurements and gene editing tools such as CRISPR-Cas will lead to detailed understanding of the biological networks at the single-cell level. The data sets from such studies are large and require new statistical and computational methods.

Metagenomics is an emerging field that holds a great promise in biomedical research. Most of the published works are still at the level of establishing association between taxa composition and microbial gene abundances with various covariates or disease states. A major challenge is to go beyond association studies and elucidate causalities. Mathematical modeling of the human gut microbiome at a genome scale is a useful tool to decipher microbe-microbe, diet-microbe and microbe-host interactions [38]. Graphical models, especially causal graphical models, provide a natural and useful tool for elucidating such causal pathways. Parallel to advances in sequencing technologies, publicly available database of experimentally elucidated metabolic pathways from all domains of life, such as MetaCyc [8], is becoming more and more complete. Currently, MetaCyc contains more than 2400 pathways from 2788 different organisms, including those involved in both primary and secondary metabolism, as well as associated metabolites, reactions, enzymes and genes. This provides important resources for analysis of microbiome and metagenomic data. Figure 21.6 presents an example of such a metabolic network. How to incorporate the metabolic networks/pathways into analysis of metagenomic data is an important area for future research.

Bibliography

[1] John Aitchison. The statistical analysis of compositional data. *Journal of Royal Statistical Society, Series B*, 44(2):139–177, 1982.

[2] Frank W. Albert and Leonid Kruglyak. The role of regulatory variation in complex traits and disease. *Nature Reviews Genetics*, 16(4):197–212, 2015.

[3] Albert-László Barabási, Natali Gulbahce, and Joseph Loscalzo. Network medicine: A network-based approach to human disease. *Nature Reviews Genetics*, 12(1):56–68, 2011.

[4] Albert-Laszlo Barabasi and Zoltan N Oltvai. Network biology: Understanding the cell's functional organization. *Nature Reviews Genetics*, 5(2):101–113, 2004.

[5] William S. Bush and Jason H. Moore. Genome-wide association studies. *PLoS Computational Biology*, 8(12):e1002822, 2012.

[6] Yuanpei Cao, Wei Lin, and Hongzhe Li. Large covariance estimation for compositional data via composition-adjusted thresholding. *arXiv:1601.04397*, 2016.

[7] J. Gregory Caporaso, Christian L. Lauber, Elizabeth K. Costello, Donna Berg-Lyons, Antonio Gonzalez, Jesse Stombaugh, Dan Knights, Pawel Gajer, Jacques Ravel, Noah Fierer et al. Moving pictures of the human microbiome. *Genome Biology*, 12(5):R50, 2011.

[8] Ron Caspi, Richard Billington, Luciana Ferrer, Hartmut Foerster, Carol A. Fulcher, Ingrid M. Keseler, Anamika Kothari, Markus Krummenacker, Mario Latendresse, Lukas A. Mueller et al. The metacyc database of metabolic pathways and enzymes and the biocyc collection of pathway/genome databases. *Nucleic Acids Research*, 44(D1):D471–D480, 2016.

[9] Min Chen, Judy Cho, and Hongyu Zhao. Incorporating biological pathways via a Markov random field model in genome-wide association studies. *PLoS Genetics*, 7(4):e1001353, 2011.

[10] David Croft, Gavin O'Kelly, Guanming Wu, Robin Haw, Marc Gillespie, Lisa Matthews, Michael Caudy, Phani Garapati, Gopal Gopinath, Bijay Jassal et al. Reactome: a database of reactions, pathways and biological processes. *Nucleic Acids Research*, 39(suppl 1):D691–D697, 2011.

[11] Lawrence A. David, Arne C. Materna, Jonathan Friedman, Maria I. Campos-Baptista, Matthew C. Blackburn, Allison Perrotta, Susan E. Erdman, and Eric J. Alm. Host lifestyle affects human microbiota on daily timescales. *Genome Biology*, 15:R89, 2014.

[12] Benjamin P. Fairfax, Peter Humburg, Seiko Makino, Vivek Naranbhai, Daniel Wong, Evelyn Lau, Luke Jostins, Katharine Plant, Robert Andrews, Chris McGee, and Julian C. Knight. Innate immune activity conditions the effect of regulatory variants upon monocyte gene expression. *Science*, 343(6175):1246949, 2014.

[13] Karoline Faust, Leo Lahti, Didier Gonze, Willem M. de Vos, and Jeroen Raes. Metagenomics meets time series analysis: unraveling microbial community dynamics. *Current Opinion in Microbiology*, 25:56–66, 2015.

[14] Timothée Flutre, Xiaoquan Wen, Jonathan Pritchard, and Matthew Stephens. A statistical framework for joint eqtl analysis in multiple tissues. *PLoS Genetics*, 9(5):e1003486, 2013.

[15] Jonathan Friedman and Eric J. Alm. Inferring correlation networks from genomic survey data. *PLoS Computational Biology*, 8(9):e1002687, 2012.

[16] Sarah A. Gagliano, Michael R. Barnes, Michael E. Weale, and Jo Knight. A Bayesian method to incorporate hundreds of functional characteristics with association evidence to improve variant prioritization. *PloS One*, 9(5):e98122, 2014.

[17] Julien Gagneur, Oliver Stegle, Chenchen Zhu, Petra Jakob, Manu M. Tekkedil, Raeka S. Aiyar, Ann-Kathrin Schuon, Dana Pe'er, and Lars M. Steinmetz. Genotype-environment interactions reveal causal pathways that mediate genetic effects on phenotype. *PLoS Genetics*, 9(9):e1003803, 2013.

[18] Eric R. Gamazon, Heather E. Wheeler, Kaanan P. Shah, Sahar V. Mozaffari, Keston Aquino-Michaels, Robert J. Carroll, Anne E. Eyler, Joshua C. Denny, GTEx Consortium, Dan L. Nicolae, Nancy J. Cox, and Hae Kyung Im. A gene-based association method for mapping traits using reference transcriptome data. *Nature Genetics*, 47(9):1091–1098, 2015.

[19] Charles Gawad, Winston Koh, and Stephen R. Quake. Single-cell genome sequencing: current state of the science. *Nature Reviews Genetics*, 17:175–188, 2016.

[20] Elin Grundberg, Kerrin S. Small, Åsa K Hedman, Alexandra C. Nica, Alfonso Buil, Sarah Keildson, Jordana T. Bell, Tsun-Po Yang, Eshwar Meduri, Amy Barrett et al. Mapping cis- and trans-regulatory effects across multiple tissues in twins. *Nature Genetics*, 44(10):1084–1089, 2012.

[21] Alexander Gusev, Arthur Ko, Huwenbo Shi, Gaurav Bhatia, Wonil Chung, Brenda W.J.H. Penninx, Rick Jansen, Eco J.C. De Geus, Dorret I. Boomsma, Fred A. Wright et al. Integrative approaches for large-scale transcriptome-wide association studies. *Nature Genetics*, 48:245–252, 2016.

[22] Xin He, Stephan J. Sanders, Li Liu, Silvia De Rubeis, Elaine T. Lim, James S. Sutcliffe, Gerard D. Schellenberg, Richard A. Gibbs, Mark J. Daly, Joseph D. Buxbaum, et al. Integrated model of de novo and inherited genetic variants yields greater power to identify risk genes. *PLoS Genetics*, 9(8):e1003671, 2013.

[23] Tianxiao Huan, Bin Zhang, Zhi Wang, Roby Joehanes, Jun Zhu, Andrew D. Johnson, Saixia Ying, Peter J. Munson, Nalini Raghavachari, Richard Wang, et al. A systems biology framework identifies molecular underpinnings of coronary heart disease. *Arteriosclerosis, Thrombosis, and Vascular Biology*, 33(6):1427–1434, 2013.

[24] Peilin Jia and Zhongming Zhao. Network-assisted analysis to prioritize GWAS results: principles, methods and perspectives. *Human Genetics*, 133(2):125–138, 2014.

[25] Minoru Kanehisa, Miho Furumichi, Mao Tanabe, Yoko Sato, and Kanae Morishima. KEGG: New perspectives on genomes, pathways, diseases and drugs. *Nucleic Acids Research*, 45(D1):D353–D361, 2017.

[26] Minoru Kanehisa and Susumu Goto. KEGG: Kyoto encyclopedia of genes and genomes. *Nucleic Acids Research*, 28(1):27–30, 2000.

[27] Minoru Kanehisa, Yoko Sato, Masayuki Kawashima, Miho Furumichi, and Mao Tanabe. KEGG as a reference resource for gene and protein annotation. *Nucleic Acids Research*, 44(D1):D457–D462, 2016.

[28] Ekta Khurana, Yao Fu, Jieming Chen, and Mark Gerstein. Interpretation of genomic variants using a unified biological network approach. *PLoS Computational Biology*, 9(3):e1002886, 2013.

[29] Gleb Kichaev, Wen-Yun Yang, Sara Lindstrom, Farhad Hormozdiari, Eleazar Eskin, Alkes L. Price, Peter Kraft, and Bogdan Pasaniuc. Integrating functional data to prioritize causal variants in statistical fine-mapping studies. *PLoS Genetics*, 10(10):e1004722, 2014.

[30] Mark D.M. Leiserson, Jonathan V. Eldridge, Sohini Ramachandran, and Benjamin J. Raphael. Network analysis of GWAS data. *Current Opinion in Genetics & Development*, 23(6):602–610, 2013.

[31] Hongzhe Li. Microbiome, metagenomics, and high-dimensional compositional data analysis. *Annual Review of Statistics and Its Application*, 2:73–94, 2015.

[32] Hongzhe Li, Zhi Wei, and John Maris. A hidden Markov random field model for genome-wide association studies. *Biostatistics*, 11(1):139–150, 2010.

[33] Wei Lin, Rui Feng, and Hongzhe Li. Regularization methods for high-dimensional instrumental variables regression with an application to genetical genomics. *Journal of the American Statistical Association*, 110(509):270–288, 2015.

[34] Li Liu, Jing Lei, Stephan J. Sanders, Arthur Jeremy Willsey, Yan Kou, Abdullah Ercument Cicek, Lambertus Klei, Cong Lu, Xin He, Mingfeng Li, et al. Dawn: a framework to identify autism genes and subnetworks using gene expression and genetics. *Molecular Autism*, 5(1):1, 2014.

[35] Mark I. McCarthy, Gonçalo R. Abecasis, Lon R. Cardon, David B. Goldstein, Julian Little, John P.A. Ioannidis, and Joel N. Hirschhorn. Genome-wide association studies for complex traits: consensus, uncertainty and challenges. *Nature Reviews Genetics*, 9(5):356–369, 2008.

[36] Michael J. McGeachie, Joanne E. Sordillo, Travis Gibson, George M. Weinstock, Yang-Yu Liu, Diane R. Gold, Scott T. Weiss, and Augusto Litonjua. Longitudinal prediction of the infant gut microbiome with dynamic Bayesian networks. *Scientific Reports*, 6(20359), 2016.

[37] Alexandra C. Nica and Emmanouil T. Dermitzakis. Expression quantitative trait loci: Present and future. *Philosophical Transactions of the Royal Society B*, 368:201220262, 2013.

[38] Cecilia Noecker, Alexander Eng, Sujatha Srinivasan, Casey M. Theriot, Vincent B. Young, Janet K. Jansson, David N. Fredricks, and Elhanan Borenstein. Metabolic model-based integration of microbiome taxonomic and metabolomic profiles elucidates mechanistic links between ecological and metabolic variation. *mSystems*, 1(1), 2016.

[39] Alexander Petersen and Hans-Georg Müller. Fréchet integration and adaptive metric selection for interpretable covariances of multivariate functional data. *Biometrika*, 103(1):103–120, 2016.

[40] Alkes L. Price, Agnar Helgason, Gudmar Thorleifsson, Steven A. McCarroll, Augustine Kong, and Kari Stefansson. Single-tissue and cross-tissue heritability of gene expression via identity-by-descent in related or unrelated individuals. *PLoS Genetics*, 7(2):e1001317, 2011.

[41] Barbara Rakitsch and Oliver Stegle. Modelling local gene networks increases power to detect trans-acting genetic effects on gene expression. *Genome Biology*, 17(1):1, 2016.

[42] Sabry Razick, George Magklaras, and Ian M. Donaldson. irefindex: a consolidated protein interaction database with provenance. *BMC Bioinformatics*, 9(1):1, 2008.

[43] Jeffry D. Sander and J. Keith Joung. CRISPR-Cas systems for editing, regulating and targeting genomes. *Nature Biotechnology*, 32:347–355, 2014.

[44] Jaeyun Sunga, Vanessa Haleb, Annette C. Merkel, Pan-Jun Kima, and Nicholas Chiab. Metabolic modeling with big data and the gut microbiome. *Applied & Translational Genomics*, in press, 2016.

[45] Damian Szklarczyk, Andrea Franceschini, Stefan Wyder, Kristoffer Forslund, Davide Heller, Jaime Huerta-Cepas, Milan Simonovic, Alexander Roth, Alberto Santos, Kalliopi P. Tsafou et al. String v10: Protein–protein interaction networks, integrated over the tree of life. *Nucleic Acids Research*, 43(D1):D447–D452, 2015.

[46] Inma Tur, Alberto Roverato, and Robert Castelo. Mapping EQTL networks with mixed graphical Markov models. *Genetics*, 198(4):1377–1393, 2014.

[47] Zhi Wei and Hongzhe Li. A Markov random field model for network-based analysis of genomic data. *Bioinformatics*, 23(12):1537–1544, 2007.

[48] Zhi Wei and Hongzhe Li. A hidden spatial-temporal Markov random field model for network-based analysis of time course gene expression data. *Annals of Applied Statistics*, 2(1):408–429, 2008.

[49] Bin Zhang and Steve Horvath. A general framework for weighted gene co-expression network analysis. *Statistical Applications in Genetics and Molecular Biology*, 4(1):1128, 2005.

[50] Zhihong Zhu, Futao Zhang, Han Hu, Andrew Bakshi, Matthew R. Robinson, Joseph E. Powell, Grant W. Montgomery, Michael E. Goddard, Naomi R. Wray, Peter M. Visscher, et al. Integration of summary data from GWAS and EQTL studies predicts complex trait gene targets. *Nature Genetics*, 48:481–487, 2016.

Index